Fundamentals of Explosive Ordnance

Explosive ordnance is used throughout the armed services to project power. Consequently, a thorough understanding of how explosive ordnance behaves regarding its effects, environmental aspects and testing methodologies, and the way that explosives function from a fundamental perspective is crucial for the safe handling of these sometimes-complex systems.

This book provides a rigorous foundation for understanding how explosives and explosive ordnance systems operate. Its scope is deliberately broad. It begins with the fundamentals of explosive materials and progresses to detonation physics, including the classical Chapman–Jouguet theory as well as Zel'dovich-von Neumann-Döring theory. The text also examines shock physics, fragmentation and penetration analysis, testing of explosives and explosive ordnance, and the use of shock tubes and blast simulators to advance our understanding of blast-wave effects. In addition, it examines the need for insensitive munitions and content on environmental considerations. Finally, it explores sacrificial claddings for blast protection and presents recent observations on the use of drones in the conflict in Ukraine.

With contributions from established researchers and practitioners across the globe, this book is written for students, engineers, defence professionals, and researchers. It offers a systematic, and sometimes in-depth introduction to the field—bridging fundamental theory with real-world ordnance behaviour. Whether used as a teaching text or a technical reference, it equips readers with the tools needed to engage with modern explosive systems.

Fundamentals of Explosive Ordnance

Edited by
Paul J. Hazell and Nicholas Kanizaj

CRC Press
Taylor & Francis Group
Boca Raton London New York

CRC Press is an imprint of the
Taylor & Francis Group, an **informa** business

Designed cover image: Shutterstock

First edition published 2027
by CRC Press
2385 NW Executive Center Drive, Suite 320, Boca Raton FL 33431

and by CRC Press
4 Park Square, Milton Park, Abingdon, Oxon, OX14 4RN

CRC Press is an imprint of Taylor & Francis Group, LLC

ISBN: [9781032863245] (hbk)
ISBN: [9781032867939] (pbk)
ISBN: [9781003529217] (ebk)

DOI: [10.1201/9781003529217]

Typeset in Times
by KnowledgeWorks Global Ltd.

Access the [Instructor and Student Resources]: *www.routledge.com/[ISBN]*

This book is dedicated to the men and women of the armed forces, entrusted with the difficult duty of using force to preserve peace and democracy in defence of their nation.

Contents

Preface

The study of explosives and explosive ordnance has a long and rich history, stretching back to the early scientific investigations of energetic materials in Europe in the thirteenth century and the descriptions of gunpowder mixtures that appeared in the military treatise of the Song Dynasty in the eleventh century. Since that time, explosives have shaped the course of human conflict, enabled large-scale engineering achievements, allowed for the mining of precious resources and provided a pathway for scientific discovery for Universities and Government laboratories around the world. From the earliest formulations of gunpowder to the sophisticated explosive ordnance of today, the field has continuously evolved in both scope and complexity.

In the twenty-first century, the nature of conflict is changing rapidly. New technologies—including the rise of autonomous systems and drones—are altering how explosive ordnance is delivered, employed, and defended against. The traditional battlefield is increasingly augmented by remotely piloted platforms, improvised explosive devices, and novel delivery mechanisms that demand both a scientific and operational understanding. Against this backdrop, the importance of mastering the fundamentals of explosive materials and their effects remains as critical as ever.

This book seeks to address those fundamentals. At its core lies the science of explosives: their chemistry, their detonation and deflagration behaviours, and the physical mechanisms by which they release energy. Understanding these principles is essential not only for safe handling and storage but also for predicting performance, effectiveness, and reliability in operational settings. Alongside the technical aspects of explosive materials themselves, this book also explores the mechanics of injury and damage caused by blasts, fragments, and thermal effects—knowledge that is vital for both military practitioners and those involved in humanitarian demining, counter-terrorism, and protection of civilian populations.

The opening chapters establish the scientific basis of energetic materials, beginning with the properties and chemistry of explosives (Chapter 1) and progressing to innovations in synthesis and manufacture (Chapter 2). The principles of detonation theory and practice (Chapter 3) are followed by the application of computational modelling to understand and predict detonation phenomena (Chapter 4), supported by a historical review of dynamic mechanical testing techniques for characterizing explosive behaviour under extreme conditions (Chapter 5).

The book then broadens in scope to consider the lifecycle and societal impacts of explosives. Chapters address the environmental consequences of explosive materials (Chapter 6), thoughts on reducing vulnerability through insensitive munitions (Chapter 7), and approaches to testing both explosives and explosive ordnance systems (Chapter 8).

Attention is next directed to blast generation and effects, encompassing the design and use of shock tubes and simulators (Chapter 9), the interaction of explosive loads with structures (Chapter 10), and the modelling of explosive effects for predictive and protective purposes (Chapter 11). Risk assessment methodologies are explored in detail, with emphasis on quantifying hazards from both blast and fragmentation (Chapter 12). This naturally leads into the human dimension, with an examination of vulnerability to blast and fragments (Chapter 13), and protective design strategies such as sacrificial claddings (Chapter 14).

Finally, the book concludes with a forward-looking perspective on the role of new technologies and delivery methods, including the growing use of drones in conflict zones, with insights drawn from the war in Ukraine (Chapter 15).

Together, these chapters form a structured and integrated reference that bridges fundamental science, experimental practice, engineering application, and contemporary operational realities, offering both students and practitioners a broad yet detailed understanding of the world of explosives and their effects. A glossary and a solutions manual are also available on the Routledge website (www.routledge.com/9781032863245).

The contents of this volume reflect a collaboration of international experts, each of whom contributes a chapter in their specialist domain. These contributions come from researchers and practitioners at leading institutions, including the Defence Science and Technology Group in Australia, Imperial College London, Cambridge University, Cranfield University, the University of New South Wales, University of Technology Sydney, Technion—Israel Institute of Technology, the University of Nottingham, Brighton and Sussex Medical School, Iqarus Gulf FZ LLC, and the University of Sheffield, among others. Each chapter has been carefully reviewed, and we are grateful to the reviewers whose expertise has ensured the scholarly rigour of this work.

Our aim in compiling this book is twofold: firstly, to provide a comprehensive foundation in the fundamentals of explosive ordnance for students, researchers, and professionals; and secondly, to situate that knowledge within the rapidly shifting context of modern warfare and security challenges. By combining a focus on the underlying science with an awareness of contemporary applications and threats, this book hopes to serve as both a reference and a guide to those who work with, study, or seek to understand the powerful and sometimes devastating role of explosives in our world.

Paul J Hazell
Nicholas Kanizaj
The University of New South Wales
Canberra, Australia
January 2026

About the Editors

Paul J. Hazell is Professor of Impact Dynamics at UNSW Canberra, Australia, and a researcher specialising in the mechanics of high-rate phenomena. His research spans shock loading, impact dynamics, and the behaviour of materials and structures under extreme loading conditions. He serves on the editorial boards of several leading journals in these fields and is the author of *Armour: Materials, Theory, and Design* (CRC Press, 2nd ed., 2022). Paul has taught courses in computational modelling, shock wave theory, impact dynamics, and protection engineering, and has received multiple awards recognising his engaging and dynamic approach to teaching. His expertise in weapons technology has also been featured in several documentaries, where he has provided expert commentary.

Nicholas Kanizaj is a Senior Lecturer in Explosive Ordnance at the School of Engineering and Technology, UNSW Canberra, Australia. He has discharged commercial research in the areas of rocket propellants, small arms propellants, explosives and pyrotechnics. As an educator, his teaching portfolio includes courses on rocket motor design, propellants, pyrotechnics, and explosive ordnance design and systems, where he combines theoretical knowledge with practical applications to prepare the next generation of specialists. His dual role as both researcher and educator allows him to bridge the gap between academic innovation and practical implementation in explosive ordnance technology, contributing to both the advancement of the field and the development of a pipeline of skilled professionals.

About the Editors

Contributors

Mathew P. Brooks
Department of Defence
Canberra, ACT, Australia

Adam S. Cumming
University of Edinburgh
Edinburgh, United Kingdom

Edward Chern Jinn Gan
University of Wollongong
Wollongong, NSW, Australia

Hezi Y. Grisaro
Technion - Israel Institute of Technology
Haifa, Israel

Mansoor Ali Khan
Iqarus Gulf FZ LLC
Dubai, UAE

Zain Ali Khan
University of Nottingham Medical School
Nottingham, United Kingdom

Nicholas Kanizaj
UNSW Canberra
Canberra, ACT, Australia

Valerian Kuznetsov
Defence Science and Technology Group
Adelaide, South Australia, Australia

Zongjun Li
UNSW Canberra
Canberra, ACT, Australia

Oleksandra Molloy
UNSW Canberra
Canberra, ACT, Australia

Federica Persico
Cranfield University
Shrivenham, Oxfordshire, United Kingdom

William Graham Proud
Imperial College London
London, United Kingdom

Alex Remennikov
University of Wollongong
Wollongong, NSW, Australia

A. D. Resnyansky
Defence Science and Technology Group
Adelaide, South Australia, Australia

S. Rigby
University of Sheffield
Sheffield, South Yorkshire, United Kingdom

Adrian Garrido Sanchis
UNSW Canberra
Canberra, ACT, Australia

Mark G. Stewart
University of Technology Sydney
Sydney, NSW, Australia

Tracey Temple
Cranfield University
Shrivenham, Oxfordshire, United Kingdom

Stephen M. Walley
Cambridge University
Cambridge, United Kingdom

Joshua T. L. Wardrop
Cranfield University
Shrivenham, Oxfordshire, United Kingdom

Ola Wattad
Technion - Israel Institute of Technology
Haifa, Israel

Caution! Please read this before reading the rest of this book

The information presented in this book, *Fundamentals of Explosive Ordnance*, is believed to be accurate and includes information that is already in the public domain. It is intended for educational and professional reference only. Explosive materials are clearly very dangerous, and neither the editors and authors nor the publisher accepts responsibility for injury, loss of life, or physical damage resulting from the misuse or misapplication of the material contained herein. The text discusses energetic materials, ordnance characteristics, and historical and theoretical aspects of explosives—none of which should be interpreted as instructions for construction, modification, handling, or employment of explosive devices, including ordnance. Readers must not attempt to replicate, test, or apply the techniques described without appropriate licences, institutional oversight, and trained supervision; doing so may be illegal and extremely dangerous. If you require practical guidance, consult qualified, licenced professionals and relevant regulations*.

* Inspired and adapted from a similar warning in: Agrawal, J. P. & Hodgson, R. D. 2006. Organic Chemistry of Explosives. John Wiley & Sons, Ltd.

1 Explosive Materials *Fundamentals*

Adrian Garrido Sanchis

1.1 INTRODUCTION: DEFINITION

Do explosives really need a presentation?

An explosive is a chemical substance or a mixture of chemical substances that, when subjected to heat, percussion, detonation, or catalysis, will release its potential energy in a very rapid (microseconds) self-sustaining exothermic decomposition, producing a large amount of heat (3000–5000°C) and pressure (gases produced expand 12,000–15,000 times the original volume).

1.2 HISTORICAL BACKGROUND

1.2.1 Origins of Explosives: The Era of Gunpowder

For more than a thousand years, since first discovered by the Chinese during the 7th century until the second half of the 19th century when the production of nitro-explosive started, black powder was the only heterogeneous mixture of compounds used as an explosive when confined, or as a propellant in guns.

Gunpowder is the oldest explosive, but there are still many unknowns about its origin. Different sources (Brown, 2010) point toward the middle of the 9th century in China by Tang dynasty alchemists in a formula contained in the *Taishang Shengzu Jindan Mijue* (The Supreme Sage Patriarch's Secret Instructions on the Golden Elixir).

In the year 1040, the Chinese literature Song dynasty text *Wujing Zongyao* (Complete Essentials from the Military Classics), written by Zeng Gongliang, mentions for the first time the composition of gunpowder, which consists of 75% potassium nitrate (the oxidizer), 15% wood charcoal acting as fuel, and 10% sulfur to lower the ignition temperature and act as fuel (Chase, 2008).

When black powder ignites (at least 300°C are required), sulfur and carbon burn rapidly using the oxygen supplied by potassium nitrate to produce a mixture of hot solid products, mainly potassium carbonate, potassium sulfate, potassium sulfide, which reduce its performance, and hot carbon gases, primarily carbon dioxide, nitrogen, and sulfur dioxide that rapidly increase the volume. This increase in volume will propel projectiles in a gun barrel, but to use it as an explosive, gunpowder needs to be confined in a borehole or sealed container so that enough pressure can build up inside the container.

Gunpowder's first recorded use was in fireworks. By the 11th century, however, its potential as a weapon was realized, leading to the development of simple bombs, fire lances, explosive arrows, and grenades. Recognizing the strategic importance of this innovation, the Chinese government centralized the production of sulfur and saltpeter, essential components of gunpowder. China relied heavily on advanced weaponry to defend against the relentless threat of Mongol invasions from Central Asia. During the following centuries, gunpowder spread both to the Orient and the West, with some sources suggesting that Mongol invasions introduced gunpowder to Europe and the Middle East between 1240 and 1280. Syrian Hassan al -Rammah wrote more than 107 gunpowder recipes in his treatise "*Al-Furusiyyah wa al-Manasib al-Harbiyya*" (The Book of Military Horsemanship and Ingenious War Devices) (Khan, 1996) and in England, Roger Bacon, a Franciscan monk who first studied gunpowder around 1260, played a notable role in its dissemination through a treatise entitled

DOI: 10.1201/9781003529217-1

"*De secretis operibus artis et naturae et de nullitate magiae*" ("On the marvelous power of art and of nature and on the nullity of magic"). In late 14th-century Europe, gunpowder production was improved by introducing the corning process, a technique that dried the powder into small granules to enhance performance by increasing the contact between the mixture of fuel and oxidizers (Kelly, 2004). During this period, the purification process for saltpeter (potassium nitrate) was also improved. They utilized wood ashes rich in potassium carbonate to remove calcium from the nitrate concentrate and employed materials such as blood and alum to clarify the mixture, significantly improving its quality and performance (Kelly, 2004). They also boiled the concentrate to remove calcium and magnesium by precipitation of their carbonates. Since potassium nitrate accounts for 75% of black powder, its production became a key and strategic resource developing different production techniques like the French and the Swiss methods based on the bacterial action during the decomposition of excreta and vegetables where the contact between putrefying material, alkaline soil, plant ashes, air, and moisture facilitates the nitrification. The French method was developed before the French Revolution. Piles containing manure with mortar (lime), wood ashes, soil, and straw were composted and protected from rain while keeping them moist with urine and frequently turned to promote aeration.

Within the pile, organic nitrogen from manure and urine was first mineralized to ammonia and ammonium ions. Microbial nitrification then oxidized ammonium to nitrate via nitrite intermediates. The acidity generated during nitrification reacted with calcium carbonate in the mortar to form soluble calcium nitrate ($Ca(NO_3)_2$).

After about a year, the decomposed material was leached with water to extract soluble calcium nitrate. The leachate was then filtered through potash dung—organic manure enriched with potassium carbonate (K_2CO_3) to convert calcium nitrate into potassium nitrate (KNO_3) through a double decomposition reaction.

$$Ca(NO_3)_2 + K_2CO_3 \rightarrow 2\ KNO_3\,(aq) + CaCO_3\,(s)$$

The Swiss method collects urine directly in a sand pit located beneath the stable (Leconte, 1862). Apart from this distinction, the remaining steps in the process are largely similar to those of the French method.

Until the discovery of nitroglycerin in the 19th century, black powder was the sole explosive used for both military and civilian applications.

1.2.2 Modern Explosives: Initial Developments

Mercury fulminate was first synthesized in the 17th century by treating mercury with nitric acid and alcohol. However, it was forgotten over time until its rediscovery in 1807, when it was proposed as a percussion initiator for black powder (Figure 1.1).

Ammonium nitrate (AN) (NH_4NO_3) was first prepared in 1654 by Glauber. In the early 19th century, Grindel and Robin introduced it as a replacement for potassium nitrate in black powder. Initially not considered an explosive, it was widely used as fertilizer grade ammonium nitrate (FGAN) with NPK rating 34-0-0 (34% nitrogen). In the early 19th century, Grindel and Robin were the first to replace potassium nitrate with ammonium nitrate (AN) in black powder formulations (Oommen and Jain, 1999).

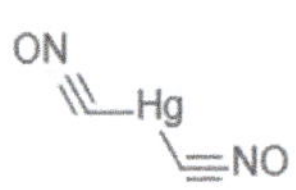

FIGURE 1.1 Mercury fulminate.

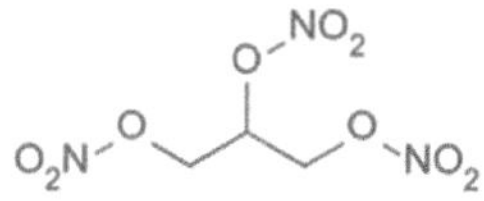

FIGURE 1.2 Nitroglycerine.

A catastrophic explosion of FGAN occurred in 1947 in the harbor of Texas City, leading to its classification as a tertiary explosive. AN is usually classified as a tertiary explosive because it is less sensitive than secondary explosives but can detonate under certain conditions, such as heating or contamination with small amounts of organic matter (Fordham, 1980).

Ammonium nitrate is typically manufactured by injecting gaseous ammonia into 40–60% nitric acid at 150°C. The resulting solution is then evaporated and converted into pilled or granular solid forms. It is the cheapest and most abundant source of oxygen available for commercial explosives, with an oxygen balance (OB) of approximately 20%. It is often combined with fuels or other explosives like nitroglycerine and TNT where the amount of AN is determined by the OB (Oommen and Jain, 1999).

In 1846, Professor Ascanio Sobrero at the University of Turin discovered liquid **nitroglycerin** (propane-1,2,3-triyl trinitrate) by adding glycerol to a mixture of sulfuric and nitric acids. He soon realized the potential of his discovery when his face was severely scarred from a nitroglycerine explosion in the 1840s, leading him to view nitroglycerin as far too dangerous for practical use. Following the accident, he strongly warned against its use in private letters and a journal article, emphasizing its extreme danger and the challenges of handling it safely (Nobel Prize Outreach, 2025) (Figure 1.2).

While nitroglycerin is primarily used in explosives and propellants, it also has medical applications. It treats heart failure and high blood pressure by relaxing blood vessels, allowing blood to flow more easily.

In 1863, Immanuel Nobel and his son Alfred Nobel developed a process for manufacturing nitroglycerin and established a small production. However, its transportation presented many risks, leading to numerous accidental explosions (The Nobel Foundation). To minimize the risk of accidental nitroglycerin explosions, Alfred Nobel invented the metal blasting cap detonator in 1864, which contained mercury fulminate. Later, in 1866, he further improved safety by stabilizing nitroglycerin (75%) through absorption into diatomaceous earth (25%) (fossilized hard-shelled algae), leading to the invention of dynamite, dynamite No. 1.

Dynamite is a general term for a wide range of explosive compositions—over a hundred mixtures—that enhance performance by partially replacing the inert base with potassium nitrate, charcoal, wood meal, potassium chlorate, magnesium carbonate, sugar and ammonium nitrate (extra dynamites). When the nitroglycerin is dissolved in a 7–8% of nitrocellulose, rather than absorbing or mixing, the new gel-like substance, which Nobel named blasting gelatin, was a highly powerful explosive. Ballistite, another Nobel's invention, was a combination of nitrocellulose, nitroglycerine, benzene, and camphor. When ammonium nitrate was added, the explosive properties of dynamite were significantly enhanced (Akhavan, 2011).

1.2.2.1 Nitrocellulose

Cellulose is the primary component of plant cell walls and is the most abundant naturally occurring organic compound. Between 1845 and 1847 C.F. Schönbein in Basel and R. Bottger in Frankfurt were able to nitrate a cellulose mixture of sulfuric and nitric acids, producing nitrocellulose. The degree of nitration is crucial in determining the application of nitrocellulose (Figure 1.3). As the nitrogen content increases, its stability decreases and is often dissolved in solvents to form a gel.

The term "nitrocellulose" does not refer to a single molecular compound but is a generic term for a family of compounds, with the nitrogen content expressed as a percentage by weight that will determine its final use as a commercial explosive for blasting purposes in torpedoes and submarine mines or as propellant when dissolved in acetone (Akhavan, 2011).

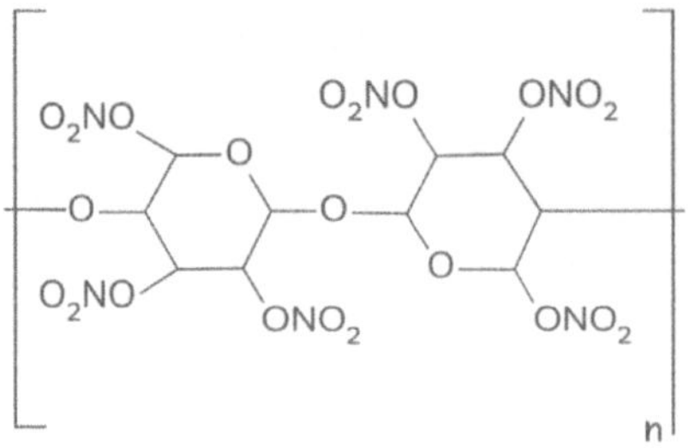

FIGURE 1.3 Nitrocellulose.

1.2.2.2 Picric Acid

Early experiments produced picric acid in 1742 by nitrating various organic substances, such as silk and natural resins. It was used until the second half of the 19th century as a fast dye for silk and wool (Woulfe, 1771). In 1885, picric acid was produced from the nitration of phenol and was found to be a suitable replacement for black powder in the filling of munitions shells, particularly in grenades and mine fillings. In 1888, Britain began manufacturing a mixture of picric acid and guncotton in Lydd, under the name Lyddite (Figure 1.4).

Lyddite remained the primary military high explosive in Britain until World War I. Dunnite, also known as Explosive D or ammonium picrate, was developed in 1906 by U.S. Army Major Beverly W. Dunn. This salt was produced by saturating an aqueous solution of picric acid with ammonia.

However, picric acid posed several problems, such as the corrosion of shells in the presence of water. Additionally, its salts were highly sensitive and prone to accidental initiation. Therefore, during World War I it was replaced by TNT by all the armies (Brown, 1998).

2,4,6-trinitrophenylmethylnitramine or tetryl was developed by Mertens in 1877 as part of his doctoral dissertation by dissolving dimethylaniline in sulfuric acid, followed by the addition of nitric and sulfuric acids (Mertens, 1877). During World War I and II, it served as a component in various explosive mixtures. Due to its high sensitivity to impact, particularly to rifle fire, tetryl was widely used in detonators and explosive booster charges, often forming the base charge of blasting caps (Figure 1.5). However, it has since been replaced by more stable, and efficient, and less toxic alternatives, such as PETN and RDX (Cooper, 1996).

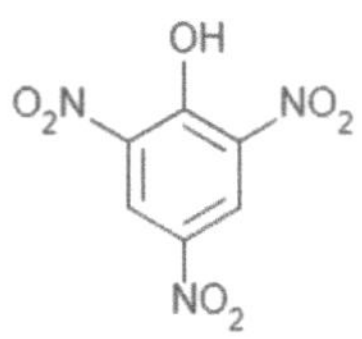

FIGURE 1.4 Picric acid.

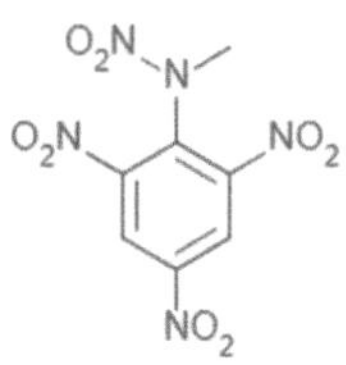

FIGURE 1.5 Tetryl.

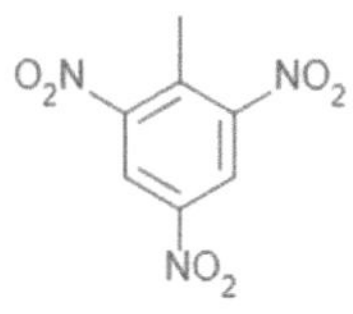

FIGURE 1.6 TNT.

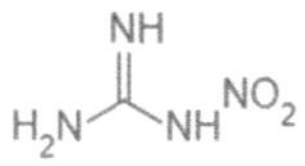

FIGURE 1.7 Nitroguanidine.

TNT (trinitrotoluene) was first synthesized by Wilbrand in 1863 and initially used as a yellow dye. Due to its high stability and difficulty to detonate, its potential as an explosive remained unknown for years. Industrial production of TNT began in Germany in 1891, and by 1899, aluminum was being mixed with TNT to improve its performance by increasing the heat of explosion. In 1902, the German Army adopted TNT as a replacement for picric acid, followed by the U.S. Army in 1912 (Akhavan, 2011). By the beginning of World War I in 1914, TNT became the standard explosive for militaries worldwide (Figure 1.6).

The use of TNT over picric acid was driven by several advantages. It was cheaper to produce and resistant to moisture. With a melting point below 100°C, TNT could be safely melted using hot water or steam, simplifying shell-filling operations. Unlike picric acid, TNT did not react with metals, reducing corrosion risks. Its stability, combined with low sensitivity to shock and friction, made it safer to handle than compounds like Lyddite.

However, TNT production during World War I was limited by the availability of toluene, which was derived from coal tar. Producing 1 ton of toluene required 600 tons of coal, leading to shortages. To address this, TNT was often mixed with other components to extend its use. Amatols, for instance, combined TNT with ammonium nitrate in varying proportions (40, 50, and 80% ammonium nitrate). Tritonal, a blend of TNT and aluminum, was also developed for enhanced explosive power. Other mixtures included Pentolite (50% TNT and 50% PETN) and Torpex (40% RDX, 40% TNT, and 20% aluminum powder) (Cooper, 1996).

Nitroguanidine, also known as Picrite, was first synthesized by Jousselin in 1877 (Figure 1.7). During World War I, the Germans utilized it as a key ingredient in bursting charges. It was often combined with ammonium nitrate and paraffin to fill trench mortar shells, demonstrating its versatility in warfare (Akhavan, 2011).

Since the 1930s, it has been used in triple-based propellants due to its unique properties, which reduce flame temperature, muzzle flash, and erosion of the gun barrel. It is characterized by a high velocity of detonation (VOD), low heat and temperature of explosion, and the production of a large volume of gases, making it an ideal component for various explosive and propellant formulations (Koch, 2019).

1.2.3 Modern Explosives: Contemporary Innovations

PETN (pentaerythritol tetranitrate), also known as penthrite, is structurally very similar to nitroglycerin. It was first synthesized in 1894, by nitrating pentaerythritol (PE) with nitric acid at temperatures under 25°C (Agrawal, 2010), commercial production did not begin until the 1930s. PETN gained widespread use during World War II (Figure 1.8).

A mixture of PETN with TNT in equal proportions (Pentolite) was used to fill hand grenades, anti-tank grenades, and detonators (Akhavan, 2011). Due to its high sensitivity to friction and

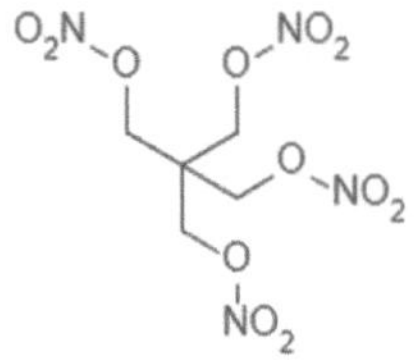

FIGURE 1.8 PETN.

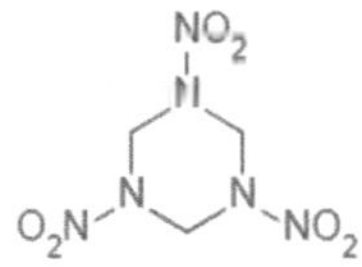

FIGURE 1.9 RDX.

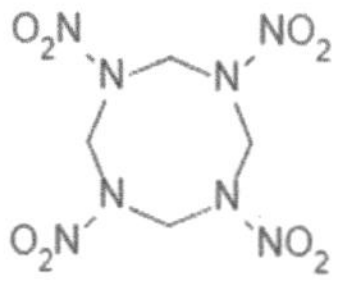

FIGURE 1.10 HMX.

impact, PETN is rarely used in its pure form. Instead, it is commonly blended with plasticizers, nitrocellulose, or synthetic rubbers to produce polymer-bonded explosives (PBXs).

When employed alone in commercial blasting caps, detonation cords, or boosters, PETN is typically desensitized by incorporating 5–10% wax. However, in military applications, PETN has largely been replaced by RDX, which offers greater stability and performance.

RDX (1,3,5-trinitroperhydro-1,3,5-triazine) (an abbreviation for "Research Department eXplosive" or "Royal Demolition eXplosive"), also known as hexogen, cyclonite, and cyclotrimethylenetrinitramine, was first synthesized in 1899 by the German chemist Henning (Figure 1.9). Its potential as an explosive was first recognized in the 1920s, but it was during World War II that its properties and methods of preparation were fully developed by two processes, the Woolwich or direct nitrolysis process and the Bachmann or combination process (Agrawal, 2010).

RDX is more thermally and chemically stable than PETN and is less sensitive to impact and friction. However, pure RDX remains highly sensitive to initiation and is often desensitized by coating its crystals with wax, oils, or grease. It is commonly used in plastic explosives, where it is mixed with materials like mineral jelly to enhance versatility. Since World War II, RDX has been widely employed in explosive mixtures with TNT, such as Cyclotols, Torpex, Composition B, and H6, offering significantly greater explosive power than picric acid and TNT (Cooper, 1996).

High Melting eXplosive (HMX) (1,3,5,7-tetranitro-1,3,5,7-tetrazocane), also referred to as **Her Majesty's eXplosive**, octogen, or cyclotetramethylenetetranitramine, is one of the most powerful chemical explosives ever manufactured, with a high VOD (Figure 1.10). Its robust thermal stability and explosive insensitivity make it suitable for demanding military applications where performance and durability are critical, for example, as a key component in detonators for nuclear weapons and solid rocket propellant (Cooper, 1996).

In 1949, it was discovered that HMX could be synthesized through the nitrolysis of RDX. This process involves dissolving RDX in a 55% nitric acid (HNO_3) solution and heating the mixture in a steam bath for approximately 6 hours (We Bachmann, 1949).

HMX is often utilized in mixtures with TNT to form **octols** or bonded with plastics to create PBXs.

Polymer-Bonded Explosives (PBXs) are formulations where plastic binders are added to powdered explosives to absorb and dissipate energy from friction and shock. They were first developed in 1952 for RDX. During the preparation process, explosive crystals are coated with binders through precipitation. These coated crystals form agglomerates known as pressing "beads," which possess good mechanical strength and can be easily pressed.

The primary purpose of incorporating a binder in PBXs is to encapsulate each explosive crystal in its own protective layer. This prevents direct contact between explosive crystals during accidental stimuli such as impact, thereby enhancing safety (Agrawal, 2010). Most PBX formulations use a rubbery polymer as the binder, often combined with a stabilizer. The introduction of binders has significantly improved safety and enabled the use of more sensitive explosives.

The most widely plasticized explosives include RDX, HMX, HNS, PETN, and TATB.

In 1960, Dr. Melvin Cook invented **water-based explosives**, which use water as a key component. Today, water-based explosives share the market with ANFO and dynamites in civilian applications, offering different formulations tailored to specific needs. These explosives are classified into two main types: water gels (or slurries) and emulsions.

- *Water-gel or slurry explosives* are single-phase systems in which all components are either dissolved or suspended in water and stabilized with cross-linking agents. Fuels and sensitizers are dispersed or suspended within the liquid medium to enhance their explosive properties. In these formulations, the oxidizer is a saturated aqueous solution of ammonium nitrate (AN) and solid oxidizers like calcium nitrate and sodium nitrate. The fuel component often includes aluminum powder, sugar, coal, fuel oil, or water-soluble fuels such as glycol and other explosive materials, including TNT and PETN that may also serve as sensitizers. Additionally, micro-balloons (hollow glass bubbles) and substances like sodium perchlorate or hexamine are sometimes incorporated to enhance shockwave propagation rate (Thurman, 2017).

  Water gels are thickened using gelling agents, such as gelatin and cross-linking compounds like guar gum, which provide structural stability. Slurries, being more liquid, are often prepared on-site and can be pumped directly into boreholes for application.

  Some water gels and slurries are classified as blasting agents because they are not cap-sensitive and lack explosive ingredients. However, others contain explosive components but remain non-cap-sensitive, while certain formulations are cap-sensitive (Cooper, 1996).
- *Emulsion explosives* were first developed in 1961 as a waterproof alternative to ANFO. These explosives are two-phase systems, fuel and oxidizer, stabilized by an emulsifier. They typically consist of dispersed droplets of an oxidizer-rich aqueous solution within a continuous fuel (oil) phase, forming a water-in-oil emulsion. Solid oxidizers can also be added to enhance performance.

  To improve sensitivity micro-balloons or chemical gassing agents are incorporated. Typical ingredients in emulsions include ammonium nitrate (AN), sodium nitrate, aluminum powder, water, sodium perchlorate, calcium nitrate, microspheres, oil, wax, emulsifiers, chemical gassing agents, and ethylene glycol dinitrate (EGDN).

  The production of emulsions involves hot concentrated solutions of ammonium nitrate, which are emulsified with surfactants (e.g., detergents) and then mixed with fuel oil, cooking oil, or wax. The small droplet size provides a large surface area in contact with the oil, allowing more intimate contact between the fuel and the oxidizer. To prevent the separation of the fuel and oxidizer, lecithin, a by-product of egg yolks or soybeans, is commonly used as an emulsifier.

1.2.4 Innovative Explosives: The Latest Advancements and Breakthroughs

TATB (triaminotrinitrobenzene), also known as 2,4,6-triamino-1,3,5-trinitrobenzene, is an aromatic explosive first produced by Adkins and Norris in 1978. TATB is extremely insensitive to initiation by shock, vibration, fire, or impact (Mitchell et al., 2002), requiring a substantial booster charge for detonation. Due to its unique safety characteristics, it is expected to replace HMX and RDX in future explosive formulations once production costs are reduced. TATB is primarily employed in applications requiring the highest safety, such as explosives in nuclear weapons (Figure 1.11). It is widely recognized as the most important insensitive high explosive for use in modern nuclear warheads due to its exceptional stability and reliability (Cooper, 1996).

HNS (hexanitrostilbene) was first synthesized in 1964 by Shipp from the U.S. Naval Ordnance Laboratory (Figure 1.12). This explosive is renowned for its resistance to heat and radiation. It is highly insensitive, being virtually unaffected by electric sparks (Shipp, 1964).

HNS is widely used in heat-resistant booster explosives and has been employed in critical applications such as stage separation in space rockets and seismic experiments on the moon.

It can be synthesized through various methods, one of which involves the oxidation of TNT dissolved in methanol using sodium hypochlorite (NaOCl) (Shipp, 1964).

NTO (5-nitro-1,2,4-triazole-3-one or nitrotriazolone), first synthesized in 1905, gained significant attention in the 1980s due to its potential as a high-performance energetic material (Figure 1.13). It is used in PBX formulations which have excellent insensitive energetic properties. NTO serves as a safer and more efficient substitute for ammonium nitrate and ammonium perchlorate in solid rocket propellants (Akhavan, 2011). Additionally, it has been developed as a replacement for traditional high explosives like RDX and HMX in composite solid propellants, making it a critical component in modern propulsion systems.

TNAZ (1,3,3-trinitroazetidine), first synthesized in 1983 at Fluorochem Inc., is a white crystalline solid recognized for its energetic properties (Figure 1.14). TNAZ is castable, making it suitable for use in gun propellants and solid rocket formulations.

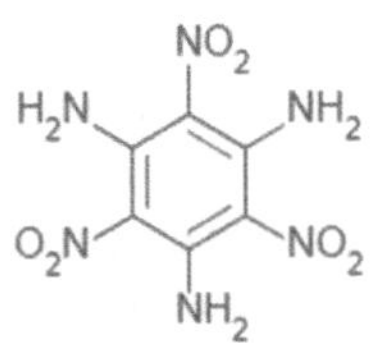

FIGURE 1.11 TATB.

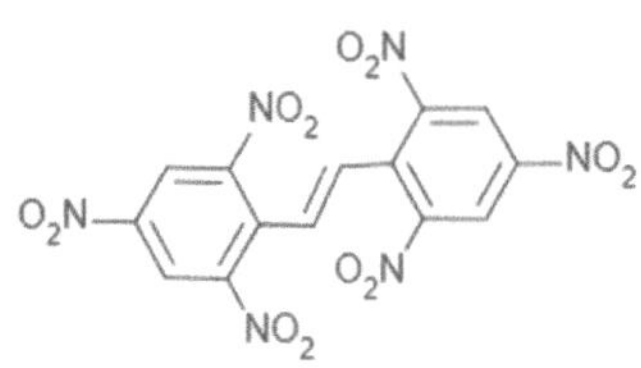

FIGURE 1.12 HNS.

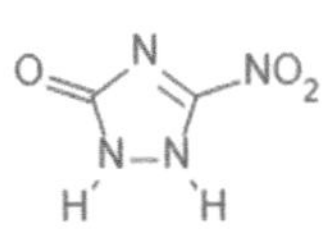

FIGURE 1.13 NTO.

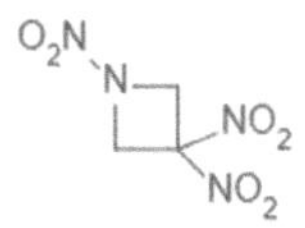

FIGURE 1.14 TNAZ.

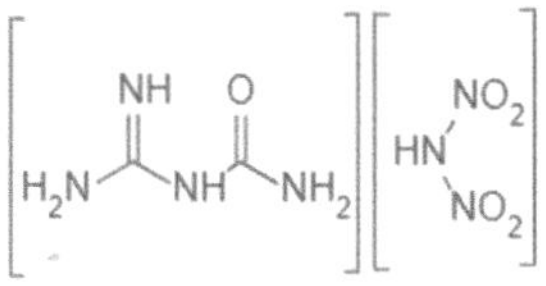

FIGURE 1.15 Fox-12.

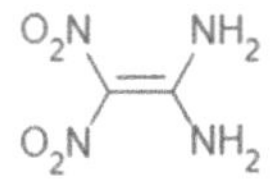

FIGURE 1.16 Fox-7.

It is able to form eutectic mixtures with TNT, tetryl, and HMX, which improves its compatibility with other energetic materials. With a melting point of 101°C, TNAZ demonstrates practical application potential. However, its adoption has been limited by high production costs and volatility (Cooper, 1996).

Strained and cage compounds represent a promising class of high-performance explosives due to their high strain energies, which result in increased density and superior detonation performance. These compounds are the focus of active research by various groups worldwide. One notable example is **CL-20**, first synthesized by Nielsen in 1991 at the Naval Air Warfare Center (NAWC) (Cooper, 1996). This compound features a strained cage structure with six nitro ($-NO_2$) groups, making it one of the most energetic high explosives available. CL-20 is utilized in tactical applications, serving both as a component in propellants and in explosive formulations.

Fox-12, also known as GUDN (Guanyl Urea Dinitramide salt), is an energetic molecule characterized by its extreme insensitivity and excellent thermal stability (Figure 1.15). Fox-12 serves as an excellent fuel in gas-generating compositions and is widely used in safety devices like airbags and belt restraint systems. Additionally, it is employed in warhead fillings for artillery, bombs, tank ammunition, and mortar rounds, highlighting its versatility and safety in both civilian and military applications.

Fox-7, also known as DADNE (1,1-diamino-2,2-dinitroethylene), was first synthesized in 1998 by the FOI Defence Research Establishment in Sweden (Figure 1.16). It is renowned for its combination of low sensitivity and high performance, making it a promising energetic material (Akhavan, 2011).

1.3 BASIC CHEMISTRY OF EXPLOSIVES

Explosives are compounds that contain a combination of oxygen, nitrogen, and oxidizable elements such as carbon and hydrogen, which act as fuels. These components enable rapid chemical reactions that release large amounts of energy. In most cases, oxygen is chemically bound to nitrogen in functional groups such as nitric oxide (NO), nitro ($-NO_2$), or nitrate ($-NO_3$), facilitating rapid oxidation. However, an important exception to this rule is the class of azides, such as lead azide (PbN_6), which contain nitrogen-rich structures (–N=N– and –N=N=N) but lack oxygen. During detonation, the bonds between nitrogen and oxygen break, allowing the free oxygen to react with fuel components, leading to the formation of stable gaseous products such as carbon dioxide (CO_2), water vapor (H_2O), and nitrogen gas (N_2) (Akhavan, 2011).

This exothermic process generates significant amounts of heat and rapidly expanding gases, which are responsible for the explosive force.

Explosives contain molecular groups which have explosive properties.

1. **Nitro (–NO_2) groups contain** the nitro group directly attached to a carbon atom. The carbon/nitrogen bond contributes to their stability, allowing them to be used safely in controlled conditions, the most well-known example is trinitrotoluene (TNT).
2. **Nitramine (N–NO_2) groups** are characterized by the nitro group directly attached to a nitrogen atom instead of carbon or oxygen, making them highly stable and energetic. Nitramines can be found in military-grade explosives and solid propellants such as RDX and HMX, both of which are used in warheads, mines, and solid rocket fuels.
3. **Nitrate ester (–O–NO_2) groups** present a (–NO_2) group bonded to an oxygen atom, creating an ester linkage that makes them highly sensitive to heat, friction, and shock. This weak bond contributes to their high explosiveness and instability, making them more sensitive than nitro compounds and nitramines. They are widely used in propellants, dynamite, and explosives. Nitroglycerin and PETN are two examples.
4. **Azo (–N=N–) and azide (–N=N=N–) groups**
   These groups appear in inorganic and organic azides, which are widely used in energetic materials, particularly as primary explosives. A notable example is lead azide ($Pb(N_3)_2$).
5. **(–NX_2) Groups**
   Where X represents a halogen, these groups are found in halogenated nitrogen compounds. Nitrogen triiodide (NI_3) is a highly sensitive explosive.
6. **Fulminate (–C≡N–O) group**
   Highly reactive compounds such as mercury fulminate ($Hg(ONC)_2$).
7. **Chlorate (–$OClO_2$) and perchlorate (–$OClO_3$) groups**
   These oxygen-rich functional groups are found in inorganic and organic chlorates and perchlorates, including potassium chlorate ($KClO_3$), potassium perchlorate ($KClO_4$), and ammonium perchlorate (NH_4ClO_4). They serve as powerful oxidizers in propellants and fireworks.
8. **Peroxide (–O–O–) and ozonide (–O–O–O–) groups**
   These groups exist in both organic and inorganic compounds, such as hydrogen peroxide (H_2O_2) and various organic peroxides. Ozonides, which contain the –O–O–O– linkage, are known for their high reactivity. For example, triacetone triperoxide (TATP) is a highly sensitive explosive used in terrorist attacks.
9. **Acetylide (–C≡C-) groups**
   These carbon–carbon triple bonds are present in acetylene (C_2H_2) and metal acetylides such as silver acetylide (Ag_2C_2) a heat- and shock-sensitive primary explosive.
10. **Metal–Carbon (M–C) bonds**
    Some organometallic compounds contain metal–carbon bonds with explosive properties.

1.4 EXPLOSIVE FORMULATIONS AND BASIC PROPERTIES

The suitability of an explosive for a specific use is determined through its properties. Some of the most important are the following:

- Sensitivity
- VOD
- Detonation pressure
- Critical diameter
- Melting temperature
- Oxygen balance
- Heat of explosion (Q)

- Volume of products of explosion
- TNT equivalent
- Explosive power

1.4.1 Sensitivity

Sensitivity measures how easily an explosive can be ignited or detonated by the formation of hotspots, small regions of 0.1–10 μm where heat is localized. These hotspots exist only for a very short duration, typically ranging from 10^{-5} to 10^{-3} seconds. During the detonation, transitions from crystalline structures and liquid phases into gaseous phases, accompanied by heat generation will occur leading to a self-sustaining combustion with an exponential increase in the rate of heat production. Explosives exhibit different types of sensitivity, with **impact sensitivity** being a critical measure of how easily an explosive can be initiated by an impact. The initiation mechanisms differ depending on whether the explosive is solid or liquid. For liquid explosives, such as nitroglycerine, adiabatic compression and heating of trapped gases play a key role. This process is governed by the relationship:

$$T_2 = T_1 \left[\frac{P_1}{P_2} \right]^{\frac{\gamma-1}{\gamma}} \tag{1.1}$$

where T_1 and T_2 are the initial and final temperatures of the trapped gas, P_1 and P_2 are the initial and final pressures inside the bubble, and γ is the ratio of specific heat capacities (C_p/C_v) where C_v is the heat capacity at constant volume and C_p is at constant pressure. The rapid compression generates hotspots, initiating the exothermic decomposition of explosive vapor and leading to the evaporation of liquid explosive from the bubble walls. For solid explosives, hotspots are formed through the heating and compression of trapped gases, as well as friction between crystal particles. These localized hotspots act as ignition sites, ultimately triggering the explosive reaction (Lovecy, 1970).

Explosives are ignited by friction when hotspots are formed by rubbing together explosive crystals. This **sensitivity to friction** is often quantified by measuring the effects when a pendulum of known weight scrapes across the explosive material, generating localized heat that can set off the material (Agrawal, 2010). The **sensitivity to heat or explosion temperature** is the temperature at which an explosive sample explodes, ignites, or decomposes within 5 seconds. This measurement is obtained by loosely loading a 0.02-g sample of the explosive into a No. 8 blasting cap, which is immersed briefly in a Wood's metal bath and tested at different temperatures (Headquarters, U.S. Army Materiel Command, 1971). **Sensitivity to initiation**, also referred to as gap sensitivity or initiability, describes the explosive's susceptibility to being initiated. This is quantified as the minimum detonating charge of mercury fulminate, lead azide, or tetryl, measured in grams, required to detonate the explosive sample (Headquarters, U.S. Army Materiel Command, 1971).

Sensitivity to bullet impact measures the explosive's reaction to the impact of a projectile. In the standard test, approximately 0.23 kg of the explosive is loaded in the same manner as it would be for actual use—cast, pressed, or in liquid form—within a 7.62-cm pipe nipple (5.08-cm inside diameter, 1.6-mm wall thickness) sealed at both ends with caps. The loaded assembly typically contains a small air space, which can be filled with a wax plug if desired. The setup is subjected to the impact of a 7.62-mm (.30-caliber) bullet fired perpendicularly to the pipe nipple's long axis from a distance of 27.4 m (Headquarters, U.S. Army Materiel Command, 1971). The results are categorized into four groups, each representing the proportion of tested samples that exhibited a specific type of reaction:

- *Group 1, Explosion*: The percentage of samples that detonated upon impact.
- *Group 2, Partial*: The percentage of samples that experienced an incomplete detonation.
- *Group 3, Burned*: The percentage of samples that ignited but did not explode.
- *Group 4, Unaffected*: The percentage of samples that showed no reaction.

High explosives are classified based on their sensitivity into three different categories. **Primary explosives** are highly sensitive materials capable of rapidly transitioning from burning to detonation. Due to their extreme sensitivity to shock, friction, electric sparks, and temperature, they are hazardous to handle. Their primary function is to initiate detonation, making them essential components in detonators used to trigger less-sensitive explosives. Common examples include lead azide, lead styphnate, and tetrazene. **Secondary explosives** are less sensitive than primary explosives and cannot be easily detonated by heat or shock. However, they are generally more powerful and require a shockwave from a primary explosive to be initiated. Some secondary explosives exhibit exceptional stability, allowing them to withstand ignition or even the impact of rifle bullets without detonating. Examples include TNT, tetryl, picric acid, nitrocellulose, nitroguanidine, ANFO, nitroglycerine, RDX, HMX, TATB, HNS, NTO, TNAZ, CL-20, and FOX-12. **Tertiary explosives** primarily consist of strong oxidizers, such as ammonium nitrate (AN, NH_4NO_3), ammonium perchlorate (AP, NH_4ClO_4), mononitrotoluene (($C_6H_4(CH_3)(NO_2)$)), and ammonium dinitramide ($[NH_4][N(NO_2)_2]$). These explosives are more difficult to initiate than secondary explosives and require a significant energy source for detonation presenting a large critical diameter. They exhibit high resistance to ignition by fire, impact, or friction (Agrawal, 2010). In their pure form, without a suitable fuel or sensitizer, tertiary explosives have relatively low explosive energy and are generally ineffective as standalone explosives (Table 1.1).

1.4.1.1 Melting Temperature

The melting temperature of an explosive is a critical property that states the feasibility of melt-casting, a widely used method in explosive manufacturing. Molten TNT, which melts at approximately 80.8°C, can incorporate higher-melting crystalline explosives such as RDX (204°C) to produce TNT-based castable formulations like Composition B and Cyclotol. Similarly, mixing TNT with HMX (276°C) results in octol formulations, while combining it with tetryl (129°C) produces tetrytol formulations. Due to its negative OB, TNT is often also mixed with oxidizers such as nitrates to enhance its performance. The melting temperature is therefore a key factor influencing the processing, formulation, and safe manufacturing of explosives (Cooper, 1996).

1.4.1.2 Critical Diameter

Critical diameter (D_c) or failure diameter (D_f) of an explosive is the minimum diameter at which detonation can still propagate through the explosive substance. For sustained detonation, the explosive must be either confined in a tube or have a cylindrical geometry to ensure that the wavefront propagates forward, rather than laterally, over a considerable distance.

The critical diameter is specific to each explosive and is influenced by several factors, including density, confinement, particle size, temperature of the unreacted explosive (Figure 1.17). As the diameter of the explosive column decreases, the detonation velocity also decreases due to energy losses at the column's sides (Cooper, 1996), see Figure 1.

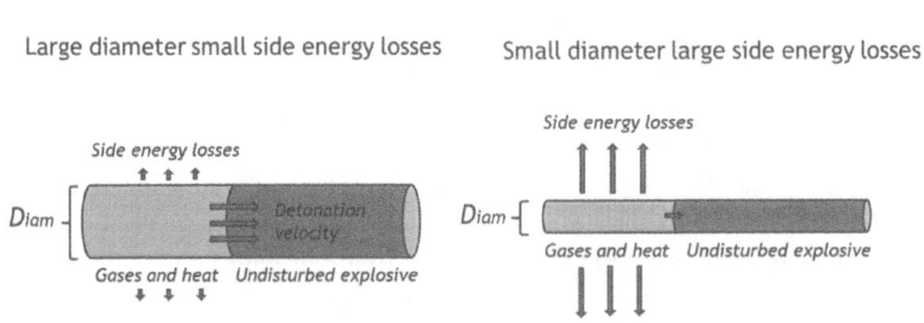

FIGURE 1.17 Influence of diameter of explosive column on the detonation velocity.

TABLE 1.1
Friction, Impact, Heat, and Bullet Impact Sensitivity of Some Primary, Secondary, and Tertiary Explosives (Headquarters, U.S. Army Materiel Command, 1971, Akhavan, 2011)

Explosive	Formula	Friction Sensitivity (N)	Impact Sensitivity (Nm)	Heat Sensitivity (°C)	Bullet Impact Sensitivity (%)	Sensitivity to Initiation (gr)
		Secondary Explosives				
Nitroglycerine	C3H5N3O9	353	0.2	222	100 explosions	N/A
Nitrocellulose (14%)	C6H7N3O11	>353	3	170	N/A	0.1 lead azide
Picric acid	C6H3N3O7	>353	7.4	300	60 partial, 40 burned	0.26 mercury fulminate 0.24 lead azide
Tetryl	C7H5N5O8	353	3	185	13 explosions, 54 partial, 10 burned, 23 unaffected	0.20 lead azide 0.25 tetryl
TNT	C7H5N3O6	>353	15	470	**N/A**	0.24 mercury fulminate 0.27 lead azide
Nitroguanidine (picrite)	CH4N4O2	>353	>49	185	100 unaffected	0.20 lead azide 0.10 tetryl
PETN	C5H8N4O12	60	3	202	100 explosions	0.17 mercury fulminate 0.03 lead azide
RDX	C3H6N6O6	120	7.5	260	100 explosions	0.19 mercury fulminate 0.05 lead azide
HMX-beta	C4H8N8O8	120	7.4	287	N/A	0.3 lead azide
TATB	C6H6N6O6	>353	50	384	N/A	0.3 lead azide
HNS	C14H6N6O12	240	50	304	N/A	N/A
		Tertiary Explosives				
Ammonium nitrate (AN)	NH4NO3	353	50	465	100 unaffected	0.20 lead azide 0.25 tetryl
Ammonium perchlorate	NH4ClO4	N/A	15	435	N/A	0.20 lead azide 0.25 tetryl
		Primary Explosives				
Mercury fulminate	HgC2N2O2	3–5	1–2	263	N/A	
Tetrazene	$C_2H_6N_{10}{\cdot}H_2O$	8	1–2	160	N/A	0.4 mercury fulminate
Lead azide	PbN6	0.1–1	2.5–4	340	N/A	
Lead styphnate	PbC6HN3O8	1.5	2.5–5	282	N/A	0.001 mercury fulminate 0.001 lead azide

1.4.1.3 Velocity of Detonation (VOD)

The VOD refers to the speed at which a detonation or shock wave propagates through an explosive material. This parameter is crucial in characterizing high explosives, as only these materials exhibit measurable VOD. The speed of detonation is directly related to the chemical reaction rate and varies significantly depending on the explosive's physical and chemical properties. Typical theoretical VOD values range from 1000 to 9500 m/s, while VOD for commercial high explosives like dynamites often exceed 2000 m/s, reaching up to 6000 m/s (Cooper, 1996) and military explosives are in a range from 6000 to 9500 m/s (Rothstein and Petersen, 1979). The VOD of an explosive is determined under conditions where the material is at its maximum density and is shaped into charge columns wider than the critical diameter (Thurman, 2017). Factors influencing detonation velocity include particle size, charge diameter, and confinement. VOD is a key factor of explosive performance, influencing fragmentation, penetration, and blast overpressure. Empirical correlations exist between an explosive's chemical composition and its VOD, providing insights into its energy release during decomposition. Several mathematical models have been developed to estimate VOD, including the Rothstein, Rothstein and Petersen, Kamlet-Jacobs, Xiong, Stein, and Keshavarz methods, each offering different predictive approaches based on thermodynamic and molecular structure analyses.

1.4.1.4 Detonation Pressure

Detonation pressure, also known as Chapman–Jouguet pressure (P_{cj}), is the peak dynamic pressure generated at the shock front of a detonation wave. It serves as a key indicator of an explosive's ability to perform work and helps classify it as either a high-brisance or low-brisance explosive. High detonation pressures are strongly associated with greater shattering (brisance) effects, making them crucial in applications such as mining, demolition, and military ordnance. The detonation pressure (P_{CJ}) in gigapascals (GPa) can be empirically calculated using the VOD (D) and the charge density (ρ). Typically, the relationship is given as:

$$P_{CJ} = \rho D^2 / 4 \tag{1.2}$$

where ρ is the charge density in g/cm³, and D is the VOD in km/s. A higher charge density and detonation velocity result in increased detonation pressure, leading to enhanced performance in practical applications (Table 1.2).

1.4.1.5 Oxygen Balance (OB)

OB represents the percentage of oxygen within an explosive, indicating whether there is sufficient oxygen for complete combustion.

When an explosive is detonated, the explosive molecule decomposes into its constituent atoms, producing various gaseous products. If complete oxidation takes place, the primary products are water (H_2O), carbon dioxide (CO_2), and nitrogen (N_2). However, if oxygen is insufficient, incomplete combustion can lead to the formation of additional products such as hydrogen (H_2), carbon monoxide (CO), and elemental carbon (C). The extent of oxidation depends on the availability of oxygen within the explosive molecule and can be quantified using the OB. It is calculated from the explosive empirical formula (*CaHbNcOd*) as the percentage of oxygen required for the complete oxidation of carbon (C) to carbon dioxide (CO_2) and hydrogen (H) to water (H_2O) (Agrawal, 2010).

$$\text{OB, \%} = \frac{\left[\boldsymbol{d} - (2\boldsymbol{a}) - (\boldsymbol{b}/2)\right] \times 1600}{\boldsymbol{MW}} \tag{1.3}$$

where ***a***, ***b***, and ***d*** are obtained from the empirical formula and ***MW*** is the molecular weight of the explosive.

TABLE 1.2
VOD, Melting Temperature, Detonation Pressure, Critical Diameter, and Theoretical Maximum Density of Some Primary, Secondary, and Tertiary Explosives (Cooper, 1996; Rothstein and Petersen, 1979)

Explosive	Formula	Mol. Weight (g/mol)	VOD (m/s)	Melting Temperature (°C)	Det. Press. (GPa)	TMD (g/cm³)	Critical Diameter (D_c) (mm)
		Secondary Explosives					
Nitroglycerine	$C_3H_5N_3O_9$	227.085	7700	13.5	23.7	1.6	0.08
Nitrocellulose (14%)	$C_6H_7N_3O_{11}$	999.405	7300	160	21.4	1.61	N/A
Picric acid	$C_6H_3N_3O_7$	229.104	7350	122.5	24.1	1.763	5.2
Tetryl	$C_7H_5N_5O_8$	287.144	7910	129	27	1.73	N/A
TNT	$C_7H_5N_3O_6$	227.13	6960	80.8	20	1.65	6–9
Nitroguanidine (picrite)	$CH_4N_4O_2$	104.07	8250	232	30	1.76	N/A
PETN	$C_5H_8N_4O_{12}$	316.135	8290	141.3	30.2	1.77	0.3
RDX	$C_3H_6N_6O_6$	222.12	8850	204	34.4	1.83	5.2
HMX	$C_4H_8N_8O_8$	296.155	9140	276	39.7	1.9	6–9
TATB	$C_6H_6N_6O_6$	258.15	7940	360	30.5	1.94	6.35
HNS-beta	$C_{14}H_6N_6O_{12}$	450.23	7120	300	22	1.74	N/A
NTO	$C_2H_2N_4O_3$	130.062	7860	270	22.2	1.44	N/A
TNAZ	$C_3H_4N_4O_6$	192.09	7300	101	24.5	1.84	N/A
CL-20	$C_6H_6N_{12}O_{12}$	438.185	9500	195	46	2.04	N/A
FOX-7	$C_2H_4N_4O_4$	148.08	8870	238	34.8	1.88	N/A
FOX-12	$C_2H_6N_4O{*}HN_3O_4$		8210	214	29.5	1.75	N/A
		Tertiary Explosives					
Ammonium nitrate (AN)	NH_4NO_3	80.043	2700	169.6	5	1.72	100
Ammonium perchlorate	NH_4ClO_4	117.49	6300	244	19.3	1.95	N/A
		Primary Explosives					
Mercury fulminate	$HgC_2N_2O_2$	284.624	4250	160	20	4.43	N/A
Tetrazene	$C_2H_6N_{10}{\cdot}H_2O$	60.06	4000	132	6.8	1.7	N/A
Silver azide	AgN_3	149.888	4000		19.2	4.8	0.025
Lead azide	PbN_6	291.24	4630	N/A	25.4	4.8	0.4–0.6
Lead styphnate	$PbC_6HN_3O_8$	450.288	5200	N/A	20.8	3.09	N/A

The **OB** quantifies whether there is sufficient oxygen for complete combustion:

- A **negative OB** occurs when there is insufficient oxygen, leading to the formation of partially oxidized products like carbon monoxide (CO) and unburnt carbon (C). This condition is associated with the release of toxic gases.
- A **positive OB** occurs when there is excess oxygen, ensuring complete oxidation of the fuel.

Understanding the OB is crucial for predicting the composition of gases released during detonation, particularly in assessing hazardous by-products like CO. Additionally, OB influences key explosive properties such as **sensitivity, detonation velocity, and explosive power**. Explosives with an OB value close to zero tend to exhibit optimal performance and higher sensitivity (Kamlet and Adolph, 1979).

1.4.1.6 Volume of Gases Produced During the Explosion

The volume of gas produced during an explosion is a key factor in determining the amount of work performed by the explosive. Gases produced expand 12,000–15,000 times the original volume at temperatures of 3000–5000°C. To enable standardized comparisons between different explosives, standard IUPAC conditions are assumed:

- *Temperature*: 273.15K (0°C)
- *Pressure*: 1 atm.

Under these conditions, one mole of gas occupies a molar volume of 22.4 dm^3, as stated by Avogadro's law. The total volume of gas V generated during detonation can be determined from the chemical decomposition equation of the explosive.

1.4.1.7 Heat of Explosion

When an explosive undergoes either burning or detonation, its chemical energy is released primarily in the form of heat. Under adiabatic conditions, this heat release is referred to as the heat of explosion (Q). For practical purposes, heat and enthalpy terms are often used interchangeably for both constant pressure and constant volume conditions. Therefore, the heat of explosion is determined by the difference between the total formation enthalpies of the explosive components and the explosion products:

$$Q = \Delta H_{detonation} = \Sigma \Delta Hf_{(products)} - \Sigma \Delta Hf_{(explosive\ components)} \tag{1.4}$$

The value of Q (in kJ/kg) is typically derived from the heat of detonation (ΔH_d) expressed in kJ/mol.

The maximum heat of explosion is observed when the OB is zero, meaning that complete oxidation occurs, leading to the most efficient energy release (Agrawal, 2010).

1.4.1.8 TNT Equivalence

TNT equivalence is a widely used normalization method for assessing the performance of an explosive relative to trinitrotoluene (TNT), which serves as the reference standard. This parameter is determined experimentally through various standardized tests, each designed to evaluate different aspects of explosive behavior. The most commonly employed methods include the plate dent test, ballistic mortar test, Trauzl test, sand crush test, and air blast test. However, these tests often yield different TNT equivalence values for the same explosive, as the measured equivalence is highly dependent on the specific test conditions and the physical properties being assessed (Cooper, 1994).

Several theoretical methods exist for calculating TNT equivalence, leading to variability in reported values, which can range by as much as 50%, with typical variations between 20 and 30%, depending on the explosive under consideration.

One of the most employed methods for determining TNT equivalence is based on the **heat of detonation**. This approach utilizes the ratio of the heat of detonation of a given explosive to that of TNT, providing a simple yet effective means of comparison. Its primary advantage lies in its straightforward implementation, making it a widely adopted metric for assessing explosive performance (Locking, 2011).

$$TNT\ Equivalence = Q\,/\,QTNT$$

TNT equivalence serves as a standardized convention for quantifying the energy released in an explosion, with trinitrotoluene (TNT) as the reference material. This metric is fundamentally based on the heat of detonation, which defines the total energy released per unit mass during detonation. The heat of detonation for TNT is 4091.61 J/g, a value closely approximating 4184 J/g (equivalent to 1 kcal/g).

This convention is particularly valuable for comparing the destructive potential of various high-energy events, including nuclear detonations and meteorite impacts, in relation to traditional explosive materials. The energy yield of nuclear weapons, for instance, is commonly expressed in kilotons or megatons of TNT equivalent, providing a standardized framework for assessing their relative impact.

Another method involves **explosive power**, which considers the ability of an explosive to perform mechanical work on its surroundings. Explosive power is directly proportional to the total volume of gases produced during detonation (V) and the thermal energy available for their expansion (Q).

$$Explosive\ \ Power = Q \times V \tag{1.5}$$

Then the TNT equivalence can be calculated by comparing the explosive power of an explosive to the explosive power of TNT.

$$TNT\ \ Equivalence = Q \times V \,/\, Q_{TNT} \times V_{TNT} \tag{1.6}$$

As a result, this approach provides a more comprehensive evaluation of an explosive's effectiveness in generating useful work (Locking, 2011) (Table 1.3).

TABLE 1.3
Heat of Explosion, Volume of Gases, Power of Explosion, Power Index, TNT Equivalence, and Oxygen Balance of Some Primary, Secondary, and Tertiary Explosives (Agrawal, 2010; Akhavan, 2011)

Explosive	Formula	Heat of Explosion (Q_v) (kJ/kg)	Volume of Gases (V) (dm³/kg)	(Power of Explosion) (kJ dm³/kg²) $Q \times V \times 10^4$	TNT Equivalence	Oxygen Balance (%)
		Secondary Explosives				
Nitroglycerine	$C_3H_5N_3O_9$	6228	715	445	1.47	3.52
EGDN	$C_2H_4N_2O_6$	6716	736	495	1.62	0
Nitrocellulose (14%)	$C_6H_7N_3O_{11}$	4313	822	354	1.17	−24.24
Picric acid	$C_6H_3N_3O_7$	3301	831	274	0.91	−43.39
Tetryl	$C_7H_5N_5O_8$	4341	820	356	1.18	−47.36
TNT	$C7H_5N_3O_6$	4092	740	303	1	−73.96
Nitroguanidine (picrite)	$CH_4N_4O_2$	2501	1076	269	0.89	−30.75
PETN	$C_5H_8N_4O_{12}$	5815	779	453	1.5	−10.12
RDX	$C_3H_6N_6O_6$	5076	908	461	1.52	−21.61
HMX	$C_4H_8N_8O_8$	5012	908	455	1.5	−21.61
TATB	$C_6H_6N_6O_6$	3498	781	273	0.9	−55.78
HNS-beta	$C_{14}H_6N_6O_{12}$	3994	746	298	0.98	−67.52
NTO	$C_2H_2N_4O_3$	2784	861	240	0.79	−24
TNAZ	$C_3H_4N_4O_6$	5891	816	481	1.59	−16.66

(Continued)

TABLE 1.3 (*Continued*)
Heat of Explosion, Volume of Gases, Power of Explosion, Power Index, TNT Equivalence, and Oxygen Balance of Some Primary, Secondary, and Tertiary Explosives (Agrawal, 2010; Akhavan, 2011)

Explosive	Formula	Heat of Explosion (Q_v) (kJ/kg)	Volume of Gases (V) (dm^3/kg)	(Power of Explosion) (kJ dm^3/kg^2) $Q \times V \times 10^4$	TNT Equivalence	Oxygen Balance (%)
CL-20	$C_6H_6N_{12}O_{12}$	5955	767	457	1.51	−10.95
FOX-7	$C_2H_4N_4O_4$	3881	908	352	1.16	−21.61
FOX-12	$C_2H_6N_4O{*}HN_3O_4$	2829	964	273	0.9	−19.13
		Tertiary Explosives				
Ammonium nitrate (AN)	NH_4NO_3	2479	980	N/A	N/A	19.99
Ammonium perchlorate	NH_4ClO_4	1972	799	N/A	N/A	34.04
		Primary Explosives				
Mercury fulminate	$HgC_2N_2O_2$	1756	215	37.77	0.12	−16.86
Lead azide	PbN6	1610	231	35.11	0.16	−5.49
Lead styphnate	$PbC_6HN_3O_8$	1853	301	55.79	0.19	−18.79

REFERENCES

Nobel Prize Outreach, 2025. *Ascanio Sobrero* [Online]. The Nobel Prize Organisation. Available: NobelPrize.org [Accessed 2025].

Agrawal, J. P. 2010. *High Energy Materials. Propellants, Explosives and Pyrotechnics*, Wiley-VCH.

Akhavan, J. 2011. *The Chemistry of Explosives*, RSC Publishing.

Brown, G. I. 1998. *The Big Bang: A History of Explosives*, Sutton Publishing.

Brown, G. I. 2010. *Explosives: History with a Bang*, The History Press Ltd.

Chase, K. 2008. *Firearms. A Global History to 1700.* Cambridge University Press.

Cooper, P. W. 1994. Comments on TNT Equivalence. *Conference: 20. International Pyrotechnics Seminar, Colorado Springs, CO (United States),25-29 Jul 1994; Other Information: PBD: [1994].* United States: Sandia National Labs., Albuquerque, NM (United States).

Cooper, P. W. 1996. *Explosives Engineering.* Wiley-VCH.

Fordham, S. 1980. Chapter 3 – Military High Explosives. *In:* Fordham, S. (ed.) *High Explosives and Propellants (Second Edition).* Pergamon.

Nobel Foundation, n.d. *The Nobel Foundation* [Online]. Available at: https://www.nobelprize.org/

Headquarters, U.S. Army Materiel Command. 1971. *Engineering Design Handbook. Explosives Series Properties of Explosives of Military Interest.* AMC Pamphlet.

Kamlet, M. J. & Adolph, H. G. 1979. The relationship of impact sensitivity with structure of organic high explosives. II. *Polynitroaromatic explosives*, 4, 30–34.

Kelly, J. 2004. *Gunpowder: Alchemy, Bombards, & Pyrotechnics: The History of the Explosive that Changed the World.* Basic Books.

Khan, I. A. 1996. Coming of Gunpowder to the Islamic World and North India: Spotlight on the Role of the Mongols. *Journal of Asian History*, 30, 27–45.

Koch, E.-C. 2019. Insensitive high explosives: IV. Nitroguanidine – Initiation & detonation. *Defence Technology*, 15, 467–487.

Leconte, J. 1862. *Instructions for the Manufacture of Saltpetre*, Evans & Cogswell.

Locking, P. 2011. The trouble with TNT equivalence. *Journal of Hazardous Materials*, 192(3), 1442–1447

Lovecy, A. L. 1970. The role of shock in the initiation of secondary explosives. *Combustion and Flame*, 15, 299–304.

Mertens, K. H. 1877. *Over nitroderivaten van dimethylaniline. Recueil des Travaux Chimiques des Pays-Bas*, 1, 109–120.
Mitchell, A. R., Coburn, M. D., Schmidt, R. D., Pagoria, P. F. & Lee, G. S. 2002. Advances in the chemical conversion of surplus energetic materials to higher value products. *Thermochimica Acta*, 384, 205–217.
Oommen, C. & Jain, S. R. 1999. Ammonium nitrate: A promising rocket propellant oxidizer. *Journal of Hazardous Materials*, 67, 253–281.
Rothstein, L. R. & Petersen, R. 1979. Predicting high explosive detonation velocities from their composition and structure. *Propellants, Explosives, Pyrotechnics*, 4, 56–60.
Shipp, K. G. 1964. Reactions of α-substituted polynitrotoluenes. I. Synthesis of 2,2′,4,4′,6,6′-hexanitrostilbene. *The Journal of Organic Chemistry*, 29, 2620–2623.
Thurman, J. T. 2017. *Practical Bomb Scene Investigation*, CRC Press.
We Bachmann, J. S. 1949. A new method of preparing the high explosive RDX1. *Journal of the American Chemical Society*, 71, 1842–1845.
Woulfe, P. 1771. Experiments to shew the nature of aurum mosaicum. *Philosophical Transactions of the Royal Society of London*, 61, 114–130.

2 Advanced Explosive Compositions

Nicholas Kanizaj

2.1 INTRODUCTION

As the saying goes, 'Prediction is very difficult, especially if it's about the future'. This is a subjective overview of the open literature regarding novel energetic materials and advanced compositions, highlighting innovative methodologies and material classes that we believe will shape the next generation of fills for explosive ordnance. It is necessarily incomplete and deliberately neglects many genuinely important innovations. For example, the use of machine learning to guide rational design of novel energetic systems, while likely to be transformative, is outside the scope of this chapter. The reader is encouraged to use this chapter as a map to begin their own explorations and, more importantly, to generate their own maps.

Achieving a robust balance between high energy output and controlled sensitivity to external stimuli (impact, friction, electrostatic discharge) is a continuous challenge. The goal is to design explosive systems that are both powerful and safe, while remaining practical and cost-effective to manufacture.

The use of heavy metals, such as lead, and the environmental impact of synthesis byproducts remain critical concerns. So-called Green Chemistry principles are increasingly guiding research to develop inherently more efficient, more environmentally benign materials and processes (Anastas and Warner, 2000). This is especially important in primary explosive manufacture, historically dominated by lead-based systems, especially lead azide and lead styphnate. Without performance or economic gains, however, it will be difficult to displace these established materials.

The drivers of modern energetic research are clear: within the parameters of economic practicality, researchers seek to enhance performance while maintaining or improving safety and environmental sustainability. The limitations of current carbon, hydrogen, oxygen, and nitrogen-containing (CHON) energetics, such as CL-20, highlight the need for new approaches that probe the periodic table.

2.2 DEFINITIONS AND PROPERTIES

Energetic materials are chemical energy storage systems capable of rapid energy release. They are kinetically metastable and thermodynamically unstable, containing both fuel and oxidiser. In contrast, reactive materials may require an external oxidiser to function. All energetic materials are reactive materials, but not all reactive materials are energetic materials. For example, diesel fuel is not considered to be an energetic material, even though upon aerobic combustion, it releases over seven times as much energy as the detonation of a military explosive such as RDX (Table 2.1).

Both energetic and reactive materials find applicability in explosive ordnance as propellants to drive projectiles, rockets, and missiles; pyrotechnics to generate thermal, spectral, or other effects and of course, as explosives to drive fragments or to deliver thermal and pressure effects.

Table 2.1 shows some typical energetic and reactive materials. Consider, for example, lead azide. This cheap, readily prepared primary explosive has a poor heat of explosion compared with secondary CHON explosives like TNT or RDX. This low performance is acceptable, since this sensitive material is used only in very small amounts to initiate the explosive train in explosive ordnance.

 DOI: 10.1201/9781003529217-2

TABLE 2.1
Selected Energetic and Reactive Materials

Material	Heat of Explosion, Gravimetric MJ/kg	Heat of Aerobic Combustion, Gravimetric MJ/kg	Heat of Aerobic Combustion, Volumetric MJ/l
Lead azide	−1.6	−2.6	−12.3
TNT	−4.6	−15.0	−24.8
RDX	−5.8	−9.4	−17.1
HMX	−5.9	−9.4	−18.0
CL-20	−6.2	−8.2	−16.8
Diesel fuel	–	−45.8	−38.2
Aluminium	–	−31.0	−83.7

The ubiquitous melt-cast aromatic nitro-explosive TNT releases over three times as much energy when burnt to completion as when it is detonated. This unburnt fuel is responsible for the sooty cloud seen upon the functioning of TNT-based formulations. The energetic nitramines also release several-fold more energy on combustion than they do during detonation. This energy difference represents lost potential performance. Driving systems towards a stoichiometric oxygen balance, however, often comes with unacceptable sensitivity profiles.

Additionally, there is a fundamental limitation to the energy content of molecular energetic materials. Their remarkable reactivities and high-power densities arise from the intimate mixing of oxidiser and fuel (on the nano-scale in molecular energetic materials). Since they contain their own oxidiser within their structures, they are *already partially oxidised*. This inherently limits their energy densities. How may this limitation be overcome?

To some extent, the energy content of explosive formulations may be enhanced by careful design incorporating the addition of fuels or oxidisers to drive non-ideal detonative processes to completion. Further, additives such as aluminium exploit both aerobic afterburn and reaction with the products of CHON explosive detonation such as water, carbon monoxide, and carbon dioxide during the detonation event (Pontalier et al., 2021).

2.3 NOVEL ENERGETIC MATERIALS

New material classes are under development to surpass the performance and safety limitations of current systems. Research into high-nitrogen content materials (Klapötke, 2022), energetic perovskites and metal–organic frameworks (MOFs) (Lai et al., 2024), and energetic ionic liquids (Bablee et al., 2025) highlights the unique advantages of particular material classes for a given application in terms of stability, energy density, and physical state. We confine our attention to solid materials that offer the ultimate in ease of synthesis: self-assembly, driven by non-covalent interactions.

Leveraging so-called reticular chemistry provides a stabilisation strategy by embedding energetic ions into MOFs or covalent organic frameworks (COFs). This allows the reduction of mechanical sensitivity, relative to simple ionic analogues, while preserving detonation performance (Lai et al., 2024).

2.3.1 Energetic Metal–Organic Frameworks (EMOFs)

MOFs have emerged as a structurally versatile class of crystalline materials, characterised by metal nodes interconnected with multidentate organic ligands. Traditionally explored for catalysis, gas storage, and sensing, they have more recently been recognised as candidate energetic materials, giving rise to energetic MOFs (EMOFs). EMOFs represent a bridge between molecular CHON energetic materials and crystalline inorganic energetic materials such as lead azide, offering both superior detonation properties and pathways towards safer, greener explosive formulations. Their tunable architectures allow integration of oxidising anions (such as nitrate, perchlorate, dinitramide) and nitrogen-rich ligands (triazoles, tetrazoles, furazans), yielding frameworks with both high energy density and controlled sensitivity, free of toxic heavy metals such as lead (Zhang and Shreeve, 2014).

The original systems that inspired EMOFs used hydrazine and related ligands to form linear (one-dimensional, 1D) coordination polymers such as nickel hydrazine perchlorate (NHP) and cobalt hydrazine perchlorate (CHP). These displayed heats of detonation comparable to CL-20 but suffered from extreme impact and flame sensitivity, rendering them impractical. Partial substitution with hydrazine carboxylate-generated 2D-layered MOFs which reduced sensitivity through structural reinforcement but at the cost of energetic output (Zhang and Shreeve, 2014).

The progression to 3D EMOFs, driven by multidentate organic ligands, marked a critical advance. In 2013 Li et al. reported various 3D EMOF systems. These frameworks were possessed of high nitrogen content (up to 53%), decomposition temperatures of 243–257°C, and impact sensitivities >20 J, significantly less sensitive than earlier hydrazine-based analogues. Importantly, functioning of their copper-based MOF reportedly exceeded the heat of detonation of CL-20 (Li et al., 2013).

A persistent challenge is reconciling high detonation performance with operational safety. Recent studies show that hydrogen bonding and solvent control can play decisive roles. For example, Chinese workers synthesised 5-(4-amino-furazan-3-yl)-1-hydroxytetrazole (AFTO)-based potassium EMOFs in hydrated and solvent-free forms. The hydrated variant ($[K(AFTO)\cdot H_2O]_n$) formed a robust hydrogen-bonded lattice with outstanding insensitivity (IS ≥ 40 J, FS ≥ 360 N), while the solvent-free framework ($[K(AFTO)]_n$) delivered superior detonation performance (detonation velocity of 8415 m/s, detonation pressure of 32.7 GPa) but slightly higher sensitivity. These results illustrate the trade-offs achievable through structural engineering of hydrogen-bonded 3D frameworks (Zhang et al., 2023).

Leveraging high sensitivity in appropriate EMOFs reveals candidates for green replacements of toxic lead azide/styphnate primaries. An illustrative example is potassium 4,8-dinitraminodifurazano[3,4-b,e]pyrazine (K-DFP). This compound exhibits detonation velocities up to 9660 m/s, rivalling or exceeding CL-20, along with high thermal stability (292°C). However, its sensitivities (impact 2 J, friction 1 N) remain in the range of primary explosives, underscoring its role as a potential high-performance lead-free initiator in advanced explosive trains (Li et al., 2018).

Several design strategies for EMOFs emerge: Multidentate nitrogen-rich backbones (triazoles, tetrazoles, furazans) improve oxygen balance and energy density, driven by the highly exothermic formation of dinitrogen upon functioning (Klapötke, 2022). Transitioning from 1D to 3D frameworks increases rigidity. It also often reduces sensitivity and enhances thermal stability, possibly by reducing the number of available 'doorway' vibrational modes leading to decomposition pathways (see, for example, Michalchuk et al., 2019). Control over intra-framework interactions offers the potential to tune the balance of sensitivity and detonation performance.

2.3.2 Energetic Perovskites

Calcium titanate occurs naturally as the mineral perovskite. It is the prototype of a class of materials of the general structure ABX_3 (Figure 2.1). In the case of calcium titanate, large sphere A represents the calcium (II) cation, small spheres B depict titanium (IV) cations, and edge units X portray oxygen anions.

Molecular halide perovskites such as methylammonium lead iodide $MAPbI_3$ (in Figure 2.1, large sphere A is methylammonium cation, small spheres B are lead (II) cations, and edge units X are

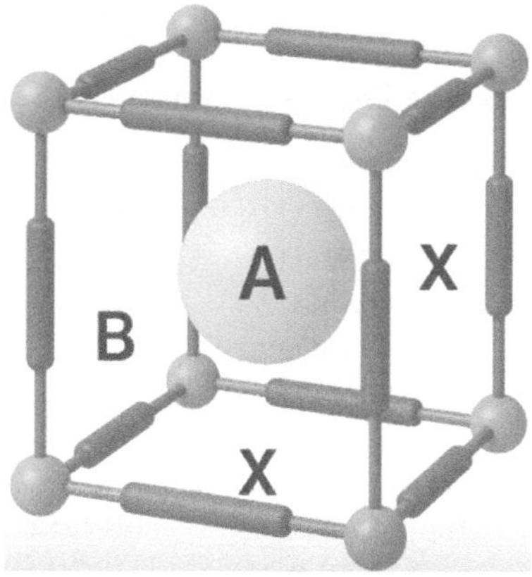

FIGURE 2.1 Idealised perovskite ABX_3 structure (author-generated schematic).

iodide anions) have revolutionised optoelectronics due to long carrier lifetimes, tunable bandgaps, and low exciton binding energies (Brenner et al., 2016; Kieslich et al., 2014).

What if we were to use oxidising anions as X species and fuel cations for B species in addition to the fuels represented by molecular cations A? We would have a dense, readily synthesised hybrid ionic-molecular structure incorporating both fuel and oxidiser. Unlike classical CHON-based explosives, perovskite energetics integrate intimately mixed yet *separate* oxidiser-rich anions and fuel-rich cations into self-assembled ABX_3 frameworks. In contrast to molecular CHON-based explosives, the molecular fuel is not already oxidised. This enables tunable properties: balancing of deflagration/detonation performance and stability (Chen et al., 2018; Gu and Xiao, 2025; Yan et al., 2025).

Energetic perovskites are hybrid organic–inorganic structures with energetic X-site oxidisers and organic amine fuel A-site cationic species, balanced by ammonium, sodium, or protonated amines at the B sites. When the B sites are also fuels (ammonia or ammonium), the energy content is maximised. Their modular self-assembly results in high densities (>2.7 g/cm^3) and favourable oxygen balances, underpinning detonation velocities and pressures comparable to nitramine explosives of military relevance but with enormously simplified synthesis.

The case of prototype material DAP-4 ($(H_2DABCO)(NH_4)(ClO_4)_3$) exemplifies this. It achieves detonation velocities of ~9 km/s and detonation pressures >30 GPa. It has a high decomposition temperature (>370°C) and activation energy, offering niche applications (Chen et al., 2018; Lai et al., 2024).

Energetic perovskites may therefore represent a transformative platform for next-generation high–energy dense materials (HEDMs). Their hybrid organic–inorganic lattice allows incorporation of oxidiser-rich X anions such as perchlorate or periodate with cationic ions derived from A fuels (e.g., 1,4-diazabicyclo[2.2.2]octane (DABCO), piperazine (Pz), ammonia) (Table 2.2).

The central design principle in energetic materials of optimising oxygen balance is well-demonstrated by reducing the amount of carbon relative to DABCO by replacement with other fuel cations. For instance, replacement with piperazine yields PAP-1 and improves oxygen balance by removing carbon atoms from the structure, while increasing the amount of oxygen in the unit cell by using the hydroxyl analogue of DABCO in DAP-O2 is a complementary strategy (Chen et al., 2018).

Already inherently tunable by formulation, as with any energetic material, energetic perovskites may obviously be formulated with other ingredients such as polymeric binders or metals. For instance, copper nanoparticles reduce DAP-4 decomposition onset by ~49°C and lower activation energy by ~80 kJ/mol (Elbasuney et al., 2024), while titanium microparticles boost combustion enthalpy (>10 kJ/g) and ignition sensitivity (Liu et al., 2025).

TABLE 2.2
Selected Perovskite Energetic Materials

Perovskite	A	B	X
DAP-1	H_2DABCO^{2+}	Na	ClO_4^-
DAP-2	H_2DABCO^{2+}	K	ClO_4^-
DAP-3	H_2DABCO^{2+}	Rb	ClO_4^-
DAP-4	H_2DABCO^{2+}	NH_4	ClO_4^-
TDPI-1	H_2DABCO^{2+}	Na	IO_4^-
TDPI-2	H_2DABCO^{2+}	K	IO_4
TDPI-4	H_2DABCO^{2+}	NH_4	IO_4^-
DPPE-1	H_2DABCO^{2+}	NH_4/Na	IO_4^-
PAP-1	H_2Pz^{2+}	Na	ClO_4^-
DAP-O2	$H_2DABCO\text{-}O^{2+}$	K	ClO_4^-

A major development in green primary explosives is represented by DPPE-1, a so-called double perovskite or $A_2BB'X_6$, where A is DABConium cation, B and B′ are the sodium and ammonium cations, respectively, and X is periodate. Like most energetic perovskites, it is prepared via one-pot aqueous synthesis; it avoids heavy metals and exhibits moderate density (2.88 g/cm³), good air/moisture/light stability, and acceptable sensitivities (impact, 3.5 J; friction, 5 N). Its ignition performance is exceptional: MPC ≤ 5 mg, rivalling lead azide while offering low-toxicity byproducts (Feng et al., 2023).

DPPE-1 demonstrates how perovskite crystallography may be leveraged for next-generation green primary explosive manufacture. It should be noted that ammonium periodate, once proposed as an oxidiser, was abandoned due to spontaneous explosive decomposition under extremely mild friction (Smith, 1951), underscoring the energetic-sensitivity tunability offered by rational perovskite design.

2.4 REACTIVE METALS AND COMPOSITES

Metal-based reactive materials (containing reactive metals and metalloids such as aluminium, boron, magnesium, or titanium) offer significant energetic advantages compared to CHON-based systems, given the high enthalpies of oxidation of these elements. Metal-based reactive materials remain central, with aluminium powders and pastes offering high reactivity for explosives, propellants, and pyrotechnics, though handling hazards and particle size effects remain critical considerations (Dreizin, 2017). Innovative applications of reactive materials span engineered micropowders, structural reactive materials, enhanced blast warheads, and operational breaching devices.

Reactive structural materials (RSMs) are now firmly established as a class of energetic composites that combine load-bearing strength with high chemical energy release. While the reader is directed to the excellent recent reviews by both Dreizin (2017) and Klapötke (2020), let us examine a thin slice of this fascinating area.

2.4.1 Reactive Metals

The metal itself, usually as an alloy, may be the reactive material. For example, pyrophoric metals such as zirconium, titanium, and lanthanide alloys exhibit much lower ignition thresholds than steel or copper, offering direct roles in reactive fragment and incendiary applications (Hillstrom, 1978). Their relatively fast aerobic reaction rates may even be sufficient for the generation of blast waves. Intimate mixing with oxidisers further increases reaction rate but often at the cost of poor sensitivity profiles.

2.4.2 Intermetallic Alloying Precursors

Reactive metal composites remain central to future energetics. Traditional aluminium, boron, magnesium, and titanium reactivity profiles are being enhanced through coatings and structural engineering. For example, leveraging the exothermic intermetallic alloying of reactive metals 'activates' metals to accelerate their combustion. Over two decades ago, Hahma showed that applying coatings of Ni, Fe, or Co to Al powders increased burn rates by up to 200× compared to untreated powders, providing scalable pathways to more reactive fuel additives (Hahma, 2005). In extension of this work, Hahma et al. (2006) demonstrated that nickel coatings catalyse aluminium nitride formation in N_2, cobalt coatings enhance ignition, and complex fluoride coatings (K_3AlF_6) suppress agglomeration and accelerate combustion in CO_2. These methods increased ignitability and burn rates several-fold compared to untreated Al.

Other workers have exploited alternative approaches to intermetallic alloying precursors. Ni-coated aluminium powders display reduced ignition delays and faster burn rates compared to bare Al, providing an avenue for engineered ignition thresholds (Dreizin, 2010).

2.4.3 (Fluorous) Thermites

The reaction of metals requires an oxidiser, such as aerobic oxygen. Using oxidisers such as less reactive metal oxides in thermites is well described by the excellent recent work by Shaw (Shaw, 2020). Fluorine is the most electronegative element but is impractical as an oxidiser in explosive ordnance. Fluoropolymers such as PTFE and PVDF as fluorine reservoirs provide superior thermal stability and chemical resistance, finding use in protective binders, insulators, and coatings relevant to energetic formulations as well as fluorous thermites in infrared countermeasures (Burgess et al., 1996; Gardiner, 2015).

Reactive metals are often coated with an inhibitory oxide layer (Altman and Pantoya, 2022). These layers tend to interfere with oxygen diffusion in the case of aerobic oxidation, limiting both reaction rate and often the extent of reaction, limiting actual performance compared with theoretical predictions. Disruption of this oxide coat is a general approach to activating reactive metals to accelerate their combustion and increase the degree of combustion. As long ago as the 1970s, gaseous hydrogen fluoride treatment of aluminium powders was shown to replace Al_2O_3 passivation layers with AlF_3, improving combustion efficiency, reducing agglomeration, and shortening ignition delays in solid propellants (Geisler et al., 1977). More readily implemented strategies to enhance metal reactivities leverage fluorinated solid materials.

Fluoropolymers in metallised composites have been shown to drastically improve ignition and burning rates and reduce agglomeration in solid propellants (Sippel et al., 2013). Computational and experimental studies of alumina surface fluorination reveal pre-ignition reactions with fluoropolymers that enhance aluminium reactivity, providing mechanistic insight into Al/PTFE combustion, and offer more general surface functionalisation strategies (Nie et al., 2020; Padhye et al., 2017).

Fluoropolymers are therefore very important as binders and oxidiser carriers. Gardiner (2015) reviewed the industrial evolution of PTFE, PVDF, fluorinated ethylene-propylene (FEP), perfluoroalkoxy alkane (PFA), Nafion and related fluoropolymers, noting their roles as binders, reactive oxidisers, and structural elastomers in explosives, propellants, and IR decoys. While PTFE is widely used, alternative melt-processable fluoropolymers such as FEP and PFA offer improved manufacturability. PFA provides higher melt strength, longer flex life, and greater thermal stability than FEP, retaining properties up to 260°C. These characteristics make PFA attractive for energetic composites where thermal resistance and durability under cyclic stress are required (International Polymer Solutions, 2021). UV-LED cured perfluoroacrylate coatings have been demonstrated with high hydrophobicity, chemical stability, and surface energy reduction (Ghazali et al., 2021). These offer potential as next-generation energetic binders suitable for the stereolithographic manufacture (SLA) of 3D-printed fluorous thermites.

The outstanding thermal stability and hydrophobicity of fluoropolymers as oxidiser reservoirs drive applications in explosive ordnance as energetic binders and as components of structural reactive materials.

2.4.4 Oxidiser Is Ubiquitous

Fuel-air explosives (FAEs) leverage the explosive dispersal of flammable materials, historically liquids, for aerobic afterburn to deliver prolonged thermal and overpressure effects. Falterman and co-workers improved classical ethylene oxide FAEs by blending with propylene oxide. This reduced vapour pressure, increased storage stability, and widened detonation limits, partially addressing the handling and reliability issues of liquid-based FAE (Falterman et al., 1979). Sullivan further demonstrated FAE canisters for experimental trials, underscoring the importance of reactive dispersal systems for testing (Sullivan, 1995).

Experimental work on droplet breakup in two-phase flows provides insight into FAE behaviour. Droplet atomisation under shock and blast loading follows regimes defined by Weber and Reynolds numbers. These processes govern evaporation rates, combustion completeness, and overpressure development in thermobaric and fuel-air systems (Gelfand, 1996).

Increased energy density is readily conferred by metallising such liquids. For example, computational simulations of Al/propylene oxide FAEs revealed optimum initiation fuel concentrations between 186 and 275 g/m³ and reliable initiation times around 250–300 ms (Ye and Zhang, 2021). Relatively simple metallised gel propellants using JP-10 and supramolecular gelators show thixotropic behaviour and stability when loaded with nano-Al, offering controllable rheology for future propulsion concepts as well as aerobic afterburn in FAE devices (Chen et al., 2017).

Entirely solid analogues of such devices are much preferred from a storage and handling perspective, as well as to maximise energy density. Predictive tools have emerged alongside experimental diagnostics. For example, structure-based models accurately estimated maximum detonation velocities of CHNOF and aluminised explosives (Keshavarz, 2012). Patented explosives suitable for high-temperature environments further demonstrate the integration of reactive metals with halogenated polymers, enabling enhanced blast effects and prolonged energy release (Lund and Braithwaite, 1995).

Experiments with shock dispersed fuel charges using aluminium, sucrose, and carbon powders revealed combustion behaviour highly dependent on confinement geometry. In small chambers, aluminium–sucrose mixtures achieved quasi-static overpressures surpassing TNT, while in larger vessels incomplete combustion limited energy release. In tunnels, combustion energy could couple back into the propagating blast, amplifying peak pressures and impulse. These findings emphasised the role of particle size, fuel selection, and geometry in performance. Thermodynamic modelling using Cheetah code has quantified aluminium combustion in shock-dispersed-fuel systems, showing near-complete combustion with heats of ~31 MJ/kg Al and validating constant-volume explosion approximations for pressure modelling (Neuwald et al., 2003; Kuhl, 2006).

By controlling the geometry of conventional dispersal charges and combining reactive metals such as aluminium with sub-stoichiometric oxidiser additives like ammonium perchlorate, dispersal of reacting clouds of fuels, primed for aerobic afterburn, is possible. Practical systems are capable of sustaining blast pressures in oxygen-limited environments and producing multi-phase flows for enhanced lethality in confined environments (Chan and Meyers, 2005; Sheridan et al., 2012). Studies on additive particle size showing finer Al/AP improve detonation behaviour (Trzciński, 2008), and thermobaric annular charges modelled with RDX-IPN-Al mixtures apparently fit Wood–Kirkwood non-ideal detonation models plus Al afterburning (Iorga et al., 2017).

In contrast to systems relying on partial confinement, open-field thermobaric effects from entirely solid systems are now able to be exploited. The geometry of practical devices is critical to their functioning. For instance, Lund and Braithwaite (1995) patented a multilayer 'jelly roll' and concentric cylinder charges where a halogenated polymer and reactive metal acceptor phase surrounds a non-metallised explosive core. Upon detonation, halogen–metal reactions sustain high temperature and extended blast pressures exceeding 200 kbar.

No less important for enhanced blast effects is the formulation. A novel polysiloxane-binder aluminised explosive (H-TBX) demonstrated catalytic enhancement of Al combustion, achieving nearly twice the open-field impulse of TNT (Kolev and Tsonev, 2020) leveraging the exothermic thermitic reaction between the metal and SiO_2 formed in-situ by polysiloxane oxidation.

2.5 ADVANCED SYNTHESIS, FORMULATION, AND PROPERTIES

Advances in flow chemistry, additive manufacturing, mechanochemistry, and spray flash evaporation provide safer and more efficient routes for scaling energetic material manufacture. For an overview of this subject, the reader is directed to the excellent recent book edited by Pantoya and co-workers (Mezger, 2017).

While additive manufacture in explosive ordnance is the subject of a standalone chapter in this book, its importance to the interplay between mesostructure and performance is illustrative. Recent advances in additive manufacturing are enabling tailored charge architectures that sustain combustion in air with or without embedded oxidisers. A notable example is the U.S. Navy's patent for additively manufactured discrete gradient charges (Castellanos et al., 2021). This approach relies on carefully layering aluminium powders of different particle sizes, bound in hydroxyl-terminated polybutadiene (HTPB) binders, and extruded by fused deposition modelling. Additive manufacturing enabled Al/PVDF structural energetic materials with tunable strength versus combustion rates, with peak burn rates near 30% Al loading (Fleck, 2021).

In addition to enabling the manufacture of complex geometries for tailored performance, additive manufacture is reshaping the manufacture of energetic composites, with extrusion, inkjet, and stereolithography enabling tailored microstructures (Muravyev et al., 2019).

Microscale formulation alters combustion behaviour. Composite Al–Mg–Zr powders produced by mechanical milling demonstrate dual-phase combustion (Al/Mg in vapour and Zr in condensed state) with microexplosions that enhance combustion efficiency up to 90%, double that of pure Al under explosive dispersal tests (Stamatis et al., 2020). Combining microscale formulation of such activated metal powders with meso-scale fabrication illustrates a transition towards safer, more powerful, and structurally integrated energetic formulations in future explosive ordnance. For instance, Dreizin's 2020 work demonstrates a scalable approach to producing spherical reactive microspheres via mechanical milling in immiscible liquid media. This process yields dense 1–100 μm microspheres of Al, B, Zr, and composite thermites (Al–CuO, Al–Fe_2O_3), significantly enhancing flowability and processability. Such microspheres potentially provide a direct feedstock for 3D-printed energetics and structural composites (Mursalat et al., 2020).

In contrast to the aggressive environment found in high-energy ball mills, resonant acoustic mixing (RAM) has emerged as a powerful alternative to ball milling for safe mechanochemical processing of sensitive materials. Michalchuk et al. (2018) show that RAM enables solvent-free, controlled co-crystal formation under reduced mechanical stress, with potential suitability for impact-sensitive explosives and propellants. This approach aligns with broader mechanochemical advances towards solvent-free, energy-efficient synthesis of energetic composites. RAM is also emerging as a solvent-free method for scalable synthesis of MOFs with energetic potential (Friščić et al., 2020).

2.6 SWITCHABLE ENERGETIC MATERIALS

By far the most exciting area in modern energetic research is the possibility of controlling deflagration and detonation rates of formulations at will. While this remains a stretch goal, there has been significant progress to date.

Electrically switchable behaviour has also been demonstrated in ammonium nitrate (AN)–epoxy composites. Zamir et al. (2021) investigated their response under elevated pressures (1–5 MPa) with applied voltages up to 250 V. A near-linear correlation between voltage and burn rate was observed, with up to a four-fold increase in combustion rate at 200 V compared to the baseline. The mechanism arises from the formation of a conductive ionic melt layer at the AN surface, enabling significant ohmic heating. Importantly, the relative influence of electrical power decreases at higher pressures, since thermal feedback from the flame dominates. Nonetheless, these findings established a pressure-dependent empirical model for predicting electrical effects on burning rate, a crucial step towards real-world rocket motor applications.

Building on earlier electrically controlled propellants (ECPs), Wang et al. (2023) reported the design of a multi-stage electrically controlled solid propellant (ECSP) using lithium perchlorate (LP) and sodium perchlorate (SP) embedded in a poly(vinyl alcohol) gel. These systems enable cyclic ignition, combustion, and extinguishment purely by adjusting applied voltage (100–200 V). Notably, the second ignition stage consistently exhibited shorter delay times and improved rate adjustability relative to the first stage due to the formation of ion concentration gradients and changes in surface chemistry during prior combustion. Combustion modelling indicated that ion transfer, electron exchange, ohmic heating, and polymer-gel flame retardancy collectively underpin the switchable behaviour.

In contrast to external circuits described above, Hahma introduced the novel concept of electrochemical combustion acceleration using short-circuited electrochemical cells embedded in the propellant matrix. When the electrolyte melts at the combustion surface, ionic conduction is established, closing the 'switch' and driving electron and ion flow between anode (magnesium) and cathode (graphite). The resulting electrochemical currents generate additional heat directly at the reaction interface, thereby accelerating combustion without the need for an external power source. This approach demonstrated burn rate increases of up to 2–3 times baseline values depending on the choice of electrolyte and conductivity enhancers (graphite flakes, ionic liquids). While the process cannot yet be switched, it represents a structurally integrated alternative to chemical catalysts for pre-programmed burn rate control (Hahma, 2024).

In a related vein, electrophoresis-based aligned nanostructure triggers represent a novel approach to wireless initiation. By dispersing and aligning carbon nanotubes or nanowires within an explosive matrix, it is possible to achieve microwave-sensitive and directionally responsive triggers, enabling efficient volumetric ignition while reducing susceptibility to accidental initiation (Imholt, 2010). Modulation of the microwave initiator offers controlled initiation of such systems to deliver tailored combustion effects.

Intriguingly, the phenomenon of ferroelectricity in energetic materials introduces the possibility of polarisation-controlled detonation rates and novel effects. For instance, the non-perovskite ferroelectric material [HDABCO]ClO_4 releases up to 6180 kJ/kg with tunable detonation velocities depending on poling state (Hu et al., 2022).

2.7 CONCLUSION

The development of advanced explosive compositions shows movement from traditional CHON-based systems towards innovative materials and formulations that balance sensitivity, performance, and safety. Emerging classes of energetic materials such as EMOFs and perovskites offer tunable architectures with the potential to replace neurotoxic lead-based primary explosives, while reactive metals and fluoropolymer composites offer tunable reactivity with increased energy densities. Advances in mechanochemistry and microstructural engineering as well as formulation methods such as additive manufacture enable the design of compositions with tailored ignition thresholds and combustion/detonation behaviours for the next generation of advanced explosive ordnance.

ACKNOWLEDGEMENT

Formatted and proof-read with the assistance of OpenAI and Google NotebookLM.

REFERENCES

Altman, I. and Pantoya, M.L. (2022) 'Comprehending metal particle combustion: A path forward', Propellants, Explosives, Pyrotechnics, 47(7), p. e202200040. https://doi.org/10.1002/prep.202200040

Anastas, P.T. and Warner, J.C. (2000) Green Chemistry: Theory and Practice, Oxford Academic, https://doi.org/10.1093/oso/9780198506980.001.0001

Bablee, A., Amarasekara, A., Gabitto, J., Bhuiyan, A. and Shamim, N. (2025) 'A comprehensive review of the thermophysical properties of energetic ionic liquids', Energies, 18(2), p. 267.

Brenner, T.M. et al. (2016) 'Hybrid organic–inorganic perovskites: Low-cost semiconductors with intriguing charge-transport properties', Nature Reviews Materials, 1, p. 15007.

Burgess, D.R., Zachariah, M.R., Tsang, W. and Westmoreland, P.R. (1996) 'Thermochemical and chemical kinetic data for fluorinated hydrocarbons', Progress in Energy and Combustion Science, 22(5), pp. 453–529. https://doi.org/10.1016/0360-1285(95)00009-7

Castellanos, J., Stamatis, D., Emery, S.B., Zamor, D.O., Gay, M.E., McDaniel, G.W. and Riggins, A.W. (2021) Process for additively manufacturing discrete gradient charges, US Patent 11,040,924. Washington, DC: USPTO.

Chan, M.L. and Meyers, G.W. (2005) Advanced Thermobaric Explosive Compositions, US Patent 6955732. Washington, DC: USPTO.

Chen, J. et al. (2017) 'Supramolecular gelled propellants for metallized systems', Fuel, 206, pp. 545–553.

Chen, S.L., Yang, Z.R., Wang, B.J., Shang, Y., Sun, L.Y., He, C.T., Zhou, H.L., Zhang, W.X. and Chen, X.M. (2018) 'Molecular perovskite high-energetic materials', Science China Materials, 61, pp. 1123–1128.

Dreizin, E.L. (2010) 'Metal-based reactive materials', Progress in Energy and Combustion Science, 36(5), pp. 527–565.

Dreizin, E.L. (2017) 'Reactive structural materials: Preparation and characterization', Advanced Engineering Materials, 20(1), p. 1700631. https://doi.org/10.1002/adem.201700631

Elbasuney, S. et al. (2024) 'Facile synthesis and decomposition kinetics of energetic molecular perovskite DAP-4 catalyzed with copper nanoparticles', Journal of Inorganic and Organometallic Polymers and Materials, 35, pp. 3633–3648.

Falterman, C.W., Bowen, J.A. and Josephson, L.H. (1979) Method for Fuel Air Explosive, US Patent 4157928. Washington, DC: USPTO.

Feng, Y. et al. (2023) 'A promising perovskite primary explosive', Nature Communications, 14, p. 7765.

Fleck, T.J. et al. (2021) 'Structural energetic properties of Al/PVDF composite materials prepared using fused filament fabrication', Propellants, Explosives, Pyrotechnics, 46, pp. 670–678.

Friščić, T., Mottillo, C. and Titi, H.M. (2020) 'Mechanochemistry for synthesis', Angewandte Chemie International Edition, 59(3), pp. 1018–1029.

Gardiner, J. (2015) 'Fluoropolymers: Origin, production, and industrial and commercial applications', Australian Journal of Chemistry, 68(1), pp. 13–22. https://doi.org/10.1071/CH14165

Geisler, R.L., Crelier, A.M. and Roberto, F.Q. (1977) Method for Improving Metal Combustion in Solid Rocket Propellants, US Patent 4017342. Washington, DC: USPTO.

Gelfand, B.E. (1996) 'Droplet breakup phenomena in flows with velocity lag', Progress in Energy and Combustion Science, 22(3), pp. 201–265. https://doi.org/10.1016/S0360-1285(96)00005-6

Ghazali, S.K., Adrus, N., Majid, R.A., Ali, F. and Jamaluddin, J. (2021) 'UV-LED as a new emerging tool for curable polyurethane acrylate hydrophobic coating', Polymers, 13, p. 487–489. https://doi.org/10.3390/polym13040487

Gu, Y. and Xiao, H. (2025) 'Energetic molecular perovskites: Design strategies and performance modulation', Crystals, 15, p. 112.

Hahma, A. (2005) Modified Metal Powder and Method of Increasing the Burn Rate and Ignitability of a Metal Powder Fuel, WO2005/121055. Geneva: WIPO.

Hahma, A. (2024) 'Electrochemical combustion acceleration', International Journal of Energetic Materials and Chemical Propulsion, 23(3), pp. 69–90. https://doi.org/10.1615/IntJEnergeticMaterialsChemProp.2024049945

Hahma, A., Gany, A. and Palovuori, K. (2006) 'Combustion of activated aluminum', Combustion and Flame, 145(3), pp. 464–480. https://doi.org/10.1016/j.combustflame.2006.01.003

Hillstrom, W.A. (1978) 'Pyrophoric metal reactions and ignition thresholds', Combustion and Flame, 31(3), pp. 309–321.

Hu, Y. et al. (2022) 'Releasing chemical energy in spatially programmed ferroelectrics', Nature Communications, 13, p. 6959.

Imholt, T.J. (2010) In-Situ and Ex-Situ Electrophoresis-Based Formation of Aligned Nanostructure Triggers for Solid Explosives, US Patent Application 20100058945. Washington, DC: USPTO.

International Polymer Solutions (2021) PFA/FEP Technical Datasheet. iPolymer.

Iorga, O.G., Orban, O., Ţigănescu, T.V., Voicu, A.E., Lupoae, M. and Iftimie, B.I. (2017) 'Experimental investigations regarding thermobaric explosive devices', MTA Review, 27(1), pp. 5–14.

Keshavarz, M.H. (2012) 'Predictive methods for detonation velocity and pressure of explosives', Journal of Hazardous Materials, 203–204, pp. 1–8.

Kieslich, G., Sun, S. and Cheetham, A.K. (2014) 'Solid-state principles applied to organic–inorganic perovskites: New tricks for an old dog', Chemical Science, 5, pp. 4712–4715.

Klapötke, T.M. (2020) 'Reactive structure materials', Engineering Science and Military Technologies, 4(2), pp. 171–177. https://doi.org/10.21608/EJMTC.2021.72401.1183

Klapötke, T.M., (2022). Chemistry of High-Energy Materials. Walter de Gruyter GmbH & Co KG.

Kolev, N.I. and Tsonev, T.T. (2020) 'Energetic blast silicates', Defence Technology, 16(4), pp. 789–797. https://doi.org/10.1016/j.dt.2020.02.002

Kuhl, A.L. (2006) Thermodynamic Model of Aluminum Combustion in SDF Explosions. 37th International Conference of ICT, Karlsruhe, Germany, June 2006.

Lai, Q., Long, Y., Yin, P., Shreeve, J.M. and Pang, S. (2024) 'Thinking outside the energetic box: Stabilizing and greening high-energy materials with reticular chemistry', Accounts of Chemical Research, 57, pp. 2790–2803.

Li, S., Wang, Y., Qi, C., Zhao, X., Zhang, J., Zhang, S. and Pang, S. (2013) '3D energetic metal–organic frameworks: Synthesis and properties of high-energy materials', Angewandte Chemie International Edition, 52, pp. 1–6.

Li, W., Wang, K., Qi, X., Jin, Y. and Zhang, Q. (2018) 'Construction of a thermally stable and highly energetic metal–organic framework as lead-free primary explosives', Crystal Growth & Design, 18, pp. 1896–1902.

Liu, Y. et al. (2025) 'Thermal decomposition and combustion performance of energetic materials/Ti systems', Korean Journal of Chemical Engineering, 42, pp. 1861–1871.

Lund, G.K. and Braithwaite, P.C. (1995) Enhanced Performance, High Reaction Temperature Explosive, US Patent 5467714. Washington, DC: USPTO.

Mezger, M.J., Tindle, K.J., Pantoya, M., Groven, L.J. and Kalyon, D.M. (eds.) (2017) Energetic Materials: Advanced Processing Technologies for Next-Generation Materials. CRC Press.

Michalchuk, A.A., Trestman, M., Rudić, S., Portius, P., Fincham, P.T., Pulham, C.R. and Morrison, C.A. (2019) 'Predicting the reactivity of energetic materials: An ab initio multi-phonon approach', Journal of Materials Chemistry A, 7(33), pp. 19539–19553.

Michalchuk, A.A.L., Hope, K.S., Kennedy, S.R., Blanco, M.V., Boldyreva, E.V. and Pulham, C.R. (2018) 'Ball-free mechanochemistry: In situ real-time monitoring of pharmaceutical co-crystal formation by resonant acoustic mixing', Chemical Communications, 54(31), pp. 4033–4036. https://doi.org/10.1039/C8CC02187B

Muravyev, N.V. et al. (2019) 'Mechanochemical routes to energetic materials', Propellants, Explosives, Pyrotechnics, 44(5), pp. 531–543.

Mursalat, M., Hastings, D.L., Schoenitz, M. and Dreizin, E.L. (2020) 'Microspheres with diverse material compositions can be prepared by mechanical milling', Advanced Engineering Materials, 22(12), p. 1901204. https://doi.org/10.1002/adem.201901204

Neuwald, P., Reichenbach, H. and Kuhl, A.L. (2003) Shock-Dispersed-Fuel Charges – Combustion in Chambers and Tunnels. 34th International ICT Conference on Energetic Materials, Karlsruhe, Germany, June 2003.

Nie, H., Pisharath, S. and Hng, H.H. (2020) 'Combustion of fluoropolymer coated Al and Al–Mg alloy powders', Combustion and Flame, 220, pp. 394–406. https://doi.org/10.1016/j.combustflame.2020.07.016

Padhye, R., Nie, H. and Hng, H.H. (2017) 'Surface fluorination effects on aluminium combustion', Combustion and Flame, 181, pp. 265–273.

Pontalier, Q., Loiseau, J., Goroshin, S., Zhang, F. and Frost, D.L. (2021) 'Blast enhancement from metalized explosives', Shock Waves, 31(3), pp. 203–230. https://doi.org/10.1007/s00193-021-00994-z

Shaw, A.P.G. (2020) Thermitic Thermodynamics: A Computational Survey and Comprehensive Interpretation of Over 800 Combinations of Metals, Metalloids, and Oxides (1st ed.). CRC Press. https://doi.org/10.1201/9781351056625

Sheridan, E.W., Hugus, G.D., Bellomo, F., Martorana, D.J. and McCoy, R.A. (2012) Thermal Enhanced Blast Warhead, US Patent 8250986. Washington, DC: USPTO.

Sippel, T.R., Son, S.F. and Groven, L.J. (2013) 'Combustion of Al/PTFE mixtures', Combustion and Flame, 160(9), pp. 1894–1902.

Smith, G.F. (1951) 'Ammonium periodate is explosive danger', Chemical & Engineering News, 29(18), p. 1770.

Stamatis, D., Wainwright, E.R., Lakshman, S.V., Kessler, M.S. and Weihs, T.P. (2020) 'Combustion of explosively dispersed Al-Mg-Zr composite particles', Combustion and Flame, 215, pp. 269–280. https://doi.org/10.1016/j.combustflame.2019.12.020

Sullivan, J.D. (1995) Fuel Air Explosive Canister, US Patent H1457. Washington, DC: USPTO.

Trzciński, W.A. (2008) 'Influence of aluminium particle size on detonation', Journal of Hazardous Materials, 157(2–3), pp. 605–612.

Wang, Z., Xie, H., Xiang, S., Ouyang, K., Bao, L., Shen, R., Ye, Y. and Zhang, W. (2023) 'Multi-stage combustion characteristics of sodium perchlorate/lithium perchlorate-based electrically controlled solid propellant', Chemical Engineering Journal, 456, p. 140958. https://doi.org/10.1016/j.cej.2022.140958

Yan, T. et al. (2025) 'Synthesis and high-pressure stability study of energetic molecular perovskite DAI-X1', Crystals, 15, p. 530.

Ye, Y. and Zhang, F. (2021) 'Numerical simulation of metallized fuel–air explosives', Defence Technology, 17(6), pp. 1783–1792.

Zamir, I., Ben-Reuven, M., Gany, A. and Grinstein, D. (2021) 'Investigation of electrically controlled ammonium nitrate–epoxy solid propellant at high pressures', Propellants, Explosives, Pyrotechnics, 46(3), pp. 477–483. https://doi.org/10.1002/prep.202000208

Zhang, J., Guo, Z., Song, Y., Hao, W., Peng, R. and Jin, B. (2023) 'Hydrogen bond and 3D frameworks to reconcile the conflict between safety and detonation performance in energetic metal–organic frameworks', Chemical Engineering Journal, 453, p. 139762.

Zhang, Q. and Shreeve, J.M. (2014) 'Metal–organic frameworks as high explosives: A new concept for energetic materials', Angewandte Chemie International Edition, 53, pp. 2540–2542.

3 Detonics
Theory and Practice

Valerian Kuznetsov

3.1 CLASSICAL MODEL OF DETONATION

3.1.1 Simple Detonation Model: CJ Model

Combustion is a general term which includes deflagration and explosion of high order. These two phenomena are very different in nature. Deflagration, also known as low-order explosion, depends on external conditions, particularly pressure, and can have different velocities of propagation. Contrarily, high-order explosion does not depend on external conditions. There is a quantitative distinction between deflagration and high-order explosion: the flame velocity is always less than the velocity of sound in the unreacted material, while the velocity of high-order explosion is always higher than the velocity of sound in the unreacted explosive.

The propagation of deflagration is facilitated by the thermal conduction of the energetic material, whereas the propagation of an explosion is characterised by a shockwave.

When certain critical conditions are met, deflagration can abruptly transition into high-order explosion. Some explosions take a form of detonation.

Detonation is the highest-order explosion which propagates at a constant velocity, which is the velocity that is maximally possible for a given explosive under the current conditions. Thus, detonation is a steady-state case of explosion where the velocity of detonation is a constant and is a fundamental characteristic of the explosive. Further in this chapter, high-order explosion is referred to as simply explosion.

Explosion is an unsteady process which, in its propagation, can either transition into detonation or decay. The unsteady stage of explosion is usually observed in the areas adjacent to the initiation point. Explosion is a general term describing both unsteady propagation and detonation which have, in principle, the same mechanism.

By the middle of last century, detonation was extensively investigated for high explosives (Baum et al. 1959). High explosives generate gas with extreme rapidity and have a shattering effect which clearly has a military application. Most of the theoretical investigations of detonation were done for gas mixtures which were the simplest detonative systems. These investigations resulted in the development of the theory of detonation, the theses of which were found suitable for the description of condensed high explosives.

The phenomenon of detonation was discovered in 1881 by French scientists Marcellin Berthelot and Paul Marie Eugène Vieille (Berthelot and Vieille 1881) and Ernest-François Mallard and Henry Louis Le Chatelier (Mallard and Le Chatelier 1881) when investigating the propagation of flames in tubes. The experiments showed that the detonation propagated at a constant velocity of 3.5–4 mm/μs in the explosive gas mix, which was above the speed of sound. From this, a hydrodynamic theory of detonation was developed. It describes all parameters of a detonation wave, such as velocity, pressure, density, temperature, and energy.

According to hydrodynamic theory, detonation is the propagation of a shockwave in an explosive material. A shockwave of sufficiently high amplitude heats up the material to a temperature of several thousand degrees which initiates a strong chemical reaction releasing energy within tens of nanoseconds. Part of the released energy is projected ahead, initiating the energy-releasing reaction. Differently from deflagration, this transfer of energy is supersonic. The released energy is

DOI: 10.1201/9781003529217-3

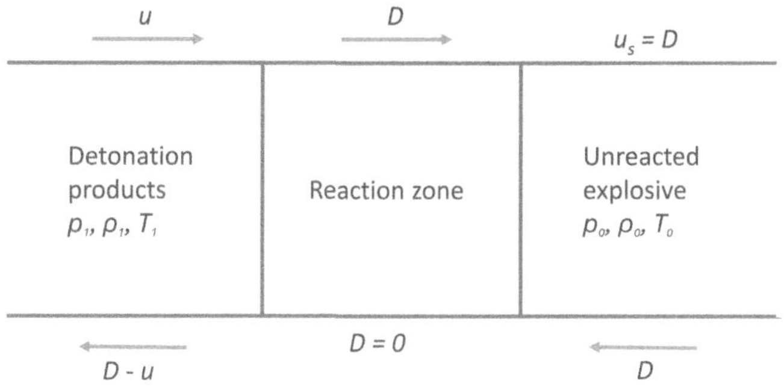

FIGURE 3.1 Structure of a detonation wave. (Based on Baum et al. (1959).)

high enough to support the propagation of the detonation wave. The wave front is where most of the energy is released, and where the matter state change occurs. The wave front has a thickness that is much smaller than the length of the explosive charge, typically tens of micrometres, and is in the direction of the wave propagation. The wave front together with the flow of the products of detonation forms the detonation wave. The detonation wave is characterised by the high pressures of 2–40 GPa in condensed explosives and 2–4 GPa for gas explosive mixtures.

Detonation is a steady process; thus, the velocity of detonation is the velocity of the shockwave in an inert matter. The propagation of a shockwave is a combination of the density jump and of the flow of the medium itself. The detonation wave possesses a more complex structure than a shockwave, and its propagation is influenced by the shockwave, the chemical reaction zone, and detonation products.

In order to describe the steady process of detonation, it is sufficient to find a correlation of the initial state of the material and of the gaseous products of reaction. The reaction zone does not need to be considered. The structure of detonation wave is shown in Figure 3.1.

A detonation wave can be described using the typical shockwave variables, such as the following:

D is the velocity of detonation which is the velocity of movement of the reaction zone;
u is the velocity of the detonation products behind the wave front;
p_0, ρ_0, T_0 are parameters of state before the chemical reaction zone;
p_1, ρ_1, T_1 are parameters of state behind the chemical reaction zone;
E_0 is the specific internal energy of unreacted explosive;
E_1 is the specific internal energy of the products of detonation;
Q is the energy of formation of the detonation products.

The relationship on both sides of the shockwave is described by the Rankine–Hugoniot conditions, and they are the following:

$$\rho_0 u_0 = \rho_1 u_1 \ (conservation\ of\ mass) \tag{3.1}$$

$$\rho_0 u_0^2 + p_0 = \rho_1 u_1^2 + p_1 \ (conservation\ of\ momentum) \tag{3.2}$$

$$E_0 + \frac{u_0^2}{2} = E_1 + \frac{u_1^2}{2} \ (conservation\ of\ energy) \tag{3.3}$$

An easy-to-understand description of the shockwave theory can be found in Cooper (1996) and in Williams (2018). A rigorous approach is presented by Zeldovich (1940).

In a coordinate system that is moving with the discontinuity, $u_0 = D$, and $u_1 = D - u$, and the Rankine–Hugoniot conditions can be rewritten as the following:

$$u = (v_0 - v)\sqrt{\frac{p_1 - p_0}{v_0 - v_1}} \tag{3.4}$$

$$D = v_0\sqrt{\frac{p_1 - p_0}{v_0 - v_1}} \tag{3.5}$$

where p is pressure and v is the specific volume $v = \frac{1}{\rho}$.

D is constant in the detonation process; thus, the latter equation can be rearranged into the linear dependence of p on v,

$$p = p_0 + \frac{(v_0 - v)}{v_0^2}D^2 \tag{3.6}$$

This is a straight line passing through the initial point (p_0, v_0) where the square of velocity D along this line is proportional to the slope. It is known as the Rayleigh line or Michelson line (Mikhelson 1893).

The equation for conservation of energy takes the form:

$$E_1 - E_0 = \frac{1}{2}(p_1 + p_0)(v_0 - v_1) + Q_v \tag{3.7}$$

and is known as the Hugoniot equation. The first term of the right-hand side is the change of the internal energy due to compression by the shockwave, and the second term is the change of energy due to reaction heat. The Hugoniot curve for detonation wave is shown in Figure 3.2.

The Hugoniot curve represents the detonation products possessing higher energy, and thus it has to be located above the Hugoniot curve of the unreacted explosive.

The detonation process is described by the region CB on the Hugoniot curve, because $D > 0$ and $u > 0$ according to Rankine–Hugoniot equations above. The region DE where $D > 0$ and $u < 0$

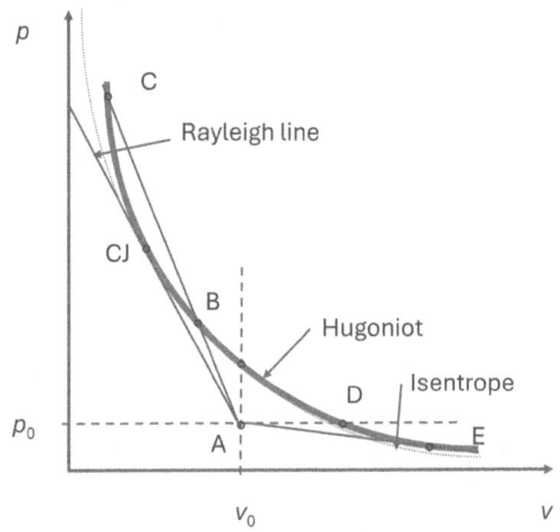

FIGURE 3.2 Hugoniot curve for a detonation wave: Chapman–Jouguet model.

corresponds to the deflagration process characterised by the reaction products moving in the direction opposite to the direction of propagation of the flame front.

The region BD does not correspond to any steady-state process, as $(p_1 - p_0) > 0$ and $(v_0 - v_1) < 0$, such that D and u are imaginary.

The Hugoniot curve can be divided into five regions (strong and weak detonations, "forbidden" zone, weak and strong deflagrations) and is described in more detail by Glassman et al. (2015).

If one draws a straight line from the point A (p_0, v_0) at an angle α, it will intersect the Hugoniot curve of the detonation products in two points. The velocity of detonation can be written as $D = v_0\sqrt{\tan\alpha}$. It appears that the same velocity of detonation can occur in two states of the reacted explosive at the wave front, which is absurd from a physical point of view.

Chapman (1899) and Jouguet (1905) proposed that the detonation process can be associated with the only state of detonation products that is indicated by the point CJ where the Rayleigh line passes through the Hugoniot adiabat at a tangent. It is obvious that tan α (and consequently the velocity of detonation) attains its minimum at the point CJ.

In a detonation wave that propagates with minimal possible velocity, the reaction zone moves with the speed of sound c relative to the reaction products. The latter moves with the particle velocity u relative to the laboratory frame of reference. Thus, the velocity of detonation is as follows:

$$D = u + c \tag{3.8}$$

Entropy of the products of detonation on the Hugoniot curve achieves its minimum in the point CJ.

It is shown in Figure 3.2 that the point CJ is where the Hugoniot curve and the isentrope intersect. While Hugoniot describes the locus of all the states which the detonation products in the shock could assume, the isentrope describes the states of the relaxed detonation products (i.e., the entropy is constant). The Rayleigh line is the common tangent to both adiabats. It is demonstrated that entropy at the front of the detonation wave increases in all the states of the detonation products beside the point CJ.

So far, the reaction zone has been neglected, assumed to have no thickness. Nevertheless, this model of detonation is sufficient for the calculation of detonation parameters.

3.1.2 A More Advanced Detonation Model: ZND Model

The description of the complete structure of detonation with propagation of shockwave in unreacted explosive was proposed by Zeldovich (1940), von Neumann (1942), and Doring (1943). The p–v diagram now incorporates the Hugoniot curve of unreacted explosive as shown in Figure 3.3.

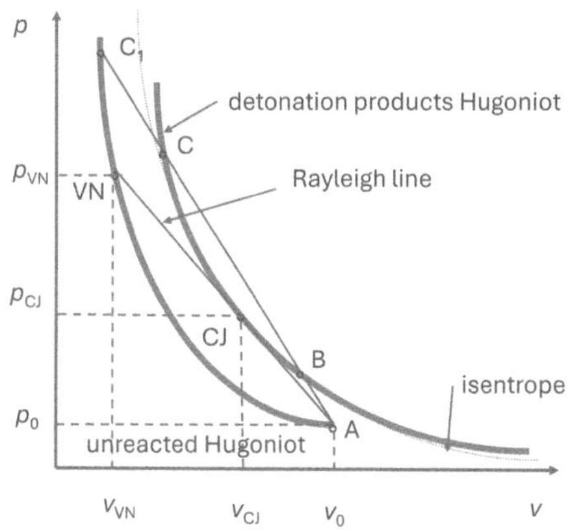

FIGURE 3.3 Formation of a steady detonation wave.

The ZND model posits that unreacted explosive is compressed at the front of the shockwave D, and the pressure reaches the maximum value termed von Neumann spike. The explosive is initiated in the reaction zone due to the rapid increase of temperature. Differently from the CJ theory, the ZND theory assumes that the chemical reaction starts only after a certain time, called induction time, and finishes over some distance called the ZND reaction zone and assumes a CJ state. For the observer at the frame of reference of the detonation front, the detonation products expand backwards from CJ plane at a velocity of $u = D - c$.

In this frame of reference, the detonation front is stationary. The detonation products in the reaction zone move at a subsonic velocity; thus, the released energy can be transported ahead and push the leading shockwave. As the velocity of the expanding products increases, it achieves the supersonic velocity relative to the detonation front, and thus at this point – CJ point – the energy of the flow cannot drive the leading shockwave anymore.

The length of the reaction zone does not depend on the geometry of the explosive charge and is often assumed to be infinitesimally thin. The shockwave and the reaction zone form a single detonation wave – a hydrodynamic discontinuity where pressure, density, and temperature jump – and a chemical reaction zone. Detonation is a steady process; thus, the von Neumann spike and reaction zone propagate through the explosive at an equal velocity. If they propagated with differing velocities, then the detonation wave would disperse and would not be steady.

The chemical reactions are complete at the CJ plane; thus, the flow behind the CJ plane expands isentropically. The expansion wave results in the decrease of pressure, density, and temperature to the ambient conditions. This expansion, or rarefaction wave, is known as Taylor–Zeldovich wave (Taylor 1950; Zeldovich and Kompaneets 1960).

This process is described graphically in Figure 3.3. The unreacted explosive (parameters p_0, v_0) is compressed by a shockwave of a velocity equal to the velocity of detonation D, to the state VN where the chemical reaction starts to occur. As the reaction takes place and the reaction heat is releasing, the matter is passing through the intermediate state between the unreacted explosive and the products of detonation. The transition from point VN to point CJ is occurring along the straight line at a constant propagation velocity and is due to the expansion of the detonation products and the release of heat. The CJ point is located on the Hugoniot curve of detonation products and is a special point where $dS = 0$ and $D_0 = u + c$. Thus, it characterises the parameters of the steady detonation wave.

The velocity of detonation wave is described by the Rankine–Hugoniot equation above which is derived only from the conservation laws of mass and of momentum and thus does not depend on the released energy Q.

This equation is applicable not only to the products of detonation but also to any intermediate state when only a part of the heat of formation Q was released. It is equally applicable to the initial state of the matter which corresponds only to shockwave compression of the initial material.

This regime can exist only if all variables of the state during the chemical reaction or after its completion are occurring along Rayleigh line $p = p_0 + \frac{D^2}{v_0^2}(v_0 - v)$. This straight line has a slope equal to $-\frac{D^2}{v_0^2}$.

If the detonation is initiated by a stronger shockwave that is faster than the velocity of detonation, then it is called "overdriven". On the unreacted Hugoniot, it is represented by the point C_1 higher than the CJ point. The transition from this state to the state where the explosive is fully consumed can occur only along the straight line AC_1. The final state of the reaction products is described by point B. However, such a detonation wave cannot be steady, as the detonation products expand with the generation of the rarefaction wave moving at a velocity of sound in these products equal to $u + c$. In other words, the chemical reactions cannot propagate faster than the velocity of sound; thus, they cannot sustain an overdriven detonation.

An overdriven detonation can be initiated only by a booster charge that has a higher velocity of detonation than the main charge. In this case, the velocity of detonation D_{CJ} is established in the charge only at some distance from the plane of initiation.

3.1.3 Calculation of Detonation Wave Parameters

A Hugoniot curve is the locus of all final equilibrium states that can be reached from an initial pressure p_0 and specific volume v_0 by a single shock process, as shown in Figure 3.3. The measurements of thermodynamic parameters, such as p, v, ρ, and E, are not required to determine a Hugoniot curve, but instead the shockwave parameters u_s and u_p could be used. For the measurement of the latter, unreacted explosive can be shock-loaded using an explosive charge or gas gun (Engelke and Sheffield 1998; Sućeska 1995). The particle velocity of explosive u_p corresponding to the loading shockwave velocity u_s can be measured, directly or indirectly. In the latter case, $u_p \approx u_{fs}/2$ where u_{fs} is free surface velocity. For various shock loadings, different final states can be achieved to produce the Hugoniot curve in the plane u_s–u_p, fitted by the expression:

$$u_s = c_0 + s u_p \tag{3.9}$$

where c_0 is the bulk speed of sound in uniaxial compression, s is the slope of Hugoniot curve, and $u_p = u_1$.

When the Rankine–Hugoniot relations are combined with the conservation laws for mass and momentum, the Hugoniot curve can be defined in the p–v plane, from the shockwave parameters,

$$p_1 - p_0 = \frac{c_0^2 (v_0 - v_1)}{\left(v_0 - s(v_0 - v_1)\right)^2} \tag{3.10}$$

The Hugoniot curves for Composition B in the u_s–u_p and p–v planes are shown in Figure 3.4.

As the detonation wave propagates through the material, it compresses the matter to a pressure known as the von Neumann pressure, $p_{VN} \approx 1.5$–$2\ p_{CJ}$. This is the high-pressure high-temperature state where the chemical reaction starts. The Hugoniot curve of the unreacted explosive and the Rayleigh line intersect at the VN point, or VN spike as shown in Figure 3.3.

The currently thought diagram of zones in a detonating explosive is presented in Figure 3.5.

The detonation front shockwave corresponds to the von Neumann spike in the ZND theory. The von Neumann spike is associated with the early ignition phase of the detonation process. There are a number of theories of how the chemical reactions occur, which determines the length of the reaction zone and how the excited constituents approach the equilibrium state CJ (Tarver 2005).

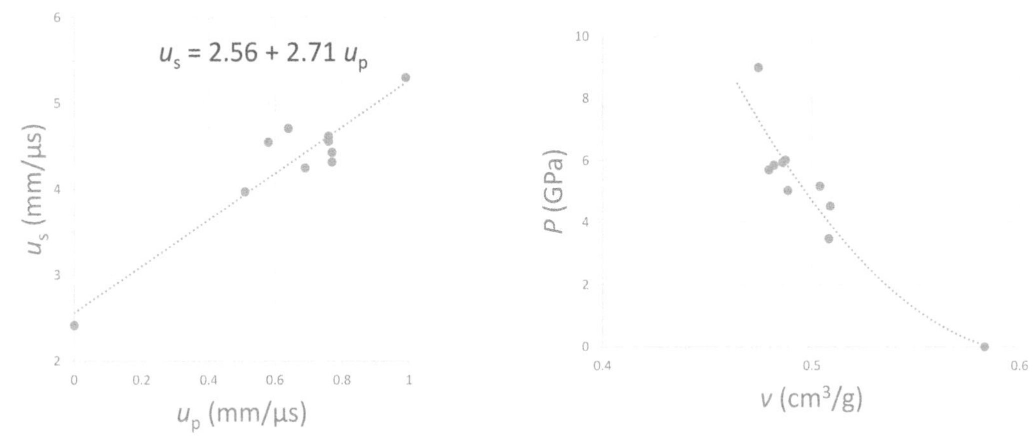

FIGURE 3.4 Plot of Hugoniot data from Marsh (1980) for Composition B in the u_s–u_p plane (left) and p–v plane (right).

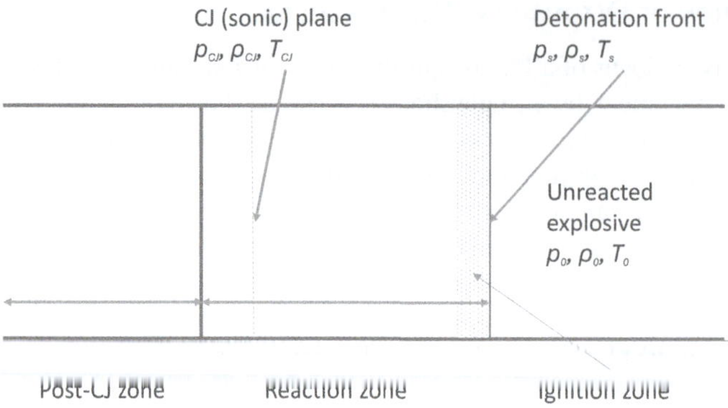

FIGURE 3.5 Structure of a detonation wave, including CJ (sonic) plane.

The chemical reaction may continue past the CJ (or sonic) plane, but the reaction products move at subsonic velocity and do not participate in driving the detonation wave.

The von Neumann spike is described theoretically and has been measured in a range of explosives. Duff and Houston (1955) reported $p_{VN} = 38.6$ GPa and $p_{CJ} = 27.2$ GPa. The authors found that the von Neumann spike disappeared after travelling 1.25 mm into aluminium plate. Tarver (2005) mentioned that a von Neumann pressure of 20 GPa in nitromethane lasted for 1–3 ns and had a CJ pressure of 13 GPa. The measurements of the von Neumann spike can be done directly using nanosecond-resolved embedded gauges, and indirectly using interferometric techniques via free surface velocity. The most popular interferometric technique is heterodyne velocimetry, also known as photon Doppler velocimetry (PDV), due to its simplicity and lower cost (Kuznetsov and Bennetts 2013 and references therein). A useful comparison of methods for the determination of detonation pressure, based on the measurement of the velocity of the interface between an explosive and flyer plate by the PDV and closing electrical contact, is reported by Pachman et al. (2018).

As shown in Figure 3.3, the pressure in the reaction zone relaxes to the CJ state which is located on the Hugoniot curve of the reaction products, also known as the Crussard curve. The Rayleigh line connects the von Neumann state and the initial state (p_0, v_0) and tangentially passes the Crussard curve in the CJ point. After the detonation products relax to the equilibrium CJ state, they expand by following the isentrope.

The Crussard curve is determined from the cylinder expansion test (Kuznetsov 2020 and references therein). The test typically involves a 1″- or 2″-diameter copper tube filled with the explosive to be tested and is initiated from one end. The acceleration of the wall expansion is measured and is then fitted to the Jones-Wilkins-Lee equation of state (EoS). The procedure is well-documented by Anastacio et al. (2019). A recent trend involves miniaturising the tube (down to a ½″ diameter), and combining it with a flyer plate test to measure detonation pressures (Maines et al. 2018).

Landau and Stanyukovich (see Baum et al. 1959) derived the equations to calculate the parameters of the detonation wave at the end of the reaction zone utilising the equation of state of strongly compressed products of detonation where the forces of repulsion between gas molecules become dominant and obtained the velocity of detonation:

$$D_{CJ} = \sqrt{\left(2\left(\gamma^2 - 1\right)Q_v\right)} \tag{3.11}$$

where Q_v is the heat of detonation at constant volume. It appears that the parameters of the CJ state do not depend on the details of the chemical reaction but on the released energy.

The index γ for Q_v = 1 cal/g was calculated and is shown below:

ρ_0 (g/cm^3)	0	0.1	0.25	0.32	0.5	0.75	1	1.25	1.7	2.25
γ	1.25	1.30	1.62	1.78	2.22	2.80	3.05	3.21	3.40	3.60

The density of condensed explosives is about 1.6 g/cm^3; thus, the value γ = 3 is often used. Pressure from detonation wave equations at the CJ plane can be written as follows:

$$p_{CJ} = \frac{\rho_0 D^2}{\gamma + 1} = \frac{\rho_0 D^2}{4} \quad for\ \gamma = 3 \tag{3.12}$$

Here, γ is the adiabatic index for a thermally perfect gas and is dependent on temperature. A calorically perfect gas is described by the adiabatic equation:

$$pv^{\gamma} = Constant \tag{3.13}$$

Here, γ is the ratio of the specific heats c_p/c_v and is constant.

3.2 INITIATION OF DETONATION

3.2.1 Initiation by a Shockwave from a Detonator

Initiation of a detonation in an explosive charge is done practically by applying a thermal or mechanical impulse, or by an immediate action of the detonator. The process of induction of detonation depends on the parameters of the initiating impulse (pressure, duration), and on the properties of the explosive to be initiated. For a successful initiation, the supplied shock energy has to concentrate within a small volume during a short duration, so that the rate of energy supply is much higher than the rate of energy dissipation due to heat conduction. The form of energy is irrelevant, as long as it exceeds the threshold value. This is known as the equivalency theorem (Koch 2021). The transition to explosion completes typically within hundreds of microseconds. If the initiating impulse was not strong enough, then initiation will not occur resulting in a localised flash or in explosion decay.

Initiation of an explosion under the influence of an initiator was first discovered by Nobel in 1864. He observed that the explosion of nitroglycerin and dynamite could be induced more easily and more completely using a blasting cap made with mercury fulminate. More potent initiators, now known as primary explosives, such as lead azide, have also been developed since. In order to initiate the less sensitive high explosives, now known as secondary explosives, boosters are used. Boosters are made from the more sensitive and powerful high explosives such as PETN and RDX. These boosters are in turn initiated by a detonator, made from the previously mentioned primary explosives. This triggering sequence, detonator to booster to secondary explosive, is known as an explosive train.

The mechanism of shockwave initiation of detonation is as follows (Baum et al. 1959). Suppose the density and compressibility of the booster and main charge are the same, so the shockwave would not change its velocity when crossing the boundary between the booster and main charge. If the velocity of the initiating shockwave is less or equal to the velocity of sound in the main charge, then the shock initiation would not occur.

Consider the attempt to initiate detonation in TNT charge of a density of 1.6 g/cm^3 by an amatol booster (90%/10% AN/TNT). If the booster diameter is close to critical (the smallest possible diameter that supports detonation), its velocity of detonation is 1600 m/s. The amatol booster will not be able to induce explosion by shock in TNT, as the sound velocity in TNT at a density of 1.6 g/cm^3 is c_0 = 1900 m/s (Baum et al. 1959).

With that said, in the example above, the detonation of the TNT can still occur because of the hot products of detonation of the booster when the deflagration transitions into detonation under favourable conditions.

If the velocity of shockwave in the main charge is below a certain critical value, then the pressure and duration of the shockwave will be insufficient to develop a self-sustaining reaction, and the detonation will not occur. This case will be accompanied by dispersion of pieces of unreacted explosive.

If the favourable conditions exist (sufficient peak overpressure of the initiating shockwave, duration of the peak, length, diameter of the charge, cladding), then the ignition of the high explosive with the transition of deflagration to detonation is possible.

Shockwave initiation of detonation in homogeneous explosives occurs through adiabatic thermal explosion (Campbell et al. 1961; Kanel et al. 2004). A shockwave from the booster transitions into the main charge, propagates through the explosive, and heats it up, so that a "slow" decomposition reaction is occurring. Heat builds up in the adiabatic reaction and accelerates the rate of decomposition. Thermal explosion develops at the front of the main charge where the compressed state occurs for the longest time. Upon conclusion of the induction time, a detonation wave is formed in the compressed material and catches up with the initiating shockwave as shown in *t*–*x* diagram (Figure 3.6). Overdriven detonation cannot be sustained, as the wave front is supersonic regarding the detonation products and slows down to D_{CJ}. For an unconfined explosive, the distance for the detonation wave to arrive from the initiation plane to the steady state is estimated to be four times the charge diameter (Miyake et al. 2023).

The size of the compressed-state initiation zone at the front of the main charge depends on the length and diameter of the booster. The larger the booster diameter, the more the contact area with the main charge is, and therefore, the larger the initiation zone is. The length of initiation zone in the beginning of the main charge is at maximum when the diameter of the booster is equal to the diameter of the main charge.

Increasing the length of the booster, too, is conducive to the increase of the initiation zone inside the main charge. The layer of the products of booster detonation prevents the rarefaction wave from propagation into the initiation zone. The thicker the layer is, the later the dissipation of the main charge products of reaction starts. The zone size reaches its maximum when the detonator diameter is about twice its length (Baum et al. 1959). For example, with the Tetryl booster of a diameter of 24.5 mm and of a density of 1.6 g/cm^3, the length of the zone (termed by the author "effective reaction zone length") was 55 mm for a 10 mm-long booster and asymptotically approached 80 mm for a 50 mm-long booster.

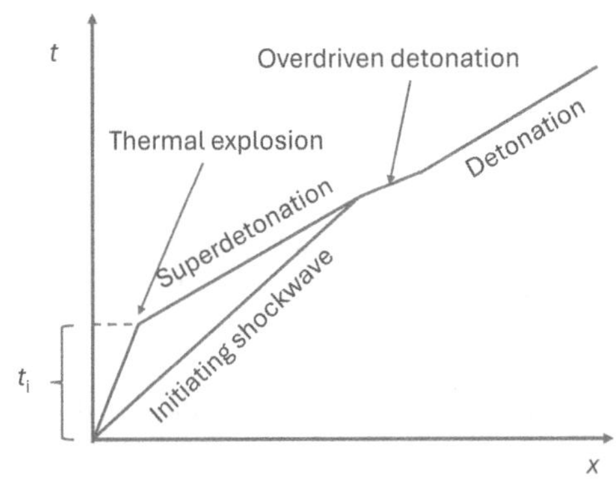

FIGURE 3.6 Time–distance diagram of shockwave initiation of detonation wave. (Adapted from Campbell et al. 1961.) The time t_i is the induction time of thermal explosion.

The characteristics of the main charge which may promote the initiation of detonation:

1. Presence of defects like pores, cracks, inclusions, grain boundaries, and crystallographic defects, also decrease in size of the explosive particles. The smaller the size of grain is, the more "hot spots" or activation centres appear per volume; thus, the faster the chemical reaction completes, and, correspondingly, the shorter the reaction zone is. This could lead to an increase in the explosive sensitivity to initiation.
2. Immersion of the booster into the main charge. This results in an increase of initiation zone and decreases the possibility for the rarefaction waves to reach the initiation zone. This happens because the booster is moved from the surface into the depth of the main charge.
3. Increase of the charge diameter, and placing it into a casing. This limits the propagation of lateral rarefaction waves into the reaction zone and increases the susceptibility of the main charge to detonation.

3.2.2 Shockwave Initiation Criteria

Detonation is often transferred from the detonating element (called donor) to the explosive charge (called acceptor) by the impact of the flyer plate. The donor charge accelerates the metal flyer plate which impacts the acceptor charge, and under the correct conditions, shock initiation occurs. The shockwave energy fluence $\frac{p^2\tau}{\rho_0 U}$ is used as an initiation criterion (Kennedy 1998). Here p is the shockwave peak overpressure driven into explosive, τ is the duration of the positive phase of the peak overpressure, ρ_0 is the density of explosive, and U is the velocity of shockwave. This initiation criterion was later reduced to $p^2\tau$. The criterion is useful for the explosive ordnance (EO) designer to devise a reliably functioning interface.

Consider the fuse FMU-139 initiating the PBXN-109 fill of the bomb BLU-129. This problem was solved numerically to investigate the effect of the gap length on initiation reliability (Lu and Ashcroft 2017). A simple Gurney analysis would suffice (Kennedy 1998). A tamped PBXN-7 booster of a length of 32 mm drives an aluminium flyer plate of a length of 2 mm over a gap of 3 mm. The shockwave energy fluence $p^2\tau$ is calculated to be about 466 $GPa^2 \cdot \mu s$. The initiating shockwave fluence of PBXN-109 has been reported to be about 8 $GPa^2 \cdot \mu s$ (Lu et al. 2016). In other words, the shockwave energy fluence from the fuse is 58 times the amount required for initiation, which is a safe fusing design approach. The large-gap sensitivity test (LGST) has shown that the replacement of the RDX-II with RDX-I (Australia-made reduced-sensitivity RDX, or RS-RDX) increased the initiation pressure from 2.5 to 5 GPa (Lochert et al. 2006). This means that the shockwave energy fluence, initiating the RS-RDX version of PBXN-109, named ARX-2014, is $(5\ GPa/2.5\ GPa)^2 = 4$ times that of the PBXN-109. Thus, one would expect that the fuse provides 58/4 = 15 times more shockwave fluence than what is needed for initiation of ARX-2014.

3.2.3 Sympathetic Detonation

A detonated explosive charge can be initiated by the detonation of another charge located at some distance away. This phenomenon was discovered in the mid-19th century and was called sympathetic detonation. The initiating explosive charge is called the donor, and the charge where the detonation is initiated is called the acceptor.

The experimental and theoretical investigation of this phenomenon is important, as the outcomes of these studies allow to determine safe distances for the storage and manufacture of explosive ordnance and can be used when constructing chains of explosive charges.

The distance of sympathetic reaction depends on many parameters: mass, density, detonation velocity of the donor, physical and chemical properties of the acceptor, casing, transmission medium, orientation of the detonating donor, and surface area of the acceptor.

Generally, the capability of an acceptor to be detonated is determined by its sensitivity, and the detonative capability of a donor is determined by the velocity of detonation, mass, and construction.

The energy of the donor can be transmitted through air by three ways: (i) by a shockwave, (ii) flow of detonation products, and (iii) thrown fragments.

Experiments showed that initiation of detonation in high-brisance high explosive (HE) without explicit deflagration is possible only if the initiating shockwave pressure is very high, at least 20–30 MPa. If the incident shockwave has a lower peak overpressure, then the acceptor first experiences a period of deflagration and then detonates. In both cases, the mechanism of detonation initiation is thermal. This explains how different parameters affect the distance of sympathetic detonation transmission. For example, with the decrease in density (increase of porosity) of the acceptor, the distance of sympathetic detonation transmission increases. The reason is that the thermal conductivity of acceptor HE decreases; thus, the heat from the reflected shockwave concentrates in a thinner area, which brings a higher increase of temperature. In addition, in a porous charge, due to breakthrough of the heated air and of the hot deflagrating products of the surface layers of the charge into the pores, the burn rate sharply increases, which facilitates the transition of deflagration into detonation. And finally, when a porous charge with low mechanical strength is subjected to a shockwave, almost adiabatic compression of air pores occurs. The pores act as "hot spots" promoting faster initiation. It was shown that heating up the "hot spots" by shock to 450–600°C leads to detonation (Baum et al. 1959). To achieve such a heating, the incident shockwave needs to have a peak overpressure of 0.8–1.5 MPa. If shockwave reflection did not happen, then the incident pressure needs to be higher, at least 3–5 MPa. The phlegmatised high-brisance explosives are less sensitive to the hot spot initiation, as the explosive particles are coated with the materials that can melt and absorb heat.

An evaluation of a sympathetic detonation test performed on the Australian "reduced-sensitivity" RDX in PBXN-109 was conducted by Lochert et al. (2006) and Lu et al. (2016). They categorised the response of the explosive fills to hazardous stimuli by the severity of the reaction according to the US Department of Defense standard MIL-STD-2105. The classifications, from least to most severe, were burning, deflagration, explosion, partial detonation, and detonation.

If the donor has a metal casing, then the transfer of energy to the acceptor also occurs by the driven fragments. A simple method to predict the conditions for sympathetic detonation of cased munitions is given by Victor (1996).

3.2.4 Shock-to-Detonation Transition

Majowicz and Jacobs (1958) and Campbell et al. (1961) first used the wedge test to study shock initiation of solid explosives. The test is named after the wedge-shaped explosive sample the base of which is exposed to the plane shocks of varied pressure. As the shock accelerates through the sample due to energy release and develops into a detonation wave, it emerges at the slanted face of the wedge and destroys the reflectivity of the aluminised coating. The light is reflected off the intact coating, and the position history of the shock is recorded by a streak camera. The streak image is recorded in the "time-distance" coordinates where the run-to-detonation part of the curve has a parabolic shape, and then the curve continues as a line representing the steady detonation. The outcome of the wedge test is a dependency between the shock compression pressure and the length of the pre-detonation portion of the wave. The longer this run up-to-detonation is, the less the explosive is sensitive to shock initiation. The schematic of the wedge test is shown in Figure 3.7.

The shock was originally produced by an explosive plane wave generator and attenuator. It was discovered, though, that the peak overpressure was quickly reduced by the rarefaction wave unloading when propagating through the explosive, and thus the initiating pressure was not well defined. The experimenters promptly switched to using a gas gun to create a long loading of constant pressure at the explosive wedge.

After the wedge test is conducted for several impact pressures, a graph of dependency of run-to-detonation distance on initiating pressures was produced under the name "Pop-plot" (Gibbs and

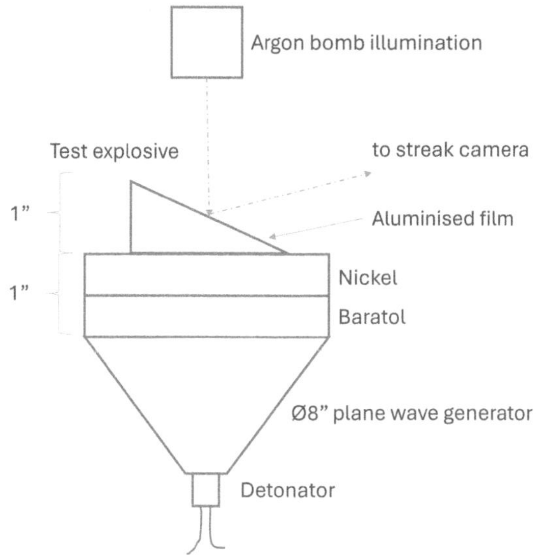

FIGURE 3.7 Arrangement of the original "wedge test" to characterise the sensitivity of explosive to shock initiation. (Adapted from Campbell et al. (1961).)

Popolato 1980) and fitted by a power function. For example, distance = 2100 * $p^{-1.34}$ for a pressure range $40 < p$ (kbar) <100.

The wedge test was conducted at DSTO utilising a 3-in. diameter explosive plane wave generator, and the results were used for the Ignition and Growth reactive flow model for modelling shock initiation (Lu et al. 2014).

3.2.5 Deflagration-to-Detonation Transition

A laminar deflagration process in an explosive can experience oscillation of pressure which would change to convective or pulsating deflagration (Koch 2021 and references therein). If deflagration pressure is increased further, this could lead to detonation. This is what is called deflagration-to-detonation transition (DDT). The transition could result in low-velocity detonation or high-velocity detonation. The development of DDT in an explosive is shown in Figure 3.8.

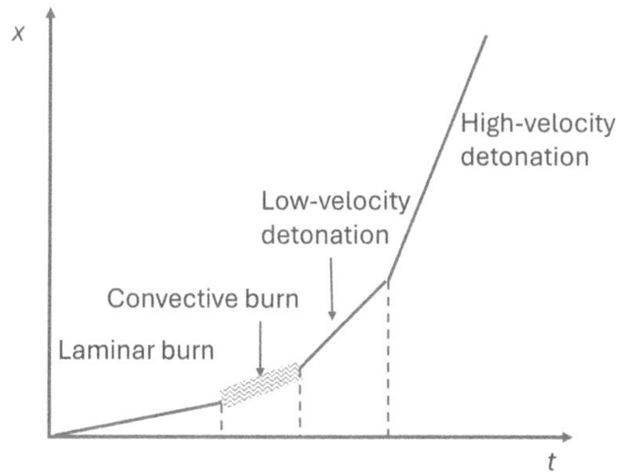

FIGURE 3.8 Distance–time diagram of deflagration-to-detonation transition (Based on Koch (2021).)

In primary explosives and oxygen-balanced HE, the deflagration wave can accelerate and turn into a detonation wave abruptly. In porous energetic materials, the deflagration can propagate by a convection mechanism. As the pressure at the deflagration front increases, it leads to an increase in burning rate, which leads to a further increase of pressure, and deflagration can transition to low-velocity detonation.

3.3 PARAMETERS OF STABLE DETONATION

3.3.1 Definition of Non-Ideal Detonation

It turns out that the one-dimensional classical model of detonation cannot completely describe the observed behaviour of explosives. The ideal velocity of detonation is independent of the reaction zone structure, and the charge diameter and confinement provide no effect on detonation properties. In reality, the ideal detonation model can provide the velocity of detonation and CJ pressure only for charges with very large diameters. In some explosives, the chemical reaction still continues after the CJ point, and this energy does not participate in driving the detonation front as was described in Section 3.1.3. However, the expanding reaction products could do the work of driving fragments or create volumetric explosive effects. This is what describes non-ideal detonation. This is where the CJ (sonic) plane does not coincide with the end of the reaction zone (Miyake et al. 2023).

Minchinton (2015) states that the fundamental and defining property of a steady non-ideal 2D detonation of a cylindrical charge is the curved detonation front, as is compared to the plane wave front of an ideal detonation. This implies that there is a pressure drop between the centre of the charge and its edge. Figure 3.9 shows the features of 2D reactive flows in non-ideal detonation.

Mader (1979) has a practical approach to separate non-ideal and ideal explosives. He defines a non-ideal explosive when there are important differences between the experimental values such as detonation pressure and velocity and the values predicted by the Baker–Kistiakowsky–Wilson (BKW) EoS. The criterion is a pressure difference of 5 GPa and a velocity difference of 0.5 mm/μs.

3.3.2 Critical Diameter and Confinement

It follows from the theory of detonation that the velocity of detonation depends on the amount of chemical energy in an explosive, known as the heat of detonation. The velocity of detonation depends on the physical properties of the explosive charge, most importantly charge diameter,

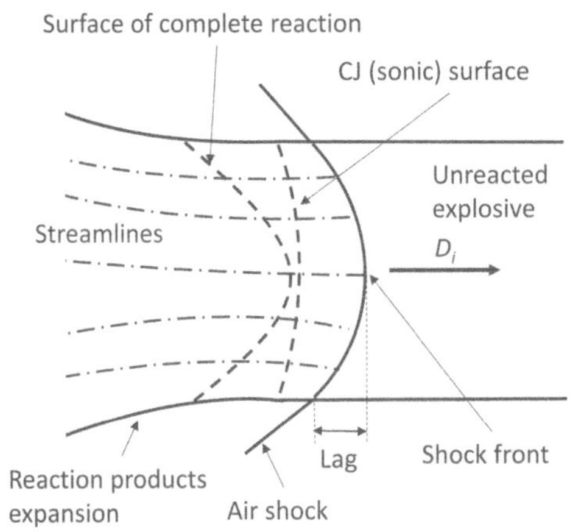

FIGURE 3.9 Phenomenology of reactive flows. It shows streamlines, detonation wave front curvature, detonation driving zone, end of reaction zone, and explosive–air interface.

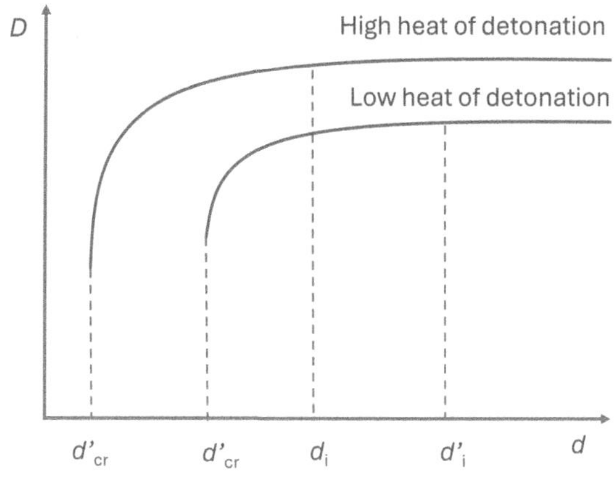

FIGURE 3.10 Dependency of the velocity of detonation on the charge diameter. (Based on Baum et al. 1959.) The top and bottom curves correspond to the HE charges with high and low heats of detonation, respectively.

density, state of matter, particle size, and casing (Baum et al. 1959). The EO designer needs to know the velocity of detonation and if it is stable. In many cases this task is reduced to the determination of the charge's critical diameter.

Investigations of the dependency of detonation velocity on charge diameter have showed that the velocity of detonation increases as the charge diameter increases. Starting with a certain diameter, d_i, the velocity of detonation does not increase anymore, as is shown in Figure 3.10. The diameter d_i corresponds to the ideal detonation velocity, or infinite diameter detonation velocity.

It has been reported that d_i is 30 mm for cast TNT and is 10 mm for pressed TNT at 1.6 g/cm^3. The "infinite" diameter is 3–4 mm for RDX at a density of 1.6 g/cm^3 (Baum et al. 1959).

If the charge diameter is decreased below d_i, the velocity of detonation decreases. At the certain diameter d_{cr} the detonation becomes unstable and decays, as shown in Figure 3.10. It is argued by Miyake et al. (2023) that ideal detonation occurs at the charge diameters $d > d_i$, and non-ideal detonation occurs at $d_{cr} < d < d_i$.

It is difficult to measure the critical diameter precisely, and the reported values can differ significantly. For example, the measured critical diameter of TNT is 2 mm (Parfenov et al. 1969) and 8–10 mm (Kubota 2007). They reported critical diameters for some common explosives: 1–1.5 mm for PETN, 0.5 mm or 1–1.5 mm for RDX, 100 mm for ammonium nitrate.

The only approach in the classical plane wave model of detonation was proposed by Khariton (1947) and reviewed within the modern framework by Leiber (1994). High pressure at the detonation wave front causes the dispersal of the detonation products from the sides of the explosive charge. The rarefaction waves appearing at the surface will enter the reaction zone and may cause the decrease of the detonation velocity due to incompleteness of the chemical reaction and reduced availability of energy to push the detonation wave.

The severity of the effect will depend on the relationship between the length of the reaction zone and the charge diameter.

Suppose t_1 is the reaction time, and t_2 is the time of propagation of the rarefaction wave, and $d/2$ is the distance to the centre of the charge. Then,

$$t_1 = \frac{l}{D}, \; and \; t_2 = \frac{d}{2c} \tag{3.14}$$

where l is the length of the reaction zone, d is the charge diameter, and c is the speed of sound in the detonation products.

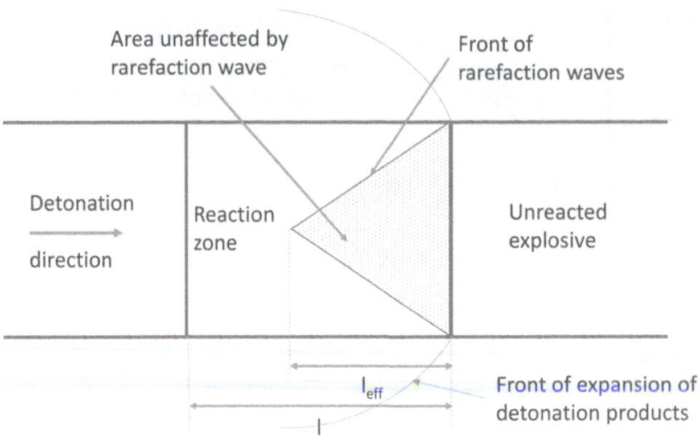

FIGURE 3.11 Detonation of a cylindrical charge. (Based on Baum et al. (1959).)

If $t_1 > t_2$, then the rarefaction wave will enter the zone of detonation products, and thus the pressure at the detonation front will drop below p_{CJ}. In other words, the detonation products will start to intensively expand before the chemical reaction completes. This is why only a part of the reaction energy will be used for driving the wave, and its velocity will be below D_{CJ}.

The position of the rarefaction wave is shown in Figure 3.11.

The velocity of detonation will depend on the ratio of the length of the zone unaffected by the rarefaction wave, l_{eff}, and of the reaction zone length l. The critical diameter of the charge, when the detonation velocity is still equal D_{CJ}, can be determined from the condition $t_1 = t_2$; thus $d_{\mathrm{cr}} = l$, considering that $c = D_{\mathrm{CJ}}/2$ (Baum et al. 1959).

In other words, the critical diameter is approximately equal to the length of the reaction zone.

The detonation theory of charges of a diameter less than the critical diameter was extensively considered by Eyring et al. (1949). The detonation velocity D, at charge diameter d, is related to the ideal detonation velocity D_i and the reaction zone length, A, by the equation:

$$D(d,\rho_0) = D_i(\rho_0) \times \left(1 - \frac{AR(\rho_0)}{d}\right) \tag{3.15}$$

This equation indicates that the detonation velocity decreases linearly with $1/d$. This formula is often used to extrapolate the critical diameter, d_{cr}, if detonation velocities are measured for a range of diameters. The theory is in good agreement with experimental outcomes; however, it lacks consideration for the changing of reaction zone length for diameters less than critical, which impacts reliability for small charge diameters.

Souers (1997) proposed a simple theory relating the decrease of the detonation velocity with the increase of the sonic reaction zone length. The size effect is described by the following:

$$\frac{u_s}{D} = 1 - \frac{\langle x_e \rangle}{\sigma R_0} \tag{3.16}$$

where R_0 is the charge radius, u_s and D the detonation velocities at R_0 and at infinite size, respectively, and σ is a function describing the extent of wall motion. The detonation wave front was suggested to have a quadratic shape. The lag distance at the edge of the cylinder, L_0, is related to the reaction zone length by $\langle x_e \rangle \approx L_0$ as shown in Figure 3.9.

The detonation wave curvature can be measured by recording the time of emergence of the detonation wave across the charge distal end. The light from adiabatic compression can be registered either by a streak camera (Sućeska 1995) or by an optical fibre timer (Lu et al. 2014).

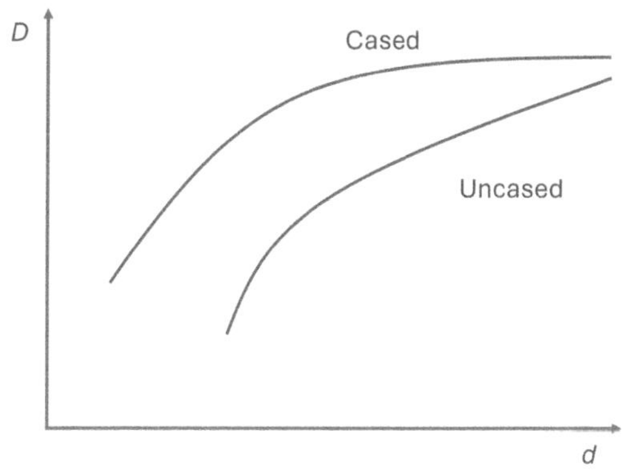

FIGURE 3.12 Change of detonation velocity D depending on diameter for uncased and cased charge, d.

If the explosive charge is enclosed in a casing, then the detonation products are confined, and the critical diameter of the charge is decreased. For example, the critical diameter of unconfined PBXW-115 (Aust.) is approximately 80 mm, and for PBXW-115 (Aust.) encased in brass tubing of 3–5 mm wall thickness, the critical diameter is 22–26 mm (Bocksteiner et al. 1994).

Casing does not affect noticeably the detonation velocity of dense HE (>1.6 g/cm^3); however, it affects significantly the detonation velocity of the porous charges of medium density (1.0–1.6 g/cm^3). The detonation velocity depends on inertial properties of the casing and its compressibility. At low density (<1 g/cm^3), the strength of casing influences the detonation velocity. The introduction of the casing allows an infinite diameter detonation velocity for small charge diameters, and detonation velocities become equal for large diameters, as is indicated in Figure 3.12.

Placing the explosive charge into a casing increases the velocity of detonation only if there is no gap between the charge and the casing. If there is a gap (few millimetres), the detonation stability may deteriorate due to the channel effect. The channel effect occurs when a shockwave propagates in the gap ahead of the detonation wave, which results in densification of the explosive ahead of the detonation front to a value close to or above the critical. This effect occurs in medium-density commercial explosives but does not occur in sensitive military HE.

One should be careful when selecting the material for casing an explosive. Metals can have a speed of sound faster than the velocity of detonation, in which case the propagating sound either supports the detonation regime or suppresses it, which will result in improvement or deterioration of explosive performance. This topic was investigated by Jackson et al. (2011) on the example of detonation of ammonium nitrate/fuel oil (ANFO) in aluminium tube.

3.3.3 Density

Secondary high explosives can be divided into two groups according to the behaviour of their granular pressed charges. For the first group which includes monomolecular high explosives such as TNT and RDX, the critical diameter of a steady detonation wave decreases with the increase of density (see Figure 3.13). In the second group of explosives, mostly mixtures with fuels, the critical diameter increases with the reduction in porosity. Examples of these explosives are ammonium nitrate, ANFO, and commercial emulsion explosives. For a detailed explanation see Sil'vestrov (2006) and references therein.

For the explosives of the first group, the detonation velocity increases monotonically with the increase of the charge diameter. For the second group, as the porosity decreases, the detonation velocity increases due to a higher heat of detonation Q since $D \sim \sqrt{Q}$. A maximum is reached at ρ_{cr} = 1.4–1.5 g/cm^3, then the detonation velocity decreases until the detonation fails at a critical diameter. The most common explanation is that the explosive is initiated thermally by the adiabatical

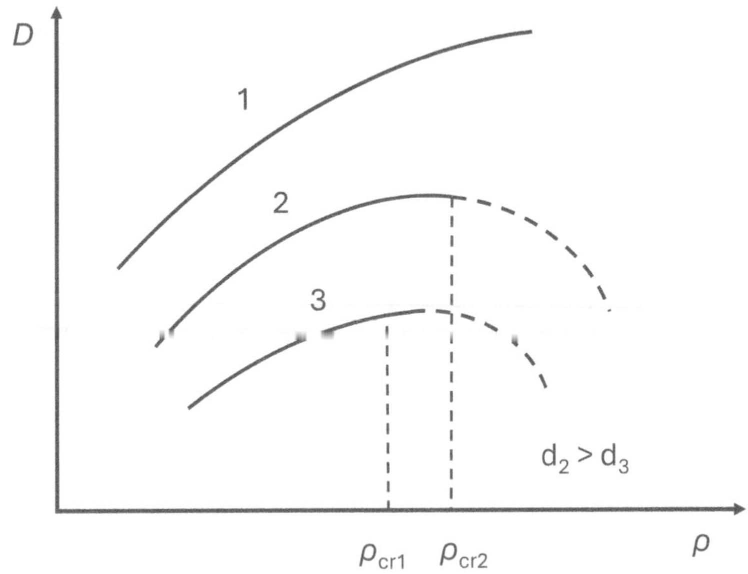

FIGURE 3.13 Dependency of velocity of detonation D on density ρ for monomolecular (1) and mixed (2, 3) types of explosives.

collapse of voids between the explosive particles, or "hot spots". The initiation period shortens until a certain pore size is reached, after which the pores cannot act as hot spots, and the explosive detonates by a different mechanism. For example, when compacted, ammonium nitrate behaves like an inert material; it absorbs energy, and thus the detonation propagation fails. If the mixed explosive includes a powerful HE (e.g., TNT or RDX), then it may compact so much that the detonation will propagate exclusively via that material, which will result in an increase of the detonation velocity.

When the diameter of an explosive charge is increased, or when the charge is placed in a casing, the critical density increases (see Figure 3.13).

The ideal detonation velocity is observed in experiments to be:

$$D_i(\rho_0) = a + b\rho_0 \tag{3.17}$$

where a and b are parameters due to the energy release in the detonation process. The ideal detonation velocity D_i is calculated from the equation of state of the detonation products (Section 3.1.3) or from experimental data by the Eyring equation mentioned above.

3.3.4 Composition and Particle Size

As the heat of detonation increases, the velocity of detonation increases, and the critical diameter decreases. For example, the heat of detonation of TNT is 3.5 MJ/kg, the velocity of detonation is 6 mm/μs and the critical diameter is 10 mm. For the more powerful RDX, those parameters are 9.5 MJ/kg, 8.4 mm/μs, and 1.5 mm (see Parfenov et al. 1969).

Critical diameter is also highly dependent on the size of the explosive particles of which an explosive consists. For example, when the TNT particle size decreases from 0.5 to 0.01 mm, the critical diameter decreases from 28 to 9 mm. When ammonium nitrate and TNT were simply mixed, the critical diameter of the resulting explosive, called amatol, was 20 mm. However, after ball-milling the mixture for 2 hours, the critical diameter decreased to 8 mm. All coarse-particle explosives have larger critical diameters than the fine-particle explosives of the same composition.

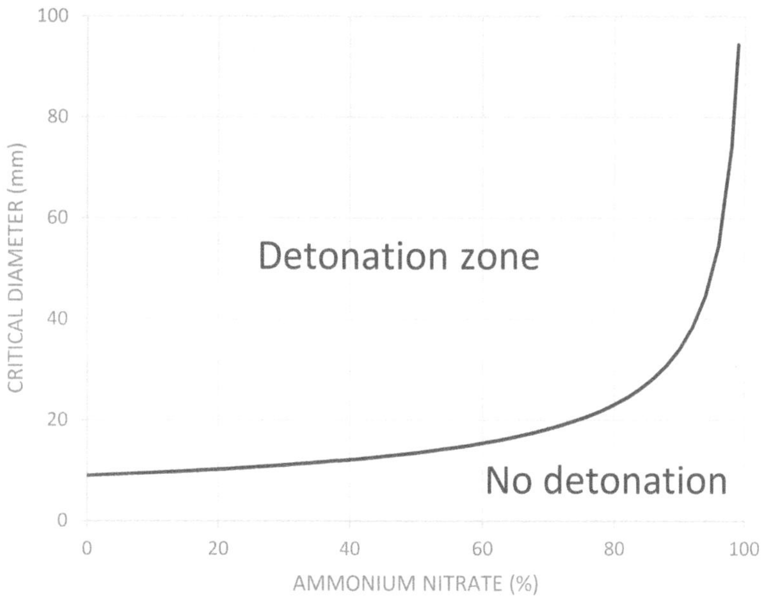

FIGURE 3.14 Critical diameter of amatol depending on the proportion of ammonium nitrate. (Adapted from Baum et al. (1959).)

The critical diameter of mixed explosives depends on their chemical composition. Figure 3.14 presents the dependency of d_{cr} on the composition of amatol. It shows that the critical diameter slowly increases with the increase of ammonium nitrate proportion from 0% to 80%, and then it increases sharply. The critical diameter for pure ammonium nitrate was found to be 80–100 mm.

The physical state of an explosive also affects its critical diameter. For example, dry granular TNT has a critical diameter of 60 mm, whereas the water-filled granular TNT charges have a critical diameter of 30 mm.

3.4 DETONATION OF EXAMPLE EXPLOSIVE SYSTEMS

3.4.1 Diluted Explosives

In 2005, the US Air Force Research Laboratory tested a system called dense inert metal explosive (DIME), also known as dense metal explosive (DME), or multiphase blast explosive (MBX) (Gibson 2008). It was an old concept covered in the US patents Nos. 3528864 and 5910638. Both are described as high explosive mixed with powdered, high-density metal such as tungsten. A representative DIME would have 75% tungsten and 25% TNT by weight.

The use of an inert metal has two effects. Firstly, it increases the density of the explosive mix, which has an advantage in applications such as warheads optimised for penetration. Secondly, it increases the blast impulse and lethality in the near field but reduces the far-field effects. Thus, it was concluded, the DIME concept was useful in urban warfare and for use in active protection systems. In 2011, the Lawrence Livermore National Laboratory team delivered a BLU-129/B bomb containing an MBX fill (Chrzanowski 2022). It had a carbon-fibre composite case that was able to penetrate concrete structures and produced no lethal fragments on detonation.

Imkhovik (2010) has reviewed the results of studies in the USA, and then in the USSR, of unusual detonation properties of mixtures of high explosives with high-density inert tungsten and lead additives. The introduction of tungsten and lead powders into RDX- and HMX-based explosives resulted in lowering the CJ pressure by 10–15 GPa measured by the free surface method compared with that by thermodynamic calculations. The measured detonation velocities turned out to be slightly higher (by 50–500 m/s) than those calculated from hydrodynamic theory. The adiabatic index of the detonation products turned out to be $\gamma = 6–8$ which was significantly higher than $\gamma = 3$ for conventional explosives, and which put this explosive in its own category.

The plate dent test showed an indentation depth higher than was predicted from test CJ pressures. There are several reasons for the observed behaviour. An increased impulse in the axial direction is due to the presence of metals which act as "internal confinement" preventing the detonation products from escaping. On the other hand, the presence of confinement allows the establishment of "underdriven" detonation in HE where the velocity relaxes immediately after the detonation front due to delayed acceleration of the metal.

It would be wrong to assume that the TNT equivalency of diluted explosives can be derived from the density alone. The correct approaches to measure the detonation velocity, pressure and energy in explosives with inert additives are discussed by Defourneaux (1994).

3.4.2 Additively Manufactured Explosives

Additive manufacturing (AM), or 3D printing of explosives, is still in its infancy. Most of the efforts thus far have been concentrated on the development of technologies to manufacture energetic materials (Muravyev et al. 2019, Zhang et al. 2022). Most research has been made into propellants and pyrotechnics due to the less-demanding safety requirement. There are fewer efforts reported in AM of high explosives. The manufactured amounts were small and not enough for proper characterisation of performance and detonation properties.

Researchers have done theoretical investigations of the detonation in large-scale charges fabricated by AM. Johnson et al. (2018) studied the blast formation due to shape variability of explosive charges, and Xiao et al. (2023) numerically demonstrated that the introduction of subsurface cavities into an explosive reduced the risk of sympathetic detonation.

Brown et al. (2023a) has reported the development of an HMX-based 3D-printed lattice of a mass of 55 g. When filled with water, the explosive lattice charge detonated and exhibited almost double the Gurney velocity. Detonation in a dry lattice failed. It was concluded that detonation propagation depended on the physical state of explosive and added water provided confinement.

Another approach taken by the researchers was the development of microscopic explosive devices. The review by Muravyev et al. (2019) discusses the development of miniaturised initiators and detonators where AM of explosives in combination with micro-electromechanical systems (MEMS) technologies would be used.

Stewart (2005) mentions two approaches taken by the designers of miniaturised detonation devices. The first was to downsize the large-scale devices (~100 mm) by 100 or 1000 times. To use the traditional large-scale design rules, the length of the reaction zone had to be scaled down correspondingly. The secondary explosives would not detonate in such systems due to their large critical diameter and long run-to-detonation distance. This meant primary explosives had to be used for the main charges. Primary explosives with reaction zones shorter than 10 μm are too dangerous to be handled safely.

The second approach is to build devices which use explosive materials for steady wave propagation at subcritical charge diameters. The detonations in these devices would be accompanied by transient effects such as run-to-detonation, corner turning, and detonation failure at subcritical diameters.

Stewart (2002) mentions that the waves propagating in charges of subcritical dimensions would experience unsteady phenomena such as a "galloping instability". This is where there is a leading shock in the tube, which has a lagging reaction zone that decays and then reignites which leads to velocity oscillations. The other phenomenon in miniaturised systems is a low detonation velocity. The low detonation velocity is supersonic but is close to the speed of sound. Its pressure is 25% of the CJ pressure. The author concluded that the wave will still be substantial enough to exert significant pressure on surrounding inert structures and will be quasistatic, which enables the design of devices.

Brown et al. (2023b) manufactured micrometre-sized strand-structured explosive charges and demonstrated the channelling effect resulting in detonation failure accompanied by an unexpected late reignition.

The review by Xue et al. (2024) summarised a significant body of work done in China. The authors mentioned TNT/nano-HMX melt-cast and CL-20/HTPB ink-based precursors as a base for new high-energy low-sensitivity explosives. AM is uniquely suitable for producing gradient and channelled structures due to its ability to vary materials and pore sizes (i.e., density). The authors mentioned the fabrication of CL-20-based explosive charges allows the control of the direction of detonation and customisation of energy release.

3.4.3 Volumetric Explosives

Volumetric explosives were mentioned first in the 1940s when the Germans attempted to detonate dispersed clouds of coal dust in air. Two-stage fuel-air explosives (FAE) were extensively used in the 1960–70s by the USA. Their applicability was relatively narrow. In Vietnam, FAE were used to clear ground from vegetation for landing helicopters. FAE were found to be mostly useful against lightweight buildings, vehicles, and personnel. In the 1980s, the USA used FAE against personnel in Lebanon and Afghanistan, and the USSR used the technology in Afghanistan and Chechnya.

In FAE, the fuel is dispersed explosively by a shockwave (Sullivan 1986), or by a gas generator via a nozzle. Liquid fuel droplets shed atomised fuel due to friction against the air and form fuel-air clouds (Zhang et al. 2010). In the second stage of the FAE operation, the fuel-air cloud is initiated by a HE charge by the shockwave and/or fireball. This results in detonation or fast deflagration of the fuel-air cloud. (Other mechanisms for the explosion effect could be diffusion, turbulent diffusion, laser-supported detonation, and photo-detonation (Myrick 2014).) Typical fuels used in FAEs are ethylene oxide, propylene oxide, isopropyl nitrate (IPN), butyl nitrite, and MAPP gas (Gelfand and Silnikov 2008; International Mine Action Standards (IMAS) 2023).

The explosive dispersal of liquid fuel results in a heterogeneous distribution of fuel in the air. The dispersal is accompanied by turbulence which makes it difficult for fuel to distribute evenly in the air. This prevents the formation of large-scale fuel-air clouds of uniform density and impedes propagation of the explosion by the detonation regime. The uniformity of the dispersed fuel is enhanced by having multiple dispersal charges as is done, for example, in the Russian ODAB-500PMV bomb. The optimal conditions for initiation of the fuel-air cloud exist only for a short period of time after fuel dispersal.

To overcome the difficulty of detonation of a heterogeneous fuel-air cloud, a very large initiating charge of HE is required. For example, 8 kg of explosive C4 was found to be necessary to initiate the detonation of a fine aluminium-air cloud which had a detonation cell size of over 0.5 m (Zhang 2009). Detonation cell size refers to a characteristic size of flow instability in a detonation wave.

In the 1970–80s, a new type of volumetric explosive appeared in the USSR. This incorporated one-stage thermobaric mixtures (termed thermobaric explosives, or TBX). Here, both the dispersal of fuel and its ignition is performed by the same HE charge. The most popular fuels have been IPN and metal fuel powders. The most well-known ammunition is the TBX warhead "Shmel" for the RPG-7. It uses a mixture of powdered magnesium and IPN as fuel. It is dispersed and initiated by a relatively large HE charge. TBX is considered to be more promising than FAE due to reduced complexity.

The disadvantage of IPN is that it can leak, vaporises easily, is toxic, and thus does not fit into the concept of insensitive munitions. In the 1990s, in order to replace TBX, the US researchers put forward a concept of enhanced blast explosive, or EBX. This is a polymer-bonded explosive (PBX) that has a fuel powder in its matrix. The fuel (typically aluminium) combusts partially during the detonation stage, thus contributing to driving the detonation wave. Aluminium reaction time can be measured by the cylinder expansion test and then optimised by changing aluminium reactivity (Anderson 2018). The fireball occupies the volume of a diameter of $d = (4...5)d_{HE}$, where d_{HE} is the diameter of the explosive charge. At the surface of the fireball, some of the incompletely reacted fuel can mix with air and release heat, thus contributing to the blast wave.

In the series FAE to TBX to EBX, the heat of combustion decreases due to the reduced use of environmental oxygen. However, the density of the detonation products and the blast are increased in the near field. EBX and some TBX compositions are efficient at driving fragmentation.

All these ammunition concepts have one idea in common: they utilise environmental oxygen for conducting the explosion, instead of the need to carry the oxidant within the warhead. This results in a decrease in the size of warhead. Furthermore, the heat of combustion of the fuels in air (20–120 MJ/kg) is 5–10 times higher than the heat of detonation of high explosives (2–25 MJ/kg).

Upon detonation, the wave front moves through the fuel-air cloud at a supersonic speed. The fuel decomposes into radicals and then recombines resulting in a release of energy which pushes the detonation wave forward. The induction time determines the length of the reaction zone. The latter should be shorter than the size of the fuel-air cloud d_c, otherwise the detonation will become unstable (Koch 2010).

When the fuel-air cloud reacts by detonation or fast deflagration, the reaction products immediately start expanding. A rarefaction wave propagates into the cloud at the speed of sound of the reaction products c_{rp}. When the reaction front reaches the distal end of the cloud, the reaction products start expanding at that end, and another rarefaction wave crosses the cloud in the opposite direction. The energy-releasing reaction in the cloud should complete within the time $t_R \approx 2d_c/c_{rp}$, so that at least some of the released heat transitions into the blast wave.

In a two-stage system, t_R is the time during which the ignition, and reaction, of a fuel-air cloud must occur. In a detonating cloud where the time of detonation propagation across the cloud is much shorter than t_R, there is sufficient time left for the afterburn of relatively large fuel particles. If the fuel-air cloud reacts by a fast deflagration mechanism, the resulting blast depends on the number of ignition sources and the rate of transition of the laminar flow into a turbulent one. In practice, this turns out to be difficult to implement, and this is why the blast performance of a fast-deflagrating composition is often lower than that of the detonating composition.

For a one-stage scheme, there are three possibilities of the functioning mechanism:

1. During the time t_R, both fuel dispersal and combustion occur. This requires a high initial velocity of dispersed fuel. The rarefaction part of the shockwave contains oxygen at high temperature and pressure which is beneficial for quick combustion of the injected fuel particles (Posey et al. 2021, Ripley et al. 2012, Ritzel et al. 2009).
2. By chemical initiation. Fuel is selected so that the combustion starts immediately after the fuel-air cloud's formation. This could be achieved by using pyrophoric fuel with a self-ignition delay. An example of such fuel is amine triethyl borane with hydrocarbons or carboranes (Gelfand and Silnikov 2008). Hypergolic approaches use bromine trifluoride, chlorine trifluoride, or perfluorohydrazine (International Mine Action Standards (IMAS) 2023). Their dispersal is accompanied by multiple-point ignition of fuel which enables volumetric explosion. In this case, the blast pressures could equal those achieved by detonative systems.
3. By the use of geometry. For example, the reflection of shockwaves in a confined or cluttered space and interaction of gas flow with environmental obstacles intensify the mixing of the fuel with air. At the same time, the unloading of blast and cooling of fuel mixture are reduced. This results in a fast afterburn, and in a high quasistatic pressure, which is comparable with that of FAE. One could look up Shchelkin spiral as an illustration of turbulence-inducing device promoting DDT.

Hahma et al. (2005) achieved efficient detonation in TBX configuration by utilising IPN and activated aluminium as fuel. The sufficiently high temperature, which was required for promoting the ignition of the metal fuel, was achieved by the addition of an oxidant such as IPN (in this case) or potassium perchlorate (Kolev et al. 2018, 2022). The gaseous products should not ignite immediately upon mixing with ambient air, and IPN was found to uniquely fulfil that requirement. Zhang et al. (2010, 2018) observed hybrid detonation in EBX charges. This is where there are multiple detonation peaks in the dispersed fuel due to different rates of reaction of the constituent fuels (zirconium and nitromethane). A critical HE charge diameter exists for enabling the reaction of metal particles in the dispersed fuel cloud.

A rarely mentioned disadvantage of aluminium as fuel is that up to 78% of heat of combustion could potentially be released in the form of light emission (González de Arrieta et al. 2023) and might not participate in driving the blast.

Notably, the heat of combustion of aluminium is about eight times the heat of detonation of TNT. This is the upper limit of TNT equivalency (also known as net explosive quantity, or NEQ) for aluminium-fuel volumetric explosive. In reality, an incomplete combustion of the metal fuel occurs; thus, some of the configurations mentioned above could achieve the NEQ of only as high as 2. Hahma et al. (2005) measured a TNT NEQ of 9.7 for his IPN/Al charges noting that TNT equivalency could not be defined easily as the thermobaric explosion source was not located in one point.

Some recent reviews of the detonation in volumetric explosives are presented by Rao (1987), Koch (2010), Trzcinski and Maiz (2015), Turker (2016), Iorga et al. (2021).

3.5 SUMMARY

Detonation is a process of propagation of a supersonic exothermic front through an energetic material. This drives a shockwave. The phenomenon of detonation was independently discovered by the French scientists Berthelot and Mallard in 1881 when investigating the propagation of flames in tubes. From this, a hydrodynamic theory of detonation was developed. Chapman (1899) and Jouguet (1905) proposed a one-dimensional model of detonation with the only state of detonation products that is indicated by the point CJ where the Rayleigh line passes through the Hugoniot adiabat at a tangent. Zeldovich (1940), von Neumann (1942), and Doring (1943) proposed the description of the complete structure of detonation with the propagation of a shockwave in an unreacted explosive and with the existence of chemical reaction zone. Practically, the Hugoniot curve of reacted explosive is determined in the cylinder expansion test. The process of initiating detonation depends on the parameters of the initiating impulse such as pressure and duration, and on the properties of the explosive to be initiated. A detonated explosive charge could be initiated by the detonation of another charge located at some distance away in a process called sympathetic detonation. A deflagration or shock in an explosive can transit into a detonation wave. A shock run-to-detonation distance is measured in the wedge test. Detonation wave cannot be sustained in an explosive charge that has a diameter less than a critical value. The detonation velocity is generally proportional to the density of explosive, and the deviations from this empirical rule are discussed. As the heat of detonation in an explosive increases, the velocity of detonation increases, and the critical diameter decreases. A high explosive with embedded heavy metal particles has an increased blast impulse and lethality in the near field, and reduced far-field effects. AM of explosives can be used to adjust the performance of charges and be a basis for miniaturised explosive devices. Volumetric explosive warheads utilise the environmental oxygen for the combustion of the carried fuel and could have up to eight times the energy density compared to warheads with conventional high explosives.

REFERENCES

Anastacio, A.C., Selesovsky, J., Künzel, M., Kucera, J., Pachman, J. (2019). "Detonation parameters of PlSEM plastic explosive", Cent. Eur. J. Energ. Mater., 16(4), pp. 487–503.

Anderson, P.E. (2018). "How and when metals react in high performance explosives", AIP Conference Proceedings 1979, 150002.

Baum, F.A., Stanyukovich, K.P., Shekhter, B.I. (1959). "Physics of explosions", in State Publishing of Physical and Mathematical Literature, Moscow (in Russian).

Berthelot, M., Vieille, P.M.E. (1881). "Sur la vitesse de propagation des phénomènes explosifs dans les gaz" ("On the velocity of propagation of explosive processes in gases"), Comptes rendus hebdomadaires des séances de l'Académie des sciences, 93, pp. 18–22.

Bocksteiner, G., Wolfson, M.G., Whelan, D.J. (1994). "The Critical Diameter, Detonation Velocity and Shock Sensitivity of Australian PBXW-115", DSTO-TR-0076.

Brown, C.B., Mueller, A.H., Sridhar, S., Lichthardt, J.P., Schmalzer, A.M., Tappan, B.C., Whitley, V.H., Hill, L.G., Lozano, E., Aslam, T.D. (March 2023a). "Switchable explosives: Performance tuning of fluid-activated high explosive architectures", Phys. Rev. Lett., 130, p. 116105.

Brown, C.B., Smilowitz, L., Remelius, D.K., Schmalzer, A.M., Tappan, B.C., Aslam, T.D., Sridhar, S., Mueller, A.H. (May 2023b). "Flash x-ray radiography analysis of detonation wave propagation in additive-manufactured high explosives", J. Appl. Phys., 133, p. 205901.

Campbell, A.W., Davis, W.C., Ramsay, J.B., Travis, J.R. (1961). "Shock initiation of solid explosives", Phys. Fluids, 4, pp. 511–521.

Chapman, D.L. (1899). Lond. Edinb.Dubl. Phil. Mag. 47, p. 90.

Chrzanowski, P. (Ed.) (2022). "LLNL 70th 1952–2022 making the impossible possible", in LLNL-AR-840123, p. 132. https://www.llnl.gov/sites/www/files/2022-09/LLNL_70th_Book_final_09.22.22.pdf

Cooper, P.W. (1996). Explosives engineering, Wiley-VCH, Inc.

Defourneaux, M. (1994). "About the misuse of the detonation velocity for the characterization of high explosives", in 25th International ICT Conference "Energetic Materials – Analysis, Characterization and Test Techniques", Karlsruhe (Germany).

Doring, W. (1943). "Uber den detonationsvorgang in gasen" (About detonation in gases), Annalen Der Physik, 435(6–7), pp. 421–436.

Duff, R.E., Houston, E. (1955). "Measurement of the Chapman-Jouguet pressure and reaction zone length in a detonating high explosive", J. Chem. Phys., 23(5), p. 1268.

Engelke, R., Sheffield, S.A. (1998). "Initiation and propagation of detonation in condensed-phase high explosives", in Davison, L. et al. (Eds.), High-Pressure Shock Compression of Solids III, Springer-Verlag New York, Inc, New York.

Eyring, H., Powell, R.E., Duffy, G.H., Parlin, R.B. (1949). "The stability of detonation", Chem. Rev., 45(1), pp. 69–181.

Gelfand, B.E., Silnikov, M.B. (2008). "Obyomnye vzryvy" (Volumetric explosions), Asterion, Saint Petersburg (in Russian).

Gibbs, T.R., Popolato, A. (1980). LASL Explosive Property Data, University of California Press, Berkeley and Los Angeles.

Gibson, N. (2008). "Hunt for high blast/low collateral damage weapons leads back to DIME/MBX", Jane's Int. Def. Rev., 41, pp.30–31.

Glassman, I., Yetter, R.A., Glumac N.G. (2015). Combustion (Fifth edition), Elsevier, p. 267.

González de Arrieta, I., Blanchard, C., Laboureur, P., Chauveau, C., Genevois, C., Rozenbaum, O., Halter, F. (2023). "Radiative properties of micron-sized Al/air premixed flames described by an effective medium core-shell formulation", Int. J. Heat Mass Transfer, 203, p. 123815.

Hahma, A., Palovuori, K., Solomon, Y. (2005). "TNT-equivalency of thermobaric explosives", 36th International Annual Conference of ICT & 32nd International Pyrotechnics Seminar "Energetic Materials. Performance and Safety".

Imkhovik, N.A. (2010). "Influence of dense inert additives (W and Pb) on detonation conditions and regime of condensed explosives", J. Energetic Mat., 28, pp. 216–230.

International Mine Action Standards (IMAS)(2023), "Fuel Air Explosive (FAE) Systems", Technical Note 09.30/04.

Iorga, O.G., Tigănescu, T.V., Sućeska, M., Rotariu, T., Marin, A., Munteanu, M., Zecheru, T., Ginghină, R.E. (2021). "Propellants Explos". Pyrotech, 46(7), pp. 1121–1135.

Jackson, S.I., Kiyanda, C.B., Short, M. (2011). "Experimental observations of detonation in ammonium-nitrate-fuel-oil (ANFO) surrounded by a high-sound-speed, shockless, aluminum confiner", Proc. Combustion Inst., 33(2), pp. 2219–2226.

Johnson, C., Mulligan, P., Williams, K., Langenderfer, M., Heniff, J. (2018). "Effect of explosive charge geometry on shock wave propagation", Proceedings of the Conference of the American Physical Society Topical Group on Shock Compression of Condensed Matter, AIP Conf. Proc. 1979, 150021.

Jouguet, E. (1905). "On the Propagation of Chemical Reactions in Gases", J. Math. Pures. Appl. Paris, 60, p. 347.

Kanel, G.I., Razorenov, S.V., Fortov, V.E. (2004). Shock-Wave Phenomena and the Properties of Condensed Matter, Springer-Verlag New York, Inc.

Kennedy, J.E. (1998). "The Gurney Model of Explosive Output for Driving Metal" (Chapter 7), in Zukas, J.A. and Walters, W.P. (Eds.), Explosive Effects and Applications, Springer-Verlag New York, Inc., New York, p. 251.

Khariton, Y.B. (1947). About Detonability of Explosives, Book 1, volume 1, USSR Academy of Science, Moscow-Leningrad (in Russian).

Koch, C.-E. (2010). "Volumetric Explosives Part 1: Fuel/Air Explosives", MSIAC Report L-165.

Koch, C.-E. (2021). High Explosives, Propellants, Pyrotechnics, Berlin/Boston, Walter de Gruyter GmbH, p. 220.

Kolev, S., Tsonev, T. (2022). "Aluminized enhanced blast explosive based on polysiloxane binder", Propellants Explos. Pyrotech., 47(2).

Kolev, S., Tsonev, T., Weinheimer, W. (2018). "Solid state fuel-air explosives", 43rd IPUSA Seminar, Fort Collins, CO, https://www.researchgate.net/publication/326985762_SOLID_STATE_FUEL-AIR_EXPLOSIVES

Kubota, N. (2007). Propellants and Explosives (Second edition), Veinhein, Wiley-VCH.

Kuznetsov, V.A. (2020). "Determination of Gurney Quantities and JWL Equation of State Parameters from the Cylinder Expansion Test", DST-Group-TR-3760, Defence Science and Technology Group, Australia.

Kuznetsov, V.A., Bennetts, S.P. (2013). "Development of Photon Doppler Velocimeter for Explosives Research", DSTO-TR-2792, Defence Science and Technology Organisation, Australia.

Leiber, C.-O. (1994). "Detonation model with spherical sources K: Critical dimension phenomena – Asymptotic considerations", In Proceedings of the 19th International Pyrotechnics Seminar, New Zealand.

Lochert, I.J., Franson, M.D., Hamshere, B.L. (2006). "Reduced Sensitivity RDX (RS-RDX) Part II: Sympathetic Reaction", DSTO-TR-1441, Defence Science and Technology Organisation, Australia.

Lu, J.P., Ashcroft, M. (2017). "Modelling of BLU warhead initiation", Parari. https://slideplayer.com/slide/17945403/

Lu, J.P., Kuznetsov, V.A., Mellen, P. (2014). "Simulation of shock initiation tests", In Proceedings of the 15th International Detonation Symposium, pp. 787–796, Office of Naval Research, Arlington, Virginia.

Lu, J.P., Lochert, I.J., Daniel, M.A., Franson, M.D. (2016). "Shock sensitivity studies for PBXN-109", Propellants Explos. Pyrotech., 41, pp. 562–571.

Mader, C.L. (1979). Numerical Modeling of Detonations, Berkeley, University of California Press.

Maines, W.R., Kittell, D.E., Hobbs, M.L. (2018). "Combined mini-cylex & disk acceleration tests in type K copper", Propellants Explos. Pyrotech., 43(5), pp. 506–511.

Mallard, E.-F., Le Chatelier, H.L. (1881). "Sur les vitesses de propagation de l'inflammation dans les mélanges gazeux explosifs" ("On the propagation velocity of burning in gaseous explosive mixtures"), Comptes rendus hebdomadaires des séances de l'Académie des sciences, 93, pp. 145–148.

Majowicz, J. M., Jacobs, S. J. (1958), "Initiation to detonation of high explosives by shocks", NAVORD Report 5710, U.S. Naval Ordnance Laboratory, White Oak, Maryland.

Marsh, S.P. (1980). LASL Shock Hugoniot Data, University of California Press, Berkeley and Los Angeles.

Mikhelson, V.A. (1893). "The normal velocity of ignition of combustible gaseous mixtures", Trans. Mosc. Univ., 10, pp. 1–92.

Minchinton, A. (2015) "On the influence of fundamental detonics on blasting practice", 11th International Symposium on Rock Fragmentation by Blasting, Sydney, NSW, Australia.

Miyake, A., Kubota, S., Kato, Y., Murata, K. (2023). "Ideal and non-ideal detonation", in Kubota, S. (Eds.), Detonation Phenomena of Condensed Explosives, Springer, Singapore.

Muravyev, N., Monogarov, K., Schaller, U., Fomenkov, I., Pivkina, A. (2019). "Progress in additive manufacturing of energetic materials: Creating the reactive microstructures with high potential of applications", Propellants Explos. Pyrotech., 44, pp. 1–30.

Myrick, J.J. (2014). "Nanoparticles, Compositions, Manufacture and Applications", US Patent Application 0227548, United States Patent and Trademark Office.

Pachman, J., Künzel, M., Němec, O., Majzlík, A. (2018). "A comparison of methods for detonation pressure measurement", Shock Waves, 28, pp. 217–225.

Parfenov, A.K., Voskoboinikov, L.M., Apin, A.I. (1969). Detonation of Coarse-Grained Explosives, LA-4199-TR, Los Alamos Scientific Laboratory of the University of California, Los Alamos, New Mexico.

Posey, J.W., Roque, B., Guhathakurta, S., Houima, R.W. (2021). "Mechanisms of prompt and delayed ignition and combustion of explosively dispersed aluminum powder", Phys. Fluids, 33, p. 113308.

Rao, A.A. (1987). "Fuel air explosives", Def. Sci. J., 37(1), pp. 23–28.
Ripley, R.C., Zhang, F., Lient, F.-S. (2012). "Acceleration and heating of metal particles in condensed matter detonation", Proc. R. Soc. A, 468, pp. 1564–1590.
Ritzel, D.V., Ripley, R.C., Murray, S.B., Anderson, J. (2009). "Near-field blast phenomenology of thermobaric explosions", P. 305, in Hannemann, K. and Seiler, F. (Eds.), 26th International Symposium on Shock Waves, Volume 1, Germany, Springer.
Sallivan, J. (1986). "Small Fuel-Air Explosive Bottle", US Statutory Invention Registration H161, United States Patent and Trademark Office. Also Sallivan, J. (1995). "Fuel-Air Explosive Canister", US Statutory Invention Registration H1457, United States Patent and Trademark Office.
Sil'vestrov, V.V. (2006). "Density dependence of detonation velocity for some explosives", Combust., Explos. Shock Waves, 42(4), pp. 472–479.
Souers, P.C. 1997). "Size effect and detonation front curvature", Propellants Explos. Pyrotech., 22(4), pp. 221–225.
Stewart, D.S. (2002). "Towards the miniaturization of explosive technology", Shock Waves, 11, pp. 467–473.
Stewart, D.S. (2005). "Miniaturization of explosive technology and microdetonics", P. 379, in W. Gutkowski, T.A. Kowalewski (Eds.), Mechanics of the 21st Century, Berlin, Springer-Verlag.
Sućeska, M. (1995). Test Methods for Explosives, Springer-Verlag New York, Inc, New York.
Tarver, C.M. (2005). "Detonation reaction zone in condensed explosives", Proc. 14th APS Topical Conference on Shock Compression of Condensed Matter, pp. 1026–1032.
Taylor, G.I. (1950). "The dynamics of the combustion products behind plane and spherical detonation fronts in explosives", Proceedings of the Royal Society of London. Series A, pp. 200–235.
Trzcinski, W.A., Maiz, L. (2015). "Thermobaric and enhanced blast explosives – Properties and testing methods", Propellants Explos. Pyrotech., 40, pp. 632–644.
Turker, L. (2016). "Thermobaric and enhanced blast explosives (TBX and EBX)", Def. Techn., 12, pp. 423–445.
Victor, A.C. (1996). "A simple method for calculating sympathetic detonation of cylindrical, cased explosive charges", Propellants Explos. Pyrotech., 21, pp. 90–99.
von Neumann, J. (1942). "Theory of detonation waves", in Progress Report OSRD 549, Prinston, NJ, Institute for Advanced Study.
Williams, F.A. (2018). Combustion Theory, CRC Press, Boca Raton, Florida.
Xiao, Z., Thornhill, R., Arora, H., Warren, R., Belblidia, F. (2023). "Impact safety improvement of high explosives through the use of internal cavity design", Propellants Explos. Pyrotech., 48(1), p. e202200152.
Xue, H., Zhou, Q., Xiao, C., Huang, G. (2024). "Perspectives on additive manufacturing for warhead applications", Defence Technology, KeAi Publishing.
Zeldovich, Y.B. (1940). "On the theory of propagation of detonation in gaseous systems", Sov. Phys. JETP, 10(5), pp. 542–568.
Zeldovich, Y.B., Kompaneets, A.S. (1960). Theory of Detonation, New York, Academic Press, p. 284.
Zhang, F. (2009). "Reaction mechanism of Al particle-air detonation", J. Propul. Power, 25, pp. 845–858.
Zhang, F. (2018). "The Julius Meszaros Lecture: Dense Metal Explosives and Energetics: Fundamentals and Beyond", in 25th International Symposium on Military Aspects of Blast and Shock, The Netherlands.
Zhang, F., Yoshinaka, A., Ripley, R. (2010). "Hybrid detonation waves in metallized explosive mixtures", In Proceedings of the 14th International Detonation Symposium, pp. 714–723, Office of Naval Research, Idaho.
Zhang, J.C., He, K., Zhang, D.W., Dong, J.D., Li, B., Liu, Y.J., Gao, G.L., Jiang, Z.X. (2022). "Three-dimensional printing of energetic materials: A review", Energ. Mater. Front., 3(2), pp. 97–108.

4 Computational Modelling of Explosive Detonation

A. D. Resnyansky

4.1 INTRODUCTION

Detonation is commonly associated with explosion or blast as a manifestation of the propagation of a shock wave, whereas a crucial element of the detonation process – a chemical reaction – is shadowed by the shock wave effects. The reaction chemically transforms an energetic composition to detonation products with the liberation of thermal energy, which is evidence of an exothermic reaction.

In general, many chemical reactions are exothermic, such as the process of brewing in winemaking. Combustion and deflagration are also chemical processes of oxidation representing exothermic reactions. For a material to be combustible, the presence of a fuel component is sufficient because an oxidizer (oxygen) can be supplied from the air. Combustion is a turbulent and slow process, whereas deflagration is a quicker process of layer-to-layer burning. The fast burning and possibility to transit to detonation at suitable conditions make deflagration more relevant to detonation. Deflagration occurs at a relatively low speed, providing sufficient time for the reaction products to expand without such violent consequences as explosion, unless the deflagrating material is confined.

Detonation phenomena were first detected in dusty gases, for example, as blasts in coal mines. An important observation that discriminates detonation from combustion was that the detonation velocities are independent of dusty gas volumes as soon as the volumes are sufficiently large. This means that for realization of the detonation regime, an explosive material must contain both fuel and oxidizer within its chemical composition. This makes explosives a special class of materials that has an oxidizer as a constituent. Heterogeneous mixtures, such as aerosols, dusty gases, and composite mixtures that give an easy access to an oxidizer, are also examples of such materials.

The main attribute of detonation as an exothermic process is a very high (supersonic, in a case of plane wave) speed of propagation. Chemical reaction at detonation creates a 'gas balloon' of gaseous products in nearly adiabatic conditions at very high pressure and temperature, followed by expansion of the products. This expansion of highly compressed hot products is capable of performing considerable mechanical work after detonation, resulting in catastrophic and noisy consequences. Detonation was first systematically described in the 19th century by Berthelot and Vieille (1881) and Mallard and Le Chatelier (1881), noting that the main cause of initiating the chemical reaction of detonation is pressure. After the initiation, the chemical reaction in a transition zone supports self-sustained propagation of detonation. Because of the resemblance of detonation to other oxidation processes, early studies interpreted this process as a "wrong" deflagration. This was one of the etymological reasons for the term 'detonation' as 'false' deflagration from French 'détonner' (to sound false) competing with another possible reason from Latin 'detonāre' (to thunder down).

This chapter primes the reader for the topic of detonation modelling. It aims to familiarize one with foundations that are needed for understanding modern relevant models by prospective modellers of reactive flows and detonation. Summarizing each section,

- The Section 4.2 provides the background to thermodynamics, with a focus on thermodynamic postulates.
- The Section 4.3 covers the chemical kinetics and thermochemistry, which should be sufficient for familiarization with the reaction equations.

DOI: 10.1201/9781003529217-4

- The Section 4.4 introduces a topic of determination of compositions at chemical equilibrium after completion of chemical reactions in the detonation process and description of applicable equations of state.
- The Section 4.5 outlines the basic elements of shock physics and continuum mechanics for substances in inert systems.
- The Section 4.6 utilizes the basics of thermodynamics, chemical kinetics, and shock physics from the previous sections for introducing the foundation detonation theories, namely, Chapman–Jouguet (CJ) and Zeldovich–von Neumann–Döring (ZND) models. These have been previously introduced in Chapter 3.
- The Section 4.7 observes several examples of using the background information on chemical kinetics, thermodynamics, and continuum mechanics for application to simulation of detonation and reactive flows in homogeneous compositions.
- The Section 4.8 applies the particulars outlined in the previous sections to the development of a model suitable for simulating detonation processes and reactive flows in homogeneous compositions.
- The Section 4.9 briefly touches on computational developments toward heterogeneous systems.
- Finally, the Section 4.10 concludes the chapter by discussing challenges and gaps in this area of research.

4.2 THERMODYNAMICS BACKGROUND

Historically, thermodynamics originated as a science about transformations between thermal and mechanical energy. After progressing further, it was found that the laws of thermodynamics are universal, i.e., they are applicable to other forms of energy. Thermodynamics deals with four types of ideal systems: (i) open systems, that can exchange both mass and energy with the environment (the exterior in relation to the system); (ii) closed systems, that cannot exchange mass, but can exchange energy; (iii) adiabatic systems, that cannot exchange energy; and (iv) isolated systems, that are both closed and adiabatic, which means that they cannot exchange mass and energy with the environment. Perfect realization of these conditions is impossible in reality for a system surrounded by an imaginary virtual boundary, through which mass and energy are inevitably exchanging with the exterior. Therefore, the ban for a mass and/or energy exchange in the system classification is replaced by a less onerous condition of essentially smaller exchange across the virtual boundary than the exchange between elements within the system.

Considerations of thermodynamic systems are possible at microscopic and macroscopic levels, resulting in statistical and phenomenological approaches. The first approach is driven by statistical analysis of particle ensembles, forming a theory of statistical thermodynamics. An alternative approach characterizes a thermodynamic system by macroscopic parameters, averaging characteristics of the system within a representative volume. These parameters can describe the system at any point omitting local microscopic fluctuations, thus characterizing overall system, which results in a theory of phenomenological thermodynamics. The present section focuses on cornerstones of the phenomenological thermodynamics.

Classical thermodynamics assumes that system states are at equilibrium, which ignores the time variable. In a real physical or chemical process, this means that a timeframe for states of the system to be considered at equilibrium, should be essentially smaller than characteristic times of the process. Extensive variables describing a state of system are measurable and obey the additivity rule for parts of the system. This means that an extensive variable of a system containing several parts is equal to the sum of the corresponding variables of the parts. Basic postulate of the classical thermodynamics presumes the existence of a parameter, called temperature T, associated with the system equilibrium, which means absence of change in the parameter throughout the system and its stationarity.

Temperature is a quantitative measure of movement of particles of a thermodynamic system. The particle movement can be typically classified as: (i) chaotic movement of particles within the system on microscopic scale that manifests itself macroscopically as an amount of heat Q which is measurable by an amount of energy transferrable to other forms of energy; and (ii) collective (ordered) movement of system particles seen as macroscopic movement of parts of the system presented as work, A. The work is a measurable amount of mechanical energy performed under the system (for example, reduction in size of a rubber balloon due to external pressure). In the origins of the research about energy transformations, the quantity A was interpreted as 'motive power', which resulted in the name 'thermodynamics' for this area of science. Heat accepted by the system is considered to be positive and the corresponding process is called endothermic. In the opposite case of heat loss, the process is called exothermic. Similarly, mechanical energy for work performed under the system and accumulated by the system is positive, whereas lost mechanical energy for work performed by the system on the environment is negative.

Transfer of energy, such as Q and A, to/from a thermodynamic system causes changes in its thermal and mechanical characteristics, affecting microscopic movement of the system's particles. As a result, the corresponding energy conversion changes an amount of energy absorbed or lost by particles of the system, quantifiable as internal energy E of the system. In addition to the thermal characteristic T of the system, it can be characterized physically by mass M, amount of substance (amount of particles) N, and volume V occupied by the system. Obviously, the variables Q, A, E, M, N, and V are extensive because they are proportional to size of the system. In contrast, temperature T is a characteristic of the system that does not depend on system's size, thus representing an intensive variable. Another intensive variable which is a characteristic of mechanical response of the system is pressure p. Depending on nature of the system, other variables, such as electrical charge, magnetic field, mechanical stress, and so on, may characterize a system's response and change its internal energy. Therefore, for an observable process of nature, independent variables, characterizing relevant physical and chemical phenomena that are critical to the system's behaviour, need to be specified before any thermodynamic analysis.

A thermodynamic system, performing or being subjected to mechanical work, changes its thermal energy (heat). First law of thermodynamics postulates that change of the full amount of heat, Q due to a work A, is balanced with an increase or expense of internal energy E of the system, which is expressed by the following formula:

$$\delta Q - \delta A = dE. \tag{4.1}$$

This formulation presumes transfer of energy within a closed system. For an isolated system, the consequence of (4.1) is $dE = 0$, which means that internal energy does not change in the system. In the framework of phenomenological thermodynamics, thermodynamic system can be characterized by functions of state with the main property to be dependent only on the current state regardless of how this state has been reached. In contrast to functions of process such as Q and A, the functions of state are not dependent on the path and history of reaching the current state of the system. A small increment of any function of process is conventionally denoted by δ, in contrast to the traditional denotation of full differential d. Mathematical formulation of changes in functions of state by small increments is represented by full differentials, because they are dependent only on initial and final states of the system. Thermodynamic paths can be rather arbitrary and the most popular (but not limited to) ones specify the following processes: isochoric at constant volume V ($dV = 0$), isobaric at constant pressure ($dp = 0$), isothermal at constant temperature ($dT = 0$), and adiabatic in the absence of heat exchange Q ($\delta Q = 0$). In classical thermodynamics, path changes in a thermodynamic system are required to be reversible and performed at equilibrium (sufficiently slow).

Specifying mechanical work performed by pressure p that results in a change of volume V of substance, the amount of work is $\delta A = pdV$ for thermodynamic processes at equilibrium. Using this, the first law of thermodynamics (4.1) takes the following form:

$$\delta Q = dE + pdV. \tag{4.2}$$

Functions of state can be related to independent thermodynamic variables via equations, such as an equation involving internal energy in the following general form $f(E, V, T) = 0$, called caloric equation of state (EOS). Another possible form of EOS, $f(p, V, T) = 0$, is called thermal EOS. In order to fully characterize a thermodynamic system, both thermal and caloric relations are needed, which represents a complete EOS. For thermodynamic processes in many gases at quasi-equilibrium, the simplest relation, connecting thermal and mechanical parameters, is known as Clapeyron's law (Clapeyron, 1834):

$$pV = NRT, \tag{4.3}$$

where $R = 8.3144$ J/(K·mol) is the gas constant and N is amount of gas measured in moles. By definition, one mole contains a fixed number, N_A, of atoms or molecules and this number $N_A \approx 6.022 \cdot 10^{23}$ associated with the atomic weight, is called Avogadro's number. The periodic table lists the atomic weights for any chemical element specified as weight in grams of one mole of substance, containing N_A atoms. The relation (4.3) is called the ideal gas law and is an example of a thermal EOS. Thermodynamic postulates will be illustrated below with molar parameters (related to one mole of substance, $N = 1$) for an ideal gas.

Historically, physicists studied mechanical processes on the assumption of constant volume (isochoric processes) for thermodynamic systems characterized by internal energy E. At the same time, chemists dealt with chemical reactions, when considering thermodynamic processes at the conditions of constant pressure (isobaric processes). Thus, usage of internal energy E is common for describing heat and work transformations in thermodynamic processes associated with mechanical changes of a system. For description of chemical reactions, energy transformations in isobaric processes are commonly analyzed with another function of state called enthalpy. Enthalpy H is defined as $H = E + pV$, resulting in a modification of the first thermodynamic law (4.2) to $\delta Q = dH - Vdp$. Obviously, for isochoric processes at $dV = 0$: $\delta Q = dE$, and for isobaric processes at $dp = 0$: $\delta Q = dH$, which explains the use of enthalpy instead of internal energy for thermodynamic analyses of chemical processes.

Capacity of a substance to accumulate heat Q is known as thermal (heat) capacity and is denoted C_x. Because Q is a function of a process, the extensive characteristic C_x depends on path of the process in thermodynamic space. Correspondingly, the subscript x is an indicator of the path of a process while keeping parameter x constant. Thus, definition of thermal capacity can be written as

$$C_x = \left(\frac{\delta Q}{dT}\right)_x. \tag{4.4}$$

By thermodynamics convention, subscripts at right parentheses shown as a list of variables, indicate that these variables are held constant, which specifies a path in quasi-equilibrium thermodynamic process. Considering several paths of common use, it is obvious that heat capacity for isothermal processes is undefined. Specifying (4.2) for isochoric and isobaric processes with internal energy and enthalpy as functions of state, the corresponding heat capacities can be defined from (4.4):

$$C_V = \left(\frac{\partial E}{\partial T}\right)_V, \quad C_p = \left(\frac{\partial H}{\partial T}\right)_p. \tag{4.5}$$

Before postulating the second law of thermodynamics, several comments can be made on the calculation of heat capacities of an ideal gas. The first law of thermodynamics can be rewritten, using the chain rule of differentiation, as follows:

$$\begin{aligned}\delta Q = dE + pdV &= \left(\frac{\partial E}{\partial T}\right)_V dT + \left(\frac{\partial E}{\partial V}\right)_T dV + pdV \\ &= \left(\frac{\partial E}{\partial T}\right)_V dT + \left[\left(\frac{\partial E}{\partial V}\right)_T + p\right] dV,\end{aligned} \tag{4.6}$$

which results in equation (4.5) for C_V from (4.4) at $V = \text{const}$. Therefore, an EOS in the form $E = E(V, T)$ is sufficient for calculation of C_V. At the same time, calculation of C_p from (4.6) at $dp = 0$ requires an additional thermal EOS, $p = p(V, T)$. Application of the ideal gas EOS (4.3) in the form $V = NRT/p$ to (4.6), results in

$$C_p = C_V + \left[\left(\frac{\partial E}{\partial V}\right)_T + p\right]\left(\frac{\partial V}{\partial T}\right)_p = C_V + \left[\left(\frac{\partial E}{\partial V}\right)_T + p\right]\frac{NR}{p}. \tag{4.7}$$

It is seen that in addition to the thermal EOS, for determination of C_p a dependence of E on V (which is a caloric EOS) is also required. In thermodynamics textbooks (e.g., see Kondepudi and Prigogine (1998)), it is common to derive a caloric EOS from the empirical Joule's law that declares independence of internal energy on volume at constant temperature for an ideal gas in isolated system, resulting in $(\partial E/\partial V)_T = 0$. From (4.7), this immediately leads to: $C_p = C_V + NR$. As a result of the Joule's law, a caloric EOS for polytropic processes (processes in substances with constant thermal capacity) in an ideal gas can be directly obtained from $(\partial E/\partial V)_T = 0$:

$$E(V,T) = C_V T + E_0. \tag{4.8}$$

To note, a caloric EOS can also be obtained from thermal EOS, using Maxwell relations. These relations can be derived after postulating the second law of thermodynamics. The derivation on the basis of the second thermodynamic law is a way of determining a caloric EOS that is an alternative to the above approach based on the Joule's law.

When considering closed systems at quasi-equilibrium, the caloric EOS (4.8) allows one to evaluate a change in heat of a polytropic gas system as follows:

$$\begin{aligned}&\frac{\delta Q}{T} = \frac{dE + pdV}{T} = \frac{C_V}{T}dT + \frac{NR}{V}dV = d\ln\left(T^{C_V}V^{NR}\right) = dS, \\ &S = \ln\left(T^{C_V}V^{NR}\right).\end{aligned} \tag{4.9}$$

Therefore, for a polytropic gas, $\delta Q/T = dS$ is satisfied, which makes S a function of state, resulting in $S(V,T) = \text{const}$ in adiabatic quasi-equilibrium processes. Thus, in this case, equation (4.9) can take the following equivalent forms:

$$T_2^{C_V} V_2^{NR} = T_1^{C_V} V_1^{NR}, \quad \frac{T_2}{T_1} = \left(\frac{V_1}{V_2}\right)^{NR/C_V}. \tag{4.10}$$

From $C_p = C_V + NR$ for an ideal gas, a more conventional representation that is popular in the modelling community can be obtained. Introducing ratio $\gamma = C_p/C_V$ that is called the polytropic index, the exponent in (4.10) is $NR/C_V = (C_p - C_V)/C_V = \gamma - 1$. This results in a condition $\gamma > 1$ and enables one to rewrite (4.10) as $TV^{\gamma-1} = \text{const}$, along any adiabatic path for a polytropic gas. Finally, from the thermal EOS (4.3), adiabatic path in (p,V)-space can be rewritten as follows:

$$pV^{\gamma} = \text{const}. \tag{4.11}$$

It will be shown later that a function of state $S(V,T)$, known as entropy and associated with the representation (4.11) for a polytropic gas, does actually exist for any system, which is a foundation of the second law of thermodynamics. Some modern theories of thermodynamics, like the theory of thermodynamic potentials by Gibbs (1928), pre-tabulate existence of entropy. In the classical phenomenological thermodynamics the second law of thermodynamics is based on a very important study by French scientist Sadi Carnot, whose only self-published manuscript (Carnot, 1897) was an analysis of heat engines in thermodynamic space. Theoretical basis of the Carnot's theory and its illustration in the (p,V)-space, which is used to date, was conducted later by Émile Clapeyron (1834).

In his manuscript, Carnot (1897) considered thermal efficiency η of heat engines for an expanding ideal gas performing work ('motive power'). For a heat ΔQ transferred to useful work, thermal efficiency η was defined as ratio of the heat loss ΔQ to the total heat Q_H lost during the expansion: $\eta = \Delta Q/Q_H$. Because heat engine requires a closed cycle in thermodynamic space to perform a useful work ΔA, the cycle needs: (i) to transform an amount of heat ΔQ to work, and (ii) to return the system to its original state for the cycle to be repeated. Before starting a cycle, a heat supply increases internal energy of the thermodynamic system reaching initial state '1' (Figure 4.1(a)), for example, by the way of contact with an appropriate heat reservoir at quasi-static conditions – heater. According to the first law of thermodynamics (4.1), the heat accumulated by the system, which results in system's temperature T_H, can be transferred to work $\Delta A_1 = Q_H$ plus an increment in internal energy. Keeping in mind that internal energy for ideal gases depends only on temperature, an important point noted by Carnot was that in agreement with the law (4.1), the only way to fully transfer heat Q_H to mechanical work ΔA_1 is to do this without any change of internal energy, which means an energy transfer at constant temperature – isothermally. According to (4.2), the work ΔA_1 of expanding gas along the path can be estimated in the (p,V)-diagram (Figure 4.1(a)) as an area $\int p dV = Q_H$ under the curve $p(V)$ (shadowed area in Figure 4.1(a)). This isothermal expansion describes a quasi-equilibrium path E_1, connecting two states '1' and '2' (Figure 4.1(a)).

For realizing a cycle, the system needs to return to its original state. Similarly to the case of isothermal expansion, the backwards path with zero loss of internal energy could be achieved along the same path in the reversed direction (from '2' to '1' along E_1 in Figure 4.1(a)), expensing the work $-\Delta A_1$ for compression. However, this hypothetical cycle does not provide any useful work, because the total work is $\Delta A_1 - \Delta A_1 = 0$ with a zero area after total integration in the (p,V)-space as the difference between the forward and reverse paths. Following this, it is obvious that the larger a difference in areas between forward expansion and reverse compression paths of a

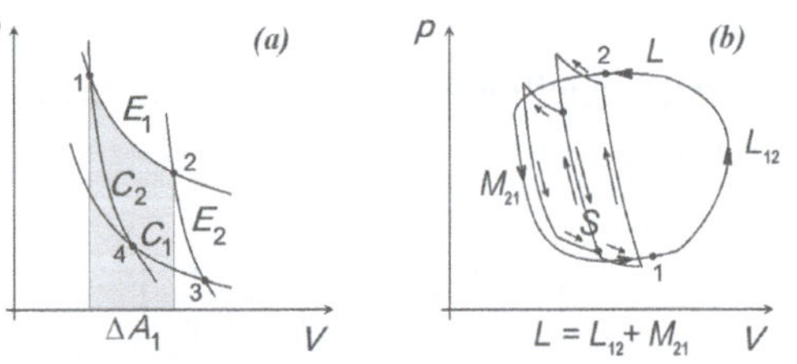

FIGURE 4.1 A Carnot cycle performing isothermal and adiabatic expansions E_1–E_2, followed by isothermal and adiabatic compressions C_1–C_2, which provides maximal thermal efficiency (a); an approximation of arbitrary cycle L by sections of Carnot cycles (b).

cycle in the (p,V)-space, the larger useful work ΔA performed by the system. When addressing this issue, another important observation by Carnot was that in order to achieve as much useful work as possible an isotherm of the reverse path (compression) should be as distant from the expansion isotherm E_1 as possible, urging arrangement of a temperature difference in the heat machine as large as possible. The only way to reduce temperature to a desirable 'cold' value T_C in the working gas participating in the cycle is to reduce internal energy by a level $-\Delta E$ providing the corresponding temperature drop. Thus, according to the first thermodynamic law (4.1), an additional forward path E_2 is required to reduce the temperature (by lowering internal energy by $-\Delta E$) and to perform extra useful work ΔA_2. At the same time, minimal heat losses can be achieved only at adiabatic expansion ($\delta Q = 0$) along a path E_2 from '2' down to '3' (Figure 4.1(a)).

By the condition of cyclic operation of heat engine, heat Q_H transferred to work $\Delta A = \Delta A_1 + \Delta A_2$ during the forward expansion (stages E_1 and E_2 of the cycle) cannot be utilized fully because the system needs to be returned to its original state. Therefore, a reverse path of the cycle starts from the largest volume at state '3' achieved during the expansion stages E_1 and E_2 (Figure 4.1(a)). Obviously, in order to avoid loss of internal energy and minimize work $\Delta A'_1$ expensed at compression of the system along the reverse path, the compression should follow isotherm C_1 (Figure 4.1(a)) at temperature $T = T_C$ (for example by the way of contact with suitable heat reservoir – cooler, such as the environment). This results in a work loss $\Delta A'_1$ identical to the heat $Q_C = |\Delta A'_1|$ gained by the working gas during the compression, according to (4.1). Similarly to the case of the adiabatic forward path E_2 corresponding to the temperature drop from T_H down to T_C, a reverse path C_2 performs full conversion of the expensed work $\Delta A'_2$ by adiabatic compression, which is required for increasing amount of heat from Q_C back to Q_H along the adiabatic path from '4' up to '1' (Figure 4.1(a)).

Following the logic of Carnot's analysis, the polytropic gas laws (4.3) and (4.11) for one mole of a polytropic gas can be utilized for isothermal and adiabatic processes, respectively. Using this, it is easy to calculate amount of work for the isothermal paths E_1 and C_1 (Figure 4.1(a)) of the Carnot cycle in the clockwise direction:

$$\begin{aligned} \Delta A_1 &= \int_{V_1}^{V_2} p dV = \int_{V_1}^{V_2} \frac{RT_H}{V} dV = RT_H \ln\left(\frac{V_2}{V_1}\right) = Q_H, \\ \Delta A'_1 &= \int_{V_3}^{V_4} p dV = \int_{V_3}^{V_4} \frac{RT_C}{V} dV = RT_C \ln\left(\frac{V_4}{V_3}\right) = -Q_C. \end{aligned} \tag{4.12}$$

Calculating pressure from (4.11) for adiabatic paths E_2 and C_2 in the forms $p = p_2V_2^\gamma/V^\gamma$ and $p = p_4V_4^\gamma/V^\gamma$, respectively, and denoting $\alpha_{23} = p_2V_2^\gamma = p_3V_3^\gamma$ and $\alpha_{41} = p_4V_4^\gamma = p_1V_1^\gamma$, it follows:

$$\Delta A_2 = \int_{V_2}^{V_3} p dV = \int_{V_2}^{V_3} \frac{\alpha_{23}}{V^\gamma} dV = \frac{p_3V_3^\gamma}{V_3^{\gamma-1}(1-\gamma)} - \frac{p_2V_2^\gamma}{V_2^{\gamma-1}(1-\gamma)} = \frac{R(T_H - T_C)}{\gamma - 1},$$

$$\Delta A'_2 = \int_{V_4}^{V_1} p dV = \int_{V_4}^{V_1} \frac{\alpha_{41}}{V^\gamma} dV = \frac{p_1V_1^\gamma}{V_1^{\gamma-1}(1-\gamma)} - \frac{p_4V_4^\gamma}{V_4^{\gamma-1}(1-\gamma)} = \frac{R(T_C - T_H)}{\gamma - 1}.$$

It is seen that gained and expensed amounts of work ΔA_2 and $\Delta A'_2$ at the adiabatic stages of the cycle are balanced $\Delta A_2 + \Delta A'_2 = 0$, resulting in the total useful work ΔA

$$\Delta A = \Delta A_1 + \Delta A_2 + \Delta A'_1 + \Delta A'_2 = \Delta A_1 + \Delta A'_1 = Q_H - Q_C.$$

This gives the efficiency coefficient $\eta = \Delta Q/Q_H = (Q_H - Q_C)/Q_H$. Noting from (4.8) that

$$\frac{V_2}{V_3} = \left(\frac{T_2}{T_3}\right)^{C_V/R} = \left(\frac{T_H}{T_C}\right)^{C_V/R}, \quad \frac{V_4}{V_1} = \left(\frac{T_4}{T_1}\right)^{C_V/R} = \left(\frac{T_C}{T_H}\right)^{C_V/R},$$

it follows $V_2/V_1 = V_3/V_4$. Using this in (4.12), the formula for thermal efficiency can be rewritten via temperatures of the heat reservoirs:

$$\eta = \frac{Q_H - Q_C}{Q_H} = \frac{T_H - T_C}{T_H}. \tag{4.13}$$

The Carnot cycle is unique. In fact, assuming that another closed cycle, utilizing the same thermal reservoirs, exists that produces more useful work, it follows that sections of this cycle adjacent to the Carnot cycle may produce work and thus accumulate extra work from nothing. To conclude this, a theorem by Carnot (1897) claims uniqueness of the cycles performing maximal work in quasi-equilibrium reversible processes between two thermal reservoirs with fixed temperatures. Another theorem by Carnot (1897) claims that (4.13) is valid for any working medium.

In order to generalize the Carnot analysis for an arbitrary cycle (cycle L in Figure 4.1(b)), equation (4.13) can be rewritten as the Clausius equality $Q_H/T_H - Q_C/T_C = 0$. Covering the cycle L by Carnot cycles, it can be noticed that for two combined adjacent Carnot cycles, energy contributions from common section of the cycles (e.g., section S in Figure 4.1(b) between two dots) cancels out. Therefore, any closed path L can be approximated by a boundary formed from exterior sections of the Carnot cycles. Difference between the external boundary of the merged Carnot cycles, approximating an area inside L, and boundary of the area itself, is tending to zero. This line of reasoning follows the proof by Clausius (1879) based on the Clausius equality, resulting in the following statement for any closed path:

$$\oint_L \frac{\delta Q}{T} = 0. \tag{4.14}$$

Statement (4.14) for a closed cycle can be rewritten for any two connected segments forming the cycle. Following this, two states '1' and '2' in thermodynamic space can be connected by any two quasi-equilibrium (reversible) paths L_{12} and M_{12} (Figure 4.1(b)) in the space and these two paths could be considered as parts of a closed cycle $L = L_{12} + M_{21}$ (the order in subscript indices corresponds to the clockwise direction for integration along respective paths). It follows from (4.14) that

$$\oint_L \frac{\delta Q}{T} = \oint_{L_{12} \cup M_{21}} \frac{\delta Q}{T} = \oint_{L_{12}} \frac{\delta Q}{T} - \oint_{M_{12}} \frac{\delta Q}{T} = 0,$$

which means that $\delta Q/T$ does not depend on path, i.e., it is a differential of a function of state and can be denoted dS. The quantity S was named entropy by Clausius (1879):

$$dS = \frac{\delta Q}{T}, \tag{4.15}$$

which proves that $S(V,T)$ defined in (4.15) is conserved along paths representing adiabatic processes, which is valid not only for polytropic gas as was derived in (4.9). This substantiates the second law of thermodynamics. A third postulate called the Nernst theorem, claims that entropy change tends to zero as absolute temperature tends to zero, which, with proper calibration, can be applied to entropy itself. This postulate is used for calibration of heats of formation in thermochemistry.

Summarizing, the first law of thermodynamics (4.2) with the use of the second law (4.15) can be rewritten for internal energy and enthalpy with differentials as the following two equations called the thermodynamic identities:

$$TdS = dE + pdV,\ TdS = dH - Vdp. \tag{4.16}$$

Thermodynamic identity involves only functions of state and full differentials of the generating functions. This relation is a basis of the Gibbs approach, where given potentials (in the present case, $E(V,S)$ and $H(p,S)$) enable one to determine dependent variables T, p and V directly from (4.16) by the chain rule differentiation, followed by separation of differentials:

$$T = \left(\frac{\partial E}{\partial S}\right)_V = \left(\frac{\partial H}{\partial S}\right)_p,\quad p = -\left(\frac{\partial E}{\partial V}\right)_S,\quad V = \left(\frac{\partial H}{\partial p}\right)_S. \tag{4.17}$$

Using different ways of differentiation of functions of state, the corresponding mathematical formalism results in a powerful tool of thermodynamics called Maxwell relations. For example, mixed second order differentiation of $E(V,S)$ with the use of (4.17) results in one of the following Maxwell relations:

$$\frac{\partial^2 E}{\partial S\,\partial V} = \frac{\partial}{\partial S}\left(\frac{\partial E}{\partial V}\right)_S = -\left(\frac{\partial p}{\partial S}\right)_V,\quad \frac{\partial^2 E}{\partial V\,\partial S} = \frac{\partial}{\partial V}\left(\frac{\partial E}{\partial S}\right)_V = \left(\frac{\partial T}{\partial V}\right)_S,$$
$$-\left(\frac{\partial p}{\partial S}\right)_V = \left(\frac{\partial T}{\partial V}\right)_S. \tag{4.18}$$

After the second law of thermodynamics ($\delta Q = TdS$) having been formulated, the earlier derived relation (4.6) with differentiation of S using the chain rule can be rewritten, separating coefficients at differentials dT and dV, as follows:

$$\left(\frac{\partial E}{\partial T}\right)_V dT + \left[\left(\frac{\partial E}{\partial V}\right)_T + p\right] dV = T\left[\left(\frac{\partial S}{\partial T}\right)_V dT + \left(\frac{\partial S}{\partial V}\right)_T dV\right],$$
$$T\left(\frac{\partial S}{\partial T}\right)_V = \left(\frac{\partial E}{\partial T}\right)_V,\quad T\left(\frac{\partial S}{\partial V}\right)_T = \left(\frac{\partial E}{\partial V}\right)_T + p.$$

Deriving a Maxwell relation for the function of state $S(V,T)$ by mixed differentiation, the second derivatives of $S(V,T)$ can be compared similarly to (4.18), which results in

$$\frac{\partial^2 S}{\partial T\,\partial V} = \frac{\partial}{\partial T}\left(\frac{\partial S}{\partial V}\right)_T = \frac{\partial}{\partial T}\left\{\frac{1}{T}\left[\left(\frac{\partial E}{\partial V}\right)_T + p\right]\right\} = \frac{\partial^2 S}{\partial V\,\partial T} = \frac{\partial}{\partial V}\left(\frac{\partial S}{\partial T}\right)_V = \frac{\partial}{\partial V}\left[\frac{1}{T}\left(\frac{\partial E}{\partial T}\right)_V\right].$$

After cancelling out the second derivatives of $E(V,T)$ in the differentiation result, this equation takes the following form:

$$-\frac{1}{T^2}\left[\left(\frac{\partial E}{\partial V}\right)_T + p\right] + \frac{1}{T}\left(\frac{\partial p}{\partial T}\right)_V = 0,\quad T\left(\frac{\partial p}{\partial T}\right)_V - p = \left(\frac{\partial E}{\partial V}\right)_T. \tag{4.19}$$

It is interesting to see that substitution of thermal EOS (4.3) for an ideal gas to (4.19) results in $(\partial E/\partial V)_T = 0$. This enables one to obtain caloric EOS from the thermal EOS and makes the Joule's law a consequence of the second thermodynamic law. It is also worth noting that with the use of the second law of thermodynamics, a Carnot cycle – represented by sections of hyperbolic curves in the (p, V)-space (Figure 4.1(a)) – takes the form of a straight rectangular in the (T, S)-space.

Equation (4.15) is valid only for processes at equilibrium. For non-equilibrium processes, this equation turns into inequality $dS > \delta Q/T$. In the case of inequality, a difference $\delta Q'/T = dS - \delta Q/T$ is positive and Clausius (1879) has called $\delta Q'$ the uncompensated heat. de Donder and Van Rysselberghe (1936) linked the uncompensated heat with non-equilibrium internal processes for a closed system with multiple components. Due to chemical reactions or other mass-exchanging processes in a system of K components, amounts of components, N_k ($k = 1, \ldots, K$), are independent variables. Hence, the uncompensated heat can be determined as $\delta Q' = \Sigma \mu_k dN_k$. For N moles of a monocomponent gas, a set of independent variables for internal energy is reduced to V, S and N. In this case, the second law of thermodynamics should take into account the uncompensated heat Q' caused by possible change in amount of substance:

$$dE = TdS - pdV + \mu dN, \tag{4.20}$$

where μ is called chemical potential. For the generating function of internal energy $E(V,S,N)$, pressure, temperature and chemical potential can be determined from (4.20) as follows:

$$T = \left(\frac{\partial E}{\partial S}\right)_{V,N}, \quad p = -\left(\frac{\partial E}{\partial V}\right)_{S,N}, \quad \mu = \left(\frac{\partial E}{\partial N}\right)_{V,S}. \tag{4.21}$$

Other thermodynamic functions, including extensive functions such as heat capacity, can also be calculated as derivatives of a function of state. For example, heat capacity for isochoric processes is an extensive parameter and can be obtained as $C_V = (\partial E/\partial T)_{V,N}$ for N moles of gas.

The nature of a thermodynamic processes under examination allows for a choice of independent variables and, consequently, a generating function of state – an important step of the examination. Two examples of the choices above illustrate this point: (i) internal energy for isochoric processes is associated with mechanical transformations, and (ii) enthalpy for isobaric processes is associated with chemical reactions. Alternative choices to the pair (V, S), as independent variables for internal energy, can be any pair of variables, including functions of state. In addition to (V, S), widely used sets of variables in the classical thermodynamics are (p, S), (S, V), and (T, p), called natural independent variables for the corresponding generating functions of state. Accordingly, thermodynamic functions can be obtained with use of a Legendre transformation, modifying internal energy to enthalpy $H = E + pV$ as considered earlier, Helmholtz free energy $F = E - TS$, and Gibbs energy $G = E + pV - TS$. Expressing internal energy E via above-defined functions H, F, and G and substituting the results for E to (4.20), alternative forms of the thermodynamic identity can be derived for these pairs of natural variables complemented with variable N for amount of substance:

$$\begin{aligned} dE &= TdS - pdV + \mu dN, \quad dH = TdS + Vdp + \mu dN, \\ dF &= -SdT - pdV + \mu dN, \quad dG = -SdT + Vdp + \mu dN. \end{aligned} \tag{4.22}$$

Thus, formulations (4.22) enable one to specify alternative generating functions $H(p,S,N)$, $F(V,T,N)$ and $G(p,T,N)$. Using these functions, dependent variables can be calculated from (4.22), similarly to the variables (4.21) calculated from the thermodynamic identity (4.20) for the generating function $E(V,S,N)$.

For closed thermodynamic systems with many components, the systems can undergo various mass and energy transformations due to internal processes associated with inter-component exchange. One of the reasons for such transformations is chemical reactions, resulting in mass exchange between the components (chemical species) and energy exchange with the environment.

For isotropic systems containing K substances A_k ($k = 1, \ldots, K$) with amounts N_k (the number of moles of A_k), internal energy with natural extensive variables takes the following form:

$$E = E(V, S, N_1, \ldots, N_K), \tag{4.23}$$

resulting in the second thermodynamic law similar to (4.20):

$$dE = TdS - pdV + \sum_{k=1}^{K} \mu_k dN_k .$$

The equilibrium requirement inherent in classical thermodynamics is overly restrictive when applied to real, time-dependent processes. This issue was a reason for an advancement of classical thermodynamics and its methods to the area of non-equilibrium (de Groot and Mazur, 1962). Even in classical thermodynamics, states in a quasi-equilibrium process are progressing in time, which is critical for fluid dynamics, detonation and shock wave phenomena and inevitably requires consideration of time. Non-equilibrium thermodynamics relaxes the requirement of process stationarity for the whole system. In order to apply methods and tools of the classical thermodynamics, two important additions to the thermodynamic postulates are made: (i) introduction of time, such that equilibrium is assumed within a timeframe where system states can be considered at local equilibrium. This is an appropriate approximation for the case when a characteristic time of a global phenomenon, involving thermodynamic subsystems, is essentially larger than characteristic times of thermodynamic processes occurring in the subsystems, which makes these subsystems be at quasi-equilibrium with respect to the global system; and (ii) local equilibrium in space, which requires applicability of the laws of thermodynamics in small subsystems of the original system, called representative volumes. This means that local gradients of parameters within the volumes are sufficiently small. Therefore, thermodynamic relations, used globally for the whole system in classical theory, can be applied to thermodynamic parameters related to a representative unit of mass, volume, or amount of substance that contain statistically large amount of particles of the substance.

Because methods of non-equilibrium thermodynamics require an analysis of small subsystems, the arsenal of classical thermodynamics needs to be adjusted by applying the classical theory to variables characterizing a system in local sense. As seen from (4.21), any intensive variable, such as p, T and μ, can be obtained by differentiation of an extensive quantity (function of state) over corresponding natural independent extensive variable. Similarly, for the local analysis, thermodynamic parameters can be 'localized' by apportioning extensive thermodynamic variables to a unit of mass, volume, or amount of subsystem. For example, for a homogenous subsystem of mass M, any extensive variable in (4.21) can be rewritten as specific intensive variable: specific internal energy $e = E/M$, specific volume $\upsilon = V/M$, specific entropy $s = S/M$, and inverse molar mass $n = N/M$. These redefinitions result in a localized formulation of the thermodynamic identity:

$$\begin{gathered} Tds = de + pd\upsilon - \mu dn, \\ T = \left(\frac{\partial e}{\partial s}\right)_{\upsilon,n}, \quad p = -\left(\frac{\partial e}{\partial \upsilon}\right)_{s,n}, \quad \mu = \left(\frac{\partial e}{\partial n}\right)_{\upsilon,s}. \end{gathered} \tag{4.24}$$

It is obvious that specific volume is a reciprocal of the average density of substance $\upsilon = V/M = 1/\rho$, which gives another representation for calculation of pressure in (4.24): $p = \rho^2 e_\rho$. Other extensive parameters can be obtained in similar fashion, for example, specific heat capacity $c_\upsilon = C_V/M$. Similarly, extensive parameters for alternative functions of state in (4.22) can also be related to a unit of mass: $h = H/M, f = F/M$, and $g = G/M$. Correspondingly, amount of substance N_k can be specified as $n_k = N_k/M = c_k/W_k$, where W_k is molar weight of component A_k and $c_k = M_k/M$ is mass concentration of the component. Alternative 'local' definitions can be obtained by apportioning extensive variables (i) to total volume or to volume increment (volume derivative for non-uniform mixtures), replacing specific variables defined above by 'densities' such as energy density E/V (or dE/dV, which is obviously linked with specific energy e as $dE/dV = (dE/dM)(dM/dV) = \rho e$) or (ii) to total amount of substance, specifying molar variables such as molar enthalpy H/N (or dH/dN).

For any process, based on the second law of thermodynamics in the local sense, non-equilibrium thermodynamics considers entropy production or dissipation function as a quantity, which specifies direction of the process. From the definition of uncompensated heat Q' by Clausius,

$$\frac{\delta Q'}{T} = dS - \frac{\delta Q}{T},$$

the total entropy increase dS can be decomposed as a sum

$$dS = d_e S + d_i S,$$

representing (i) external entropy increase $d_e S = \delta Q/T$ in the system due to effects external to the system, and (ii) internal 'uncompensated' part $d_i S = \delta Q'/T$ due to internal non-equilibrium processes inside the system.

This leads to an important attribute of non-equilibrium thermodynamics associated with time t, namely, entropy production for the 'uncompensated' heat characterized by

$$\sigma' = \frac{d_i S}{dt}.$$

which is energetically associated with the dissipation function Φ':

$$\Phi' = \frac{\delta Q'}{dt} = T\frac{d_i S}{dt} = T\sigma'.$$

Such defined dissipation function should be non-negative due to the second law of thermodynamics.

The uncompensated heat $\delta Q'$ due to a chemical reaction can be specified in the local sense ($\delta q' = \delta Q'/M$) for a system with K chemical species as $\delta q' = (\partial e/\partial n_k)\, dn_k = \mu_k\, dn_k$. Here, component amounts N_k for each specie in the mixture are variables of specific internal energy $e(\upsilon, s, n_1, \ldots, n_K)$, where its partial derivative over amount of substance, $\mu_k = \partial E/\partial N_k = \partial e/\partial n_k = W_k(\partial e/\partial c_k)$ at all remaining parameters fixed, is called chemical potential. Using analogues of the chemical potentials in the case of mass concentration parameters: $\Lambda_k = \partial e/\partial c_k$, the thermodynamic identity (4.24) for K components can be rewritten as

$$de = Tds - pd\upsilon + \sum_{k=1}^{K}\left(\frac{\partial e}{\partial c_k}\right)dc_k. \tag{4.25}$$

The thermodynamic identities (4.22) can be rewritten, similarly to (4.25), for alternative choices of natural independent variables related to the corresponding specific functions of state: specific enthalpy $h(p, s, N_1, \ldots, N_K) = e + p\,\upsilon$, specific free energy $f(\upsilon, T, N_1, \ldots, N_K) = e - T\,s$, and specific Gibbs potential $g(p, T, N_1, \ldots, N_K) = e + p\,\upsilon - T\,s$, as follows:

$$\begin{aligned} &dh = Tds + \upsilon dp + \sum_{k=1}^{K}\left(\frac{\partial e}{\partial c_k}\right)dc_k, \quad df = -sdT - pd\upsilon + \sum_{k=1}^{K}\left(\frac{\partial e}{\partial c_k}\right)dc_k, \\ &dg = -sdT + \upsilon dp + \sum_{k=1}^{K}\left(\frac{\partial e}{\partial c_k}\right)dc_k. \end{aligned} \tag{4.26}$$

To remind, the extensive variables apportioned to mass, while introducing specific variables above, could be apportioned to any extensive quantity such as amount of substance N (molar variables) or volume V (density variables), depending on nature of the process.

For localized specific variables, general principles of the non-equilibrium thermodynamics (de Groot and Mazur, 1962), are based on a local entropy production $\sigma = ds/dt$ related to dissipation function $\Phi = \delta q'/dt$ as $\Phi = T\sigma$. The dissipation function represents a total product of J_k and X_k, where irreversible generalized fluxes J_k are caused by generalized forces X_k:

$$T\frac{ds}{dt} = \Phi = \sum_{k=1}^{K} X_k J_k, \quad J_k = \frac{dq_k}{dt}, \tag{4.27}$$

where q_k are generalized coordinates. Forces X_k causing the fluxes (flows of some quantities q_k in time) are commonly gradients of a transfer potential Ψ: $X_k = -\nabla\Psi$. For example, forces X_k such as gradients of internal energy in (4.25) are causing flow in the coordinates presented as a change in mass concentration c_k (for example, mass fluxes of components due to phase transitions or chemical reactions), resulting in the following contribution to the dissipation function for kth component: $X_kJ_k = -(\partial e/\partial c_k) dc_k/dt$. For a reversible process in an adiabatic system $\sigma = d_is/dt = 0$, however, for a system with internal energy sensitive to parameter c_k, where these generalized coordinates are managed by constitutive equations $dc_k/dt = -\varphi_k$, internal dissipation is possible with $\Phi = \Sigma X_kJ_k = \Sigma(\partial e/\partial c_k)dc_k/dt = \Sigma\Lambda_k\varphi_k$.

In summary, this section briefly described postulates of classical thermodynamics and major statements of non-equilibrium thermodynamics, followed by consequential relations. These relationships are the basis for detonation research, including chemistry- and physics-related aspects addressed in subsequent sections.

4.3 CHEMICAL REACTIONS AND THERMAL EFFECTS

The critical point of the detonation chemistry is a chemical transformation of the explosive material into a gaseous mixture of detonation products. The transformation is accompanied by a release of a large amount of energy via an exothermic chemical reaction. This transformation requires an analysis of the chemical reaction process in order to determine: (i) what is timeframe of the process of energy release; (ii) how much energy is released; (iii) what is final composition of the products, and (iv) what are physical parameters of the products after the transformation. The first two items of these questions will be addressed in the present section.

In general, simplest reactions, performing a single molecular rearrangement, called elementary reactions, involve 1, 2 or, less likely, 3 molecules, corresponding to monomolecular, bimolecular, or termolecular reactions, respectively. Typical reactions such as those at combustion or detonation, transform a chemical composition of reactants to that of products, superposing from tens up to hundreds of elementary reactions with highly different reaction rates. An analysis of a complete set of reactions can be very difficult, therefore, researchers frequently consider only one (global), or a very limited set of complex reactions transforming a source composition of reactants to final products. The corresponding effective reaction equations are derived either from empirical data or from an analysis of a system of elementary reactions for determination of reaction parameters and rates within a reaction mechanism. An example of an elementary reaction is the hydroxyl radical reaction $OH + H_2 \rightarrow H_2O + H$. In contrast, the hydrogen explosion $2H_2 + O_2 \rightarrow 2H_2O$ is an example of a global reaction which contains dozens of elementary reactions, such as the hydroxyl radical reaction.

Firstly, a single reaction involving K substances of a system is considered, which can be formulated as follows:

$$\sum_{i=1}^{R} \nu_i' A_i \rightleftarrows \sum_{j=1}^{P} \nu_{j+R}' A_{j+R}, \tag{4.28}$$

here A_i and A_j are chemical symbols representing species for R reactants and P reaction products ($R + P = K$), and ν'_i, ν'_j are corresponding stoichiometric coefficients. For example, for the hydrogen explosion reaction drawn above, $A_1 = H_2$, $A_2 = O_2$, $A_3 = H_2O$, $\nu'_1 = 2$, $\nu'_2 = 1$, $\nu'_3 = 2$. To note, sets of reactants and products may contain common elements. Double arrow in (4.28) indicates that the reaction may potentially go in the both directions: from reactants to products (forward reaction) and in the opposite direction (reverse reaction). If W_i and W_j are the corresponding molecular weights of reactants and products, and the substances are taken in a ratio such that reactant species are completely transformed to products during the reaction, then the balance reaction condition (the conservation law) can be written down as

$$W_0 = \sum_{i=1}^{R} \nu'_i W_i = \sum_{j=1}^{P} \nu'_{j+R} W_{j+R}. \tag{4.29}$$

For components of the system involved in reaction (4.28), amount of kth substance is N_k. As a result, the corresponding mass is $M_k = W_k N_k$ and total mass of the system $M = \Sigma M_k$, where summations over k, i and j in this subsection are from 1 to K, 1 to R and 1 to P, respectively. Current average molecular weight W could be calculated as $W = M/N$, where N is current amount of moles in the mixture $N = \Sigma N_k$. For kth substance, partial density ρ_k^p is defined as ratio of the corresponding mass M_k to the mixture volume V: $\rho_k^p = M_k/V$, which results in $\rho = \Sigma \rho_k^p$ ($\rho = M/V$). Therefore, current mass concentration of kth substance in the mixture is defined as $c_k = \rho_k^p/\rho = M_k/M$ ($k = 1, \ldots, K$) and this set has only $K - 1$ independent variables because $\Sigma c_k = 1$. With definition of molar concentration $x_k = N_k/N$, average molecular weight of the mixture can be obtained as $W = \Sigma M_k/N = \Sigma W_k N_k/N = \Sigma x_k W_k$. In the case of known mass concentrations, $1/W = N/M = \Sigma N_k/M = (\Sigma M_k/W_k)/M = \Sigma c_k/W_k$. The reaction formula (4.28) is frequently rewritten as one-sided relation, which, for the forward reaction, results in the following:

$$\sum_{k=1}^{K} \nu_k A_k \rightarrow 0, \tag{4.30}$$

where stoichiometric coefficients in (4.30) are taken in accordance with the convention of negative coefficients for reactants and positive for products: $\nu_i = -\nu'_i$ ($i = 1, \ldots, R$), $\nu_{j+R} = \nu'_j$ ($j = 1, \ldots, P$). In this formulation, possible common species in reactants and products on both sides in (4.28) are numbered differently. Therefore, for a component A_i, the corresponding coefficients ν'_i in reactants need to be subtracted from those for the same component in the products. It is clear that the mass conservation law (4.29) can be rewritten in the following form for reaction (4.30):

$$\sum_{k=1}^{K} \nu_k W_k = \sum_{i=1}^{R} \nu_i W_i + \sum_{j=1}^{P} \nu_{j+R} W_{j+R} = -W_0 + W_0 = 0. \tag{4.31}$$

With a reaction progressing, masses M_k and component amounts N_k change in time. In order to characterize the reaction process, the law of definite proportions (Prigogine and Defay, 1954) introduces a parameter called the reaction extent ξ'. This law claims that a change in N_k is proportional to the change in parameter ξ' multiplied by the corresponding stoichiometric coefficient ν_k: $dN_k = \nu_k d\xi'$. This means that the reaction extent shows how many units of the substance involved in the reaction have been expensed in reactants and generated in products (ξ' is measured in the mole units).

Similarly to a definition by Zemansky and Dittman (1997), the reaction extent parameter can be nondimensionalized (denoted by ξ). Assuming fully balanced reaction, the parameter is obtained from ξ' by division over a coefficient N_{00}, calculated as ratio of the total mass M over total molecular weight W_0 ($N_{00} = M/W_0$) of the components involved in the reaction. Molecular weight, W_0, is equal to the weight of reactants before and after the reaction due to the mass conservation law (4.29), assuming the reactants are fully transformed to the products of reaction. Thus, equations for these reaction extent parameters are

$$d\xi = \frac{W_0}{M}\frac{dM_1}{\nu_1 W_1} = \ldots = \frac{W_0}{M}\frac{dM_K}{\nu_K W_K}, \quad d\xi' = \frac{dN_1}{\nu_1} = \ldots = \frac{dN_K}{\nu_K}, \quad \xi = \frac{\xi'}{N_{00}}. \tag{4.32}$$

The definition (4.32) means that masses and amounts of the reactant species decrease because $\nu_i < 0$ and masses of the product species increase because $\nu_j > 0$, proportionally to the number of reaction units. Equation (4.32) can also be rewritten for the mass concentration parameters in the following form

$$dc_k = \frac{\nu_k W_k}{W_0} d\xi, \quad k = 1,\ldots,K. \tag{4.33}$$

In case of complete transformation of reactants to reaction products, initial concentrations of reactants are $c_i^0 = -\nu_i W_i/W_0$ ($\nu_i < 0$, $i = 1, \ldots, R$) and for products: $c_j^0 = 0$ ($j = R + 1, \ldots, K$). Correspondingly, for the final state after the reaction: $c_i = 0$ ($i = 1, \ldots, R$) and $c_j = \nu_j W_j/W_0$ ($j = R + 1, \ldots, K$). In general, the concentration parameters are not independent ($\Sigma c_k = 1$) and can be calculated using the parameter ξ as follows:

$$c_i = c_i^0 + \frac{\nu_i W_i}{W_0}\xi, \quad c_j = \frac{\nu_j W_j}{W_0}\xi, \quad i = 1,\ldots,R, \quad j = R+1,\ldots,K, \tag{4.34}$$

which results in $\Sigma c_i^0 = 1$ at $\xi = 0$ (the summation over i is from 1 to R) from the mass conservation (4.31) for reactants and $c_j^0 = 0$ ($j = R + 1, \ldots, K$) for products. At the end of the reaction (at $\xi = 1$): $c_i = 0$ ($i = 1, \ldots, R$) for reactants and from the mass conservation (4.31) $\Sigma c_j = 1$ for products (the summation over j is from $R + 1$ to K).

For illustration, an example of hydrogen explosion in the stoichiometric hydrogen-air mixture can be considered, using the reaction representation (4.28). Taking air as a mixture of approximately 20% oxygen and 80% nitrogen by molar concentration, the chemical balance equation for one mole of oxygen in the hydrogen-air mixture is

$$2H_2 + O_2 + 4N_2 \rightarrow 2H_2O + 4N_2,$$

which gives $R = 3$, $\nu'_1 = 2$, $\nu'_2 = 1$, $\nu'_3 = 4$, and $P = 2$, $\nu'_4 = 2$, $\nu'_5 = 4$. Reformulating this and using one-sided reaction representation (4.30):

$$-2H_2 - O_2 - 4N_2 + 2H_2O + 4N_2 \rightarrow 0 \tag{4.35}$$

with $K = 5$, it follows $\nu_1 = -2$, $\nu_2 = -1$, $\nu_3 = -4$, $\nu_4 = 2$, $\nu_5 = 4$. To note, nitrogen (N_2) does not participate in the reaction, which is expressed by $\nu_3 + \nu_5 = 0$, however, this component is a part of the total mass M in the mixture of reactants and products. Initial amounts of the components are $N_1^0 = 2$, $N_2^0 = 1$, $N_3^0 = 4$, $N_4^0 = 0$, $N_5^0 = 0$ and final amounts (after the reaction having completed): $N_1^f = 0$, $N_2^f = 0$, $N_3^f = 0$, $N_4^f = 2$, $N_5^f = 4$. With the present component molar weights in g/mol: $W_1 = 2$, $W_2 = 32$, $W_3 = 28$, $W_4 = 18$, and $W_5 = 28$, the total molar mass of the reactants: $W_0 = 148$ g/mol, initial masses of the components of the mixture before reaction are $M_1^0 = 2W_1 = 4$ g, $M_2^0 = W_2 = 32$ g, $M_3^0 = 4W_3 = 112$ g, $M_4^0 = M_5^0 = 0$, and final masses: $M_1^f = M_2^f = M_3^f = 0$, $M_4^f = 2W_4 = 36$ g, $M_5^f = 4W_5 = 112$ g. Therefore, total mass of the mixture $M = \Sigma M_i^0 = \Sigma M_i^f = 148$ g, initial mass concentrations $c_1^0 = 0.027$, $c_2^0 = 0.216$, $c_3^0 = 0.757$, $c_4^0 = 0$, $c_5^0 = 0$, and final mass concentrations: $c_1^f = c_2^f = c_3^f = 0$, $c_4^f = 0.243$, $c_5^f = 0.757$. Thus, mass concentrations are calculated from (4.33) as:

$$c_1 = c_1^0(1 - \xi), \; c_2 = c_2^0(1 - \xi), \; c_3 = c_3^0(1 - \xi)c, \quad c_4 = c_4^f\xi, \quad c_5 = c_5^f\xi.$$

As seen from this solution, A_3 and A_5 denote the same substance (N_2), with total mass concentration for this component $c_3 + c_5 = c_3^0 = c_5^f = \text{const}$, that is not changing during the reaction.

The reaction extent parameter ξ' is a variable of reaction equation, which is changing in time with certain velocity J. The reaction equation can be written (de Donder and Van Rysselberghe, 1936) for single reaction as follows:

$$\frac{d\xi'}{dt} = J. \tag{4.36}$$

This equation can be rewritten for N_k and mass concentrations c_k, using (4.33):

$$\frac{dN_k}{dt} = \nu_k J, \quad \frac{dc_k}{dt} = \frac{\nu_k W_k}{M} J. \tag{4.37}$$

Instead of substance amount N_k, chemists prefer molar concentration (molar densities) in mol/cm^3, denoted as $[A_k] = N_k/V = \rho_k^p/W_k$ that are interchangeable in this section with $\beta_k = [A_k]$. For a fixed volume V, a reaction equations for $[A_k]$ can be derived from (4.37)

$$\frac{d[A_k]}{dt} = \nu_k \, J/V = \nu_k J_0 .$$

Stoichiometric coefficients of reactants, such as a and b in an elementary reaction $a\mathrm{A} + b\mathrm{B} \rightarrow \mathrm{C}$, characterize how many molecules of substances A and B are involved in the reaction, forming product C. The integer parameters a and b are called reaction orders to components A and B, and the sum $a + b$ is called total reaction order (the complete molality). For monomolecular reactions of dissociation (A → products) the total reaction order is 1, for bimolecular (A + B → products or A + A → products) – 2, for termolecular (A + B + C → products, or 2A + B → products, or A + A + A → products) – 3. The number of molecule interactions controls reaction velocity that is proportional to a product of reactant concentrations. In the case of interaction of multiple molecules of the same sort when a or b are more than 1, multiplication of the same reactant concentrations a or b times results in the corresponding power exponents. Thus, reaction rate for an elementary reaction with stoichiometric coefficients a and b of the reactants is proportional to $[A]^a[B]^b$. Temperature is an important characteristic to favour molecule interactions and to activate reaction, which is a key contributor to additional proportionality coefficient k. This is physics-based and called the reaction rate constant. The reaction constant is a measurable function known for a large number of substances (Chase, 1998) from experiments and is usually described by Arrhenius law

$$k_r = AT^b \exp\left(-\frac{E}{RT}\right), \tag{4.38}$$

where A is pre-exponential coefficient, b is exponent index, and E is activation energy. Parameters A, b, and E are empirical, where E is treated as an energetic barrier for the reaction to start.

Summarizing, in a case of analysis of a full kinetic mechanism for K components subjected to R elementary reactions with reaction constants k_r shown above the reaction arrow, the system of reactions can be put in writing as follows:

$$\sum_{k=1}^{K} \nu_{kr}^{(r)} A_k \xrightarrow{k_r} \sum_{k=1}^{K} \nu_{kr}^{(p)} A_k, \quad r = 1,\ldots,R, \tag{4.39}$$

where some of the stoichiometric coefficients for reactants, $\nu^{(r)}{}_{rk}$, and products, $\nu^{(p)}{}_{rk}$, can be zero, because sums on the both sides contain all, reactant and product, components of the reaction. As discussed above, rate of the forward rth reaction in (4.39) depends on concentrations of reactants and it is proportional to k_r and to molar concentrations of the reactants $\beta_k = [A_k]$ in mol/cm^3 with exponents equal to the reaction orders. Thus, the kinetic laws for reaction extent of each reaction and for molar concentration of each component are

$$\frac{d\xi_r'}{dt} = J_r, \quad J_r = k_r\left(v_{rk}^{(p)} - v_{rk}^{(r)}\right)\prod_{s=1}^{K}(\beta_s)^{v_{rs}^{(r)}}, \quad r = 1,\ldots,R,$$
$$\frac{d\beta_k}{dt} = \sum_{r=1}^{R} J_r, \quad k = 1,\ldots,K, \tag{4.40}$$

where J_r is velocity of rth reaction. For R multiple reactions with all components written on the left side: $\Sigma\nu_{rk}A_k \to 0$, ($r = 1, \ldots, R$), change in component masses during the reactions can be calculated from $M_k = W_kN_k$:

$$dM_k = W_k\sum_{r=1}^{R} v_{rk}d\xi_r'.$$

Then, from $\Sigma W_k\nu_{rk} = 0$ obtained from (4.31) for rth reaction, the mass conservation law for each of the reactions results in conservation for all reactions:

$$\sum_{k=1}^{K}\left(\sum_{r=1}^{R} W_k v_{rk} d\xi_r'\right) = \sum_{r=1}^{R}\left(\sum_{k=1}^{R} W_k v_{rk}\right) d\xi_r' = 0.$$

For an example of elementary termolecular reaction A + 2B → C with reaction constant k, reaction equations for component concentrations take the following form:

$$\frac{d[\mathrm{A}]}{dt} = -k[\mathrm{A}][\mathrm{B}]^2, \quad \frac{d[B]}{dt} = -2k[\mathrm{A}][\mathrm{B}]^2, \quad \frac{d[\mathrm{C}]}{dt} = k[\mathrm{A}][\mathrm{B}]^2.$$

A full kinetic mechanism includes all elementary reactions, which results in a large number of reaction equations. On the other hand, a global reaction formulates only a single relation between initial reactants and final products, ignoring intermediate products. In the reaction equation for a global reaction, reaction constant k is an effective parameter calibrated for relevant range of thermodynamic parameters. For example, a combination of the following two, monomolecular (dissociation) and bimolecular, elementary reactions results in a global reaction with reaction products, cancelling out an intermediate product C:

$$\mathrm{A} \xrightarrow{k_1} \mathrm{B} + \mathrm{C}, \quad \mathrm{C} + \mathrm{D} \xrightarrow{k_2} \mathrm{E} \quad \Rightarrow \quad \mathrm{A} + \mathrm{D} \xrightarrow{k_{eff}} \mathrm{B} + \mathrm{E},$$

where k_1 and k_2 above arrows show the reaction rate constants for corresponding elementary reactions. Applying (4.40) to specification of reaction velocities and keeping in mind that the intermediate product [C] is created in the first reaction and expended in the second, the reaction equations for component concentrations of the elementary reactions are

$$\frac{d[\mathrm{A}]}{dt} = -k_1[\mathrm{A}], \quad \frac{d[\mathrm{B}]}{dt} = k_1[\mathrm{A}], \quad \frac{d[\mathrm{C}]}{dt} = k_1[\mathrm{A}] - k_2[\mathrm{C}][\mathrm{D}],$$
$$\frac{d[\mathrm{D}]}{dt} = -k_2[\mathrm{C}][\mathrm{D}], \quad \frac{d[\mathrm{E}]}{dt} = k_2[\mathrm{C}][\mathrm{D}].$$

For the case of constant rates k_1 and k_2, the equations for [A], [B] and [C] − [D] can be easily solved (for example, $[B](t) = [B]_0 + [A]_0[1 - \exp(-k_1 t)]$). Expressing, for example, [C] = [D] +([C] − [D])(t), the equations for [D] and [C] can be solved as well, followed by solution of the last equation for [E]. Thus, effective rate k_{eff} and exponents α and β for the reactant concentrations can be evaluated, resulting in the following equations describing the global reaction:

$$-\frac{d[\mathrm{A}]}{dt} = -\frac{d[\mathrm{D}]}{dt} = \frac{d[\mathrm{B}]}{dt} = \frac{d[\mathrm{E}]}{dt} = J_0, \quad J_0 = k_{eff}[\mathrm{A}]^{\alpha}[\mathrm{D}]^{\beta}.$$

It should be noted that the reaction velocity J_0 for the global reaction can now contain exponents α and β for [A] and [D], and the exponents can be non-integer and non-positive. When specifying effective reaction rates, the use of process-based assumptions and information about reaction process is customary. For example, for the dominating influence of reaction rates of slow elementary reactions, the Bodenstein method of quasi-stationary concentrations for various elementary reactions (Donoghue, 1974; Semenoff, 1939; etc.) is common in textbook content (Warnatz et al., 2006). Use of reaction equations for global reactions reduces the number of equations and saves significantly computational resources.

Chemical kinetics and associated inter-component mass exchange in closed systems brings specifics into representation of thermodynamic laws. Multiple components of chemical reaction require an EOS for multi-component mixture that takes the form (4.23) for caloric EOS. Thermal EOS (4.3) for an ideal gas can be formulated for multi-component mixture and for each component in the following forms:

$$pV = \sum_{k=1}^{K} N_i RT, \quad p_i V = N_i RT, \tag{4.41}$$

where N_i are molar amounts of the components and p_i are partial pressures. Caloric and thermal equations of state (4.23) and (4.41) can be reformulated for component masses $M_i = W_i N_i$, resulting in

$$E = E(V, S, M_1 \ldots, M_K), \quad p_i V = \frac{M_i}{W_i} RT. \tag{4.42}$$

With the possibility of internal dissipation due to change in component amounts, N_k, or component masses, M_k, the thermodynamic identity takes the following forms:

$$TdS = dE + pdV - \sum_{k=1}^{K} \mu_k dN_k, \quad TdS = dE + pdV - \sum_{k=1}^{K} \Lambda_k dM_k, \tag{4.43}$$

where functions μ_k and Λ_k are called the chemical potentials and can be calculated from (4.43) as follows:

$$\mu_k = \left(\frac{\partial E}{\partial N_k}\right)_{V,S,N_{i,i\neq k}}, \quad \Lambda_k = \left(\frac{\partial E}{\partial M_k}\right)_{V,S,M_{i,i\neq k}}. \tag{4.44}$$

As formulated in the previous section, relations (4.42)–(4.44) can be rewritten in terms of specific variables apportioned to total mass M of the system (or, alternatively, to N or V).

Using the definitions of chemical potentials (4.44), uncompensated heat $M\delta q$ for the whole system (non-locally) with a single chemical reaction (4.30) and conventional variable for reaction extent, ξ', is reduced to

$$M\delta q = \sum_{k=1}^{K} \mu_k dN_k = \sum_{k=1}^{K} \nu_k \mu_k d\xi' = -A' d\xi', \tag{4.45}$$

where A' is called the chemical affinity, following traditional definition by de Donder and Van Rysselberghe (1936). In the framework of non-equilibrium thermodynamics (de Groot and Mazur, 1962), the affinity is a generalized force that is driving chemical reaction. For a localized consideration with nondimensionalized extent of reaction ξ (4.32), comparison of the whole uncompensated heat (4.45) with uncompensated heat related to a mass unit of the system

$$\begin{gathered}\delta q = \sum_{k=1}^{K} \Lambda_k dc_k = \frac{1}{W_0} \sum_{k=1}^{K} \nu_k W_k \Lambda_k d\xi = -Ad\xi, \\ A'd\xi' = N_{00} A'd\xi = MAd\xi, \quad A' = \frac{M}{N_{00}} A = W_0 A,\end{gathered} \tag{4.46}$$

enables one to relate traditional definition of affinity A' based on the molar extent of reaction ξ' to the definition of affinity A based on the nondimensionalized extent of reaction ξ.

Summarizing, the thermodynamic identities taking into account chemical reactions, follow representations (4.24) and (4.26) with the affinity A driving reaction at the rate characterized by reaction extent ξ:

$$\begin{gathered}de = Tds - pd\upsilon - Ad\xi, \quad dh = Tds + \upsilon dp - Ad\xi, \\ df = -sdT - pd\upsilon - Ad\xi, \quad dg = -sdT + \upsilon dp - Ad\xi.\end{gathered}$$

It is interesting to note that Gibbs energy G is the only generating function of state with the pair of independent intensive variables p and T, which is commonly used when reactions are running at constant pressure and temperature. Condition of equilibrium in a closed thermodynamic system with chemical reactions is formulated as a condition of achieving maximal entropy. Thus, at equilibrium, the entropy does not change: $ds = 0$, which, for reacting systems, results in a minimum for Gibbs energy. Following this and taking relations (4.46) scaled by W_0, the condition of equilibrium for chemically reactive systems is

$$A' = AW_0 = -\sum_{k=1}^{K} \nu_k W_k \Lambda_k = -\sum_{k=1}^{K} \nu_k \mu_k = 0. \tag{4.47}$$

Chemical affinity A and Gibbs energy are critical functions managing the reaction direction and identifying composition of reaction products at equilibrium.

In order to estimate the energy released during reaction, it is necessary to find out how strong the molecular bonds are and what energy can be released or absorbed during dissociation or formation of a substance. Formation of compound substances can be considered as a product of a reaction from basic source substances, which is accompanied by release or absorption of heat. From the first law of thermodynamics (4.2), amount of heat transferred to/from the system enables one to assess change of internal energy at isochoric processes (V = const) because $dE = dQ_V$. Thus, in this case Q_V is a function of state and the heat transferred in the process from initial ('0') to current state is

$$Q_V = \int_0 dQ = \int_0 dE = E - E_0 = \Delta E. \tag{4.48}$$

Similarly, for isobaric processes (p = const): $dH = dQ_p$, resulting in

$$Q_p = \int_0 dQ = \int_0 dH = H - H_0 = \Delta H. \tag{4.49}$$

Correspondingly, from the definitions of full thermal capacity for these processes:

$$dE = C_V dT, \quad dH = C_p dT. \tag{4.50}$$

For exothermic processes $\Delta H < 0$ (for isobaric processes) or $\Delta E < 0$ (for isochoric processes), stating that internal energy or enthalpy is released from the system to the environment. In formulas of the following section, extensive variables will be treated as molar ones.

Thus, a change of internal energy or enthalpy can be obtained from (4.50) by measuring amount of heat from the change in system's temperature in the processes such as chemical reactions. Germain Hess observed (Hess, 1840) that change of heat (such as enthalpy ΔH in isobaric processes) for a substance obtained as a product of different chemical reactions is the same regardless of the reactions used for formation of the substance. For any substance, enthalpy of formation at standard conditions ($T = T_0 = 298.15°K$, $p = p_0 = 0.1$ MPa) is denoted $\Delta_f H^0_{298}$ (superscript '0' relates enthalpy to $p = p_0$). A convention is accepted that for pure substances (atomic substances in a stable state) standard enthalpies of formation are zero. Therefore, as a result of a reaction leading to formation of substances of the products, enthalpy change $\Delta_R H^0$ for the products (marked by subscript 'R' for the corresponding change due to reaction) can be obtained from contributions of the enthalpies of formation $\Delta_f H^0$ of components in the reaction, according to the Hess's law (Hess, 1840; Prigogine and Defay, 1954). This means that enthalpy of formation of a substance is the amount of heat to expend in order to form the substance from the constituting components. Thus, any reaction can be characterized by the heat of reaction and Hess's law can be written for the reaction (4.30) as follows:

$$\Delta_R H^0 = \sum_{k=1}^{K} \nu_k \Delta_f H_k^0, \tag{4.51}$$

where $\Delta_f H^0{}_k$ is standard enthalpy of formation for substance A_k. Considering combustion as a reaction of oxidation, the heat of reaction at standard conditions $\Delta_c H^0$ can also be calculated from standard heats of combustion $\Delta_c H^0{}_k$. If a substance was formed as a result of the reaction, then this value from (4.51) provides enthalpy of formation for the substance. From the definition of enthalpy of formation for a substance it follows that it is positive when heat is expended for the formation of the substance and negative when energy is released during the process of formation. The latter occurs in exothermic processes such as combustion. From the definitions of thermodynamic potentials for isothermal-isobaric and isothermal-isochoric processes, the corresponding energy changes during the reaction are

$$\Delta G = \Delta H - T\Delta S, \; \Delta F = \Delta E - T\Delta S. \tag{4.52}$$

Thus, for example, for Gibbs energy it follows from (4.51) that for reaction (4.30),

$$\Delta_R G = \sum_{k=1}^{K} \nu_k \Delta_f G_k.$$

In thermochemical calculations, enthalpy or energy changes are commonly required at high temperatures, which can be obtained through using and combining of the thermodynamic relations written for molar variables in (4.4), (4.16), $(\delta Q/dT)_p = (\partial H/\partial T)_p = T(\partial S/\partial T)_p$, and $(\partial(G/T)/\partial T)_p = (\partial(H/T)/\partial T)_p - (\partial S/\partial T)_p$, as follows:

$$\left(\frac{\partial(\Delta H)}{\partial T}\right)_p = \Delta C_p^0, \quad \left(\frac{\partial(\Delta S)}{\partial T}\right)_p = \frac{\Delta C_p^0}{T}, \quad \left(\frac{\partial\left(\Delta G/T\right)}{\partial T}\right)_p = -\frac{\Delta H}{T^2}, \tag{4.53}$$

where $\Delta C^0_p = \Sigma\nu'_j C^0_{p,j} - \Sigma\nu'_i C^0_{p,i} = \Sigma\nu_k C_{p,k}$, $C_{p,k}$ are molar capacities (in J/(mol·grad)) of the components. Dependencies $C_p(T)$ are available for a large number of elements and chemical compositions from empirical tables (Chase, 1998) in a general form

$$C_{p,k}(T) = a_k + b_k T + c_k T^2 + c'_k/T^2 + d_k T^3, \tag{4.54}$$

In order to obtain a change ΔC^0_p in (4.53), changes Δa, Δb, Δc, $\Delta c'$, and Δd during the reaction are required, replacing the corresponding coefficients in (4.54). Specifically, $\Delta a = \Sigma\nu'_j a_j - \Sigma\nu'_i a_i = \Sigma\nu_k a_k$ is a difference of the coefficient a obtained for products from that for reactants. Finally, enthalpy change at given temperature T can be found by integration of (4.53) with the use of enthalpy (4.51) at standard conditions, resulting in Kirchhoff's equation:

$$\Delta_R H^0(T) = \Delta_R H^0 + \int_{T_0}^{T} \Delta C^0_p(T')dT',$$

where thermal capacities are taken from (4.54), employing the coefficient changes in ΔC^0_p as per (4.54) in agreement with the reaction stoichiometry. Here, $\Delta_R H^0$ is standard molar enthalpy of reaction at standard temperature obtained from (4.51) with enthalpies of formation available in JANAF tables (Chase, 1998). Similarly, change of molar Gibbs energy at elevated temperature can be obtained from (4.53) by integration. In a case when ΔG^0_{298} is not known and ΔS^0_{298} is known, ΔS^0_T and ΔG^0_T can be found by integration from (4.53).

Using $V = (\partial G/\partial p)_{N,T} = N(\partial\mu/\partial p)_{N,T}$ for a monocomponent ideal gas, chemical potential can be easily found from the EOS (4.3), written down as $NRT/p = N(\partial\mu/\partial p)_{N,T}$, by integration over p:

$$\mu = \mu^0(T, p_0) + RT\ln\left(\frac{p}{p_0}\right). \tag{4.55}$$

For multi-component mixtures, a chemical potential calculated from the Gibbs energy is

$$\mu_k = \left(\frac{\partial G}{\partial N_k}\right)_{p,T,N_{s,s\neq k}}, \tag{4.56}$$

which can be obtained explicitly for ideal gases from the second law of thermodynamics, using the complete EOS $p_i = N_i RT/V$, $E_i = E_{i0} + C_{V,i}(T - T_0)$ and $C_{p,i} = C_{V,i} + N_i R$, where $C_{p,i}$ and $C_{V,i}$ are full thermal capacities for N_i moles of gas:

$$dS_i = \frac{C_{V,i}dT}{T} + p_i\frac{dV}{T} = \frac{C_{V,i}dT}{T} + \frac{p_i}{T}\left(\frac{RTdN_i}{p_i} + \frac{N_i RdT}{p_i} - \frac{N_i RTdp_i}{(p_i)^2}\right) =$$
$$= \frac{C_{p,i}dT}{T} + RdN_i - \frac{N_i Rdp_i}{p_i}.$$

After integration, this equation gives $S_i(T,p_i)$. Substituting it into $G = \Sigma N_i(E_i + p_i V - TS_i)$ and differentiating G over N_i with remaining parameters fixed in (4.56), a relationship similar to (4.55) can be obtained for the chemical potential of ith component:

$$\mu_i = \mu_i^0(T, p_0) + RT\ln\left(\frac{p_i}{p_0}\right). \tag{4.57}$$

For non-ideal gases, to keep consistency with the ideal gas representation (4.55), chemical potential can be formulated in a similar way

$$\mu = \mu^0(T, p_0) + RT\ln\left(\frac{f(p,T)}{f(p_0,T_0)}\right) = \mu^0(T, p_0) + RT\ln(a(p,T)), \tag{4.58}$$

where a function called fugacity f is used. The function f is introduced for convenience, because in conditions close to the normal conditions of rarefied gas satisfying EOS (4.3), the fugacity tends to p and equation (4.58) converges to (4.55). Non-ideal systems, specifically condensed systems, employ representation (4.58) with a function $a(p,T)$ called activity as a measure of deviation from the ideal state. Fugacity can be calculated by comparing molar Gibbs energies for ideal (G_{id}) and non ideal (G_{re}) gases in an isothermal process. Integration of $dG = Vdp$ from zero to current value p for a difference between the representations of non-ideal (4.58) and ideal (4.55) gases, gives:

$$G_{re} - G_{id} = \int_0^p \left[V(p) - \frac{RT}{p}\right]dp = RT\ln\left(\frac{f}{p}\right). \tag{4.59}$$

If V is known as a function of p and T, then it can be obtained from $V = (\partial G/\partial p)_T$ and (4.59):

$$RT\left(\frac{\partial \ln f}{\partial p}\right)_T = V.$$

In case of p represented as a function of V and T, substitution of the derivative $(\partial \ln f/\partial p)_T = (\partial \ln f/\partial V)_T\,(\partial V/\partial p)_T$ to the last equation gives

$$\left(\frac{\partial \ln f}{\partial V}\right)_T = \frac{V}{RT}\left(\frac{\partial p}{\partial V}\right)_T. \tag{4.60}$$

The ideal gas EOS (4.3) is adequate for gases at pressures below tens of atmospheres, but it manifests notable deviation from experimental data at higher pressures. When progressing toward higher pressures, molecule volume (covolume), b, and molecular repulsion, caused by bringing molecules closer to each other, need to be taken into account. The simplest EOS, expressing the repulsion by a pressure term a/V^2 and the covolume, is known as the Van der Waals EOS and takes the following form for one mole of gas:

$$\left(p + \frac{a}{V^2}\right)(V - b) = RT. \tag{4.61}$$

Because this EOS (4.61) is a simple generalization of the ideal gas EOS, it keeps many properties of the ideal gas EOS, which can be applied to thermodynamic analysis. When relying on representation (4.55) of the chemical potential for an ideal gas, an equation for molar fugacity can be derived from (4.60) as follows:

$$\left(\frac{\partial \ln f}{\partial V}\right)_T = -\frac{V}{(V-b)^2} + \frac{2a}{RTV^2}.$$

Noting, that from (4.61) $p(V - b) \to RT$ at $V \to \infty$, this equation can be integrated

$$\ln f = \ln\left(\frac{RT}{V-b}\right) + \frac{b}{V-b} - \frac{2a}{RTV},$$

which can fully specify the fugacity after expressing V via p and T from (4.61).

For condensed components, a standard pressure chemical potential can be found at any temperature from molar enthalpy $\partial(\mu/T)/\partial T = -h/T^2$ by direct integration

$$\frac{\mu(p_0,T)}{T} = \frac{\mu(p_0,T_0)}{T_0} - \int_{T_0}^{T} \frac{h_0(p_0,T)}{T^2} dT, \tag{4.62}$$

where initial $\mu(p_0,T_0) = \Delta_f H^0(T_0) - T_0 \Delta S_0(T_0)$ and $h_0(p_0,T)$ can be calculated from tabular heat capacities, using formulas (4.54) above, along with the reference entropy change ΔS_0 available in tables (Chase, 1998). After that, chemical potential at any pressure can be found from (4.57) and (4.58) for gaseous systems. For condensed systems, functions of state at a higher than standard pressure can be extended, using $dE = TdS - pdV$. This equation, differentiated over p at T = const with application of a Maxwell relation in the form $-T(\partial S/\partial p)_T = (\partial V/\partial T)_p$, to $dG = -SdT + VdP$, gives

$$\left(\frac{\partial E}{\partial p}\right)_T = T\left(\frac{\partial S}{\partial p}\right)_T - p\left(\frac{\partial V}{\partial p}\right)_T = -T\left(\frac{\partial V}{\partial T}\right)_p - p\left(\frac{\partial V}{\partial p}\right)_T.$$

Then, the corresponding enthalpy derivative

$$\left(\frac{\partial H}{\partial p}\right)_T = \left(\frac{\partial(E+pV)}{\partial p}\right)_T = -T\left(\frac{\partial V}{\partial T}\right)_p - p\left(\frac{\partial V}{\partial p}\right)_T + V + p\left(\frac{\partial V}{\partial p}\right)_T = V - T\left(\frac{\partial V}{\partial T}\right)_p$$

enables one to calculate an enthalpy change at higher pressure via integration. If $V(p,T)$ is known from a thermal EOS, a thermal capacity at higher pressure can also be calculated by integration of $(\partial C_p/\partial p)_T = -T(\partial^2 V/\partial T^2)_p$. For volume $V = (\partial G/\partial p)_T$ reduced to the molar volume V_m, the integration provides chemical potential $\mu = (\partial G/\partial N)_{p,T}$ for condensed components at a given pressure

$$\mu(p,T) = \mu(p_0,T) + \int_{p_0}^{p} V_m(p,T)\,dp.$$

Finally, the use of enthalpies of formation in Hess's law enables one to determine explosion heats, when considering detonation as a reaction. First, calculation of enthalpy of formation is illustrated by an example of a combustion reaction. The chemical formula for many typical energetic materials is $C_aH_bN_cO_d$, resulting in the following global reaction of combustion (letters 's', 'l' and 'g' in parentheses refer to solid, liquid and gaseous aggregate states):

$$\begin{aligned} C_aH_bN_cO_d(s) + \left(a + \frac{b}{4} - \frac{d}{2}\right)O_2(g) \rightarrow aCO_2(g) + \\ + \frac{b}{2}H_2O(l) + \frac{c}{2}N_2(g). \end{aligned} \tag{4.63}$$

In the case of measured molar energy of combustion $\Delta_c E$, when comparing with contributions from gaseous components, heats of combustion $\Delta_c H$ for condensed (liquid and solid) components of the reaction can be neglected. Therefore, due to negligible volume change of the condensed components, heat of combustion is proportional to a molar change of gaseous components, which follows from the general equation for enthalpy of reaction and the ideal gas EOS presented as $\Delta_R H = \Delta_R E + p\Delta_R V = \Delta_R E + \Delta N \cdot RT$. Therefore, at normal conditions ($T = 298°K$, $p = 0.1$ MPa, $RT \approx 2.48$ kJ/mol), the heat of combustion can be obtained from (4.63):

$$\Delta_c H = \Delta_c E + \Delta N(g) RT = \Delta_c E + (d/2 - b/4 + c/2) \cdot 2.48 \text{ kJ/mol}$$

Heats of combustion for many substances are available, e.g., from Rouse (1976).

For determination of oxygen balance Ω of the CHNO-system explosives (see Keshavarz and Klapötke (2020)), the main question is how many oxygen atoms are required for a complete reaction. Assuming that such systems are fully converted to CO_2, H_2O and N_2 as to (4.63), oxygen can be in excess (positive balance) or in deficit (negative balance). For an explosive with the general formula $C_aH_bN_cO_d$, oxygen is balanced in the formula with d atoms of oxygen if $d - 2a - b/2 = 0$ as follows from (4.63) for the combustion reaction. Thus, the balance in % is defined as follows:

$$\Omega = \Omega_{CO_2} = [d - 2a - b/2]1600/W,$$

where W is molecular weight of explosive in g/mol and a coefficient of 16 (multiplied by 100% in the formula) is the molar mass of O_2. For example, for RDX with chemical formula $C_3H_6N_6O_6$, and molecular weight of 222 g/mol, the oxygen deficiency is

$$\Omega = -(3 \cdot 1600 / 222)\% \approx -21.6\%.$$

This means that for complete reaction three extra atoms of oxygen are required in the explosive compound to achieve balance.

At explosion, for majority of explosives, H_2O and CO are generated in the products in agreement with the explosion reaction suggested by Cowan and Fickett (1956), following the Kistiakowsky–Wilson rules (Akhavan, 2011). Hence, the explosion reaction takes the form

$$\begin{aligned} C_aH_bN_cO_d(s) \rightarrow \frac{c}{2}N_2(g) + \frac{b}{2}H_2O(g) + \\ + \left(d - \frac{b}{2}\right)CO(g) + \left(a - d + \frac{b}{2}\right)C(s). \end{aligned} \tag{4.64}$$

In case of a higher oxygen content (for example, nitroglycerine with positive oxygen balance) a modified Kistiakowsky–Wilson rule (Kamlet and Jacobs, 1968), suggests forming H_2O and CO_2 in reaction products:

$$\begin{aligned} C_aH_bN_cO_d(s) \rightarrow \frac{c}{2}N_2(g) + \frac{b}{2}H_2O(g) + \\ + \left(\frac{d}{2} - \frac{b}{4}\right)CO_2(g) + \left(a - \frac{d}{2} + \frac{b}{4}\right)C(s). \end{aligned} \tag{4.65}$$

Heat of formation of the products can be determined from (4.63), using Hess's law. Standard heats of formation of pure substances O_2 and N_2 are zero. Heat of combustion of explosive $\Delta_c H$ is available from tables (Rouse, 1976). Therefore, heat of formation of explosive $\Delta_f H_{HE}$ can be calculated from (4.63) as

$$\Delta_f H_{HE} = a\Delta_f H_{CO_2} + (b/2)\,\Delta_f H_{H_2O} - \Delta_c H$$

As a result, specific heat of explosion Q is calculated (Keshavarz and Klapötke, 2020) as

$$Q = -\left[\sum v_i \Delta_f H_i - \Delta_f H_{HE}\right] / W, \tag{4.66}$$

where the stoichiometric coefficients of reactions (4.64) or (4.65) are used according to Hess's law in the first term of (4.66), depending on the high explosive (HE) oxygen balance.

For example, for RDX, $C_3H_6N_6O_6$, with $\Delta_c H = -2.1$ MJ/mol, $\Delta_f H_{CO_2} = -393.5$ kJ/mol, and $\Delta_f H_{H2O} = -285.8$ kJ/mol, it follows $\Delta_f H_{RDX} = (-3{\cdot}393.5 - 3{\cdot}285.8 + 2100)$ kJ/mol = 62.1 kJ/mol. Finally, keeping in mind a negative oxygen balance for RDX, zero heats of formation for N_2 and C and enthalpies of formation for vapour H_2O and CO_2: $\Delta_f H_{H_2O(g)} = -248.3$ kJ/mol, $\Delta_f H_{CO} = -110.5$ kJ/mol, heat of formation for the products, required by reaction (4.66), can be calculated from (4.64)

$$\sum v_i \Delta_f H_i = (b/2)\Delta_f H_{H_2O} + (d - b/2)\Delta_f H_{CO} = -1.0764 \text{ MJ/mol}$$

and from (4.66), specific heat of explosion of RDX:

$$Q_{RDX} = \left[-\left(-1076.4 - 62.1\right)\text{kJ/mol}\right] / \left(222 \text{ g/mol}\right) = 5.1 \text{ kJ/g}.$$

Summarizing, this section outlined fundamental principles of chemical kinetics and thermochemistry, including specification of reaction rates, thermodynamic relations required for evaluating balance in components of reactions, and estimates of fuel-oxidizer balance in explosives and heat of explosion in the detonation process.

4.4 EQUATIONS OF STATE AND EQUILIBRIUM COMPOSITIONS

The second law of thermodynamics (4.20), (4.22) extended to the case of many components provides the following conditions of equilibrium for a closed system, $dE_{S,V} = dH_{S,p} = dF_{T,V} = dG_{T,p} = \Sigma\mu_k dn_k = 0$. For each natural pair of variables, change in the corresponding generating function of state informs about direction of reaction, reaching a minimum at equilibrium. Therefore, for a chemical reaction at constant T and p, the corresponding relation is $dG_{T,p} = 0$ that can be expressed as the following condition for affinity: $\Sigma\nu_i\mu_i = 0$. As noted in the previous section, a more realistic EOS is likely to be required for specification of chemical potential, because the ideal gas EOS may not be sufficient for characterization of chemical compositions at local equilibrium after detonation reaction. This characterization would enable one to evaluate detonation performance and to simulate behaviour of HEs and their products.

Equations of state for real gases are routinely based on generalizations of the ideal gas EOS as a virial expansion series:

$$\frac{pV}{NRT} = 1 + \frac{B}{V} + \frac{C}{V^2} + \ldots,$$

where B, C, … are virial coefficients responsible for inter-molecular interaction. For example, if expanding $1/(V - b)$ in (4.61) as the Taylor series, the van der Waals EOS can be represented as the virial expansion series with temperature correction for B. Temperature corrections for virial coefficients result in the following generalization of the virial series EOS by Jones and Miller (1948):

$$\frac{pV}{NRT} = 1 + \frac{A(T)}{V} + \frac{B(T)}{V^2} + \frac{C(T)}{V^3} + \ldots \quad . \tag{4.67}$$

An approximation of the series expansion (4.67) by analytical functions, reduces it to an equation in the form suggested by Becker (1922)

$$\frac{pV}{NRT} = 1 + X\exp(X), \quad X = \frac{k}{V},$$

where k is covolume. Consideration of temperature in this representation is achieved by taking $X = k/(VT^{1/3})$ by Kistiakowsky and Wilson (1941), which generalizes the last EOS and represents it as (e.g., see Mader (1979)):

$$\frac{pV}{NRT} = 1 + X\exp(bX), \quad X = \frac{k}{VT^{a}}. \tag{4.68}$$

This EOS is a good approximation for a number of explosives. For example, Cowan and Fickett (1956) modified equation (4.68) to avoid thermodynamic extremities by correction of T in X and representing covolume for a mixture of constituents in HE products as follows:

$$\frac{pV}{NRT} = 1 + X\exp(bX), \quad X = \sum \frac{k_k x_k}{V(T+\theta)^a}, \tag{4.69}$$

which is known as semi-empirical Becker–Kistiakowsky–Wilson (BKW) EOS for gaseous products.

An EOS that is simple in structure employs another virial expansion series in pressure (Jones and Miller, 1948), while taking temperature into account within the ideal gas approximation:

$$\frac{pV}{NRT} = 1 + b\,p + c\,p^2 + d\,p^3 + \ldots \,. \tag{4.70}$$

Equations of state for condensed substances exploit a wide variety of thermodynamically consistent representations used in hydrocode calculations. A relatively simple empirical polynomial EOS aiming at the high temperature range was employed by Cowan and Fickett (1956):

$$p = p_1(\xi) + a(\xi)T + b(\xi)T^2, \tag{4.71}$$

where $\xi = V_0/V$ and p_1, a, and b are rational polynomial functions. A simple way of using a mathematically convenient and thermodynamically consistent approximation is to generalize the polytropic gas EOS, $e = p/[\rho(\gamma - 1)]$, which represents the EOS in specific variables with polytropic index γ as defined in (4.11). An example of such generalization is the stiffened gas EOS (Menikoff, 2007)

$$e = \frac{p + \gamma p_0}{\rho(\gamma - 1)}. \tag{4.72}$$

This EOS is very popular amongst users and developers of numerical methods employing Riemann solvers (Godunov et al., 1979; Toro, 2009), because it enables one to analytically calculate Riemann invariants (see the next section) used in a variety of numerical schemes.

Due to a small change of elastic characteristics of solids with specific volume, simple EOS generalizations in polynomial form can be obtained, using an approximation of constant elastic moduli. One of the moduli, bulk modulus B_0, is related to the corresponding sound velocity c_0 as $B_0 = \rho_0 c_0^2$. Sound velocity, c, in thermodynamic interpretation is associated with velocity of propagation of a change in coordinated particle movement of a substance via inter-atomic compression. This response, related to EOS characteristics in the (p, υ)-space, will be discussed in the subsequent section. For example, the bulk elastic modulus, representing response of a substance to omnidirectional compression in adiabatic system, is proportional to $-p_\rho = c^2$ (see the next section). Thus, for a constant bulk modulus B_0, internal energy after integration of B_0 could be approximated as a polynomial function of specific volume $\upsilon = 1/\rho$ as follows:

$$e = e_0 + B_0\left(\upsilon_0 - \upsilon + \upsilon \ln\frac{\upsilon}{\upsilon_0}\right).$$

Similarly, for a linear dependence of bulk modulus with pressure $B = B_0 + B_1 p$, internal energy $e(\upsilon)$ can be obtained via the integration of a thermodynamic relation $B = -\upsilon(\partial p/\partial \upsilon)_T$, resulting in $p(\upsilon) = B_0[(\upsilon_0/\upsilon)B_1 - 1]/B_1$, followed by another integration of $e_\upsilon = -p$. This approach was first used by Murnaghan (1944) with the assumption of the elastic modulus to be a linear polynomial in pressure, which results in well-known Birch–Murnaghan EOS that was expressed via 'strain' $\varepsilon = ½[(\upsilon_0/\upsilon)^{⅔} - 1]$ (Birch, 1947).

An empirical extension of the Murnaghan EOS on a high temperature zone for deformations of solids taking the shear modulus effect into account was described by Godunov et al. (1974). For small elastic deformations, the deformation tensor is involved via first two invariants of the tensor (Godunov and Romenskii, 2003), with the first one represented by material density and the second – by invariant d for deviatoric part of elastic strain ε_{ij} (noted by prime), resulting in $d = ½\varepsilon'_{ij}\varepsilon'_{ij}$. In a compact form, this EOS can be written in the following approximation proposed by Dorovskii et al. (1983):

$$\begin{aligned} e\left(\rho, e_{ij}, s\right) &= \frac{c_0^2}{2\alpha^2}\left(\delta^{\alpha} - 1\right)^2 + 2\rho_0 b_0^2 \delta^{\beta+2} d + \\ &+ c_\upsilon T_0 \delta^{\Gamma}\left[\exp\left(s/c_\upsilon\right) - 1\right] + e_0 - \frac{p_0}{\rho}, \end{aligned} \tag{4.73}$$

where $\delta = \rho/\rho_0$. In this EOS, ρ is current density, ρ_0 and T_0 are reference density and temperature, α and β are power exponents characterizing density variations of bulk and shear moduli, Γ is the Grüneisen coefficient, c_υ is specific heat capacity, e_0 is an adjustment constant, p_0 is ambient initial pressure, and reference sound velocities c_0 and b_0 are related to the bulk and shear moduli, B_0 and G_0, as $B_0 = \rho_0 c_0^2$, $G_0 = \rho_0 b_0^2$.

A universal approach, making EOS for condensed substances consistent with the ideal gas EOS, is a representation of equations of state in the Mie–Grüneisen form:

$$p = p_c(\rho) + p_T(\rho, e), \quad p_T(\rho, e) = \rho\Gamma\left[e - e_c(\rho)\right]. \tag{4.74}$$

This representation decouples material response to a 'cold' part p_c, such as the Birch–Murnaghan EOS, when temperature influence is negligible, and a 'thermal' part p_T. The latter extends the response to conditions when the material manifests behaviour of liquid or even gaseous substance at extreme temperatures with possible convergence to the ideal gas response. For example, EOS (4.73) is in agreement with the Mie–Grüneisen representation (4.74). Here, the Grüneisen parameter Γ corresponds to $\gamma - 1$ for the polytropic gas EOS. In order to derive EOS as a function of specific internal energy, the representation (4.74) needs to be complemented with caloric EOS such as $e - e_c = c_\upsilon T$ or, alternatively, a caloric EOS can be obtained by integration of (4.19) rewritten in specific variables:

$$e(T, \upsilon) = e\left(T, \upsilon_{ref}\right) + \int_{\upsilon_{ref}}^{\upsilon}\left[T\left(\frac{\partial p}{\partial T}\right)_{\upsilon} - p\right] d\upsilon,$$

which completes the EOS by a dependence for specific internal energy

$$e(\rho,s) = e_c(\rho) + \rho\Gamma e_T(\rho,s). \tag{4.75}$$

The major assumption of this EOS is that the reference ('cold') part is dependent only on density, which is considered to be a reasonable approximation below the Debye temperature (Kittel, 2004) for a variety of condensed materials. The corresponding reference curve can be an isentrope or Hugoniot (see the next section). The Murnaghan EOS can be employed as a high-pressure approximation, taking it as a polynomial in density.

Wide range equations of state were derived for condensed substances under high-pressure and high-temperature shock loading. For these EOSs, thermal effects were taken into account both within a response representable by the Mie–Grüneisen approximation and within an extreme response of solid substances in the plasma state beyond limits of this approximation, which was addressed by a number of studies (Lozano et al., 2023; Trunin, 1998; Zharkov and Kalinin, 1971).

An application of the Mie–Grüneisen representation (4.74) and (4.75) to describe the detonation products has been fruitful for the derivation of a wide-spread Jones–Wilkins–Lee (JWL) EOS. This is very popular in hydrocode simulations and was first described by Jones and Miller (1948), Wilkins et al. (1964), and Lee et al. (1968). This EOS was calibrated (Dobratz and Crawford, 1985) by experimental data on isentropic expansion of detonation products for a large number of explosives. Referring to the following isentropic curves in the representation (4.74) and (4.75) with the Grüneisen parameter selected as $\Gamma = \omega$:

$$p_{ref}(\eta) = Ae^{-R_1\eta} + Be^{-R_2\eta} + \frac{C}{\eta^{\omega+1}}, \quad e_{ref}(\eta) = \frac{A}{R_1}e^{-R_1\eta} + \frac{B}{R_2}e^{-R_2\eta} + \frac{C}{\omega\eta^{\omega}} + e_0,$$

the JWL EOS can be derived, following a formal analysis by Segletes (2018):

$$p(\rho,e) = A\left(1 - \frac{\omega}{R_1\eta}\right)e^{-R_1\eta} + B\left(1 - \frac{\omega}{R_2\eta}\right)e^{-R_2\eta} + \frac{\omega}{\eta}(e - e_0), \tag{4.76}$$

here A, B, R_1, R_2, ω and e_0 are empirical constants, $\eta = \rho_{0s}/\rho$, and ρ_{0s} is density of unreacted explosive.

Focusing on equilibrium conditions, it is convenient to select T and p as independent variables for chemical reactions, which results in the condition of equilibrium: $dG_{T,p} = 0$. As a result, change in the Gibbs potential is an indicator of the direction of chemical reaction: for $dG < 0$ the reaction is running in the forward direction, otherwise, in the reverse direction. In the terms of chemical potentials for reaction (4.28), the local equilibrium condition is equivalent to the condition of zeroing out chemical affinity: $\Sigma\nu_k\mu_k = 0$.

For a single reaction ($\Sigma\nu_k A_k \to 0$) involving K components of an ideal gas, the equilibrium condition $\Sigma\nu_k\mu_k = 0$ with the use of (4.57) takes the following form:

$$\sum_{k=1}^{K} \nu_k\mu_k = \sum_{k=1}^{K}\left(\nu_k\mu_k^0 + \nu_k RT\ln\frac{p_k}{p_0}\right) = 0.$$

This equation is reduced to the following

$$K_0 = \prod_{k=1}^{K}\left(\frac{p_k}{p_0}\right)^{\nu_k} = \exp\left(-\frac{\sum_{k=1}^{K}\nu_k\mu_k^0}{RT}\right), \quad K_0 = \prod_{k=1}^{K} p_k^{\nu_k}, \quad \Delta_R G^0 = \sum_{k=1}^{K}\nu_k\mu_k^0, \tag{4.77}$$

where K_0 is called the thermodynamic equilibrium constant for the chemical reaction, and $\Delta_R G^0$ is called the standard molar Gibbs energy of reaction. Traditional equilibrium constant K_p has the dimension of $p^{\Delta\nu}$, where $\Delta\nu = \Sigma\nu_k = \Sigma\nu'_j - \Sigma\nu'_i$ (index j refers to products and i – to reactants), which gives

$$K_p = \prod_{k=1}^{K} p_k^{\nu_k} = K_0 p_0^{\Delta\nu}, \quad Q = \prod_{k=1}^{K} p_k^{\nu_k},$$
$$\Delta_R G = \sum_{k=1}^{K} \nu_k \mu_k = \Delta_R G^0 + RT \ln Q, \tag{4.78}$$

Q is called the reaction quotient. An equation derived from (4.77) as $\ln K_0 = \ln K_p = -\Delta G_R^0/(RT)$ and (4.78) results in

$$\Delta_R G = \sum_{k=1}^{K} \nu_k \mu_k = -RT\left(\ln K_p - \ln Q\right),$$

which clearly indicates that large K_p and small Q favour the reaction. The parameters $\Delta_R G^0$ and Q control attainment of equilibrium at $\Delta_R G = 0$. To note, Q is generally calculated with use of partial pressures out of equilibrium.

Using molar fractions $x_k = N_k/N = p_k/p$, the equilibrium constant can be calculated from (4.78) as follows:

$$K_0 = \prod_{k=1}^{K}\left(\frac{p_k}{p_0}\right)^{\nu_k} = \prod_{k=1}^{K}\left(\frac{p_k}{p}\right)^{\nu_k}\left(\frac{p}{p_0}\right)^{\Delta\nu} = \prod_{k=1}^{K} x_k^{\nu_k}\left(\frac{p}{p_0}\right)^{\Delta\nu} = K_x\left(\frac{p}{p_0}\right)^{\Delta\nu}.$$

Similarly, for molar concentrations $C_k = N_k/V = p_k/RT$ (introduced in the previous section as β_k), the corresponding equilibrium constant is

$$K_C = \prod_{k=1}^{K} C_k^{\nu_k} = \prod_{k=1}^{K}\left(\frac{p_k}{RT}\right)^{\nu_k} = \prod_{k=1}^{K}\left(\frac{p_k}{p_0}\right)^{\nu_k}\left(\frac{p_0}{RT}\right)^{\Delta\nu} = K_0\left(\frac{p_0}{RT}\right)^{\Delta\nu}.$$

It is interesting to note that $\Delta_R G^0$ does not specify unconditionally if a reaction is likely to progress and in which direction. For example, for a reaction A + B → C with $p_0 = 0.1$ MPa and with pressure of the mixture during the reaction $p = 0.1$ MPa, $Q = p_C/(p_A p_B) = x_C/(x_A x_B)$. It is seen from (4.78) that small Q (for example, due to a small concentration of products) makes $\Delta_R G$ negative, which favours the reaction even at positive $\Delta_R G^0$.

In general, for multiple reactions $r = 1, \ldots, R$, which can progress in both, forward and reverse directions with rates $k_{f,r}$ and $k_{r,r}$, respectively, the reaction (4.39) can be written in the following general form:

$$\sum_{k=1}^{K} \nu_{kr}^{(r)} A_k \underset{k_{r,r}}{\overset{k_{f,r}}{\rightleftarrows}} \sum_{k=1}^{K} \nu_{kr}^{(p)} A_k, \quad r = 1,\ldots,R. \tag{4.79}$$

Equilibrium in this process is achieved when a generating function (the Gibbs energy for the case of p and T as independent variables) reaches minimum, or reaction rates for the components are equal. The reaction velocities are specified as

$$\frac{d\xi_r}{dt} = J_r = k_{f,r}\prod_{k=1}^{K}[\mathrm{A}_k]^{\nu_{kr}^{(p)}} - k_{r,r}\prod_{k=1}^{K}[\mathrm{A}_k]^{\nu_{kr}^{(r)}}.$$

For ith component A_i, the rate of production of the component

$$\frac{d[A_i]}{dt}=\sum_{r=1}^{R}k_{f,r}\left(v_{ri}^{(p)}-v_{ri}^{(r)}\right)\prod_{k=1}^{K}[A_k]^{v_{rk}^{(r)}}-\sum_{r=1}^{R}k_{r,j}\left(v_{ri}^{(p)}-v_{ri}^{(r)}\right)\prod_{k=1}^{K}[A_k]^{v_{rk}^{(p)}}.$$

It is seen that the equilibrium condition for kth reaction $J_k = 0$, can be formulated as follows:

$$k_{f,r}\prod_{k=1}^{K}[A_k]^{v_{kr}^{(p)}}=k_{r,r}\prod_{k=1}^{K}[A_k]^{v_{kr}^{(r)}},\quad K_{C,r}=k_{f,r}/k_{r,r}=\prod_{k=1}^{K}[A_k]^{v_{kr}^{(r)}-v_{kr}^{(p)}}$$

with reaction rates $k_{f,r}$ and $k_{r,r}$ for the forward and reverse rth reaction. This provides the equilibrium conditions, which can be used for assessment of the reaction rates.

Equilibrium constants can be calculated from $\Delta_R G^0$. For example, for a stoichiometry balanced reaction $A + 2B \rightarrow C$, with three moles of reactants in a closed system at temperature T and pressure $p = 0.1$ MPa, $\Delta G_{fA}{}^0$, $\Delta G_{fB}{}^0$ and $\Delta G_{fC}{}^0$ are calculated from the molar heat capacity dependencies $C_p(T)$ in (4.54) or can be taken directly from the JANAF tables (Chase, 1998). Following this, standard molar Gibbs energy of reaction is calculated as $\Delta_R G^0 = \Delta_f G_C{}^0 - \Delta_f G_A{}^0 - 2\Delta_f G_B{}^0$ and the equilibrium constant can be determined as $K_p = \exp[-\Delta_R G^0/(RT)]$. Denoting a reacted fraction of the component A as α, the corresponding fraction of reacted B is 2α and C $- \alpha$. As the next step, in the case if the mixture is at equilibrium after having reached this fraction, the corresponding fractions of the components A, B and C are $1 - \alpha$, $2 - 2\alpha$ and α, respectively with a total reacted fraction of $3 - 2\alpha$. As a result, the molar concentrations at equilibrium are $x_1 = (1 - \alpha)/(3 - 2\alpha)$ for A, $x_2 = (2 - 2\alpha)/(3 - 2\alpha)$ for B and $x_3 = \alpha/(3 - 2\alpha)$ for C. Then, the equilibrium constant is $K_x = x_3/(x_1 x_2{}^2) = \alpha(3 - 2\alpha)^2/(2 - 2\alpha)^2 = K_p p^{\Delta\nu}$. Using K_p determined earlier, the concentrations at equilibrium can be found as a solution of the last algebraic equation for α.

In general, for a closed system with initial molar amounts N_{k0} and total initial amount $N_0 = \Sigma N_{k0}$, taking α as the extent of reaction for the substances that has reached equilibrium, the amounts of the substances at equilibrium are $N_k = N_{k0} + \nu_k\alpha$, which results in the total amount $N = \Sigma N_k = \Sigma(N_{k0} + \nu_k\alpha) = N_0 + \Delta\nu\cdot\alpha$. Thus, equilibrium concentrations x_k and equilibrium constant K_p for reaction (4.30) are

$$x_k=\frac{N_k}{N}=\frac{N_{k0}+v_k\alpha}{N_0+\Delta v\alpha},$$
$$K_p=p_0^{\Delta v}\prod_{k=1}^{K}\left(\frac{N_{k0}+v_k\alpha}{N_0+\Delta v\alpha}\right)^{v_k}=\exp\left(-\frac{\Delta_R G^0}{RT}\right).\tag{4.80}$$

Solving this equation, the reaction extent α provides equilibrium concentrations after determination of $\Delta_R G^0$.

Similar logic is followed by popular numerical codes, which analyze cases of multiple reactions. A preliminary phase of general flowchart of relevant codes is to specify possible components involved in reactions that should be linear independent. Formulas of possible chemical reactions can be written down as a system such that

$$\sum_{k=1}^{K}v_{kr}A_k\rightarrow 0,\quad r=1,\ldots,R.\tag{4.81}$$

As a first step, the set of reactions needs to be specified on the basis of balance of elements representing the formula of a substance under investigation with atomic balance for N atoms: $A_k = \{N_{k1}, N_{k2}, \ldots, N_{kN}\}$. For K substances, consisting of N atoms the column of all of them can be presented as a product of a $K \times N$ matrix $\{N_{kn}\}$ and an N-vector representing the column of N atoms. Then the balance conservation laws for R reactions (4.81) are

$$\sum_{k=1}^{K} \nu_{kr} N_{kn} = 0, \quad n = 1,\ldots,N, \quad r = 1,\ldots,R. \tag{4.82}$$

Relations (4.82) is a system of linear equations for R sets of variables $\{\nu_{1r}, \ldots, \nu_{Kr}\}(r = 1, \ldots, R)$. It is seen that the number of independent reactions is determined from rank r_0 of the matrix $\{N_{kn}\}$. These independent solutions give stoichiometric coefficients and specify corresponding independent reactions with R solutions, where $R = K - r_0$. This enables one to determine mathematically a solution.

For example, for RDX explosive with formula $A_1 = C_3H_6N_6O_6$ with $N = 4$ atoms {C, H, N, O}, the corresponding formula for A_1 can be represented as a product of row {3, 6, 6, 6} by the column {C, H, N, O}. A system for these four atoms can be considered with the following eight components {H_2O, C, N_2, CO_2, O_2, CO, H_2, NO}. For these components ($K = 8$), the matrix $\{N_{kn}\}$ takes the form

$$\begin{Bmatrix} H_2O \\ C \\ N_2 \\ CO_2 \\ O_2 \\ CO \\ H_2 \\ NO \end{Bmatrix} = \begin{bmatrix} 0 & 2 & 0 & 1 \\ 1 & 0 & 0 & 0 \\ 0 & 0 & 2 & 0 \\ 1 & 0 & 0 & 2 \\ 0 & 0 & 0 & 2 \\ 1 & 0 & 0 & 1 \\ 0 & 2 & 0 & 0 \\ 0 & 0 & 1 & 1 \end{bmatrix} \begin{Bmatrix} C \\ H \\ N \\ O \end{Bmatrix}. \tag{4.83}$$

It is seen that rank of the matrix is $r_0 = 4$, hence the number of linear independent solutions is $R = K - r_0 = 4$. Thus, from (4.83) the system of linear independent equations for stoichiometric coefficients is

$$\nu_{2r} + \nu_{4r} + \nu_{6r} = 0,\ 2\nu_{1r} + 2\nu_{7r} = 0,\ 2\nu_{3r} + 2\nu_{8r} = 0, \nu_{1r} + 2\nu_{4r} + 2\nu_{5r} + \nu_{6r} + \nu_{8r} = 0.$$

For example, one of the independent solutions for the system is $\{\nu_{1r}, \ldots, \nu_{8r}\}$= {2, 0, 0, 0, −1, 0, −2, 0}, which corresponds to the reaction $2H_2O - O_2 - 2H_2 \rightarrow 0$.

Assuming that R independent reactions have been specified, for each of the reactions, standard Gibbs energy of rth reaction $\Delta_R G_r^0$ can be found at a given T for gaseous components and at given p and T for condensed components. This may formally allow one to build a system of algebraic equations similar to (4.77) and solve it. It should be noted that partial pressure in formulation (4.77) may need to be replaced by activity for solid components and fugacity for non-ideal gases. However, this way is cumbersome to proceed. Therefore, the common approach is to minimize the Gibbs or free energy, depending on a process that involves the reactions. Minimization is conducted subject to the atomic balance conditions that generate a set of limiting conditions appended to the optimization process with the method of Lagrange multipliers.

Numerical codes for determining properties of selected HEs, specifically, condensed explosives, involve balancing the detonation products thus enabling one to employ the CJ theory (introduced in Chapter 3 and Section 4.6) and examine explosive performance. For example, the 'BKW Fortran' code by Mader (1967) calculates an equilibrium composition for a pre-given set of components for a given formula of explosive. Further, the code calculates molar concentrations x_k, molar Gibbs energies G_k and enthalpies H_k, using BKW EOS for the products and EOS (4.74) for condensed components. The calculation is subject to balancing chemical elements for possible set of reactions plus satisfying total conservation of the elements. Internal energy calculations are followed by Hugoniot temperature calculation, which is used to calculate a CJ state, using the Hugoniot and CJ conditions as described in subsequent sections. For entropy calculation, isentropic state through to the CJ state can be determined, concluding with the Hugoniot curve calculation. A large variety of codes (Cowperthwaite and Zwisler, 1973; Fried and Souers, 1994; Gordon and McBride, 1976; Heuze et al., 1985; Nichols and Ree, 1990) employ the same idea with variations in choice of equations of state, explosives for analysis, product compositions, additional analyses for the CJ and Hugoniot states, etc.

To summarize, this section presented several selected thermodynamic functions of state and their expressions via equations of state. The described equations of state are typically used in shock physics and detonation research. A scheme of determination of equilibrium compositions, using chemical equilibrium relations for components of reaction, was also outlined in the section. Equations of state are key relations for the specification of substances, and used extensively in detonation modelling addressed in subsequent sections.

4.5 SHOCK TRANSITIONS IN INERT MEDIA

Observable flow of material systems at detonation and shock loading is a phenomenological process because it involves a coordinated movement of large ensembles of particles within a representative volume. Such a volume is sufficiently small to consider it as a point in space for the phenomenological description. On the other side, it contains sufficiently large amount of particles, allowing one to ignore individual movement of the particles in a statistical sense and can be considered as a thermodynamic system in the local sense. Therefore, thermodynamic equilibrium within a representative volume is achievable even if the phenomenological flow is not stationary, because a timescale of flow in the volume is vastly smaller than a characteristic time of the observable flow (local stationarity).

Material flows presented as closed systems under phenomenological consideration, are subject to conservation laws. On the other hand, thermodynamic equilibrium in the local sense enables one to consider flow parameters within the framework of non-equilibrium thermodynamics. Thus, a flow of material can be described with a system of conservation laws closed by thermodynamic relations for specific variables and possible kinetics and/or constitutive equations if the material behaviour is affected by internal processes. Thermodynamic parameters are subject to the thermodynamic laws, thus, relating them to each other by an EOS. A generating function of state, such as specific internal energy $e(\rho, s)$, enables one to calculate pressure and temperature using (4.24).

The processes, accompanied by propagation of shock or detonation waves, cannot be considered as continuous smooth flows. Solutions for a flow with discontinuities cannot be obtained from systems of differential equations describing the flow because derivatives cannot be calculated at the points of discontinuity. Therefore, in order to find how mechanical, physical and chemical characteristics of a closed material system evolve within the flow space, mass, momentum, and energy conservation laws should be taken in integral form:

$$\begin{gathered}\frac{d}{dt}\int_{\Omega}\rho dV = 0, \quad \frac{d}{dt}\int_{\Omega}\rho\mathbf{u}\,dV = -\oint_{\partial\Omega}p\,\mathbf{n}\,dS, \\ \frac{d}{dt}\int_{\Omega}\rho\left(e+\frac{\mathbf{u}^2}{2}\right)dV = -\oint_{\partial\Omega}p\mathbf{u}\cdot\mathbf{n}\,dS.\end{gathered} \tag{4.84}$$

These integral equations express conservation of mass within a fluid volume, conservation of momentum as the equality of the rate of total momentum within the volume to an external force acting on the volume boundary, and conservation of energy as equality of the rate of total energy to a power acting on the volume boundary. The fluid volume $\Omega(t)$ represents a closed system of the continuum, consisting of the same material particles moving in time and surrounded by the boundary surface $\partial\Omega$. It is called a fluid ('Lagrangian') volume, because the constituting particles are moving with the particle velocities and does not depend on observer's position.

The discontinuity points in space can be commonly considered as a smooth surface. A fluid volume can be selected in a vicinity of the discontinuity surface with a thickness in the direction normal to the surface, tending to zero, while holding a finite area along the surface. Integrating over this volume, the jump conditions (relationships for changes in the flow parameters across the discontinuity surface) can be obtained for the flow parameters in (4.84) in general three-dimensional case. For the sake of simplicity, one-dimensional case with the space variable x is considered, which reduces (4.84) to three equations with velocity u directed along the axis x. Such a jump surface, imitating a shock wave or another discontinuity with zero thickness of the zone (the front) where parameters undergo sudden change, is moving with velocity D. A fluid volume containing the front can be chosen with a thickness of Δx. Assuming constant parameters ahead of the shock front (denoted with subscript '0') and behind the front (with subscript '1') and taking $\Delta x \to 0$, integration of (4.84) over the selected volume gives dependencies, linking the flow parameters before and after the jump. They are called the Rankine–Hugoniot (RH) relationships (Courant and Friedrichs, 1976):

$$\begin{gathered}\left[\rho(u-D)\right]=0,\quad \left[\rho u(u-D)+p\right]=0,\\ \left[\rho(u-D)\left(e+{u^2}/{2}\right)+pu\right]=0,\end{gathered} \tag{4.85}$$

here $[a] = a_1 - a_0$. The simplest solution, satisfying the first equation in (4.85), is $D = u$. This condition physically means that the discontinuity is moving with particle velocity u, and it is called a contact discontinuity because the velocity u is the same on both sides of the jump. Cancelling out the corresponding term in the second momentum equation results in equal pressures on both sides of the jump. With velocity and pressure identical on both sides, the energy RH condition in (4.85) is satisfied automatically. Thus, the contact jump conditions can be formulated as follows:

$$D=u,[u]=0,\left[p\right]=0. \tag{4.86}$$

To note, density and entropy (as well as temperature and internal energy) can undergo non-zero changes subject to the pressure continuity (4.86) and fulfilment of the EOS.

In a case of non-zero mass flow through the jump surface, denoted by $F = \rho_0 (u_0 - D)$, the RH relations (4.85) describe conditions of a strong discontinuity called a shock wave and with a jump surface as the shock front. Parameters with subscript '0' refer to a state ahead of the front and subscripts for parameters at state behind the front are omitted. The state ahead of the front can be taken at rest, assigning $u_0 = 0$. The mass flow F is used in order to obtain particle and discontinuity velocities:

$$\frac{u}{D}=1-\frac{\upsilon}{\upsilon_0},\quad D=\mp\upsilon_0\sqrt{\frac{p-p_0}{\upsilon_0-\upsilon}},\quad F=\pm\sqrt{\frac{p-p_0}{\upsilon_0-\upsilon}}. \tag{4.87}$$

It should be noted that for the case when state '0' is behind the front, a negative sign is assigned to D and positive to F, corresponding to a shock wave going in the left direction with respect to the flow, or, in short, to a left-going shock wave. For a right-going wave, the choice of sign is opposite. Eliminating D and u with the use of mass and momentum jump conditions presented by (4.87), the RH condition for total energy from (4.85) can be formulated as follows:

$$H(p,\upsilon;p_0,\upsilon_0)=e(p,\upsilon)-e(p_0,\upsilon_0)-\frac{1}{2}(p+p_0)(\upsilon_0-\upsilon)=0. \tag{4.88}$$

This equation is known as the Hugoniot equation and it specifies a locus of all possible states behind the front in the (p, υ)-plane related to state '0' ahead of the front, presented by a point (p_0, υ_0) called the Hugoniot centre.

The Hugoniot equation is specified with an EOS, which is illustrated below with use of the polytropic gas EOS (4.11). For molar weight of a gas W, polytropic index γ can be found from (4.11) via relation $R/W = c_p - c_\upsilon = (\gamma - 1)c_\upsilon$. Thus, the EOS takes the form:

$$p/\rho = (\gamma - 1)c_\upsilon T, \tag{4.89}$$

where an internal specific energy function (the caloric EOS) was pre-tabulated in (4.8):

$$e = c_\upsilon T. \tag{4.90}$$

Internal energy is specified by an arbitrary additive that is taken zero in (4.8). Using the following equations in partial derivatives obtained from (4.24), relations (4.89) and (4.90) result in

$$p = \rho^2\left(\frac{\partial e}{\partial \rho}\right)_s = \rho(\gamma - 1)e, \quad e = c_\upsilon\left(\frac{\partial e}{\partial s}\right)_\rho = c_\upsilon T,$$

which provides the following EOS representation

$$e = \frac{p}{\rho(\gamma - 1)}, \quad \frac{p}{p_r} = \left(\frac{\rho}{\rho_r}\right)^\gamma \exp\left(\frac{s - s_r}{c_\upsilon}\right),$$
$$\frac{T}{T_r} = \left(\frac{\rho}{\rho_r}\right)^{\gamma - 1} \exp\left(\frac{s - s_r}{c_\upsilon}\right). \tag{4.91}$$

Here, parameters with subscript 'r' relate to a reference state.

Using this EOS for derivation of jump conditions, the contact conditions take the form $[\rho T] = 0$ or $[\rho^\gamma \exp(s)] = 0$. For the case of a strong discontinuity with an EOS $e = e(p, \upsilon)$, a curve described by the Hugoniot equation (4.88) in (p, υ)-space is called Hugoniot curve with the centre at (p_0, υ_0). The Hugoniot curve connects the state '0' at (p_0, υ_0) ahead of the wave with states (p, υ) at any point $p > p_0$ behind the shock front and the corresponding specific volume υ is calculated from (4.88). For a gas flow with the polytopic gas EOS, providing $e - e_0 = (p\upsilon - p_0\upsilon_0)/(\gamma - 1)$ where $e_0 = e(p_0, \upsilon_0)$, the Hugoniot equation can be obtained analytically in two equivalent forms:

$$\frac{p}{p_0} = \frac{(\gamma + 1)\upsilon_0 - (\gamma - 1)\upsilon}{(\gamma + 1)\upsilon - (\gamma - 1)\upsilon_0}, \quad \frac{\upsilon}{\upsilon_0} = \frac{(\gamma - 1)p + (\gamma + 1)p_0}{(\gamma + 1)p + (\gamma - 1)p_0}. \tag{4.92}$$

It can be easily found that asymptotes of the hyperbola (4.92) are limited by lines $\upsilon_a = \upsilon_0(\gamma - 1)/(\gamma + 1)$ and $p_a = -p_0(\gamma - 1)/(\gamma + 1)$ as shown in Figure 4.2(a). The line $\upsilon = \upsilon_a$ gives the value of maximal compression of polytropic gas by shock wave.

Conservation laws (4.84) for homogeneous gaseous and condensed systems describe time evolution of mass, momentum and total energy. The Gauss–Ostrogradsky theorem enables one to move derivatives in (4.84) inside the integrals as follows:

$$\frac{d}{dt}\int_\Omega f dV = \int_\Omega \frac{\partial f}{\partial t} dV + \oint_{\partial\Omega} f \mathbf{u} \cdot \mathbf{n} dS. \tag{4.93}$$

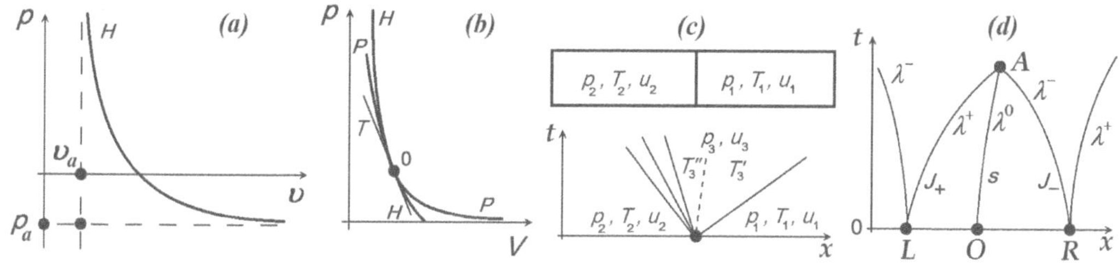

FIGURE 4.2 Hugoniot curve H for a polytropic gas with asymptote lines (a); Hugoniot 'H' and Poisson 'P' curves passing through the centre '0' (b); Riemann problem illustrated with an (x,t)-diagram of the flow process (c); the characteristic cone for a solution at 'A' influenced by initial data from the area 'LR' (d).

This general formula (4.93) can be applied to equations of the system (4.84) for smooth variables. As a result, the corresponding system of partial differential equations in an Eulerian coordinate system $\{x_i\}$, $i = 1, 2, 3$ (laboratory system that is fixed and associated with the observer), can be obtained from (4.84) by transfer of the time derivatives inside integrals via application of the Gauss–Ostrogradsky theorem (4.93), which is legitimate for smooth flows and for any fluid volume Ω. Integrands in the transformed equations can be equated, which gives a system of five differential equations called the Euler model (Courant and Friedrichs, 1976) in absence of external sources of mass, force and energy:

$$\frac{\partial \rho}{\partial t}+\frac{\partial \rho u_k}{\partial x_k}=0, \quad \frac{\partial \rho u_i}{\partial t}+\frac{\partial \rho u_i u_k}{\partial x_k}+\frac{\partial p}{\partial x_i}=0,$$
$$\frac{\partial \rho\left(e+u^2/2\right)}{\partial t}+\frac{\partial\left[\rho\left(e+u^2/2\right)u_k+p u_k\right]}{\partial x_k}=0, \tag{4.94}$$

where u_i ($i = 1, 2, 3$) are components of the velocity vector, $u^2 = u_i u_i$ is the squared velocity magnitude (the summation rule is assumed for silent indices).

System (4.94) describes non-stationary smooth continuous flows, however, the number of variables (ρ, u_i, p, e) in the system exceeds the number of equations by one. In order to close the system, an extra relation – EOS, such as $p = \rho(\gamma - 1)e$ from (4.90) and (4.91) for a polytropic gas – is required, which presumes local equilibrium.

Differentiating $e = e(\rho, s)$ with the chain rule and expanding conservation equations (4.94), the system can be rewritten in one-dimensional case as follows:

$$\frac{\partial \rho}{\partial t}+u\frac{\partial \rho}{\partial x}+\rho\frac{\partial u}{\partial x}=0, \quad \frac{\partial u}{\partial t}+u\frac{\partial u}{\partial x}+\frac{1}{\rho}\frac{\partial p}{\partial x}=0, \quad \frac{\partial s}{\partial t}+u\frac{\partial s}{\partial x}=0. \tag{4.95}$$

It will be shown below that system (4.95) and hence (4.94) can be transformed to a system of transport equations, thus manifesting itself as a system of hyperbolic equations. This means that a response to stimuli in the material system is transmitted with a finite velocity.

It is convenient to introduce trajectory (material) derivative $d/dt = \partial/\partial t + u\,\partial/\partial x$ which is a derivative along the Lagrangian path for any particle (x, t) in a flow with the space coordinate $x(t)$ specified as a solution of equation $dx/dt = u(x, t)$. Denoting $c = (p_\rho)^{1/2}$ with $p(\rho, s)$ differentiated under adiabatic conditions and taking the last equation of (4.95), $ds/dt = 0$, into account, the continuity equation (the mass conservation law) can be rewritten as $dp/dt = c^2 d\rho/dt$ and system (4.95) as

$$\frac{dp}{dt}+\rho c^2\frac{\partial u}{\partial x}=0, \quad \frac{du}{dt}+\frac{1}{\rho}\frac{\partial p}{\partial x}=0, \quad \frac{ds}{dt}=0. \tag{4.96}$$

It is seen that for smooth gas flows, entropy does not change along the particle trajectory. Expressing density and c via pressure and entropy with the help of EOS, a function $I = \int dp/(\rho c)$ can be introduced, which reduces the system (4.96) to

$$\frac{dI}{dt} + c\frac{\partial u}{\partial x} = 0, \quad \frac{du}{dt} + c\frac{\partial I}{\partial x} = 0, \quad \frac{ds}{dt} = 0.$$

An equivalent form of this system, obtained by summation and subtraction of the first two equations from each other, is

$$\frac{d(u+I)}{dt} + c\frac{\partial(u+I)}{\partial x} = 0, \quad \frac{d(u-I)}{dt} - c\frac{\partial(u-I)}{\partial x} = 0, \quad \frac{ds}{dt} = 0 \tag{4.97}$$

The representation (4.97) shows that in addition to constant entropy s along the trajectory $dx/dt = u = \lambda^0$, combinations $J_+ = u + I$ and $J_- = u - I$ are conserved along curves, that are called the characteristic paths defined by equations $dx/dt = u + c = \lambda^+$ and $dx/dt = u - c = \lambda^-$. The combinations J_+ and J_- are called the Riemann invariants (Toro, 2009). Therefore, a solution for a smooth flow at any point $A\{x_c, t_c\}$ can be found by drawing the trajectory, λ^0, and characteristic paths, λ^+ and λ^-, from this point backwards down to three corresponding points of the x-line at $t = 0$: $A\{x^0, 0\}$, $R\{x^+, 0\}$ and $L\{x^-, 0\}$, respectively, as shown in Figure 4.2(d). Assuming that J_+, J_- and s are known at $t = 0$ due to initial conditions, these invariants and entropy can be found at $t = t_c$ in the form $J_+(x_c, t_c) = J_+(x^+, 0)$, $J_-(x_c, t_c) = J_-(x^-, 0)$ and $s(x_c, t_c) = s(x_0, 0)$. This method of calculation of smooth flows is called the method of characteristics. This solution is sufficient for determination of all flow parameters at the point (x_c, t_c). Specifically, pressure, density and temperature can be found from entropy s and function I provided that an EOS is given.

Obviously, due to the conservation of entropy along trajectories in smooth flows, the trajectory curves transfer entropy from a reference state throughout the flow. For a polytropic gas with EOS (4.91) the condition of adiabaticity takes the form similar to (4.11):

$$p\upsilon^\gamma = p_r\upsilon_r^\gamma, \tag{4.98}$$

which represents an isentropic curve or Poisson's adiabat for a polytropic gas. In summary, a substance can be compressed by a shock wave, satisfying a discontinuous solution obtained with the use of integral conservation laws (4.84), and by an isentropic compression wave, satisfying a smooth solution obtained with the use of the same conservation laws in differential form (4.94) or (4.95).

It is seen that c represents speed in the transport equations (4.97). Therefore, a value of c actually characterizes the velocity of propagation of a compression response with respect to the flow. In case of a weak (acoustic) response, c represents the velocity of sound. The entropy equation $ds = 0$ can be expressed via p and ρ as $dp = c^2 d\rho + p_s ds$, using the definition of c, hence system (4.96) can be rewritten in the following form:

$$\frac{dp}{dt} + \rho c^2\frac{\partial u}{\partial x} = 0, \quad \frac{du}{dt} + \frac{1}{\rho}\frac{\partial p}{\partial x} = 0, \quad \frac{dp}{dt} - c^2\frac{d\rho}{dt} = 0. \tag{4.99}$$

For the acoustic (linear) approximation in vicinity of the solution $p = p_0$, $c = c_0$, $\rho = \rho_0$, $u = u_0$, after a regrouping similarly to (4.97), system (4.99) with the function $I = \int dp/(\rho c) = p/(\rho_0 c_0)$ takes the following form alike

$$\frac{d^+}{dt}\left(u + \frac{p}{\rho_0 c_0}\right) = 0, \quad \frac{d^-}{dt}\left(u - \frac{p}{\rho_0 c_0}\right) = 0, \quad \frac{d^0}{dt}\left(p - c_0^2\rho\right) = 0. \tag{4.100}$$

Here, $d^+/dt = \partial/\partial t + (u_0 + c_0)\,\partial/\partial x$, $d^-/dt = \partial/\partial t + (u_0 - c_0)\,\partial/\partial x$ and $d^0/dt = \partial/\partial t + u_0\,\partial/\partial x$. It is seen that the last equation in (4.100) expresses the conservation of entropy because for $p = p(\rho, s)$: $dp = p_\rho d\rho + p_s ds$ and in the acoustic approximation: $dp = c_0^2 d\rho + p_s ds$. Hence, for normal gases obeying $p_s > 0$, the condition $d(p - c_0^2\rho) = 0$ is equivalent to $ds = 0$, which means conservation of entropy along $dx/dt = u_0$. Conservation of the Riemann invariants $u \pm p/(\rho_0 c_0)$ along lines $dx/dt = u_0 \pm c_0$ means that the transport equations are restricting change in a solution inside a zone expanding at velocities c_0 and $-c_0$ with respect to a gas flow with velocity u_0. It is obvious, therefore, that c_0 (c, in the general quasi-linear case) represents the sound velocity, and the product $\rho_0 c_0$ is called the acoustic impedance. As a result, line $dx/dt = \lambda^0$ is called the contact characteristic and lines $dx/dt = \lambda^\pm$ – sound characteristics. The sound characteristics λ^+ and λ^- form a characteristics cone with the cone base LR enveloping initial data required for determination of solution at the cone apex A (Figure 4.2(d)).

A case of compressible gas dynamics, reducible to algebraic formulas, is the case of 'polytropic' gas with EOS (4.91). In this case, $c^2 = (\partial p/\partial \rho)_s = p_r(\rho/\rho_r)^{\gamma-1}\exp[(s - s_r)/c_v] = \gamma p/\rho$ and for the isentropic flows Riemann invariants can be analytically obtained. Specifically, p and ρ can be expressed via c as

$$\frac{\rho}{\rho_r} = \left(\frac{c}{c_r}\right)^{\frac{2}{\gamma-1}}, \quad \frac{p}{p_r} = \left(\frac{c}{c_r}\right)^{\frac{2\gamma}{\gamma-1}}, \quad \frac{dp}{\rho c} = \frac{p_r}{\rho_r c_r^2}\frac{2\gamma}{\gamma-1}dc. \tag{4.101}$$

Therefore, from $c_r^2 = \gamma p_r/\rho_r$, the integral I introduced above can be written down in canonical form in the (p, u)-space with a reference to state '0' ahead of the wave:

$$I = \int \frac{dp}{\rho c} = \int \frac{2}{\gamma-1}dc = \frac{2}{\gamma-1}c,$$
$$I = \int \frac{dp}{\rho c} = \frac{2c_0}{\gamma-1}\int d\left(\frac{p}{p_0}\right)^{\frac{\gamma-1}{2\gamma}} = \frac{2c_0}{\gamma-1}\left(\frac{p}{p_0}\right)^{\frac{\gamma-1}{2\gamma}}. \tag{4.102}$$

For continuous flows with a constant state ahead of the smooth wave, a solution for the left-going wave is obtained with the use of the Riemann invariant J_- that is conserved along λ^--characteristics, whereas the λ^+-characteristics are straight lines and vice versa for the right-going continuous wave. Entropy is conserved along the contact characteristics, therefore, continuous solutions with a constant state ahead of the wave are isentropic. In summary, for a characteristic fan, specifying a rarefaction wave, initiating at point $\{x_0, 0\}$ in the (x, t)-space, a self-similar solution for a polytropic gas takes the following form for u and c:

$$u \pm \frac{2}{\gamma-1}c = \text{const}, \quad \frac{dx}{dt} = \frac{x - x_0}{t} = u \mp c. \tag{4.103}$$

Lower and upper signs should be chosen for the left- and right-going waves, respectively. Remaining parameters of the flow can be found from EOS (4.91), keeping in mind that the flow is isentropic.

Temperatures in a polytropic gas, compressed in the following two ways – by a shock wave and by a smooth compression wave – up to the same specific volume, can be compared with the use of the polytropic gas EOS in the form

$$T/T_0 = e/e_0 = (p/p_0)(\upsilon/\upsilon_0). \tag{4.104}$$

Then, denoting $\xi = \upsilon_0/\upsilon$ and comparing p/p_0 from (4.92) with p/p_0 from (4.98) at the same specific volume, temperature at the shock compression could be easily compared with that at the isentropic compression using (4.104), which is expressed as $[(\gamma+1)\xi-(\gamma-1)]/[(\gamma+1)-(\gamma-1)\xi]\geq\xi^{\gamma}$ at $\xi\leq 1$ (the compression branch). This means that temperature and pressure under shock loading (curve H in Figure 4.2(b)) are higher than those under isentropic compression with the same density (curve P in Figure 4.2(b)). It is also seen that curves H and P intersect at $x = 1$ (the Hugoniot centre '0'). Moreover, their derivatives are equal, which means that Hugoniot and Poisson curves have the same tangent line T at the intersection point '0'. From EOS (4.91) for temperatures at the same density, it follows that entropy behind the shock front increases. In fact, for gases with an EOS $p(v, s)$, having the following properties of normal gas

$$p>0,\ p_{v}<0,\ p_{vv}>0,\ p_{s}>0,\ e_{p}>0, \tag{4.105}$$

it can be proved that the Hugoniot and Poisson curves intersect at the Hugoniot centre and are tangent to each other at the centre. A line, connecting the centre with any point (p, v) of the Hugoniot at $p > p_0$ has slope $(p - p_0)/(v - v_0)$ specifying a shock velocity from (4.87), which is called the Rayleigh line (also introduced by Michelson (1893)). From the thermodynamic identity $de = Tds - pdv$ and (4.88), differential along the Hugoniot is

$$\begin{aligned} dH &= Tds - pdv - \frac{v_0 - v}{2}dp + \frac{p+p_0}{2}dv = \\ &= Tds + \frac{v - v_0}{2}\left(\frac{dp}{dv} - \frac{p-p_0}{v-v_0}\right)dv. \end{aligned} \tag{4.106}$$

It is seen that along the Rayleigh line: $dH = Tds$. It follows from the above that slope of the Rayleigh line reaches the tangent line (Figure 4.2(b)) with slope $-p_v$, at v tending to v_0, which results in $D \rightarrow c_0$ from (4.87) because of $\rho p_{\rho} = -vp_v$. Therefore, in case of a weak shock wave, thermodynamic states are close to the Hugoniot centre and change in entropy tends to zero because of the closeness to the Poisson adiabat, while the corresponding shock wave velocity approaches sound velocity. It is obvious that entropy $s > s_0$ for points above P (Figure 4.2(b)) and $s < s_0$ for points below P. Therefore, from the position of the Poisson adiabat P below Hugoniot H at $p > p_0$ and P above H – at $p < p_0$, Tsemplen's theorem follows, which bans Hugoniot transitions at $p < p_0$ (rarefaction shock waves). From this, it also follows that for a one-dimensional case, state of gas ahead of the shock front is supersonic (with respect to the gas flow), and state of gas behind the shock front is subsonic. In general multi-dimensional case, this is valid for shock waves undergoing velocity change in the direction normal to the front surface – normal shocks.

The RH conditions and their consequences from the above enable the determination of the parameters behind the shock front in a case, when all parameters on one side of the front are pre-given and one of the parameters on the other side, or a shock velocity D, is given. In the case when p is given, v can be found from (4.92) (or (4.88) in the general case) and viceversa. For temperature or internal energy, equation (4.91) needs to be used. Because the mass and momentum jump conditions result in a quadratic equation for F, which suggests two possible signs for the mass flow as seen in (4.87), a shock velocity D has two options of choosing a sign associated with direction of travel of the corresponding shock wave in respect to the flow (left- or right-going waves). Therefore, in the case of a given shock velocity, statement of Tsemplen's theorem is used to specify direction of the shock wave propagation. Specifically, sound velocity is calculated on the side, where all parameters are given. From the comparison of the calculated sound velocity with D, it follows that the shock wave should be directed toward the side with the supersonic flow. Pressure and specific volume before or after the front can be found by using two equations (4.87) and (4.88) after identification of the shock wave direction and selection of the corresponding signs in (4.87).

In the (p, u)-space, the momentum conservation condition transformed with use of the mass conservation relation (4.85) can be rewritten in the following form

$$p = p_0 + \rho_0 (D - u_0)(u - u_0) \tag{4.107}$$

or, in the form, where shock velocity D has been excluded by subtraction of the two representations of D in (4.87) calculated with respect to the state ahead of the wave and behind the wave:

$$(u - u_0)^2 = (p - p_0)(\upsilon_0 - \upsilon). \tag{4.108}$$

As seen in Figure 4.2(b), it follows from (4.108) that Hugoniot states obey one of the two conditions: (i) $p < p_0$; $\upsilon > \upsilon_0$; or (ii) $p > p_0$; $\upsilon < \upsilon_0$. Using the Hugoniot equation (4.88) (specified as (4.92) for a polytropic gas) in the form $\upsilon = \upsilon(p)$, equation (4.108) gives two branches corresponding to the left- and right-going waves.

Hugoniot and Poisson curves rewritten in the (p, u)-space are widely used for analysis of wave interactions. In fact, the (p, u)-analysis is established for well-known Riemann problem statement. The Riemann problem in one-dimensional case utilizes a schematic shown in Figure 4.2(c), which involves volumes of gases in a shock tube at constant states (known also as a Sod problem (Sod, 1978) for the shock tube analysis) separated by a diaphragm with a gas on the right side in state '1' and on the left side – in state '2'. The Riemann problem analyses the gas flow after removal of the diaphragm. Specifically, after having two material systems with different constant parameters brought into contact, left- and right-going waves are formed as shown in Figure 4.2(c) in the (x, t)-diagram. An illustration in Figure 4.2(c) shows a fan of a rarefaction wave (a family of λ^--characteristics separating gases at constant states) as the left-going wave and a line representing trajectory of a shock wave with slope $dx/dt = D$ as the right-going wave. These waves are separated by a contact interface (the dashed line with slope $dx/dt = u$). According to the contact jump condition (4.86), pressure and volume on both sides of the interface are at the same (p, u)-state '3' as shown in Figure 4.2(c). However, these states separated by the contact interface may not be identical, even allowing two different substances occupying spaces corresponding to the states '1' and '2'. Therefore, as shown in Figure 4.2(c), temperatures may be different (T_3'' on the left and T_3' on the right). However, because in the (p, u)-space the condition at the contact interface is represented by a single point, an analysis in this space is very convenient. Any wave going in a given direction can be represented by a two-branched curve: Hugoniot curve 'H' at pressures above p_0 and Poisson adiabat 'P' at $p < p_0$, the both expressed in the (p, u)-variables, as shown in Figure 4.3(a). Because of the smooth connection at the Hugoniot centre, the corresponding curve composed from the two branches in the (p, u)-space can be formally considered as a single combination curve (see Figure 4.3(a)). The curve shown in Figure 4.3(a) corresponds to the case of increasing pressure while particle velocity is decreasing behind the wave front, which is the case for a left-going wave.

In the case of a substance (regardless of its aggregate state) at state '1' placed on the left side from contact interface, a locus of states behind the front is the combination curve for a left-going wave

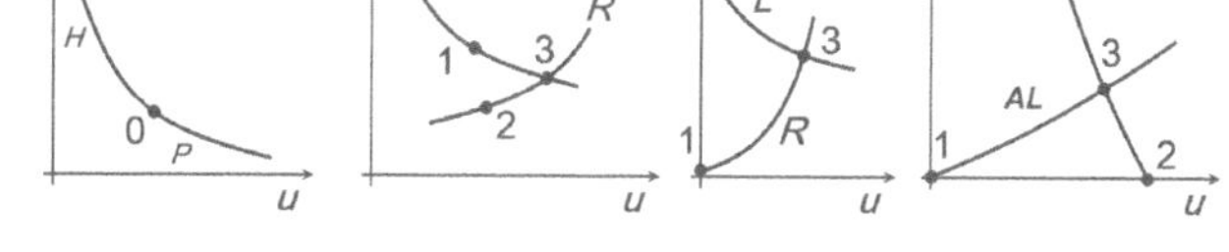

FIGURE 4.3 Combination curve for the Riemann problem in the (p, u)-space (a); wave interaction illustrated by intersection of the combination curves (b); the Riemann problem diagram for gases in a shock tube (c); illustration of the impedance mismatch method for a problem of collision of metallic plates (d).

(similarly to the case in Figure 4.3(a)). For another substance at state '2' on the right, a locus of states behind the front is the combination curve R for a right-going wave with initial states for pressure and particle velocity as shown in Figure 4.3(b). State at the contact interface between these substances is determined from the (p, u)-diagram as state '3', which must be an intersection of the combination curves because the state '3' should be simultaneously behind the right-going wave with state '2' ahead of the front (curve R) and behind the left-going wave with state '1' ahead of the front (curve L). It is seen that states '1' and '3' are separated by a rarefaction wave because of $p_3 < p_1$ and states '2' and '3' – by a shock wave because of $p_3 > p_2$, which demands a configuration with the left-going rarefaction wave and the right-going shock wave with pressure p_3 ($p_1 > p_3 > p_2$) and velocity $u_3 > \max\{u_1, u_2\}$ after the waves.

An analysis of gas flows in the (p, u)-space originated the Riemann solver technique (Toro, 2009), first realized in the Godunov numerical scheme (Godunov, 1959). This solver can be represented in an analytic form for the case of a polytropic gas. For an analysis schematized in Figure 4.3(c), the corresponding illustration in Figure 4.2(c) describes a shock tube set-up with gases being at rest on both sides of the diaphragm with the gas on the left side being at a higher pressure. For these two gases at state '1' on the right and '2' on the left, a solution should combine a left-going rarefaction wave and right-going shock wave. For rarefaction waves, a solution in the (p, u)-space is simply the transmission of Riemann invariants along sound characteristics:

$$u \pm I(p) = \text{const}, dx / dt = \lambda^{\pm}. \tag{4.109}$$

However, the integral $I(p)$ needs to be expressed solely via p, which can be done for the case of polytropic gas, using (4.102). Therefore, in this case, a relation connecting states '2' and '3', is $u_3 + I(p_3) = u_2 + I(p_2)$ that corresponds to curve L in Figure 4.3(c) and can be explicitly written down for the left-going wave (4.103) as follows:

$$u_3 + \frac{2c_0}{\gamma-1}\left(\frac{p_3}{p_0}\right)^{\frac{\gamma-1}{2\gamma}} = u_2 + \frac{2c_0}{\gamma-1}\left(\frac{p_2}{p_0}\right)^{\frac{\gamma-1}{2\gamma}}. \tag{4.110}$$

For the right-going shock wave, $F < 0$ in (4.87), which gives the following equation for the Hugoniot section of the curve R from (4.92):

$$u_3 - u_1 = \sqrt{2\upsilon_1}\frac{p_3 - p_1}{\sqrt{(\gamma+1)p_3 + (\gamma-1)p_1}}. \tag{4.111}$$

Subtraction of equation (4.111) from (4.110) results in an algebraic equation for p_3, which can be solved numerically, followed by calculation of u_3 from (4.110) or (4.111). Temperature T_3' behind the shock front can be calculated with the use of (4.104), and temperature T_3'' – from the EOS after determination of a constant entropy in the isentropic flow on the left side from the contact interface.

For condensed materials, a modification of the ideal gas EOS called the 'stiffened gas' EOS (Menikoff, 2007) and applied to the numerical algorithms by Godunov et al. (1979), takes the form:

$$e = \frac{p - c_0^2(\rho - \rho_0)}{(\gamma-1)\rho}. \tag{4.112}$$

This EOS is convenient, when analytical Hugoniots and Riemann invariants are desirable for use in the Riemann solvers for condensed substances, which closely follow relationships (4.110) and (4.111) with a slight modification of the solver's formulae. It is seen that the polytropic gas EOS (4.91) and EOS (4.112) are of the same Mie–Grüneisen type (4.74) and (4.75) with $\Gamma = \gamma - 1$, $e_c = 0$, $p_c = 0$ for EOS (4.91) and $e_c = -c_0^2/(\gamma - 1)$, $p_c = -\rho_0 c_0^2$ for EOS (4.112).

Reformulation of the jump condition (4.107) based on an empirical dependence $D = D(u)$ is very popular in the applications that employ the (p, u)-analysis for condensed materials. The $D(u)$-dependence is determined from shock wave experiments, usually, in the linear form $D = a + bu$, where a and b are empirical constants. When assuming strong waves, pressure ahead of the front is frequently neglected ($p_0 = 0$), and material ahead of the front is taken at rest ($u_0 = 0$). As a result, equations (4.107) with centre at the origin can be rewritten as:

$$p = \rho_0(a + bu)u. \tag{4.113}$$

For condensed materials, the rarefaction section of a combination curve (represented by (4.110) in the case of polytropic gases) is frequently approximated by extension of the parabola (4.113) over the range of rarefaction below the centre with the same formula even for strong waves. For left-going waves the parabola (4.113) is usually mirrored. For waves, originating from a point on the Hugoniot section and propagating in the opposite direction through the same material, a combination curve described by the same parabolic function with centre at this point can be mirrored as well. Moreover, a rarefaction curve for the same material with centre at a point on the Hugoniot curve can be reflected or mirrored with the same Hugoniot curve as an approximation. Because the technique is based on the use of curves represented by (4.113) in the (p, u)-space, shock velocity in (4.113) is sometimes approximated by the speed of sound in case of moderate loads. Therefore, slopes of the two-branch combination lines in the linear case are simply acoustic impedances, which generates a class of impedance matching techniques (van Thiel et al., 1977).

For example, for an analysis of impact of an aluminium target ($\rho_{AL} = 2.712$ g/cm^3, $D_{AL} = 5.38$ km/s + 1.34 u) by a copper plate ($\rho_{CU} = 8.924$ g/cm^3, $D_{CU} = 3.91$ km/s + 1.51 u) with a velocity of $U_0 = 2$ km/s, pressure and velocity at the impact interface (the aluminium-copper interface) have to be taken equal. Using the parabolic curves (4.113), as shown in Figure 4.3(d), the corresponding equations are

$$p = \rho_{AL} D_{AL}(u)u = \rho_{CU} D_{CU}(U_0 - u)(U_0 - u),$$
$$\rho_{AL}(5.38 + 1.34u)u = \rho_{CU}(3.91 + 1.51(2 - u))(2 - u),$$

which results in $u \approx 1.377$ km/s, $p \approx 19.594$ GPa. Approximating the empirical parabolic curves by straight lines with slopes at the centres determined from bulk sound velocity at normal conditions, the acoustic impedances can be matched in the (p, u)-space. Calculating bulk sound velocities from longitudinal and shear sound velocities (Marsh, 1980) for these materials as $c_{AL} = 5.234$ km/s and $c_{CU} = 3.927$ km/s, the impedance mismatch in the (p,u)-space gives $p = \rho_{AL} c_{AL} u = \rho_{CU} c_{CU}(U_0 - u)$, resulting in $u \approx 1.423$ km/s, $p \approx 20.197$ GPa. It is seen that an error of linearization for the impedance mismatch method is on the order of 3% for these moderately strong shock waves. It should be noted that the impedance mismatch method, applied to processes involving strong rarefaction waves, may result in errors that are not as moderate as in the example above.

In summary, fundamental solutions and jump conditions across shock discontinuity were described in the section. Schematic approaches, used for calculation of states connected by shock transitions and rarefaction waves, were also outlined. These RH conditions and elementary analytical solutions for smooth flows are important elements for derivation of classical detonation solutions considered in the next section.

4.6 CLASSICAL DETONATION THEORIES

An analysis of mechanical and energetic effects in detonation processes was conducted for the first time by Chapman (1899) and Jouguet (1905) and is known as the CJ theory. The analysis is based on the shock physics and thermochemistry foundations from previous sections, which can be directly applied to the case of detonation. The process of detonation in the CJ theory is simplified by assumption that entire energy of the chemical reaction is liberated within the shock wave zone (exothermic jump) and a thickness of the zone is negligible. Thus, the CJ theory assumes that heat liberated at the detonation reaction is released instantaneously, and detonation wave propagates simultaneously with the shock wave initiating the reaction.

As a first step, for assessment of liberated energy in the detonation process, the exothermic reaction within the shock front is presented as a global single-stage reaction

$$R \rightarrow P,$$

where component R (reactants), representing solid or gaseous HE, is transformed to component P (products), representing gaseous detonation products. Equations of state for these components are denoted $e_{0s}(p, v)$ for the unreacted explosive mixture (source reactants) and $e_{0g}(p, v)$ for the gaseous products. The products are formed with a released heat (enthalpy of reaction) $\Delta_R H^0$ obtained as a sum of molar enthalpies of formation of components in the reaction. It follows from (4.51): $\Delta_R H^0 = \Sigma\, \nu_k \Delta_f H^0{}_k$ for k from 1 to K for all components (in the present case, $K = 2$). For the above reaction, $\nu_1 = -1$, $\nu_2 = 1$, $\Delta_R H^0 = \Delta_f H^0{}_P - \Delta_f H^0{}_R$ (subscript 'R' at H^0 refers to reactants and 'P' – to products), and the enthalpy change has a negative value because the reaction is exothermic: $\Delta_R H^0 = -Q'$. The difference $Q' = \Delta_f H^0{}_R - \Delta_f H^0{}_P = -\Delta_R H^0$ is called the heat of explosion and specific heat of explosion $Q = -\Delta_R h_0 = -\Delta_R H^0/W$ can be calculated from (4.65). Positive Q means an increase in specific internal energy of products after completion of the detonation reaction by Q, expressed as $e_{0g}(p, v) - e_{0s}(p, v) = h_{0,g}(p, v) - h_{0,s}(p, v) = Q_{p,v} > 0$ from equations (4.48) and (4.49). Thermodynamic paths within the framework of conventional thermochemistry are routinely taken at constant p and T (for enthalpy) or v and T (for internal energy). In contrast to the thermochemistry analysis, heat at the detonation reaction is assessed at constant p and v, because within a timeframe of the detonation process, the internal chemical reaction is fast, making the process nearly adiabatic and occurring within a limited volume. In the present case, because the reaction is faster than exchange with the environment, the assumption of an isolated thermodynamic system is used, which includes reactants and products.

Detonation products are considered to be gaseous, therefore, it is assumed that an EOS for the products is the polytropic gas EOS with effective polytropic index γ:

$$pv = (\gamma - 1)c_v T\,.$$

In general, this index can be introduced as a ratio

$$\gamma = (h - \Delta_R h_0)/(e - \Delta_R h_0),$$

which, from the definition of enthalpy, gives: $\gamma e - \gamma\Delta_R h_0 = e + pv - \Delta_R h_0$, resulting in

$$e = \frac{p}{(\gamma - 1)\rho} + Q, \tag{4.114}$$

where $Q = -\Delta_R h_0$ and $\rho = 1/v$. It is seen that this EOS is a modification of the polytropic gas EOS (4.91) with exothermic heat release.

By illustration of the CJ theory in a one-dimensional case, it is convenient to use the polytropic gas EOS for a source component and the modified EOS (4.114) for the products. The Hugoniot equation for exothermic jump modifies the energy conservation RH condition (4.88) to the following

$$e_{0g}(p, v) - e_{0s}(p_0, v_0) = \frac{1}{2}(p + p_0)(v_0 - v). \tag{4.115}$$

The Hugoniot equation for the source reactant with initial unchanged chemical composition

$$e_{0s}(p, v) - e_{0s}(p_0, v_0) = \frac{1}{2}(p + p_0)(v_0 - v), \tag{4.116}$$

is identical to the Hugoniot H $(p, v; p_0, v_0)$ considered in the previous section. For EOS of the products, the relation (4.115) for the exothermic jump can be formally rewritten as

$$e_{0g}(p, v) - e_{0g}(p_0, v_0) = \frac{1}{2}(p + p_0)(v_0 - v) + Q, \tag{4.117}$$

which can be denoted by D $(p, v; p_0, v_0)$. Four sectors in the (p, v)-plane for curve H $(p, v; p_0, v_0)$ can be identified with respect to the Hugoniot centre (p_0, v_0) for the unreacted component R. It can be shown that curve D is located above curve H and does not intersect it because this would contradict the condition of positive energy release Q in (4.117). From the Hugoniot analysis, it follows that the Hugoniot H occupies sectors II and IV (Figure 4.4(a)). It is seen that curve D in sector II comes to line $v = v_0$ and in sector IV – to line $p = p_0$ because $e_p > 0$ and $Q > 0$. Inverting functions H and D in equations (4.116) and (4.117) as $p^H(v)$ and $p^D(v)$, respectively, it is seen that at any v, $p^D(v) > p^H(v)$. From (4.87), the slope of Rayleigh lines is proportional to the squared shock velocity, which results in:

$$\frac{dp}{dv} = \frac{p - p_0}{v - v_0} = -\frac{D^2}{v_0^2}. \tag{4.118}$$

It can be shown that a Rayleigh line (dashed line in Figure 4.4(b)) may intersect $D(p, v; p_0, v_0)$ at no more than two points P_1 and P_2 or at a single point C that is the point of tangency to the Rayleigh line touching D (the solid line in Figure 4.4(b)). Because of an upper position of the curve D with respect to H, the curve D can only go through sectors I, II, and IV. However, Rayleigh lines to points of D in sector I have negative slopes and cannot correspond to any shock velocity, hence outlining unachievable D in this sector by dashed section of the curve (Figure 4.4(a)). Therefore, only branches of the curve $D(p, v; p_0, v_0)$ in sector II, denoted as $D_{II}(p, v; p_0, v_0)$, and in sector IV, denoted as $D_{IV}(p, v; p_0, v_0)$, have physical sense.

Therefore, solutions, specifying exothermic shock waves with specific heat of explosion Q, are possible only for the Rayleigh lines intersecting reactive Hugoniot D in sectors II and IV. In general, only two tangency points, touching the Rayleigh lines drawn from the Hugoniot centre (p_0, v_0) to $D(p, v; p_0, v_0)$, exist in sectors II and IV. The tangency point C in sector II is associated with the minimal shock velocity, because slopes of other Rayleigh lines crossing D_{II} are larger than the slope of the tangent line, and B – with the maximal jump velocity in sector IV, because slopes of other Rayleigh lines crossing D_{IV} are smaller.

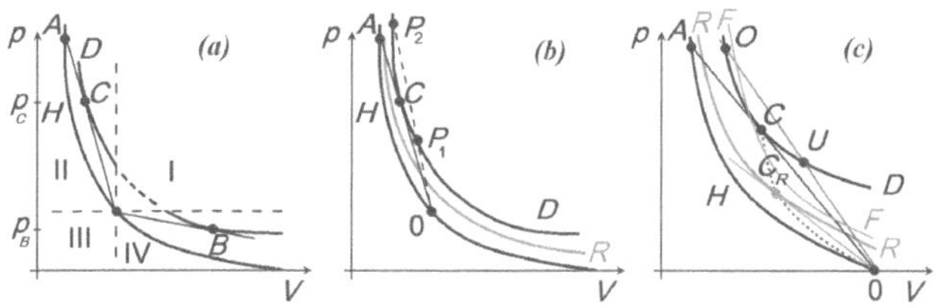

FIGURE 4.4 A schematic of unreacted Hugoniot H and reactive Hugoniot D for exothermic jump (a); the schematic complemented with partially reacted Hugoniot R (b); detailed view of Hugoniots H, D and R with Rayleigh lines illustrating the CJ, underdriven and overdriven regimes (c).

The jump velocity in sector II is supersonic because it corresponds to a shock wave with the Rayleigh line intersecting curve H at point A. For a polytropic gas with sound velocity c_0 ahead of the shock front ($c_0^2 = \gamma p_0/\rho_0$) and the shock velocity obtained from (4.87) plus the Hugoniot (4.92), it follows:

$$\frac{(D-u_0)^2}{c_0^2} = \frac{(\gamma-1)p_0 + (\gamma+1)p}{2\gamma p_0} > 1, \quad p > p_0,$$

which results in $|D - u_0| > c_0$. For any Rayleigh line in sector IV, absolute value of the line slope equal to $\rho_0^2(D-u_0)^2$, is smaller than the slope of the tangent line equal to $\rho_0^2 c_0^2$, which results in $|D - u_0| < c_0$. A solution for section $D_{II}(p, v; p_0, v_0)$ in sector II corresponds to detonation (supersonic) and for branch $D_{IV}(p, v; p_0, v_0)$ in sector IV – to deflagration (subsonic).

For points (p, v) of $D_{II}(p, v; p_0, v_0)$ at $p > p_C$ and for points (p, v) of $D_{IV}(p, v; p_0, v_0)$ at $p > p_B$, the jump velocity is subsonic $D < u + c$. Further, for points (p, v) at $p < p_C$ in sector II and at $p < p_B$ in sector IV, the jump velocity is supersonic: $D > u + c$. Indeed, tracing entropy along the Hugoniot, it is seen from (4.106) that entropy reaches a maximum between the intersection points and $ds = 0$ at the tangency point. For normal gas $(\partial s/\partial p)_v > 0$, from $dp/dv = (p - p_0)/(v - v_0)$ at the point of maximum and Maxwell relation $(\partial s/\partial v)_p = -(\partial s/\partial p)_v(\partial p/\partial v)_s$, a derivative along the Rayleigh line is:

$$\frac{ds}{dv} = \left(\frac{\partial s}{\partial v}\right)_p + \left(\frac{\partial s}{\partial p}\right)_v \frac{dp}{dv} = \left(\frac{\partial s}{\partial v}\right)_p + \left(\frac{\partial s}{\partial p}\right)_v \frac{p-p_0}{v-v_0} = -\left(\frac{\partial s}{\partial p}\right)_v \left(\frac{\partial p}{\partial v}\right)_s + \left(\frac{\partial s}{\partial p}\right)_v \frac{p-p_0}{v-v_0} =$$

$$= \left(\frac{\partial s}{\partial p}\right)_v \left[\frac{p-p_0}{v-v_0} - \left(\frac{\partial p}{\partial v}\right)_s\right] = \left(\frac{\partial s}{\partial p}\right)_v \left[\frac{p-p_0}{v-v_0} - \frac{c^2}{v^2}\right] = \left(\frac{\partial s}{\partial p}\right)_v \left[\frac{(D-u)^2}{v^2} - \frac{c^2}{v^2}\right],$$

which proves that at $ds = 0$: $D - u = \pm c$. The tangency point C gives the condition of self-sustaining detonation $D = u + c$, known as the CJ condition. The stationary value of entropy on $D(p, v; p_0, v_0)$ reaches minimum at C and maximum at B. Thus, entropy increment ahead of the self-sustaining detonation wave is less than that in deflagration wave, which ensures more efficiency for a detonation heat machine, when comparing with combustion.

In summary, at the tangent points C and B, the jump velocity with respect to flow velocity is equal to the local sound velocity:

$$u + c = D. \tag{4.119}$$

In contrast to detonation, the critical role of diffusion and thermal conductivity is higher at deflagration, which makes description of the deflagration regime (sector IV) not as good as description of the detonation regime with the CJ theory. For points (p, v) in sector II at $p > p_C$ (points O in Figure 4.4(c)), a regime is called the overdriven detonation and it is supersonic with respect to the flow ahead of the front. For points in the same sector at $p < p_C$ (points U in Figure 4.4(c)), a supersonic regime is called the underdriven detonation. Obviously, for the supersonic regime, disturbances behind the front of the overdriven detonation (due to rarefaction in the wake of detonation wave) have higher velocity and are overtaking the front, thus gradually reducing parameters at the front to those of the self-sustained detonation wave, and decaying the wave for the underdriven detonation.

Using the CJ theory for a known mixture, a classic set-up for determining the detonation wave parameters requires calculation of the detonation velocity D, and parameters p, u, ρ and T behind the detonation front. This assumes that the state ahead of the front, specified by p_0, u_0 and ρ_0, is given. In the previous section, a set-up for determination of parameters of the shock wave in inert materials with a given state ahead of the front, required one extra parameter behind the front for closing the system. This requirement of an extra parameter for closing the system is replaced by the CJ condition (4.119) in the present case of an exothermic jump. For illustration of the solution process, it can be taken that $u_0 = 0$, negligible p_0 ($p_0 = 0$), and the same polytropic index γ for an EOS describing the substance for both, reactants and products. As a result, the system of equations for determining the flow parameters includes: (i) the Hugoniot for exothermic jump representing detonation wave:

$$e(p,\upsilon) - e(p_0,\upsilon_0) = \frac{1}{2}(p + p_0)(\upsilon_0 - \upsilon) + Q, \tag{4.120}$$

(ii) and (iii) two equations, obtained from (4.87) and (4.108) for the mass and momentum conservation across the shock front

$$D = \upsilon_0\sqrt{\frac{p - p_0}{\upsilon_0 - \upsilon}}, \quad u = (\upsilon_0 - \upsilon)\sqrt{\frac{p - p_0}{\upsilon_0 - \upsilon}}, \tag{4.121}$$

(iv) complete EOS for a polytropic gas specifying the substance before and after the detonation wave:

$$e(p,\upsilon) = \frac{p\upsilon}{\gamma - 1}, \quad e = c_\upsilon T, \quad c^2 = \gamma p\upsilon, \tag{4.122}$$

and, finally, (v) the CJ condition (4.119): $u + c = D$. Therefore, five equations for variables p, υ, D, u and e are written down. Temperature and sound velocity are linked with the parameters above by relations (4.122).

When solving the system of equations, the CJ condition can be replaced by a combination of the equations (4.120) and (4.122) with the CJ condition, resulting in

$$D - u = \upsilon\sqrt{\frac{p - p_0}{\upsilon_0 - \upsilon}} = c, \quad \frac{p - p_0}{\upsilon_0 - \upsilon} = \frac{c^2}{\upsilon^2} = \frac{\gamma p}{\upsilon}. \tag{4.123}$$

For the case of negligible pressure ahead of the detonation front $p_0 \ll p$, the Hugoniot equation (4.120) and the modified CJ condition (4.123) provide the following

$$\frac{p\upsilon}{\gamma - 1} = \frac{1}{2}p(\upsilon_0 - \upsilon) + Q, \quad \frac{p}{\upsilon_0 - \upsilon} = \frac{\gamma}{\upsilon},$$

which gives specific volume (or density $\rho = 1/\upsilon$) and pressure behind the front:

$$\upsilon = \frac{\gamma\upsilon_0}{\gamma + 1}, \quad p = \frac{2(\gamma - 1)Q}{\upsilon_0}. \tag{4.124}$$

From p and υ obtained in (4.124), detonation velocity, temperature, velocity and sound velocity at the CJ point can be calculated from (4.122) and (4.123):

$$D = \sqrt{2(\gamma^2 - 1)Q}, \quad T = \frac{2\gamma(\gamma - 1)Q}{c_\upsilon(\gamma + 1)},$$
$$u = \sqrt{\frac{2(\gamma - 1)Q}{\gamma + 1}} = \frac{D}{\gamma + 1}, \quad c = \sqrt{\frac{\gamma^2 2(\gamma - 1)Q}{\gamma + 1}} = \frac{\gamma D}{\gamma + 1}. \tag{4.125}$$

If p_0 is not negligible, which is usually the case for gaseous explosive compositions with moderate p_{CJ} (4.124), then the equations for determining CJ parameters can be conveniently reformulated, using the sound velocity of gas ahead of the front from $c_0{}^2 = \gamma p_0 \upsilon_0$. Introducing variables $\Delta p = p - p_0$ and $\Delta \upsilon = \upsilon_0 - \upsilon$, and rewriting $\Delta p/\Delta \upsilon = D^2/\upsilon_0{}^2$ for the CJ condition $D - u = \upsilon(\Delta p/\Delta \upsilon)^{1/2} = c$, the RH conditions take the following form

$$\Delta \upsilon = \frac{\upsilon_0}{\gamma+1}\left(1-\frac{c_0^2}{D^2}\right), \quad \Delta p = \frac{D^2}{\upsilon_0(\gamma+1)}\left(1-\frac{c_0^2}{D^2}\right).$$

Following this, the Hugoniot equation (4.120) is reduced to the following $D^2(1 - c_0{}^2/D^2)^2 = 2(\gamma^2 - 1)Q$, which provides the CJ velocity solution as a result

$$D = \sqrt{\frac{(\gamma^2-1)Q}{2} + c_0^2} + \sqrt{\frac{(\gamma^2-1)Q}{2}}, \quad u = \frac{D}{\gamma+1}\left(1-\frac{c_0^2}{D^2}\right). \tag{4.126}$$

Behind the detonation front, flow is represented by a smooth isentropic rarefaction wave. For example, for a wave propagating in a shock tube with closed left end, the gas flow behind the right-propagating detonation wave is represented as a flow between the left end of the tube and the detonation front. For an ideal gas flow described by the Euler model (4.94), a solution for the right-going rarefaction wave is described by Riemann invariants. The right boundary condition for a smooth solution in the shock tube at $t > 0$ is provided by the CJ solution (4.125) and (4.126), and the left boundary condition is a requirement of resting gas $u_R = 0$ due to the closed end of the shock tube. Subscript 'R' refers to the state of gas at rest in the shock tube zone adjacent to the left wall. From (4.103), the corresponding solution takes the following form for a polytropic gas:

$$u - \frac{2c}{\gamma-1} = u_{CJ} - \frac{2c_{CJ}}{\gamma-1} = -\frac{D}{\gamma-1}, \quad \frac{x}{t} = u_{CJ} + D_{CJ}.$$

For this self-similar solution, a rarefaction fan originates at $t = 0$ from the point $x = 0$ simultaneously with the detonation wave. Therefore, the general solution with the boundary conditions above, is:

$$u - \frac{2c}{\gamma-1} = u_{CJ} - \frac{2c_{CJ}}{\gamma-1} = -\frac{D}{\gamma-1}, \quad \frac{x}{t} = u_{CJ} + D_{CJ}. \tag{4.127}$$

Using (4.125), the transmitted Riemann invariant is $J_- = u_{CJ} - 2c_{CJ}/(\gamma - 1) = -D/(\gamma - 1)$. From the left boundary condition $u_R = 0$ and (4.127), resulting in $u_R = 0 = (2c_R - D)/(\gamma - 1)$, it follows that sound velocity c_R in the zone of rest should be equal to $D/2$. Therefore, the equation for sound λ^+-characteristic, separating the zone of constant solution at rest from a smooth flow with varying parameters, is $x/t = u_R + c_R = D/2$. In the zone between this characteristic and detonation front, a solution for varying flow is calculated directly from (4.127), which results in

$$\begin{aligned} u &= \left\{\frac{2}{\gamma+1}\frac{x}{t} - \frac{D}{\gamma+1}, \ \text{at}\ \frac{x}{t} > \frac{D}{2};\ \ 0,\ \text{at}\ \frac{x}{t} < \frac{D}{2}\right\}, \\ c &= \left\{\frac{\gamma-1}{\gamma+1}\frac{x}{t} + \frac{D}{\gamma+1}, \ \text{at}\ \frac{x}{t} > \frac{D}{2};\ \ c_R,\ \text{at}\ \frac{x}{t} < \frac{D}{2}\right\}. \end{aligned} \tag{4.128}$$

Using the condition of isentropic flow $p/\rho^\gamma = p_{CJ}/(\rho_{CJ})^\gamma$ and a relation for the sound velocity in a polytropic gas $p/\rho = c^2/\gamma$, pressure and density can be calculated from the sound velocity, c, obtained by equation (4.128). Further, temperature can be obtained from the thermal EOS (4.122). Results of the calculation are plotted in Figure 4.5, where curves '1', representing the solution, are drawn at

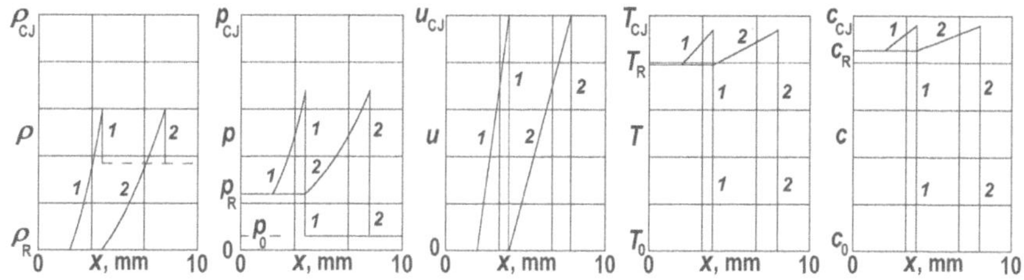

FIGURE 4.5 A solution of the Taylor wave propagation for density (ρ), pressure (p), particle velocity (u), temperature (T) and sound velocity (c).

some time $t = t_1$, and curves '2' – at $t = 2t_1$. A solution for the flow behind detonation front is known as the Taylor wave solution (Taylor, 1950).

The next analysis advances the CJ theory toward refining the detonation wave structure due to the chemical reaction occurring behind the shock front. This analysis employs information from the previous section about chemical kinetics managing reactions. Detonation products are formed in the reaction process initiated by shock wave, propagating in an inert explosive composition, which is managed by the Hugoniot H. After completion of the reaction, the CJ condition is achieved at the point C on reactive Hugoniot D. As the Hugoniot analysis, schematizing the CJ theory in Figure 4.4(c), shows, the Rayleigh line to the tangency point C of the reactive Hugoniot D, intersects the reactant Hugoniot H at a point A. Thus, because the reaction is not completed instantaneously, a shock transition along the Rayleigh line '0A', firstly loads the unreacted explosive to a point A (called the von Neumann (VN) point or the point of 'chemical peak') with a pressure p_{VN} higher than p_{CJ} at the point C. Secondly, a transformation from this precursory load up to the level p_{VN} down to p_{CJ} requires a more thorough consideration of the detonation front, which includes an analysis of the transition zone of a finite size with parameters varying from the VN point A down to the CJ point C. This analysis was first undertaken by Zeldovich (1940), von Neuman (von Neumann) (1942) and Döring (1943), resulting in a theory known as ZND theory. This theory was introduced in Chapter 3.

As a first step of the analysis, a system of equations of the Euler model in one-dimensional case complemented with simple chemical kinetics with rates ω_k for K components, is written as follows:

$$\begin{gathered}
\frac{\partial \rho_k}{\partial t} + \frac{\partial \rho_k u}{\partial x} = \omega_k, \quad k = 1,\ldots,K, \quad \sum_{k=1}^{K} \omega_k = 0, \\
\frac{\partial \rho u}{\partial t} + \frac{\partial\left(\rho u^2 + p\right)}{\partial x} = 0, \quad \rho = \sum_{k=1}^{K} \rho_k, \\
\frac{\partial \rho E}{\partial t} + \frac{\partial\left[\rho E u + p u\right]}{\partial x} = 0, \quad E = e + {u^2}/{2}.
\end{gathered} \tag{4.129}$$

EOS formulations in this section employ ρ_k that are partial densities: $\rho_k = \rho c_k$. The system can be closed by the ideal gas EOS in the form:

$$e = \sum_{k=1}^{K} c_k h_k(T) - \frac{p}{\rho},$$

here h_k are enthalpies of components that can be specified as described in Section 4.3.

Simplest analysis of the ZND theory is based on assumptions similar to the CJ theory that uses two components ($K = 2$), connected by a reaction $R \to P$, where R denotes reactants and P – reaction products.

Mass concentration of the reactants, denoted as c_R, is initially 1 and reaches zero, when the reaction has completed. For convenience, variable z is interchangeable with c_R ($z = c_R$), resulting in mass concentration for the products $c_P = 1 - z$. Assuming a constant specific heat capacity c_p = const, specific enthalpy $h = e - p/\rho$ can be calculated by integration over temperature. After substitution to e, total specific energy takes the following form

$$E = \frac{u^2}{2} + zh_R(T) + (1-z)h_P(T) - \frac{p}{\rho} =$$

$$= \frac{u^2}{2} + z\left[c_p(T - T_0) + \Delta_f h_{0,R}\right] + (1-z)\left[c_p(T - T_0) + \Delta_f h_{0,P}\right] - \frac{p}{\rho} =$$

$$= \frac{u^2}{2} + c_p T + z\left(\Delta_f h_{0,R} - c_p T_0\right) + (1-z)\left(\Delta_f h_{0,P} - c_p T_0\right) - \frac{p}{\rho} =$$

$$= \frac{u^2}{2} + c_p T + zQ - \frac{p}{\rho} = \frac{u^2}{2} + \frac{p}{\rho(\gamma - 1)} + zQ.$$

In this formula, the total heat effect is concentrated in the reactants as specific heat of explosion $Q = \Delta_f h_{0,R} - c_p T_0$ and specific heat of formation of the products $\Delta_f h_{0,P} = c_p T_0$. Relations for specific heat capacities from Section 4.3 and relationship $c_p = \gamma c_v$ are used in this derivation. From (4.89) and (4.90), resulting in $c_p T - p/\rho = \gamma c_v T - p/\rho = \gamma e - p/\rho = \gamma p/[\rho(\gamma - 1)] - p/\rho = p/[\rho(\gamma - 1)]$, the formula above for E can be obtained.

For the case of $K = 2$ with denotation of ω replacing ω_1, system (4.129) is reduced to

$$\frac{\partial \rho}{\partial t} + \frac{\partial \rho u}{\partial x} = 0, \quad \frac{\partial \rho z}{\partial t} + \frac{\partial \rho z u}{\partial x} = \omega,$$
$$\frac{\partial \rho u}{\partial t} + \frac{\partial(\rho u^2 + p)}{\partial x} = 0, \tag{4.130}$$
$$\frac{\partial \rho E}{\partial t} + \frac{\partial\left[\rho E u + p u\right]}{\partial x} = 0, \quad E = e + {}^{u^2}\!/_2,$$

where the EOS closing (4.130) is $e = p/[\rho(\gamma - 1)] + zQ = e' + zQ$. To note, the energy conservation law in (4.130) could be written down in the following equivalent form:

$$\frac{\partial \rho\left(e' + {}^{u^2}\!/_2\right)}{\partial t} + \frac{\partial\left[\rho\left(e' + {}^{u^2}\!/_2\right)u + pu\right]}{\partial x} = -Q\omega,$$

where an alternative EOS $e' = p/[\rho(\gamma - 1)]$ for inert polytropic gas is used and the chemical reaction effect is interpreted as an external energy source. Using system (4.130), the first two equations enable one to specify the chemical kinetics equation along a particle trajectory as

$$\rho\frac{dz}{dt} = \omega, \quad \omega = -A\rho^a z \exp\left(-\frac{E_a}{RT}\right), \tag{4.131}$$

where the dependence for ω is taken as an Arrhenius kinetics for the reaction, using (4.37) and (4.38). The exponent a with respect to ρ depends on reaction order and a discussion about this parameter for global reactions can be found in Section 4.3.

The system of equations (4.130) can be used for full-scale simulations of reactive processes. However, structure of the reaction zone can be analyzed easily in detail if a problem set-up for this system is reduced to one-dimensional case with two states separated by the reaction zone. Specifically, the state ahead of the wave is on Hugoniot H at the VN point A and a state behind the front on the reactive Hugoniot D is at the CJ point C (see Figure 4.3(c)). The structure of the reaction zone can be analyzed in more detail if a stationary solution is sought. For analysis of a stationary solution, the coordinate system $x' = x - Dt$ is introduced that is linked with the shock front. Particle velocity in this coordinate system x' is $v = D - u$, resulting in the particle velocity ahead of the wave: $v_0 = D$ (at $x < 0$). The coordinate system in (4.130) is replaced by x' that is denoted back to x for simplicity. Energy conservation law from (4.130) in the stationary case is written as:

$$v\frac{d}{dx}\left(\rho\frac{v^2}{2}+\rho e+p\right)+\left(\rho\frac{v^2}{2}+\rho e+p\right)\frac{dv}{dx}=0.$$

Mass, momentum, and kinetic equations of system (4.130) can be rewritten as follows:

$$\begin{aligned}&\frac{d\rho}{dx}=-\frac{\rho}{v}\frac{dv}{dx},\quad \frac{dp}{dx}=-\rho v\frac{dv}{dx},\\&\frac{dz}{dx}=-\frac{A\rho z}{v}\exp\left(-\frac{E_a}{RT}\right),\quad \frac{d}{dx}\left(\frac{v^2}{2}+h\right)=0.\end{aligned}\tag{4.132}$$

Using EOS in the form $h = \gamma p/[\rho(\gamma - 1)] + zQ$, the last equation in (4.132) can be expanded to

$$\frac{dh}{dx}+v\frac{dv}{dx}=\frac{\gamma}{\gamma-1}\left(\frac{1}{\rho}\frac{dp}{dx}-\frac{p}{\rho^2}\frac{d\rho}{dx}\right)+Q\frac{dz}{dx}+v\frac{dv}{dx}=0,$$

and substituting dp/dx and $d\rho/dx$ from (4.132), the last equation takes the following form

$$\frac{dv}{dx}=-\frac{Qv(\gamma-1)\gamma}{c^2\left(1-v^2/c^2\right)}\frac{dz}{dx}.\tag{4.133}$$

The system of equations in (4.132) and (4.133) for z, p, ρ and v enables one to obtain a solution if the initial data at $x = 0$ ahead of a reaction zone (just behind the shock front) are given, while x progressing up to the point where v is sufficiently close to c, which corresponds to the end of the reaction zone approaching the CJ condition ($u + c = D$) at the point C (Figure 4.4(b)). The initial data corresponds to the VN point A, which can be found from the unreacted (reactant) Hugoniot H with $D = D_{CJ}$.

For the example of a hydrogen explosion (see Section 4.3) initiated in a stoichiometric hydrogen-air mixture (4.35), containing 20% oxygen and 80% nitrogen in molar concentrations, the chemical reaction for approximately one mole of the mixture is

$$0.28572{\cdot}H_2+0.14286{\cdot}O_2+0.57144{\cdot}N_2\rightarrow 0.28572{\cdot}H_2O+0.57144{\cdot}N_2.$$

Average molar weight of the mixture is $W_{mix} \approx 21$ g/mol. In agreement with data by Gamezo et al. (2008), the following parameters of the mixture are specified: $p_0 = 1$ bar, $T_0 = 293$°K, $\rho_0 = 8.7345{\cdot}10^{-4}$ g/cm^3, $\gamma = 1.17$. Constants of the Arrhenius kinetics are $A = 6.85{\cdot}10^{12}$ cm^3/(g·s), $E_a = 46.37\ RT_0$, and heat of explosion is $Q = 43.28\ RT_0/W_{mix}$. Initial data are taken at the VN spike, which can be calculated from the corresponding Hugoniot state in unreacted gas. For specification of parameters at the point A (Figure 4.4(c)), slope of the Rayleigh line is obtained from the shock velocity (4.126), corresponding to the CJ condition for exothermic jump. From relations for: (i) the

CJ condition (4.123) $\Delta p/\Delta \upsilon = (c_{CJ}/\upsilon_{CJ})^2 = D^2/\upsilon_0^2$, ii) the Hugoniot condition for unreacted gas (4.92) in variables Δp, $\Delta \upsilon$: $\Delta p/\Delta \upsilon = 2\gamma p_0/[2\upsilon_0 - (\gamma + 1)\Delta \upsilon]$, and iii) $c_0^2 = \gamma p_0 \upsilon_0$, it can be found

$$\Delta \upsilon = \frac{2\upsilon_0}{\gamma+1}\left(1-\frac{c_0^2}{D^2}\right), \quad \Delta p = \frac{2\gamma p_0}{\gamma+1}\left(\frac{D^2}{c_0^2}-1\right).$$

Using the momentum jump condition for particle velocity $u = (\Delta p \Delta \upsilon)^{1/2}$, and the mass jump condition in terms of relative velocity $v = D - u$, parameters at the VN point are calculated as follows:

$$u_{VN} = \frac{2D}{\gamma+1}\left(1-\frac{c_0^2}{D^2}\right), \quad p_{VN} = \frac{2\rho_0 D^2}{\gamma+1} - \frac{\gamma-1}{\gamma+1}p_0,$$

$$v_{VN} = D\left[1-\frac{2}{\gamma+1}\left(1-\frac{c_0^2}{D^2}\right)\right], \quad \rho_{VN} = \rho_0\frac{D}{v_{VN}}.$$

A numerical solution for the system of ordinary differential equations (4.131) and (4.132) by the fourth order Runge–Kutta method, using the initial data at the VN point from the above, is shown in Figure 4.6. The solution is calculated until a point in the close vicinity to the CJ state. As seen in Figure 4.6, width of the chemical zone for this example is approximately 3.3 mm. It is seen that two zones can be roughly identified: (i) ignition delay with temperature T close to T_{VN}, followed by (ii) ignition of the mixture with T close to T_{CJ}. Physically, the first zone is associated with dissociation of molecules and generation of active zones. After accumulation of the active zones, an exothermic reaction occurs with major heat release, reaching the CJ state. This two-zone scheme by Korobeinikov and Levin (1969) was used in a number of models simplifying the reaction zone and forming a group of models for detailed analysis of the detonation processes (e.g., see Guirguis et al. (1986); Hishida et al. (2009); Ng et al. (2005)). Several cases of using this approach in applications will be described in the next section.

To note briefly, thermodynamic stability of the stationary solution depends on the activation energy, resulting in a more stable solution for smaller activation energies as discovered by Shchelkin (1966). Several comments can be made on the mutual relationship between the unreacted and reactive Hugoniots. Another family of reactive Hugoniots can be specified by amount of liberated energy. For an intermediate value of heat of explosion $Q_R < Q$, the corresponding Hugoniots R are located between the reactant Hugoniot H and the product Hugoniot D with the heat of explosion Q, as shown in Figure 4.4(b). The CJ points C_R on these Hugoniots form a locus of states that constitute a curve progressing from the Hugoniot centre '0' on H to the CJ point C on D as shown by dotted curve in Figure 4.4(c). States on a Hugoniot R assume that the detonation reaction has completed for corresponding Q_R thus resulting in the curve R being a product Hugoniot. In

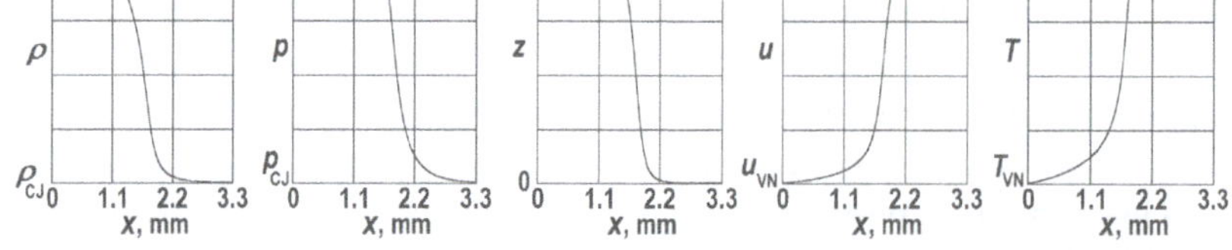

FIGURE 4.6 Structure of the reaction zone calculated as a stationary solution of the reactive flow system simulating an explosion of a hydrogen-air mixture.

fact, one more family of Hugoniots can be derived, while tracking gradual transformation of initial unreacted composition to final products via intermediate compositions during the reaction $R \rightarrow P$. This set of the shock response curves is represented by Hugoniots F (Figure 4.4(c)), each of them drawn at an intermediate (frozen) composition of the reaction (at a fixed reaction extent). These curves, being mixture Hugoniots and positioned above the reactant Hugoniot H, may intersect the product Hugoniot D.

Summarizing, the classical detonation theories, including basic solutions of the CJ and ZND theories, were described in the present section and illustrated with examples. Fundamental CJ relationships widely used in detonation applications were also derived. These relationships are a traditional benchmark solution, employed for the validation of advanced models and computer codes, which will be used and discussed in subsequent sections.

4.7 DETONATION MODELLING OF HOMOGENEOUS EXPLOSIVES

The information presented in previous sections has covered several topics related to detonation modelling. In order to simulate the detonation process in its entirety, all elements of this information need to be incorporated into a comprehensive mathematical model. Next, the model requires a calibration to processes and substances of interest, followed by validation of the model.

Models used in modern hydrocodes for simulating detonation and reactive flows in homogeneous mixtures are mainly based on formulations similar to the ZND theory presented by system (4.129). Steps of this basic approach include: (i) detecting flow parameters within the framework of non-equilibrium thermodynamics, as shown in Section 4.2; (ii) specification of kinetic equations and reaction rates for parameters of mass concentrations or substance amounts of the components of a reacting mixture; alternatively, construction of physics-driven single- or multi-stage kinetics sharing the ZND-type approach as illustrated in Section 4.6; (iii) calculation of the heat of reaction in order to specify liberated heat in an exothermic reaction (such as the heat of explosion) for its subsequent use in an EOS; specification of chemical potentials used for chemical affinity, which is driving the reaction, as described in Section 4.3; and specification of the CJ parameters related to the reactive components under examination for a benchmark test; (iv) determination of products' composition with the use of a thermochemical code; and utilization of this information for determination of reaction parameters and specification of components in EOS, as described in Section 4.4; and (v) use of the shock physics theory outlined in Section 4.5 for specification of shock wave parameters; and derivation of governing equations underlying a reactive flow model, such as the Euler model employed for description of gaseous detonation products and explosive reactants with negligible strength.

As an illustration of applying these principles to a global chemical reaction, examples of set-ups for simulation of combustion of gaseous mixtures are considered below. The first example considers a schematic reaction, representing a part of the combustion reaction of propane considered by Levin et al. (2021) in an oxygen atmosphere, written down as the following global reaction for four components:

$$A_1 + 5A_2 \underset{k_r}{\overset{k_f}{\rightleftarrows}} 4A_3 + 3A_4,$$

where $A_1 = C_3H_8$ (propane), $A_2 = O_2$ (oxygen), $A_3 = H_2O$ (vapour water) and $A_4 = CO_2$ (carbon dioxide). In this reaction, k_f and k_r are reaction rate constants, obtained empirically for the forward and reverse reactions. For the flow parameters to be specified in the local sense, instead of molar concentrations $\beta_k = [A_k]$, it is convenient to deal with partial densities ρ_k that are related to the molar concentrations as $[A_k] = \rho_k/W_k$, where W_k (k =1, 2, 3, 4) are molar weights of the components. Reaction rates ω_k for the present four components A_k and kinetic equations (various forms of kinetic equations are listed in Section 4.3) can be specified, using (4.37) for the forward reaction, as

$$
\begin{aligned}
&J_0 = \frac{J}{V}, \quad \omega_k' = \frac{\omega_k}{W_k} = \nu_k J_0 = \frac{\rho}{M}\nu_k J, \\
&J_0 = \frac{\omega_k'}{\nu_k} = \frac{\omega_k}{\nu_k W_k} = -\frac{\omega_1}{W_1} = -\frac{\omega_2}{5W_2} = \frac{\omega_3}{4W_3} = \frac{\omega_4}{3W_4}, \\
&\frac{d[A_k]}{dt} = \omega_k', \quad \frac{d\xi'}{dt} = J = \frac{M}{\rho}J_0, \quad \frac{d\xi}{dt} = \frac{W_0}{\rho}J_0, \\
&\rho\frac{dc_k}{dt} = \omega_k = W_k\omega_k' = W_k\nu_k J_0.
\end{aligned}
\tag{4.134}
$$

Here, ν_k are stoichiometric coefficients, ξ' and ξ are conventional and nondimensionalized reaction extent parameters, and the rates ω_k are responsible for production of mass of kth component per unit time. For the present reaction, relations of the rates ω_k to reaction velocity J_0 are specified with the following Arrhenius kinetics:

$$
J_0 = k_f[A_1]^\alpha[A_2]^\beta = AT^\delta\left(\frac{\rho_1}{W_1}\right)^\alpha\left(\frac{\rho_2}{W_2}\right)^\beta \exp\left(-\frac{E_a}{RT}\right),
$$

where E_a is activation energy, and reaction constants A, δ, α, and β are empirical parameters. In the case of reaction running in both directions, the reaction equations take the form:

$$
\frac{\rho}{\nu_k W_k}\frac{dc_k}{dt} = J_0 = k_f[A_1]^\alpha[A_2]^\beta - k_r[A_3]^\gamma[A_4]^\varepsilon, \quad k = 1,2,3,4.
$$

The second example considers combustion of fuel $C_aH_bO_c$ in an oxygen atmosphere with oxidizer O_2. The reaction, similarly to (4.63), can be written as follows:

$$
C_aH_bO_c + \left(a + \frac{b}{4} - \frac{c}{2}\right)O_2 \rightarrow aCO_2 + \frac{b}{2}H_2O.
$$

Similar to the first example, the reaction equation for this case can take the form of (4.134) with reaction rates $\omega'_k = \nu_k J_0$ that are determined via reactant concentrations entering J_0 as

$$
J_0 = AT^\delta \exp\left(-\frac{E_a}{RT}\right)[C_aH_bO_c]^\alpha[O_2]^\beta.
$$

Similarly, molar concentrations can be replaced by partial densities via $\rho_k = W_k[A_k]$ with $W_1 \approx 12a + b + 16c$ g/mol for the $C_aH_bO_c$ component and $W_2 \approx 32$ g/mol for the O_2 component.

Summarizing, chemistry-driven kinetics can be directly applied to both single- and multi-reaction cases. In general, a system of ordinary differential equation of the chemical kinetics takes the form (4.40) for rates J_r and molar concentrations β_k. This system can be used in an alternative form, shown in (4.134), for mass concentrations with reaction rates $\omega_k = W_k\omega'_k$.

The Euler model for reactive flows with single reaction is expressed by a system of equations similar to system (4.129) used for the general ZND case:

$$\frac{\partial\rho}{\partial t}+\frac{\partial\rho u_s}{\partial x_s}=0,\quad \frac{\partial\rho_k}{\partial t}+\frac{\partial\rho_k u_s}{\partial x_s}=\omega_k,\quad k=1,\dots,K,$$
$$\frac{\partial\rho u_s}{\partial t}+\frac{\partial(\rho u_s u_m+p\delta_{sm})}{\partial x_m}=0,\quad \rho=\sum_{k=1}^{K}\rho_k,\quad s=1,2,3, \tag{4.135}$$
$$\frac{\partial\rho\left(h-{p}/{\rho}+{u^2}/{2}\right)}{\partial t}+\frac{\partial\rho u_s\left(h+{u^2}/{2}\right)}{\partial x_s}=0,\quad u^2=\sum_{m=1}^{3}u_m u_m,$$

utilizing reaction rates ω_k similar to those from the above.

For gaseous products in the examples above, complete EOS is specified via empirical heat capacities as functions of temperature in the following form

$$p=\sum_{k=1}^{K}p_k=\sum_{k=1}^{K}\frac{\rho_k R}{W_k}T=\sum_{k=1}^{K}\frac{\rho c_k R}{W_k}T=\frac{\rho RT}{W_{\text{mix}}}=\rho R_m T,$$
$$h=\sum_{k=1}^{K}c_k h_k(T), \tag{4.136}$$

where $1/W_{\text{mix}}=\Sigma c_k/W_k$ and $R_m=R/W_{\text{mix}}$. Specific enthalpies $h_k=H_{k,\text{mol}}/W_k$ can be found from $dH_{k,\text{mol}}=C_{pk,\text{mol}}(T)dT$, resulting in

$$H_{k,\text{mol}}(T)=\int_{T_0}^{T}C_{pk,\text{mol}}(T')dT'+\Delta_f H^0_{k,\text{mol}},$$

where $\Delta_f H^0{}_{k,\text{mol}}$ are molar enthalpies of formation of the components. Molar capacities $C_{pk,\text{mol}}(T)$ as functions of T in the form (4.54) can be taken from NASA or JANAF tables (Chase, 1998). Specific heat capacities are calculated as $c_{\upsilon k}=C_{\upsilon k,\text{mol}}/W_k$ and effective polytropic index $\gamma(T)$ can be obtained as follows:

$$\gamma=\frac{C^{\text{mix}}_{p,\text{mol}}(T)}{C^{\text{mix}}_{\upsilon,\text{mol}}(T)}=\frac{C^{\text{mix}}_{\upsilon,\text{mol}}(T)+R}{C^{\text{mix}}_{\upsilon,\text{mol}}(T)}=1+\frac{R}{C^{\text{mix}}_{\upsilon,\text{mol}}(T)}=1+R\frac{1}{W_{\text{mix}}\sum_{k=1}^{K}c_k c_{\upsilon k}(T)}.$$

The system of equations (4.135) is a typical system for modelling reactive flows within single-velocity approximation for the case of single reaction (e.g., see Oran and Boris (2001)). The system can be solved by finite difference or finite volume methods (e.g., numerical schemes by Lax and Wendroff (1964), von Neumann and Richtmyer (1950), MacCormack (1969), Godunov (1959)). Type of the system of partial differential equations for the Euler model (4.94), (4.135) is hyperbolic, which means that all characteristics of the system are real (e.g., λ^+-, λ^-- and λ^0-characteristics). In numerical realizations of models described by hyperbolic systems, including the Euler model, an effect of a zone of flow at current time-step spreads by trajectories of sound characteristics of the system to a zone of influence. The characteristics are erected from points of the zone at the current time-step spreading to the target area formed by projections of the characteristic trajectories in space at the next time-step (Figure 4.2(d)). This imposes a limitation on a size of time-step, limited by a time increment, which is the time required to spread beyond the adjacent cells by either λ^+- or λ^--characteristics. Ratio of this time increment to cell size is called the Courant number, which is a coefficient determined by the velocities of sound characteristics in order to prevent expanding of the zone of influence by more than the cell size. This limitation formally agrees with the stability condition of explicit numerical schemes for hyperbolic systems.

However, the Courant limitation for the Euler model (4.94) is not sufficient for reactive flow simulations, because in this case the Euler model is coupled with reaction equations such as (4.40). The kinetics (4.40) for R reactions form a system of ordinary differential equations that may have dispersed eigenvalues determined from the corresponding reaction rates. Possible large rates relate to small reaction times associated with fast elementary or global reactions that may be much smaller than the corresponding time of propagation of sound. Eigenvalues of the system of reaction equations enforce extremely tough restrictions on stability for explicit methods approximating the reaction equations, which makes inevitable using implicit schemes for calculation of reaction flows.

In general, typical approach for the reactive flow modelling is to split a numerical scheme into two successive routines, which is known as the physical splitting into hydrodynamic and reactive stages. If a solution (a set of independent variables) at current time t^n is denoted by U^n then at the next time cycle $t^{n+1} = t^n + \Delta t$, increased by time-step increment Δt, the solution advances to U^{n+1}. The first stage of the splitting algorithm ignores chemical reactions and applies an operator L of the hydrodynamic numerical scheme for the system of partial differential equations (4.135) with zeroed right-hand sides representing chemical reaction rates to a solution at the current step. This process of numerical solution can be schematically presented as

$$U^{n+1}_{1/2} = L\langle U^n \rangle.$$

The first stage usually employs conventional explicit numerical schemes such as those in (Lax and Wendroff, 1960, 1964; Richtmyer and Morton, 1967), Flux-Corrected Transport (FCT) (Boris and Book, 1973), as well as schemes based on Riemann solvers (Toro, 2009) that were originated from the Godunov method (Godunov, 1959), a family of high-order upwind schemes (Kolgan, 1972; van Leer, 1974) and others derived herefrom, such as Total Variation Diminishing (TVD) scheme (Colella and Woodward, 1984; Harten, 1983) and ARbitraty high-order with DERivatives (ADER) schemes (Titarev and Toro, 2002).

The second stage of the splitting algorithm calculates chemical reaction equations, such as the equation for parameter z in system (4.130), equations for reaction rates (4.40), or equations for molar or mass concentrations (4.134). As seen in (4.130) or (4.135), the reaction equations are formulated along particle trajectories of the hydrodynamic flow and calculated at completion of the first stage. Mathematically, this means that the reactive stage solves a system of ordinary differential equations of chemical kinetics using an implicit scheme:

$$\frac{dU}{dt} = M\langle U \rangle, \quad U^{n+1} - U^{n+1}_{1/2} = M\langle U^{n+1} \rangle \Delta t, \tag{4.137}$$

where M is an operator, representing reaction rates of the reaction equations. The operator M linearized on a solution of the finite difference scheme can be actualized as a matrix that may have a wide-spread spectrum of eigenvalues. In particular, a chemical reaction system, such as (4.40), may have large reaction rates for fast reactions, which makes it a stiff system of ordinary differential equations. The corresponding rates of chemical reactions may have relaxation times much smaller than the time-step increment for the hydrodynamic stage allowable by the Courant number. Two factors need to be taken into account for the second (reactive) stage:

1. Stability of numerical methods for solving ordinary differential equations, which requires an implicit method schematically written down in (4.137). This requirement is presently a standard, because stability conditions for explicit methods may require, when comparing with the Courant time-step, many orders smaller size of the time-step dictated by reaction rates. A large family of methods starting from Newton–Raphson up to multi-stage Runge–Kutta and Gear's methods (Butcher, 2008; Gear, 1971; Rosenbrock, 1963) are used for solving these equations; and
2. Accuracy of numerical methods for calculating the reactive system (4.137). A time-step size acceptable for stability may not be sufficiently small for achieving desired accuracy, because solutions, like those considered in the examples of Section 4.3, have high-frequency features that may be missed. Therefore, at the second stage (4.137), higher-order accuracy methods for stiff ordinary differential equations (e.g., the methods by Rosenbrock (1963) and Gear (1971)) or subcycling within a time increment Δt allowable by the Courant condition, need to be used for achieving desirable accuracy.

When having achieved required accuracy and stability of solutions at the both stages, a superposition of these two solutions may provide a satisfactory solution for a coupled system such as (4.135). Issues in numerical methods used for the reactive flow modelling were reviewed by Oran and Boris (1981).

Thorough realization of the reactive flow modelling algorithms was critical for analysis of non-ideal detonation regimes. Detonation front instabilities were analyzed by Shchelkin (1966) soon after discovery of regular features associated with circulation of disturbances across the wave propagation direction. These regular instabilities are presently classified as elements of cellular and spinning detonation waves. The cellular or multi-front detonation structure was related to the propagation of detonation in channels and tubes. The cellular detonation was first observed with soot foils placed on side walls of shock tubes (Bull et al., 1982) and by the Schlieren method (a photography technique sensitive to the density variation in a gas flow) in the 1950s and 60s (Bdzil et al., 2002; Mitrofanov and Soloukhin, 1965; Voytsekhovskiy et al., 1963). The spinning regime, which is essentially a three-dimensional structure of a spiraling wave, was first observed at the end of shock tube also by soot traces and photography in works by Voytsekhovskiy (1959) and Voytsekhovskiy et al. (1963).

Numerous studies employed the methods described above for simulation of reactive flows representing combustion and detonation, using a global reaction analysis (Gamezo et al., 2008; Levin et al., 2021), a reduced mechanism (O'Connor et al., 2006), or multiple elementary reactions (Konnov, 2008; Levin et al., 1995; Oran and Boris, 1981; Warnatz et al., 2006). A system of reaction equations describing the full mechanism for a hydrogen-air system has been used by Levin et al. (1995). The problem of ignition of a mixture in a gradient zone from high down to low pressure, promoting overdriven detonation, revealed conditions of ignition and transition to detonation. The same chemistry-driven model was used by Levin et al. (2021) in two-dimensional case for a numerical analysis of cell structure in a detonation wave propagating in shock tube. This model was also used for simulation of the cellular and spinning detonation regimes in a propane burning mixture (Levin et al., 2016). Similar to the kinetics used in the example before, a detailed analysis by Wackerle et al. (1978) also used the Arrhenius kinetics for study of initiation of PBX-9501.

A physics-driven approach helps to construct reaction kinetics on the grounds of physical behaviour within the reaction zone. This can be convenient for a description and calibration of global reactions when details of elementary reactions are not critical for the analysis. For example, the ZND simulation in Section 4.6 identified two clear features in the chemical peak zone: (i) ignition delay and (ii) reaction with major heat release. Experimentally, the delay time for various conditions of gaseous detonation is determined by the shock tube tests. The shock tubes used for this task were instrumented with pressure gauges mounted at the closed end of the tube opposite to the end where detonation was initiated. Another method employed for determining the delay time is the Schlieren (contrast photography) technique (Oppenheim and Soloukhin, 1973; Saytzev and Soloukhin, 1961). When using the pressure measurement technique, a gauge is recording the time of arrival of the shock wave, t_s, followed by the spike time t_i due to ignition. By variation of the shock loading conditions, the ignition time can be obtained as a function of pressure and temperature. For a time difference $\tau = t_i - t_s$ called the induction time, $1/\tau$ is proportional to reaction rate, which helps to estimate $\ln \tau \sim E_a/(RT)$ within a given temperature range. This can be used for evaluating the reaction velocity (4.131) and activation energy, E_a. The induction time function $t_{ind}(t)$ at selected constant pressure, temperature and density, having applied for a time period t_i decreases when temperature is increasing, which can be managed by the Arrhenius kinetics as formulated by Korobeinikov and Levin (1969):

$$t_{ind}(t) = kp^{-n_1}\rho^{-l_1}\exp\left(\frac{E_1}{RT}\right), \tag{4.138}$$

where k, n_1, l_1 are empirical constants and E_1 is activation energy. At pressure and temperature evolving in time t, the induction time $t_{ind}(t)$ needs to be integrated over time at nearly constant pressure and temperature within the integration time increment. In order to conduct the integration, fraction of the induction time (fraction of the shortfall to ignition) α is introduced, such that $d\alpha = -dt/t_{ind}(t)$. If $t_{ind} = \tau = \text{const}$, then the result of integration of α from the time of arrival of shock wave at $t = t_s$ up to the start of ignition (the end of induction time) at $t = t_i = t_s + \tau$ leads to the integral $\int dt/t_{ind}(t) = \int dt/\tau$ equal to 1. This principle was applied by Korobeinikov and Levin (1969) to non-constant $t_{ind}(t)$, such that

$$\int_{t_s}^{t_s+\tau} \frac{dt}{t_{ind}(t)} = 1, \tag{4.139}$$

which, at constant parameters of flow in (4.139), is reduced to $\tau = t_{ind}$. It is obvious that $\alpha = 0$ at the end of induction ($t = t_i = t_s + \tau$) corresponds to the start of ignition. In differential form, the particle derivative equation $d\alpha/dt = -1/t_{ind}(t)$ written down for fraction α of the induction time with initial condition $\alpha(t_s) = 1$, takes the following form

$$\frac{d\alpha}{dt} = -k_1 p^{n_1}\rho^{l_1}\exp\left(-\frac{E_1}{RT}\right) = f_1(p,\rho,T), \tag{4.140}$$

which informs on the end of induction (start of ignition), until α reaching zero during integration of (4.140) in a flow. Major heat release starts at the second stage, resulting in an increase of mass concentration c of the reaction products for simplest reaction

$$R \rightleftarrows P.$$

Because this is a global reaction, mass concentration of reactants is $c = M_R/(M_R + M_P)$ with masses of reactants M_R and products M_P. Evolution of c in the corresponding forward and reverse reactions can be described by the following reaction equation

$$\begin{aligned}\frac{dc}{dt} &= -\omega_f + \omega_r = -k_2 p^{n_2}\rho^{l_2}c^{m_2}\exp\left(-\frac{E_2}{RT}\right) + \\ &+ k_3 p^{n_3}\rho^{l_3}c^{m_3}\exp\left(-\frac{E_2+Q}{RT}\right) = f_2(p,\rho,c,T),\end{aligned} \tag{4.141}$$

here k_i, l_i, and n_i are empirical constants, power exponents m_i are also empirical or can be obtained from an analysis of reaction mechanism (see Tsuboi et al. (2009)), E_2 is activation energy and Q is heat of reaction. It is seen that at some time, the forward and reverse reactions may equilibrate when $f_2 = 0$ for some parameters of flow. As a result, the two-stage kinetics can be a part of a model for the simulation of gaseous detonation, which includes conservation laws (4.135), the ideal gas EOS (4.136), and the following kinetic equations (4.140) and (4.141):

$$\frac{\partial \rho\alpha}{\partial t}+\frac{\partial \rho u\alpha}{\partial x}=\rho f_1(p,\rho,T),\quad \frac{\partial \rho c}{\partial t}+\frac{\partial \rho uc}{\partial x}=\rho f_2(p,\rho,c,T).$$

This model by Korobeinikov and Levin (1969) enables one to calibrate the kinetics and validate the model against independent experiments, which provides a tool for describing many important phenomena. Among them are a simulation of the development of cellular detonation (Taki and Fujiwara, 1978) and an analysis of efficiency of the rotating detonation process for possible rotating detonation engines (Levin et al., 2021). A similar two-stage approach was used for a cell structure analysis in the spinning detonation by Vasil'ev and Nikolaev (1976). The Korobeinikov–Levin (KL) approach, used in the examples of gaseous detonation above, agrees with the two-stage nature of energy release in the chemical reaction zone of condensed explosives. Several kinetics consistent with the KL approach and applied to the simulation of detonation in condensed explosives are surveyed below.

For a condensed explosive, the cellular structure was observed by Urtiew et al. (1970) in channels filled with Nitromethane and simulated by Guirguis et al. (1986) with a kinetics suggested by Shchelkin (1959). Similar kinetics for the mass concentration of products by Lobanov et al. (1982) based on different physical assumptions, employed the two-stage approach in the following form:

$$\frac{dc}{dt}=f_k(\rho_1,\rho_2)c^{a_k}(1-c)^{b_k}\,p\exp\left(-\frac{E_k}{RT}\right),\quad k=1,2,$$

where ρ_1 and ρ_2 are densities of solid reactants and gaseous products. At the induction stage $a_1 = 2/3$, $b_1 = 0$ and at the second stage of major energy liberation $a_2 = 0$, $b_2 = 2/3$, and the activation energy at the second stage is three times higher than at the first stage, $E_2 = 3E_1$. This model was used for simulating shock-initiated detonation of a solid HE at high velocity impact (Titov et al., 1981).

An empirical approach to describing reaction kinetics is the fitting of parameters of a polynomial, representing the reaction rate, to available experiments. One of the kinetics of this type is a popular ignition and growth (I&G) model by Lee and Tarver (1980) that uses a polynomial dependence of mass concentration c for reaction products:

$$\frac{dc}{dt}=I(1-c)^x\left(\frac{\rho}{\rho_0}-1\right)^r+G(1-c)^x c^y p^z, \tag{4.142}$$

where x, r, y, z, I and G are constants that have been specified for a number of condensed explosives, ρ_0 is initial explosive density and ρ is current density. First term in (4.142) is responsible for ignition delay (induction stage), relying on the hot spot mechanism. This mechanism is initiated due to void collapse and material shearing along grain interfaces, which presumes a heterogeneous composition of solid HE. It should be noted, however, that the HE substance is considered as a homogeneous mixture with heterogeneities hidden in the kinetics (4.142), while manifesting these meso-scale effects phenomenologically. The first stage is active when mass concentration is less than a certain threshold, assuming slow release of energy with a relatively low reaction velocity. The second term in (4.142) is responsible for coalescence and fast growth of active zones and this term is activated when c is above a certain concentration. A modification of the I&G model by Tarver (2021) includes an additional term similar to the last term in (4.142) but with different power exponents, that are involved when c is less than a certain limit. This term is associated with reaction 'growth' in this modification followed by reaction 'completion' performed by the last term in (4.142) at c above a certain level. Parameters of the reaction velocity were calibrated numerically by Vandersall et al. (2018) to experimental pressure profiles from detonation experiments, employing embedded manganin gauges.

In paper by Bdzil et al. (1999), the Euler model utilized specific internal energy $e(p, \upsilon)$, obeying the additivity rule for gaseous products and solid HE components. As one of the modelling options, it was assumed that pressure and temperature were at equilibrium between the components. The reaction rate was calibrated by Bdzil et al. (1999) to velocity gauge data (Sheffield et al., 1998) for the I&G model (Tarver and McGuire, 2002) managed by the I&G kinetics. After the calibration, the kinetics were initially used in the Baer–Nunziato (BN) model (Baer and Nunziato, 1986; Nunziato and Walsh, 1980), implemented and validated by Schmitt et al. (2002). It is seen that the I&G model (4.142) couples two stages, similarly to those observed in the ZND modelling of reaction zone, which agrees with the two-zone logic, separating the induction stage from ignition stage. A multi-stage model with four stages, enhancing the orginal I&G approach with two additional developing growth and burn stages and employing specific combinations of exponents a and b in a polynomial $c^a(1 - c)^b$, was used by Desbiens (2017).

An example of a truly polynomial-type kinetics is the Forest Fire kinetics (Forest, 1978; Forest and Mader, 1979) that was employed by Mader and Forest (1976) and Mader (1979) for analysis of detonation in condensed explosives. Assuming that mass concentration of products c is a parameter of an EOS $e = e(p, \upsilon, c)$, this kinetics is taken in the following form:

$$\frac{dc}{dt} = (1-c)\exp\left[\sum_{k=1}^{n} a_k p^k\right], \tag{4.143}$$

at $p_{\min} < p < p_{CJ}$, where the polynomial index n may reach a maximum of 14. Mader (1979) calibrated this kinetics by effective utilizing of Hugoniots of reacted explosives and empirical detonation runtime (the distance when detonation velocity becomes stationary after impact initiation).

It should be noted that when size of a chemical reaction zone is on the order of a fraction of millimetre, a large number of numerical cells is required for resolving structure of the zone in detonation wave. In order to alleviate this requirement and resolve the reaction zone, numerical algorithms aimed at increasing space accuracy are being developed for tracking discontinuities at detonation. An example of such algorithm is the level set method (Aslam et al., 1996) that was used to enhance space resolution in vicinity of a moving discontinuity surface $\varphi(x_1, x_2, x_3, t) = 0$ governed by the equation

$$\partial\varphi / \partial t + D\mathbf{n}\cdot\nabla\varphi = 0,$$

where $\nabla_k = \partial/\partial x_k$ is denotation of the gradient derivative, $\boldsymbol{n}$ is external unit vector normal to the discontinuity surface representing a detonation wave, and D is detonation velocity in the normal direction. This tracking method was used to highlight a discontinuity feature and to provide a high resolution locally rather than to increase spatial accuracy for the whole problem. A simplified version of the Forest Fire kinetics (4.143) with single p^n-term and two-step Arrhenius law, representing a thermally neutral induction step plus exothermic ignition step with an extensive energy release, was used by Bdzil et al. (2002) in an Euler model. This numerical realization was equipped with the level set algorithm for detonation front tracking (Bdzil et al., 2002). Similar representation of a simplified Forest Fire kinetics with $n = 1$ was used by Aslam and Bdzil (2002) for an analysis of shape of the detonation front, which enabled the authors to conclude that the detonation wave was getting flatter when increasing confinement.

The BN model (Baer and Nunziato, 1986) was implemented in the CTH hydrocode by Hertel et al. (1995) for the Deflagration-to-Detonation Transition (DDT) simulation. The CTH implementation by Schmitt et al. (2002) employed a level set method developed by Sethian (1999) and was complemented with a viscoelastic Maxwell-type model, classified as ViscoElastic Plasticity (VEP) model in the CTH material model database for simulation of condensed materials. Schmitt et al. (2002) used the CTH hydrocode equipped with the BN model for a DDT simulation of cook-off test.

In summary, this section has illustrated an application of the physics and chemistry basics from previous sections for realization of kinetic formulations that are available in the literature for modelling reactive flows and detonation. An example of a step-by-step construction of a model for simulation of detonation and reactive flows will be conducted in the next section.

4.8 MODEL DEVELOPMENT FOR REACTIVE FLOWS AND DETONATION

This section demonstrates a flowchart for constructing and using a model for simulating reactive flows and detonation. The step-by-step procedure for model construction concludes with numerical examples. While using physically sensible parameters, the examples do not calibrate and validate kinetic and constitutive equations of the model, leaving this task to experiment-driven studies. The present approach utilizes information from previous sections, attempting to link specific subject areas with the corresponding steps of the model construction. Another focus of the procedure is unification of formulation, aimed at presenting the model as a set of conservation laws complemented with constitutive and kinetic equations that describe evolution of relevant parameters along flow trajectories. It should be noted that this approach is limited to the single-velocity approximation of material flow. Severity of these limitations will be discussed at the end of the section.

The present formulation follows the framework of non-equilibrium thermodynamics (de Groot and Mazur, 1962). Variables of components in the traditional formulation (4.135) and (4.136) include only mass or molar amounts of the components, ignoring distribution of component's volumes in the composition. This is acceptable for gaseous mixtures because of their homogeneity and component compliance at compression, but may not be sufficient for compositions containing condensed or density dissimilar constituents. Correct account for a dissipation in processes accompanied by compaction of the compositions, requires an adequate description of individual constituents because of heterogeneity at the meso-level. For example, the role of a dissipation mechanism under compaction for a granular bed HE was considered by Kober and Menikoff (1999). If volume concentration parameters of the components are specified, this information is sufficient to specify actual specific volumes of components, and thus calculate all component parameters when equations of state of the components are given. Many models take into account the component compaction, including the model by Baer and Nunziato (1986). However, when attempting to unify the formulation for provision of a single system of conservation laws, the majority of models do not specifically address this point, which makes implementation of such models difficult in commercial and research hydrocodes.

The present formulation describes a mixture of components A_k ($k = 1, \ldots, K$) that are characterized by specific variables obeying thermodynamic laws introduced in Section 4.2. Actual densities ρ_k complemented with mass and volume concentrations of kth component, $c_k = M_k/M$ and $\theta_k = V_k/V$, enable one to relate actual and partial densities, $\rho_k{}^p$, to density of the mixture, ρ, as follows:

$$\rho_k{}^p = \rho_k\theta_k = \rho c_k, \rho = \Sigma\rho_k{}^p, \rho_k = \rho\, c_k/\theta_k, k = 1,\ldots,K. \tag{4.144}$$

A homogenization approach (Resnyansky and Bourne, 2004; Romenski et al., 2007) enables one to specify parameters of components via averaged variables of mixture and auxiliary concentration and exchange parameters (component specifiers). This makes it possible to replace conservation laws and constitutive equations for individual components, such as those in the BN model (Baer and Nunziato, 1986) and models alike, by a single system of conservation laws for parameters of the mixture complemented with constitutive and kinetic equations for the concentration and exchange parameters.

For a closed monocomponent system representing a solid substance, on top of two independent thermodynamic parameters (e.g., p and T), state of the system at local equilibrium is affected by additional parameters. Depending on nature of the system, the parameters may significantly vary in number and describe material response by a variety of rheological theories for relevant materials.

Not going beyond the case of isotropic solid materials, components in the solid state for irreversible deformations can be described by constitutive equations for effective elastic deformations using the Maxwell-type theory by Godunov and Romenskii (2003). This description requires six extra parameters which are responsible for stress relaxation, such as an effective tensor of small elastic deformations, presented by the Almansi tensor, ε_{ij}. The number of parameters can be reduced to five if omnidirectional elastic deformation, represented by trace of the tensor tr(ε_{ij}), is expressed via density. Then, shear deformations can be characterized by symmetric deviatoric tensor $\varepsilon'_{ij} = \varepsilon_{ij} - \text{tr}(\varepsilon_{ij})/3$ with five independent elements. It should be noted that effective elastic deformations in the model by Godunov and Romenskii (2003) differ from conventional elastic deformations in the way that they take into account irreversible processes and can be directly used for calculation of stresses (for details, see Godunov and Romenskii (2003)). In the present case, small elastic deformations are used in a weighted dimensionalized form (Resnyansky, 2010) $e_{ij} = \varepsilon'_{ij}/\rho = \upsilon\varepsilon'_{ij}$ defined at local thermodynamic equilibrium in the units of specific volume. For a non-local system considered in classical thermodynamics, the weighted small elastic deformations would actually be extensive variables.

Setting independent thermodynamic variables that fully specify a function of state, specific internal energy of kth component can be written as $e_k = e_k(\rho_k, e_{ij}^{(k)}, s_k)$, $k = 1, \ldots, K$ and $i, j = 1, 2, 3$ are coordinate indices for tensors of stress and small elastic deformations. To homogenize the system, each independent thermodynamic parameter of a component needs to be represented via corresponding averaged parameter of the mixture and parameters-specifiers of the component. For example, the component density ρ_k is represented in (4.144) via averaged density of the mixture, ρ, and specifiers that are mass and volume concentrations of the component, c_k and θ_k. Thus, the following arguments of specific internal energy, small elastic deformations $e_{ij}^{(k)}$ and specific entropy s_k, also require similar specification rules. Specification of these parameters may be subject to a certain freedom. For example, the reactive flow formulation (4.135) and (4.136) is suitable for calculating average density of the mixture and partial densities (mass concentrations) of the components, which enables one to describe flow of the mixture as a whole but does not allow specification of individual components, unless volume concentrations are introduced.

Definition of component specifiers can be made with the use of meso-mechanical assumptions, where the freedom of choice is restricted by thermodynamic and conservation conditions. In particular, the conditions of conservation are $\rho = \Sigma\rho_k^p = \Sigma\theta_k\rho_k$, $e = \Sigma c_k e_k$, $\rho e_{ij} = \Sigma c_k \rho e_{ij}^{(k)}$, and $s = \Sigma c_k s_k$. Inter-component imbalances and exchanges within the system are actually governed by constitutive and kinetic equations, which are calibrated to empirical data and may be highly sensitive to material microstructure. When direct exchange data are unavailable, judgement on reasonability of the choice can only be made from indirect comparison of the phenomenological description with experiment. An option to relate component parameters $e_{ij}^{(k)}$ and s_k to the corresponding mixture parameters is suggested in the present case as follows:

$$e_{ij}^{(k)} = \frac{\theta_k}{c_k} e_{ij} + \frac{\lambda_{ij}^{(k)}}{c_k}, \quad s_k = \frac{s}{Kc_k} + \frac{\chi_k}{c_k}, \quad k = 1, \ldots, K, \tag{4.145}$$

where component specifiers $\lambda_{ij}^{(k)}$ and χ_k are responsible for mechanical work mismatch and heat exchange between the components. Here, for the parameters-specifiers c_k, θ_k, χ_k, and $\lambda_{ij}^{(k)}$, each set has $K - 1$ independent parameters because of the conservation conditions: $\Sigma c_k = 1$, $\Sigma\theta_k = 1$, $\Sigma\chi_k = 0$, and $\Sigma\lambda_{ij}^{(k)} = 0$ (the summation is from 1 to K). Definition (4.145) is one of the simplest choices, defining component specifiers as $\lambda_{ij}^{(k)} = c_k e_{ij}^{(k)} - \theta_k e_{ij}$ and $\chi_k = c_k s_k - s/K$, which is consistent with the conservation conditions. Choice of parameters-specifiers may significantly affect values of individual components, and to a less degree – characteristics of the mixture as a whole. For example, alternative options for specifying heat exchange were considered by Resnyansky et al. (2023) and Resnyansky (2023a). Parameters-specifiers for s_k were selected in a way that temperatures in a representative volume are energetically balanced, tending to the relation at equilibrium, such that $T = \Sigma c_k c_{\upsilon k} T_k / \Sigma c_k c_{\upsilon k}$ for components with constant heat capacities $c_{\upsilon k}$.

Internal energy is an extensive variable of the global system, therefore specific internal energy, e, is subject to the additivity rule $e = \Sigma c_k e_k$. After specifications of arguments for specific internal energy of the mixture from (4.144) and (4.145), e is determined as follows

$$\begin{aligned} e\left(\rho, e_{ij}, s, c_1, \theta_1, \lambda_{ij}^{(1)}, \chi_1, \ldots, c_K, \theta_K, \lambda_{ij}^{(K)}, \chi_K\right) &= \sum_{k=1}^{K} c_k e_k\left(\rho_k, e_{ij}^{(k)}, s_k\right) = \\ &= \sum_{k=1}^{K} c_k e_k\left(\rho \frac{c_k}{\theta_k}, \frac{\theta_k}{c_k} e_{ij} + \frac{\lambda_{ij}^{(k)}}{c_k}, \frac{s}{K c_k} + \frac{\chi_k}{c_k}\right), \end{aligned} \tag{4.146}$$

with $K - 1$ independent parameters-specifiers for each set as mentioned in the comment to definitions (4.145).

Assuming that inter-component exchanges are slower than thermodynamic processes inside components, the components are at quasi-equilibrium, which enables one to use the thermodynamic identity within each component:

$$T_k ds_k = de_k + p_k d\left(\frac{1}{\rho_k}\right) - s_{ij}^{(k)} de_{ij}^{(k)}, \quad k = 1, \ldots, K.$$

This allows one to determine dependent thermodynamic variables of the components, such as pressure, temperature and deviatoric stresses, from EOSs of the components, $e_k = e_k(\rho_k, e_{ij}^{(k)}, s_k)$, by separating coefficients at differentials of independent variables:

$$\begin{aligned} &p_k = (\rho_k)^2 \left(\frac{\partial e_k}{\partial \rho_k}\right)_{s_k, e_{ij}^{(k)}}, \quad T_k = \left(\frac{\partial e_k}{\partial s_k}\right)_{\rho_k, e_{ij}^{(k)}}, \\ &s_{ij}^{(k)} = \left(\frac{\partial e_k}{\partial e_{ij}^{(k)}}\right)_{\rho_k, s_k}, \quad k = 1, \ldots, K. \end{aligned} \tag{4.147}$$

Finally, for a multi-component system, the thermodynamic identity applied to the mixture includes uncompensated heat due to inter-component interactions causing internal mass exchange, inter-component momentum and heat exchange, and inter-component transfer of mechanical work due to shear deformations. This results in the following thermodynamic identity for the mixture:

$$\begin{aligned} T ds = de + p d\left(\frac{1}{\rho}\right) - s_{ij} de_{ij} - \\ - \Lambda_k dc_k - \Pi_k d\theta_k - Q_{ij}^{(k)} d\lambda_{ij}^{(k)} - \Psi_k d\chi_k, \end{aligned} \tag{4.148}$$

which gives relationships for dependent variables of the mixture similar to (4.147):

$$\begin{aligned} &p = \rho^2 \frac{\partial e}{\partial \rho}, \quad T = \frac{\partial e}{\partial s}, \quad s_{ij} = \frac{\partial e}{\partial e_{ij}}, \\ &\Lambda_k = \frac{\partial e}{\partial c_k}, \quad \Pi_k = \frac{\partial e}{\partial \theta_k}, \quad Q_{ij}^{(k)} = \frac{\partial e}{\partial \lambda_{ij}^{(k)}}, \quad \Psi_k = \frac{\partial e}{\partial \chi_k}. \end{aligned} \tag{4.149}$$

Here, for each of the relationships in (4.149), independent parameters not involved in differentiation are held constant, omitting lists of fixed parameters outside parentheses displayed in (4.147).

For pressure p in a gaseous mixture of components with chemical concentrations x_k, partial pressure is defined as $p_k^p = p \cdot x_k$ and it represents pressure of kth component of the mixture occupying the whole volume in the case of no remaining components. For an ideal gas mixture with EOS: $p = \rho R_0 T$, $R_0 = (\gamma - 1)c_\upsilon$, density of the mixture from (4.144) is $\rho = \Sigma\rho_k^p = \Sigma\theta_k\rho_k$, where R_0 is assumed to be the same for all components. Substituting ρ as the sum above, the ideal gas EOS for the mixture takes the form $p = \rho R_0 T = \Sigma\theta_k\rho_k R_0 T$. Because ρ_k is actual density of kth component, $p_k = \rho_k R_0 T$ is actual pressure in the component occupying volume V_k if no other components are present. Denoting $p_k^p = \theta_k p_k$ and focusing on $V_k = \theta_k V$ it can be suggested that p_k^p represents partial pressure for the case of chemically unspecified components for the whole volume, where $p = \Sigma p_k^p = \Sigma\theta_k p_k$.

Dependent thermodynamic parameters in (4.148) can be directly calculated from (4.149), using the chain rule (as showed in detail for the calculation of pressure below) for differentiation of the function of state, e, taken from (4.146):

$$
\begin{aligned}
& p = \rho^2 \frac{\partial}{\partial \rho}\left(\sum_{k=1}^{K} c_k e_k\right) = \sum_{k=1}^{K} \rho^2 c_k \left(e_k\right)_{\rho_k} \frac{c_k}{\theta_k} = \sum_{k=1}^{K} \theta_k p_k , \\
& s_{ij} = \frac{\partial e}{\partial e_{ij}} = \sum_{k=1}^{K} \theta_k s_{ij}^{(k)}, \quad \Lambda_m = \frac{\partial e}{\partial c_m} = e_m + \frac{p_m}{\rho_m} - s_{ij}^{(m)} e_{ij}^{(m)} - T_m s_m , \\
& T = \frac{\partial e}{\partial s} = \frac{1}{K}\sum_{k=1}^{K} T_k , \quad \Pi_m = \frac{\partial e}{\partial \theta_m} = -\frac{p_m}{\rho} + s_{ij}^{(m)} e_{ij} , \\
& Q_{ij}^{(m)} = \frac{\partial e}{\partial \lambda_{ij}^{(m)}} = s_{ij}^{(m)}, \quad \Psi_m = \frac{\partial e}{\partial \chi_m} = T_m .
\end{aligned}
\tag{4.150}
$$

It is seen from this derivation that the definition of partial pressure, analyzed in the case of ideal gas in the previous paragraph, is applicable to any material mixture. Similar to pressure, deviatoric stresses s_{ij} in the mixture are volume averaged over the stresses in components, and partial stresses $s_{ij}^{(k)p}$ can be introduced in the same way. It is also seen from (4.150) that, when comparing roles of deviatoric stresses and pressure, 'deformations' e_{ij} play the role similar to that of specific volume $\upsilon = 1/\rho$. To detail, specific volume from (4.144) for kth component $\upsilon_k = 1/\rho_k$ is related to 'partial' counterpart υ_k^p as $\upsilon_k^p = 1/\rho_k^p = 1/(\rho_k\theta_k) = \upsilon_k/\theta_k$. Similarly, partial weighted elastic deformations can be determined as $e_{ij}^{(k)p} = e_{ij}^{(k)}/\theta_k$. To summarize, these definitions of partial elastic deformation $e_{ij}^{(k)p}$ and density ρ_k^p from (4.144) for kth component, should be complemented with definitions of partial pressure and deviatoric stresses, p_k^p and $s_{ij}^{(k)p}$, that are related to thermodynamic pressure and stresses, p_k and $s_{ij}^{(k)}$, as $p_k^p = \theta_k p_k$ and $s_{ij}^{(k)p} = \theta_k s_{ij}^{(k)}$.

Derivatives in (4.150) are calculated under the assumption of independent component specifiers. In reality, from the conservation conditions, specifiers for one of the components (for example, Kth one) must be expressed via the $K - 1$ component that can be declared to be truly independent parameters. Thus, denoting Σ' as the sign of summation from 1 up to $K - 1$, dependent specifiers can be obtained as: $c_K = 1 - \Sigma' c_k$, $\theta_K = 1 - \Sigma'\theta_k$, $\chi_K = -\Sigma'\chi_k$ and $\lambda_{ij}^{(K)} = -\Sigma'\lambda_{ij}^{(k)}$. Therefore, when calculating derivatives over the component specifiers in (4.150), the corresponding Kth term should be summed up for each of the functions Λ_k, Π_k, $Q_{ij}^{(k)}$ and Ψ_k $(k \neq K)$ with coefficient –1, because the derivative of Kth specifier over any corresponding independent one is –1 ($\partial c_K/\partial c_k = -1$ at any $k \neq K$, and so on).

In the case of chemical reaction (4.30), mass production of kth component is related to the reaction rate obeying equation (4.33), as

$$\frac{dc_k}{dt} = \frac{\nu_k W_k}{W_0} \frac{d\xi}{dt}, \tag{4.151}$$

where, as defined in (4.32), ξ is nondimensionalized reaction extent variable (conventional reaction extent variable ξ' is in mole units), W_0 is total molecular mass of reactants before reaction, ν_k are stoichiometric coefficients of reaction (4.30) and W_k are molecular masses of components. If the total mass of the substance is M and $N_{00} = M/W_0$, then from (4.35) and (4.36) it follows:

$$dN_k = \nu_k d\xi', \quad d\xi = \frac{d\xi'}{N_{00}}, \quad dM_k = W_k dN_k, \quad dc_k = \frac{\nu_k W_k}{W_0} d\xi, \quad \mu_k dN_k = \nu_k \mu_k d\xi'.$$

Conventional chemical affinity $A' = -\Sigma \nu_k \mu_k$ is related to the global thermodynamic system with internal energy E of the whole system. In the local sense, chemical affinity $A = -(\Sigma \nu_k W_k \Lambda_k)/W_0$ is related to a representative volume with specific internal energy e under the assumptions of non-equilibrium thermodynamics. Comparing (4.45) with (4.46), the affinities A' and A are linked by the following relations:

$$A' = -\frac{\partial E}{\partial \xi'}, \quad A = -\frac{\partial e}{\partial \xi}, \quad A' = W_0 A, \quad \sum_{k=1}^{K} \nu_k \mu_k = \sum_{k=1}^{K} \nu_k W_k \Lambda_k.$$

The affinity-driven reaction production replaces applicable contributions in (4.148) driven by chemical potentials Λ_k from (4.150) as follows:

$$\begin{gathered} Tds = de + pd\left(\frac{1}{\rho}\right) - s_{ij} de_{ij} + A d\xi - \Pi_k d\theta_k - Q_{ij}^{(k)} d\lambda_{ij}^{(k)} - \Psi_k d\chi_k, \\ A = -\sum_{k=1}^{K} \frac{\nu_k W_k}{W_0} \Lambda_k. \end{gathered} \tag{4.152}$$

Relationships for dependent variables and governing equations for models with mass production of components, such as phase transition, can be derived in a way similar to (Resnyansky, 2023a). The present case, dealing with the modelling of detonation accompanied by a reaction in progress, will be illustrated for a case of $K = 2$, where only two components are involved in the reaction:

$$R \rightarrow P, \tag{4.153}$$

where R denotes explosive reactants and P – detonation products. In this case, the mass production of components is linked with the progress of a chemical reaction, representing detonation of a solid HE. Similar to the analysis with the ZND model in Section 4.6, the mixture contains two components: solid HE reactants R (component 1) and gaseous detonation products P (component 2). For reaction (4.153), this gives $\nu_1 = -1$, $\nu_2 = 1$, $W_1 + W_2 = W_0$, and from the mass conservation $\nu_1 W_1 + \nu_2 W_2 = 0$, it follows $W_1 = W_2 = W_0/2$, hence $\nu_1 W_1 = -W_0/2$. Thus, constitutive equation (4.151) takes the following form

$$\frac{dc}{dt} = -\frac{1}{2}\frac{d\xi}{dt} = -\varphi, \tag{4.154}$$

where initial mass concentration of reactants c is close to 1 for the detonation reaction (4.153). At $K = 2$, chemical potential $\Lambda_2 = -\Lambda_1$ from (4.150) and $A = -(\nu_1 W_1/W_0)\Lambda_1 - (\nu_1 W_1/W_0)\Lambda_2 = ½(\Lambda_1 - \Lambda_2)$

from (4.152). Therefore, specifiers from (4.144) and (4.145) and dependent variables from (4.150) reduced to the case $K = 2$, are defined as follows:

$$\begin{gathered}
\rho_1 = \rho\frac{c}{\theta}, \quad \rho_2 = \rho\frac{1-c}{1-\theta}, \quad p = \theta p_1 + (1-\theta)p_2, \\
s_1 = \frac{s}{2c} + \frac{\chi}{c}, \quad s_2 = \frac{s}{2(1-c)} - \frac{\chi}{1-c}, \\
e_{ij}^{(1)} = \frac{\theta}{c}e_{ij} + \frac{\lambda_{ij}}{c}, \quad e_{ij}^{(2)} = \frac{1-\theta}{1-c}e_{ij} - \frac{\lambda_{ij}}{1-c}, \\
T = \frac{T_1 + T_2}{2}, \quad s_{ij} = \theta s_{ij}^{(1)} + (1-\theta)s_{ij}^{(2)}, \quad Q_{ij} = s_{ij}^{(1)} - s_{ij}^{(2)}, \\
\Psi = T_1 - T_2, \quad \Lambda = \frac{\partial e}{\partial c} = \Lambda_1 - \Lambda_2 = \\
= e_1 + \frac{p_1}{\rho_1} - s_{ij}^{(1)}e_{ij}^{(1)} - T_1 s_1 - e_2 - \frac{p_2}{\rho_2} + s_{ij}^{(2)}e_{ij}^{(2)} + T_2 s_2, \\
A = \frac{1}{2}\Lambda, \quad \Pi = -\frac{p_1 - p_2}{\rho} + \left(s_{ij}^{(1)} - s_{ij}^{(2)}\right)e_{ij},
\end{gathered} \tag{4.155}$$

where c is calculated from the reaction extent variable ξ in (4.154), using a solution similar to (4.34). In (4.155), inter-dependencies of any second specifier with the first one were taken into account (e.g., $c_1 = c$ and $c_2 = 1 - c$) as commented earlier to (4.150). This results in dependent variables Λ, Π, Q_{ij} and Ψ be presented as differences of the paired variables, such as $\Lambda = \Lambda_1 - \Lambda_2$ for the derivative of e over c, and so on. For derivatives calculated in (4.155) over each independent parameter, it is presumed that remaining parameters are fixed, omitting explicit inventory of fixed parameters outside parentheses, similarly to (4.149).

For deriving the governing equations of a closed thermodynamic system containing a K-component mixture, the single-velocity approximation is taken, assuming the same velocities u_i ($i = 1, 2, 3$) for each of the components. Considering kth component as a subsystem of the mixture, balance equations can be obtained from the conservation laws for a subsystem with source exchange terms. The subsystems are characterized by partial variables, whereas the source terms are responsible for mass, momentum and energy exchange between the components. In the case of a solid component, the corresponding balance system needs to be complemented with constitutive equations for effective elastic deformations, obeying the Maxwell-type theory (Godunov and Romenskii, 2003). As a result, a set of balance equations for the component subsystems takes the following form:

$$\begin{gathered}
\frac{\partial \rho_k^p}{\partial t} + \frac{\partial \rho_k^p u_m}{\partial x_m} = m_k, \quad \frac{\partial \rho_k^p u_i}{\partial t} + \frac{\partial\left(\rho_k^p u_i u_m + p_k^p \delta_i^m - s_{im}^{(k)p}\right)}{\partial x_m} = n_k^i, \\
\frac{\partial \rho_k^p\left(e_k + {u^2}/{2}\right)}{\partial t} + \frac{\partial\left[\rho_k^p\left(e_k + {u^2}/{2}\right)u_m + p_k^p u_m - s_{mj}^{(k)p}u_j\right]}{\partial x_m} = \\
= l_k + u_m n_k^m - \frac{u^2}{2}m_k, \quad k = 1,2, \\
\frac{\partial \rho_k^p e_{ij}^{(k)p}}{\partial t} + \frac{\partial \rho_k^p u_k e_{ij}^{(k)p}}{\partial x_k} - \frac{1}{2}\left(\frac{\partial u_i}{\partial x_j} + \frac{\partial u_j}{\partial x_i} - \frac{2}{3}\frac{\partial u_k}{\partial x_k}\delta_{ij}\right) = -\varphi_{ij}^{(k)},
\end{gathered} \tag{4.156}$$

where δ_i^m are elements of the unit tensor (the Kronecker delta, $\delta_i^m = 1$ at $i = m$, $\delta_i^m = 0$ at $i \neq m$), $u^2 = u_i{\cdot}u_i$ is the squared velocity magnitude and $\varphi_{ij}^{(k)} = \rho_k^p e_{ij}^{(k)p}/\tau_k = \rho_k e_{ij}^{(k)}/\tau_k$. Function τ_k describes relaxation of deviatoric (shear) stresses $s_{ij}^{(k)}$ that are proportional to deviatoric elastic deformations for many solid substances with a coefficient of shear modulus G_k, such that $s_{ij}^{(k)} = G_k\varepsilon_{ij}^{(k)} = G_k\rho_k e_{ij}^{(k)}$. For a closed system of the two components, the balance conditions in (4.156) require $m_1 + m_2 = 0$, $n^i_1 + n^i_2 = 0$ ($i = 1, 2, 3$) and $l_1 + l_2 = 0$. As a result, summation of the balance laws (4.156) over k from 1 to 2 gives the conservation laws of mass, momentum and energy for the mixture complemented with constitutive equations for elastic deformations and component specifiers. Similar equations have been derived by Resnyansky (2023a, 2023b) for multiple phases with components of a mixture subject to phase transitions and heat exchange.

As an illustration, reaction rate φ in (4.154) can be selected to be proportional to a rate with the Arrhenius kinetic $A_0\rho c{\cdot}\exp[-E_a/(RT)]$. With this choice, a system of governing equations for a mixture in the case of $K = 2$ is transformed to the following system of conservation laws

$$\frac{\partial \rho}{\partial t} + \frac{\partial \rho u_s}{\partial x_s} = 0, \quad \frac{\partial \rho u_i}{\partial t} + \frac{\partial(\rho u_i u_s - s_{is})}{\partial x_s} + \frac{\partial p}{\partial x_i} = 0,$$
$$\frac{\partial \rho\left(e + {u^2}/{2}\right)}{\partial t} + \frac{\partial\left[\rho\left(e + {u^2}/{2}\right)u_s - s_{is}u_i + pu_s\right]}{\partial x_s} = 0, \tag{4.157}$$

complemented with a system of constitutive equations for small elastic deformations and equations for parameters-specifiers:

$$\frac{\partial \rho e_{ij}}{\partial t} + \frac{\partial \rho u_s e_{ij}}{\partial x_s} - \frac{1}{2}\left(\frac{\partial u_i}{\partial x_j} + \frac{\partial u_j}{\partial x_i} - \frac{2}{3}\frac{\partial u_k}{\partial x_k}\delta_{ij}\right) = -\rho\phi_{ij},$$
$$\frac{\partial \rho\lambda_{ij}}{\partial t} + \frac{\partial \rho\lambda_{ij}u_s}{\partial x_s} = -\rho\pi_{ij}, \quad \frac{\partial \rho c}{\partial t} + \frac{\partial \rho c u_s}{\partial x_s} = -\rho\varphi, \tag{4.158}$$
$$\frac{\partial \rho\theta}{\partial t} + \frac{\partial \rho\theta u_s}{\partial x_s} = -\rho\psi, \quad \frac{\partial \rho\chi}{\partial t} + \frac{\partial \rho\chi u_s}{\partial x_s} = -\rho\omega.$$

For the present case of chemical reaction (4.153), the constitutive equation for mass concentration c is replaced by the chemical reaction kinetics (4.154), resulting in

$$\frac{\partial \rho\xi}{\partial t} + \frac{\partial \rho\xi u_s}{\partial x_s} = 2\rho\varphi. \tag{4.159}$$

The conservation laws (4.157) can be closed with equations of state for the components $e_k = e_k(\rho_k, e_{ij}^{(k)}, s_k)$ substituted into (4.146), which provides a dependence for specific internal energy of the mixture.

The thermodynamic identity (4.152) for the case of chemical reaction is obtained with replacement of the chemical potentials Λ_k (4.150) driving the mass productions dc_k, by affinity $A = -(\partial e/\partial\xi) = ½\Lambda$ driving the reaction progress $d\xi$. Rates of the constitutive equations (4.158) can be selected in the following general form for K components:

$$\rho\phi_{ij} = \sum_{k=1}^{K}\theta_k\varphi_{ij}^{(k)}, \quad \rho\pi_{ij}^{(m)} = \theta_m\left(\varphi_{ij}^{(m)} - \sum_{k=1}^{K}\theta_k\varphi_{ij}^{(k)}\right).$$

For the present case of two components ($K = 2$), this choice of rates is reduced to

$$\rho\phi_{ij} = \theta\varphi_{ij}^{(1)} + (1-\theta)\varphi_{ij}^{(2)}, \quad \rho\pi_{ij} = \theta(1-\theta)\left(\varphi_{ij}^{(1)} - \varphi_{ij}^{(2)}\right).$$

The condition of correctness for the model (4.157) and (4.158) under the assumptions of non-equilibrium thermodynamics requires a non-negative entropy production, resulting in dissipation. Using the mass and momentum conservation laws from system (4.157), the energy conservation law can be reduced to the following

$$\rho\frac{de}{dt} + p\frac{\partial u_i}{\partial x_i} - s_{ij}\frac{\partial u_i}{\partial x_j} = 0. \tag{4.160}$$

After application of the chain rule of differentiation to e from (4.146) and substitution of the corresponding derivatives from (4.149), equation (4.160) can be rewritten as

$$\frac{p}{\rho}\left(-\rho\frac{\partial u_i}{\partial x_i}\right) + s_{ij}\left(\frac{\partial u_i}{\partial x_j} - \rho\phi_{ij}\right) + \rho T\frac{ds}{dt} - \rho A\frac{d\xi}{dt} +$$

$$+\rho\Pi\frac{d\theta}{dt} + \rho Q_{ij}^{(k)}\frac{d\lambda_{ij}^{(k)}}{dt} + \rho\Psi\frac{d\chi}{dt} + p\frac{\partial u_i}{\partial x_i} - s_{ij}\frac{\partial u_i}{\partial x_j} = 0.$$

It is seen that the shear work and power terms cancel out, and from this equation the dissipation function takes the following form

$$\begin{aligned} \rho\Phi = \rho T\frac{ds}{dt} &= \rho\phi_{ij}s_{ij} + \rho\Lambda\varphi + \rho\Pi\psi + \\ &+ Q_{ij}^{(k)}\theta(1-\theta)\left(\varphi_{ij}^{(1)} - \varphi_{ij}^{(2)}\right) + \rho\Psi\omega = \\ &= \theta s_{ij}^{(1)}\varphi_{ij}^{(1)} + (1-\theta)s_{ij}^{(2)}\varphi_{ij}^{(2)} + \rho\Lambda\varphi + \rho\Pi\psi + \rho\Psi\omega. \end{aligned} \tag{4.161}$$

For checking non-negativity of the dissipation function Φ in (4.161), the first two terms, representing viscoelastic work, are specified with the choice of $\varphi_{ij}^{(k)}$ in (4.156), which gives a non-negative contribution $\Sigma\theta_k s_{ij}^{(k)}\rho_k e_{ij}^{(k)}/\tau_k = \Sigma\theta_k s_{ij}^{(k)} s_{ij}^{(k)}/(2G_k\tau_k)$. The last three terms are non-negative with the following choice of rates in equations (4.158) for the parameter-specifiers, where the mass production can be replaced by the reaction rate from (4.159) for the case of the chemical reaction:

$$\varphi = \Lambda\varphi_0, \quad \psi = \Pi\psi_0, \quad \omega = \Psi\omega_0. \tag{4.162}$$

With the choice (4.162), the dissipation function (4.161) can be rewritten as follows

$$\begin{aligned} \Phi = T\frac{dS}{dt} &= \\ &= \Lambda^2\varphi_0 + \Pi^2\psi_0 + \Psi^2\omega_0 + \theta\frac{\left(s_{ij}^{(1)}s_{ij}^{(1)}\right)^2}{2G_1\tau_1} + (1-\theta)\frac{\left(s_{ij}^{(2)}s_{ij}^{(2)}\right)^2}{2G_2\tau_2}, \end{aligned} \tag{4.163}$$

which satisfies the thermodynamic requirement of non-negative entropy production with arbitrary non-negative functions φ_0, ψ_0, and ω_0 (silent i and j are the summation indices from 1 up to 3).

The rate of chemical reaction φ, with the above-mentioned Arrhenius kinetics for φ_0, can be written as follows

$$\varphi = \Lambda \varphi_0 = \Lambda A_0 \rho c \exp\left(-\frac{E_0}{RT}\right), \tag{4.164}$$

where the coefficient Λ is a key contributor to the reaction behaviour. This coefficient, as seen from (4.155), is actually proportional to the Gibbs energy. The function Λ manages reaction, enforcing it: (i) to be at equilibrium when Λ is zero, (ii) to run the reaction in forward direction at $\Lambda > 0$, and (iii) in reverse direction at $\Lambda < 0$.

A constitutive equation for volumetric compression, assuming co-existence of solid and gaseous components, is managed by the following kinetics (Resnyansky, 2010):

$$\psi = \Pi \psi_0 = \Pi \theta(1-\theta)\left(\theta - \theta_p(p)\right), \quad \theta_p(p) \le \theta, \tag{4.165}$$

where $\theta_p(p)$ characterizes a compaction limit. Specification of this function for the present case of gas-solid mixture (gaseous products and solid reactants) can be conducted in the way described comprehensively for porous materials in the paper by Weckert and Resnyansky (2022).

In the two-component case, $l_1 = -l_2$ in (4.156), and heat exchange rate ω in (4.158) can be taken to be proportional to $h\,(T_1 - T_2)$, as suggested by Nigmatulin (1991) and used by Resnyansky (2008) for provision of ω. This is achieved automatically with Ψ from (4.155) and ω from (4.162), which satisfies the correctness condition (non-negative dissipation), if taking ω_0 as

$$\omega_0 = \frac{h}{\rho}\left(\frac{1}{T_1} + \frac{1}{T_2}\right), \tag{4.166}$$

where h is the heat conduction coefficient. Specification of ω_i for a case of three phases was demonstrated in paper by Resnyansky et al. (2011).

The model developed above was implemented by Resnyansky et al. (2011) in the CTH hydrocode (Schmitt et al., 2017). An algorithm (Resnyansky, 2012), employed earlier for implementation of a two-phase model (Resnyansky, 2010) in an earlier version of the CTH code, was advanced by Resnyansky (2023a) and used for implementation of the two-phase model in a recent version of the code (Schmitt et al., 2017), which was applied to the detonation model in the present section.

Several detonation modelling examples are considered below, using a simplest form (4.164) of the Arrhenius kinetics. Objective of these calculations is to demonstrate that the model developed along the lines of the step-by-step construction procedure above is capable of producing a sensible numerical analysis of a detonation or reactive flow process. In order to adequately describe empirical data, several more steps, supplementing the model formulation, are required: (i) selection and calibration of equations of state for the components; (ii) formulation of constitutive equations and parameter calibration for these equations; (iii) specification of exchange rates and assessment of relevant parameters via material characterization; and (iv) development of chemical kinetics that take into account elements of reaction mechanism required for desired resolution of the chemical transition zone, and calibration of the kinetic parameters. A brief description of these steps was outlined in the previous sections.

The first example simulates initiation of a solid explosive upon impact by a 3 mm thickness steel plate, initially occupying a one-dimensional space from 0.1 to 0.4 cm within a fixed Eulerian mesh from 0 up to 2.4 cm. The flyer-plate collides against a 18 mm thick block of explosive, occupying initially a space from 0.4 up to 2.2 cm, with an impact velocity of 500 m/s. This example is calculated within one-dimensional set-up, using EOS parameters that roughly approximate Trinitrotoluene (TNT) explosive.

For an assessment of heat of explosion, thermochemical equations, obtained at the end of Section 4.3, are applied to a TNT explosive with chemical formula $C_7H_5N_3O_6$. For this substance, the

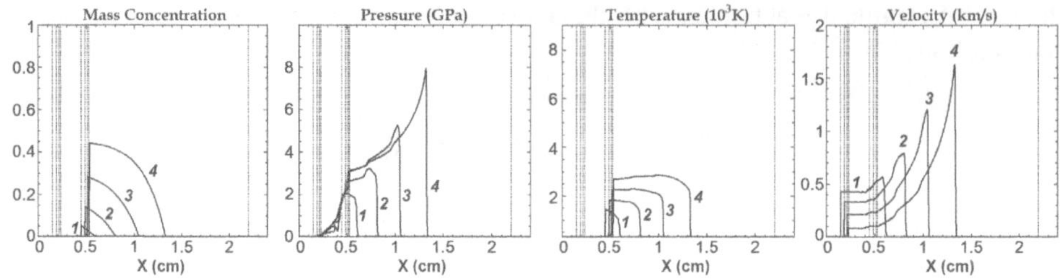

FIGURE 4.7 One-dimensional calculation of the detonation wave propagation with a small pre-exponential coefficient: Mass concentration of reactive products along with pressure, temperature and velocity profiles.

numbers of atoms in this formula are $a = 7$, $b = 5$, $c = 3$, $d = 6$, which results in the following molecular mass of the explosive $W_{TNT} = 12a + 1b + 14c + 16d \approx 227$ g/mol and the oxygen deficiency number $\Omega = -74\%$. Using tabular data (Chase, 1998) for the TNT heat of combustion: $\Delta_c H_{TNT} = -3.41$ MJ/mol, heat of formation from (4.63) is calculated similarly to the example in Section 4.3: $\Delta_f H_{TNT} = a\, \Delta_f H_{CO2} + b/2\, \Delta_f H_{H2O} - \Delta_c H_{TNT} \approx 59$ kJ/mol. Because TNT is an oxygen deficient explosive, heat of formation of products for reaction (4.64) is $\Sigma N_i \Delta_f H_i = (b/2)\, \Delta f H_{H2O} + (d - b/2)\, \Delta_f H_{CO} \approx -1.0075$ MJ/mol. Therefore, from (4.66), specific heat of explosion is: $Q_{TNT} = -(-1007.5 - 59)/227$ kJ/g $\approx$ 4.7 kJ/g. Thus, an estimate of $Q \approx 5$ kJ/g will be used for specifying an EOS required for reaction components in this exothermic process.

In the present examples, EOS for the solid component (component 1) is taken as a Murnaghan EOS in the Mie–Grüneisen form (4.73). Bulk and shear modulus in (4.73) and the corresponding power exponents can be calculated from known sound velocities and Hugoniot data (van Thiel et al., 1977). Gaseous products (component 2) are described by the polytropic gas EOS (4.91) modified with (4.114) for the exothermic reaction, taking the heat of explosion $Q = 5$ kJ/g from the above estimate. These EOS approximations might not be accurate at the extremities but are sufficient for an illustration of the approach. The viscoelastic response of the condensed constituent (component 1) with initial density of 1.65 g/cm^3 is described here by the Maxwell-type rheology (Godunov and Romenskii, 2003). Rate sensitivity of the yield stress response in the constitutive equations of viscoelasticity (4.156) is taken into account via setting two yield limits (40 and 60 MPa) at two different strain rates (10^{-2} and 10^3 1/s, respectively). Using these data, a procedure described by Resnyansky (2012), specifies the relaxation time τ_1 in (4.156). The Grüneisen parameters of $\Gamma_1 = 1.55$ and 0.4 (the polytropic index $\gamma_2 = 1.4$) are taken for the condensed and gaseous components, respectively.

First numerical example with results shown in Figure 4.7 was calculated using a pre-exponential coefficient that was 20 times smaller than that used in the second simulation (the corresponding results shown in Figure 4.8). Curves are labelled by numbers that correspond to the time in

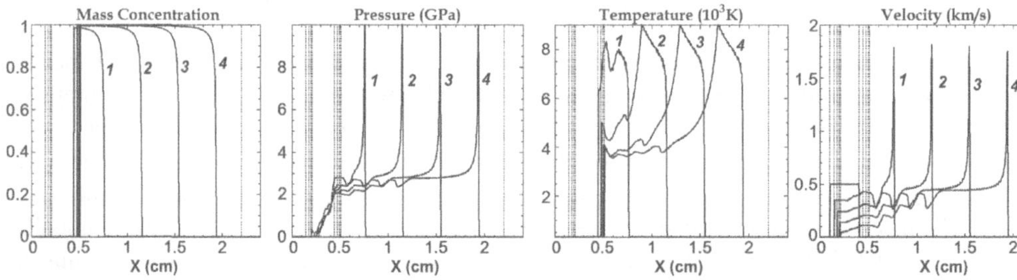

FIGURE 4.8 One-dimensional calculation of the detonation wave propagation with a large pre-exponential coefficient: Mass concentration of reactive products along with pressure, temperature and velocity profiles.

microseconds after impact for the corresponding profiles. As shown earlier, mass concentration of the gaseous component, representing detonation products, is directly linked with nondimensionalizied reaction extent variable ξ. It is seen that the reaction in the first case has not been completed, which results in a mixture behind the detonation front that shares gaseous products with a large portion of the solid component. In this case, thermodynamic states behind the detonation front may partially correlate with states on a 'frozen' Hugoniot F (Figure 4.4(c)). On the contrary, the reaction has almost fully completed in the second case. Obviously, in the first case the Taylor wave solution was not realized, and while the reaction was gradually developing, the temperature peak remained below the ignition level as seen in Figure 4.7.

From the numerical solution plotted in Figure 4.8, the velocity of detonation wave, D, can be estimated to be close to $D \approx 3.9$ km/s. Referring to the classical CJ theory, the detonation velocity, D, and pre-given heat of explosion $Q = 5$ kJ/g could be used for evaluation of an 'effective' polytropic index from (4.125) as $\gamma = (D^2/(2Q) + 1)^{1/2} \approx 1.54$. This effective γ takes an intermediate value between the polytropic indices of $\gamma_1 = \Gamma_1 + 1 = 2.55$ and $\gamma_2 = 1.4$ for the present two components with the chosen EOSs. Using the isentropic condition $(p/p_{CJ})^{\gamma-1} = (c/c_{CJ})^{2\gamma}$ in the Taylor wave, pressure can be evaluated from the classical CJ theory (4.128) in the zone of rest as $p \approx 3.2$ GPa, if the values of D, Q and γ from the above are used. This pressure value correlates with the corresponding calculated pressure observed in Figure 4.8, for the case of completed detonation reaction. It should be noted that the classical CJ theory appreciably simplifies detonation process, therefore, direct comparison of the numerical results with characteristics provided by the CJ theory is unwarranted.

Material parameters, employed in the constitutive equations and equations of state for the simulation presented in Figure 4.8, were also used in two-dimensional modelling examples of shock-initiated detonation. The first example simulates initiation of a HE charge by a plate impact in an axisymmetric set-up. The problem considers collision of a cylindrical 54 mm diameter 3 mm thick steel flyer plate (FP) against a 50 mm diameter 18 mm thick block of the TNT HE with a velocity of impact of 500 m/s in the axial (Y) direction. The calculation results are shown in Figure 4.9 at times of 2 and 4 μs after impact, displaying shapes of the plate and charge that initially occupy a space in Y-direction from −3 mm up to 0 for the plate and from 0 up to 18 mm – for the HE charge.

Calculated distributions of the mass concentration of HE products, which is directly related to the reaction extent and pressure in the plate-target assembly, are complemented with the corresponding profiles along the symmetry axis in Figure 4.4. It is seen that the mass concentration and pressure profiles along the symmetry axis are similar to the corresponding profiles in Figure 4.8, including the same detonation velocity and pressure in the zone of rest of the Taylor wave. However, rarefaction waves from free boundaries at the lateral sides of the HE block cause incomplete reaction due to transverse relief waves, as seen in Figure 4.9.

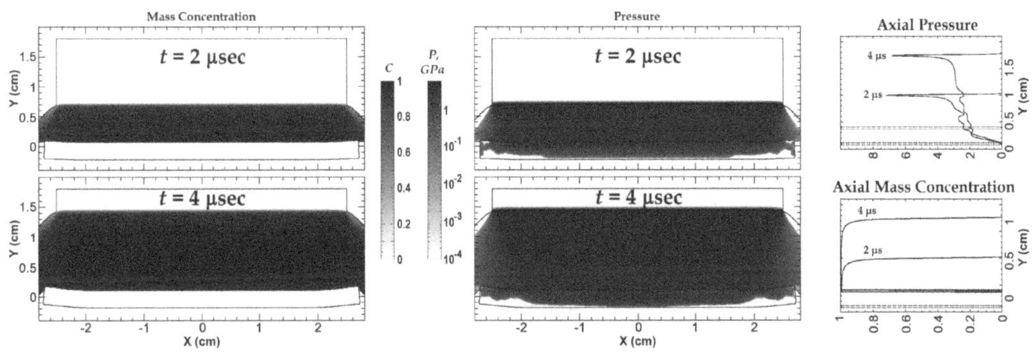

FIGURE 4.9 Two-dimensional calculation of collision of a steel flyer plate against a block of high explosive.

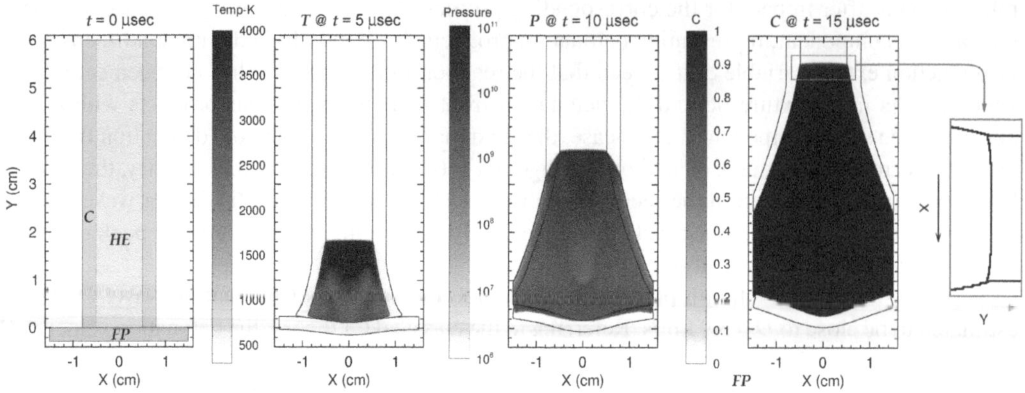

FIGURE 4.10 Calculations of initiation of a long cylindrical HE charge due to high-velocity impact by flyer plate (*FP*) in the axisymmetric set-up.

Boundary effects can be better observed in the next example of impact against a cylindrical HE charge, calculated in an axisymmetric setup, which is presented in Figure 4.10 at $t = 0$. This set of plots illustrates a case of detonation initiation in a long charge due to high-velocity impact. Target is a narrow 10 mm diameter 60 mm long rod of explosive (*HE*) encased in a 3 mm wall thickness plastic cylinder (*C*). The charge is subject to impact by a 30 mm diameter 2.8 mm thick steel *FP* with a velocity of 500 m/s in the axial direction. Numerical results are shown in Figure 4.10 at $t = 5$ μs (temperature), 10 μs (pressure) and 15 μs (mass concentration of detonation products). The present calculation demonstrates the wave diffraction effects due to wave interactions with the cylinder wall *C*, resulting in propagation of transverse relief waves, which cause curvature of an initially planar detonation wave. A detailed view of the front curvature is displayed in the zoomed-in fragment of the last frame. To achieve an adequate description of shape of the head wave during detonation propagation throughout the length of a long *HE* charge, laborious calibration of chemical kinetics and a high accuracy numerical analysis are required. In the present calculations, which use a relatively low mesh resolution, the calculation results cannot be considered sufficiently accurate for the purposes of validation.

Calculations by Mader (1979), employing an Arrhenius kinetics for modelling detonation in a homogeneous explosive, could be referred to as an example displaying an edge effect on shape of the detonation wave. These calculations of shock initiation of a confined charge demonstrate the wall guide effect on the propagating detonation wave, resulting in reverse curvature. An adequate modelling of the detonation front curvature requires high spatial resolution, and the curvature feature of the front has been confidently observed in calculations by Mader (1979) with the Forest Fire kinetics. A high accuracy study of detonation propagation in a long HE charge was conducted by Aslam and Bdzil (2002) with detailed calibration of the I&G reaction kinetics. The calculation of wave propagation followed the calibration, while focusing on the detonation wave zone with use of a tracking algorithm (Aslam et al., 1996) for achieving a high resolution in vicinity of the detonation front.

In summary, efforts in model development and calibration for detonation and reactive flows need to be focused on the following topics: (i) model formulation, model correctness and EOS formulations within the framework of non-equilibrium thermodynamics; (ii) development of constitutive equations and chemical kinetics, using a rheology of irreversible processes in condensed substances and an analysis of chemical mechanism in reactive components; (iii) thermochemical analysis, aimed at specifying the equations of state and chemical kinetics equations; and (iv) characterization of the system components by specification of the model parameters. The present section has demonstrated that successive application of these focus points, required for simulation of detonation processes, enables one to design and utilize a model that may incorporate effects of chemical reactions to the phenomenological behaviour of detonation systems subject to the shock wave loading.

4.9 MODELLING DETONATION IN HETEROGENEOUS EXPLOSIVES

Gaseous detonating mixtures is an example of a homogeneous explosive with 'heterogeneous' chemical composition. Exothermic reactions in the gaseous explosives typically transform gaseous reactants to products in the same aggregate state. Chemical transformation of a solid explosive to gaseous products, similar to that considered in the previous section, occurs while the condensed reactant components co-exist with products in the gaseous aggregate state, which makes the combination of reactants and products a heterogeneous mixture. During a numerical analysis, this composition is frequently considered as a homogenized multi-component mixture under the condition of inter-component velocity equilibrium.

Explosives that are classified as truly heterogeneous could be represented by energetic materials with components that are initially in dissimilar aggregate states or have contrasting properties, such as porous condensed explosives, composite explosive mixtures with density dissimilar components, liquid fuels with bubbles, dusty gaseous explosive compositions, and explosive aerosols. The need for the description of the shock response of a heterogeneous explosive in its unreacted state, which is represented by a Hugoniot H (Figure 4.4), should anticipate an analysis of a response of the explosive after the detonation initiation. Urgency of the corresponding shock response description of unreacted explosives has initiated a variety of studies producing Hugoniots of heterogeneous inert explosives. The shock response of heterogeneous multi-component mixtures, including those with different aggregate states, is frequently assessed with the additivity rule applied to Hugoniots, which was used in one of the first publications by Dremin and Karpukhin (1960). This rule has been successfully applied to a Hugoniot analysis by Bakanova et al. (1972) for a large number of alloys. Specifying a mixture as monocomponent substance is also not uncommon. In order to examine such a substance, mixture Hugoniots are frequently employed as a basis for construction of EOS within the Mie–Grüneisen approach (4.74) and (4.75).

Shock loaded heterogeneous systems, containing pores filled with gas (e.g., the air), manifest a temperature increase up to extreme values even at moderate loads (Williamson, 1990). As a result, when loaded by shock waves, materials with high porosities may respond in an abnormal way, seen as an increase of Hugoniot pressure with decreasing density. This abnormality cannot be described by traditional equations of state for homogeneous monocomponent materials satisfying the conditions of 'normal' substance (4.105). In order to fill this gap within a monocomponent analysis, EOS descriptions available in the literature attempt to incorporate high-temperature effects in the EOSs by additional non-linear temperature terms (e.g., see (4.67) and (4.71)) in commonly used equations of state (Trunin, 1998). Other more specific ways of adjusting the Hugoniot description to the abnormal response of porous mixtures are to utilize a temperature-variable Grüneisen parameter (Cao et al., 2018) and to vary maximum compression ratios (Kinelovskii and Maevskii, 2016).

For porous condensed HEs, the additivity rule is applicable only to a very limited range of porosities. In addition, this approach typically ignores multi-phase nature of porous materials by replacing the gaseous phase with voids. However, some studies (e.g., see Voskoboinikov et al. (1967)), while utilizing the additivity rule, have considered the air in pores as a constituent. Thermodynamic inter-component equilibrating of parameters under shock loading was discussed by Krueger and Vreeland (1991) and possible metastable states in heterogeneous systems with the air were analyzed by Trofimov (1967). Condensing a mixture of several phases or components to a monocomponent substance can be performed with the homogenization procedure used in the previous section. This makes it possible to consider a heterogeneous composition, at the same time uncoupling response of individual components. This can be done with the use of the approach employed for two-component reactants-products mixture under the assumption of inter-component velocity equilibrium.

When approximating a multi-phase inert mixture as a monocomponent substance, design of Hugoniots for the mixture may help in deriving equations of state for the substance, for example, by using the constructed Hugoniots as reference curves in the Mie–Grüneisen EOS representation (4.74) and (4.75). The present line of action realized below utilizes the two-component approach, when selecting equations of state for components and relaxing some conditions of inter-component equilibrium for thermodynamic variables. In addition, states of the components before and after the shock front are assumed to be constant. Conventional jump conditions in the form $\rho(D - u) = \rho_0 D$ and $p - p_0 = \rho_0 Du$, representing mass and momentum conservation across the shock wave moving with velocity D, are complemented with the RH equation

$$e - e_0 - \frac{1}{2}(p + p_0)(1/\rho_0 - 1/\rho) = 0, \tag{4.167}$$

where parameters with subscript '0' refer to the state ahead of the shock wave, e is the specific internal energy, ρ is the density related to specific volume as $\upsilon = 1/\rho$, p is pressure, and u is particle velocity. For an EOS $e = e(p, \upsilon)$, equation (4.167) can be rewritten in the form $p = p^H(\upsilon)$. The RH conditions can be applied either to components of a substance or to the mixture. This is because the conservation laws are the same for both cases. This is valid when inter-component exchanges and production rates are negligible within the shock transition zone, which is the case for metastable states. For a case when inter-component mass exchange is slower than the shock transition (in the vicinity of the shock front), the conservation law for mass concentration, obtained from (4.158) in a one-dimensional case with zero mass production rates, takes the form

$$\frac{\partial \rho c}{\partial t} + \frac{\partial \rho c u}{\partial x} = 0. \tag{4.168}$$

Noting the mass flow through the shock front $F = \rho_0 (u_0 - D)$, the jump condition for the flux of mass concentration derived directly from (4.87) and (4.168) is $[F \cdot c] = 0$. Because $F \neq 0$ for strong shocks and since $[F] = 0$ by mass conservation, this immediately implies $[c] = 0$, which declares that parameter-specifier c obeying (4.168) does not actually change across the shock wave.

Considering a shock loaded porous material as an adiabatically isolated system without mass exchange between gas in the pores and solid matrix, inter-component thermodynamic heat and momentum fluxes are still possible, resulting in a contribution to the dissipation function. This means that phase pressures and temperatures can be at inter-component non-equilibrium during the shock wave compression. Pressure and velocity equilibrium are more likely during the shock transition as discussed by Resnyansky (2008) and Krueger and Vreeland (1991), which may result in absence of momentum exchange. However, as discussed by Resnyansky (2008, 2010, 2016), inter-component heat exchange takes more time than momentum exchange. Numerical simulations by Resnyansky (2008) have demonstrated that pressure equilibrium occurs within a pre-compaction precursor wave well before the main shock front. Therefore, a case of pressure non-equilibrium and temperature equilibrium is physically very unlikely because the mechanical equilibrium ($p_1 = p_2 = p$) is likely to reach faster than temperature equilibrium. Using this implicit pressure equilibrium assumption, an analysis by Nikolaevskii (1969) comes to the conclusion of impossibility of thermal equilibrium along Hugoniots for individual components. However, this analysis misses the option of equilibrating pressure and temperature within the shock transition, which questions validity of individual Hugoniots but raises the possibility of a Hugoniot for mixture.

Thus, a Hugoniot analysis of a mixture can be schematized in three utmost regimes: (i) independent pressures and temperatures in components; (ii) absence of heat exchange between the components, resulting only in pressure equilibrium; and (iii) complete exchange, resulting in inter-component temperature equilibrium in addition to the pressure equilibrium. For simplicity, any kinetic processes during the inter-component exchanges are ignored, reducing the consideration to the cases of (1) independent parameters p and T (case (i)), (2) keeping independent temperatures in the components after the shock propagation with, however, inter-component pressure equilibrium (case (ii)), and (3) instantaneous reaching of a state of complete inter-component equilibrium (case (iii)). Depending on the above options of equilibrium between the components, the following Hugoniots can be analyzed: (1) at non-equilibrium (independent parameters) of p and T (*NE* Hugoniot), (2) at equilibrium of p only (*PE* Hugoniot), and (3) at equilibrium of both p and T (*PTE* Hugoniot).

In the present analysis, a porous material is represented as a solid matrix (component 1) containing gaseous pores (component 2). For this two-phase mixture ($K = 2$), using (4.155), pressure and specific volume are linked with those of the components, resulting in

$$p = cp_1\frac{\upsilon_1}{\upsilon} + (1-c)p_2\frac{\upsilon_2}{\upsilon}, \quad \upsilon = c\upsilon_1 + (1-c)\upsilon_2, \tag{4.169}$$

where volume concentrations are written from (4.144) as $\theta_k = c_k\upsilon_k/\upsilon$. As discussed above, parameter c (mass concentration of the condensed component) is considered to be continuous within the shock transition zone.

First option of *NE* Hugoniot considers inter-component non-equilibrium, which means independent shock transitions for the components. Because of the velocity equilibrium, pressures and specific volumes of components sustain independent Hugoniots, resulting in equations

$$p_1 = p_1^H(\upsilon_1), \quad p_2 = p_2^H(\upsilon_2). \tag{4.170}$$

Because mass and volume concentrations at local equilibrium are constant across the shock front, the system of equations (4.169) and (4.170) contains four equations for variables p_1, p_2, υ_1 and υ_2, which can be expressed as functions of p and υ. Rewriting the RH conservation of mass for each of the components and for the mixture under the condition of velocity equilibrium, it follows

$$\frac{u-D}{u_0-D} = \frac{\upsilon_1}{\upsilon_{01}} = \frac{\upsilon_2}{\upsilon_{02}} = \frac{\upsilon}{\upsilon_0},$$

where $\upsilon_0 = c\upsilon_{01} + (1-c)\upsilon_{02}$. Thus, the equation for specific volume (4.169) gives

$$\upsilon_1 = (\upsilon_{01}/\upsilon_0)\upsilon, \upsilon_2 = (\upsilon_{02}/\upsilon_0)\upsilon. \tag{4.171}$$

Substituting (4.171) to (4.170) and using (4.169) for pressure, *NE* Hugoniot with p and T at non-equilibrium is expressed as follows

$$\begin{aligned} p^{NE}(\upsilon) &= c\frac{\upsilon_{01}}{\upsilon_0}p_1^H(\upsilon_1) + (1-c)\frac{\upsilon_{02}}{\upsilon_0}p_2^H(\upsilon_2) = \\ &= c\frac{\upsilon_{01}}{\upsilon_0}p_1^H\left(\frac{\upsilon_{01}}{\upsilon_0}\upsilon\right) + (1-c)\frac{\upsilon_{02}}{\upsilon_0}p_2^H\left(\frac{\upsilon_{02}}{\upsilon_0}\upsilon\right). \end{aligned} \tag{4.172}$$

This Hugoniot dependence $p^{NE}(\upsilon)$ presumes independent shock transformations for each of the mixture components. Comparison of the *NE* Hugoniot with available experimental data for porous materials has shown (Resnyansky, 2008) that non-equilibrium for both, pressure and temperature, is unlikely. It seems that thermodynamic conditions provided by the shock transition are usually sufficient at least for equilibrating pressure between the components. Attempts to merge individual Hugoniots of components by the additivity rule to a Hugoniot of mixture were undertaken in the literature, starting from an early paper by Dremin and Karpukhin (1960) and further, in a number of papers surveyed by Trunin (1998). However, it appears that these studies implicitly presume equilibrium of pressure between components, which is analyzed next.

An analysis of *PE* Hugoniot under the assumption of pressure equilibrium, requires specification of equations of state for components. The first component is taken to be a solid material described by the stiffened gas EOS (4.72) and the second component – a gaseous substance described by the polytropic gas EOS (4.91):

$$e_1 = \frac{p_1 + \gamma_1 p_{0s}}{\gamma_{1m}} \upsilon_1 + e_{01}, \quad e_2 = \frac{p_2 \upsilon_2}{\gamma_{2m}} + e_{02}. \tag{4.173}$$

In these formulas, it is denoted $\gamma_{km} = \gamma_k - 1$, $\gamma_{kp} = \gamma_k + 1$, $(k = 1, 2)$, p_{0s} is the stiffening constant for the solid component and e_{01}, e_{02} are inter-phase adjustment constants of specific internal energies. For these equations of state, the component Hugoniots (4.170) used as inverse functions. Hugoniot for the condensed component is taken as a generalization of (4.92) for the stiffened gas EOS, and for the gaseous component – the polytropic gas Hugoniot, as follows:

$$\upsilon_1 = \upsilon_{01} \frac{(p_0 + p_{0s})\gamma_{1p} + (p_1 + p_{0s})\gamma_{1m}}{(p_1 + p_{0s})\gamma_{1p} + (p_0 + p_{0s})\gamma_{1m}}, \quad \upsilon_2 = \upsilon_{02} \frac{p_0\gamma_{2p} + p_2\gamma_{2m}}{p_2\gamma_{2p} + p_0\gamma_{2m}}. \tag{4.174}$$

Applying the additivity rule (4.169) for specific volume to (4.174) under the assumption of pressure equilibrium $p_1 = p_2 = p$, Hugoniot $\upsilon^{PE}(p)$ can be obtained

$$\upsilon^{PE}(p) = c\upsilon_{01} \frac{(p_0 + p_{0s})\gamma_{1p} + (p + p_{0s})\gamma_{1m}}{(p + p_{0s})\gamma_{1p} + (p_0 + p_{0s})\gamma_{1m}} + (1-c)\upsilon_{02} \frac{p_0\gamma_{2p} + p\gamma_{2m}}{p\gamma_{2p} + p_0\gamma_{2m}}. \tag{4.175}$$

Finally, the *PE* Hugoniot $p^{PE}(\upsilon)$ can be obtained by inversion of the function (4.175).

The last equilibrium option specifies *PTE* Hugoniot, where both pressure and temperature are at inter-component equilibrium ($p_1 = p_2 = p$ and $T_1 = T_2 = T$). Because variables p and υ are not natural for specific internal energy e, therefore, equations of state (4.173) should be complemented with the corresponding caloric EOSs for temperatures:

$$T_1 = \frac{p_1 + \gamma_1 p_{0s}}{c_{\upsilon 1}\gamma_{1m}} \upsilon_1, \quad T_2 = \frac{p_2 \upsilon_2}{c_{\upsilon 2}\gamma_{2m}},$$

where $c_{\upsilon 1}$ and $c_{\upsilon 2}$ are specific heat capacities for the solid and gaseous components, respectively. Using the additivity rule for specific internal energy and substituting (4.173) to (4.167), a Hugoniot for the mixture can be derived. Complementing it with the caloric EOSs under the assumptions of inter-component pressure and temperature equilibrium and the additivity rule for specific volume from (4.169), the following system can be obtained:

$$c\frac{p + \gamma_1 p_{0s}}{\gamma_{1m}} \upsilon_1 + (1-c)\frac{p\upsilon_2}{\gamma_{2m}} - e_0 = \frac{1}{2}(p + p_0)(\upsilon_0 - \upsilon),$$

$$\frac{1}{c_{\upsilon 1}} \frac{p + \gamma_1 p_{0s}}{\gamma_{1m}} \upsilon_1 = \frac{1}{c_{\upsilon 2}} \frac{p\upsilon_2}{\gamma_{2m}}, \quad \upsilon = c\upsilon_1 + (1-c)\upsilon_2.$$

The last two equations provide expressions $\upsilon_1 = \upsilon_1(p, \upsilon)$ and $\upsilon_2 = \upsilon_2(p, \upsilon)$. Substituting them into the first equation enables one to obtain $p = p^{PTE}(\upsilon)$. This Hugoniot can be rewritten in an inverse analytical form as a rational dependence $\upsilon = \upsilon^{PTE}(p)$:

$$\upsilon^{PTE}(p) = \frac{\left[\frac{1}{2}(p + p_0)\upsilon_0 + e_0\right](c_{\upsilon m}p + c_{\upsilon 2m}p_{0s})}{p(c_{\upsilon s}p + c_{\upsilon g}p_{0s}) + \frac{1}{2}(p + p_0)(c_{\upsilon m}p + c_{\upsilon 2m}p_{0s})}. \tag{4.176}$$

Detailed formulas for constants e_0, $c_{\upsilon s}$, $c_{\upsilon g}$, $c_{\upsilon m}$ and $c_{\upsilon 2m}$ are specified by Resnyansky (2016). Behaviour of the *PTE* Hugoniot can be more easily analyzed for a simplified approximation, when neglecting pressure ahead of the shock wave ($p_0 = 0$). Then, expression (4.176) is reduced to $\upsilon_0{}^{PTE}(p)$ as follows

$$\upsilon_0^{PTE}(p) = \frac{\left(\frac{1}{2}p\upsilon_0 + e_0\right)(c_{\upsilon m}p + c_{\upsilon 2m}p_{0s})}{p(c_{\upsilon sm}p + c_{\upsilon gm}p_{0s})}.$$

A Hugoniot analysis for the *NE*, *PE*, and *PTE* options of inter-component equilibrium with equations of state in general form was conducted by Resnyansky (2017), and a numerical analysis utilizing a general Mie–Grüneisen EOS (4.75) for these three options are presented in a paper by Resnyansky (2008).

For the conventional Hugoniot dependence of a monocomponent substance represented as $p = p^H(\upsilon)$ or $\upsilon = \upsilon^H(p)$, relation $(\partial\upsilon^H/\partial p) < 0$ is always satisfied due to the thermodynamic limitations of normal materials (4.105) for corresponding EOS. However, in the present two-component case, it was shown by Resnyansky (2016) that derivative $\partial\upsilon_0{}^{PTE}(p)/\partial p$ is a rational function with a numerator proportional to $Ap^2 - Bp - C$ and a positive denominator. As Resnyansky (2016) showed for highly porous materials, A and C can be positive, which makes discriminant of the quadratic polynomial positive too. Hence, in this case, the numerator is negative at $p \to 0$ and positive at $p \to \infty$, which ensures existence of a value υ_b, such that the denominator and, consequently, derivative $\partial\upsilon_0{}^{PTE}(p)/\partial p$ turns to zero at $\upsilon = \upsilon_b$ (Figure 4.11(a)). This transition to abnormality occurs at certain combinations of pressure and specific volume within a specific range of porosities that is related to the mass concentration parameter (Resnyansky, 2016).

Thus, in a case of a multi-component mixture, when a relation between pressure and specific volume at shock loading is not under the requirement to follow the Hugoniot derived from a monocomponent EOS, the condition $(\partial\upsilon^H/\partial p) > 0$ is thermodynamically admissible for components satisfying

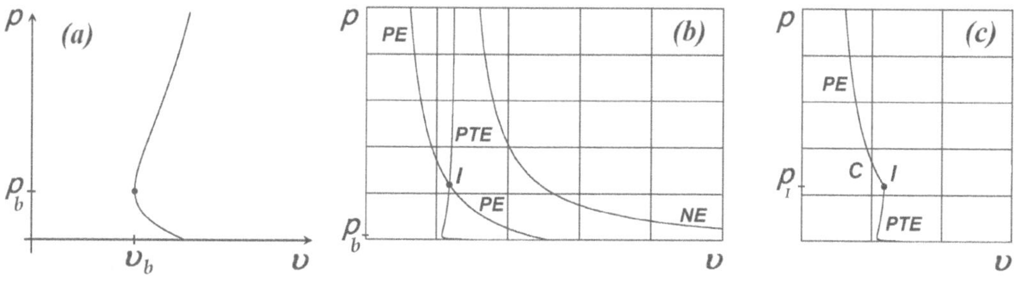

FIGURE 4.11 A PTE Hugoniot schematic for highly porous substance (a); Hugoniot schematics for three equilibrium options (b); Schematic of assembly of a composite Hugoniot $\upsilon^C(p)$ from the pressure-temperature equilibrium Hugoniot $\upsilon^{PTE}(p)$ and pressure equilibrium Hugoniot $\upsilon^{PE}(p)$ (c).

the restrictions of a normal material (4.105). It should be noted that within the multi-component consideration, the abnormality condition can be achieved only for a *PTE* Hugonot.

The value of υ_b was determined by Resnyansky (2016) as the following

$$\upsilon_b = \frac{2e_0}{\rho_{0s}} \frac{c_{\upsilon sm}}{c_{\upsilon gm}} \frac{c_{\upsilon s} c_{\upsilon 2m} + c_{\upsilon m}\left(c_{\upsilon s} + c_{\upsilon m}\right)}{\left(c_{\upsilon m} - c_{\upsilon 2m}\right)\left(c_{\upsilon s} + c_{\upsilon m}\right)},$$

where $c_{\upsilon m} - c_{\upsilon 2m} = c\gamma_{1m} c_{\upsilon 1} > 0$, which means that υ_b can always be calculated when the condition of positive discriminant is satisfied. In comparison with typical shock loading levels, pressure p_b at $\upsilon = \upsilon_b$ is quite low, as shown in Figure 4.11(b). The study by Resnyansky (2016) showed that major contribution to the value of υ_b is made by the part υ_{1b} from solid component. When neglecting mass of gaseous component ($c \approx 1$), the corresponding abnormality condition for specific volume is reduced to $\upsilon_0 > \upsilon_{01}\, \gamma_{1p}/\gamma_{1m}$, which is equivalent to $m > \gamma_{1p}/\gamma_{1m}$, where $\upsilon_{01} = 1/\rho_{01}$ is initial specific volume of the solid component and m is porosity $m = \upsilon_0/\upsilon_{01}$. The traditional analysis by Zeldovich and Raiser (1967) considers porous materials as monocomponent homogeneous substances described by the Mie–Grüneisen EOS (4.74) and (4.75) with Grüneisen parameter Γ. Within the framework of this analysis, denominator in the corresponding Hugoniot is an expression being proportional to $(\gamma + 1)/(\gamma - 1) - \upsilon_0/\upsilon$ (similar to the Hugoniot (4.92) for a polytropic gas) or to $(\Gamma + 2)/\Gamma - \upsilon_{01}/\upsilon$ for the Mie–Grüneisen EOS with initial specific volume υ_{01}, which specifies maximal compression ratio of the substance when the denominator reaching zero (as a reminder, $\Gamma = \gamma - 1$). The analysis by Zeldovich and Raiser (1967) extends the conventional consideration over an area of larger initial specific volumes $\upsilon_0 = m\upsilon_{01} > \upsilon_{01}$ for porous substances, which results in a negative denominator of the Hugoniot dependence at $m > (\Gamma + 2)/\Gamma$ and leads to abnormality with a singularity at $m = (\Gamma + 2)/\Gamma$. In contrast, the present abnormality condition $m > \gamma_{1p}/\gamma_{1m}$, when neglecting mass of the gaseous component, is thermodynamically consistent for the components without any singularities at the transition and abnormality points. This condition is identical to the condition $m > (\Gamma + 2)/\Gamma$ obtained by Zeldovich and Raiser (1967) for monocomponent substances.

It should be noted that the analysis by Zeldovich and Raiser (1967) does not consider inter-component interaction and refers the Hugoniot condition to a condensed material as a monocomponent substance, violating the maximal compression ratio in the Hugoniot dependence. In contrast to the analysis by Zeldovich and Raiser (1967), the maximal compression ratios are not violated in the present two-component case. Gaseous and solid Hugoniots (4.174) have their own maximal compression ratios (e.g., volume υ_a in Figure 4.2(a)). Meso-mechanical analysis by Resnyansky (2016) showed that intersection point I (Figure 4.11(b)) describes a state, where compression of the gaseous component reaches nearly maximum, corresponding to pore collapse. This point specifies a critical pressure in the mixture behind the shock front, above which, time of the shock transition is not sufficient for component temperatures to equilibrate. On the contrary, for a shock loading below this pressure, the time is sufficient for temperature equilibration. During this loading time, (i) gaseous pores are being adiabatically compressed and heated up to extreme temperatures; (ii) the pores are not fully collapsed at this loading pressure, therefore, interstitial gas after the compression still occupies a residual volume that has a low resistance to expansion of condensed phase; and (iii) the condensed component heated up by the compressed gas is expanding towards the residual volume. Because of the dominating contribution from mass concentration of the solid component to specific volume in (4.169) at high porosity, an increase in the volume due to the matrix expansion manifests itself as abnormality. This explains visible abnormal expansion of the mixture on the section of *PTE* Hugoniot above pressure p_b (Figure 4.11(a)) and below pressure p_I at point I (Figure 4.11(b)). Therefore, the point I specifies a transition state from the *PTE* to *PE* regime.

Summarizing, a composite Hugoniot of a porous mixture can be constructed from (i) a branch of *PTE* Hugoniot at pressures below p_I, and (ii) a branch of *PE* Hugoniot at pressures above p_I (Figure 4.11(c)). Selection of these branches in correspondence with actual material response follows the principle of minimal specific volumes among all possible Hugoniot states, which is represented by the following simple expression assembled from the Hugoniot branches

$$\upsilon^{C}(p) = \min\{\upsilon^{PTE}(p), \upsilon^{PE}(p)\}. \tag{4.177}$$

It should be noted that regardless of conventional or abnormal behaviour of the *PTE* Hugoniots, transition from the *PTE* Hugoniot, $\upsilon^{PTE}(p)$, to *PE* Hugoniot, $\upsilon^{PE}(p)$, occurs at the threshold pressure p_I, which will be demonstrated in examples below.

It should be mentioned that examples overriding the transition assumptions, which may result in a single *PE* or *PTE* Hugoniot or in a shift (up or down) of the transition pressure, might be possible. For example, structure of a porous material, composed from large solid particles surrounded by significant gaseous pockets, may prevent achieving temperature equilibrium even at slow rates of loading at shock pressures below p_I. This may result in an extension of *PE* Hugoniot branch, $\upsilon^{PE}(p)$, on $\upsilon^{C}(p)$ below the threshold pressure. On the other hand, a porous material with ultra-dispersed particles in the nanometre size range may respond to shock by achieving pressure-temperature equilibrium even at very strong loading, corresponding to pressures above p_I. In this case, a *PTE* Hugoniot contribution, $\upsilon^{PTE}(p)$, may extend over $\upsilon^{C}(p)$ above the threshold pressure.

Examples of composite Hugoniots $\upsilon^{C}(p)$ calculated for porous inert TNT are shown in Figure 4.12. In order to calculate contributing Hugoniot branches, $\upsilon^{PTE}(p)$ from (4.176) and $\upsilon^{PE}(p)$ from (4.175), thermo-mechanical constants of the EOSs need to be specified. Parameters for the gaseous substance (component 2 – the air) are taken from the polytropic gas EOS used earlier. Parameters for condensed constituent (component 1 – solid TNT) are taken from data by Erkman and Edwards (1974), specifically, $c_{\upsilon 1} = 0.71$ J/(g·K), $\Gamma_1 = 0.737$, $\rho_{01} = 1.65$ g/cm^3, $c_{01} = 2.7$ km/s, where the Grüneisen parameter is taken from paper by Fortov and Dremin (1975). Available experimental data (Erkman and Edwards, 1974; van Thiel et al., 1977) for shock loaded porous inert TNT compositions were used for comparison with calculated Hugoniots. The present illustration covers data with porosities $m = 1$, 1.27 and 1.68, where $m = \rho_{01}/\rho_0$, and ρ_0 is average density of the porous HE. As seen in Figure 4.12, threshold pressure p_I for the composite Hugoniots is approximately $p_I \approx 2$ GPa, corresponding to intersection of Hugoniots $\upsilon^{PTE}(p)$ with Hugoniots $\upsilon^{PE}(p)$ at $m = 1.27$ and 1.68. Thus, composite Hugoniots $\upsilon^{C}(p)$, shown by solid curves in Figure 4.12, are combined from the branches of contributing Hugoniots with specific volume taking minimal value, according to (4.177). Evidence of abnormality can be noticed at $m = 1.68$ for both the calculations and experiments.

Some explosives, like porous and bubbly mixtures, can be analyzed under single-velocity approximation. However, for the case of solid or liquid particles within a gaseous carrier, an adequate model requires knowledge of velocities for both, gaseous and condensed, components. General approach extends the Euler or Navier-Stokes governing systems of equations to the two-velocity case. This is achieved by combination of the corresponding systems of balance governing equations for each of the components, coupled by mass, moment and energy exchange via right-hand terms,

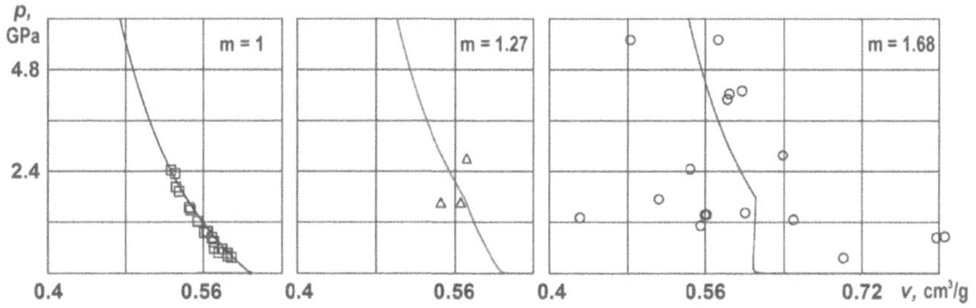

FIGURE 4.12 Composite Hugoniots of TNT inert explosive (solid curves) at different porosities compared with the corresponding experimental data (points).

which results in an extended system. An example of such a combination for two components is a modified system of eight equations (4.156), where four equations for first component are coupled with the corresponding equations for second component by right-hand exchange terms. In contrast to (4.156), the extended system requires velocities in the component's systems to be related to the corresponding component, relaxing the assumption of inter-component velocity equilibrium. For examples of fuel particles in air, the major contributing factors to substance delimitation in the corresponding models are EOS, exchange rates and energy release kinetics for a condensed component, such as liquid droplets or solid particles. A large number of studies employ the extended systems, e.g., models by Baer and Nunziato (1986) and Romenski et al. (2007).

One of the challenges for extended systems is the role of independent velocities of components in an analysis. To note, the component velocities may equilibrate due to momentum exchange, however, corresponding relaxation times can be larger than a characteristic time of the flow. In one-dimensional case, the extended system can be reduced to the 'unified' formulation (Resnyansky, 2006), consisting of a single system of conservation laws, complemented with material derivative equations consisting of chemical kinetics and constitutive equations for internal parameters and parameters-specifiers, including ones for the velocity disequilibrium, as discussed in the previous section. Unfortunately, examples of system unification are unknown in two- and three-dimensional cases. Other approaches, such as velocity splitting to barycentric and diffusive parts (Chinnayya et al., 2004; Nigmatulin, 1991) are worthwhile to explore, which, however, may require some limitations on rates for the split velocity components. Similar to the use of the additivity rule (e.g., see Trunin (1998)) for condensed substances, assumptions of inter-component equilibrating for pressure and temperature are common in analysis of heterogeneous explosives (e.g., see Nigmatulin (1991)). A relevant extended system (Nigmatulin, 1991) was used in numerical simulations of cellular detonation of sprays (aerosols) by Zhdan and Prokhorov (2000) and dusty gaseous mixtures by Briand et al. (2010). The same approach analyzed the burning of Al particles in air (Tropin and Lavruk, 2022), whereas a more sophisticated model by Ruggirello et al. (2012), which exploits the extended system by Nigmatulin (1991) and is complemented with specialized kinetics, was used for layer-by-layer burning of Al particles. Specifying a particle-gas mixture with an Arrhenius kinetics as an example, the corresponding extended system included two mass, momentum and energy balance systems for the components, complemented with inter-component mass, momentum and energy exchange terms (Fedorov et al., 2008):

$$
\begin{aligned}
&\frac{\partial \rho_k}{\partial t} + \frac{\partial \rho_k u_s^{(k)}}{\partial x_s} = \pm J, \quad k = 1,2, \\
&\frac{\partial \rho_k u_i^{(k)}}{\partial t} + \frac{\partial\left(\rho_k u_i^{(k)} u_s^{(k)} + p_k \delta_i^k\right)}{\partial x_s} = \pm\left(-f_i + J u_i^{(2)}\right), \\
&\frac{\partial \rho_k E_k}{\partial t} + \frac{\partial\left[\rho E_k u_s^{(k)} + p_k u_s^{(k)}\right]}{\partial x_s} = \pm\left(q - f_s u_s^{(2)} + J E_2\right),
\end{aligned}
\tag{4.178}
$$

where q describes inter-component heat exchange similar to that chosen in (4.162) and (4.166), f_i is responsible for momentum exchange characterized by a drag force and J – mass exchange that is managed by an Arrhenius kinetics. Shape of the wave front for cellular detonation was analyzed by Fedorov and Khmel (2005) with this approach. Extended systems can be complemented with constitutive equations responsible for internal processes in mixture, such as droplet breakup (Zhdan, 1977).

Systems for nearly condensed substances with a possibility of compaction (e.g., granular beds) usually employ extended systems similar to the BN model (Baer and Nunziato, 1986; Romenski et al., 2007). As a rule, a set of independent thermodynamic parameters in these systems includes volumetric concentration for characterizing momentum exchange at compaction (Saurel et al., 2014). For example, the deflagration-to-detonation two-phase model by Baer and Nunziato (1986) for a granular explosive has been used by Baer et al. (1996). Similarly, internal processes such as burning of bed particles and change in the grain surface area during reaction should be managed by additional constitutive equations (Ruggirello et al., 2012). Models alike may take into account both multi-component and multi-phase nature of mixture (Chinnayya et al., 2002). Within the single-velocity approximation of the work by Chinnayya et al. (2002), a mixture, containing N phases, has volume concentration α_n for nth phase ($n = 1, \ldots, N$) that contains N_n chemical species with mass fraction $Y_{m,n}$ of mth specie ($m = 1, \ldots, N_n$). Equations for $\rho_n\alpha_n$ are equivalent to conventional equations for partial densities due to definition (4.144). However, constitutive equations for α_n require specification of compaction rates, representing momentum exchange via pre-specified drag forces. Similarly, addressing heat exchange, the corresponding rates are required for managing inter-component balance of thermal energy. Finally, kinetic equations for mass concentrations $Y_{m,n}$ require reaction rates (e.g., Vielle's law for solid propellants). On the preliminary stage of the study by Chinnayya et al. (2002), phase transitions were not considered.

Non-ideal heterogeneous explosives that are characterized by dramatically different reaction rates may require introduction of an additional reaction stage dealing with intermediate products. For example, Miller (1995) considered a two-stage global reaction for an explosive (HE) with Aluminium additives (AA) schematized with the following chemical formula

$$R \to DP \to CP.$$

The formula splits the global detonation reaction into two stages: (i) first rapid reaction transforming reactants R to a mixture of detonation products DP, representing fully reacted HE component, with aluminium particles AA subject to subsequent afterburning; and (ii) a relatively slow reaction, transforming the particle-products mixture DP to gaseous combustion products CP, by afterburning of the AA component. In this study, conventional Euler model system under the assumptions of inter-component pressure and temperature equilibrium is complemented with kinetic equations for the reactions governed by two reaction progress variables: (i) fraction λ of the detonated HE component managed by a three-term I&G reaction kinetics (Ignition, Growth of hot spots, and rapid energy-release reaction); and (ii) fraction α of the slowly burning AA component that has not been transformed fully at the first detonation stage, managed by a pressure sensitive burning kinetics with a rate proportional to $p^{1/6}(1 - \alpha)^{1/2}$. As a result, total volume of the substance is presented as $V = (1 - \lambda)V_{HE} + (1 - \alpha)\lambda V_{DP} + \alpha V_{CP}$, following the additivity rule with two mass concentrations representing the corresponding reaction extent parameters.

Summarizing, the present section briefly outlined modelling approaches to a Hugoniot analysis and reactive flow simulation, related to heterogeneous explosive mixtures. For these explosives, relevant hydrocodes utilize models formulated with the extended systems, calculating particle velocities for each of the components. These models require more significant computational resources compared to models for homogeneous explosives.

4.10 CONCLUSION

The chapter has provided an indepth review of major areas relevant to the numerical simulation of the detonation processes and reactive flows. Background information, required for (i) development of models and codes, (ii) informed usage of the codes, and (iii) handling numerical results produced by the detonation modelling software, was outlined, including basics of non-equilibrium thermodynamics, thermochemistry, chemical kinetics and shock physics. Classical detonation theories founded on this basic information, specifically the CJ and ZND detonation models, were summarized and illustrated with examples. Solutions of these classical models can serve as a benchmark for development of modern detonation and reactive flow theories and for analysis of robustness of relevant software. A flowchart of a step-by-step procedure for development of relevant thermodynamically consistent models was provided. The chapter information, while advancing along the flowchart steps, is comprehensively employed for construction of a model. This procedure represents an example of a model formulation suitable for a unified hydrocode implementation process. The model formulation is based on principles of non-equilibrium thermodynamics, while employing a homogenization technique for description of mixture. Derivation of constitutive and kinetic equations specifying mixture components is supported by a meso-mechanical analysis. A model implementation in the CTH hydrocode was used in order to obtain detonation modelling solutions for a generic homogeneous explosive and to compare the results with solutions of the benchmark theories. Possible techniques of phenomenological analysis for describing heterogeneous explosives as substances with a reduced number of components are outlined. Use of the extended systems of conservation laws for simulation of multi-component explosive substances is discussed at the end of the chapter.

The present-day gaps, concerning numerical modelling aspects for detonation analysis, are nested in several disciplines and specific areas of applied science dealing with this topic. In particular, these areas include physics-related subjects, addressing the areas of chemical reactions and advanced numerical methods specific to detonation modelling. Special subjects in question are associated with inter-disciplinary topics of physics and chemistry dealing with numerical implementations and with topics sharing physics and chemical kinetics.

In regard to the physical aspects of available models, the phenomenological analysis of detonation lacks detailed consideration of material structure in condensed substances. Specifically, a meso-mechanical analysis of inter-component mass, momentum and energy exchange may be required for simulation of heterogeneous and non-ideal explosives. Information about inter-component interaction is critical for specification of momentum and heat exchange rates, which is required for possible application to multi-scale model design and use of these kinetic equations in the extended systems of balance laws. Incorporation of thermodynamics, rheology and kinetics for condensed components of explosive mixtures into phenomenological consideration requires a thorough examination of internal physical and mechanical processes for individual components. Analysis of these processes is important for derivation of constitutive and kinetic equations for condensed components that describe chemical reactions, irreversible rate-sensitive deformations, compaction and heat release. Conversion of structurally heterogeneous components to a homogenized substance for an analysis on the phenomenological level is impossible without closing these gaps. Mathematical description of the shock loading and detonation processes presumes a use of wide-range equations of state and phase transitions kinetics. Research in this area of extreme loads requires more attention to the issues associated with thermodynamic equilibrium in the local sense, which is frequently limited to quasi-static approximations in thermodynamic space. These issues are of particular importance, when interpreting data for fast processes within the zone of shock wave transition, in particular, analyzing dynamic wave structure through effective realization of advanced kinetics in the detonation modelling.

Development of chemical kinetics for exothermic reactions at detonation eventuates in specification of the mass exchange rates for kinetics associated with chemical reactions. Additional studies are required for elaboration of the mechanisms of reduction of elementary reactions to a compact set of global reactions. In doing so, highly relevant features in non-ideal explosives that are essential for a phenomenological description of the detonation effects require an analysis of influence of the kinetic effects on phenomenological behaviour. Chemical transformation of a solid explosive to gaseous products raises the issue of multi-scale reactions, which includes (i) spatial splitting requiring micro- and meso-scale analyses for homogenization of heterogeneous mixtures with co-existing solid and gaseous components, and (ii) temporal splitting of physical and chemical processes in reactive substances, specifically in heterogeneous and non-ideal explosives, subjected to a range of chemical reactions, transforming the components to intermediate products with a variety of reaction rates. The issue of equilibrium in the shock-induced reactions, where reaction times are on the order of the duration of shock wave propagation through the shock transition zone, also requires more attention, specifically in condensed and non-ideal explosives with extended reaction zones. A coupling of physical and chemical processes, through analysis of the effects of physical stimuli on mechanisms of chemical reactions and sensitivity of physical behaviour to chemical processes, also requires more attention of researchers in the future. In particular, determination of parameters of chemical kinetics for detonation reactions using tailored tests, followed by validation of full-scale modelling results against independent experimental data, is of high importance.

Numerical aspects of the detonation modelling traditionally focus on spatial resolution of the chemical peak zone by tracking algorithms and enhanced temporal resolution of the chemical reaction zone. The chemical reaction solutions are obtained by calculation of stiff ordinary differential equations, which consumes vast computer resources. The issue, concerning utilization of available and emerging reactive flow models in research and commercial codes, is associated with the model suitability for universal implementation process. This can be addressed with advanced model formulations and representation of governing, constitutive and kinetic equations in a unified form agreeable with thermodynamics, physics and chemistry of detonation processes and reactive flows.

Problems, raising awareness of the detonation modelling community, include an analysis of detonation initiation due to high energy stimuli with special attention to sympathetic detonation. Specifically, (i) elaboration of model ingredients, represented by parameters of constitutive and reaction equations, that are subject to mechanical characterization and explosive sensitivity tests; and (ii) design of independent sensitivity experiments for subsequent validation of numerical results, are still urgent tasks. Other problems of high awareness include studies of structure of the detonation waves in regimes deviating from self-sustained one, an analysis of the chemical peak structure, and description of turbulence effects in non-ideal gas detonation. Several other speciality tasks attracting attention of detonation modellers include investigation of shape of the detonation front, three-dimensional structure of detonation wave, efficiency of the detonation process with controlled release of energy such as rotating detonation engine, etc. In summary, with further progress of computational resources, more detailed and tailored processes related to detonation are to be the focus of future modelling efforts.

REFERENCES

Akhavan, J. 2011. *The Chemistry of Explosives*. 3rd ed. Cambridge, UK: Royal Society of Chemistry.

Aslam, T. D. and Bdzil, J. B. 2002. Numerical and Theoretical Investigations on Detonation-Inert Confinement Interactions. In: *Proceedings Twelfth International Detonation Symposium*. Arlington, VA: Office of Naval Research.

Aslam, T. D., Bdzil, J. B. and Stewart, D. S. 1996. Level set methods applied to modeling detonation shock dynamics. *Journal of Computational Physics*, 126(2), 390–409.

Baer, M. R., Hertel, E. S. and Bell, R. L. 1996. Multidimensional DDT modeling of energetic materials. *AIP Conference Proceedings*, 370, 433–436.

Baer, M. R. and Nunziato, J. W. 1986. A two-phase mixture theory for the deflagration-to-detonation transition (DDT) in reactive granular materials. *International Journal of Multiphase Flow*, 12(6), 861–889.

Bakanova, A. A., Dudoladov, I. P. and Sumulov (Sutulov), Y. N. 1972. Compliance with the additivity rule for a number of alloys under shock compression. *Journal of Applied Mechanics and Technical Physics*, 13(6), 903–907.

Bdzil, J., Aslam, T. and Short, M. 2002. DSD Front Models: Nonideal Explosive Detonation in ANFO. In: *Proceedings Twelfth International Detonation Symposium*. Arlington, VA: Office of Naval Research.

Bdzil, J. B., Menikoff, R., Son, S. F., Kapila, A. K. and Stewart, D. S. 1999. Two-phase modeling of deflagration-to-detonation transition in granular materials: A critical examination of modeling issues. *Physics of Fluids*, 11(2), 378–402.

Becker, R. 1922. Stoßwelle und Detonation. *Zeitschrift für Physik*, 8, 321–362 (in German).

Berthelot, M. and Vieille, P. M. E. 1881. Sur la vitesse de propagation des phénomènes explosifs dans les gaz (On the velocity of propagation of explosive processes in gases). *Comptes rendus hebdomadaires des séances de l'Académie des sciences*, 93, 18–22.

Birch, F. 1947. Finite elastic strain of cubic crystals. *Physical Review*, 71(11), 809–824.

Boris, J. P. and Book, D. L. 1973. Flux-corrected transport I, SHASTA, a fluid transport algorithm that works. *Journal of Computational Physics*, 11(1), 38–69.

Briand, A., Veyssiere, B. and Khasainov, B. A. 2010. Modelling of detonation cellular structure in aluminium suspensions. *Shock Waves*, 20(6), 521–529.

Bull, D. C., Elsworth, J. E. and Shuff, P. J. 1982. Detonation cell structures in fuel/air mixtures. *Combustion and Flame*, 45, 7–22.

Butcher, J. C. 2008. *Numerical Methods for Ordinary Differential Equations.* New York: John Wiley and Sons.

Cao, X., Chen, J., Yu, Y. and Wu, Q. 2018. An equation of state for abnormal expansion of shocked porous materials. *Journal of Applied Physics*, 124(21), 215903.

Carnot, S. 1897. Reflections on the Motive-Power of Heat, and on Machines Fitted to Develop That Power. In: *Reflections on the Motive Power of Heat. From the Original French. Accompanied by An Account of Carnot's Theory by Sir William Thomson Lord Kelvin.* New York: John Willey and Sons (Translated from Carnot, S. 1824. *Réflexions sur la puissance motrice du feu et sur les machines propres à développer cette puissance.* Paris, Bachelier), pp. 37–126.

Chapman, D. L. 1899. On the rate of explosion in gases. *The London, Edinburgh, and Dublin Philosophical Magazine and Journal of Science, Series 5*, 47(284), 90–104.

Chase, M. W., Jr. 1998. *NIST-JANAF Thermochemical Tables 2 Volume-Set.* 4th ed. College Park, MD: American Institute of Physics.

Chinnayya, A., Daniel, E. and Saurel, R. 2002. A New Concept for the Modelling of Detonation Waves in Multiphase Mixtures. In: *Proceedings Twelfth International Detonation Symposium.* Arlington, VA: Office of Naval Research.

Chinnayya, A., Daniel, E. and Saurel, R. 2004. Modelling detonation waves in heterogeneous energetic materials. *Journal of Computational Physics*, 196(2), 490–538.

Clapeyron, E. 1834, Mémoire sur la puissance motrice de la chaleur. *Journal de l'École polytechnique*, 14, 153–190 (in French).

Clausius, R. 1879. *The Mechanical Theory of Heat*, Browne, W. R. (tr.), London: Macmillan and Co.

Colella, P. and Woodward, P. R. 1984. The piecewise parabolic method (PPM) for gas-dynamical simulations. *Journal of Computational Physics*, 54(1), 174–201.

Courant, R. and Friedrichs, K. 1976. *Supersonic Flow and Shock Waves.* New York: Springer-Verlag.

Cowan, R. D. and Fickett, W. 1956. Calculation of the detonation properties of solid explosives with the Kistiakowsky-Wilson equation of state. *Journal of Chemical Physics*, 24(5), 932–939.

Cowperthwaite, M. and Zwisler, W. H. 1973. *TIGER Computer Program Documentation.* Stanford Research Institute Publication Number 2106. Menlo Park, CA: Stanford Research Institute.

Desbiens, N. 2017. Modeling of the jack rabbit series of experiments with a temperature based reactive burn model. *AIP Conference Proceedings*, 1793, 040034.

Dobratz, B. M. and Crawford, P. C. 1985. *LLNL Explosives Handbook: Properties of Chemical Explosives and Explosive Simulants. Change 2.* Lawrence Livermore National Library Technical Report UCRL-52997-Chg. 2. Livermore, CA: Lawrence Livermore National Laboratory, University of California.

de Donder, T. and Van Rysselberghe, P. 1936. *Thermodynamic Theory of Affinity: A Book of Principles.* Oxford, UK: Oxford University Press.

de Groot, S. R. and Mazur, P. 1962. *Nonequilibrium Thermodynamics.* Amsterdam: North Holland.

Donoghue, E. 1974. Interpretation of phenomenological rate constants in the Bodenstein approximation. *Journal of Chemical Physics*, 61, 2180–2188.

Döring, W. 1943. Über Detonationsvorgang in Gasen (On detonation processes in gases). *Annalen der Physik*, 43 (6–7), 421–436 (in German).

Dorovskii, V. N., Iskol'dskii, A. M. and Romenskii, E. I. 1983. Dynamics of impulsive metal heating by a current and electrical explosion of conductors. *Journal of Applied Mechanics and Technical Physics*, 24(4), 454–467.

Dremin, A. N. and Karpukhin, I. A. 1960. Method of determining the shock adiabats of dispersed substances, *Zhurnal Prikladnoi Mekhaniki i Tekhnicheskoi Fiziki*, 1(3), 184–188 (in Russian).

Erkman, J. O. and Edwards, D. J. 1974. *Computed and Experimental Hugoniots for Unreacted Porous High Explosives.* Naval Ordnance Laboratory Technical Report NOLTR 74-213. White Oak, Silver Spring, MD: Naval Surface Weapons Center.

Fedorov, A. V. and Khmel, T. A. 2005. Numerical simulation of formation of cellular heterogeneous detonation of aluminum particles in oxygen. *Combustion, Explosion and Shock Waves*, 41(4), 435–448.

Fedorov, A. V., Khmel, T. A. and Kratova, Y. V. 2008. Shock and detonation wave diffraction at a sudden expansion in gas–particle mixtures. *Shock Waves*, 18(4), 281–290.

Forest, C. A. 1978. *Burning and Detonation*. Los Alamos Scientific Laboratory Report LA-7245. Los Alamos, NM: Los Alamos Scientific Laboratory, University of California.

Forest, C. A. and Mader, C. L. 1979. Numerical modeling of the deflagration-to-detonation transition. *Symposium (International) on Combustion*, 17(1), 35–41.

Fortov, V. E. and Dremin, A. N. 1975. Semiempirical equation of state of trinitrotoluene. *Doklady Physical Chemistry*, 222(1), 463–466 (trans. from Doklady Akademii Nauk SSSR, 1975, 222(1), 162–165).

Fried, L. E. and Souers, P. C. 1994. *CHEETAH: A Next Generation Thermo-Chemical Code*. Lawrence Livermore National Laboratory Report UCRL-ID-117240. Livermore, CA: University of California, Lawrence Livermore National Laboratory.

Gamezo, V. N., Ogawa, T. and Oran, E. S. 2008. Flame acceleration and DDT in channels with obstacles: Effect of obstacle spacing. *Combustion and Flame*, 155(1–2), 302–315.

Gear, C. W. 1971. The automatic integration of ordinary differential equations. *Communications of the ACM*, 14(3), 176–179.

Gibbs, J. W. 1928. Equilibrium of Heterogeneous Substances (1876). In: *Collected Works, V. 1, Thermodynamics*. New York: Longmans.

Godunov, S. K. 1959. Finite difference method for numerical computation of discontinuous solutions of the equations of fluid dynamics. *Matematičeskij sbornik*, 47(89), 271–306 (in Russian) (translation: I. Bohachevsky, hal-01620642).

Godunov, S. K., Kozin, N. S. and Romenskii, E. I. 1974. Equations of state of the elastic energy of metals in the case of a nonspherical strain tensor. *Journal of Applied Mechanics and Technical Physics*, 15(2), 246–250.

Godunov, S. K. and Romenskii, E. I. 2003. *Elements of Continuum Mechanics and Conservation Laws*. New York: Kluwer Academic Publ.

Godunov, S. K., Zabrodine, A., Ivanov, M., Kraiko, A. and Prokopov, G. 1979. *Résolution numérique des problèmes multidimensionnels de la dynamique des gaz. Moscou: Editions Mir* (in French). https://archive.org/details/godounov-zabrodine-ivanov-kraiko-prokopov-resolution-numerique-des-problemes-mul

Gordon, S. and McBride, B. J. 1976. *Computer Program for Calculation of Complex Chemical Equilibrium Compositions, Rocket Performance, Incident and Reflected Shocks, and Chapman-Jouguet Detonations*. Report NASA SP-273. Cleveland, OH: NASA Lewis Research Centre.

Guirguis, R., Oran, E. S. and Kailasanath, K. 1986. Numerical simulations of the cellular structure of detonations in liquid nitromethane—Regularity of the cell structure. *Combustion and Flame*, 65(3), 339–365.

Harten, A. 1983. High resolution schemes for hyperbolic conservation laws. *Journal of Computational Physics, Journal of Computational Physics*, 49(3), 357–393.

Hertel, E. S., Jr., Bell, R. L., Elrick, M. G., Farnsworth, A. V., Kerley, G. I., McGlaun, J. M., Petney, S. V., Silling, S. A., Taylor, P. A. and Yarrington, L. 1995. CTH: A Software Family for Multi-Dimensional Shock Physics Analysis. In: Brun, R. and Dumitrescu, L.Z. (eds.) *Shock Waves @ Marseille I*. Berlin, Heidelberg: Springer.

Hess, H. 1840. Recherches thermochimiques. *Bulletin de l'Académie impériale des sciences de St.-Pétersbourg*, 7(18), 257–272.

Heuze, O., Bauer, P., Presles, H. N. and Brochet, C. 1985. The Equation of State of Detonation Products and Their Incorporation into the QUATUOR Code. In: *Proceedings Eighth Symposium (International) on Detonation*. White Oak, Silver Spring, MD: Naval Surface Weapons Center.

Hishida, M., Fujiwara, T. and Wolanski, P. 2009. Fundamentals of rotating detonations. *Shock Waves*, 19, 1–10.

Jones, H. and Miller, A. R. 1948. The detonation of solid explosives: The equilibrium conditions in the detonation wave-front and the adiabatic expansion of the products of detonation. *Proceedings of the Royal Society of London. Series A, Mathematical and Physical Sciences*, 194(1039), 480–507.

Jouguet, É. 1905. Sur la propagation des réactions chimiques dans les gaz (On the propagation of chemical reactions in gases). *Journal de mathématiques pures et appliquées, Série 6*, 1, 347–425 (in French).

Kamlet, M. J. and Jacobs, S. J. 1968. Chemistry of detonations. I. A simple method for calculating detonation properties of C–H–N–O explosives. *Journal of Chemical Physics*, 48(1), 23–35.

Keshavarz, M. H. and Klapötke, T. M. 2020. *Energetic Compounds. Methods for Prediction of Their Performance*. 2nd ed. Berlin: De Gruyter.

Kinelovskii, S. A. and Maevskii, K. K. 2016. Estimation of the thermodynamic parameters of a shock-wave action on high-porosity heterogeneous materials. *Technical Physics*, 61(8), 1244–1249.

Kistiakowsky, G. B. and Wilson, E. B., Jr. 1941. *Prediction of Detonation Velocities of Solid Explosives*. Office of Scientific Research and Development Report OSRD-69. VA: National Defense Research Committee.
Kittel, C. 2004. *Introduction to Solid State Physics*. 8th ed. New York: John Wiley and Sons, Inc.
Kober, E. and Menikoff, R. 1999. *Compaction Waves in Granular HMX*. LANL Report LA-13546-MS. Los Alamos, NM: Los Alamos National Laboratory.
Kolgan, V. P. 1972. Application of the minimum-derivative principle in the construction of finite-difference schemes for numerical analysis of discontinuous solutions in gas dynamics. *Uchenie Zapiski TsAGI*, 3(6), 68–77 (in Russian) (English translation: 2011. *Journal of Computational Physics*, 230(7), 2384–2390).
Kondepudi, D. and Prigogine, I. 1998. *Modern Thermodynamics: From Heat Engines to Dissipative Structures*. Chichester, UK: John Wiley & Sons.
Konnov, A. A. 2008. Remaining uncertainties in the kinetic mechanism of hydrogen combustion. *Combustion and Flame*, 152(4), 507–528.
Korobeinikov, V. P. and Levin, V. A. 1969. Strong explosion in a combustible gas mixture. *Fluid Dynamics*, 4(6), 30–32.
Krueger, B. R. and Vreeland, T., Jr. 1991. A Hugoniot theory for solid and powder mixtures. *Journal of Applied Physics*, 69(2), 710–716.
Lax, P. D. and Wendroff, B. 1960. Systems of conservation laws. *Communications on Pure and Applied Mathematics*, 13, 217–237.
Lax, P. D. and Wendroff, B. 1964. Difference schemes for hyperbolic equations with high order of accuracy. *Communications on Pure and Applied Mathematics*, 17, 381–398.
Lee, E. L., Horning, H. C. and Kury, J. W. 1968. *Adiabatic Expansion of High Explosives Detonation Products*. The University of California Radiation Laboratory Report UCRL-50422, Livermore, CA: Lawrence Radiation Laboratory.
Lee, E. L. and Tarver, C. M. 1980. Phenomenological model of shock initiation in heterogeneous explosives. *Physics of Fluids*, 23(12), 2362–2372.
Levin, V. A., Manuylovich, I. S. and Markov, V. V. 2016. Numerical simulation of spinning detonation in circular section channels. *Computational Mathematics and Mathematical Physics*, 56(6), 1102–1117.
Levin, V. A., Manuylovich, I. S. and Markov, V. V. 2021. Rotating detonation waves in annular gap with variable stagnation pressure. *Shock Waves*, 31, 651–663.
Levin, V. A., Markov, V. V. and Osinkin, S. F. 1995. Initiation of detonation in hydrogen – Air mixture by explosion of a spherical TNT charge. *Combustion, Explosion and Shock Waves*, 31(2), 207–210.
Lobanov, V. F., Karakhanov, S. M. and Bordzilovskii, S. A. 1982. Transient processes in the impact initiation of TNT. *Combustion, Explosion and Shock Waves*, 18(3), 342–349.
Lozano, E., Cawkwell, M. J. and Aslam, T. D. 2023. An analytic and complete equation of state for condensed phase materials. *Journal of Applied Physics*, 134, 125102.
MacCormack, R. W. 1969. *The Effect of Viscosity in Hypervelocity Impact Cratering*. AIAA Paper No. 69-354. Reston, VA: American Institute of Aeronautics and Astronautics.
Mader, C. L. 1967. *Fortran BKW: A Code for Computing the Detonation Properties of Explosives*. Los Alamos Scientific Laboratory Report LA-3704. Los Alamos, NM: Los Alamos Scientific Laboratory, University of California.
Mader, C. L. 1979. *Numerical Modeling of Detonations*. Berkeley, CA: University of California Press.
Mader, C. L. and Forest, C. A. 1976. *Two-Dimensional Homogeneous and Heterogeneous Detonation Wave Propagation*. Los Alamos Scientific Laboratory Report LA-6259. Los Alamos, NM: Los Alamos Scientific Laboratory, University of California.
Mallard, E.-F. and Le Chatelier, H. L. 1881. Sur les vitesses de propagation de l'inflammation dans les mélanges gazeux explosifs (On the propagation velocity of burning in gaseous explosive mixtures). *Comptes rendus hebdomadaires des séances de l'Académie des sciences*, 93, 145–148.
Marsh, S. P. 1980. *LASL Shock Hugoniot Data*. Berkeley, CA: University of California.
Menikoff, R. 2007. Empirical EOS for Solids. In: Horie, Y. (ed.) *Shock Wave Science and Technology Reference Library, 2, Solids 1*. New York: Springer.
Michelson, V. A. 1893. On the normal ignition velocity of explosive gaseous mixtures. *Scientific Transactions of Imperial Moscow University on Mathematics and Physics*, 10, 1–93.
Miller, P. J. 1995. A reactive flow model with coupled reaction kinetics for detonation and combustion in non-ideal explosives. *MRS Online Proceedings Library*, 418, 413–420.
Mitrofanov, V. V. and Soloukhin, R. I. 1965. The diffraction of multifront detonation waves. *Soviet Physics Doklady*, 9(12), 1055–1058.
Murnaghan, F. D. 1944. The compressibility of media under extreme pressures. *Proceedings of the National Academy of Sciences of the United States of America*, 30(9), 244–247.

Ng, H. D., Radulescu, M. I., Higgins, A. J., Nikiforakis, N. and Lee, J. H. S. 2005. Numerical investigation of the instability for one-dimensional Chapman-Jouguet detonations with chain-branching kinetics. *Combustion Theory and Modelling*, 9(3), 385–401.

Nichols, A. L. and Ree, F. H. 1990. *CHEQ 2.0 User's Manual.* Lawrence Livermore National Laboratory Report UCRL-MA-106754. Livermore, CA: University of California, Lawrence Livermore National Laboratory.

Nigmatulin, R. I. 1991. *Dynamics of Multiphase Media.* New York: Hemisphere.

Nikolaevskii, V. N. 1969. Hydrodynamic analysis of shock adiabats of heterogeneous mixtures of substances. *Journal of Applied Mechanics and Technical Physics*, 10(3), 406–411.

Nunziato, J. W. and Walsh, E. K. 1980. On ideal multiphase mixtures with chemical reactions and diffusion. *Archive for Rational Mechanics and Analysis*, 73, 285–311.

O'Connor, P. D., Long, L. N. and Anderson, J. B. 2006. *The Direct Simulation of Detonations.* AIAA Paper No. 2006-4411. Reston, VA: American Institute of Aeronautics and Astronautics.

Oppenheim, A. K. and Soloukhin, R. I. 1973. Experiments in gasdynamics of explosions. *Annual Review of Fluid Mechanics*, 5(1), 31–58.

Oran, E. S. and Boris, J. P. 1981. Detailed modelling of combustion systems. *Progress in Energy and Combustion Science*, 7(1), 1–72.

Oran, E. S. and Boris, J. P. 2001. *Numerical Simulation of Reactive Flow.* 2nd ed. Cambridge, UK: Cambridge University Press.

Prigogine, I. and Defay, R. 1954. *Chemical Thermodynamics.* London: Longmans.

Resnyansky, A. D. 2006. *A Thermodynamically Complete Model for One-Dimensional Two-Phase Flows with Heat Exchange.* DSTO Technical Report DSTO-TR-1862. Edinburgh, Australia: Defence Science and Technology Organisation.

Resnyansky, A. D. 2008. Constitutive modeling of Hugoniot for a highly porous material. *Journal of Applied Physics*, 104(9), 093511.

Resnyansky, A. D. 2010. Constitutive modeling of shock response of phase-transforming and porous materials with strength. *Journal of Applied Physics*, 108(08), 083534.

Resnyansky, A. D. 2012. *CTH Implementation of a Two-Phase Material Model With Strength: Application to Porous Materials.* DSTO Technical Report DSTO-TR-2728. Edinburgh, Australia: Defence Science and Technology Organisation.

Resnyansky, A. D. 2016. Two-zone Hugoniot for porous materials. *Physical Review B*, 93, 054103.

Resnyansky, A. D. 2017. An equation of state consistent with the Hugoniot abnormality. *AIP Conference Proceedings*, 1793, 050003.

Resnyansky, A. D. 2023a. *Multi-Phase Prototype Models in Standard Hydrocode Framework for Reactive Materials.* DSTG Technical Report DSTG-TR-4039. Edinburgh, Australia: Defence Science and Technology Group.

Resnyansky, A. D. 2023b. Numerical Modelling of a Material Reactivity at High-Velocity Impact. In: Hao H. and Wu C. (eds.) *Proceedings of the 11th International Symposium on Impact Engineering (ISIE2023).* Perth, Australia: Curtin University.

Resnyansky, A. D. and Bourne, N. K. 2004. Shock-wave compression of a porous material. *Journal of Applied Physics*, 95(4), 1760–1769.

Resnyansky, A. D., Bourne, N. K. and Brown, E. N. 2023. Fracture of PC and PMMA cylinders in the Taylor tests. *AIP Conference Proceedings*, 2844, 400002.

Resnyansky, A. D., Bourne, N. K., Millett, J. C. F. and Brown, E. N. 2011. Constitutive modeling of shock response of polytetrafluoroethylene. *Journal of Applied Physics*, 110(3), 033530.

Richtmyer, R. D. and Morton, K. W. 1967. *Difference Methods for Initial-Value Problems.* New York: Interscience.

Romenski, E., Resnyansky, A. D. and Toro, E. F. 2007. Conservative hyperbolic formulation for compressible two-phase flow with different phase pressures and temperatures. *Quarterly of Applied Mathematics*, 65(2), 259–279.

Rosenbrock, H. H. 1963. Some general implicit processes for the numerical solution of differential equations. *The Computer Journal*, 5(4), 329–330.

Rouse, P. E., Jr. 1976. Enthalpies of formation and calculated detonation properties of some thermally stable explosives. *Journal of Chemical Engineering Data*, 21(1), 16–20.

Ruggirello, K. P., DesJardin, P. E., Baer, M. R., Kaneshige, M. J. and Hertel, E. S. 2012. A reaction progress variable modeling approach for non-ideal multiphase explosives. *International Journal of Multiphase Flow*, 42, 128–151.

Saurel, R., Le Martelot, S., Tosello, R. and Lapébie, E. 2014. Symmetric model of compressible granular mixtures with permeable interfaces. *Physics of Fluids*, 26(12), 123304.

Saytzev, S. and Soloukhin, R. 1961. Study of combustion of an adiabatically-heated gas mixture. *Symposium (International) on Combustion*, 8(1), 344–347.

Schmitt, R. G., Crawford, D. A., Harstad, E. N., Hensinger, D. M. and Ruggirello, K. P. 2017. *CTH User's Manual and Input Instructions, Version 12.0.* Albuquerque, NM: Sandia National Laboratories.

Schmitt, R. G., Taylor, P. A. and Hertel, E. S. 2002. Application of a Multiphase Mixture Theory with Coupled Damage and Reaction to Energetic Material Response, In: *Proceedings Twelfth International Detonation Symposium.* Arlington, VA: Office of Naval Research.

Segletes, S. B. 2018. *An Examination of the JWL Equation of State.* US Army Research Laboratory Technical Report ARL-TR-8403. Aberdeen Proving Ground, Maryland: Weapons and Materials Research Directorate, US Army Research Laboratory.

Semenoff, N. 1939. On the kinetics of complex reactions. *Journal of Chemical Physics*, 7(8), 683–699.

Sethian, J. A. 1999. *Level Set Methods and Fast Marching Methods, Evolving Interfaces in Computational Geometry, Fluid Mechanics, Computer Vision, and Materials Science.* 2nd ed. Cambridge, UK: Cambridge University Press.

Shchelkin, K. I. 1959. Two cases of unstable combustion. *Soviet Physics JETP*, 9(2), 416–420.

Shchelkin, K. I. 1966. Instability of combustion and detonation of gases. *Soviet Physics Uspekhi*, 8(5), 780–797.

Sheffield, S. A., Gustavsen, R. L., Hill, L. G. and Alcon, R. R. 1998. Electromagnetic Gauge Measurements of Shock Initiating PBX9501 and PBX9502 Explosives. In: *Proceedings Eleventh International Detonation Symposium.* Arlington, VA: Office of Naval Research.

Sod, G. A. 1978. A survey of several finite difference methods for systems of nonlinear hyperbolic conservation laws. *Journal of Computational Physics*, 27(1), 1–31.

Taki, S. and Fujiwara, T. 1978. Numerical analysis of two-dimensional nonsteady detonations. *AIAA Journal*, 16(1), 73–77.

Tarver, C. M. 2021. Ignition and growth modeling of shock initiation using embedded particle velocity gauges in the plastic bonded explosive LX-14. *Journal of Energetic Materials*, 39(4), 494–505.

Tarver, C. M. and McGuire, E. M. 2002. Reactive Flow Modeling of the Interaction of TATB Detonation Waves with Inert Materials. In: *Proceedings Twelfth International Detonation Symposium.* Arlington, VA: Office of Naval Research.

Taylor, G. I. 1950. The dynamics of the combustion products behind plane and spherical detonation fronts in explosives. *Proceedings of the Royal Society of London. Series A, Mathematical and Physical Sciences*, 200(1061), 235–247.

Titarev, V. and Toro, E. 2002. ADER: Arbitrary high order Godunov approach. *Journal of Scientific Computing*, 17(1–4), 609–618.

Titov, V. M., Lobanov, V. F., Bordzilovsky, S. A. and Karakhanov, S. M. 1981. Investigation of Transient Processes at Initiation of High Explosives. In: *Proceedings Seventh Symposium (International) on Detonation.* White Oak, Silver Spring, MD: Naval Surface Weapons Center.

Toro, E. F. 2009. *Riemann Solvers and Numerical Methods for Fluid Dynamics.* 3rd ed. Berlin, Heidelberg: Springer-Verlag.

Trofimov, V. S. 1967. Thermodynamic investigation of the shock adiabat. *Combustion, Explosion and Shock Waves*, 3(4), 352–358.

Tropin, D. A. and Lavruk, S. A. 2022. Physicomathematical modeling of attenuation of homogeneous and heterogeneous detonation waves by clouds of water droplets. *Combustion, Explosion and Shock Waves*, 58(3), 327–336.

Trunin, R. F. 1998. *Shock Compression of Condensed Materials.* Cambridge, UK: Cambridge University Press.

Tsuboi, N., Asahara, M., Hayashi, A. and Koshi, M. 2009. Effects of Detailed Chemical Reaction Model on Detonation Simulations. In: Hannemann, K. and Seiler, F. (eds,) *Shock Waves.* Berlin, Heidelberg: Springer.

Urtiew, P. A., Kusubov, A. S. and Duff, R. E. 1970. Cellular structure of detonation in nitromethane. *Combustion and Flame*, 14(1), 117–122.

van Leer, B. 1974. Towards the ultimate conservative difference scheme. II. Monotonicity and conservation combined in a second-order scheme. *Journal of Computational Physics*, 14(4), 361–376.

van Leer, B. 1979. Towards the ultimate conservative difference scheme. V. A second-order sequel to Godunov's method. *Journal of Computational Physics*, 32(1), 101–136.

van Thiel, M., Shaner, J. and Salinas, E. 1977. *Compendium of Shock Wave Data.* Lawrence Livermore National Laboratory Report UCRL-50108. Livermore, California: University of California, Lawrence Livermore National Laboratory.

Vandersall, K. S., DeHaven, M. R., Strickland, S. L., Tarver, C. M., Springer, H. K. and Cowan, M. R. 2018. Shock initiation experiments with ignition and growth modeling on the HMX-based explosive LX-14. *AIP Conference Proceedings*, 1979, 100045.

Vasil'ev, A. A. and Nikolaev, Y. A. 1976. Model of the nucleus of a multifront gas detonation. *Combustion, Explosion and Shock Waves*, 12(5), 667–674 (1976).

von Neuman, J. 1942. *Theory of Detonation Waves*. Progress Report to Division B National Defense Research Committee of OSRD-549. Washington, DC: Office of Scientific Research and Development, National Defense Research Committee, Division B.

von Neumann, J. and Richtmyer, R. D. 1950. A method for the numerical calculation of hydrodynamic shocks. *Journal of Applied Physics*, 21(3), 232–237.

Voskoboinikov, I. M., Afanasenkov, A. N. and Bogomolov, V. M. 1967. Generalized shock adiabat for organic liquids. *Combustion, Explosion and Shock Waves*, 3(4), 359–364.

Voytsekhovskiy, B. V. 1959. A steady detonation, *Doklady Akademii Nauk SSSR*, 129(6), 1254–1256 (in Russian).

Voytsekhovskiy, B. V., Mitrofanov, V. V. and Topchiyan, M. E. 1963. *The Structure of a Detonation Front in Gases*. Novosibirsk, Russia: Siberian Branch of the USSR Academy of Sciences (in Russian) (English translation: 1966. Wright–Patterson AFB Report FTD-MT-64-527, Ohio: Foreign Technology Division, Wright–Patterson Air Force Base).

Wackerle, J., Rabie, R. L., Ginsberg, M. J. and Anderson, A. B. 1978. *A Shock Initiation Study of PBX-9404*. LASL Report LA-UR-78-1219. Los Alamos, NM: Los Alamos Scientific Laboratory.

Warnatz, J., Maas, U. and Dibble, R. W. 2006. *Combustion: Physical and Chemical Fundamentals, Modeling and Simulation, Experiments, Pollutant Formation*. 2nd ed. Berlin Heidelberg: Springer-Verlag.

Weckert, S. A. and Resnyansky, A. D. 2022. Experiments and modelling for characterisation and validation of a two-phase constitutive model for describing sands under explosive loading. *International Journal of Impact Engineering*, 166, 104234.

Wilkins, M. L., Squier, B. and Halperin, B. 1964. *The Equation of State of PBX 9404 and LX04-01*. The University of California Radiation Laboratory Report UCRL-7797. Livermore, CA: Lawrence Radiation Laboratory.

Williamson, R. L. 1990. Parametric studies of dynamic powder consolidation using a particle-level numerical model. *Journal of Applied Physics*, 68(3), 1287–1296.

Zeldovich, Y. B. 1940. K Teorii Rasprostranenia Detonatsii v Gasoobraznykh Sistemakh. *Zhurnal Experimentalnoi i Teoreticheskoi Fiziki,* 10, 542–568 (in Russian), (translation: Zeldovich, Y. B. 1950. *On the Theory of the Propagation of Detonation in Gaseous Systems*. NACA Technical Memorandums Report NACA-TM-1261).

Zeldovich, Y. B. and Raiser, Y. P. 1967. *Physics of Shock Waves and High Temperature Hydrodynamic Phenomena*. New York: Academic Press.

Zemansky, M. W. and Dittman, R. H. 1997. *Heat and Thermodynamics*. 7th ed. Boston: McGraw-Hill Inc.

Zharkov, V. N. and Kalinin, V. A. 1971. *Equations of State for Solids at High Pressures and Temperatures*. New York: Consultants Bureau.

Zhdan, S. A. 1977. Calculation of heterogeneous detonation taking into account deformation and breakdown of fuel droplets. *Combustion, Explosion and Shock Waves*, 13(2), 217–221.

Zhdan, S. A. and Prokhorov, E. S. 2000. Calculation of the cellular structure of detonation of sprays in an H2-O2 system. *Combustion, Explosion and Shock Waves*, 36(6), 777–784.

5 Historical Reasons for Determining the High-Rate Mechanical Properties of Solid Explosives and Propellants

Stephen M. Walley

5.1 BACKGROUND

Just like anything else on a battlefield, explosive ordnance can be subjected to impact by bullets or fragments travelling at up to 1.5 km/s (Curtis et al., 2006, 2010; Rosenberg and Dekel, 2020). Other situations where explosives can be subjected to rapid deformation include vibration during transport (Perry and Walley, 2022) and low-speed impact due to being accidentally dropped when being loaded onto or off a vehicle (Xue et al., 2024; Xiao et al., 2025). This is a particular issue in naval docks where upward-pointing spikes (spigots) may be present on the quayside (Hughes et al., 2015).

The deformation caused by terminal ballistic impact is three-dimensional in nature (Figure 5.1) (Hörz et al., 1994; Wang et al., 2022; Guo et al., 2023; Liu et al., 2024). However, most techniques for studying the high-rate deformation of materials generate stress and strain information in only one dimension. Such techniques include dropweights (Figure 5.2) (Lieb and Rocchio, 1983; Gazonas, 1991), Kolsky (split-Hopkinson) pressure bars (Figure 5.3) (Kolsky, 1949; Chen and Song, 2011; Othman, 2018) and shocks produced by the planar impact of flat flyer plates (Figure 5.4) (Braithwaite et al., 2012; Boddy et al., 2014), laser pulses (Gottfried, 2014, 2015; Dlott, 2022) or the wedge test (Figure 5.5) (Murri et al., 1982; Ichiyama et al., 2025).

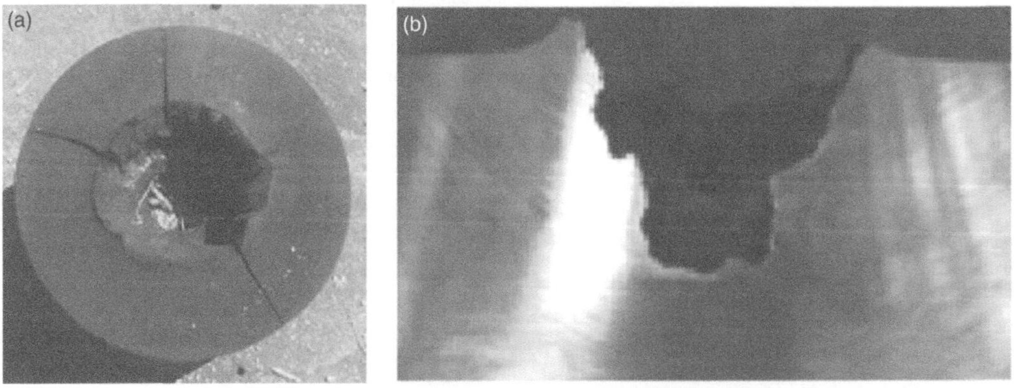

FIGURE 5.1 (a) Top view of a crater produced by the impact on a steel cylinder (120 mm diameter, 100 mm height) of a reactive shaped charge jet generated from a pressed cone of a 73.5 wt.% PTFE 26.5 wt.% aluminium powder mix (Guo et al., 2019). (b) Cross-section of the same crater (Guo et al., 2023).

DOI: 10.1201/9781003529217-5

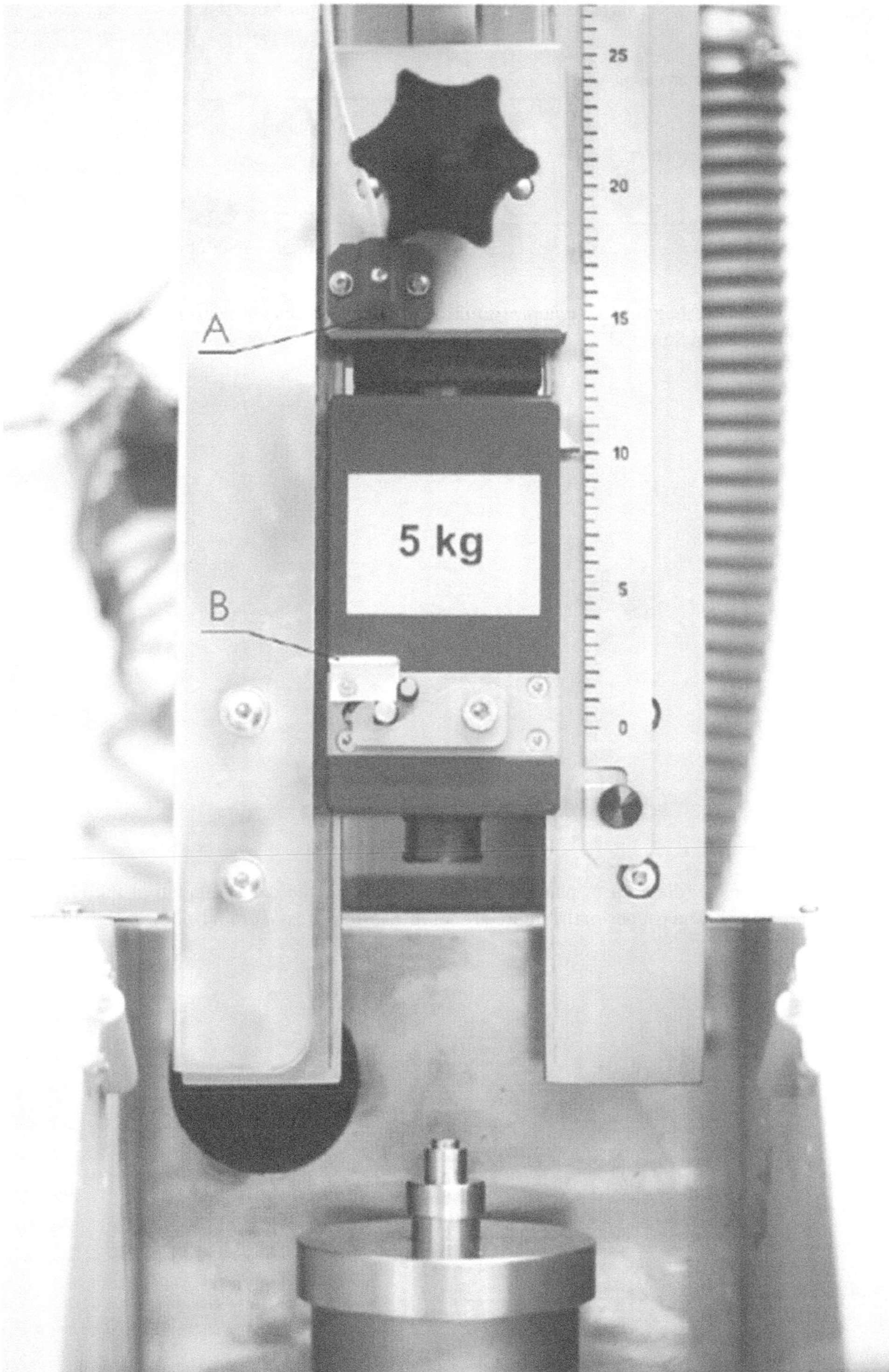

FIGURE 5.2 Photograph of the OZM BAM Fall Hammer which has a PDV probe (A) aimed at a retroreflector (B) attached to the dropweight. The PDV probe allows the velocity of the dropweight to be measured. (From Künzel et al. (2024).)

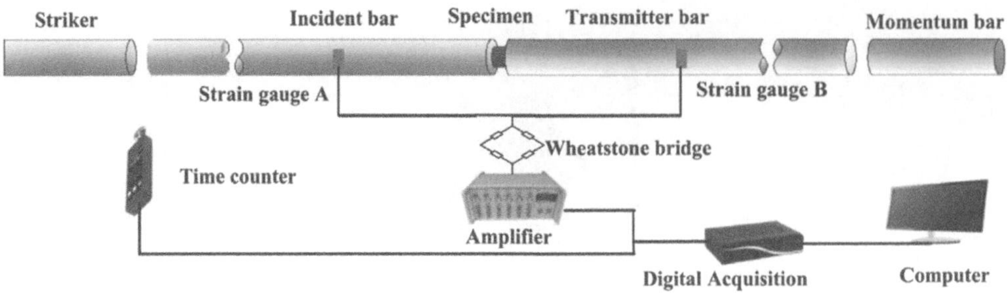

FIGURE 5.3 Schematic diagram of a compression Kolsky bar apparatus for measuring the high-rate properties of a propellant. (From Sun et al. (2015).)

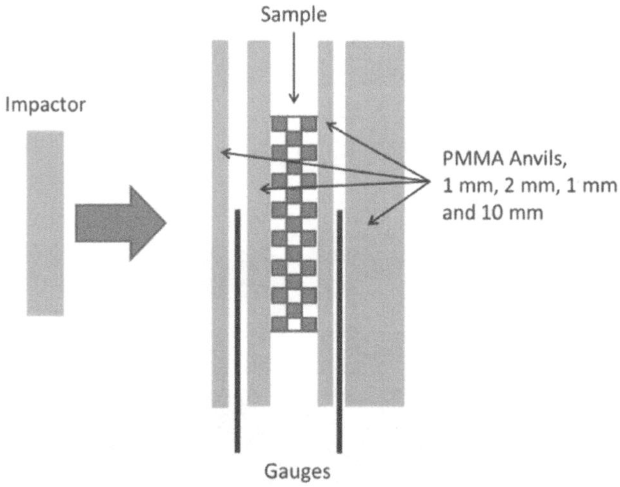

FIGURE 5.4 Schematic diagram of planar flyer plate impact for the determination of the Hugoniot of a double-base propellant. The gauges in the figure are for measuring stress. (From Boddy et al. (2014).)

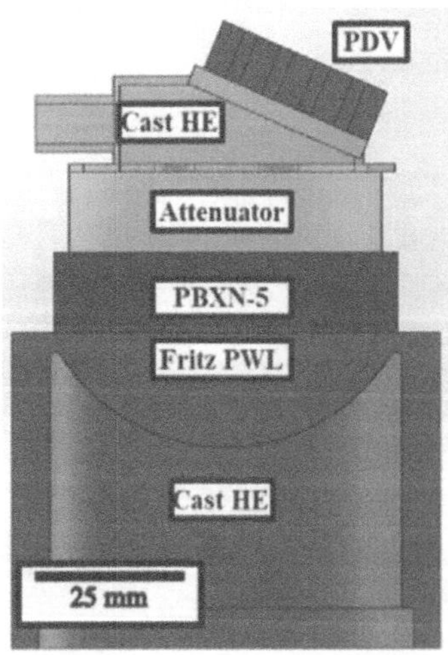

FIGURE 5.5 Schematic diagram of a modern version of the wedge test for a high explosive (HE) specimen. (From Ichiyama et al. (2025).)

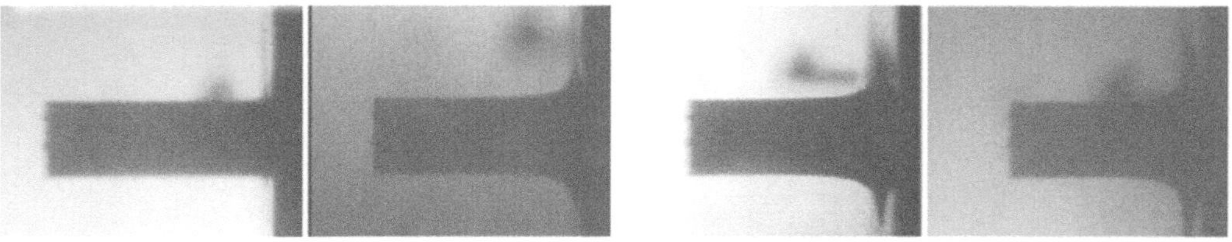

FIGURE 5.6 Consecutive images from a high-speed photographic sequence of a classic Taylor impact test performed on a rod fabricated from a mixture of PTFE and titanium powders. Impact velocity 406 m/s, left to right. Interframe time 2 μs. Original dimensions: 7.5 mm diameter, 19.8 mm long. (From Shen (2012).)

Impact techniques that produce deformation in more than one dimension include the Taylor impact test (Figure 5.6) (Taylor, 1948; Walley and Couque, 2024), the Brazilian indirect tension test (Figure 5.7) (Hondros, 1959; Jun-Ling et al., 2012; Kendall and Siviour, 2015; Zhou et al., 2015; Parab et al., 2016) and the Steven test (Figure 5.8) (Chidester et al., 1995; Gruau et al., 2009; Picart et al., 2023). The Brazilian test was in fact originally developed in Japan during World War 2 for obtaining the tensile strength of concrete (Akazawa, 1943). Although a summary of Akazawa's paper and PhD thesis was made available in 1953 in a more widely understood language (French) (Akazawa, 1953), the summary was published in a later issue of the *RILEM Bulletin* than the one where two Brazilian authors (hence the name of the technique) published two articles on the topic (Carneiro and Barcellos, 1953; Barcellos, 1953). To this day, Akazawa's priority has not been acknowledged in the name of the method.

FIGURE 5.7 Selected frame from a high-speed photographic sequence of a high-rate Brazilian test carried out in a Kolsky bar. Time from start of deformation: 149 μs. A crack can be seen developing in the middle of the specimen parallel to the direction of loading. (From Jun-Ling et al. (2012).)

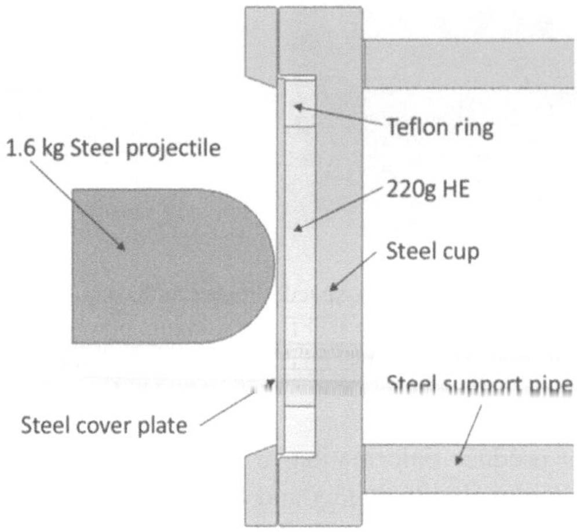

FIGURE 5.8 Schematic diagram of the Steven Test for high explosives (HE). (From Reaugh et al. (2020).)

The Brazilian test started to be used in the 1970s to obtain information on the low-rate tensile strength of polymer-bonded explosives (PBXs) (Johnson, 1979; Palmer et al., 1993). The two main advantages of the indirect tension method over conventional tensile loading for PBXs are that (i) discs are much easier to make than tensile dog-bones and (ii) they are considerably smaller so that if initiation does occur, the consequences are less severe. The Steven test is one of the tests that has been developed to study low-velocity (tens of metres per second) initiation of high explosives (HEs) to impacts that include shear (such as transportation accidents).

Optical methods (Figure 5.9) (Hung and Hovanesian, 1972; Barker and Fourney, 1976; Grantham et al., 2004; Siviour et al., 2009; Chiang and Mao, 2016) and X-ray techniques (Figures 5.10–5.12)

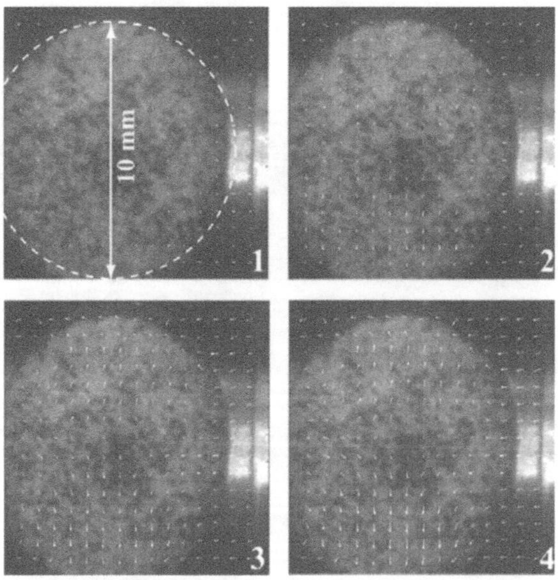

FIGURE 5.9 High-speed photographic sequence (20 μs interframe time) of the deformation of a high-rate Kolsky bar Brazilian test of a polymer-bonded sugar, an inert simulant of a polymer-bonded explosive (PBX). The arrows indicate the magnitude and direction of the movement of speckles on the surface of the disc. The speckles are the sugar crystals at the surface. (From Grantham et al. (2004).)

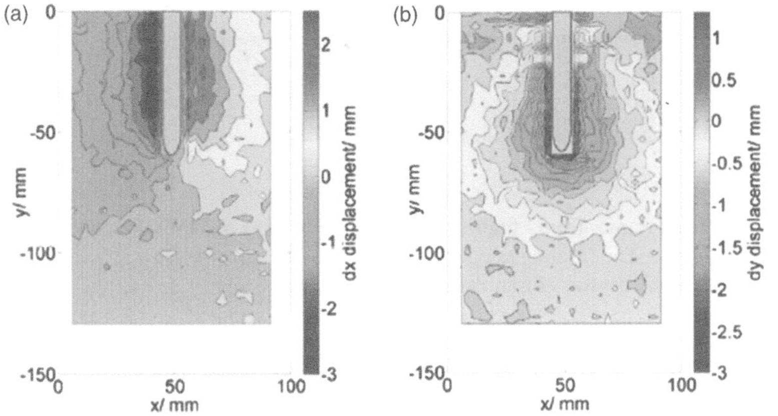

FIGURE 5.10 Colour-coded maps of (a) the lateral and (b) the longitudinal displacements of a thin layer of 1-mm diameter lead shot (less than 100% coverage) 450 μs after impact produced by the normal ballistic impact on a cylinder of dry sand of an ogive-nosed rod (10 mm diameter, 100 mm long) at 200 m/s. The displacements were measured using a flash X-ray. (From Collins et al. (2011).)

FIGURE 5.11 Enhanced X-ray image of a layer of 1-mm diameter lead shot before impact. Estimated coverage: 30–40%. (From Collins et al. (2011).)

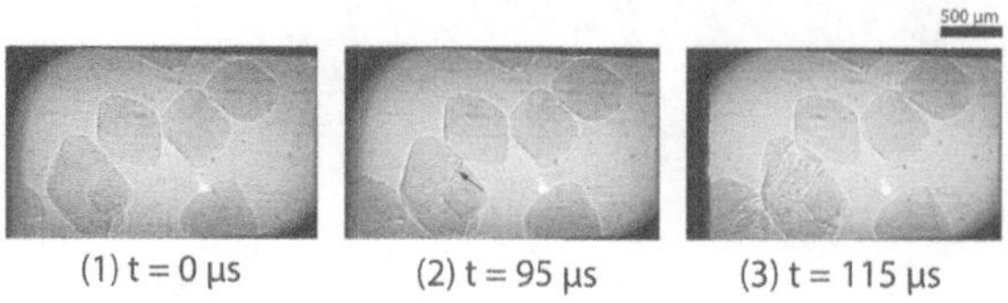

FIGURE 5.12 Selected frames from a high-speed X-ray sequence of an HMX-based PBX being deformed in a compression Kolsky bar. The framing rate was 200,000 fps. The arrow in frame (2) indicates the position of tensile fracture in an HMX crystal. (From Parab et al. (2016).)

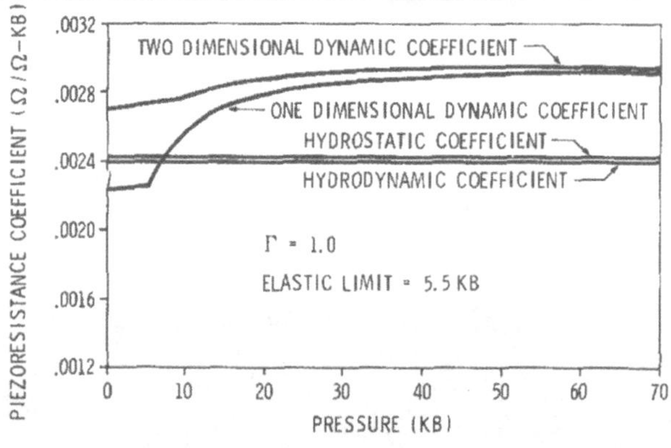

FIGURE 5.13 Piezoresistance coefficient of manganin (a metal alloy often used in gauges to measure stress) as a function of pressure for one- and two-dimensional loading. (From Barsis et al. (1970).)

(Collins et al., 2011; Addiss et al., 2015; Parab et al., 2016; Paulson et al., 2020; Wagner et al., 2020) can give quantitative information about surface and internal deformation in more than one direction, but it remains challenging to measure dynamic multiaxial forces (or stresses) for the reasons shown in Figure 5.13 (Barsis et al., 1970) (see also Cosculluela and Forquin (2022)).

5.2 HISTORICAL DEVELOPMENT OF DETONATING EXPLOSIVES

Only a brief summary will be given here since the history of the development of detonating explosives can be found in two books (Brown, 1998, 2010).

Guncotton was developed as a blasting agent in the 1840s (Schoenbein, 1847; Combes, 1848) followed by nitroglycerine in the 1860s (Mayer, 1868a, 1868b; Eissler, 1884). These chemicals differ from the previous and widely used energetic mixture (gunpowder) in that the fuel (carbon) and oxidiser (nitro groups) are contained in every single molecule (Figure 5.14) allowing the energy to be released on a sufficiently short timescale to sustain a shock (or detonation) wave.

Although gunpowder (alternatively called black powder) can produce impressive explosions (Majendie, 1877; Reitsma, 2001), very few studies (all Russian) have been performed into its shock and detonation properties (Belyaev and Kurbangalina, 1964; Foteenkov et al., 1982; Zanegin and Karachinskii, 2001; Subbotin et al., 2009; Ermolaev et al., 2010). The only authors I have been able to find who claimed to have measured the detonation velocity of black powder are Belyaev and Kurbangalina (1964). Early in their paper they state, "There is no direct experimental proof that black powder can be detonated. However, we may not conclude that this is altogether impossible, although there is no doubt that the capacity of black powder for detonation is low". Later they state, "We attempted to excite the detonation of black powder in strong steel tubes". One result of their experiments was that they found a constant and reproducible velocity of propagation of the

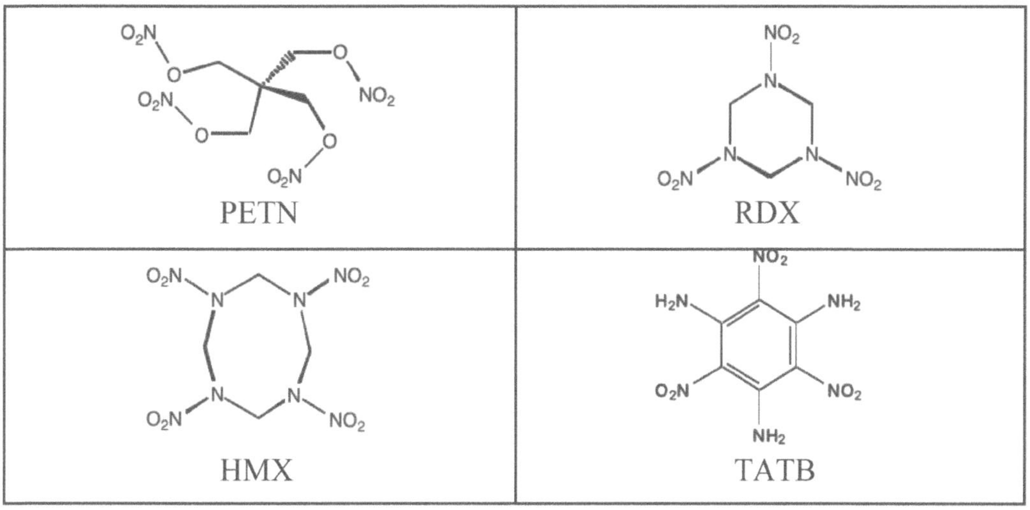

FIGURE 5.14 Chemical drawings of the explosive molecules that are the energetic chemicals contained within most of the PBXs discussed in this chapter.

explosive reaction of 1300 m/s "at a certain distance from the initiating charge" which, they said, "is undoubtedly that of detonation", being a similar value to that of ammonium nitrate (1500 m/s). However, Zanegin and Karachinskii (2001) stated the following: "...because of the original use of gun powders as propellants, their detonation properties have been inadequately studied, and literature data on this issue are controversial and, sometimes, are not confirmed by practice".

A few years later still Ermolaev et al. (2010) commented that "It is extremely difficult to detonate black powder, even using a powerful booster detonator". This is because even though the fuel (charcoal) and oxidiser (potassium nitrate) powders are intimately mixed, powders cannot easily be ground finer in scale than a few microns (Kendall, 1978; Hagan, 1981). And since the burning velocity in gunpowder even at 50 atmospheres pressure is only around 10 m/s (Blackwood and Bowden, 1952), the burning front takes about 0.1 ms to cross a 1 mm-sized grain. There are also additional times required to disrupt the grains, mix them and for them to begin to react. For this reason, gunpowder was and is mostly used as a propellant (as its name suggests).

Ermolaev et al. (2010) went on to summarise the Russian studies thus,

> Being a precursor of modern composite propellants and explosive compositions, black powder, nevertheless, differs from them drastically, demonstrating rather unusual, mostly inexplicable behavior during deflagration and detonation. First, in contrast to explosive compositions based on ammonium nitrate and ammonium perchlorate, black powder exhibits no deflagration-to-detonation transitions. Even in long, strong tubes, the burning velocity does not exceed 400-440 m/s [Foteenkov et al. (1991)]. Upon shock initiation, two different processes with stable characteristics are observed: a combustion wave propagating at a velocity of ~400 m/s (as in the case of thermal initiation!) and a combustion wave propagating at a velocity of 1100-1300 m/s, which was termed detonation (Belyaev and Kurbangalina, 1964). Experiments with charges placed into long, thin-walled, low-strength tubes demonstrated the unique ability of black powder to sustain a fast burning in a quasi-steady pulsating regime with periodic local ruptures of the shell.

Methods for quantifying the sensitiveness of energetic chemicals or explosive formulations to compressive or frictional impact were developed early in the 20th century (Robertson, 1921a; Copp et al., 1948; Napier et al., 1948; Bawn and Rotter, 1956; Koenen et al., 1961; Boyars and Levine, 1968; Trimborn, 1970; Walley et al., 2006, 2015). But with the exception of some studies on cordite (a mixture of nitrocellulose and nitroglycerine) during World War 2 (Eley and Pepper, 1947;

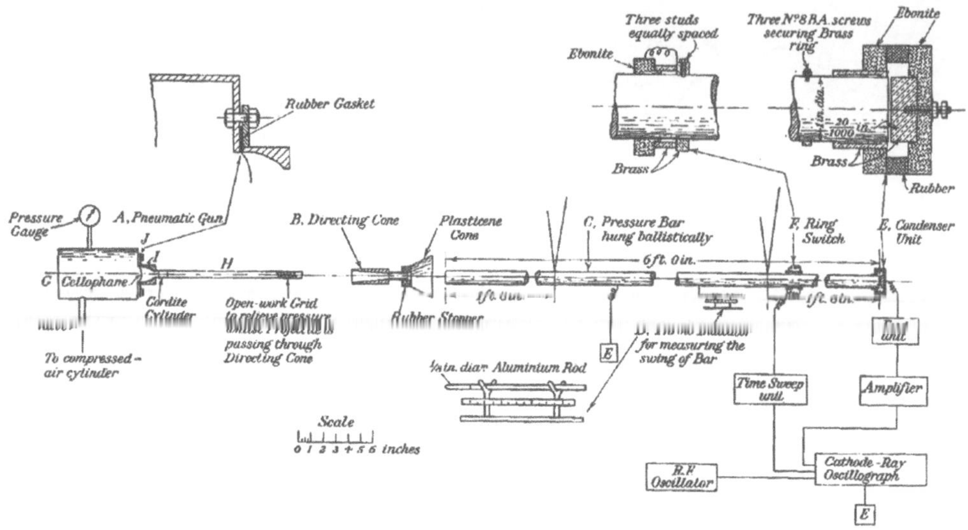

FIGURE 5.15 Schematic diagram of the apparatus used by Taylor and colleagues during World War 2 to measure the high-rate mechanical properties of cordite. (From Taylor and Davies (1958).)

Ward, 1956; Taylor and Davies, 1958), there was little interest in measuring the high-rate mechanical properties of explosives or solid propellants until the mid- to late 1960s (Hoge, 1967, 1970). Ward's (1956) study on cordite also made use of oscillatory loading.

Note that up to the late 1950s, the 'strength' of an explosive did not refer to its mechanical properties. Rather it was a measure of the amount of damage it could do to other materials when initiated (see, for example, Shepherd (1956)).

During World War 2, Taylor and Davies (1958) found using a single Hopkinson pressure bar (Hopkinson, 1914b) (Figures 5.15 and 5.16) that the dynamic mechanical strength of cordite (a

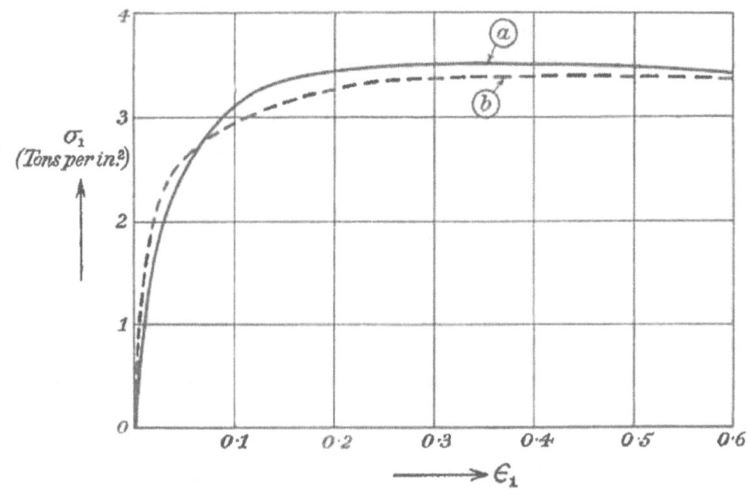

FIGURE 5.16 Graph showing the relation between the true stress and the strain for cordite subjected to impact using the apparatus shown in Figure 5.15. The two curves (a) and (b) are for the same data but calculated using two different formulas for the stress. For more details, see Taylor and Davies (1958).

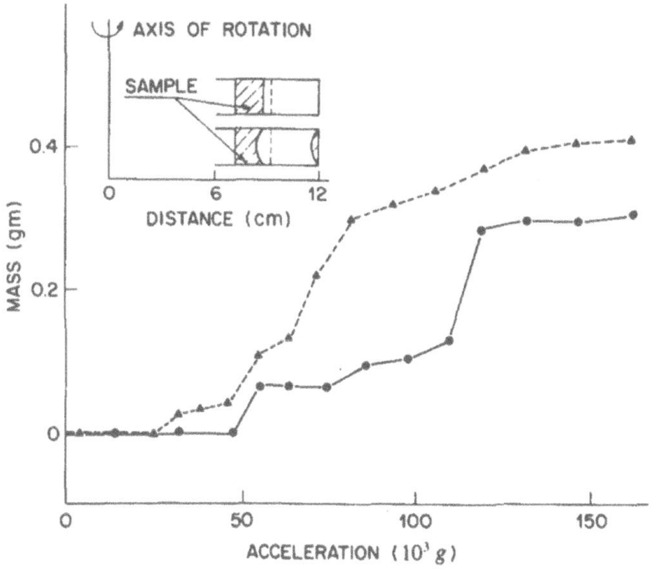

FIGURE 5.17 Plot of fractured mass of melt-cast trinitrotoluene (TNT) (solid triangles) and Composition B (solid circles) as a function of acceleration during gun launch of a shell containing explosive. The inserted diagram schematically shows the effect of acceleration on the explosive in the shell: upper drawing, no acceleration; lower drawing, sample under high acceleration. (From Lanzerotti and Sharma (1981).)

propellant) could be briefly (10–100 μs) up to seven times its static strength. Unlike for World War 1 (Robertson, 1921a, 1921b; Landon and Quinney, 1923), I have not so far found any document which discusses the purpose of carrying out such measurements or the uses to which the results were put (Walley, 2020).

Due to continuing improvement in numerical simulation methods (Church et al., 1999, 2014), there is an increasing desire to apply these techniques to explosives as materials in their own right (Walley et al., 2010, 2018). This chapter will therefore concentrate on the high-rate testing of PBXs which were originally developed at the Los Alamos National Laboratory in the early 1950s (James and Smith, 1952; Land, 1961; Wilkins et al., 1965). The aim is to obtain data to improve the quality of their constitutive relations so that their response to ballistic impact (Wang et al., 2022; Bi et al., 2023; Rollin and Ressler, 2024), rapid acceleration during gun launch (*ca.* 10^6 m/s^2) (Figure 5.17) (Lanzerotti and Sharma, 1981; Moxnes and Ødegårdstuen, 1998; Church et al., 2005; Carlucci et al., 2013; Ma et al., 2018; Walley et al., 2018; Benfante et al., 2025; Sandusky and Felts, 2025) and impact of ordnance on a target (Church et al., 2005; Liu et al., 2015; Xu et al., 2020, 2021; Bi et al., 2023; Wei et al., 2023) can be more realistically modelled.

As can be seen in Figure 5.18, a PBX is a highly filled (*ca.* 95%) particulate composite containing a wide range of explosive crystal sizes and shapes. The crystals are held together using a low-strength polymer binder acting as an adhesive (Yang, 2008; Yeager et al., 2010). As neither the particle shape nor size distributions are usually known, they are often modelled as spheres (Ge et al., 2021) rather than the irregularly shaped particles they actually are.

5.3 DROPWEIGHTS

For explosives and their formulations, dropweights are mostly used to determine their relative impact sensitivity (Walley et al., 2006, 2015). There are, however, problems with the statistical

FIGURE 5.18 Optical micrograph of polished and stained PBX 9501. The crystals are β-HMX; the binder is Estane 5703. (From Rae et al. (2002).)

interpretation of the results (Dixon and Mood, 1948; Neyer, 1991, 1994a, 1994b, 1997; Chao and Fuh, 2003; Alouaamari et al., 2008; Preston et al., 2012, 2016; Nance, 2017; Cummock et al., 2021; Muhr et al., 2022; Christensen et al., 2024; Künzel et al., 2024; Turpeinen et al., 2025). However, dropweights have been used in the past (although not nowadays) to measure the high-rate mechanical properties of solid propellants (Figure 5.19) (Lieb and Rocchio, 1983; Gazonas, 1991).

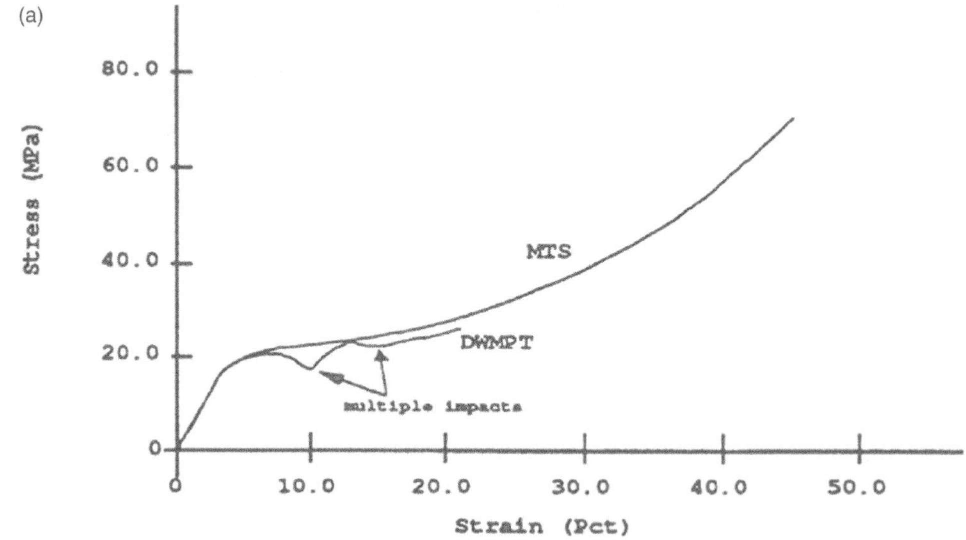

FIGURE 5.19 Comparison between stress–strain curves obtained for (a) JA2 solid propellant and (b) M30 solid propellant at a strain rate of 250 s^{-1} using an MTS servohydraulic mechanical testing machine and a dropweight (DWMPT). (From Gazonas (1991).) *(Continued)*

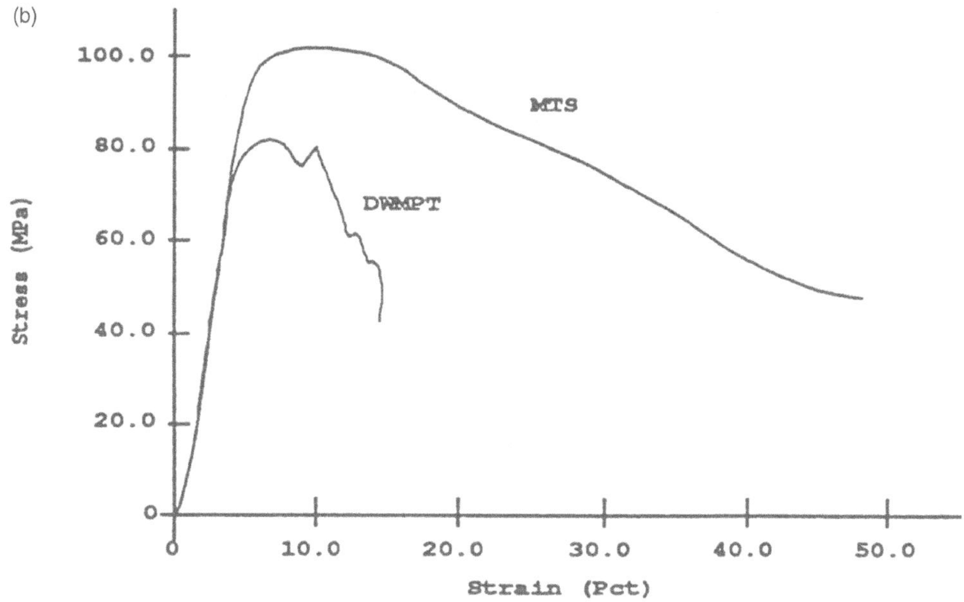

FIGURE 5.19 (*Continued*)

Gazonas used a servohydraulic mechanical testing machine to determine the rate sensitivity of the mechanical properties of three propellants (Figure 5.20). However, he also made a comparison between stress–strain curves obtained at a strain rate of about 250 s^{-1} using both a servohydraulic testing machine and a dropweight (Figure 5.19). The two techniques agreed well for JA2 (Figure 5.19a) but poorly for the other propellants he studied (Figure 5.19b). He attributed the good agreement for JA2 to its high ductility and the poor agreement for the other two to their brittle response.

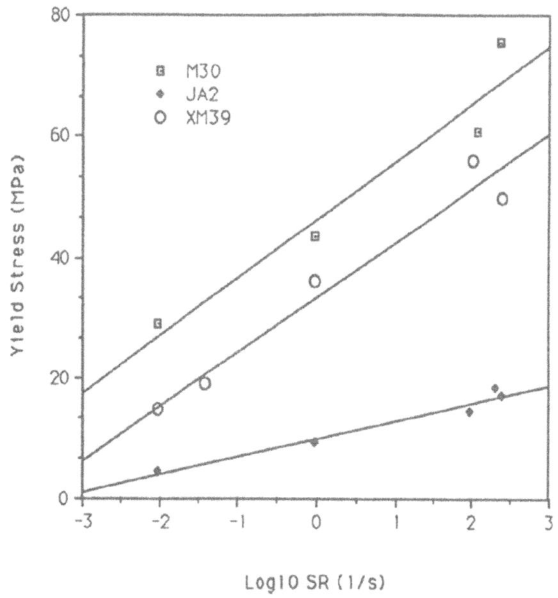

FIGURE 5.20 Plot of the yield stress as a function of $\log_{10}$(strain rate) for three solid propellants obtained using an MTS servohydraulic testing machine. (From Gazonas (1991).)

5.3.1 Multiaxial High-Rate Tests

Most of the techniques discussed in this chapter subject specimens to uniaxial deformation whereas Jiao et al. (1985) stated that in the 'real world' "...detonation and shock wave phenomena are not really one-dimensional, but generally two or three-dimensional. To measure the velocity of such shock waves is difficult". The main reason is that forces and strains have traditionally been measured using stress and strain gauges which can only be calibrated for one-dimensional deformation (Rosenberg et al., 1980, 1983). If, for example, a stress gauge is subjected to multiaxial deformation its change in resistance will be substantially different to the one-dimensional case (Barsis et al., 1970; Rosenberg et al., 1981) (Figure 5.13). Thus, two- or three-dimensional tests such as the Steven Test (Figure 5.8), the Susan Test and spigot intrusion which are all mainly used for accidental ignition studies at relatively low velocities (tens of metres per second) (Chidester et al., 2000; Ma et al., 2013a, 2014; Wen et al., 2014; Liu et al., 2021; Xiao et al., 2025) will not be considered further in this chapter. In order to numerically model 3D tests, one-dimensional high-rate mechanical data obtained using the Kolsky bar has to be used (Xiao et al., 2020; Picart et al., 2023). This only works if the material is isotropic and remains so during the impact event.

5.3.2 The Kolsky Bar

Although as already mentioned in the introduction to this chapter, an energetic material (cordite) was one of the first substances to have its high-rate compressive strength measured using a single Hopkinson pressure bar (Taylor and Davies, 1958), it can be seen from Figure 5.21 that it wasn't until the late 1980s that the Kolsky (or split Hopkinson) pressure bar began to be widely used to obtain the high-rate mechanical properties of energetic materials (a term encompassing PBXs, solid propellants, pyrotechnics, reactive metal-polymer mixes and reactive metal alloys). Up until the time of writing (October 2025), there have been only about 360 publications on this topic compared

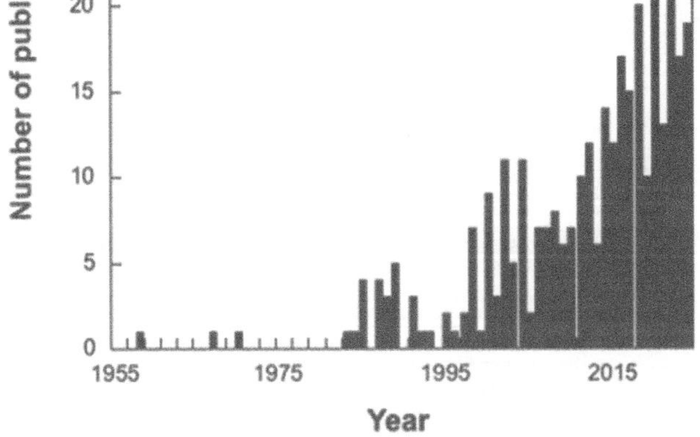

FIGURE 5.21 Bar chart showing the number of papers published each year up to 2024 where a Kolsky bar was used to measure the high-rate mechanical properties of energetic materials (PBXs, solid propellants, pyrotechnics, reactive metal-polymer mixes, reactive metal alloys).

to about 5300 for metals. The Kolsky bar has also been used to study energy release mechanisms in energetic materials (Ho, 1992; Yang et al., 2022; Ravindran et al., 2023; Williamson and Morley, 2023; Wu et al., 2023). This chapter will mostly consider the high-rate properties of PBXs containing the secondary explosive molecules shown in Figure 5.14.

The first high-rate results for a PBX measured using a Kolsky bar were obtained in the late 1960s by Hoge in both compression (Figures 5.22 and 5.23) (Hoge, 1967) and direct tension (Figure 5.24) (Hoge, 1970). Note that Hoge reported that deflagration occurred in some of his compression Kolsky bar experiments "sufficient to scorch the faces of the bars". It can be seen from Figures 5.22–5.24 that strain rate has a strong effect on the mechanical properties of the PBXs that Hoge studied.

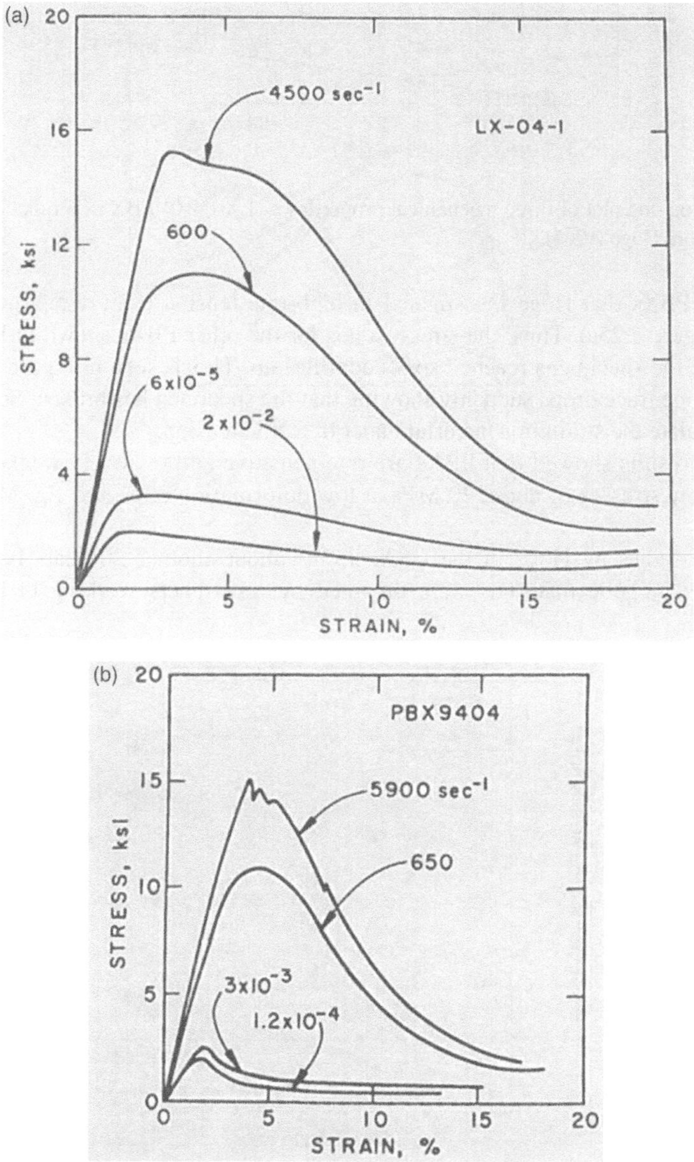

FIGURE 5.22 Compressive stress–strain curves for (a) LX-04-01 and (b) PBX 9404 PBXs at four different strain rates. To two significant figures, 1 ksi is 6.9 MPa. (From Hoge (1967).)

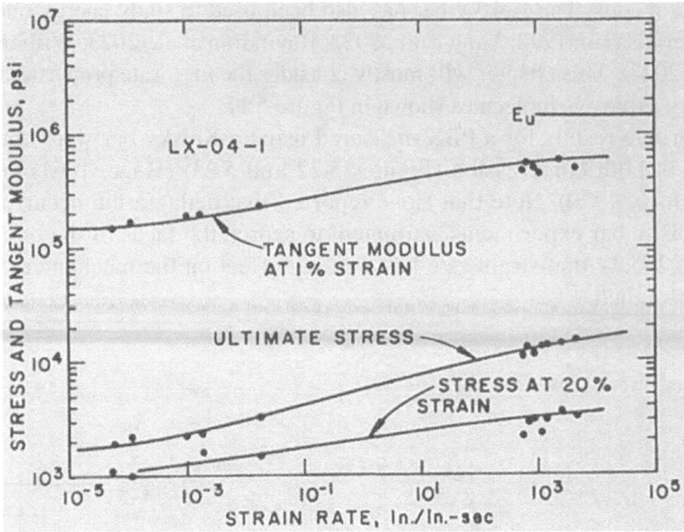

FIGURE 5.23 log–log plot of three mechanical properties of LX-04-01 PBX as a function of strain rate in compression. (From Hoge (1967).)

Of the four PBXs that Hoge investigated in high-rate tension (Figure 5.24), only RX-08-AZ was ductile (Figure 5.25a). Thus, the stress values for the other PBXs shown in Figure 5.24 were obtained before the specimens reached stress equilibrium. This is seen in Figure 5.25b in that the lower oscilloscope trace drops suddenly showing that the specimen has broken. Such traces cannot be used to calculate the strain in a material under dynamic tension.

These early results showed that PBXs are rate-sensitive and relatively weak substances with compressive flow stresses of about 20 MPa at low deformation rates and 100 MPa at high rates of strain.

After these studies by Hoge in the USA, it was about another 20 years before more high-rate results for energetic materials were obtained by researchers working in France on PBXs

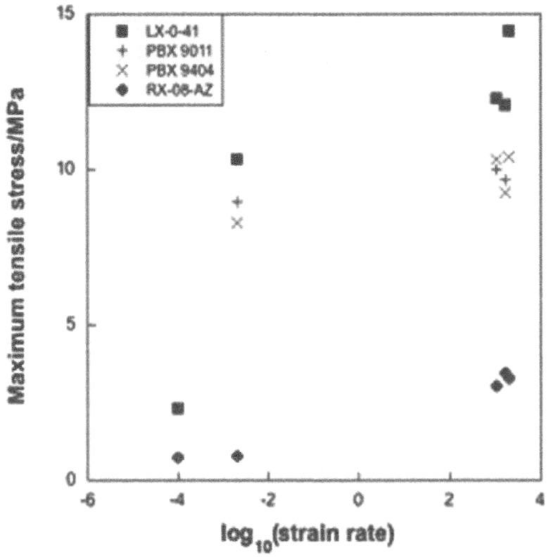

FIGURE 5.24 Semi-log plot of the maximum tensile stress supported by four PBXs as a function of strain rate. (Data tabulated in Hoge (1970).)

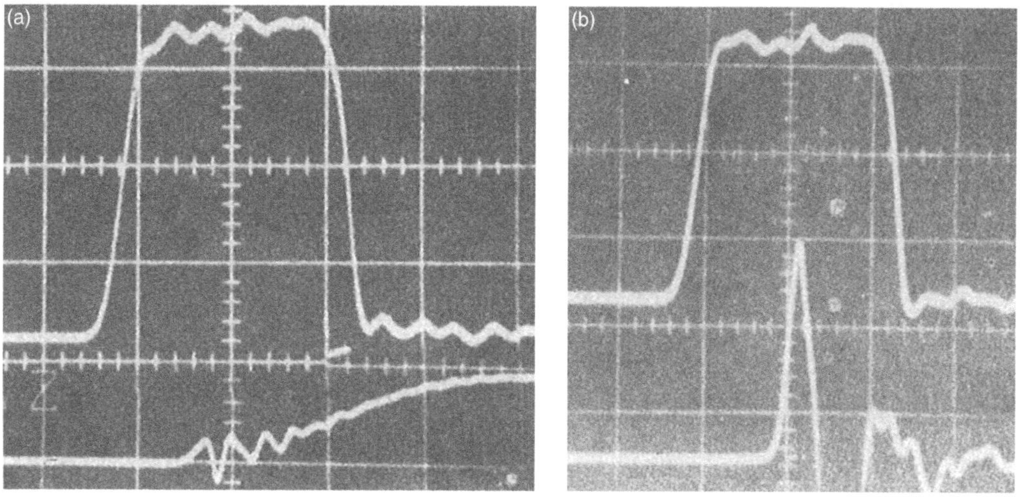

FIGURE 5.25 Photographs of oscilloscope voltage traces obtained from tensile Kolsky bars used to load (a) RX-08-AZ and (b) PBX 9404. In both cases, the upper trace was recorded from the input bar, and the lower trace was recorded from the output bar. The timescale on the oscilloscope was 20 μs/cm (the squares on the oscilloscope screen were one centimetre in size). (From Hoge (1970).)

(Figure 5.26) (Bonthoux and Girodon, 1985; Quidot et al., 1985; Quidot, 1988) and in Australia on solid propellants (Figure 5.27) (Fong, 1985; Ho and Fong, 1987a, 1987b). Figure 5.27 shows that there is both a strong temperature and strain rate dependence of the yield (or flow) stress for the solid propellant that Ho and Fong investigated. Quidot (1988) did not investigate the temperature dependence of the high-rate mechanical properties for a PBX but showed that both the modulus and maximum stress strongly depended on strain rate (Figure 5.26).

Ho and Fong (1987b) found that their stress–strain rate data could be fitted by equation (5.1):

$$\sigma_y = K_1 + K_2 \log\left(\dot{\varepsilon} a_T\right) \tag{5.1}$$

where σ_y is the stress, $\dot{\varepsilon}$ is the strain rate, T is the temperature, K_1 and K_2 are constants (K_2 characterises the temperature and strain rate sensitivity of the propellant) and a_T is a shift factor. As far as

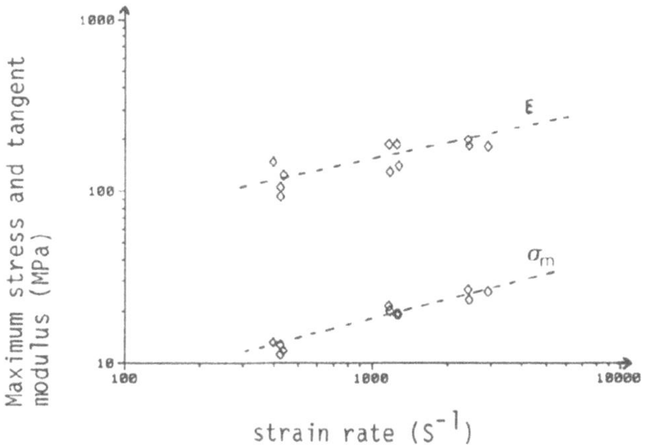

FIGURE 5.26 log–log plot of tangent modulus (E) and maximum stress (σ_m) of a PBX as a function of strain rate. Data obtained using a Kolsky bar. (From Quidot (1988).)

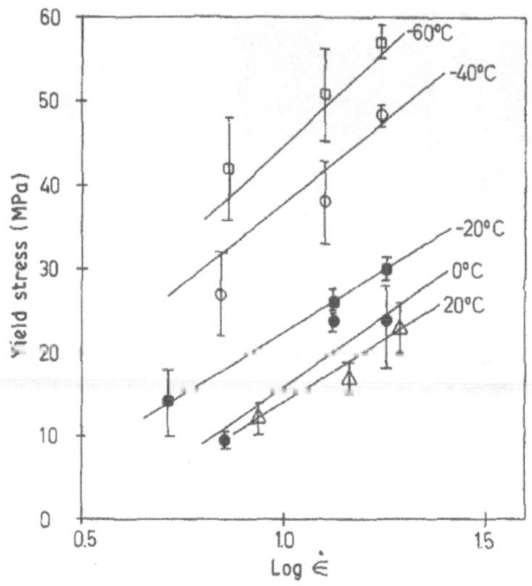

FIGURE 5.27 Plot of the compressive yield stress against log($\dot{\epsilon}$) for an ammonium perchlorate/GAP propellant at various temperatures. Data obtained using a Kolsky bar. (From Ho and Fong (1987b).)

I know, this was the first time that time–temperature superposition had been applied to the high-rate mechanical/fracture properties of an energetic material, although Heijboer (1968) and Vincent (1972) had previously found a relation between loss peaks of inert polymers and their impact fracture toughness.

An equation for the stress buildup σ_p in a polymer (high-density polyethylene) was derived by Briscoe and Nosker (1984):

$$\sigma_p = \sigma_0\left(1-\alpha^{c_p t/l_0}\right) \tag{5.2}$$

where α is the elastic wave reflection coefficient between the steel input bar and the polymer. A plot of this function for $\sigma_0 = 40$ MPa, $\alpha = 0.93$, $c_p = 1.6$ km/s and $l_0 = 2.0$ mm is given in Figure 5.28a.

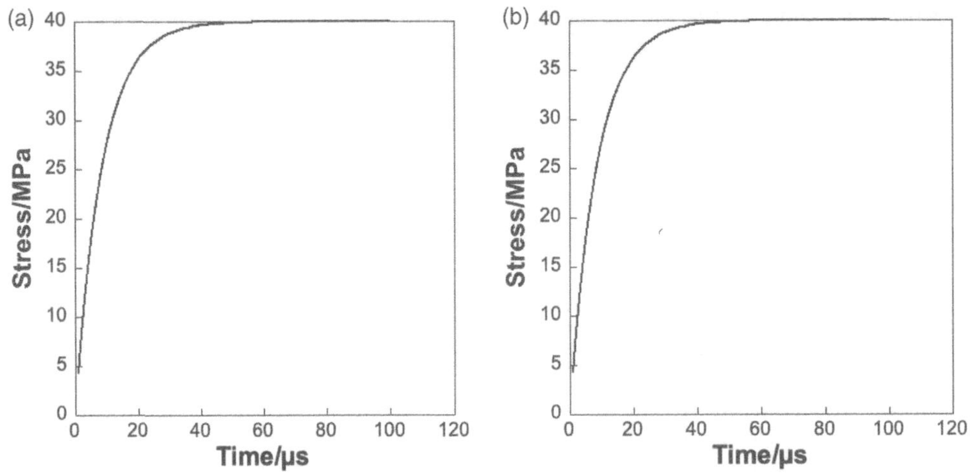

FIGURE 5.28 Plots of the stress buildup in (a) a 2-mm thick and (b) a 1-mm thick polyethylene specimen being deformed in a steel Kolsky bar according to the formula published by Briscoe and Nosker (1984) (see equation (5.2)).

It can be seen that it takes around 80–100 μs for a 2-mm thick soft polymer specimen to reach stress equilibrium in a steel Kolsky bar system. Since typical input stress pulses in laboratory compression Kolsky bar systems are of similar duration, the data will not meet the stress equilibrium condition for validity (Gray III, 2000). Equation (5.2) shows that there are two ways of reducing the time for stress equilibrium to be achieved: (a) reduce the specimen thickness l_0 (Figure 5.28b) or (b) reduce the reflection coefficient α by using low impedance Kolsky bars such as a magnesium alloy (Gray III et al., 2000) or a polymer (Gary et al., 1995; Zhao et al., 1997).

Another topic of interest to the researchers carrying out high-rate studies of PBXs and solid propellants in the 1980s was their impact fracture properties. For example, Fong et al. (1983) investigated the high-rate fracture properties of nitroguanidine propellant grains at four different temperatures (−40°C, −10°C, +5°C, +20°C) using both a dropweight and a Kolsky bar. They found that propellants containing small nitroguanidine particles were less resistant to fracture in their lab experiments, although the reverse was found to be the case during acceleration of a shell in a gun barrel. They attributed the difference in behaviour to a distinction between 'resistance to fracture' and 'resistance to fragmentation', but they did not explain what they meant by these terms or why they differed.

Figure 5.29 from a recently published paper demonstrates a problem common to most investigations of the high-rate mechanical properties of materials in that since low-strain rate mechanical testing machines are common and relatively inexpensive and Kolsky bars do not cost much to develop or to buy, mechanical data is often only obtained at low and high strain rates. But for full constitutive model development, data is required at intermediate strain rates (Walley and Field, 1994; Porter, 1995; Porter and Gould, 2009). However, servohydraulic machines for obtaining intermediate strain rate information are expensive and hence uncommon. Alternative less expensive oscillatory methods for obtaining mechanical data in the strain rate range 10^{-1}–10^{2} s^{-1} have therefore been developed (Figure 5.30) (Williamson et al., 2008; Picart et al., 2009; Le et al., 2010).

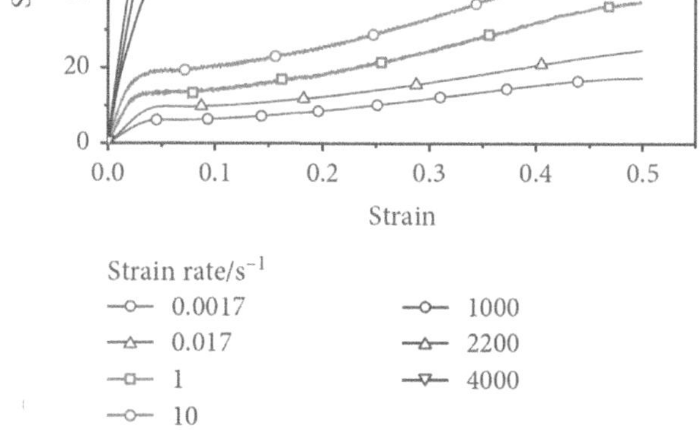

FIGURE 5.29 Low- and high-strain rate compressive stress–strain curves for a CMDB propellant. (From Zhang et al. (2023).)

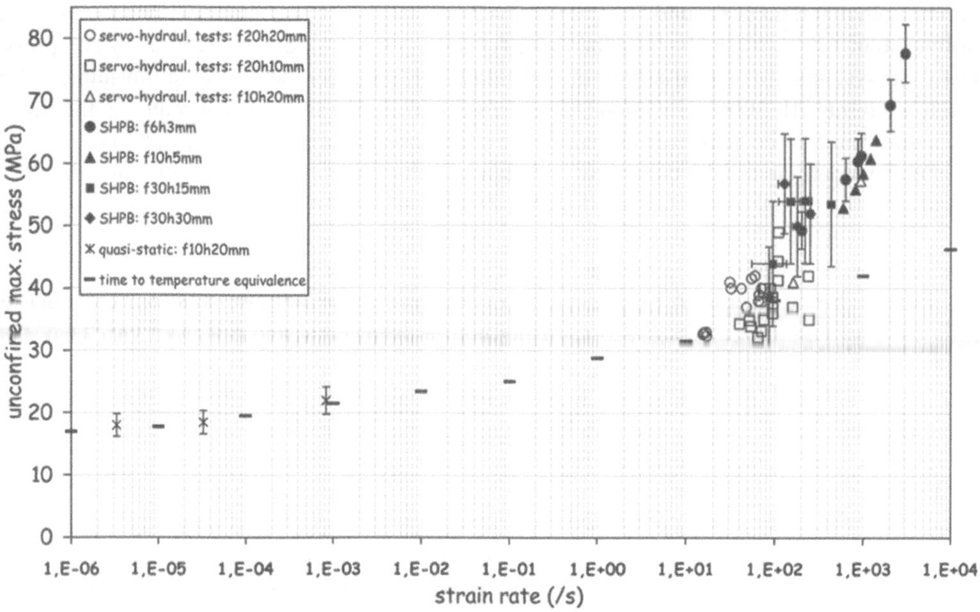

FIGURE 5.30 Unconfined uniaxial compressive stress versus log (strain rate) for a pyrotechnic at 23°C. (From Picart et al. (2009).)

5.4 SHOCK-LOADING

Hugoniots of a wide range of explosives studied up to 1980 may be found in Gibbs and Popalato (1980) and Marsh (1980). As Hugoniot's theory of shocks can be found in Davison and Graham (1979) and Gathers (1994) (as well as many other places), his analysis (Hugoniot, 1888, 1889, 1998) will not be reproduced here. I will just quote Brinkley Jr. and Seely (1969) who state that a Hugoniot is a line of points (p_1, v_1) "that are compatible with the requirement that mass, momentum and energy be conserved across a propagating discontinuity separating the states (p_1, v_1) from a specified initial state (p_0, v_0), where p is the pressure and v is the specific volume" i.e. a Hugoniot is a line connecting the states that can be reached in a one-dimensional shock experiment.

Every substance has its own Hugoniot, which is therefore one of its intrinsic properties. Note that in any given 1D shock experiment, p and v do not pass through the points on the Hugoniot line but rather go straight from the initial state to the final state i.e. they follow the so-called Rayleigh line (Rayleigh, 1910). The first published studies of the Hugoniots of secondary explosives were performed in the Soviet Union in the late 1950s (Figures 5.31 and 5.32) using the sort of apparatus shown schematically in Figure 5.33. As can be seen in Figure 5.32, an alternative way of presenting a material's shock response is to plot the shock velocity, u_s, against the particle velocity, u_p (the velocity at which particles of the shocked material move after the shock wave has set them in motion).

That Soviet researchers were probably the first to perform such studies can be deduced from the following two quotes:

> A considerable amount of work has been published on the study of shock wave compression of various materials (Goranson et al., 1955; Walsh and Christian, 1955; Walsh and Rice, 1957; Walsh et al., 1957). The shock sensitive nature of high explosives has, for the most part, limited work on these materials to a study of the detonation wave [*Figure 5.32b*]. However, the relatively low sensitivity observed for liquid TNT (trinitrotoluene) ... suggested the feasibility of studying this material. The pressure range over which measurements on the undetonated material can be made, and the widespread applications of TNT, further suggested the liquid as a logical candidate for extensive study.
>
> **(Garn, 1958)**

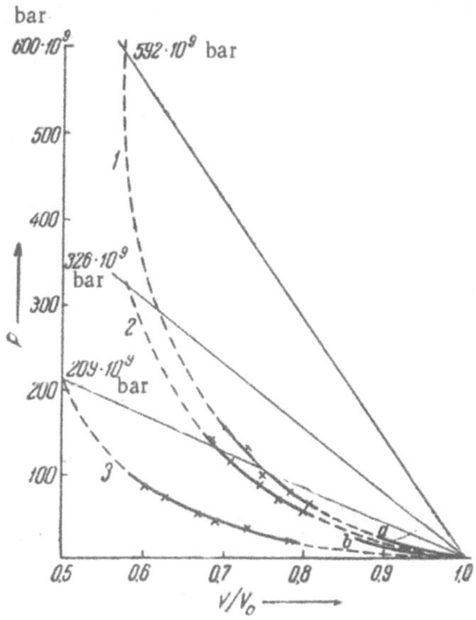

FIGURE 5.31 The first known published plot of the Hugoniot of three secondary explosive chemicals showing the relationship between p and v/v_0. for (1) hexogen (RDX), (2) trinitrotoluene (TNT), (3) nitromethane. The dashed lines are extrapolations of the experimental Hugoniot curves. (From Ilyukhin et al. (1960).)

> In recent years, great use has been made of dynamic methods of investigating the compressibility of various inert materials (Altshuler et al., 1958b; Altshuler et al., 1958a; Walsh and Christian, 1955). The absence of any corresponding work on condensed explosive materials is due to their high sensitivity to shock. The relatively low sensitivity to shock of cast trinitrotoluene, crystalline hexogene, and nitromethane has made it possible to measure their Hugoniot curves under conditions where detonation is absent.
>
> **(Ilyukhin et al., 1960)**

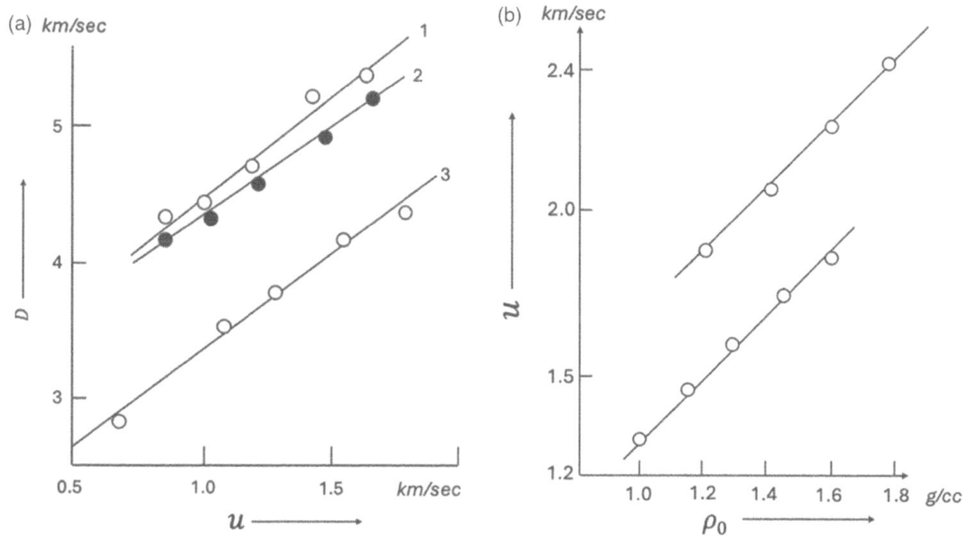

FIGURE 5.32 (a) Relationship between the velocity of the shock wave and the mass velocity. Key: (1) for hexogen (RDX); (2) for trinitrotoluene (TNT); (3) for nitromethane. (b) Experimental plot of the mass velocity, u, on the detonation wave front against initial density, ρ_0 of trotyl (TNT) (lower line) and hexogen (RDX) (upper line). ((a) Redrawn from Ilyukhin et al. (1960). (b) Redrawn from Dremin and Pokhil (1959).)

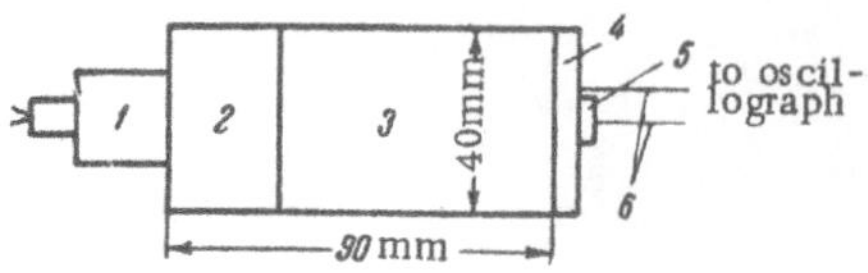

FIGURE 5.33 Schematic diagram of the method used by Soviet researchers in the late 1950s to measure the velocity of shock waves in unreacted explosive substances. Key: (1) detonator and intermediate explosive charge, (2) explosive lens used to generate a planar shock, (3) explosive charge used to create a shock, (4) metal plate, (5) explosive substance whose shock properties were under investigation, (6) electrical signal wires to measure the shock wave velocity by a contact method. (From Ilyukhin et al. (1960).)

Figures 5.34 and 5.35 present the first published Hugoniot plots (dating from the mid-1960s) from American authors for the explosives TNT, Composition B and TATB.

Table 5.1 summarises the studies into the shock properties of PBXs and solid propellants published between 1966 and 1989, all except one being by American and Soviet researchers:

5.5 SPALL

Since there have not been very many experimental studies of spall produced in explosives and propellants by shock-loading (Seaman et al., 1985; Seaman and Erlich, 1991; Golubev et al., 1995;

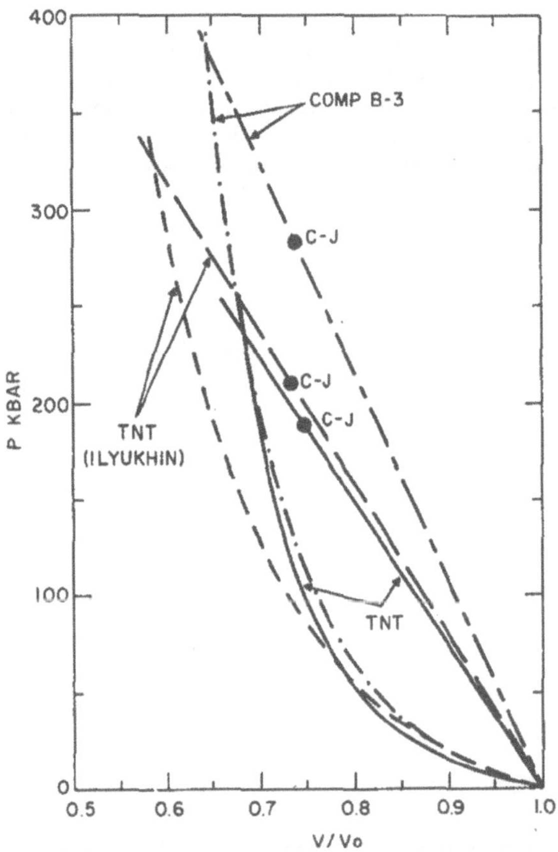

FIGURE 5.34 Shock Hugoniots for non-reacting TNT and Composition B-3. (From Coleburn and Liddiard (1966).)

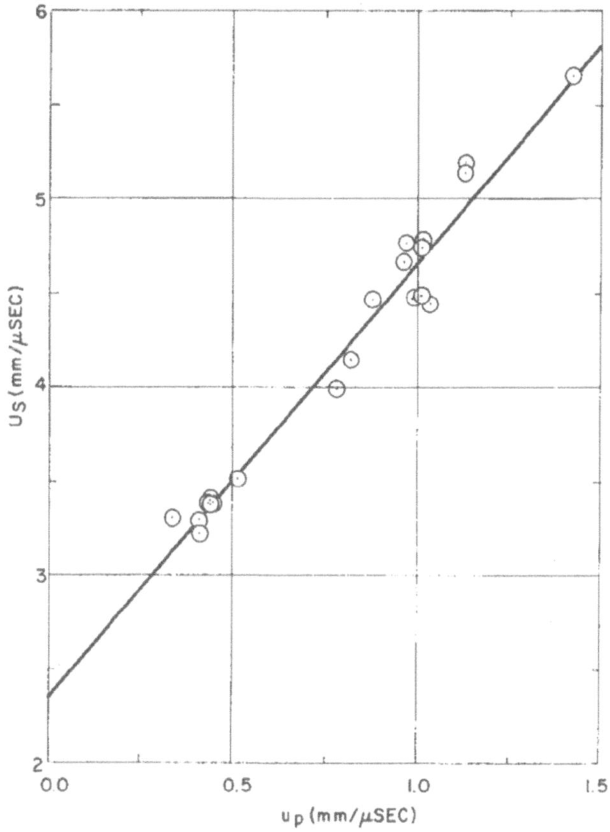

FIGURE 5.35 Plot of the shock wave velocity, u_s, versus particle velocity, u_p, of TATB. (From Coleburn and Liddiard (1966).)

TABLE 5.1
Publications Up to 1989 About the Shock Properties of Explosives and Solid Propellants

Year	Energetic Material	Country	References
1966	Pentolite, Composition B (and B-3), TNT, Octol, Baratol, Nitrosol	USA	Boyle et al. (1966)
1966	TNT, TNB, TNA, DATB, TATB, Composition B-3, LX-04-0, LX-04-1, PBX 9404-03, RDX/TNT/Al/wax	USA	Coleburn and Liddiard (1966)
1967	TNT	USSR	Vasilev et al. (1967)
1969	RDX	USA	Brinkley Jr. and Seely (1969)
1969	TNT	USA	Wasley and Walker (1969)
1970	TNT, Tritonal, PBX 9404, H-6	USA	Boyle et al. (1970)
1970	HNS, PBX 9404, NONA, RDX, PETN	USA	Roth (1970)
1971	TNT, Tetryl	USSR	Dremin et al. (1971)
1975	PETN, TATB, DATB	USA	Bauer (1975)
1975	Propellant	USA	Kuo and Summerfield (1975)
1976	Lead azide	USA	Davies et al. (1976)
1976	Porous TNT	USA	Erkman and Edwards (1976)

(*Continued*)

TABLE 5.1 (*Continued*)
Publications Up to 1989 About the Shock Properties of Explosives and Solid Propellants

Year	Energetic Material	Country	References
1980	Baratol, Composition B, Cyclotol, DATB, HMX, nitroguanidine, Octol, PBXs (various), PETN, RDX, TATB, Tetryl, TNT	USA	Gibbs and Popalato (1980)
1980	Baratol, Composition B, HMX, PETN, RDX, TATB, Tetryl, TNT, nitroguanidine, PBXs (various)	USA	Marsh (1980)
1982	Propellant	USA	Murri et al. (1982)
1982	Techniques paper	USA	Vorthman (1982)
1984	Composition B, PBXW-109(I)	USA	Lemar et al. (1984)
1985	Hexogen (RDX)	P.R. China	Jiao et al. (1985)
1985	Composition B	USA	Lemar et al. (1985)
1988	PBX 9502	USA	Dick et al. (1988)
1989	TNT	USSR	Balinets and Gogulya (1989)

Ostrik and Petrovskii, 1995; Golubev and Pogorelov, 1999; Huang and Ding, 1999; Golubev and Pogorelov, 2000; Novikov, 2000; Batsanov et al., 2004; Golubev, 2007; Helte and Lundgren, 2014; Zhang et al., 2019), a brief introduction to the phenomenon is given making use of published studies of metals and polymers.

Spall is the detachment of fragments from a material due to a blow. It has been used for thousands of years to shape stones for tools and weapons (Speth, 1972; Dibble and Whittaker, 1981). The first physical explanation (in terms of stress waves) for spall produced by an explosive in contact with a ship steel was given by Bertram Hopkinson in the years leading up to World War 1 (Figure 5.36) (Hopkinson, 1914a, 1914b, 1921a). Studies on 3D spall in metals produced by the high-velocity impact of non-planar projectiles (Figure 5.37) have been published by (Rinehart and Pearson, 1954; Rinehart, 1964; Wilson, 1967; Bertholf et al., 1975; Belov et al., 1997; Zhou et al., 2000; Fortov et al., 2006; Grady, 2006).

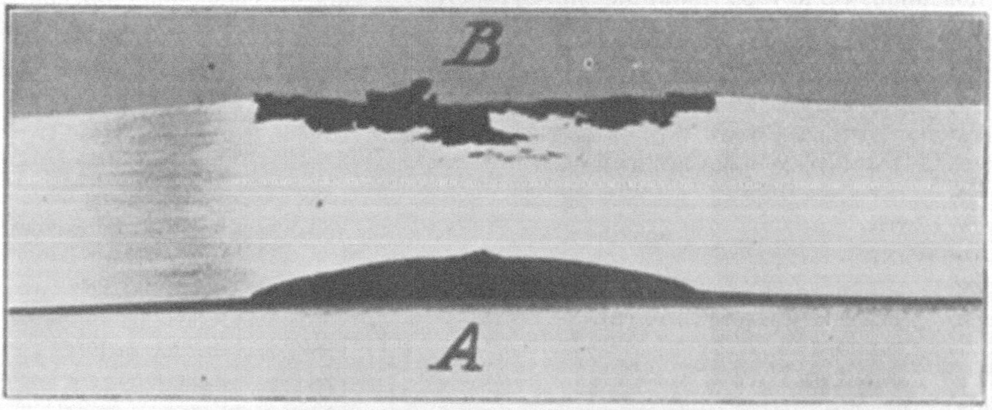

FIGURE 5.36 Section through a 19-mm thick mild steel plate which had been subjected to the detonation of guncotton in contact with side A. Spall occurred on the other side (side B). (From Hopkinson (1921b).)

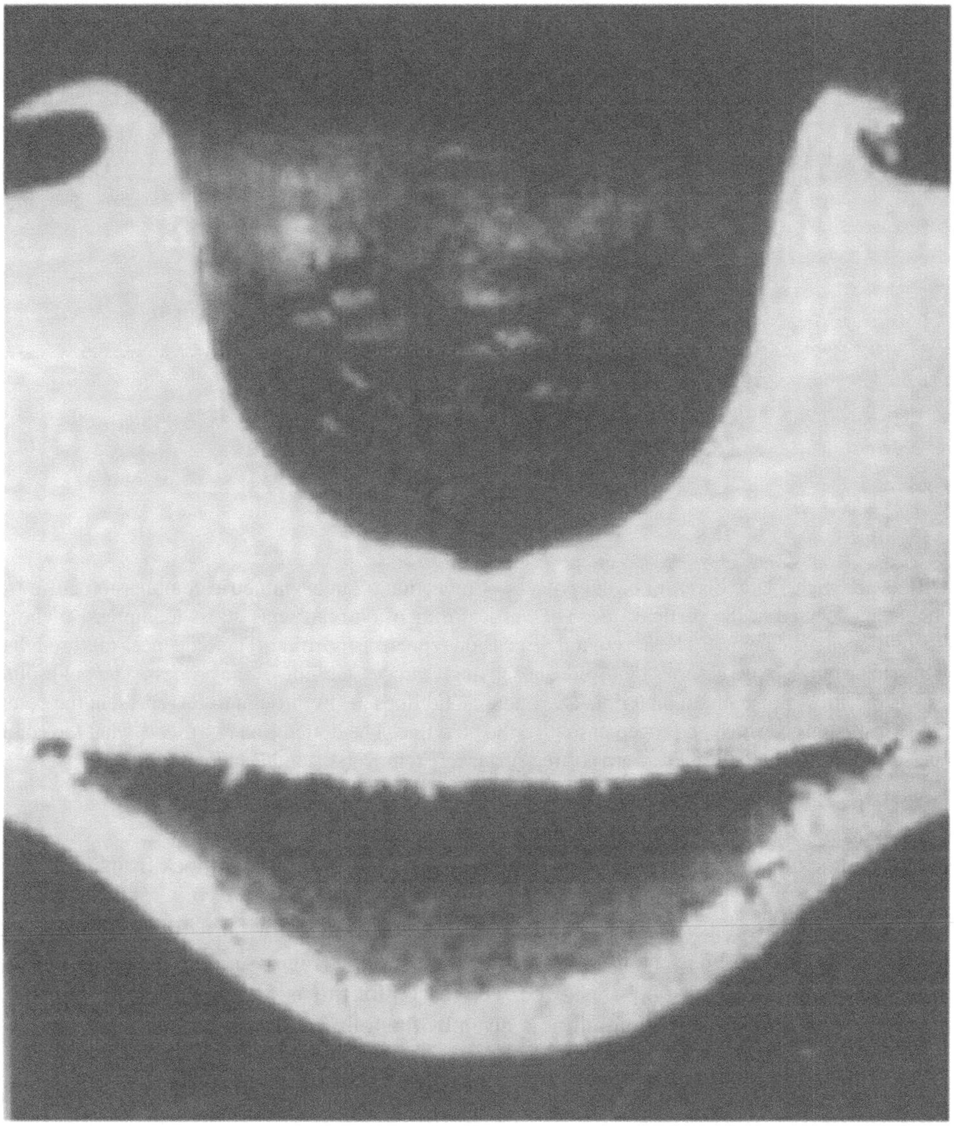

FIGURE 5.37 Spall produced by the reflection of a three-dimensional shock wave at the rear surface of a 12.5-mm thick aluminium alloy plate subjected on the upper side to the normal impact of a 3.175-mm diameter glass sphere at about 6 km/s. (From Eftis et al. (2003).)

Rinehart and Pearson (1954) outlined the main differences between spall and fracture under static loading as follows:

> There are marked differences between the fracture phenomena associated with impulsive loads and those associated with static loads.
>
> ...
>
> The most characteristic feature of fracturing under impulsive loading is that the load will almost always generate a well defined and reproducible fracture pattern within the specimen. The geometry of the pattern is often complex; yet it can be, in most cases, unambiguously associated with the transient events that lead to its formation. Indeed, random break-up or fracturing that must be treated statistically seems to occur more frequently under static loading than under impulsive loading.

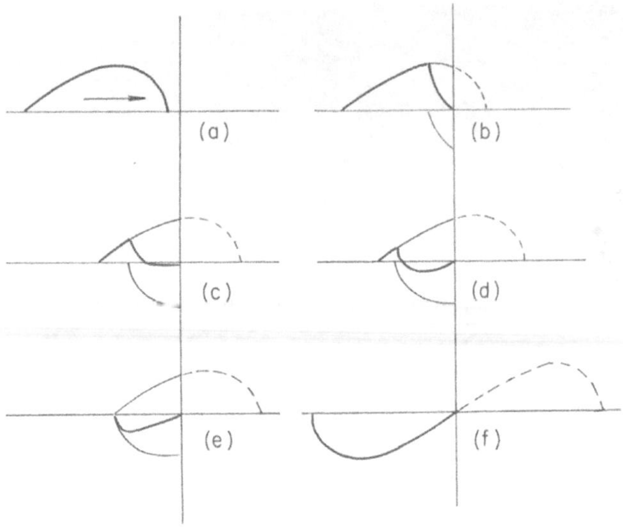

FIGURE 5.38 Schematic diagram of the process of reflection of an asymmetric compression pulse from a free boundary located at the vertical line. The vertical dimension represents stress (compressive above the horizontal line, tensile below). The horizontal dimension represents position. The sequences (a) through to (f) show the shape of the pulse at times that were chosen to give understanding of the process. The bold solid line represents the stress pulse in the material. The lighter solid lines in (b) through to (e) represent the compression and tensile pulses which add together to give the bold line. The dotted line represents what the compression pulse would have done if there were no free surface. (From Kolsky and Rader (1968).)

The reasons spalls form in a solid due to the reflection of a compressive shock from a free surface are given below.

An elastic compression wave in a solid that is normally incident on a free surface is totally reflected back into that solid as a tensile wave of the same shape (Figure 5.38). This happens because the free surface of a solid must be stress-free. The reason for this is that the reflection coefficient R of elastic waves at a material discontinuity is given by the following:

$$R = \frac{Z_1 - Z_2}{Z_1 + Z_2},\ Z_2 > Z_1 \tag{5.3}$$

where $Z\ (=\sqrt{E\rho})$ is the acoustic (or mechanical) impedance of the solid, E is Young's modulus and ρ is the density (Kennelly and Kurokawa, 1921). If $Z_1 = 0$ then $R = -1$. Strictly speaking, the surface of a solid will only be stress-free if the solid is in contact with a vacuum (for which Z is identically zero). However, since the values for the dynamic modulus and density of a 'typical' PBX are 5 GPa (Banerjee and Adams, 2004) and 1900 kg/m^3, respectively, while those for air at the Earth's surface are *ca.* 140 kPa and 1.225 kg/m^3 giving values for the mechanical impedances of *ca.* 3 MPa.s^{-1}m for a PBX and 0.4 kPa.s^{-1}m for air, R for the PBX/air boundary is −0.9997.

Spall will only occur sometime between Figure 5.38b and 5.38f if the tensile stress in the bottom left quadrant exceeds the dynamic fracture strength of the material. The position of the spall fracture depends on the shape of the compressive pulse. For as Kolsky (1959) pointed out,

> The maximum tensile stress, which is equal to the maximum compressive stress in the incident pulse, first occurs at a distance less than, or equal to, half the pulse length from the surface. For a steep-fronted pulse, it occurs at exactly half the pulse length, while for a symmetrical pulse… it occurs at a quarter of the pulse length.

FIGURE 5.39 Hopkinson fracture produced in a 0.5-in (12.7 mm) diameter polystyrene rod. (From Kolsky and Rader (1968).)

Spall in transparent polymers was studied by Kolsky from the late 1940s onwards (Figures 5.39 and 5.40) (Kolsky and Shearman, 1949; Kolsky, 1953, 1974; Kolsky and Rader, 1968). As well as reproducing the classic Hopkinson fracture in a rod of polystyrene (Figure 5.39), Kolsky also studied spall in square, cylindrical and conical Perspex specimens (Figure 5.40) produced by the detonation of 0.3–0.5 g lead azide charges.

Novikov (2000) stated in connection with spall in explosives that

> At present there is voluminous experimental material available in the literature for nonreactive materials that allows the behavior of constructions loaded with shock waves to be reliably predicted. Data on the spalling strength of solid heterogeneous [high explosives] are fairly scanty which is undoubtedly associated with the specificity of tests of these, so-called energetic, materials.
>
> As follows from studies of spalling of metals, partial fracture of the material is observed at a stress level lower than the level corresponding to complete spalling destruction. Most interesting for practical purposes is a situation when pulse tensile stresses induce in the material first visible zones of incipient fracture, because the destruction zones rapidly grow with a further, even insignificant, increase in the tensile stress.

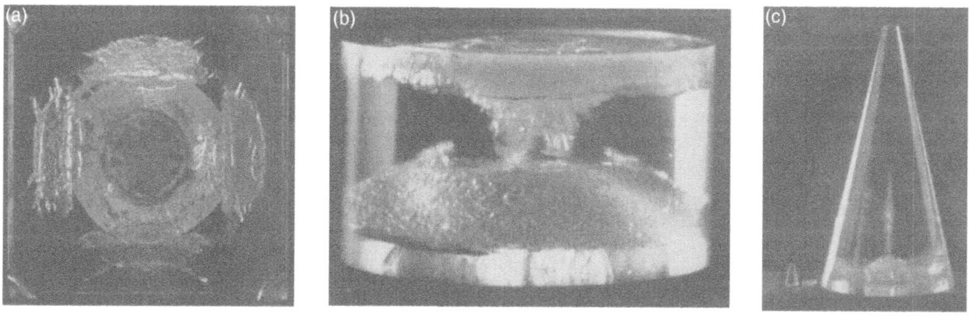

FIGURE 5.40 (a) Top view of spall produced in a square Perspex specimen by the detonation of 0.3 g of lead azide at the centre of the top surface; (b) side view of spall produced in a cylinder of Perspex by the detonation of 0.3 g of lead azide at the centre of the top surface; (c) side view of spall produced in a cone of Perspex by the detonation of 0.4 g of lead azide at the centre of the bottom surface. (From Kolsky (1953). See also (Kolsky and Shearman, 1949; Kolsky and Rader, 1968; Kolsky, 1974).)

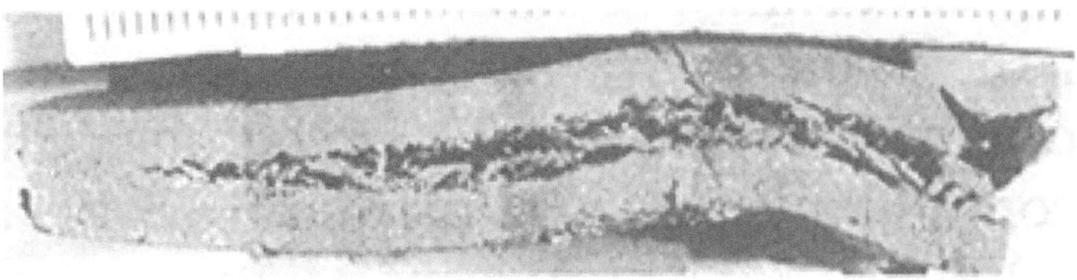

FIGURE 5.41 First reported spall produced by shock-loading an energetic material, a propellant, in a wedge configuration. (From Murri et al. (1982).)

The first studies of spall in energetic materials were performed in the 1980s by Seaman and co-workers for a propellant (Murri et al., 1982; Seaman et al., 1985; Seaman and Erlich, 1991) (Figures 5.41 and 5.42) and in 1995 for a PBX (Golubev et al., 1995). Seaman et al. (1985) derived a value for the dynamic fracture toughness of the propellant they were studying of 1.28 $MPa.m^{0.5}$.

There have been few experimental studies of this phenomenon since. One reason there have been so few experimental studies was recently given by Ge et al. (2021):

> Because of the instability and chemical reactions of PBX, the fracture phenomenon is hard to examine in experiments. Investigations have mainly focused on spallation that occurs during shock compression and tensile stages of metals. The effects of microstructures and impact velocity on spall fracture of PBX in shock loading are not well understood.

Apart from reasons of safety, Ge et al. (2021) also stated that "Computer simulation can provide detailed spatial stress and temperature distributions, and these properties are essential in appreciating the shock propagation properties in PBX under different loading regimes".

5.5.1 Taylor Impact Test

It has been known since at least 1885 that if a rod of material impacts a solid surface, only the impact end deforms (Kick, 1885b). Figure 5.43 is the first representation of this phenomenon that I have found. The figure is a drawing showing the deformation of rods of porcelain paste (built up in horizontal and vertical layers) that had been dropped onto a solid surface from two different heights

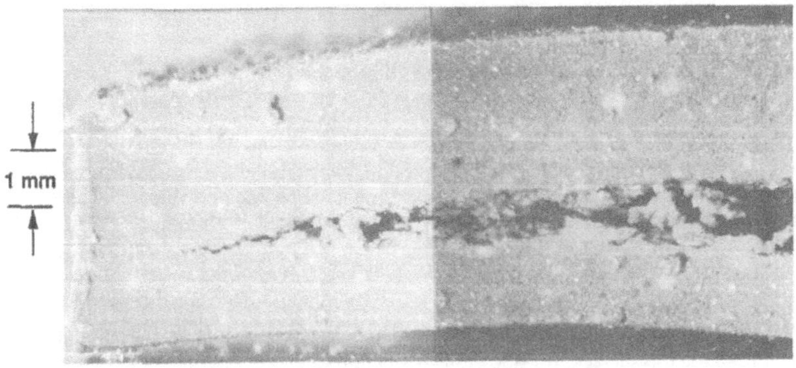

FIGURE 5.42 Optical micrograph of cross-section of a propellant shocked by a thin PMMA flyer plate at 145 m/s. (From Seaman and Erlich (1991).)

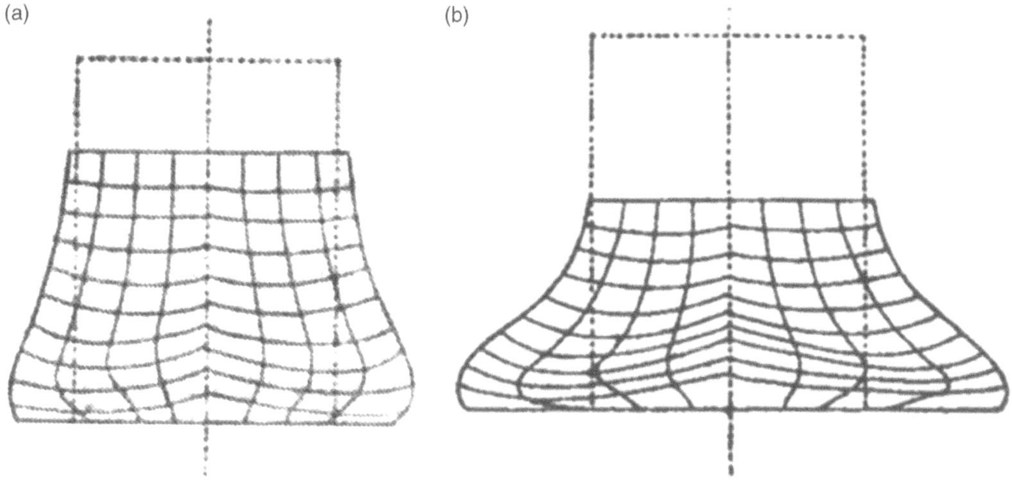

FIGURE 5.43 Drawings comparing deformed and undeformed cylinders of porcelain paste after impact on a solid surface at (a) 4.9 m/s and (b) 6.3 m/s. (From Kick (1885b, page 95).)

(5 and 8 m) i.e. at about 4.9 and 6.3 m/s. The text that follows is a description of the experiment in the original German, along with a simplified translation into English. The figure numbers in Kick's text have been changed to correspond with those in this chapter.

> Die Stauchung tritt einseitig stets ein, wenn ein in der Hand gehaltener oder aufgehängter Cylinder plastischen Materials gegen eine seiner Endflächen in der Achsenrichtung Stössen ausgesetzt wird; desgleichen, wenn ein cylindrischer Körper dieser Art gegen eine feste Wand geschossen oder geschleudert wird, oder wenn er in der Achsenrichtung auf eine feste Unterlage auffällt. Hiervon überzeugt man sich am leichtesten, wenn man einen Cylinder aus weicher Porzellanmasse auf einen geeignet geformten Fallklotz und letzteren auf den Amboss fallen lässt, wobei die Porzellanmasse sich staucht. Ist dieselbe entsprechend horizontal und vertical geschichtet, so erhält man die in Figur [5.43] dargestellte Schichtenverwerfung. Figur [5.43a] ist bei geringerer Fallhöhe (5m), Figur [5.43b] bei grösserer Fallhöhe (8m) erhalten. (When a cylinder of plastic material axially impacts a solid wall, deformation always occurs at the impacting end. The easiest way to convince yourself of this is to attach a cylinder made of soft porcelain paste to a suitably shaped block and let the latter fall onto an anvil. The cylinder of paste will become deformed. If it is layered horizontally and vertically, the shapes shown in Figure [5.43] are obtained. Figure [5.43a] was obtained for a fall height of 5m and Figure [5.43b] for a fall height of 8m).

The first known application of the rod impact technique was carried out by French military researchers at the beginning of the 20th century to study the impact properties of metals (de Maupeou, 1902; Régnauld, 1927). However, these researchers did not perform (or at least did not publish) an analysis and so they did not report dynamic strengths for the metals they investigated. The first (approximate) mathematical analysis of the deformation produced in a rod by impact was performed by G.I. Taylor during World War 2 to allow the high-rate compressive strengths of metals to be estimated (Taylor, 1946, 1948). Hence, the technique later became known as the Taylor impact test (Erlich, 1985; Walley and Couque, 2024). However, since a more accurate technique of obtaining such information based on the Hopkinson pressure bar (Hopkinson, 1914b; Landon and Quinney, 1923) was under development around the same time and by several of the same people (Taylor, 1946; Kolsky, 1949; Walley, 2020), the Taylor impact has not often been used since World War 2 to obtain the high-rate properties of materials.

The first application of Taylor's impact test to an explosive was reported by Napadensky et al. (1970). They calculated the deformation (Figure 5.44) and the temperature rise (Figure 5.45) that

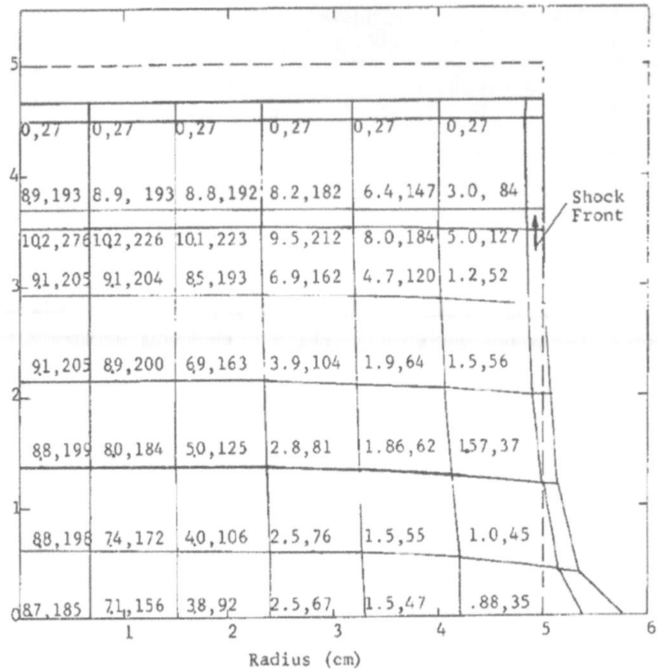

FIGURE 5.44 Calculation of the deformation of a right circular cylinder (5 cm diameter, 5 cm long) of TNT 15.7 μs after impacting a rigid, lubricated target at 210 m/s. The pairs of numbers at each node give the pressure (kbar) and temperature (°C). (From Napadensky et al. (1970).)

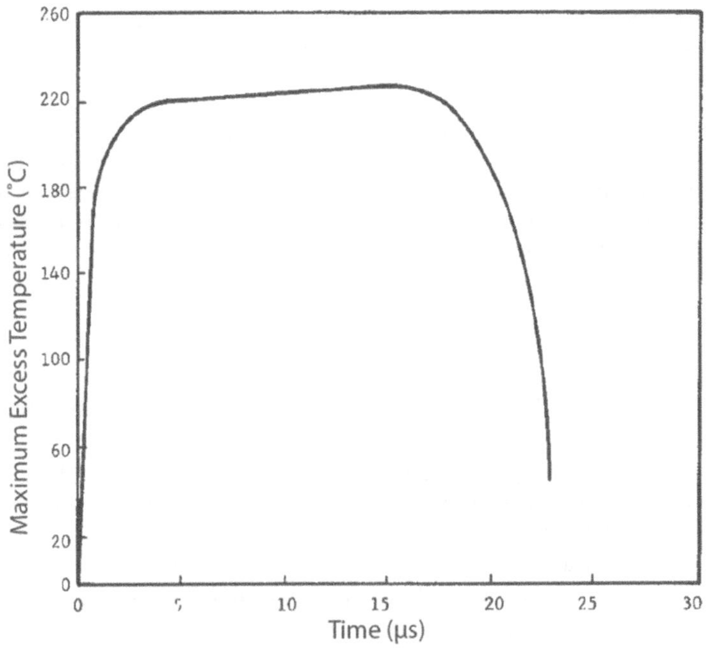

FIGURE 5.45 Calculated maximum rise in temperature (°C) on the cylinder axis as a function of time (μs) for a 5 cm diameter, 5 cm long right circular cylinder of TNT impacting a rigid lubricated target at 210 m/s. (From Napadensky et al. (1970).)

FIGURE 5.46 Recovered cylinder of PBXN-109 after being subjected to a reverse Taylor impact test at 33 m/s. The dimensions of the specimen were not given. (From Quidot and Chabin (2007).)

would occur if a 5 cm diameter, 5 cm long cylinder of TNT were to be fired at 210 m/s at a rigid lubricated surface. However, there is no indication in their article that they actually performed this experiment.

Unlike metals for which deformation beyond the elastic limit is permanent, the large deformation of viscoelastic materials such as polymers, PBXs and solid propellants is mostly recoverable (see Figure 5.46 and Bernatskii and Rabinovich (1964)). For this reason, Taylor impact testing of such materials requires the deformation during impact to be determined either by high-speed photography (Briscoe and Hutchings, 1976) or flash X-radiography (Figure 5.47) (Fugelso et al., 1982).

Firing a rod sufficiently accurately so that it deforms symmetrically about its axis for the whole duration of a classic Taylor impact test is difficult (Ma et al., 2013b). For this and other reasons, the reverse Taylor impact test was developed in the early 1980s (Gust, 1982; Murri et al., 1982). In this test, a rod of the material of interest is held stationary and is then impacted by a moving anvil, usually a metal or ceramic plate carried on the front of a sabot (Figure 5.48). High-speed photography is then essential whatever the nature of the target material as the specimen is usually destroyed during the experiment unless a great deal of care is taken.

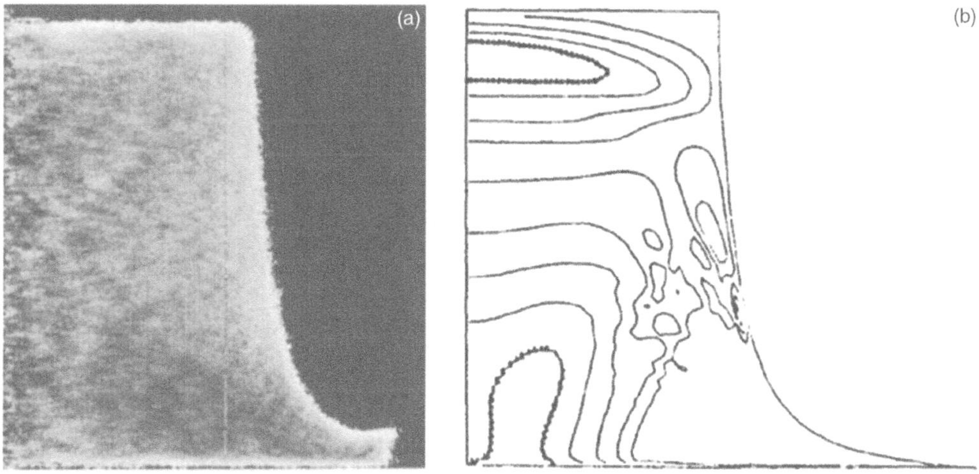

FIGURE 5.47 (a) Flash radiograph of a mock PBX cylinder 4.8 μs after impact at 677 m/s (original dimensions: 8.85 mm diameter, L/D = 1.2). (b) Density contours (0.033 separation) for the same cylinder 6.0 μs after impact. (From Fugelso et al. (1982).)

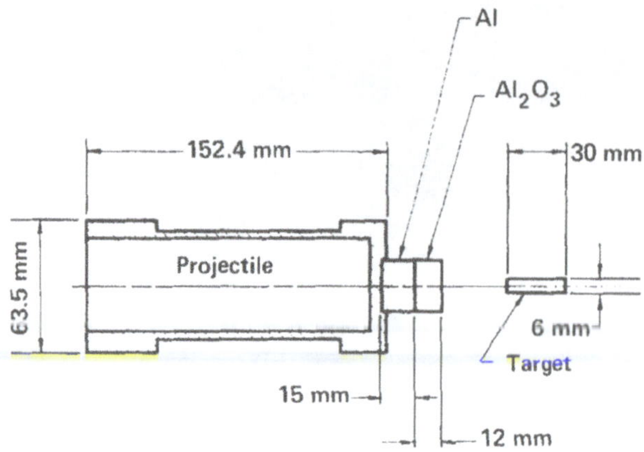

FIGURE 5.48 Schematic diagram of the first reported experimental reverse Taylor impact test setup (for studying copper). (From Gust (1982).)

One of the first simulations of reverse Taylor impact testing of an energetic material was performed by Murri et al. (1982) for a cylinder of propellant (see Figure 5.49a). One of their aims was to determine where the cylinder went into tension (Figure 5.49b) and hence how cracks formed within the cylinder during impact.

Quidot (1988) used the classic Taylor impact test to study the effect of impact velocity (and hence strain rate) on the dynamic fragmentation of a PBX (Figure 5.50). He also compared the experimentally determined fragment size with the predictions of three different fragmentation theories (Figure 5.51). He stated that Rittinger's theory was closest to experiment, although as can be seen from Figure 5.51, there is not much difference between the three theories.

Seaman and Erlich (1991) reported radial cracks formed in a propellant specimen on which a reverse Taylor impact test had been performed (Figure 5.52). They attributed these cracks to the tensile stress arising from the rarefaction waves that propagated from the free outer cylindrical surface towards the cylinder axis.

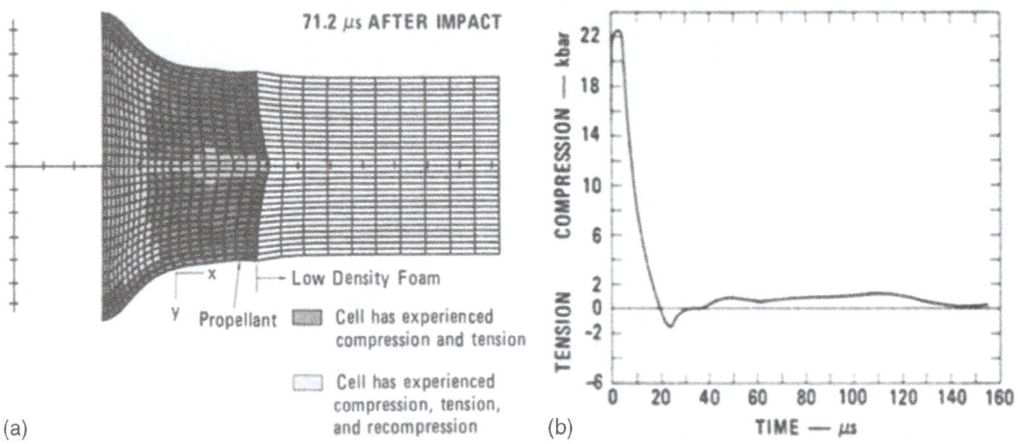

FIGURE 5.49 (a) Two-dimensional numerical simulation of the impact of a 3-in (76 mm) diameter, 3-in (76 mm) long propellant cylinder being impacted axially by a steel cylinder (i.e. reverse Taylor impact test) 71.2 μs after impact. (b) Calculated stress/kbar in a cell in the centre of the cylinder, 1.35 in (34.3 mm) from the impact surface as a function of time. (From Murri et al. (1982).)

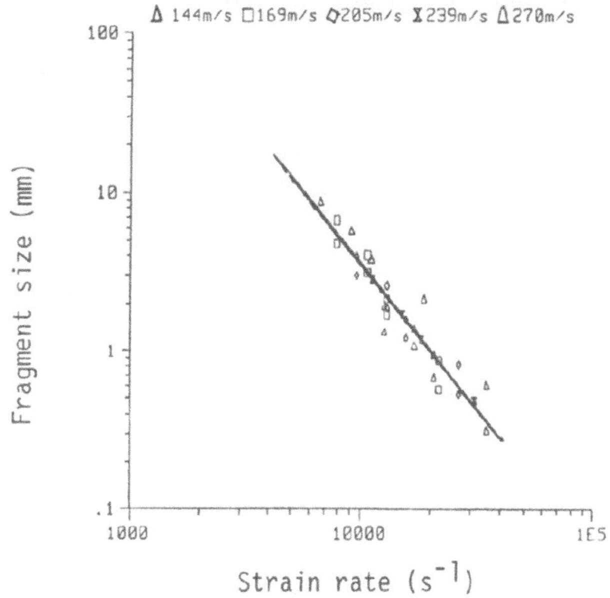

FIGURE 5.50 Plot of fragment size as a function of strain rate for a PBX. Data obtained using a classic Taylor impact test (symbol key to impact velocity given at the top of the figure). (From Quidot (1988).)

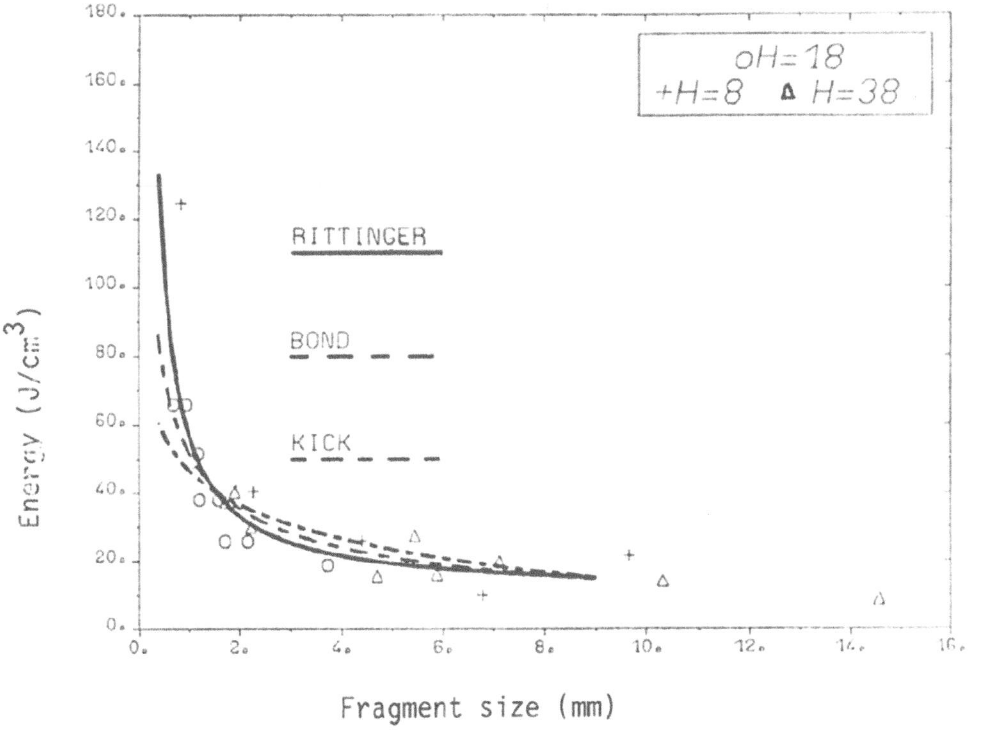

FIGURE 5.51 Comparison of the predictions of fragment size for a classic Taylor impact test of a PBX as a function of impact energy for three different fragmentation models: Rittinger (von Rittinger, 1867), Bond (Bond, 1952), and Kick (Kick, 1885a). All specimens were 18 mm in diameter. Three different lengths (H) were used: 8 mm (+), 18 mm (o), 38 mm (Δ). (From Quidot (1988).)

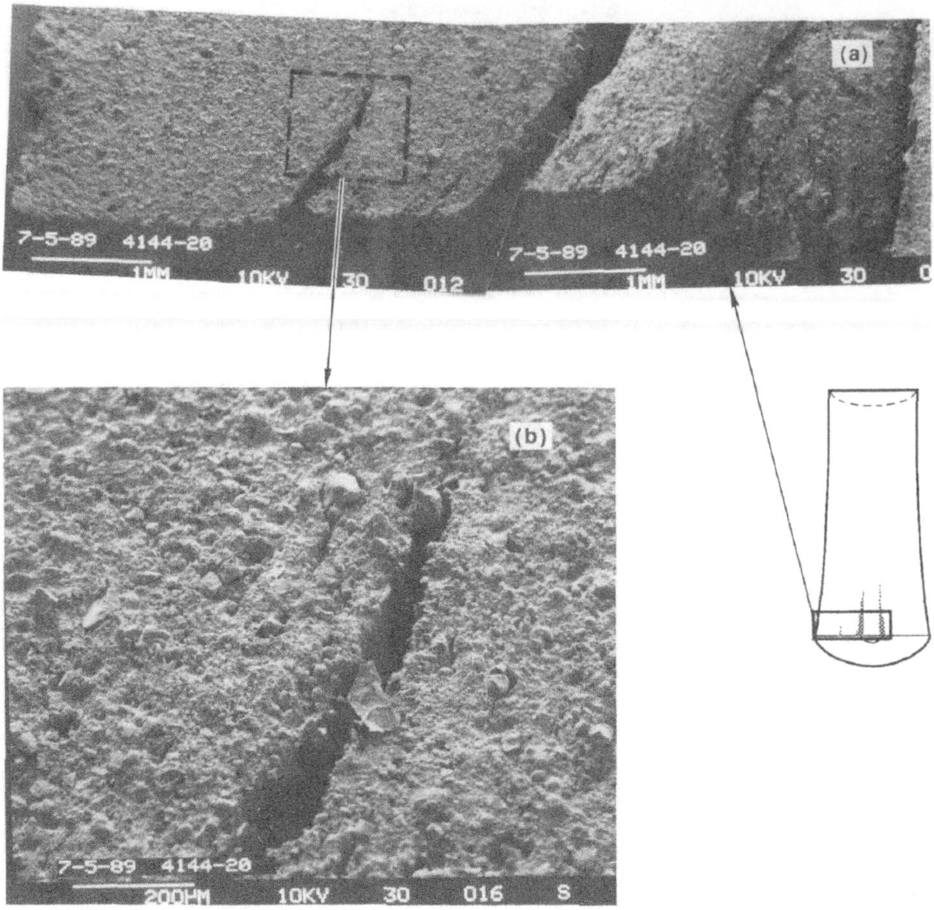

FIGURE 5.52 Scanning electron micrographs of the cracked region near the impact end of a reverse Taylor impact propellant cylinder (original dimensions: 30.6 mm long, 15.2 mm diameter) performed at 117 m/s. The drawing shows the region of the cylinder from where the micrographs were obtained. (From Seaman and Erlich (1991).)

5.6 SUMMARY

The techniques developed since World War 2 (the Kolsky bar, the Taylor impact test, plate impact shock-loading) for measuring and studying the high-rate mechanical properties of materials began to be applied to PBXs and solid propellants from about the mid-1960s. This chapter takes the story of the application of these techniques to PBXs and solid propellants up to the end of the 1980s. This date was chosen as being when constitutive models began to be sufficiently accurate to allow impact events to be modelled to a reasonable level of accuracy, at least for relatively straightforward materials such as metals and polymers. A follow-on review bringing the story of high-rate mechanical properties of explosives and solid propellants up to the present day is planned for the future.

ACKNOWLEDGEMENT

The author thanks the referee for the care and attention with which they read my typescript. The chapter is much improved as a result.

REFERENCES

Addiss, J.W., Collins, A.L., Walley, S.M. & Proud, W.G. 2015. The use of digital speckle radiography to investigate the internal flow fields during the ballistic penetration of sand. *In:* Iskander, M., Bless, S. & Omidvar, M. (eds.) *Rapid Penetration into Granular Media: Visualizing the Fundamental Physics of Rapid Earth Penetration*. pp. 229–265, Amsterdam: Elsevier. https://doi.org/10.1016/B978-0-12-800868-3.00007-9

Akazawa, T. 1943. New test method for evaluating internal stress due to compression of concrete: The splitting tension test (in Japanese). *J. Jpn. Soc. Civil Eng.*, 29, 777–787.

Akazawa, T. 1953. Méthode pour l'essai de traction de bétons (Tensile testing method for concrete). *RILEM Bulletin*, 16, 11–23.

Alouaamari, M., Lefebvre, M.H., Perneel, C. & Herrmann, M. 2008. Statistical assessment methods for the sensitivity of energetic materials. *Propell. Explos. Pyrotech.*, 33, 60–65. https://doi.org/10.1002/prep.200800210

Altshuler, L.V., Krupnikov, K.K. & Brazhnik, M.I. 1958a. Dynamic compressibility of metals under pressures from 400,000 to 4,000,000 atmospheres. *Sov. Phys. JETP*, 7, 614–619.

Altshuler, L.V., Krupnikov, K.K., Ledenev, B.N., Zhuchikhin, V.I. & Braznik, M.I. 1958b. Dynamic compressibility and equation of state of iron under high pressure. *Sov. Phys. JETP*, 7, 606–614.

Balinets, Y.M. & Gogulya, M.F. 1989. Emission from shocked pressed TNT. *Sov. J. Chem. Phys.*, 5, 382–387.

Banerjee, B. & Adams, D.O. 2004. On predicting the effective elastic properties of polymer bonded explosives using the recursive cell method. *Int. J. Solids Struct.*, 41, 481–509. https://doi.org/10.1016/j.ijsolstr.2003.09.016

Barcellos, A. 1953. Correlation entre la résistance à la traction et la résistance à la compression des bétons (Correlation between the tensile and compressive strengths of concrete). *RILEM Bulletin*, 13, 109–125.

Barker, D.B. & Fourney, M.E. 1976. Displacement measurements in the interior of 3D bodies using scattered light speckle patterns. *Exp. Mech.*, 16, 209–214. https://doi.org/10.1007/BF02329270

Barsis, E., Williams, E. & Skoog, C. 1970. Piezoresistivity coefficients in manganin. *J. Appl. Phys.*, 41, 5155–5162. https://doi.org/10.1063/1.1658638

Batsanov, S.S., Gavrilkin, S.M., Gordopolov, A.Y. & Gordopolov, Y.A. 2004. Spalling phenomena in shock-recovery capsules during shock compression of inert and reactive mixtures. *Combust. Explos. Shock Waves*, 40, 605–611. https://doi.org/10.1023/B:CESW.0000041414.23884.48

Bauer, R.J. 1975. An analysis of small scale gap test sensitivity data using porosity theory and nonreactive shock Hugoniots. White Oak, MD, Naval Surface Warfare Center. Report no. NSWC/WOL/TR 75-67.

Bawn, C.E.H. & Rotter, G. (eds.) 1956. *Science of Explosives*, London: Her Majesty's Stationery Office.

Belov, G.V., Golubev, V.K. & Yutkina, N.A. 1997. Damage of a steel plate from hypervelocity impact. *Combust. Explos. Shock Waves*, 33, 100–102. https://doi.org/10.1007/BF02671860

Belyaev, A.F. & Kurbangalina, R.K. 1964. Realisation of detonation conditions for black powder. *Russ. J. Phys. Chem.*, 38, 309–310.

Benfante, A., Samuels, P., Kasabwala, P. & Tashjian, N. 2025. Comparison of TNT ignition due to cavity collapse in gun and setback simulator tests. *In:* Mason, M.H. (ed.) *Proceedings 17th International Detonation Symposium*. pp. 498–503, Arlington, VA: Office of Naval Research.

Bernatskii, A.D. & Rabinovich, A.L. 1964. Strain in certain crosslinked polymers. *Polym. Sci. USSR*, 6, 1166–1175. https://doi.org/10.1016/0032-3950(64)90383-1

Bertholf, L.D., Buxton, L.D., Thorne, B.J., Byers, R.K., Stevens, A.L. & Thompson, S.L. 1975. Damage in steel plates from hypervelocity impact. 2: Numerical results and spall measurement. *J. Appl. Phys.*, 46, 3776–3783. https://doi.org/10.1063/1.322163

Bi, C., Guo, X., Wang, A.H., Weng, G.J., Qu, K.P., Shen, F. & Zhu, L.L. 2023. Strain-rate-dependent cohesive zone modelling of charge damage behavior when a projectile penetrates multilayered targets. *Acta Mech.*, 234, 2869–2887. https://doi.org/10.1007/s00707-023-03541-2

Blackwood, J.D. & Bowden, F.P. 1952. The initiation, burning and thermal decomposition of gunpowder. *Proc. R. Soc. Lond. A*, 213, 285–306. https://doi.org/10.1098/rspa.1952.0127

Boddy, R.L., Taylor, N.E., Braithwaite, C.H., Williamson, D.M., Jardine, A.P., Gould, P., Cullis, I.G. & Johnson, R. 2014. Application of group interaction modelling to the shock Hugoniots of double-base propellants. *In:* Pachman, J., Selesovsky, J. & Matyas, R. (eds.) *Proceedings 17th Seminar on New Trends in Research of Energetic Materials*. pp. 96–102, Pardubice, Czech Republic: University of Pardubice.

Bond, F.C. 1952. The third theory of comminution. *Trans. AIME*, 220, 484–494.

Bonthoux, F. & Girodon, A. 1985. Adaptation d'un dispositif du type Hopkinson-Kolsky à l'étude du comportement mécanique de matériaux explosibles en température (Adaptation of a Hopkinson-Kolsky type device to the study of the mechanical behavior of explosive materials at temperature). *J. Phys. France Colloq.*, 46(C5), 639–644. https://doi.org/10.1051/jphyscol:1985582

Boyars, C. & Levine, D. 1968. Drop-weight impact sensitivity testing of explosives. *Pyrodynamics*, 6, 53–77.

Boyle, V.M., Jameson, R.L. & Sultanoff, M. 1966. Determination of shock Hugoniots for several condensed phase explosives. *In: Proceedings Fourth Symposium (International) on Detonation*. pp. 241–247, Washington DC: Office of Naval Research.

Boyle, V.M., Smothers, W.G. & Ervin, L.H. 1970. The shock Hugoniots of unreacted explosives. *In: Proceedings Fifth Symposium (International) on Detonation*. pp. 251–258, Arlington, VA: Office of Naval Research.

Braithwaite, C.H., Chapman, D.J., Church, P.D., Gould, P.J. & Williamson, D.M. 2012. Hugoniot of a reactive metal powder mixture. *In:* Kontis, K. (ed.) *Proceedings 28th International Symposium on Shock Waves. Volume 1*. pp. 153–158, Berlin: Springer. https://doi.org/10.1007/978-3-642-25688-2_23

Brinkley, S.R., Jr. & Seely, L.B. 1969. Construction of the Hugoniot curve and calculation of the Chapman-Jouguet points for general equations of state. *Combust. Flame*, 13, 506–510. https://doi.org/10.1016/0010-2180(69)90090-X

Briscoe, B.J. & Hutchings, I.M. 1976. Impact yielding of high density polyethylene. *Polymer*, 17, 1099–1102. https://doi.org/10.1016/0032-3861(76)90013-6

Briscoe, B.J. & Nosker, R.W. 1984. The influence of interfacial friction on the deformation of high density polyethylene in a split Hopkinson pressure bar. *Wear*, 95, 241–262. https://doi.org/10.1016/0043-1648(84)90140-6

Brown, G.I. 1998. *The Big Bang: A History of Explosives*, Stroud: Sutton Publishing.

Brown, G.I. 2010. *Explosives: History with a Bang*, Stroud: Sutton Publishing.

Carlucci, D.A., Frydman, A.M. & Cordes, J.A. 2013. Mathematical description of projectile shot exit dynamics: Set-forward. *Trans. ASME: J. Appl. Mech.*, 80, 031501. https://doi.org/10.1115/1.4023335

Carneiro, F.L.L.B. & Barcellos, A. 1953. Résistance à la traction des bétons (The tensile strength of concrete). *RILEM Bulletin*, 13, 103–108.

Chao, M.T. & Fuh, C.D. 2003. Why wild values occur in pyrotechnic sensitivity analysis. *Propell. Explos. Pyrotech.*, 28, 216–218. https://doi.org/10.1002/prep.200300008

Chen, W. & Song, B. 2011. *Split Hopkinson (Kolsky) Bar: Design, Testing and Applications*, New York: Springer. https://doi.org/10.1007/978-1-4419-7982-7

Chiang, F.P. & Mao, L. 2016. Evolution of speckle photography: From macro to nano and from 2D to 3D. *In:* Jin, H., Yoshida, S., Lamberti, L. & Lin, M.-T. (eds.) *Advancement of Optical Methods in Experimental Mechanics*. pp. 71–77, New York: Springer. https://doi.org/10.1007/978-3-319-22446-6_9

Chidester, S.K., Green, L.G. & Lee, C.G. 1995. A frictional work predictive method for the initiation of solid high explosives from low-pressure impacts. *In:* Short, J.M. & Tasker, D.G. (eds.) *Proceedings 10th International Detonation Symposium*. pp. 786–792, Arlington, VA: Office of Naval Research.

Chidester, S.K., Tarver, C.M., DePiero, A.H. & Garza, R.G. 2000. Single and multiple impact ignition of new and aged high explosives in the Steven impact test. *AIP Conf. Proc.*, 505, 663–666. https://doi.org/10.1063/1.1303560

Christensen, D., Novik, G.P. & Unneberg, E. 2024. Estimating sensitivity with the Bruceton method: Setting the record straight. *Propell. Explos. Pyrotech.*, 49, e202400022. https://doi.org/10.1002/prep.202400022

Church, P., Cornish, R., Cullis, I., Gould, P. & Lewtas, I. 2014. Using the split Hopkinson pressure bar to validate material models. *Phil. Trans. R. Soc. A*, 372, 20130294. https://doi.org/10.1098/rsta.2013.0294

Church, P.D., Andrews, T. & Goldthorpe, B. 1999. A review of constitutive model development within DERA. *In:* Jerome, D.M. (ed.) *Structures under Extreme Loading Conditions (PVP 394)*. pp. 113–120, New York: American Society of Mechanical Engineers.

Church, P.D., Townsley, R., Bezance, T., Proud, W.G., Grantham, S.G., Bourne, N.K. & Millett, J.C.F. 2005. Simulation of precise setforward and setback experiments. *Int. J. Impact Eng.*, 32, 80–91. https://doi.org/10.1016/j.ijimpeng.2005.07.011

Coleburn, N.L. & Liddiard, T.P. 1966. Hugoniot equations of state of several unreacted explosives. *J. Chem. Phys.*, 44, 1929–1936. https://doi.org/10.1063/1.1726963

Collins, A.L., Addiss, J.W., Walley, S.M., Promratana, K., Bobaru, F., Proud, W.G. & Williamson, D.M. 2011. The effect of rod nose shape on the internal flow fields during the ballistic penetration of sand. *Int. J. Impact Eng.*, 38, 951–963. https://doi.org/10.1016/j.ijimpeng.2011.08.002

Combes, M. 1848. Sur la manière d'employer le pyroxyle dans l'exploitation des mines (On the use of guncotton in blasting). *Compte Rendu Séances Académie Sciences*, 26, 61–64.

Copp, J.L., Napier, S.E., Nash, T., Powell, W.J., Skelly, H., Ubbelohde, A.R. & Woodward, P. 1948. The sensitiveness of explosives. *Phil. Trans. R. Soc. Lond. A*, 241, 197–203. https://doi.org/10.1098/rsta.1948.0012

Cosculluela, A. & Forquin, P. 2022. Damage in armor ceramics subjected to high-strain-rate dynamic loadings: The spherical expansion shock wave pyrotechnic test. *In:* Voyiadjis, G.Z. (ed.) *Handbook of Damage Mechanics: Nano to Macro Scale for Materials and Structures*. pp. 609–638, New York: Springer. https://doi.org/10.1007/978-1-4614-8968-9_81-1

Cummock, M.R., Casey, A.D. & Son, S.F. 2021. The effect of the chosen distribution form on reaction probability estimates from drop-weight impact results. *J. Energ. Mater.*, 39, 377–398. https://doi.org/10.1080/07370652.2020.1799115

Curtis, J.P., Mills, J.T. & Lee, P.R. 2010. Response of explosives to hypervelocity impact and shearing. *Adv. Chem. Res.*, 4, 243–267.

Curtis, J.P., Mills, J.T., Lynch, N.J., Cornish, R. & Lee, P.R. 2006. An analytical model for high velocity impact on confined explosive configurations. *Int. J. Impact Eng.*, 33, 210–218. https://doi.org/10.1016/j.ijimpeng.2006.09.088

Davies, F.W., Zimmerschied, A.B., Bogardt, F.G. & Avrami, L. 1976. The Hugoniot and shock initiation threshold of lead azide. *J. Chem. Phys.*, 64, 2295–2302. https://doi.org/10.1063/1.432537

Davison, L. & Graham, R.A. 1979. Shock compression of solids. *Phys. Rep.*, 55, 255–379. https://doi.org/10.1016/0370-1573(79)90026-7

de Maupeou, G.L. 1902. Action comparée des forces sur les solides: Invariables, élastiques, déformables (Comparisons of the action of forces on solids: Invariant, elastic, deformable). *Bull. Assoc. Technique Maritime*, 13, LVIII–LIX, 271–365.

Dibble, H.L. & Whittaker, J.C. 1981. New experimental evidence on the relation between percussion flaking and flake variation. *J. Archaeol. Sci.*, 8, 283–296. https://doi.org/10.1016/0305-4403(81)90004-2

Dick, J.J., Forest, C.A., Ramsay, J.B. & Seitz, W.L. 1988. The Hugoniot and shock sensitivity of a plastic bonded TATB explosive PBX 9502. *J. Appl. Phys.*, 63, 4881–4888. https://doi.org/10.1063/1.340428

Dixon, W.J. & Mood, A.M. 1948. A method for obtaining and analyzing sensitivity data. *J. Amer. Statistical Assoc.*, 43, 109–126. https://doi.org/10.1080/01621459.1948.10483254

Dlott, D.D. 2022. Laser pulses into bullets: Tabletop shock experiments. *Phys. Chem. Chem. Phys.*, 24, 10653–10666. https://doi.org/10.1039/D2CP00418F

Dremin, A.N., Koldunov, S.A. & Shvedov, K.K. 1971. Shock-initiated detonation of bulk-density charges. *Combust. Explos. Shock Waves*, 7, 87–92. https://doi.org/10.1007/BF00748921

Dremin, A.N. & Pokhil, P.F. 1959. The detonation wave parameters of trotyl, hexogen, nitroglycerin and nitromethane. *Proc. Acad. Sci. USSR: Phys. Chem. Sect.*, 128, 839–841.

Eftis, J., Carrasco, C. & Osegueda, R.A. 2003. A constitutive-microdamage model to simulate hypervelocity projectile-target impact, material damage and fracture. *Int. J. Plast.*, 19, 1321–1354. https://doi.org/10.1016/S0749-6419(02)00036-0

Eissler, M. 1884. *The Modern High Explosives Nitro-Glycerine and Dynamite: Their Manufacture, Their Use, and Their Application to Mining and Military Engineering*, New York: Wiley.

Eley, D.D. & Pepper, D.C. 1947. The viscoelastic behaviour of cordite. 1: Plastic flow in compression at high stresses. *Trans. Farad. Soc.*, 43, 559–568. https://doi.org/10.1039/TF9474300559

Erkman, J.O. & Edwards, D.J. 1976. Computed and experimental Hugoniots for unreacted porous high explosives. *In:* Edwards, D.J. (ed.) *Proceedings Sixth Symposium (International) on Detonation*. pp. 766–776, Arlington, VA: Office of Naval Research.

Erlich, D.C. 1985. Rod impact (Taylor) test. *In: Metals Handbook. Volume 8* (Ninth Edition). pp. 203–207, Metals Park, OH: American Society of Metals.

Ermolaev, B.S., Belyaev, A.A., Viktorov, S.B., Slepstov, K.A. & Zharikova, S.U. 2010. Nonideal regimes of deflagration and detonation of black powder. *Russ. J. Phys. Chem. B*, 4, 428–439. https://doi.org/10.1134/S1990793110030103

Fong, C.W. 1985. Crack initiation in perforated propellants under high strain rate impact conditions. *Propell. Explos. Pyrotech.*, 10, 91–96. https://doi.org/10.1002/prep.19850100402

Fong, C.W., Warren, R.C. & Odgers, S.G. 1983. The effect of nitroguanidine particle size and orientation on the fracture toughness of M30 propellant and the relationship with propellant grain fracture during gun firings. *J. Ballistics*, 7, 1832–1889.

Fortov, V.E., Lebedev, E.F., Luzganov, S.N., Kozlov, A.V., Medin, S.A., Parshikov, A.N., Politschook, V.P. & Shurupov, A.V. 2006. Railgun experiment and computer simulation of hypervelocity impact of lexan projectile on aluminum target. *Int. J. Impact Eng.*, 33, 254–263. https://doi.org/10.1016/j.ijimpeng.2006.09.008

Foteenkov, V.A., Korotkov, A.I., Ermolaev, B.S. & Sulimov, A.A. 1982. Propagation of convective combustion in poured-density explosives. *Combust. Explos. Shock Waves*, 18, 250–252. https://doi.org/10.1007/BF00789627

Foteenkov, V.A., Sulimov, A.A. & Korotkov, A.I. 1991. Explosion development in charges of black powder. *In: Proceedings 16th International Pyrotechnics Seminar*. pp. 121–130, Sundyberg: Royal Swedish Academy of Sciences.

Fugelso, E., Jacobson, J.D., Karpp, R.R. & Jensen, R. 1982. Radiographic study of impact in polymer-bonded explosives. *AIP Conf. Proc.*, 78, 607–612. https://doi.org/10.1063/1.33283

Garn, W.B. 1958. Determination of the unreacted Hugoniot for liquid TNT. *J. Chem. Phys.*, 30, 819–822. https://doi.org/10.1063/1.1730051

Gary, G., Klepaczko, J.R. & Zhao, H. 1995. Generalization of split Hopkinson bar technique to use viscoelastic materials. *Int. J. Impact Eng.*, 16, 529–530. https://doi.org/10.1016/0734-743X(94)90166-I.

Gathers, G.R. 1994. *Selected Topics in Shock Wave Physics and Equation of State Modeling*, Singapore: World Scientific.

Gazonas, G.A. 1991. The mechanical response of M30, XM39, and JA2 propellants at strain rates from 10^{-2} to 250 s^{-1}. Aberdeen, MD, Ballistic Research Laboratory. Report no. BRL-TR-3181.

Ge, S.Y., Yuan, S., Yang, G.L., Zhu, Q., Li, S.B. & Dou, Y.S. 2021. Stress wave propagation and fracture formation in HMX-estane polymer bonded explosive under impact loading: A mesoscale numerical simulation. *J. Appl. Phys.*, 129, 165104. https://doi.org/10.1063/5.0040824

Gibbs, T. & Popalato, A. 1980. *LASL Explosive Property Data*, Berkeley, CA: University of California Press.

Golubev, V. 2007. Influence of conditions of shock wave loading on spall fracture of samples of four high explosive compositions. *In:* Ottis, J. & Krupka, M. (eds.) *Proceedings 10th Seminar on New Trends in Research of Energetic Materials*. pp. 616–621, Pardubice: University of Pardubice.

Golubev, V.K., Novikov, S.A. & Sobolev, Y.S. 1995. Effect of heating on the cleavage fracture of certain explosive compounds. *Combust. Explos. Shock Waves*, 31, 617–619. https://doi.org/10.1007/BF00743814

Golubev, V.K. & Pogorelov, A.P. 1999. Effect of loading conditions and initial state on the spall fracture of an explosive RDX-based compound. *J. Appl. Mech. Tech. Phys.*, 40, 791–795. https://doi.org/10.1007/BF02468460

Golubev, V.K. & Pogorelov, A.P. 2000. Effect of loading conditions and the initial state on the spalling of an octogen-based high explosive. *Strength Mater.*, 32, 136–140. https://doi.org/10.1007/BF02511671

Goranson, R.W., Bancroft, D., Burton, B.L., Blechar, T., Houston, E.E., Gittings, E.F. & Landeen, S.A. 1955. Dynamic determination of the compressibilities of metals. *J. Appl. Phys.*, 26, 1472–1479. https://doi.org/10.1063/1.1721933

Gottfried, J.L. 2014. Influence of exothermic chemical reactions on laser-induced shock waves. *Phys. Chem. Chem. Phys.*,16, 21452–21466. https://doi.org/10.1039/C4CP02903H

Gottfried, J.L. 2015. Laboratory-scale method for estimating explosive performance from laser-induced shock waves. *Propell. Explos. Pyrotech.*, 40, 674–681. https://doi.org/10.1002/prep.201400302

Grady, D.E. 2006. Comparison of hypervelocity fragmentation and spall experiments with Tuler-Butcher spall and fragment size criteria. *Int. J. Impact Eng.*, 33, 305–315. https://doi.org/10.1016/j.ijimpeng.2006.09.064

Grantham, S.G., Siviour, C.R., Proud, W.G. & Field, J.E. 2004. High-strain rate Brazilian testing of an explosive simulant using speckle metrology. *Meas. Sci. Technol.*, 15, 1867–1870. https://doi.org/10.1088/0957-0233/15/9/025

Gray, G.T, III. 2000. Classic split-Hopkinson pressure bar testing. *In:* Kuhn, H. & Medlin, D. (eds.) *ASM Handbook. Vol. 8: Mechanical Testing and Evaluation*. pp. 462–476, Materials Park, OH: ASM International. https://doi.org/10.31399/asm.hb.v08.a0003296

Gray, G.T., III, Idar, D.J., Blumenthal, W.R., Cady, C.M. & Peterson, P.D. 2000. High- and low-strain rate compression properties of several energetic material composites as a function of strain rate and temperature. *In:* Short, J.M. & Kennedy, J.E. (eds.) *Proceedings 11th International Detonation Symposium*. pp. 76–84, Arlington, VA: Office of Naval Research.

Gruau, C., Picart, D., Belmas, R., Bouton, E., Delmaire-Sizes, F., Sabatier, J. & Trumel, H. 2009. Ignition of a confined high explosive under low velocity impact. *Int. J. Impact Eng.*, 36, 537–550. https://doi.org/10.1016/j.ijimpeng.2008.08.002

Guo, H.-G., Su, C.-H., Cai, Y.-Q., He, S., Yu, Q.-B. & Wang, H.F. 2023. Reactive jet density distribution effect on its penetration behavior. *Defence Technol.*, 24, 190–202. https://doi.org/10.1016/j.dt.2022.03.001

Guo, H.-G., Zheng, Y.-F., Yu, Q.-B., Ge, C. & Wang, H.-F. 2019. Penetration behavior of reactive liner shaped charge jet impacting thick steel plates. *Int. J. Impact Eng.*, 126, 76–84. https://doi.org/10.1016/j.ijimpeng.2018.12.005

Gust, W.H. 1982. High impact deformation of metal cylinders at elevated temperatures. *J. Appl. Phys.*, 53, 3566–3575. https://doi.org/10.1063/1.331136

Hagan, J.T. 1981. Impossibility of fragmenting small particles: Brittle-ductile transition. *J. Mater. Sci.*, 16, 2909–2911. https://doi.org/10.1007/BF02402857

Heijboer, J. 1968. Dynamic mechanical properties and impact strength. *J. Polym. Sci. C: Polym. Symp.*, 16, 3755–3763. https://doi.org/10.1002/polc.5070160716

Helte, A. & Lundgren, J. 2014. Spall initiation of confined explosives. *In:* Ames, R.G. & Boeka, R.D. (eds.) *Proceedings 28th International Symposium on Ballistics*. pp. 247–254, Lancaster, PA: DesTech.

Ho, S.Y. 1992. Impact ignition mechanisms of rocket propellants. *Combust. Flame*, 91, 131–142. https://doi.org/10.1016/0010-2180(92)90095-7

Ho, S.Y. & Fong, C.W. 1987a. Correlation between fracture properties and dynamic mechanical relaxations in composite propellants. *Polymer*, 28, 739–744. https://doi.org/10.1016/0032-3861(87)90222-9

Ho, S.Y. & Fong, C.W. 1987b. Temperature dependence of high strain-rate impact fracture behaviour in highly filled polymeric composite and plasticized thermoplastic propellants. *J. Mater. Sci.*, 22, 3023–3031. https://doi.org/10.1007/BF01086508

Hoge, K.G. 1967. The behavior of PBXs under dynamic compressive loads. *Appl. Polym. Symp.*, 5, 19–39.

Hoge, K.G. 1970. Dynamic tensile strength of explosive materials. *Explosivstoffe*, 18, 39–41.

Hondros, G. 1959. The evaluation of Poisson's ratio and the modulus of materials of a low tensile resistance by the Brazilian (indirect tension) test with particular reference to concrete. *Australian J. Appl. Sci.*, 10, 243–268.

Hopkinson, B. 1914a. The effects of the detonation of gun-cotton. *Trans. North-East Coast Inst. Eng. Shipbuilders*, 30, 199–217.

Hopkinson, B. 1914b. A method of measuring the pressure produced in the detonation of high explosives or by the impact of bullets. *Phil. Trans. R. Soc. Lond. A*, 213, 437–456. https://doi.org/10.1098/rspa.1914.0008

Hopkinson, B. 1921a. The effects of the detonation of gun cotton. *In:* Ewing, J.A. & Larmor, J. (eds.) *The Collected Scientific Papers of Bertram Hopkinson*. pp. 461–474, Cambridge: Cambridge University Press.

Hopkinson, B. 1921b. The pressure of a blow. *In:* Ewing, J.A. & Larmor, J. (eds.) *The Scientific Papers of Bertram Hopkinson*. pp. 423–437, Cambridge: Cambridge University Press.

Hörz, F., Cintala, M., Bernhard, R.P. & See, T.H. 1994. Dimensionally scaled penetration experiments: Aluminum targets and glass projectiles 50 μm to 3.2 mm in diameter. *Int. J. Impact Eng.*, 15, 257–280. https://doi.org/10.1016/S0734-743X(05)80017-X

Huang, F. & Ding, J. 1999. Dynamic fracture of solid propellant. *In:* Zhang, S., Zhao, Y., Li, C., Zhou, Z. & Sun, J. (eds.) *Theory and Practice of Energetic Materials. 3*. pp. 118–123, Chengdu: Sichuan Publishing House of Science and Technology.

Hughes, C.T., Curtis, J.P., Reaugh, J.E. & Jones, A.G. 2015. Explosive response to pinch arising from low speed spigot impact. *In:* Carney, J. & Maienschein, J. (eds.) *Proceedings 15th International Detonation Symposium*. pp. 683–692, Arlington, VA: Office of Naval Research.

Hugoniot, P.H. 1888. Mémoire sur la propagation du mouvement dans les corps et spécialement dans les gaz parfaits. 1. *J. École Polytechnique*, 57, 3–97.

Hugoniot, P.H. 1889. Mémoire sur la propagation du mouvement dans les corps et spécialement dans les gaz parfaits. 2. *J. École Polytechnique*, 58, 1–125.

Hugoniot, P.H. 1998. On the propagation of motion in bodies and in perfect gases in particular. *In:* Johnson, J.N. & Chéret, R. (eds.) *Classic Papers in Shock Compression Science*. pp. 161–358, Berlin: Springer. https://doi.org/10.1007/978-1-4612-2218-7_7

Hung, Y.Y. & Hovanesian, J.D. 1972. Full-field surface-strain and displacement analysis of three-dimensional objects by speckle interferometry. *Exp. Mech.*, 12, 454–460. https://doi.org/10.1007/BF02328815

Ichiyama, R.T., Scarpelli, N.P., Stanisha, M., Smith, J.L., Anderson, P.E. & Oxley, J.C. 2025. Simulations and experiments of photon Doppler velocimetry diagnosed wedge tests. *Propell. Explos. Pyrotech.*, 50, e12046. https://doi.org/10.1002/prep.12046

Ilyukhin, V.S., Pokhil, P.F., Rozanov, O.K. & Shvedova, N.S. 1960. Measurement of the shock adiabatics of cast trinitrotoluene, crystalline hexogene, and nitromethane. *Soviet Phys. Dokl.*, 5, 337–340.

James, E. & Smith, L.C. 1952. Plastic bonded RDX. Los Alamos, NM, Los Alamos National Laboratory. Report no. LA-UR-98-3473.

Jiao, Q., Ding, J. & Liang, Y.-M. 1985. Hugoniots and reaction rates from EMV gauge measurements and Lagrange analysis. *In:* Short, J.M. (ed.) *Proceedings Eighth Symposium (International) on Detonation*. pp. 83–88, White Oak, MD: Naval Surface Weapons Center.

Johnson, H.D. 1979. Mechanical properties of LX-10-1 evaluated with the diametric disc test. Amarillo, TX, Mason and Hanger-Silas Mason Co. Inc., Report no. MHSMP-79-08

Jun-Ling, L.I., Hua, F.U., Duo-Wang, T.A.N., Fang-Yun, L.U. & Rong, C.H.E.N. 2012. Fracture behaviour investigation into a PBX. *Strain*, 48, 463–473. https://doi.org/10.1111/j.1475-1305.2012.00842.x.

Kendall, K. 1978. The impossibility of comminuting small particles by compression. *Nature*, 272, 710–711. https://doi.org/10.1038/272710a0

Kendall, M.J. & Siviour, C.R. 2015. Experimentally simulating high rate composite deformation in tension and compression: PBX simulant. *J. Dyn. Behav. Mater.*, 1, 114–123. https://doi.org/10.1007/s40870-015-0018-2

Kennelly, A.E. & Kurokawa, K. 1921. Acoustic impedance and its measurement. *Proc. Amer. Acad. Arts Sci.*, 56, 3–42. https://doi.org/10.2307/20025828

Kick, F. 1885a. *Das Gesetz der Proportionalen Widerstände und seine Anwendungen* (The Law of Proportional Resistances and Its Applications), Leipzig: Verlag von Arthur Felix.

Kick, F. 1885b. Uber das Verhalten verschiedener Materialien bei gleicher Einwirkung äusserer Kräfte auf congruente Probestücke und über den Einfluss der Geschwindigkeit der Formänderungen auf den Widerstand oder die Beziehungen von Druck und Stoss (The behaviour of test pieces of various materials of the same shape subjected to the same external forces, the influence of the rate of deformation on their resistance, or the relationships between pressure and impact). *In: Das Gesetz der Proportionalen Widerstände und seine Anwendungen (The Law of Proportional Resistances and Its Applications).* pp. 75–113, Leipzig: Verlag von Arthur Felix.

Koenen, H., Ide, K.H. & Swart, K.-H. 1961. Sicherheitstechnische Kenndaten explosionsfähiger Stoffe. 1: Prüfverfahren der Bundesanstalt für Materialprüfung (BAM) Berlin-Dahlem (Safety-related characteristics of explosive Materials. 1: Test procedure of the Federal Institute for Materials Testing, Berlin-Dahlem). *Explosivstoffe*, 9, 4–13, 30–42.

Kolsky, H. 1949. An investigation of the mechanical properties of materials at very high rates of loading. *Proc. Phys. Soc. Lond. B*, 62, 676–700. https://doi.org/10.1088/0370-1301/62/11/302

Kolsky, H. 1953. Fractures produced by stress waves. *In: Stress Waves in Solids*. pp. 183–196, Oxford: Clarendon Press.

Kolsky, H. 1959. Fractures produced by stress waves. *In:* Averbach, B.L., Felbeck, D.K., Hahn, G.T. & Thomas, D.A. (eds.) *Fracture*. pp. 281–296, New York: John Wiley.

Kolsky, H. 1974. Wave propagation effects and fracture. *Inst. Phys. Conf. Ser.*, 21, 199–214.

Kolsky, H. & Rader, D. 1968. Stress waves and fracture. *In:* Liebowitz, H. (ed.) *Fracture: An Advanced Treatise. 1: Microscopic and Macroscopic Fundamentals*. pp. 533–569, New York: Academic.

Kolsky, H. & Shearman, A.C. 1949. Investigation of fractures produced by transient stress waves. *Research*, 2, 384–389.

Künzel, M., Kucera, J., Smid, J. & Horkel, J. 2024. Verification of impact energy delivered by a dropweight. *Central Eur. J. Energ. Mater.*, 21, 171–187. https://doi.org/10.22211/cejem/190439

Kuo, K.K. & Summerfield, M. 1975. High speed combustion of mobile granular solid propellants: Wave structure and the equivalent Rankine-Hugoniot relation. *In: Proceedings 15th Symposium (International) on Combustion*. pp. 515–527, Pittsburgh, PA: The Combustion Institute.

Land, W.E. 1961. Plastic bonded explosives rocket warhead. *United States Patent no.* 2,998,772.

Landon, J.W. & Quinney, H. 1923. Experiments with the Hopkinson pressure bar. *Proc. R. Soc. Lond. A*, 103, 622–643. https://doi.org/10.1098/rspa.1923.0084

Lanzerotti, M.Y.D. & Sharma, J. 1981. Brittle behavior of explosives during high acceleration. *Appl. Phys. Lett.*, 39, 455–457. https://doi.org/10.1063/1.92737

Le, V.D., Gratton, M., Caliez, M., Frachon, A. & Picart, D. 2010. Experimental mechanical characterization of PBXs. *J. Mater. Sci.*, 45, 5802–5813. https://doi.org/10.1007/s10853-010-4655-5

Lemar, E.R., Forbes, J.W. & Watt, J.W. 1984. Unreacted Hugoniot and shock initiation studies of composition B-3 and PBXW-109(I). *In:* Asay, J.R., Graham, R.A. & Straub, G.K. (eds.) *Shock Waves in Condensed Matter – 1983*. pp. 539–541, New York: Elsevier.

Lemar, E.R., Forbes, J.W., Watt, J.W. & Elban, W.L. 1985. Unreacted Hugoniot of composition B-3 for stresses of 0-16 kbar. *J. Appl. Phys.*, 58, 3404–3408. https://doi.org/10.1063/1.335758

Lieb, R.J. & Rocchio, J.J. 1983. Standardization of a dropweight mechanical properties tester for gun propellants. Aberdeen, MD, Ballistics Research Laboratory. Report no. ARBRL-TR-02516

Liu, R., Wang, X.J., Chen, P.W., Kang, G., Zhu, S.P. & Guo, Y.S. 2021. The role of tension-compression asymmetrical microcrack evolution in the ignition of PBXs under low-velocity impact. *J. Appl. Phys.*, 129, 175108. https://doi.org/10.1063/5.0046011

Liu, W., Wu, L.Z., Sun, X.X., Shen, R.Q. & Ye, Y.H. 2015. PTFE energy absorber of explosive initiator under high-g loading. *Adv. Eng. Res.*, 2, 425–428. https://doi.org/10.2991/iea-15.2015.104

Liu, W.B., Duan, Z.P., Liu, Y., Zhou, T.T. & Huang, F.L. 2024. Numerical simulation of the damage and ignition responses of high explosives under low-velocity impact using the SPH method. *Eng. Anal. Bound. Elem.*, 166, 105830. https://doi.org/10.1016/j.enganabound.2024.105830

Ma, D., Chen, P., Dai, K. & Zhou, Q. 2014. Specimen size effect of explosive sensitivity under low velocity impact. *J. Phys.: Conf. Ser.*, 500, 052026. https://doi.org/10.1088/1742-6596/500/5/052026

Ma, D.Z., Chen, P.W., Zhou, Q. & Dai, K.D. 2013a. Ignition criterion and safety prediction of explosives under low velocity impact. *J. Appl. Phys.*, 114, 113505. https://doi.org/10.1063/1.4821431

Ma, W.H., Wang, N.F., Sui, X., Li, S.P. & Xie, K. 2018. Thermal safety of a gun-launched missile's solid rocket motor under conditions of high environmental temperature and overloaded forces. *Propell. Explos. Pyrotech.*, 43, 1277–1286. https://doi.org/10.1002/prep.201800050

Ma, X., Zhang, D.Z., Giguere, P.T. & Liu, C. 2013b. Axisymmetric computation of Taylor cylinder impacts of ductile and brittle materials using original and dual domain material point methods. *Int. J. Impact Eng.*, 54, 96–104. https://doi.org/10.1016/j.ijimpeng.2012.11.001

Majendie, V.D. 1877. A list of some remarkable and important explosions of gunpowder which are recorded as having occurred previous to 1866. *Min. Proc. R.A. Inst.*, 9, 246–250.

Marsh, S.P. 1980. High explosives, high-explosive simulants, and propellants. *In: LASL Shock Hugoniot Data.* pp. 591–651, Berkeley, CA: University of California Press.

Mayer, J. 1868a. Nitroglycerine: Its claims as a new industrial agent. 1. *J. Franklin Inst.*, 86, 40–45. https://doi.org/10.1016/0016-0032(68)90352-9

Mayer, J. 1868b. Nitroglycerine: Its claims as a new industrial agent. 2. *J. Franklin Inst.*, 86, 119–130. https://doi.org/10.1016/0016-0032(68)90180-4

Moxnes, J.F. & Ødegårdstuen, G. 1998. Prediction of deformation of the powder within the nose cap of an MP-projectile during setback and spin. *In:* van Niekerk, C. (ed.) *Proceedings 17th International Symposium on Ballistics. Volume 2.* pp. 57–66, San Antonio, TX: International Ballistics Society.

Muhr, M., Klapötke, T.M. & Holl, G. 2022. Sensory monitoring of drop hammer experiments with multivariate statistics. *Propell. Explos. Pyrotech.*, 47, e202200025. https://doi.org/10.1002/prep.202200025

Murri, W.J., Curran, D.R. & Seaman, L. 1982. Fracture model for high energy propellant. *AIP Conf. Proc.*, 78, 460–464. https://doi.org/10.1063/1.33305

Nance, D.V. 2017. Analysis of sensitivity experiments: An expanded primer. Eglin, FL, Air Force Research Laboratory. Report no. AFRL-RW-EG-TP-2017-001.

Napadensky, H.S., Eichler, T.V., Kot, C.A. & Zaker, T.A. 1970. Deformation of a cylinder of explosive material in unconfined impact. *In: Proceedings Fifth Symposium (International) on Detonation.* pp. 313–320, Arlington, VA: Office of Naval Research.

Napier, S.E., Powell, W.J. & Ubbelohde, A.R. 1948. The sensitiveness of initiators to friction: Apparatus for measuring relative sensitiveness to grazing from with or without grit. *Phil. Trans. R. Soc. Lond. A*, 241, 265–272. https://doi.org/10.1098/rsta.1948.0017

Neyer, B.T. 1991. Sensitivity testing and analysis. *In: Proceedings 16th International Pyrotechnics Seminar.* pp. 87–106, Sundyberg: Royal Swedish Academy of Sciences.

Neyer, B.T. 1994a. A D-optimality-based sensitivity test. *Technometrics*, 36, 61–70. https://doi.org/10.1080/00401706.1994.10485401

Neyer, B.T. 1994b. How to learn more from sensitivity tests. *In: Proceedings 15th Symposium on Explosives and Pyrotechnics.* Paper 13, Oaks, PA: Franklin Applied Physics.

Neyer, B.T. 1997. $\mu \pm k\sigma$ analysis of various threshold tests. *In: Proceedings 16th Symposium on Explosives and Pyrotechnics.* Paper 13, Oaks, PA: Franklin Applied Physics.

Novikov, S.A. 2000. Mechanical properties of solid high explosives under intense shock loading. *Chem. Phys. Rep.*, 18, 1955–1968.

Ostrik, A.V. & Petrovskii, V.P. 1995. Two-dimensional spalling fracture of polymeric cylindrical constructions. *Chem. Phys. Rep.*, 14, 2–9.

Othman, R. (ed.) 2018. *The Kolsky-Hopkinson Bar Machine: Selected Topics*, Berlin: Springer. https://doi.org/10.1007/978-3-319-71919-1

Palmer, S.J.P., Field, J.E. & Huntley, J.M. 1993. Deformation, strengths and strains to failure of polymer bonded explosives. *Proc. R. Soc. Lond. A*, 440, 399–419. https://doi.org/10.1098/rspa.1993.0023

Parab, N.D., Roberts, Z.A., Harr, M.H., Mares, J.O., Casey, A.S., Gunduz, I.E., Hudspeth, M., Claus, B., Sun, T., Fezzaa, K., Son, S.F. & Chen, W.W. 2016. High speed X-ray phase contrast imaging of energetic composites under dynamic compression. *Appl. Phys. Lett.*, 109, 131903. https://doi.org/10.1063/1.4963137

Paulson, S.C., Roberts, Z.A., Sorensen, C.J., Kerschen, N.E., Harr, M.H., Parab, N.D., Sun, T., Fezzaa, K., Son, S.F. & Chen, W.W. 2020. Observation of damage during dynamic compression of production and low-defect HMX crystals in Sylgard binder using X-ray phase contrast imaging. *J. Dyn. Behav. Mater.*, 6, 34–44. https://doi.org/10.1007/s40870-019-00225-8

Perry, J.I. & Walley, S.M. 2022. Vibration-induced heating of energetic materials: A review. *J. Dyn. Behav. Mater.*, 8, 162–177. https://doi.org/10.1007/s40870-021-00322-7

Picart, D., Biessy, M. & Brigolle, J.L. 2009. Intermediate strain rate characterization of a plastic bonded explosive composition. *In*: Dyckmans, G. (ed.) *Proceedings 9th International Conference on the Mechanical and Physical Behaviour of Materials under Dynamic Loading (DYMAT 2009)*. pp. 343–348, Les Ulis: EDP Sciences. https://doi.org/10.1051/dymat/2009048

Picart, D., Tchikladze, N. & Camus, S. 2023. Mechanical behaviour of pressed HMX and TATB-based PBXs during low-velocity impact. *J. Dyn. Behav. Mater.*, 9, 345–364. https://doi.org/10.1007/s40870-023-00390-x

Porter, D. 1995. *Group Interaction Modelling of Polymer Properties*, New York: Marcel Dekker.

Porter, D. & Gould, P.J. 2009. Predictive nonlinear constitutive relations in polymers through loss history. *Int. J. Solids Struct.*, 46, 1981–1993. https://doi.org/10.1016/j.ijsolstr.2009.01.012

Preston, D.N., Brown, G.W., Sandstrom, M.M., Pollard, C.J., Warner, K.F., Remmers, D.L., Phillips, J.J., Shelley, T.J., Reyes, J.A., Hsu, P.C. & Reynolds, J.G. 2016. Small-scale safety testing of ammonium nitrate and mixtures. *Propell. Explos. Pyrotech.*, 41, 9–13. https://doi.org/10.1002/prep.201500124

Preston, D.N., Brown, G.W., Skidmore, C.B., Reardon, B.L. & Parkinson, D.A. 2012. Small-scale explosives sensitivity safety testing: A departure from Bruceton. *AIP Conf. Proc.*, 1426, 713–716. https://doi.org/10.1063/1.3686378

Quidot, M. 1988. Dynamic fragmentation of compact energetic materials. *In:* Chiem, C.Y., Kunze, H.-D. & Meyer, L.W. (eds.) *Impact Loading and Dynamic Behaviour of Materials*. pp. 609–614, Oberursel: DGM Informationsgesellschaft mbH.

Quidot, M., Bertrand, A. & Bonthoux, F. 1985. Étude de la fragmentation de matériaux energétiques compacts (Study of the fragmentation of compact energetic materials). *J. Phys. France Colloq.*, 46(C5), 425–433. https://doi.org/10.1051/jphyscol:1985553

Quidot, M. & Chabin, P. 2007. Constitutive model and equation of state of unreacted PBXN109. *In:* Peiris, S. & Doherty, R. (eds.) *Proceedings Thirteenth International Detonation Symposium*. pp. 559–565, Arlington, VA: Office of Naval Research.

Rae, P.J., Goldrein, H.T., Palmer, S.J.P., Field, J.E. & Lewis, A.L. 2002. Quasistatic studies of the deformation and failure of β-HMX based polymer bonded explosives. *Proc. R. Soc. Lond. A*, 458, 743–762. https://doi.org/10.1098/rspa.2001.0894

Ravindran, S., Gupta, V. & Kidane, A. 2023. Mechanism of hotspot formation in energetic materials under mild impact loading. *J. Dyn. Behav. Mater.*, 9, 375–383. https://doi.org/10.1007/s40870-023-00396-5

Rayleigh. 1910. Aerial plane waves of finite amplitude. *Proc. R. Soc. Lond. A*, 84, 247–284. https://doi.org/10.1098/rspa.1910.0075

Reaugh, J.E., Vandersall, K.S., Jones, A.G. & Curtis, J.P. 2020. Computer simulations to study the post-ignition violence of HMX explosives in the Steven test. *In: Proceedings 16th International Detonation Symposium*. pp. 830–837, Arlington, VA: Office of Naval Research.

Régnauld, P. 1927. La fragilité de l'acier (The fragility of steel). *Rev. Métall. Mémoires*, 24, 509–515. https://doi.org/10.1051/metal/192724090509

Reitsma, H.J. 2001. The explosion of a ship, loaded with black powder, in Leiden in 1807. *Int. J. Impact Eng.*, 25, 507–514. https://doi.org/10.1016/S0734-743X(00)00067-1

Rinehart, J.S. 1964. Fracturing by spalling. *Wear*, 7, 315–329. https://doi.org/10.1016/0043-1648(64)90224-8

Rinehart, J.S. & Pearson, J. 1954. Fracturing under impulsive loading. *In: Behavior of Metals under Impulsive Loads*. pp. 123–145, Cleveland, OH: American Society of Metals.

Robertson, R. 1921a. Some properties of explosives. *J. Chem. Soc. Trans.*, 119, 1–29. https://doi.org/10.1039/CT9211900001

Robertson, R. 1921b. Some war developments of explosives. *Nature*, 107, 524–527. https://doi.org/10.1038/107524a0

Rollin, B. & Ressler, J.J. 2024. Numerical investigation of a high explosive response to a fragment impact. *AIP Conf. Proc.*, 3066, 610007. https://doi.org/10.1063/12.0028527

Rosenberg, Z. & Dekel, E. 2020. Terminal Ballistics (Third Edition), Berlin: Springer. https://doi.org/10.1007/978-3-030-46612-1

Rosenberg, Z., Partom, Y. & Keren, B. 1983. Calibration of commercial manganin stress gauges under static uniaxial strain conditions. *J. Appl. Phys.*, 54, 2824–2826. https://doi.org/10.1063/1.332315

Rosenberg, Z., Partom, Y. & Yaziv, D. 1981. The response of manganin gauges shock loaded in the 2D straining mode. *J. Appl. Phys.*, 52, 755–758. https://doi.org/10.1063/1.328758

Rosenberg, Z., Yaziv, D. & Partom, Y. 1980. Direct measurement of strain in plane impact experiments by a longitudinal resistance gauge. *J. Appl. Phys.*, 51, 4790–4798. https://doi.org/10.1063/1.328311

Roth, J. 1970. Shock sensitivity and shock Hugoniots of high-density granular explosives. *In: Proceedings Fifth Symposium (International) on Detonation*. pp. 219–230, Arlington, VA: Office of Naval Research.

Sandusky, H.W. & Felts, J.E. 2025. Survivability of TNT-based explosives with dynamically collapsing cavities. *In:* Mason, M.H. (ed.) *Proceedings 17th International Detonation Symposium*. pp. 432–442, Arlington, VA: Office of Naval Research.

Schoenbein, P. 1847. On the discovery of guncotton. *Philos. Mag. (Ser. 3)*, 31, 7–12. https://doi.org/10.1080/14786444708645774

Seaman, L., Curran, D.R. & Murri, W.J. 1985. A continuum model for dynamic tensile microfracture and fragmentation. *Trans. ASME: J. Appl. Mech.*, 52, 593–600. https://doi.org/10.1115/1.3169106

Seaman, L. & Erlich, D.C. 1991. Fracture in rocket propellant. *J. Phys. IV France*, 1(C3), 557–564. https://doi.org/10.1051/jp4:1991378

Shen, Y.B. 2012. The chemical and mechanical behaviors of polymer/reactive metal systems under high strain rates. PhD thesis, Georgia Institute of Technology.

Shepherd, W.C.F. 1956. Strength of high explosives and effects due to shape. *In:* Bawn, C.E.H. & Rotter, G. (eds.) *Science of Explosives*. pp. 290–406, 533–534, London: Her Majesty's Stationery Office.

Siviour, C.R., Grantham, S.G., Williamson, D.M., Proud, W.G. & Field, J.E. 2009. Novel methods of material properties at high rates of strain using speckle metrology. *Imaging Sci. J.*, 57, 326–332. https://doi.org/10.1179/174313109X454783

Speth, J.D. 1972. Mechanical basis of percussion flaking. *Am. Antiq.*, 37, 34–60. https://doi.org/10.2307/278884

Subbotin, S.G., Proskurin, A.V. & Melnikova, A.Y. 2009. Long-term shock loading of solids by a combined charge with explosive initiation. *J. Appl. Mech. Tech. Phys.*, 50, 547–552. https://doi.org/10.1007/s10808-009-0073-3

Sun, C.X., Xu, J.S., Chen, X., Zheng, J., Zheng, Y. & Wang, W.Q. 2015. Strain rate and temperature dependence of the compressive behavior of a composite modified double-base propellant. *Mech. Mater.*, 89, 35–46. https://doi.org/10.1016/j.mechmat.2015.06.002

Taylor, G.I. 1946. The testing of materials at high rates of loading. *J. Inst. Civil Eng.*, 26, 486–519. https://doi.org/10.1680/ijoti.1946.13699

Taylor, G.I. 1948. The use of flat ended projectiles for determining yield stress. 1: Theoretical considerations. *Proc. R. Soc. Lond. A*, 194, 289–299. https://doi.org/10.1098/rspa.1948.0081

Taylor, G.I. & Davies, R.M. 1958. The mechanical properties of cordite during impact stressing. *In:* Batchelor, G.K. (ed.) *The Scientific Papers of Sir Geoffrey Ingram Taylor. Vol. 1: Mechanics of Solids*. pp. 480–495, Cambridge: Cambridge University Press.

Trimborn, F. 1970. Mechanische Messungen am großen Fallhammer der Bundesanstalt für Materialprüfung (BAM) (Mechanical measurements on the large impact Testing apparatus of the Federal Institute for Materials Testing). *Explosivstoffe*, 18, 49–56.

Turpeinen, T., Ahponen, J. & Vesterinen, A. 2025. A practical comparative investigation of Bruceton and 3POD2.0 statistical testing methods on friction and impact sensitivity testing. *J. Energ. Mater.*, https://doi.org/10.1080/07370652.2024.2425102

Vasilev, V.A., Bolkhovitinov, L.G. & Khristoforov, B.D. 1967. Dynamic compressibility of powdered TNT. *Combust. Explos. Shock Waves*, 3, 371. https://doi.org/10.1007/BF00741690

Vincent, P.I. 1972. A correlation between critical tensile strength and polymer cross-sectional area. *Polymer*, 13, 558–560. https://doi.org/10.1016/0032-3861(72)90116-4

von Rittinger, P.R. 1867. *Lehrbuch der Aufbereitungskunde in ihrer neuesten Entwicklung und Ausbildung Systematisch Dargestellt (Textbook of Processing Science in Its Latest Developments and Training Systematically Presented)*, Berlin: Ernst and Korn.

Vorthman, J.E. 1982. Facilities for the study of shock-induced decomposition of high explosives. *AIP Conf. Proc.*, 78, 680–684. https://doi.org/10.1063/1.33249

Wagner, K.B., Keyhani, A., Boddorff, A.K., Kennedy, G., Montaigne, D., Jensen, B.J., Beason, M., Zhou, M. & Thadhani, N.N. 2020. High-speed X-ray phase contrast imaging and digital image correlation analysis of microscale shock response of an additively manufactured energetic material simulant. *J. Appl. Phys.*, 127, 235902. https://doi.org/10.1063/5.0003525

Walley, S.M. 2020. Highways and byways in the history of high rate mechanical testing. *J. Dyn. Behav. Mater.*, 6, 113–158. https://doi.org/10.1007/s40870-020-00237-9

Walley, S.M. & Couque, H. 2024. The Taylor impact experiment. *In:* Hokka, M. (ed.) *Dynamic Behavior of Materials: Fundamentals, Material Models, and Microstructure Effects*. pp. 219–267, Amsterdam: Elsevier. https://doi.org/10.1016/B978-0-323-99153-7.00007-4

Walley, S.M. & Field, J.E. 1994. Strain rate sensitivity of polymers in compression from low to high strain rates. *DYMAT J.*, 1, 211–228.

Walley, S.M., Field, J.E., Biers, R.A., Proud, W.G., Williamson, D.M. & Jardine, A.P. 2015. The use of glass anvils in drop-weight studies of energetic materials. *Propell. Explos. Pyrotech.*, 40, 351–365. https://doi.org/10.1002/prep.201500043

Walley, S.M., Field, J.E. & Greenaway, M.W. 2006. Crystal sensitivities of energetic materials. *Mater. Sci. Technol.*, 22, 402–413. https://doi.org/10.1179/174328406X91122
Walley, S.M., Siviour, C.R., Drodge, D.R. & Williamson, D.M. 2010. High-rate mechanical properties of energetic materials. *JOM*, 62(1), 31–34. https://doi.org/10.1007/s11837-010-0007-9
Walley, S.M., Taylor, N.E. & Williamson, D.M. 2018. Temperature and strain rate effects on the mechanical properties of a polymer-bonded explosive. *EPJ Spec. Top.*, 227, 127–141. https://doi.org/10.1140/epjst/e2018-00060-6
Walsh, J.M. & Christian, R.H. 1955. Equation of state of metals from shock wave measurements. *Phys. Rev.*, 97, 1544–1556. https://doi.org/10.1103/PhysRev.97.1544
Walsh, J.M. & Rice, M.H. 1957. Dynamic compression of liquids from measurements of strong shock waves. *J. Chem. Phys.*, 26, 815–823. https://doi.org/10.1063/1.1743414
Walsh, J.M., Rice, M.H., McQueen, R.G. & Yarger, F.L. 1957. Shock-wave compressions of twenty-seven metals: Equations of state of metals. *Phys. Rev.*, 108, 196–216. https://doi.org/10.1103/PhysRev.108.196
Wang, Z.J., Qiang, H.F., Geng, B. & Geng, T.J. 2022. Numerical simulation of fragment impacting solid rocket motors. *AIP Adv.*, 12, 055204. https://doi.org/10.1063/5.0088412
Ward, A.G. 1956. The mechanical properties of solid propellants. *In:* Bawn, C.E.H. & Rotter, G. (eds.) *Science of Explosives.* pp. 793–814, 1052–1053, London: Her Majesty's Stationery Office.
Wasley, R.J. & Walker, F.E. 1969. Dynamic compressive behavior of a strain-rate sensitive, polycrystalline, organic solid. *J. Appl. Phys.*, 40, 2639–2648. https://doi.org/10.1063/1.1658047
Wei, Y.M., Xu, W.Z., Guo, F.W., Tan, X.P., Zheng, X., Zhao, L.Y., Zhao, Y.L. & Wang, J.Y. 2023. Experimental and numerical simulation investigation on dynamic fracture behavior of PBXs under different high-g impact conditions. *Int. J. Impact Eng.*, 181, 104735. https://doi.org/10.1016/j.ijimpeng.2023.104735
Wen, Y.-S., Dai, X.-G., Han, Y. & Xiang, Y. 2014. Reaction characteristic for various scale explosive under mild impact. *J. Energ. Mater.*, 32, S41–S50. https://doi.org/10.1080/07370652.2013.820808
Wilkins, M.L., Squier, B. & Halperin, B. 1965. Equation of state for detonation products of PBX 9404 and LX04-01. *In: Proceedings Tenth Symposium (International) on Combustion.* pp. 769–778, Pittsburgh, PA: The Combustion Institute. https://doi.org/10.1016/S0082-0784(65)80220-X.
Williamson, D.M. & Morley, O.J. 2023. A split Hopkinson pressure bar investigation of impact-induced reaction of HMX, RDX and PETN. *J. Dyn. Behav. Mater.*, 9, 384–401. https://doi.org/10.1007/s40870-023-00404-8
Williamson, D.M., Siviour, C.R., Proud, W.G., Palmer, S.J.P., Govier, R., Ellis, K., Blackwell, P. & Leppard, C. 2008. Temperature-time response of a polymer bonded explosive in compression (EDC37). *J. Phys. D: Appl. Phys.*, 41, 085404. https://doi.org/10.1088/0022-3727/41/8/085404
Wilson, M.P.W. 1967. The deformation of metals by hypervelocity impact: The kinetics of twinning. PhD thesis, Cambridge.
Wu, Y., Fan, Y.N., Chen, X.X., Wen, J.J., Wang, Q.L. & Huang, J.Y. 2023. Low/intermediate speed impact-induced ignition and damage of a novel high-energy solid propellant. *Propell. Explos. Pyrotech.*, 48, e202300056. https://doi.org/10.1002/prep.202300056
Xiao, H., Duan, Z.P., Liu, W.B., Bai, Z.L. & Huang, F.L. 2025. Experimental and numerical study on safety of shelled high explosives during drop. *J. Appl. Phys.*, 137, 225903. https://doi.org/10.1063/5.0270917
Xiao, Y.C., Fan, C.Y., Wang, Z.J. & Sun, Y. 2020. Coupled mechanical-thermal model for numerical simulations of PBXs under low-velocity impacts. *Propell. Explos. Pyrotech.*, 45, 823–832. https://doi.org/10.1002/prep.201900345
Xu, W.Z., Guo, F.W., Liang, X., Yan, T.L., Xu, Y., Deng, J., Li, Y.X. & Wang, J.Y. 2021. Dynamic response properties of polymer bonded explosives under different excitation by deceleration. *Mater. Des.*, 206, 109810. https://doi.org/10.1016/j.matdes.2021.109810
Xu, W.Z., Liang, X., Li, H., Deng, J., Gao, F.W., Li, Y.X., Yan, T.L. & Wang, J.Y. 2020. Experiment study on the influencing factors of mechanical response of HMX-based PBXs in the high-g deceleration environments. *J. Energ. Mater.*, 39, 33–47. https://doi.org/10.1080/07370652.2020.1750507
Xue, H.J., Wu, Y.Q. & Yang, K. 2024. Multiple mesoscopic hotspot mechanisms model and its applications in drop impact hazard simulations. *J. Energ. Mater.*, 42, 747–771. https://doi.org/10.1080/07370652.2023.2168311
Yang, K., Duan, H.Z., Wu, Y.Q., Hou, X. & Huang, F.L. 2022. Impact-induced deformation and ignition related to localized viscous shear flow heating for high-ductility composite energetic materials. *Mater. Des.*, 223, 111185. https://doi.org/10.1016/j.matdes.2022.111185
Yang, L. 2008. Surface polarity of β-HMX crystal and the related adhesive forces with adhesive binder. *Langmuir*, 24, 13477–13482. https://doi.org/10.1021/la802494b

Yeager, J.D., Dattelbaum, A.M., Orler, E.B., Bahr, D.F. & Dattelbaum, D.M. 2010. Adhesive properties of some fluoropolymer binders with the insensitive explosive TATB. *J. Colloid Interface Sci.*, 352, 535–541. https://doi.org/10.1016/j.jcis.2010.08.063

Zanegin, I.V. & Karachinskii, S.I. 2001. Detonation and physicochemical properties of cannon powders. *Combust. Explos. Shock Waves*, 37, 563–566. https://doi.org/10.1023/A:1012345104370

Zhang, Y.M., Wang, N.F., Chen, X.X., Wang, R., Bai, L., Dang, J.F. & Wu, Y. 2023. Investigation of the effect of axial gap on the mechanical response of a cartridge-loaded CMDB propellant grain under vibration loads. *Int. J. Aerosp. Eng.*, 2451808. https://doi.org/10.1155/2023/2451808

Zhang, Y.P., Liu, H., Yang, Z., Li, Q.K. & He, Y.H. 2019. Shock-induced hot spot formation and spalling in RDX containing a cube void. *ACS Omega*, 4, 8031–8038. https://doi.org/10.1021/acsomega.9b00589

Zhao, H., Gary, G. & Klepaczko, J.R. 1997. On the use of a viscoelastic split Hopkinson pressure bar. *Int. J. Impact Eng.*, 19, 319–330. https://doi.org/10.1016/S0734-743X(96)00038-3

Zhou, J.S., Zhen, L., Yang, D.Z. & Li, H.T. 2000. Macro- and microdamage behaviors of the 30CrMnSiA steel impacted by hypervelocity projectiles. *Mater. Sci. Eng. A*, 282, 177–182. https://doi.org/10.1016/S0921-5093(99)00754-6

Zhou, Z., Dai, K., Chen, J., Liu, Y. & Chen, P. 2015. Experimental study on dynamic tensile properties of PBXs. *In:* Carney, J. & Maienschein, J. (eds.) *Proceedings 15th International Detonation Symposium*. pp. 356–362, Arlington, VA: Office of Naval Research.

6 Environmental Impact of Explosive Materials Through Life

Joshua T. L. Wardrop, Federica Persico, and Tracey Temple

6.1 INTRODUCTION

The environmental impact of explosives extends far beyond their immediate destruction during conflicts and the production during manufacturing, with long-lasting effects on ecosystems, soil quality, air, and water contamination that often go unnoticed after their use. Explosives have been extensively used since their invention during the 9th century, with increasing safety/environmental incidents since the discovery of gunpowder (Skelly, 2022). Research into the environmental impact of explosives, considering their manufacture, use, and disposal, only began in the 1970s, following the increased awareness of the potential effects of anthropogenic activities on the ecosystems and the introduction of environmental legislation.

Explosive manufacturing is a significant source of emissions (to air, water, and ground). Industrial wastewater (known as effluent) from processing is one of the largest consistent streams. This water is likely to be contaminated with various nitrates and chemical by-products in addition to heavy metals. The most concerning risk is that this water may contaminate local domestic water supplies leading to potential public health concerns. However, this source (or waste stream) is relatively consistent in this industry and wastewater treatment methods are well established (Bhanot et al., 2020; Fawcett-Hirst et al., 2021). Air emissions include nitrogen oxides (NO_x), sulphur oxides (SO_x), ammonia (NH_3), volatile organic compounds (VOCs), and particulate matter (PM), with local-scale impacts on air pollution and environmental health and wider potential global contributions to global warming and ozone depletion (Sandham et al., 2013). Other sources of emissions from manufacturing include spills and leaks from equipment (i.e. loss of containment). Preventing these accidental emissions requires effective training and risk assessment procedures (Barbosa et al., 2018; Organisation for Economic Co-operation and Development (OECD), 2023), regular maintenance of equipment (U.S. Department of Labor, Occupational Safety and Health Administration (OSHA), 2017), and additional containment (i.e. chemically resistant bunds) (National Institute for Occupational Safety and Health (NIOSH), 2024).

Ground (or soil) pollution arises from manufacturing activities typically from direct spills or deposition of airborne or waterborne contaminants. It is therefore acknowledged that established explosive manufacturing sites typically have significant amounts of ground pollution. While emissions from industrial activities are increasingly regulated through environmental permits, especially in jurisdictions like the European Union (EU), less centralised enforcement regimes in other parts of the world present consistency challenges (Cardesa-Salzmann, 2012; Engel & Saleska, 2005).

Beyond manufacturing, the impact of explosives continues to be a major concern during their use. The detonation of explosives remains the most common way to release energetics in the environment. This chemical process is never fully complete, and this renders an unpredictable fate and transport of any residual explosive material released into an ecosystem (Hristov & Hristov, 2017). Research has shown that explosives during use can be transported through multiple environmental pathways, such as air, water, and soil. Literature reported that for many training ranges, ecosystem recovery in case of explosives and heavy metal contamination can take up to 42 years

DOI: 10.1201/9781003529217-6

(Jones & Schmitz, 2009; Zentelis et al., 2017). However, the complexity of the different ecosystems, together with the variability of environmental conditions worldwide, complicates the assessment of the extent and magnitude of the impacts of energetics. It is well documented that the use of traditional explosives, such as 2,4,6-trinitrotoluene (TNT) and 1,3,5-trinitroperhydro-1,3,5-triazine (RDX), has led to contamination of soil and groundwater(Clausen et al., 2004; Hewitt et al., 2005; Walsh et al., 2014). Moreover, as new insensitive high explosives (IHEs), such as 1,2,4-triazol-5-one (NTO) and 2,4-dinitroanilisole (DNAN), are increasingly being introduced into service, there is a rising concern regarding the need for more stable energetics being introduced in the environment. There is currently limited knowledge about their environmental impact (Persico et al., 2023).

Finally, disposal of explosive materials poses a significant environmental challenge. The disposal of end-of-life munitions and explosives, whether through open burning (OB) or inadequate storage, can exacerbate contamination risks in soil, water, and air (Christian et al., 2023; Vučićević et al., 2022). Efforts to develop more sustainable disposal methods, including environmentally safe decommissioning and recycling strategies, are crucial in mitigating these long-term impacts.

The aim of this chapter is to analyse the environmental impact of explosives throughout their lifecycle and to explore potential "green alternatives", focusing on future advancements and sustainable solutions.

6.2 ENVIRONMENTAL CONSIDERATIONS OF EXPLOSIVE MANUFACTURING PROCESSES

Generally, the manufacturing phase of the product lifecycle converts raw materials into a desired output. It often includes a wide range of industrial processes throughout the product's supply chain with various impacts on the environment (Duflou et al., 2012; Sandham et al., 2013). Once the product is produced, much of the potential environmental impact has already happened. Specific features of the design have consequences for in-service and disposal phases of the lifecycle, most of these are irreversibly embedded into the product at the manufacturing stage, for example the use of hazardous materials and the practicality of recycling are all determined at the earliest stages.

Figure 6.1 shows a simplified explosives manufacturing process highlighting environmental considerations. Whilst every manufacturing process will be unique and may have different stages, generally all methods will require inputs in the form of additional resources (energy, raw materials consumables, etc.). Each activity has undesired outputs, "waste" (including by-products from production and any emissions from the activities), while the desired output, "the product", progresses through the manufacturing system (or supply chain). During the manufacturing

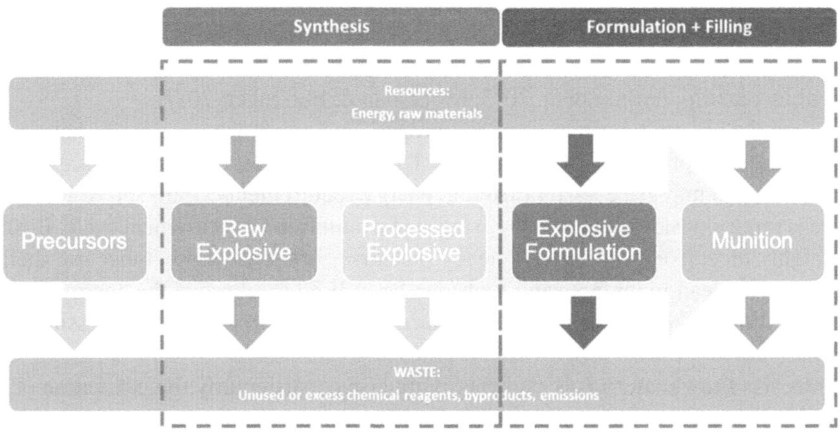

FIGURE 6.1 A simplified representation of a multiple stage explosives manufacturing process.

process, an individual may only have influence over a specific area of a given process; however a holistic consideration of all elements is beneficial to maximise sustainability throughout (Herrmann et al., 2014).

When considering the environmental or sustainability concerns of any manufacturing activity, energy efficiency is often a priority (Duflou et al., 2012; Abdul Rashid et al., 2008, Aflaki et al. (2013)). Excessive energy usage is wasteful by its nature, ultimately contributing to the unnecessary use of non-resources (e.g. fossil fuels, uranium, or other materials), and is also the principal cause of greenhouse gas emissions (Dones et al., 2004; Khan et al., 2014). Renewable energy sources are worth considering due to their low operational energy costs. However, the building, decommissioning, and disposal of renewable infrastructure can be expensive and generate unavoidable waste. Therefore, minimising energy waste throughout the entire lifecycle of energy generation is essential, as all methods carry inherent environmental costs (De Wild-Scholten, 2013; Raugei et al., 2012; Turconi et al., 2013). Hence, why, understanding energy consumption of a manufacturing process is essential for effective environmental and economic management. There are methods, such as techno-economic analysis (TEA), which can be used for estimating the technical and economic performance of a product, process, or service; or the lifecycle assessment (LCA) which focuses on the environmental impact of the same throughout its lifecycle which are currently aiding in addressing environmental impacts holistically (Ferreira et al., 2016; Reardon, 1995). Moreover, it is now increasingly expected for all companies to monitor their greenhouse gas emissions, including explosive manufacturing and other activities (e.g. carbon accounting) using various techniques including the Greenhouse Gas Protocol (World Resources Institute and World Business Council for Sustainable Development (WRI & WBCSD), 2004) and ISO 14064 (International Organization for Standardization (ISO), 2018) by counting and differentiating their emissions into three categories Scope 1, 2, and 3 emissions.[1]

Scope 1 emissions: Are those directly made by an organisation. For example, the running boilers or business vehicles.

Scope 2 emissions: Are those made indirectly by the business activity. Examples include the buying of electricity or heating/cooling of buildings.

Scope 3 emissions: Are often viewed as more complicated. They include all emissions associated with both up and down the value chain. This prevents the offsetting indirect emissions to suppliers. Examples include all emissions from suppliers' activities, including waste treatment/disposal.

This accounting approach is designed to be comprehensive. Whilst not directly due to manufacturing activities, it is important to consider the energy consumption of general facility usage, for lighting, heating/cooling, IT systems etc. Often these are essentially fixed costs not directly dependent on the production volume. Facilities management should consider the energy efficiency of all systems within a facility (Aflaki et al.,2013; Moynihan & Barringer, 2017).

Considering the energy requirement of any process equipment is necessary to have a full understanding of the energy efficiency of any process. Different manufacturing techniques (e.g. pressing, casting, and mixing) may have vastly different energy requirements (Ibbotson & Kara, 2018), and this feature is rarely considered at the early stages of a munition's procurement cycle. Businesses are normally highly receptive to energy efficiency incentives due to a direct impact on their financial costs. Additionally, lean manufacturing tools can be utilised to improve the energy efficiency of a process by limiting wasteful activities (e.g. unnecessary transport and overprocessing) (Herrmann et al., 2014).

The synthesis of raw materials is also energy intensive, particularly the generation of "energetic materials" which are by definition stores of chemical energy, while waste management activities often include additional energy requirements. Thus, manufacturers have a potential risk of unintentionally offsetting the energy requirement to other areas of the supply chain. This highlights the need to consider a whole lifecycle approach to energy usage (which should be accounted for within the Scope 3 emissions).

Just in time manufacturing is a lean production strategy to create products to meet demand, not in advance of need by aligning production with demand and raw material supply. Lean manufacturing it aims to increase efficiency and reduce "waste". This philosophy can be used in a manufacturing settling, limiting unnecessary inventory of product/materials, over-production, over-processing, excessive transportation, and defective products (Chen et al., 2020; Green et al., 2019). Although often applied for economic reasons, environmental justifications include the limiting of resource depletion, emissions of pollutants, and energy conservation. While this approach offers advantages in terms of waste and cost reduction and potential improvements in quality of the products, reduced stocks of material present risks of delays if immediate access to specific energetic materials are required (e.g. during war time).

6.2.1 Eco-Toxicological Risks from Material Selection

The toxicological properties of substances used in explosives manufacturing need to be considered both in terms of the formulation and any processes, tooling and equipment. Inorganic hazardous materials including lead, mercury, cadmium, chromium, and even some radioactive substances have all been used extensively in explosive systems. Organic hazardous substances such as TNT, RDX, nitroglycerin (NG), per- and polyfluoroalkyl substances (PFAS), and chemical warfare agents have all been linked to detrimental impacts on human health and the environment (Lima et al., 2011; Panieri et al., 2022). However, it is important to note that other explosives, such as Tetryl, are considered more toxic than RDX or TNT, while NG, although hazardous, is not particularly toxic and is even used medically to treat angina. When PFAS is found in explosives, it typically originates from its use as a plasticiser in plastic-bonded explosives, for example, in HNS/Teflon mixtures (90/10). Additionally, for energetic materials, detonation residues or other by-products may need to be considered. To reduce the risk posed responsible manufacturers use a hierarchy of controls-based approach to hazardous materials (Figure 6.2).

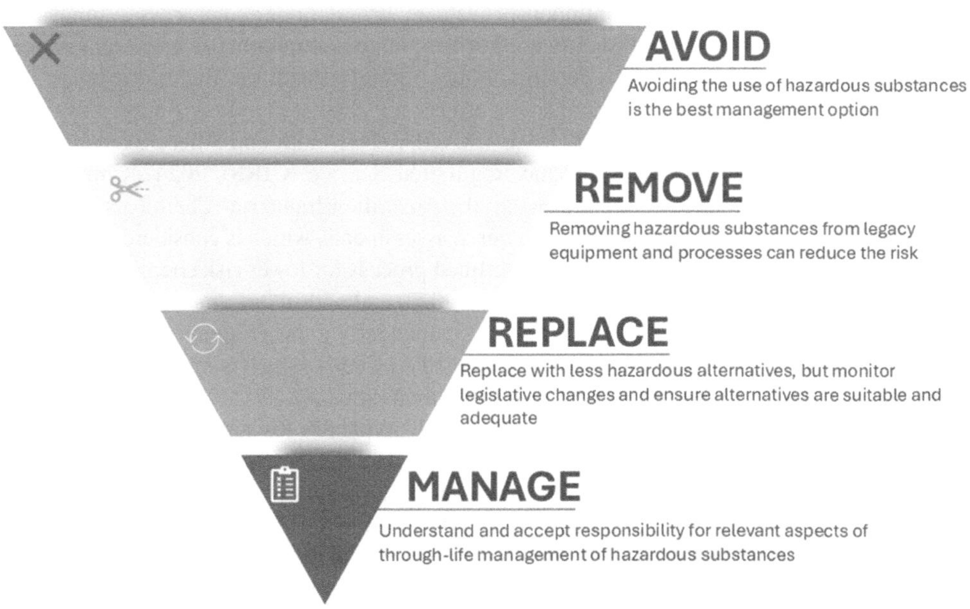

FIGURE 6.2 A suggested hierarchy of controls approach to the use of hazardous materials.

When designing a product or process, the usage of hazardous materials should be avoided where possible with the use of less hazardous alternatives where required. From an energetic materials standpoint, potential substitutions aimed at minimising environmental impact include the following:

1. *Lead-free primary explosives*: Replacing traditional lead-based compounds such as lead azide and lead styphnate with environmentally friendly alternatives can reduce toxicity and environmental contamination.
2. *Phthalate-free plasticisers*: Utilising plasticisers like bis(2,2-dinitropropyl) acetal (BDNPA) offers less toxic options compared to conventional phthalate-based plasticisers.
3. *IHEs*: Adopting IHEs, such as triaminotrinitrobenzene (TATB), enhances safety by reducing sensitivity to impact, heat, and friction, thereby mitigating the risk of unintended detonations.

Where replacement is not an option, the through life implications of these materials must be fully understood and mitigated against.

While alternative synthetic routes and sustainable feedstock may be applied to current raw materials used in energetic formulations, there is increasing pressure (cultural, economic, and legislative) to improve the human health and environmental impact of chemical substances necessitating the innovation of less hazardous alternatives.

The EU REACH (Registration, Evaluation, Authorisation, and Restriction of Chemicals) legislation aims to protect humans and the environment from the risks posed by particularly harmful chemicals and to promote the use of less hazardous alternatives. In terms of explosives manufacturing, since August 2023 di-/poly-isocyanates, which are used to cure poly-alcohols to form polyurethane thermosetting binders, have been subjected to additional regulations concerning their use and supply [Annex XVII]. Additionally, the antioxidant AO2246 used in polyurethane binders utilised in PXBs and some propellants has been classified as a "Substance of Very High Concern", increasing the scrutiny of its use and obligations to inform downstream users and to search for alternatives. Some substances notably various chromium VI and lead compounds, dinitrotoluene (DNT), and tris(2-chloroethyl) phosphate (TCEP) have been added to the Authorisation List [Annex XIV], requiring specific authorisation for use beyond a defined "sunset date". Overall, the impact of REACH requires manufacturers to continuously evaluate formulations and processes, innovating to find alternative materials or provide significant justification to continually use hazardous materials. Whilst compliance is technically only required when trading in/with the EU, it is seen as a gold-standard internationally. Global harmonisation efforts for chemical industry necessitate similar frameworks to maintain trade with the EU, see UK-REACH (Health and Safety Executive (HSE), 2022), K-REACH (Chemical Inspection and Regulation Service (CIRS), 2024), China REACH etc. Other similar systems include the following: the Australian Industrial Chemicals Introduction Scheme (AICIS); a risk based approached with tiered assessments, which is considered more flexible and less burdensome than REACH with a streamlined process for lower risk chemicals (Australian Industrial Chemicals Introduction Scheme (AICIS), 2023) and within the United States of America: The Toxic Substances Control Act (TSCA) of 1976 amended by the Lautenberg Chemical Safety Act in 2016 (U.S. Environmental Protection Agency (EPA), 2024), which is more risk focussed with the burden of proof often lying with regulators rather than industry.

PFAS is a highly topical class of synthetic chemicals, typically resistant to heat, oil, and water sometimes termed "forever chemicals" (Panieri et al., 2022). PFAS have been used extensively in both industrial and consumer products since the 1940s (e.g. firefighting foams, non-stick items, and water-repellent products). Polytetrafluoroethylene (PTFE) known under the brand Teflon is perhaps the most recognisable PFAS. Specifically for energetics Viton A is used as a binder in formulations such as PBXN-5 and PBXN-7. Increased environmental contamination of PFAS (notably water, soil, and bio accumulation) has raised concern alongside links with human health impacts such as increased cancer risk, immune system suppression, endocrine disruption, and liver/kidney damage.

In 2017 DuPont and Chemours settled a circa $921M Lawsuit regarding water contamination in West Virgina and Ohio. Increased regulatory and scientific interest has highlighted PFAS concerns recently incentivising the development of safer alternatives and methodologies for environmental remediation.

6.3 INSERVICE CONSIDERATIONS OF EXPLOSIVE MATERIALS

During the First (1914–1918) and Second (1939–1945) World Wars, research into the development of explosives increased moving from NG to TNT, RDX, pentaerythritol tetranitrate (PETN), and octahydro-1,3,5,7-tetranitro-1,3,5,7-tetrazocine (HMX) (Chatterjee et al., 2017). The use of these explosives has been associated with different environmental issues, from residues left on the soil (Hewitt et al., 2005), explosives dissolved in groundwater systems (Bordeleau et al., 2008), oceans (Maser & Strehse, 2020), and emissions such as carbon monoxide, mixed nitrogen oxides (NO_x), and VOCs to air (Szostak & Cleare, 2000), which varies depending on the type of explosives and the amount used. From a chemical perspective, explosive substances are generally categorised by their chemical composition and structural molecular groups for example: (1) nitro compounds (e.g. TNT), (2) nitrate esters (e.g. pentaerythritol tetranitrate or PETN), (3) nitramines (e.g. RDX, HMX), (4) derivatives of chloric and perchloric acids, and (5) azides (Chatterjee et al., 2017).

Extensive research has examined the effects that TNT, RDX, HMX, and their degradation products have on the environment and the health effects associated with their use. It was discovered that TNT has carcinogenic and mutagenic effects at high concentrations in humans (Nipper et al., 2001) and it is also genotoxic in fish, microorganisms, and mammals (Inouye et al., 2009; Koske et al., 2019; Rosen & Lotufo, 2007). TNT, as many other explosive substances and mixtures (Persico, Coulon, et al., 2022), tends to bind and behave differently depending on the organic matter content in the soil, enhancing or diminishing (high organic content will retain the explosive in the soil more and also increase the formation of degradation products) their fate and transport in this matrix or the production of degradation products (Dontsova, 2018; Lynch et al., 2002). The impact of TNT and RDX has been considered in different training ranges, where it was discovered that at high concentration their use has a negative impact on both the earthworm population (Lachance et al., 2004; Vila et al., 2007, 2008) and plant growth (Vila et al., 2007).

This has been established after many studies defined a close link between the use of energetic materials during training and environmental issues in the area. One example is from a study conducted at the Cold Lake Air Weapons Range (CLAWR) in Canada, where air-to-air and air-to-ground trainings with live bombs, dummy bombs, rockets, OB, and open detonation (OD) take place (Bordeleau et al., 2008). Evidence showed that depending on the type of training that was happening in the area a higher or lower explosive contamination was found (Figure 6.3). Where live bombing was taking place, TNT and RDX contamination were, respectively, 500 and 100 times higher compared to where dummy bombs were fired. HMX contamination was higher where anti-tank weapons were used, or energetic materials contain RDX, like C-4 (HMX being an impurity of RDX). Metal contamination was found across the board with high concentrations in soil, surface water, and groundwater highlighting a greater concern as contaminants are more likely to transport in wider areas. Moreover, contamination was also detected in areas farther from the range where it was most likely that residues were transported by the wind in nearby lakes.

In the United States, the Department of Defense (DoD) established that the first cause of environmental concern from propellants, explosives, and pyrotechnics (PEP) on their training areas was from the combustion products during OB and OD procedures. It was established that during OD 98.5% of the carbon in the PEP was converted to carbon dioxide (96.9%) or carbon monoxide (1.6%) and 1.2% of the nitrogen was converted to nitrous oxides (NO_x) (Cook & Spillman, 2000), all contaminants that are either contributing to climate change or ozone depletion.

Bordeleau et al. (2013) explored the environmental impact of double-based propellant residues on training ranges due to the left-over materials of explosives or rocket fuels mainly made of NG and

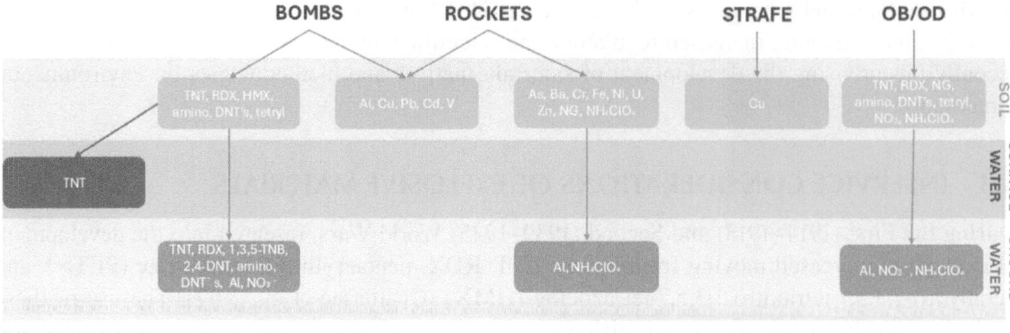

FIGURE 6.3 Contaminants in soils, groundwater, and surface water related to military training activities. The diagram illustrates the types and pathways of contaminants resulting from four types of training activities: Bombing, rocket firing, strafing, and open burning/open detonation (OB/OD). Contaminants include both organic explosives (e.g., TNT, RDX, HMX, tetryl, and DNTs) and inorganic elements or ions (e.g., metals, nitrate, and ammonium perchlorate). Arrows indicate the potential migration of these substances into surface water and groundwater. (Source: Adapted from Bordeleau et al. (2008).)

nitrocellulose (NC). Double-base propellants are usually used in small-arm ammunition typically that contains up to 84% NC, with 10% NG, a stabiliser, and up to 6% filler compounds (Lotufo et al., 2013). These substances can dissolve and migrate from soil (leachate) to surface water bodies to groundwater systems following a rainfall or a snowmelt infiltrating the water table (Bordeleau et al., 2013). It was discovered NG, if left undisturbed on the soil surface can go photolysis, significantly degrading in the environment. Depending on the length of exposure to the sun and the particle size of the materials degradation of NG varies, although it tends to be slower compared to other traditional explosives, such as RDX. In similar environmental conditions, it is bound to NC, which slows down the degradation process. In rodent studies, NG showed moderate acute toxicity with LD50 values ranging from 500 to over 1,100 mg/kg, caused toxic effects in organs with repeated exposure, and reduced fertility at high doses, with estimated NOAEL and LOAEL values at 3 and 32 mg/kg/day, respectively; limited data is currently available for birds, and none for reptiles or amphibians with low toxicity issues defined in aquatic organisms (Lotufo et al., 2013; Williams et al., 2015).

Lastly, other explosives, such as the nitrate ester PETN, have been extensively used on training ranges as it is considered less toxic than energetic materials, such as RDX. Compared to other traditional explosives, PETN has a short half-life of 2.4 days meaning that it tends to degrade faster, although it is more persistent compared to RDX (Thiboutot et al., 2015). PETN can slightly decompose forming alkoxy radicals which can interact with other molecules in the environment, especially in water, triggering autocatalytic reactions. However, research from Thiboutot et al. (2015) showed that, when in use, PETN does not have a high deposition rate compared to RDX and TNT. It should also be noted that TNT can persist in the environment for several years particularly when deposited as partially detonated fragments or low-order residues, as documented at long-used military training ranges (Hewitt et al., 2005),

As new safety requirements advance, IHEs were developed to create explosive ordnance that were insensitive to unintentional stimuli. Hence why the military started to develop polymer-bonded explosives (PBX), which are not melt-castable and not currently suitable for artillery rounds (TNT is still used as a melt-castable binder) and are homogeneous mixtures of a solid explosive bound by a polymer matrix such as hydroxyl terminated polybutadiene (HTPB). They are used in conjunction with other explosives such as TATB and RDX entrapped in a polymer with PBX coating the explosive rendering it insensitive (Kumar et al., 2010). PBX formulations are resistant to leaching due to their insoluble binders that protect the explosive material; however, they can expand with heat and degrade under UV light, constantly and slowly releasing the explosive in the environment (Kumar et al., 2017).

Persico et al. (2022) have shown that a 155-mm insensitive munition (IM) filled with an IHE mixture (NTO (53%), DNAN (32%), and RDX (15%) deposits slightly higher explosives residues compared to the values recorded for legacy (RDX/TNT) filled munitions. This deposition can cause concern as the chemical components of the IHE have different properties. For example, DNAN, compared to TNT, has a higher toxicity in plants. On the other hand NTO, while less toxic compared to TNT, is highly mobile with a concentration of 10 mg/L can discolour water, which is considered, under multiple UK and European regulation, pollution (Tennant et al., 2019). Moreover, the fate and transport of the explosive vary depending on the release location. Pristine soil, compared to soil already contaminated in a training area, is more likely to retain the energetic compounds, whereas soil in a training range has an increased likelihood of allowing these compounds to reach the groundwater system and cause environmental issues (Persico et al., 2023).

Christian et al. (2023) showed that other explosives, such as 2,4,6-triamino-1,3,5-trinitrobenzene (TATB), which is a nitroaromatic compound with high thermal stability (Dobraz, 1995; Watt et al., 2006; Yan et al., 2020), tend to remain primarily at the soil surface and it is transported at a slow rate with a low probability of developing toxic compounds. Although, TATB is of a bright yellow colour, which changes when exposed to light for too long, especially when it sits undisturbed on the surface for a great amount of time. The colour variations are classified as pollutants as they can create toxic by-products having implications on the ecosystem.

6.4 DISPOSAL OF EXPLOSIVE MATERIALS

According to the United Nations guidelines, disposal is the last step of an explosive's ordnance lifecycle, in which demilitarisation and/or destruction techniques can be applied to eliminate either obsolete or surplus stock, and with this, the risks associated to gradual decomposition of energetic materials (IATG, 2021). OB and OD, which consist of positioning explosives ordnance on the ground and then igniting to detonation/combustion through high explosives, liquid fuel, or accelerants, respectively (Duijm, 2002), have been globally considered not only as suitable, but also straightforward and cost-effective disposal techniques (Snellinx et al., 2002).

Currently, there is no specific legislation that covers safe disposal of ammunition worldwide, with only the London Convention of 1972 that bans disposal of ammunition at sea. Relevant legislation only covers broad principles of pollution. For example, relevant pieces of legislation include the EU's Water Framework Directive 2000/60/EC, the Water Resource Act 1991, the Environmental Protection Act 1990, the Environment Act 2021 in the United Kingdom, and the United States Clean Water Act 1972. Beneath the level of legislation, the International Ammunition Technical Guideline (IATG) specifies the need for national authorities to set the standards governing environmental protection within national borders with uncertainty on how much this principle has been applied (Evans & Duncan, 2020).

Selection of the "best" disposal technique is a multi-criteria decision problem which needs to consider the environment, safety of personnel, costs, and capacity (Duijm & Markert, 2002).

OB, OD, closed detonation, fluidised bed combustion, rotary kiln, and mobile furnaces are among the available options to dispose of energetic materials. However, each of these solutions has pros and cons (Table 6.1) which will play a role in deciding the best available technique, depending on who is responsible for the clean-up and any environmental issues that may occur.

At the beginning of the 1990s concerns were raised regarding current disposal options for energetic materials. For example, the Netherlands in 1991, discovered that OB exceeded safety thresholds for different compounds, e.g. HCl, Cl_2, SO_x, NO_x, from the disposal of traditional explosives (Van Ham & Bartelds, 1992). More recent, studies have also suggested that disposal techniques such as OB, whilst economically viable, are not always eco-friendly, leaving explosives residues behind. A study from Christian et al. (2023) explored the efficiency of OB to dispose of TATB trialling different methods, such as adding different waste materials (e.g. wood and paper waste), settings, and accelerants (e.g. Kerosene) to the burn. It was discovered that TATB does not readily

TABLE 6.1
Multi-Criteria Analysis of Six Different Disposal Options, Based on Safety, Environment, Costs and Capacity, with Consideration of the Pro and Cons Involved for Each Method

Disposal option	Safety	Environment	Costs	Capacity
Open burning	**Pro** Simple set up **Cons** Manual preparation Uncontrolled burning	**Pro** No need for extensive infrastructure **Cons** Debris and deposition on soil Ashes and VOC in the air	**Pro** Minimal technology **Cons** Long-term damage can lead to clean up costs	**Pro** Can dispose of large quantities **Cons** Not available for all explosives
Open detonation	**Pro** Reduced need for handling hazardous materials **Cons** Manual preparation Potential accidents can occur	**Pro** No need for extensive infrastructure **Cons** Debris and deposition on soil Noise pollution	**Pro** Costs only for detonators **Cons** Long-term damage can lead to clean up costs	**Pro** Efficient for large quantities **Cons** Difficult to control
Closed detonation	**Pro** Contained environment **Cons** Manual preparation	**Pro** Reduced emission to air **Cons** Noise pollution Waste from filters	**Pro** Cost effective for small operations **Cons** Higher costs due to set up and safety requirements	**Pro** Suitable for small ammunition **Cons** Not scalable for large volumes
Fluidised bed combustion	**Pro** Reduced risk of detonation **Cons** Requires handling pre-processing	**Pro** No wastewater produced **Cons** Ashes production	**Pro** Moderate cost if facilities are available **Cons** Higher than OB-OD as it needs infrastructure	**Pro** Effective for TNT **Cons** Not effective for some explosives
Rotary kiln	**Pro** Controlled environment reduces explosion risk **Cons** Needs carefully monitoring and management	**Pro** Reduced NO_x emission **Cons** NO_x reduced compared to OB and OD but still present	**Pro** Moderate costs if integrated with existing waste facilities **Cons** Higher costs compared to the other method	**Pro** Suitable for large scale operations **Cons** Less mobile compared to other methods
Mobile furnaces	**Pro** Contained combustion **Cons** Transport of hazardous waste	**Pro** Emissions are controlled **Cons** Disposal of ash and residues still necessary	**Pro** Lower costs compared to large infrastructure **Cons** Costs for transportation and operation costs	**Pro** Flexible system that can be used for more explosives type **Cons** Limited capacity compared to the Rotary Kiln and the Fluidised bed combustion

burn in almost any condition and while not leaving any visual traces of explosive in the environment high concentration of the energetic were still found during analysis. In both the Netherlands and Germany, OB and OD have been banned since the 2000s, and, currently, the NATO's Support and Procurement Agency prohibits OB and OD in industrial demilitarisation contracts, though it remains permitted elsewhere (Evans & Duncan, 2020).

As research developed different technologies, some disposal methods have moved from OB and OD to rotary kiln, although studies suggested that NO_x emissions from this technique are still high, with some suggesting that in the specific environmental conditions a fluidised bed incinerator is more efficient and has lower emissions (Cho et al., 2020). However, in both cases, these techniques can be costly and require movement of explosives to the appropriate infrastructure.

6.5 TOWARDS GREENER EXPLOSIVES: INNOVATIONS FOR SUSTAINABILITY

The wider chemical industry is undergoing a sustained transition away from fossil-derived feedstocks toward renewable, bio-based sources, driven by decarbonisation targets, resource security, and advances in industrial biotechnology and green chemistry (Clomburg et al. 2017, Lee et al. 2019, Kircher 2025). Biomass-derived intermediates (e.g. sugars, lignocellulosics, and bio-derived aromatics) are increasingly used as drop-in or functional replacements for petrochemical building blocks within established manufacturing value chains (Kohil et al. 2019, Lopez et al. 2023,). Energetic materials represent a comparatively small and specialised sub-sector of the chemical industry, but one that is largely dependent on organic molecular frameworks derived from conventional aromatic precursors. As such, the same bio-based feedstock strategies already demonstrated at scale in the wider sector can, in principle, be adapted for use in energetics manufacture. This presents a renewable source for raw materials, often with a significantly lower carbon footprint as any CO_2 emissions from their combustion (or indeed, detonation), and is offset against the CO_2 absorbed in the growth of the biomass. A biosynthetic pathway for converting glucose into phloroglucinol, a precursor for the insensitive TATB, has been established using *Escherichia coli* (Achkar et al., 2005). Frost (2010) then demonstrated this biosynthetic phloroglucinol could be converted into both TNT and TATB. The TNT synthesis is particularly interesting as it eliminates the use of mixed acid nitrating agents and thus the red water waste stream, and the suggested overall reaction is described below:

$$C_6H_3(OH)_3 + CH_3I + 3NH_2OH + 3HNO_3 \rightarrow C_6H_2(CH_3)(NO_2)_3 + 6H_2O + HI$$

Biomass feedstock can be diverse, including waste materials (e.g. food waste or by-products), which could potentially be a valuable resource reducing dependence on imported fossil fuels. Feedstock availability and consistency is perhaps the greatest concern of using bio-based reagents for explosives manufacturing; however, other highly regulated chemical industries such as pharmaceuticals are considering these routes (Espro et al., 2021; Kumar et al., 2022).

Moreover, international legislation is now paving the way for the development of new "green" energetic materials which has seen improvement in the development of metal-free (Huynh et al., 2006; Tang et al., 2019) to carbon-free explosives (Bondarchuk, 2020). Research is currently focussing on making these explosives materials, especially primary explosives, eco-friendly during the entire lifecycle of the product. For example, Yi et al. (2018) developed a sulphate-based nano-thermite n-Al/$CuSO_4 \cdot 5H_2O$ to directly detonate RDX as a metal-free alternative to conventional nanothermites. This compound is more stable compared to metal oxide-based materials, and because of its improved properties, such as significantly higher heat release, it indicates a potential as a low-toxicity initiating or ignition component. Bondarchuk (2020) developed two novel carbon-free energetic compounds: 1,4,2,3,5,6-dioxatetrazinane (DOTZ) and its 2,5-dinitro derivative (DNDOTZ). DOTZ has a better detonation performance compared to traditional and new explosives with only N_2 and H_2O as detonation products potentially representing one of the most balanced sustainable options currently present. DNDOTZ has a weaker detonation performance although it can enhance explosives with high negative oxygen balance, such as TNT, TNB, and TATB, and is a very effective solid oxidant, which outperforms gaseous oxygen. Tang et al. (2019) developed a metal-free poly-azido compound, 3,6-bis-[2-(4,6-diazido-1,3,5-triazin-2-yl)-diazenyl]-1,2,4,5-tetrazine, with great detonation performance which can substitute current traditional secondary explosives. In the same study, a primary explosive, namely 3,6-bis-[2-(4,6-diazido-1,3,5-triazin-2-yl)-hydrazinyl]-1,2,4,5-tetrazine, was used demonstrating an initiation capacity for RDX comparable to the current lead azide. As research into green explosives is very recent, studies are currently mainly focussing on minimising the impact during manufacturing and the development of energetics which does not release metals and carbons in the environment as these represented in the past have been the main cause of concern in training areas.

6.6 TOWARDS MORE SUSTAINABLE EXPLOSIVES MANUFACTURING: TWELVE PRINCIPLES OF GREEN CHEMISTRY

The manufacturing of explosives in many ways is a typical chemical manufacturing process with some additional implications due to the nature of energetic materials (e.g. toxicology and explosive risk), as mentioned in Section 2.1. Most literature on sustainable chemical processes is applicable to the manufacture of explosives. The twelve principles of green chemistry (Anastas & Warner, 1998 is a widely accepted framework for achieving more sustainable chemical processes (Figure 6.4).

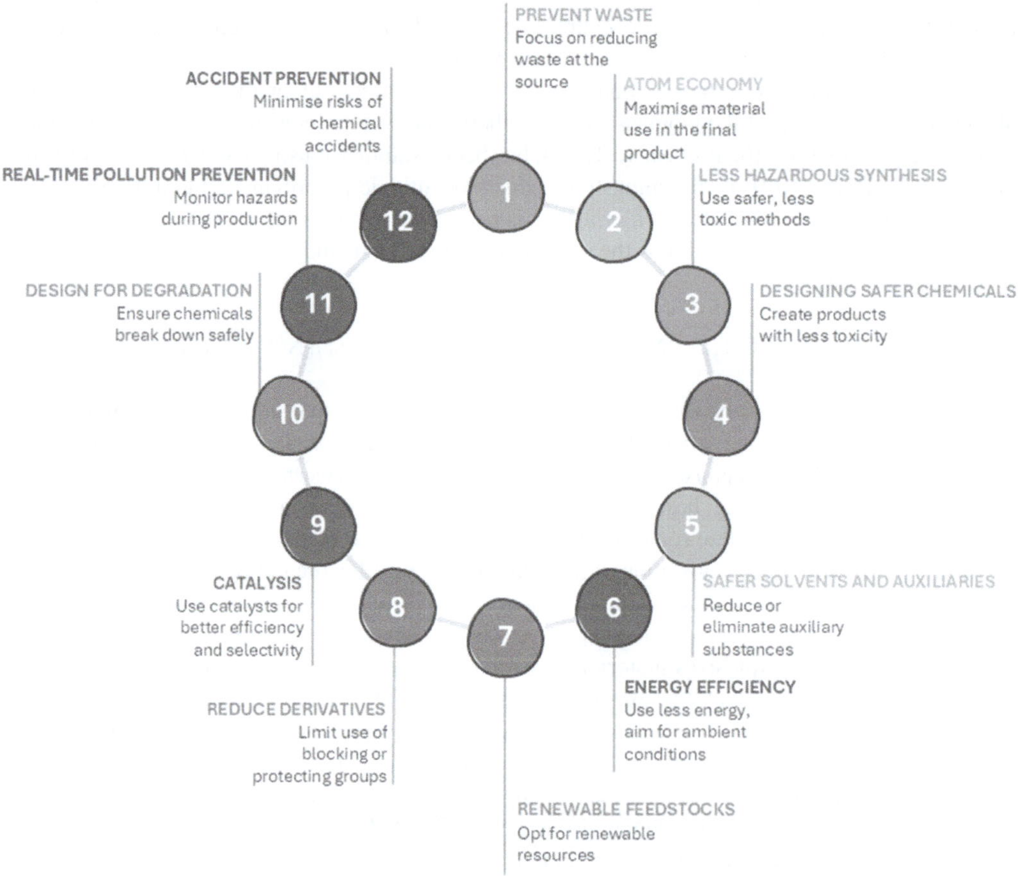

FIGURE 6.4 The twelve principles of green chemistry (Anastas & Warner, 1998).

Explosive manufacturing processes can be broadly categorised into batch and continuous production routes, with batch manufacturing being the predominant industrial approach. Batch production offers flexibility in production, typically lower capital investment, versatility, and greater control of production volumes. However, batch processes also typically require more downtime between production cycles for cleaning and reconfiguration activities. Large volume of waste material, particularly from vessel washing activities, is characteristic of batch processes and it is worth noting that smaller batch sizes often result in a greater proportion of waste material being generated. Continuous processes run with minimal downtime, leading to higher production rates and better efficiency, despite requiring detailed production planning to prevent over manufacture of material.

Additive manufacturing (AM) of explosives is also being explored as an alternative to conventional explosive manufacturing processes however this remains at a relatively early stage of technical development, with most studies still confined to laboratory-scale demonstrations and proof-of-concept applications (Zong et al., 2024). By contrast, AM of polymers, metals, and ceramics is comparatively mature, and its environmental performance has been extensively examined using life-cycle assessment approaches (Kellens et al., 2017a; Kellens et al., 2017b). This broader literature shows that many AM processes are energy-intensive, often requiring one to two orders of magnitude more energy than conventional manufacturing routes such as machining or injection moulding (Yoon et al. 2014, Kellens et al., 2017).

6.6.1 Atom Economy Concept

The second principle of green chemistry, atom economy (AE), is a useful concept to assess the efficiency of a chemical reaction in terms of the efficiency of materials used to form the desired product. Introduced in Trost (1991), it is a calculated ratio of molecular weight of desired product to the total molecular weight of all reactants. AE is a theoretical value only requiring an overall chemical equation and relevant molecular masses of for a given reaction.

Assuming the simple reaction

$$X + Y \rightarrow P + U$$

Where

X + Y are Reactants

P is the desired product

U is an undesired by-product

$$Atom\ Economy\ (\%) = \frac{Molecular\ weight\ of\ desired\ products\ (P)}{Molecular\ weight\ of\ all\ reactants\ (X+Y)} \times 100$$

A higher AE infers a more efficient reaction with fewer by-products (P) and waste. This represents a useful tool to critique the sustainability of a chemical synthesis in terms of resource efficiency. As the theoretical AE has a fixed value for each specific reaction, the AE should be considered during the design phase of a process when selecting synthetic pathways. In later design stages AE should be considered holistically with chemical yield to identify the most efficient process.[2]

$$Chemical\ Yield\ (\%) = \frac{Actual\ Yield}{Theoretical\ Yield} \times 100$$

Chemical yield is more traditionally taught as a measurement of the efficiency of a reaction (or process); a higher yield means more product is obtained from the reaction. This contrasts with AE

TABLE 6.2
The Atom Economy of TNT Synthesis via the Nitration of Toluene

	Substance	Molecular Mass (M_r)
Desired Product (P)	TNT, $C_7H_5(NO_2)_3$	227 g/mol
Reactants (X + Y)	Toluene, C_7H_8	92 g/mol
	Nitric Acid, HNO_3	(3 × 63) = 189 g/mol

which focuses on how efficiently the reactants (X + Y) are incorporated into the desired product (U). Ideally, a process would have a high AE and yield to maximise the efficiency of reactants.

6.6.1.1 TNT via Nitration of Toluene

In industry TNT is typically synthesised in a three-step nitration process using sulphuric (H_2SO_4) and nitric acid (HNO_3). The nitric acid is consumed, whereas the sulphuric acts as a catalyst and thus is not considered here.

$$C_7H_8 + 3HNO_3 \rightarrow C_7H_5\left(NO_2\right)_3 + 3H_2O$$

$$Atom\ Economy\ (\%) = \frac{227\ g/mol}{92+189\ g/mol} \times 100$$

The AE (Tab. 6.2) of this reaction (80.8%) is relatively high demonstrating a high percentage of reactants being incorporated into the TNT product. It is also worth noting that the only by-product is water, the formation of which reduces the AE. Additionally, in practice less stable isomers of TNT are removed via a sulfidation process as red water; this increases the proportion of the desired product but lowers the actual yield of reaction.

6.6.1.2 RDX via the Hale Process

The original process for generating RDX was developed by Hale at Picatinny Arsenal in 1925 (Hale, 1925), in which hexamine is nitrated using 98–100% nitric acid, generating RDX yields of 68%. The original formulated reaction suggested by Hale:

$$C_6H_{12}N_4 + 3\ \mathrm{HNO_3} \rightarrow C_3H_6O_6N_6 + 3\ \mathrm{HCHO} + NH_3$$

$$Atom\ Economy\ (\%) = \frac{222\ g/mol}{140+189\ g/mol} \times 100$$

The AE (Tab. 6.3) of the Hale Process is relatively high (67.5%) demonstrating efficiency in converting the raw materials into product. However, Bachmann and Sheehan (1949) highlight that a large excess of nitric acid is required and indeed the formula provided by Hale (1925) appears to

TABLE 6.3
The Atom Economy of RDX Synthesis via the Hale Process

	Substance	Molecular Mass (M_r)
Desired Product (P)	RDX, $C_3H_6N_6O_6$	222 g/mol
Reactants (X + Y)	Hexamine, $C_6H_{12}N_4$	140 g/mol
	Nitric Acid, HNO_3	(3 × 63) = 189 g/mol

TABLE 6.4
The Atom Economy of RDX Synthesis via the Nitration of Hexamine

	Substance	Molecular Mass (M_r)
Desired Product (P)	RDX, $C_3H_6N_6O_6$	222 g/mol
Reactants (X + Y)	Hexamine, $C_6H_{12}N_4$	140 g/mol
	Nitric Acid, HNO_3	(10 × 63) = 630 g/mol

be an oversimplification of the overall reaction. A more comprehensive evaluation of the nitration of hexamine to RDX is provided (Luo et al., 2002):

$$C_6H_{12}N_4 + 10\ \mathrm{HNO_3} \rightarrow C_3H_6N_6O_6 + 3\ \mathrm{CH_2}\left(\mathrm{ONO_2}\right)_2 + \mathrm{NH_4NO_3} + 3\ \mathrm{H_2}O$$

$$Atom\ Economy\ (\%) = \frac{222\ g/mol}{140+630\ g/mol} \times 100$$

This more robust evaluation of the nitration of hexamine (Tab. 6.4) has a far lower AE (28.8%) highlighting the inefficiency in practice of such a synthesis. The synthesis is relatively simple however requiring only two reagents (hexamine and nitric acid) and reducing the dependence on particular raw materials. The by-products are methanediol ($CH_2(OHO_2)_2$), ammonium nitrate (NH_4NO_3), and water. Methanediol (also known as formaldehyde monohydrate) is not very stable and easily decomposes into water and formaldehyde (CH_2O); it is a carcinogen with eco-toxic effects.

6.6.1.3 RDX Bachmann Process

The Bachmann process (Bachmann & Sheehan, 1949) is a typically utilised continuous flow process for the generation of two moles of RDX:

$$C_6H_{12}N_4 + 4\ \mathrm{HNO_3} + 2\ \mathrm{NH_4NO_3} + 6\left(\mathrm{CH_3CO}\right)_2 O \rightarrow 2\ C_3H_6N_6O_6 + 12\ \mathrm{CH_3COOH}$$

$$Atom\ Economy\ (\%) = \frac{2\ (222)\ g/mol}{140+252+160+612\ g/mol} \times 100$$

The AE of the Bachmann process (Tab 6.5) is still relatively low (38.1%) demonstrating inefficiency in converting the raw materials into product. The yield of RDX is high circa. overall, 85–90% compared to 68% for the Hale process. Additionally, the Bachmann process yields both RDX and HMX. It is also worth noting that the continuous flow nature of this synthesis process offers many advantages over batch manufacture, notably from a sustainability perspective due to

TABLE 6.5
The Atom Economy of RDX Synthesis via the Bachmann Process

	Substance	Molecular Mass (M_r)
Desired Product (P)	RDX, $C_3H_6N_6O_6$	222 g/mol
Reactants (X + Y)	Hexamine, $C_6H_{12}N_4$	140 g/mol
	Nitric Acid, HNO_3	(4 × 63) = 252 g/mol
	Ammonium Nitrate, NH_4NO_3	(2 × 80) = 160 g/mol
	Acetic anhydride, $(CH_3CO)_2O$	(6 × 102) = 612 g/mol

more precise control of reagents, a lower requirement for vessel washing, and reduced inventory (of both reactants and product). Other environmental considerations include the following: the reactant, acetic anhydride, is reactive and volatile potentially contributing to air emissions and the process generates a large proportion of an acetic acid (CH_3COOH) by-product which could potentially be recycled but needs to be managed.

AE presents a simple method to quantify the resource efficiency of chemical synthesis. The AE is a purely theoretical value specific to a given reaction. Arguably the greatest benefit of AE is that changes the perspective of a chemist evaluating a reaction to consider what material is used and what material is wasted, thus incentivising the development of more efficient synthetic routes.

However, because AE only considers a single reaction, the impact of the other lifecycle stages (for both raw materials and undesired by-products) is not considered thus limiting the concept to a simple tool to help quantitatively evaluate reactions. Additionally, some by-products may have value to other customers, while others may be extremely benign (e.g. water). Understanding the by-products generated from a process is essential when planning any form of manufacturing.

AE has been shown to be useful in improving the synthesis of the pharmaceutical ibuprofen (Murphy, 2018). Traditional synthesis of ibuprofen (AE, 44%) was complex, requiring multiple stages and toxic reagents, and generated significant volumes of waste, including numerous by-products (hydrochloric acid, ammonia, acetic acid, and aluminium trichloride hydrate). A more efficient method (the Boots–Hoechst–Celansese, BHC process) was developed, simplifying the process, reducing waste, and increasing the yield of desired product (AE, 75%). Indeed, the only by-product from the BHC process is acetic acid, which could potentially be reused in other processes.

6.6.2 Enhancing the Environmental Performance of In-Service Impacts

As mentioned throughout this chapter the manufacturing, use and disposal of conventional explosives results in the release of toxic residues, which contaminate soil and water sources, creating significant environmental challenges. Transitioning to green explosives is imperative to mitigate these adverse effects while ensuring their efficacy in industrial and military contexts. Currently, most of the research sector has focussed on developing explosives free of metals, as contamination from lead-based projectiles has been in the past a huge issue that has compromised the land and its productivity. Developments are mostly around primary explosives, which can be initiated through small stimuli, which are evidenced as a source of contamination as they might contain lead-based materials (Oyler, 2014). To overcome this issue, researchers have developed several lead-free primary explosives which can lead to reduced environmental contamination while maintaining the explosive performance expected during combat. Silver azide has been one of the primary explosives investigated, for both detonators and primer, to be discussed as a more environmentally friendly solution. Although this solution has been around for a while, the costs of silver represented a drawback for its production (Oyler, 2014). Huynh et al. identified six criteria that an explosive must have to be identified as "green" and evaluated the efficiency for environmental protection for cat $[Fe^{II}(NT)_3(H_2O)_3]$, $cat_2[Fe^{II}(NT)_4(H_2O)_2]$, $cat_3[Fe^{II}(NT)_5(H_2O)]$, and $cat_4[Fe^{II}(NT)_6]$ with (cat: cation and NT = 5-nitrotetrazolato-N^2) identifying more chemically stable and environmentally friendly solutions (Huynh et al., 2006; Talawar et al., 2009). Tang et al. (2019) developed the 3,6-bis-[2-(4,6-diazido-1,3,5-triazin-2-yl)-hydrazinyl]-1,2,4,5-tetrazine which is another lead-free primary explosive, while Toku Kokuloku et al. (2024) synthetised nine different environmentally friendly, high-dense, thermostable, and safe organic energetic materials.

6.7 REDUCING ENVIRONMENTAL HARM OF DISPOSAL PRACTICES

It has been proven that repurposing ammunitions reduces primary energy uses and minimises the environmental impact compared to traditional disposal options (Ferreira et al., 2019) although it is only in recent years that technologies have been developed to ensure that organisations meet environmental objectives and targets set by international legislation.

Metal recycling of the casing of the ammunition is potentially the fastest solution currently available to minimise material use. Alverbro et al. (2009) analysed the environmental impacts associated with different disposal options available for a grenade. Static kiln and combination treatment were compared to OD to determine the best available environmental disposal option where the latter had been identified to have the highest environmental impact. However, it was suggested that if the metals are recycled before the detonation occurs, the generic environment practices can be improved, although still lower compared to the other disposal option presented. In Athens, at the Portland cement clinker production centre, Kolovos (2006) studied how to substitute ordinary raw cement mix with waste ammunition materials (WAM). This process incorporated metals from WAM, such as lead, copper, zinc, tin, and antimony, in traditional raw materials such as limestone and clay which are traditionally used to prepare clinker. This mixture also improves the strength of the cement decreasing the environmental impact associated with disposing of the ammunition. Different organisations worldwide have also started to run campaigns to safely dispose of ammunition and recycle both the metals and the energetics. In Norway and Sweden, military companies recycle over 90% of their ammunition during demilitarisation processes, spanning from small calibre convention ammunition to aircraft bombs and missile systems. This approach is used to repurpose the materials, including the energetics, for the civilian market.

The impact of ammunition goes beyond the disposing of it and repurposing of the materials. Passon et al. (2023) identified 58 activities involved with the demilitarisation process, spanning from the energy used to recycle the materials, the transport from the different infrastructure, and the removal of the energetics from the ammunition. This research identified that for almost every activity involved in the demilitarisation process, energy and water usage are the main embodied impacts, with emitted aspects that span from release of CO_2 in the environment, to waste by-products (e.g. ashes) and contaminated water. In the same paper, Passon et al. (2023) compared two scenarios and the environmental impact of the demilitarisation of a 105-mm High Explosive M1 ammunition. The first scenario involved OB and OD, while the second scenario involved mechanical and kiln-based destruction highlighting the impact that these activities have on climate change, aquatic toxicity, and human health and how these impacts can be diminished by the recycling opportunities. There are also many examples of processes of recovery of explosives from the casing for recyclable purposes, as research tend to discuss specifically only metal recovery. For example, TNT and RDX can be removed from the casing by melting the explosives, induction heating, ammonia wash out, and microwave washout. The same can be achieved with PBX by cryogenic washout, high pressure water jet, and supercritical solvent extraction (Akhavan, 2004).

6.8 SUMMARY AND CONCLUSIONS

This chapter has focused on the environmental impacts of explosive materials throughout their lifecycle, particularly focusing on the manufacturing, in-service, and disposal stages. It highlights a need for a holistic approach to mitigate against these various impacts while maintaining functional requirements and alludes to advances in green chemistry and other practices to further reduce the environmental impact often in line with other similar industries.

The manufacturing of explosives is particularly resource intensive; indeed, the purpose is to generate "energetic materials" (i.e. those with chemically stored energy). Various multistage processing activities require energy and generate by-products, air, and industrial effluent emissions, and the use of various hazardous materials (particularly eco-toxic materials) generates the potential contamination of soil, water, and air. Green chemistry principles, such as maximising the AE of synthetic routes and a reduced reliance on hazardous materials, provide opportunities for reducing the environmental impact of the manufacturing process and explosive materials through life.

The in-service phase introduces additional environmental concerns, with the deposition of eco-toxic residues such as RDX, TNT, and emerging energetic materials like DNAN and NTO. These traditional substances (RDX and TNT) are known for their persistence within environments and have the potential to bioaccumulate, presenting challenges for environmental remediation and any potential ecosystem recovery. While IHEs are gaining greater attention (e.g. DNAN, NTO, and TATB) due to their improved safety features, their long-term behaviour within the environment requires further study. Advances in alternative formulations such as lead-free primary explosives and bio-based energetic materials present promising avenues for minimising contamination from hazardous materials in the future.

Traditional disposal practices previously have included the dumping of unrequired munitions at sea. However, in recent years disposal practises have been dominated by OB and OD. Whilst simple and cost effective, these methods lack detailed control resulting in the release of by-products into the environment. Emerging technologies such as fluidised bed combustion and rotary kilns provide more advanced techniques offering reduced emissions though at higher operational and infrastructural costs. The integration of recycling and repurposing strategies includes the reclaiming of energetic materials and metal recycling highlights the potential for a circular economy approach. This demonstrates clear environmental and economic benefits; however, such practises remain in their infancy.

The transition towards "greener explosives" necessitates a multifaceted approach combining innovations across chemistry and engineering, regulatory enforcement, and industrial collaboration. It will become increasingly necessary to reduce eco-toxic materials within explosive products. Emerging technologies in bio-based chemistry, additive manufacturing, and more sustainable formulations hold the potential to reshape the field of energetic materials. A full product lifecycle approach should be considered in the development of new materials. Simultaneously, initiatives to recycle and repurpose obsolete munitions underscore the importance of adopting circular economy principles in the defence sector, both from an environmental perspective and to safeguard access to a strategically essential resource.

While progress in this field to date has been significant, the explosives industry has historically lagged similar chemical industry (e.g. FMCG, synthetic polymers, and pharmaceuticals) due to the necessity for meeting robust defence requirements. Achieving true sustainability for explosives requires addressing the interdependencies established at the design phase (i.e. material selection) leading to manufacturing, usage, and disposal. Collaboration across academia, industry, and governments will be vital in overcoming these challenges and ensuring that future advancements remain in step with environmental goals through the life of a program (i.e. product life). By integrating innovative practices in line with increasingly stringent regulatory measures, the explosives industry can reduce its environmental impact while maintaining its critical roles for defence and wider industry.

NOTES

1. The terms Scope 1, 2, and 3 emissions originates from the Green House Gas Protocol of 2001 and is the basis of GHG reporting in many nations worldwide.
2. In practice a low yield reaction with high AE could result greater resource usage.

REFERENCES

Abdul Rashid, S. H., Evans, S., & Longhurst, P. (2008). A comparison of four sustainable manufacturing strategies. *International Journal of Sustainable Engineering*, *1*(3), 214–229. https://doi.org/10.1080/19397030802513836

Achkar, J., Xian, M., Zhao, H., & Frost, J. W. (2005). Biosynthesis of phloroglucinol. *Journal of the American Chemical Society*, *127*(15), 5332–5333. https://doi.org/10.1021/ja042340g

Aflaki, S., Kleindorfer, P. R., & deMiera Polvorinos, V. S. (2013). Finding and implementing energy efficiency projects in industrial facilities. *Production and Operations Management*, *22*(3), 503–517. https://doi.org/10.1111/j.1937-5956.2012.01377.x

Akhavan, J. (2004). *The Chemistry of Explosives* (3rd ed.). Royal Society of Chemistry.

Alverbro, K., Björklund, A., Finnveden, G., Hochschorner, E., & Hägvall, J. (2009). A life cycle assessment of destruction of ammunition. *Journal of Hazardous Materials, 170*(2–3), 1101–1109. https://doi.org/10.1016/j.jhazmat.2009.05.092

Anastas, P. T., & Warner, J. C. (1998). *Green Chemistry: Theory and Practice.* Oxford University Press.

Australian Industrial Chemicals Introduction Scheme (AICIS). (2023). *About AICIS.* Retrieved from https://www.industrialchemicals.gov.au/

Bachmann, W. E., & Sheehan, J. C. (1949). A new method of preparing the high explosive RDX. *Journal of the American Chemical Society, 71*(5), 1842–1845. https://doi.org/10.1021/ja01173a092

Barbosa, J., Carvalho, F., & Santos, R. (2018). Risk assessment methodologies applied to chemical process industries: A review. *Safety Science, 103*, 147–161. https://doi.org/10.1016/j.ssci.2017.11.017

Bhanot, P., Celin, S. M., Sreekrishnan, T. R., Kalsi, A., Sahai, S. K., & Sharma, P. (2020). Application of integrated treatment strategies for explosive industry wastewater—A critical review. *Journal of Water Process Engineering, 35*, 101232. https://doi.org/10.1016/j.jwpe.2020.101232

Bondarchuk, S. V. (2020). Modeling of explosives: 1,4,2,3,5,6-dioxatetrazinane as a new green energetic material with enhanced performance. *Journal of Physics and Chemistry of Solids, 142*, 109458. https://doi.org/10.1016/j.jpcs.2020.109458

Bordeleau, G., Martel, R., Ampleman, G., & Thiboutot, S. (2008). Environmental impacts of training activities at an air weapons range. *Journal of Environmental Quality, 37*(2), 308–317. https://doi.org/10.2134/jeq2007.0197

Bordeleau, G., Martel, R., Ampleman, G., & Thiboutot, S. (2013). Photolysis of RDX and nitroglycerin in the context of military training ranges. *Chemosphere, 93*(1), 14–19. https://doi.org/10.1016/j.chemosphere.2013.04.048

Cardesa-Salzmann, A. (2012). Constitutionalising secondary rules in global environmental regimes: Non-compliance procedures and the enforcement of multilateral environmental agreements. *Journal of Environmental Law, 24*(1), 103–132.

Chatterjee, S., Deb, U., Karmakar, S., & Sikder, A. K. (2017). Environmental fate of high explosives: A review. *Chemosphere, 184*, 438–451. https://doi.org/10.1016/j.chemosphere.2017.06.035

Chemical Inspection and Regulation Service (CIRS). (2024). *About K-REACH.* Retrieved from https://kreach.cirs-group.com/en/k-reach/background

Chen, P.-K., Lujan-Blanco, I., Fortuny-Santos, J., & Ruiz-de-Arbulo-López, P. (2020). Lean manufacturing and environmental sustainability: The effects of employee involvement, stakeholder pressure, and ISO 14001. *Sustainability, 12*(18), 7258. https://doi.org/10.3390/su12187258

Cho, S., Park, C., Lee, J., Lyu, B., & Moon, I. (2020). Finding the best operating condition in a novel process for explosive waste incineration using fluidized bed reactors. *Computers & Chemical Engineering, 142*, 107054. https://doi.org/10.1016/j.compchemeng.2020.107054

Christian, O., Spencer, M., Ladyman, M., Persico, F., Gutierrez-Carazo, E., Kadansky, E., & Temple, T. (2023). Persistence of 2,4,6-triamino-1,3,5-trinitrobenzene in the environment. *Environmental Research, 239*, 117378. https://doi.org/10.1016/j.envres.2023.117378

Clausen, J., Robb, J., Curry, D., & Korte, N. (2004). A case study of contaminants on military ranges: Camp Edwards, Massachusetts, USA. *Environmental Pollution, 129*(1), 13–21. https://doi.org/10.1016/j.envpol.2003.10.002

Clomburg, J. M., Crumbley, A. M. & Gonzalez, R. (2017). Industrial biomanufacturing: The future of chemical production. *Science, 355*(6320), aag0804. https://doi.org/10.1126/science.aag0804.

Cook, D. S., & Spillman, E. (2000). Military training ranges as a source of environmental contamination. *Federal Facilities Environmental Journal, 11*(2), 27–37. https://doi.org/10.1002/ffej.3330110205

De Wild-Scholten, M. J. (2013). Energy payback time and carbon footprint of commercial photovoltaic systems. *Solar Energy Materials and Solar Cells, 119*, 296–305. https://doi.org/10.1016/J.SOLMAT.2013.08.037

Dobraz, B. M. (1995). *The insensitive high explosives (TATB): Development and characterization – 1888 to 1994.*

Dones, R., Heck, T., & Hirschberg, S. (2004). *Greenhouse Gas Emissions from Energy Systems: Comparison and Overview* (B. Gschwend, Ed.). Switzerland.

Dontsova, K., Taylor, S., & Pesce-Rodriguez, R. (2018). Dissolution of NTO, DNAN and insensitive munitions formulations and their fates in soils. SERP Project ER-2220. Retrieved from https://apps.dtic.mil/sti/pdfs/AD1057612.pdf

Duflou, J. R., Sutherland, J. W., Dornfeld, D., Herrmann, C., Jeswiet, J., Kara, S., … & Kellens, K. (2012). Towards energy and resource-efficient manufacturing: A processes and systems approach. *CIRP Annals, 61*(2), 587–609. https://doi.org/10.1016/j.cirp.2012.05.002

Duijm, N. J. (2002). Hazard analysis of technologies for disposing explosive waste. *Journal of Hazardous Materials*, *90*(2), 123–135. https://doi.org/10.1016/S0304-3894(01)00357-0

Duijm, N. J., & Markert, F. (2002). Assessment of technologies for disposing explosive waste. *Journal of Hazardous Materials*, *90*(2), 137–153. https://doi.org/10.1016/S0304-3894(01)00358-2

Engel, K. H., & Saleska, S. R. (2005). Contextual environmental federalism. *NYU Environmental Law Journal*, *14*, 108–150.

Espro, C., Paone, E., Mauriello, F., Gotti, R., Uliassi, E., Bolognesi, M. L., … & Luque, R. (2021). Sustainable production of pharmaceutical, nutraceutical, and bioactive compounds from biomass and waste. *Chemical Society Reviews*, *50*(20), 11191–11207. https://doi.org/10.1039/D1CS00524C

Evans, R., & Duncan, A. (2020). Disposal of explosive ordnance and environmental risk mitigation. *The Journal of Conventional Weapons Destruction*, *24*(1).

Fawcett-Hirst, W., Temple, T. J., Ladyman, M. K., & Coulon, F. (2021). A review of treatment methods for insensitive high explosive contaminated wastewater. *Heliyon*, *7*(7), e07438. https://doi.org/10.1016/j.heliyon.2021.e07438

Ferreira, C., Ribeiro, J., Almada, S., Rotariu, T., & Freire, F. (2016). Reducing impacts from ammunitions: A comparative life-cycle assessment of four types of 9mm ammunitions. *Science of the Total Environment*, *566–567*, 34–40. https://doi.org/10.1016/j.scitotenv.2016.05.005

Ferreira, C., Ribeiro, J., Clift, R., & Freire, F. (2019). A circular economy approach to military munitions: Valorization of energetic material from ammunition disposal through incorporation in civil explosives. *Sustainability*, *11*(1), 255. https://doi.org/10.3390/su11010255

Frost, J. W. (2010). *Manufacture of TATB and TNT from Biosynthesized Phloroglucinols*. Draths Corporation. https://apps.dtic.mil/sti/tr/pdf/ADA581260.pdf

Green, K. W., Inman, R. A., Sower, V. E., & Zelbst, P. J. (2019). Impact of JIT, TQM and green supply chain practices on environmental sustainability. *Journal of Manufacturing Technology Management*, *30*(1), 26–47. https://doi.org/10.1108/JMTM-01-2018-0015

Hale, G. C. (1925). The nitration of hexamethylenetetramine. *Journal of the American Chemical Society*, *47*(11), 2754–2763. https://doi.org/10.1021/ja01688a017

Health and Safety Executive (HSE). (2022). *UK REACH explained*. Retrieved from https://www.hse.gov.uk/reach/about.htm

Herrmann, C., Schmidt, C., Kurle, D., et al. (2014). Sustainability in manufacturing and factories of the future. *International Journal of Precision Engineering and Manufacturing-Green Technology*, *1*, 283–292. https://doi.org/10.1007/s40684-014-0034-z

Hewitt, A. D., Jenkins, T. F., Walsh, M. E., Walsh, M. R., & Taylor, S. (2005). RDX and TNT residues from live-fire and blow-in-place detonations. *Chemosphere*, *61*(6), 888–894. https://doi.org/10.1016/J.CHEMOSPHERE.2005.04.058

Hristov, H. P., & Hristov, H. I. (2017). Environmental impact of energetics on test ranges. *Propellants, Explosives, Pyrotechnics*, *42*(1), 84–89. https://doi.org/10.1002/prep.201600157

Huynh, M. H. V., Coburn, M. D., Meyer, T. J., & Wetzler, M. (2006a). Green primary explosives: 5-Nitrotetrazolato-N^2-ferrate hierarchies. *Proceedings of the National Academy of Sciences*, *103*(27), 10322–10327. https://doi.org/10.1073/pnas.0604241103

IAGT. (2021). *Demilitarization and destruction of conventional ammunition*. IATG 10.10:2021[E]. Retrieved from https://data.unsaferguard.org/iatg/en/V3_IATG-10.10_en.pdf

Ibbotson S. M. & Kara S (2018). A framework for determining the life-time energy consumption of a product at the concept design stage. *Procedia CIRP*, *69*, 704–709. https://doi.org/10.1016/j.procir.2017.11.052

International Organization for Standardization (ISO). (2018). *ISO 14064-1:2018: Greenhouse gases*. Retrieved from https://www.iso.org/standard/66453.html

Jones, H. P., & Schmitz, O. J. (2009). Rapid *Recovery of Damaged Ecosystems. PLoS ONE*, *4*(5), e5653. https://doi.org/10.1371/journal.pone.0005653

Kellens, K Mertens, R., Paraskevas, D., Dewulf, W., & Duflou, J. R. (2017a). Environmental impact of additive manufacturing processes: Does AM contribute to a more sustainable way of part manufacturing? *Procedia CIRP*, *61*, 582–587. https://doi.org/10.1016/j.procir.2016.11.153

Kellens, K., Baumers, M., Gutowski, T. G., Flanagan, W., Lifset, R., & Duflou, J. R. (2017b). Environmental dimensions of additive manufacturing: Mapping application domains and lifecycle phases. *Journal of Industrial Ecology*, *21*(S1), S49–S68. https://doi.org/10.1111/jiec.12629

Khan, M. A., Khan, M. Z., Zaman, K., & Naz, L. (2014). Global estimates of energy consumption and greenhouse gas emissions. *Renewable and Sustainable Energy Reviews*, *29*, 336–344. https://doi.org/10.1016/j.rser.2013.08.091

Kircher, M. (2025). Chemical production based on biomass—Potential and limits. *Biomass*, *5*(1), 8. DOI: https://doi.org/10.3390/biomass5010008

Kohli, K., Prajapati, R. & Sharma, B. K. (2019). Bio-based chemicals from renewable biomass for integrated biorefineries. *Energies*, *12*(2), 233. DOI: https://doi.org/10.3390/en12020233

Kolovos, K. G. (2006). Waste ammunition as secondary mineralizing raw material in Portland cement production. *Cement and Concrete Composites*, *28*(2), 133–143. https://doi.org/10.1016/j.cemconcomp.2005.10.001

Koske, D., Goldenstein, N. I., & Kammann, U. (2019). Nitroaromatic compounds damage the DNA of zebrafish embryos (Danio rerio). *Aquatic Toxicology*, *217*, 105345. https://doi.org/10.1016/j.aquatox.2019.105345

Kumar, A. S., Rao, V. B., Sinha, R. K., & Rao, A. S. (2010). Evaluation of plastic bonded explosive (PBX) formulations based on RDX, aluminum, and HTPB for underwater applications. *Propellants, Explosives, Pyrotechnics*, *35*(4), 359–364. https://doi.org/10.1002/prep.200800048

Lachance, B., Renoux, A. Y., Sarrazin, M., Hawari, J., & Sunahara, G. I. (2004). Toxicity and bioaccumulation of reduced TNT metabolites in the earthworm *Eisenia andrei* exposed to amended forest soil. *Chemosphere*, *55*(10), 1339–1348. https://doi.org/10.1016/j.chemosphere.2003.11.049

Kumar, M., Ladyman, M. K., Mai, N., Temple, T., & Coulon, F. (2017). Release of 1,3,5-trinitroperhydro-1,3,5-triazine (RDX) from polymer-bonded explosives (PBXN-109) into water by artificial weathering. *Chemosphere*, *169*, 604–608. https://doi.org/10.1016/j.chemosphere.2016.11.107

Kumar, V., Bansal, V., Madhavan, A. et al. (2022). Active pharmaceutical ingredient (API) chemicals: A critical review of current biotechnological approaches. *Bioengineered*, *13*(2), 4309–4327. https://doi.org/10.1080/21655979.2022.2031412

Lee, S. Y., Kim, H. U., Chae, T. U., Cho, J. S., Kim, J. W., Shin, J. H., et al. (2019). A comprehensive metabolic map for production of bio-based chemicals. *Nature Catalysis*, *2*, 18–33. DOI: https://doi.org/10.1038/s41929-018-0212-4

Lima, D. R. S., Bezerra, M. L. S., Neves, E. B., & Moreira, F. R. (2011). Impact of ammunition and military explosives on human health and the environment. *Reviews on Environmental Health*, *26*(2), 101–110. https://doi.org/10.1515/reveh.2011.014

Lopez, G., Keiner, D., Fasihi, M., Koiranen, T. & Breyer, C. (2023). From fossil to green chemicals: Sustainable pathways and new carbon feedstocks for the global chemical industry. *Energy & Environmental Science*, *16*, 2879–2909. DOI: https://doi.org/10.1039/D3EE00478C

Lotufo, G. R., Rosen, G., Wild, W., & Carton, G. (2013). *Summary Review of the aquatic toxicology of munitions constituents.* Environmental Laboratory. www.erdc.usace.army.mil.http://acwc.sdp.sirsi.net/client/default.

Luo, U.-M., Lin, S.-H., Chang, J.-G., & Huang, T.-H. (2002). Evaluations of kinetic parameters and critical runaway conditions in the reaction system of hexamine-nitric acid to produce RDX in a non-isothermal batch reactor. *Journal of Loss Prevention in the Process Industries*, *15*(2), 119–127. https://doi.org/10.1016/S0950-4230(01)00027-4

Lynch, J. C., Brannon, J. M., & Delfino, J. J. (2002). Dissolution rates of three high explosive compounds: TNT, RDX, and HMX. *Chemosphere*, *47*(7), 725–734. https://doi.org/10.1016/S0045-6535(02)00035-8

Maser, E., & Strehse, J. S. (2020). "Don't Blast": blast-in-place (BiP) operations of dumped World War munitions in the oceans significantly increase hazards to the environment and the human seafood consumer. *Archives of Toxicology*, *1*, 3. https://doi.org/10.1007/s00204-020-02743-0

Moynihan, G. P., & Barringer, F. L. (2017). Energy efficiency in manufacturing facilities: Assessment, analysis, and implementation. *InTech*. https://doi.org/10.5772/64902

Murphy, M.A. (2018). Early industrial roots of green chemistry and the history of the BHC Ibuprofen process invention and its quality connection. *Foundations of Chemistry*, *20*, 121–165. https://doi.org/10.1007/s10698-017-9300-9

National Institute for Occupational Safety and Health (NIOSH). (2024). *Engineering controls.* Centers for Disease Control and Prevention. Retrieved December 20, 2024, from https://www.cdc.gov/niosh/engcontrols/

Nipper, M., Carr, R. S., Biedenbach, J. M., Hooten, R. L., Miller, K., & Saepoff, S. (2001). Development of marine toxicity data for ordnance compounds. https://doi.org/10.1007/s002440010253

Organisation for Economic Co-operation and Development (OECD). (2023). *OECD Guiding Principles for Chemical Accident Prevention, Preparedness and Response* (3rd ed). https://doi.org/10.1787/162756bf-en

Oyler, K. D. (2014). Green primary explosives. *Green Energetic Materials*, *2014*, 103–132.

Pallvi, B., Sreekrishnan, T. R., & S. Mary Celin. (2020). Application of integrated treatment strategies for explosive industry wastewater: A critical review. *Journal of Water Process Engineering*, *35*, 101232. https://doi.org/10.1016/j.jwpe.2020.101232

Panieri, E., Baralic, K., Djukic-Cosic, D., Buha Djordjevic, A., & Saso, L. (2022). PFAS molecules: A major concern for human health and the environment. *Toxics*, *10*(2), 44. https://doi.org/10.3390/toxics10020044

Passon, B. C., Galante, E. B. F., & Ogliari, A. (2023). A systematic approach to assist in life-cycle assessment of ammunition demilitarization process: A case study with the 105-mm HE M1 ammunition. *The International Journal of Life Cycle Assessment*, *28*(4), 398–428. https://doi.org/10.1007/s11367-023-02149-x

Persico, F., Temple, T., Ladyman, M., Gilroy-Hirst, W., Guiterrez-Carazo, E., & Coulon, F. (2022). Quantitative environmental assessment of explosive residues from the detonation of insensitive high explosive filled 155 mm artillery shell. *Propellants, Explosives, Pyrotechnics*, *47*(3). https://doi.org/10.1002/prep.202100220

Persico, F., Coulon, F., Ladyman, M., López, C. F., & Temple, T. (2023). Evaluating the effect of insensitive high explosive residues on soil using an environmental quality index (EQI) approach. *Science of the Total Environment*, *869*. https://doi.org/10.1016/J.SCITOTENV.2023.161797

Raugei, M., Fullana-i-Palmer, P., & Fthenakis, V. (2012). The energy return on energy investment (EROI) of photovoltaics: Methodology and comparisons with fossil fuel life cycles. *Energy Policy*, *45*, 576–582. https://doi.org/10.1016/J.ENPOL.2012.03.008

Reardon, P. T. (1995). Life-cycle assessment (LCA) methodology applied to energetic materials. *Los Alamos National Laboratory*. Retrieved from https://www.osti.gov/servlets/purl/114498

Rosen, G., & Lotufo, G. R. (2007). Toxicity of explosive compounds to the marine mussel, Mytilus galloprovincialis, in aqueous exposures. *Ecotoxicology and Environmental Safety*, *68*(2), 228–236. https://doi.org/10.1016/j.ecoenv.2007.03.006

Sandham, L. A., Van der Vyver, F., & Retief, F. P. (2013). The performance of environmental impact assessment in the explosives manufacturing industry in South Africa. *Journal of Environmental Assessment Policy and Management*, *15*(03), 1350014. https://doi.org/10.1142/S1464333213500130

Skelly, D. (2022). Storage of military ammunition and explosives—Risk evaluation and hazard management. In T. Temple & M. Ladyman (Eds.), *Challenges in Risk Analysis for Science and Engineering Development of a Common Language* (1st ed., Vol. 1, pp. 3–1-3.47). IoP Publishing.

Snellinx, Z., Nepovím, A., Taghavi, S., Vangronsveld, J., Vanek, T., & Van der Lelie, D. (2002). Biological remediation of explosives and related nitroaromatic compounds. *Environmental Science and Pollution Research*, *9*(1), 48–61.

Szostak, R., & Cleare, K. (2000). Emissions related to munitions firing: A case study of nitrogen oxides, volatile organic compounds, and energetic residue from detonable munitions. *Federal Facilities Environmental Journal*, *11*(3), 87–104. https://doi.org/10.1002/ffej.3330110309

Talawar, M. B., Sivabalan, R., Mukundan, T., Muthurajan, H., Sikder, A. K., Gandhe, B. R., & Rao, A. S. (2009). Environmentally compatible next generation green energetic materials (GEMs). *Journal of Hazardous Materials*, *161*(2–3), 589–607. https://doi.org/10.1016/j.jhazmat.2008.04.011

Tang, J., Chen, D., Zhang, G., Yang, H., & Cheng, G. (2019). A "green" primary explosive: Design, synthesis, and testing. *Synlett*, *30*(08), 885–892. https://doi.org/10.1055/s-0037-1611696

Tennant, M., Chew, S. C., Krämer, T., Mai, N., McAteer, D., & Pons, J. F. (2019). Practical colorimetry of 3-nitro-1,2,4-triazol-5-one. *Propellants, Explosives, Pyrotechnics*, *44*(2), 198–202. https://doi.org/10.1002/prep.201800239

Thiboutot, S., Brousseau, P., & Ampleman, G. (2015). Deposition of PETN following the detonation of seismoplast plastic explosive. *Propellants, Explosives, Pyrotechnics*, *40*(3), 329–332. https://doi.org/10.1002/prep.201500019

Toku Kokuloku Jr, L., Djam Miensah, E., Gu, A., Chen, K., Wang, P., Gong, C., Jiao, Y., Chen, K., & Yang, Y. (2024). A comparative adsorption study of activated carbon and Fe-modified activated carbon for trinitrotoluene removal. *Journal of the Taiwan Institute of Chemical Engineers*, *161*, 1876–1070. https://doi.org/10.1016/j.jtice.2024.105519

Trost, B. M. (1991). The atom economy—A search for synthetic efficiency. *Science*, *254*(5037), 1471–1477. https://doi.org/10.1126/science.1962206

Turconi, R., Boldrin, A., & Astrup, T. (2013). Life cycle assessment (LCA) of electricity generation technologies: Overview, comparability and limitations. *Renewable and Sustainable Energy Reviews*, *28*, 555–565. https://doi.org/10.1016/J.RSER.2013.08.013

U.S. Department of Labor, Occupational Safety and Health Administration (OSHA). (2017). *Process safety management for storage facilities* (OSHA 3909-03). Retrieved from https://www.osha.gov/sites/default/files/publications/OSHA3909.pdf

U.S. Environmental Protection Agency (EPA). (2024). *Summary of the Toxic Substances Control Act*. Retrieved from https://www.epa.gov/laws-regulations/summary-toxic-substances-control-act

Van Ham, N., & Bartelds, H. (1992). *Environmentally acceptable disposal of munition and explosives.* Retrieved from https://apps.dtic.mil/sti/citations/ADA506937

Vila, M., Lorber-Pascal, S, & Laurent, F. (2008). Phytotoxicity to and uptake of TNT by rice. *Environmental Geochemistry and Health*, 30, 199–203. https://doi.org/10.1007/s10653-008-9145-1

Vila, M., Mehier, S., Lorber-Pascal, S., & Laurent, F. (2007). Phytotoxicity to and uptake of RDX by rice. *Environmental Pollution, 145*(3), 813–817. https://doi.org/10.1016/j.envpol.2006.05.009

Vučićević, N., Vuković, M., & Papić, M. (2022). Open burning and open detonation of explosives. Prediction of pollutant emissions. *Environment Protection Engineering, 48*(1). https://doi.org/10.37190/epe220102

Walsh, M. R., Walsh, M. E., Ramsey, C. A., Thiboutot, S., Ampleman, G., Diaz, E., & Zufelt, J. E. (2014). *Energetic residues from the detonation of IMX-104 insensitive munitions. Propellants, Explosives, Pyrotechnics, 39*(2), 243–250. https://doi.org/10.1002/prep.201300095

Watt, D., Deschambault, E., & Touze, P. (2006). Insensitive munitions (IM): A key aspect of improved munitions safety. *Hanneng Cailiao, 14*(5), 323–329.

Williams, M. A., Sample, B. E., & Leach, G. J. (2015). Wildlife toxicity assessment for nitroglycerin. In *Wildlife Toxicity Assessments for Chemicals of Military Concern* (pp. 239–255). Elsevier. https://doi.org/10.1016/B978-0-12-800020-5.00013-2

World Resources Institute and World Business Council for Sustainable Development (WRI & WBCSD). (2004). *The Greenhouse Gas Protocol: A Corporate Accounting and Reporting Standard* (Revised ed.). Retrieved from https://ghgprotocol.org

Yan, T., Yang, H., Yang, C., Yi, Z., Zhu, S., & Cheng, G. (2020). *An advanced and applicable heat-resistant explosive through controllable regiochemical modulation.* https://doi.org/10.1039/d0ta09158h

Yi, Z., Ang, Q., Li, N., Shan, C., Li, Y., Zhang, L., & Zhu, S. (2018). Sulfate-based nanothermite: A green substitute of primary explosive containing lead. *ACS Sustainable Chemistry & Engineering, 6*(7), 8584–8590. https://doi.org/10.1021/acssuschemeng.8b00522

Yoon, HS., Lee, JY., Kim, HS. et al. (2014). A comparison of energy consumption in bulk forming, subtractive, and additive processes: Review and case study. *International Journal of Precision Engineering and Manufacturing-Green Technology, 1*, 261–279. https://doi.org/10.1007/s40684-014-0033-0

Zentelis, R., Banks, S., Roberts, J. D., Dovers, S., & Lindenmayer, D. (2017). Managing military training-related environmental disturbance. *Journal of Environmental Management, 204*, 486–493. https://doi.org/10.1016/j.jenvman.2017.09.029

Zong, H., Wang, S., Ren, H., & others. (2024). Energetic materials in 3D: An in-depth exploration of additive manufacturing techniques. *International Journal of Advanced Manufacturing Technology, 133*, 3059–3080. https://doi.org/10.1007/s00170-024-13937-6

7 Introduction to Insensitive Munitions and the Reduction of Vulnerability

Adam S. Cumming

7.1 INTRODUCTION

Munitions have always been risky. Their use involves the rapid release of energy which, while aimed at an effect, can and does pose risks to users/operators. As weapons became more effective, the risks increased to the point that they were considered unacceptable, in part through accidents. This led to the development of Insensitive or Reduced Vulnerability Munitions.

Insensitive Munitions (IM) now play an important part of the programmes within all NATO (North Atlantic Treaty Organization) nations and beyond. They are used to improve safety and reduce accidents through the application of an understanding of both the likely threats and the technology available to both assess those threats and counter them, thus managing the vulnerability of munitions. Proper operation requires a systematic approach based on the understanding of vulnerability and reaction. This has been developing over the last 45 years, and these studies have produced a workable approach, though one that is continuing to develop through improved understanding of both risks and materials.

This understanding has led to an IM STANAG and its implementation in most NATO nations and beyond and has led to the formation of an advice group, the Munitions Safety Information Analysis Center (MSIAC) in NATO which while supporting its subscribers has also promoted the development and application of IM knowledge more generally. Common agreement on terms and science with the understanding of properties and common understanding in application is needed to minimise risks especially while undertaking joint operations.

NATO has ratified STANAG 4439 (IM) (1) and this is part of the policy of most NATO member nations. It is now recognised as an important approach to meeting the requirement worldwide and has therefore become a normal consideration in munition design.

What do we mean by IM?

STANAG 4439 defines IM as

> … munitions which reliably fulfil their performance, readiness and operational requirements on demand, but which minimize the probability of inadvertent initiation and severity of subsequent collateral damage to weapon platforms, logistic systems and personnel when subjected to unplanned stimuli….

7.1.1 History

There have always been accidents with munitions. For example:

James II, King of Scots, was killed by an exploding cannon at the siege of Roxburgh Castle in 1460.

The sinking of UK battlecruisers at Jutland in WWI, worsened by the vulnerability of the shell fill used at the time.

DOI: 10.1201/9781003529217-7

Accidents in the Gulf of Tonkin during the Vietnam War. USS Forrestal and USS Enterprise.
Ammunition depots exploding in UK and elsewhere, sometime caused by fires which led to sympathetic reaction.

IM policies are the result of experience in the last 60 years and have largely been led by the US, particularly the US Navy. The reasons for this include experience in the Vietnam War and later experiences in the Gulf and elsewhere (2).

However, in many places there remains a higher acceptance of risk. Those who subscribe to IM policies do not regard this level as acceptable and subscribe to the management of reduced risk through research on Insensitive Systems and the continual development of IM policy.

The increased effectiveness and power of many weapons have changed the balance of risk. Accidental damage can therefore be more catastrophic to both the munition owner and facilities and may also prevent successful prosecution of any mission. It puts operators at risk. However, at the same time understanding of both risks and their causes has improved which permits better risk management. This has of course not eliminated the hazards, which remain intrinsic.

The development of IM assessments complements the standard hazard tests as listed in the UN Recommendations on the Transport of Dangerous Goods (3). A series of tests were defined and listed in the STANAG and associated documents. They were aimed at assessing the response to a range of likely threats. These are shown below with the acceptable response.

Threat	Passing Criterion
Fast Cook-off	Type V Reaction
Slow Cook-off	Type V Reaction
Bullet Impact	Type V Reaction
Fragment Impact	Type V Reaction
Sympathetic Reaction	Type III Reaction

Threats and IM acceptable responses
The responses are generally defined as

Type I/II	Detonation or Partial Detonation
Type III	Explosion
Type IV	Deflagration or Propulsive Reaction
Type V	Burning Reaction

These tests and their analysis continue to evolve with the acquisition of data and improved understanding. This does mean that an IM 'fingerprint', the description produced by a series of tests, may need reassessment in the light of improved understanding.

7.2 DEVELOPMENT

- Vulnerability affects many aspects:
  - Manufacture
  - Transport
  - Storage and handling
  - Logistics
  - Use

Vulnerability is affected by the whole system; it is not merely the energetic materials, which are the main reactive component. It is also the interaction of these energetic materials with other components. It is therefore possible to introduce vulnerability into any system through poor design choices. However, this can and should be managed by proper understanding of the risks and compensating for them (4).

7.3 PROGRESS

Following experiences in the 1960s, NATO members realised that the technology and understanding probably existed to begin to reduce and manage the risks. As stated above, this initiative was US-led. It led to an agreement to create a Memorandum of Understanding (MOU) to form firstly a pilot project, and once this had demonstrated its worth, the NATO Insensitive Munitions Information Center (NIMIC) which later became MSIAC. (Munitions Safety Information Analysis Center)

In parallel there were UK/US/Ca/Aus studies on IM which produced draft protocols for assessment which were later transferred to NIMIC under the MOU produced for information sharing. France also contributed significantly to this development (5).

The Ammunition Safety Group in NATO was created to assist in the coordination and understanding of tests. This is a task it still pursues, often involving nations not normally connected with NATO (6).

7.4 ENERGETIC MATERIALS AND IM

As energetic materials are the source of most of the hazards, it is worth considering them separately:

- *Ingredients and Compositions*
  - *Performance and Safety are related*: Greater performance means greater risk, but only for a given set of ingredients! *Changing ingredients/formulation constituents can modify this balance.*
  - Several developments have taken place in the last 20 years. These offered new options to assist in IM development.
  - The aim is to provide well-designed and well-understood compositions to meet IM and other service requirements.
  - Existing ingredients can be modified and used/processed in different ways.
  - Improved measurement and modelling of sensitivity and initiation improve the risk assessment and help in choosing ingredient options.
  - Composite rocket propellant technology gave rise to Polymer-Bonded eXplosives (PBX), particularly in cast or extruded forms.
  - PBX contain solid filler in a rubbery matrix, the energetic loading meeting most performance goals (>80%), less in propellants. The proportion can be 95% in pressed compositions.
  - High solids loading means that the most desirable mechanical properties are not fully obtained.
  - It is sometime assumed that the performance of an Insensitive High Explosive (IHE) will be lower than that of more traditional materials. This is inaccurate but the design may need more care in order to achieve the required output.
  - *However, testing results indicate that cast PBX are overall less vulnerable than melt-cast alternatives, and performance targets can be met. There are of course exceptions and limitations.*
  - *It has been established that in some cases novel technology has raised the threshold for violent response. The energy remains, and provided the trigger lies beneath that threshold a benign response is likely. However, any response is likely to be more energetic as a result.*
- *Hazard Assessment and Modelling*
  - Modelling and mitigation are significant parts of the approach to IM systems.

  - Existing tests replicate accidents and are hard to transfer into models.
  - Research is aimed at diagnostic small scale tests providing data for validated models.
  - Linked with NATO groups and to assessment and revision of IM requirements.
  - *The aim should be to provide SUFFICIENT information for a systems assessment.*
  - *This requires that likely threats are identified with sufficient detail to perform a Threat Hazard Assessment (THA).*
- *Initiation Systems*
  - A critical part of the system
  - Effective low-vulnerability systems are essential. They can form a trigger for accidental events.
  - Modelling and prediction of initiation is important.
  - Electrical, mechanical, and laser systems are being investigated. Effective laser systems are available for strategic systems and could be applied elsewhere. Flyer plate technology and exploding bridge wires are also options.
- *Systems Integration*
  - Proper use must be made of energetic materials.
  - Systems design and incorporation is crucial if the best results are to be obtained.
  - Compositions should be designed to meet system requirements, and systems designed to make the best use of available technology.
  - Performance and mitigation methods need consideration to ensure a cost-effective solution.
  - *Full Systems Integration is essential for maximum benefit.*

IM have been produced and fielded. The concept has become standard requirement for new systems and for retrofitting older systems during a mid-life update.

They may now be better described as Reduced Vulnerability Munitions as they are still energetic and their insensitivity is based on the understanding of their behaviour, the threats to which they will be exposed, and the environment in which they are manufactured, stored, transported, and used. The understanding is developing with new tools produced based on research. This is clear when reviewing the literature (7,8).

When reports and papers are published, it is clear that there is always more work proposed. That is the nature of research and indeed a response to changes in understanding of both the technology and the environment as well as of the changing threats.

The example of XDT (Unknown to Detonation Transition) is an indication of this since removing the possibility of immediate initiation only revealed another mechanism for reaction – this has been shown by delayed detonation in a damaged material where the damage introduced the possibility of reaction (9).

It must be remembered that munitions are by their nature risky; they are designed to function and release energy in a useful way. However, there will always be the risk of improper functioning. The aim is to manage the risks as efficiently as possible.

There are several aspects of this, but the most important one is to understand the risks. That understanding must cover all the aspects that affect vulnerability throughout the lifetime of the munition. This is covered in the THA mentioned above. Without that understanding management is very much more difficult, though the level of understanding need only be sufficient to manage risk properly. However, this always needs review in the light of new understanding and an appreciation of new threats. This understanding can stimulate the development of new techniques for risk management. It is equally important to be sufficiently aware of the interdependences of the various aspects so that the various components can be combined in a suitable manner.

Thus, new ingredients and new formulations need to be considered together with their processing and performance, as well as their vulnerability. Ageing also needs to be considered since reactions continue within the chemical energetic system, and this may change the vulnerability signature. All need to be related to the systems in which the components are combined.

This requires an assessment of developing tools, provision of accurate data, empirical testing and development of new tools. It cannot all be done at once, and technical areas develop at different rates. Any theoretical approach must be validated, and, in some cases, a more empirical approach may be all that is possible and indeed this can be all that is needed, provided the assessment is as realistic as possible.

Therefore, research must be a mixture of empirical and theoretical work. That balance will change, as noted above, and must be subject to review.

7.5 RESEARCH PRIORITIES

If an iterative programme of improved understanding is to be obtained, then the end user needs must be understood as well as the science and technology directed to meeting those needs. They may not immediately answer all those needs but should offer improvement over the existing or previous position, so that the risk is managed and understood in ways relevant to the user. Therefore, the aims should be the following:

1. To be able to provide advice on vulnerability throughout the life cycle of a munition.
2. To be able to predict the characteristics of a munition system sufficiently well to allow proper fielding and management.
3. To be able to design reduced vulnerability systems meeting performance, life and environmental requirements.
4. To understand ignition well enough to design an optimised system.
5. To be able to predict the effect of ageing on vulnerability.
6. To provide intelligent assessments of IM characteristics when faced with unexpected or novel threats.

Different areas of the user community will have different needs. The balance of the priorities will therefore change depending on circumstances. Often systems have been accepted for service in a relatively narrowly defined environment and use outside of that environment can produce risks through temperature exposure (for example).

One issue has been and remains is the need to manage the vulnerability of munitions in urban terrain and in conflicts such as Iraq and Afghanistan where the environment results in new risks.

One of these newer problem areas is that of munition disposal. While this may seem to be peripheral to the main aspects of IM Technology, it is an area of increasing importance.

The use of IM systems has produced problems for disposal in the field, as existing tools for disposal are less adequate for the purpose. The existing tools are designed for disrupting more traditional explosives but are much less effective with materials that have a reduced vulnerability. They assume a level of response that, while typical of older materials, will probably not be produced by reduced-vulnerability materials.

This area is one that is growing in importance as a priority. Disruption is opening the munition in such a way as to make it unlikely to function and allow for access. This will, however, produce residues which need to be handled. The disruption itself may leave the energetic materials more sensitive, producing risks to those engaged in clean-up. There is also a security issue that can drive disposal as residues may be misused.

Residual materials also have a high risk of an environmental impact, and this also requires to be understood and its effect both measured and mitigated. Residual materials will include explosives such as TNT; HMX; and RDX; initiation systems and materials as well as contaminated components. In conflict zones, residues will be left and these need to be managed. Designing munitions to minimise this impact is important but limits the options in terms of explosive materials available for future systems.

All these factors are part of the changing IM landscape.

7.6 AREAS OF ACTIVE WORK

There are several apparently discrete areas where research is in progress, with different emphases in different countries, based in part on their local priorities. In some cases, they contribute to international collaboration in NATO or in other forums. However, these should be seen as the building blocks for an integrated approach to providing validated IM technology and a clear understanding of the direction of future research.

It is hard to get these disparate pieces to fit together. There is a danger of the view being 'stove piped' with limited awareness of broader technical issues. It is therefore hard to incorporate new options either from within the academic world or from technology developments in non-munitions areas. However, while it challenges the status quo and may need investment, it is not impossible. A UK example of this is the use of materials technology to provide improved modelling tools using the Porter–Gould Model (see Porter and Gould, 2003) (20,21) described in several papers at the International Ballistics Symposium and the Detonation Symposium.

In short, integration is an issue and there will be a continuing need to seek novel solutions to enable the whole technology area to meet future needs. This also requires active technology watching to ensure that the best novel solutions are developed.

It is important to examine the various strands of activity in overview before looking in more detail at the research. Many of these are already joined in some way, and one of the roles of MSIAC is to foster this integration amongst firstly their members and secondly in NATO through the Ammunition Safety Group.

The outline research topics are discussed in the following sections.

7.6.1 Detonics

It includes connecting microscale detonation properties to the macroscale via studies of the meso scale.

The ability to understand and predict behaviour remains an issue. It is clear from the reported results that there are more complications than simple Chapman–Jouget (CJ) theory suggests (see Chapters 3 and 4 for an overview of CJ Theory). The challenge remains to provide enough information to support a satisfactory analysis for use. Some of the complications are outlined below.

7.6.2 Testing

It is used to provide data for both empirical and semi-quantitative assessments as well as to sustain and direct modelling. Many tests simulate accidents and may therefore give false confidence in the result. An example of this is the experience with damaged material. This is known as XDT (unknown to Detonation Transition). It is now understood that a negative response can be followed by an event which is a result of damage to the energetic material and a delayed ignition. It was first observed with PBX which in contrast to melt-cast explosives did not react immediately. Simple application of the existing tests as a pass/fail would not show that the materials had been damaged sufficiently to produce risk.

The original NATO STANAG tests for IM are now seen to have limitations and they are currently being examined for updating or replacement. It is essential to understand what is being tested and why a test is performed. It is important to avoid a routine approach to testing where the result is disconnected from the risk but merely a process to be followed. The STANAG tests were produced in an attempt to use the best scientific information at the time to produce a series of tests to provide the best possible advice. Most of the existing national tests which are used to help assess the vulnerability of a munition are based on accidents which produced a violent response. These tests mimic the accident in a controlled way. For example, fragment attack and fuel fire clearly attempt to assess what will happen in these circumstances. Others like oblique impact mimic the effect of falling from a vehicle and being subjected to ground friction. They provide useful vulnerability information, but the data are limited.

7.6.3 Prediction of Response

How do we refine the STANAG Tests to provide more useful information? What is required in sufficient analytical information to support modelling and provide the means to predict response more accurately. This would provide information on the hot spot evolution and the growth of the reaction, rather than observing the macroscopic event. Studies of mesoscopic detonation processes are part of the (24). The understanding has moved on and so the tests need to change and evolve too. MSIAC is promoting studies of this, as are groups in NATO nations.

There is the same need to understand what is being tested and why a test is performed. Part of the general aim is to move from reliance on large-scale tests to the intelligent use of small-scale diagnostic tests with validation, if required, by full-scale testing. It is likely that full scale will be required in some ways for the foreseeable future. The cross-over between IM testing and more normal hazard testing has been recognised and future testing needs to provide useful information for a range of applications (6,11,22,23).

This is far from easy and is one of the possible uses of the studies now going on into the behaviour of energetic ingredients at high pressures. Even knowing the equation of state in those conditions would be helpful. That understanding should also help to design either crystals or compositions with a more benign response.

It will be very hard to produce something energetic capable of a benign response in such a harsh environment, but mitigation and minimisation may well be possible.

The questions remain:

- Can we provide data to populate models and refine them?
- Can we use small-scale tests to predict full-scale responses?
- Can we connect detonic-scale events to munition design?

7.6.4 Hazard Analysis

It helps in assessing the material hazard properties.

Linked with detonics and testing, it must now include examination of the toxicity of the materials being considered and the lifetime and compatibility requirements. Much of this should be covered as part of the review of the system and its components.

7.6.5 Threat Analyses

This must include new and changing threats and also be capable of modelling and predicting the likely scenarios.

The data provided will support the direction of research in order to support risk management both present and future. It is iterative since new materials and new concepts may change the threat priorities. For example, the likelihood of significant quantities of unreacted reduced sensitivity (RS) explosives becoming available will need to be considered.

7.6.6 Modelling

It helps in predicting effects and properties.

While the development of models and software can be considered in isolation, it is more usually linked with detonics and testing as the model provides a means for taking the theory and using it to predict the results. Often applying the theory and the modelling approach will define the data needs and therefore the experiments that are needed to provide that data, so that the models more closely approach reality. The aim should be to move to predictive models and away from curve-fitting approaches which often have limited value and can lose their validity with new understanding or indeed with new materials.

As noted below, modelling is becoming more important in the design of new or improved materials/ingredients. This area also includes predicting the mechanical properties and the changes with time as ageing can affect the IM signature. This is an important part of assessing the long-term suitability for service of a system.

Recent work has shown that it may be possible to predict sensitivity and use this understanding in munition design (10).

7.6.7 Ingredients

Ingredients which should provide improved options for formulations and include both existing and new options.

One of the factors affecting the existing response of materials is that the technology used in their manufacture is now obsolete. It is possible to produce ingredients whose size, shape, and size distribution are much more carefully controlled.

The area includes not just new materials such as novel nitrogen-only materials which have been worked on (12,13). Ingredients such as FOX-7, NTO, DNAN have been and are being investigated but have been shown to have their own problems (14). However, there has also been the development of improved versions of existing materials.

The best known of these is RS RDX (RS-RDX). The history of RS-RDX is interesting. I-RDX® was 'discovered' by SNPE in the 1990s. Other manufacturers identified their materials as also meeting an RS requirement – called RS-RDX. Several features appeared to define RS-RDX:

- RS-RDX crystals are high-quality crystals.
- RS-RDX crystals have high density.

Most are made by direct nitration, not the Bachman process. This reduces the amount of HMX present, and which makes them less sensitive. Work published by Chemring in Norway indicated that this was not the whole story, however. Cast PBX formulation (N-109) based on RS-RDX has less shock sensitivity than that based on standard RDX. Chemring in Norway have also produced RS-HMX (18,19). These and others are the result of improved understanding of crystallisation, and of the methods to apply that to produce materials closer to ideal, with fewer impurities and defects present. Thus, the improvement in crystal quality and prediction is offering new option for managing the vulnerability properties of ingredients.

Another area of development is the study of polymorphs where the species can exist in different polymorphic forms (14). These forms can exhibit different safety characteristics (such as HMX, for example) as well as differences in crystal density, for example CL20. The temperature at which forms change is an important parameter as this may well affect the safety of any formulation.

Beyond this there are designer species now being studied such as nanomaterials and co-crystals. Understanding the properties of crystals makes it possible to produce such materials with the possible option for designing an ingredient with specific properties.

Solid ingredients are only part of the solution for the binders, and other ingredients need consideration too. The origin of PBX and composite propellants illustrates the early stages of that process, with the use of rubbery material to give more ideal mechanical properties enabling the delocalisation of the shock. Such inert materials could not contribute to the energy of the system, though the common binder hydroxyl-terminated polybutadiene (HTPB) acts as a fuel in propellants. The solid loading was therefore too high to optimise the mechanical properties. This led to the preparation of energetic binders which would contribute to the output.

This approach seemed promising but has not been adopted significantly due to cost and the changes in the priorities for operations. If there is a need, they may be employed more extensively. There is, however, a problem with their cure chemistry as they use isocyanates to produce the polymer, as indeed does HTPB. Isocyanates are highly toxic and so research work is in progress to produce alternatives.

This last point indicates one of the new priorities affecting the search for new options. An assessment of toxicity is required, in the EU to meet REACH (Registration, Evaluation, Authorisation, and Restriction of Chemicals) requirements and in the US to meet EPA (Environment Protection Agency) regulations. Toxicity is another factor to consider.

The application of artificial intelligence and machine learning may have a very significant impact in all the areas noted in modelling above. All these offer new options for formulation.

7.6.8 Formulations

It provides new options for IM systems.

Well-understood insensitive formulations are needed for

- Warheads
- Gun systems
- Rocket systems
- Ignition
- Pyrotechnics

In general terms there is a relation between safety and performance. Increasing the explosives output tends to reduce safety. There are ways to manage this through improving the quality of the materials and their properties.

As mentioned in the introduction, one of the first routes to reduced vulnerability was by the use of rubbery binders to produce improved mechanical properties which reduce the vulnerability of the compositions to impact (for example). This had been developed for composite rocket propellants (7,8).

One of the features in the last 20 years has been the drawing together of energetics research strands with the enhancement of the technology through the synergy of that process of integration.

The use of energetic binders was seen as a way of maintaining performance but has so far failed to make a significant mark. New options for melt cast such as 2,4-Dinitroanisole (DNAN) or Guanylurea dinitramide (GUDN) appeared to offer some promise as alternatives to more traditional routes. However, they have problems with dimensional stability of heat cycling (DNAN) and other issues. Some of this in the case of DNAN is probably due to polymorphic shifts.

While explosives have the highest energy and may appear to offer the highest risk, the rocket motor is larger and if ignited does significant damage, even leading to detonation in some cases. Therefore, the same approaches are being used, and the same problems investigated.

Newer materials must also meet increasingly stringent environmental constraints which are affecting their introduction. They also have their own intrinsic limitations.

7.6.9 Components

Components are an important part of system construction and integration for IM.

Energetics are not the only technology to consider. Their interaction and compatibility with other materials need assessment, and those other materials can add to the solution or if misused or wrongly chosen can provide a route to vulnerability. This has been examined in an international programme (11).

The approach to system integration has changed only slowly over the last 20/30 years. This is in part because of the need to ensure minimum risk. Much of that risk assessment is based on assumptions made in the past and only slowly evolving to meet user's changing requirements. Mitigation techniques are being used as part of a systems approach (15).

Modern materials and understanding are changing those assumptions, and these changes are providing the munition designer with choices that would not have been possible ten years ago. However, the impact of these changes still needs to be carefully evaluated especially with respect to time and the impact of ageing. This is one area where modelling and AI/ML may have a significant impact.

7.6.10 Processing

Anything needs to be made.

This is often the constraint that prevents new materials and new approaches being adopted. Novel approaches such as Resonant Acoustic Mixing (RAM) or twin-screw extrusion offer some advantages. Recent research on RAM mixing offers promise for the manufacture of suitable compositions. It allows for the easy processing of difficult materials, and it may allow proper and effective handling of nanomaterials.

7.6.11 Performance

Predictable performance is essential to ensure that we can meet the requirements of the military users.

Whatever is produced and offered needs to meet the performance requirements for the system. This has often been quoted as a problem with IM, in which it was deemed to require a performance reduction. This has been shown to be untrue, but it must be demonstrated that the system fully meets the requirements as well as the IM need. Other constraints will operate too, such as life, environment, and disposability.

7.6.12 Cost

Cost assessment is needed to ensure we can afford the proposed solution.

Whatever options are considered, the cost needs to be one factor in that consideration. Often, however, true costs only become apparent after research and development investment. New options in development will cost more and the cost can only reduce if those materials and systems are developed to a manufacturing level. Cost awareness is necessary but should not be the only factor.

Whole Life Cost Assessment (WLA) is one area now being used to assess real cost. This considers all parameters that affect value throughout the life of the munition system. This can give a better understanding of real cost and provide information to justify new approaches (16).

7.7 CONCLUSIONS

Overall, IM has developed continuously over decades and while it is apparently no longer a major issue, it remains an important part of both research and development. The requirement is a normal one for all NATO nations, and others, which leads to continued investment and changes. Since it was introduced other requirements such as those for dealing with residues and disposal generally have been developed, as has a need to manage toxicity and environmental impact.

Risks have been reduced and new technology fielded. This is likely to continue. One feature is the re-engineering of existing materials to improve their behaviour through the application of new understanding and approaches. That is likely to continue.

Therefore, IM will remain a system requirement, while evolving to respond to changes in requirements and technology (17).

Various reviews have indicated the areas where research work was thought to be needed, while building on existing expertise, experience, and needs. It is worth concluding with a version of that list as it indicates where gaps are seen, and which are also seen to be capable of resolution with research. It also shows the research interests of those involved; for like any list it is partial.

- Phenomenology
  - Understanding underlying mechanisms for reactivity
- Boosters
- Energetic Materials and their contribution to sensitivity
  - Particle size/particle packing
  - Crystal morphology/crystal shape
  - Crystal quality
  - Phase changes/phase stabilisation
- Mitigation
  - Ageing of IM solutions
- Sub-scale to full-scale testing comparisons

Most of these have been reflected earlier, though others such as processing do not appear. Research is needed, and new approaches can provide new solutions and new options.

Developments in all these areas are reported regularly, both at conferences, and in the literature. *Propellants, Explosives Pyrotechnics* is one journal worth following, as is the *Journal of Energetic Materials*. Work is presented at the annual ICT Conference in Germany and at the NDIA – both conferences held regularly. There are other meetings, such as the Detonation Symposium, and New Trends in Energetic Materials, which is held annually in Pardubice and aimed at younger scientists, where information is shared and discussed.

All these conferences produce awareness of developments and provide opportunities for technology watch.

REFERENCES

1. NATO STANAG 4439, Edition X (2010); "Policy for Introduction and Assessment of Insensitive Munitions", March 2010; AOP-39, Edition 3, "Guidance on the Assessment and Development of Insensitive Munitions".
2. M Till, K Kosar (2006); "Formation of the US Navy's Munition Reaction Evaluation Board (MREB)", Proceedings from the 2006 Insensitive Munitions & Energetic Materials Technology Symposium, Bristol, UK.
3. UN Recommendations on the Transport of Dangerous Goods, Manual of Tests and Criteria (2015); 6th Edition.
4. IJ Powell (2016); Insensitive Munitions – Design principles and technology developments; Propellants, Explosives and Pyrotechnics; 41(3), 409–413. Wiley-VCH.
5. S Bordachar, P Lamy (2012); "A New Impulse for the French MURAT (IM) Policy", Proceedings from the 2012 Insensitive Munitions & Energetic Materials Technology Symposium, 14-17 May 2012, Las Vegas, NV, USA.
6. C Jacq, K Tomasello, E Baker, M Sharp (2019); Review and Update of Insensitive Munitions Test Procedures; 2019 Insensitive Munitions & Energetic Materials Technology Symposium, 21-24 October 2019, Seville, Spain, NDIA.
7. P Lamy, CO Leiber, AS Cumming, M Zimmer (1996); "ASNR LTTP on Insensitive High Explosives", Proceedings of the 27th ICT Conference, Karlsruhe, ICT, 25–28, June 1996.
8. AS Cumming et al (2007); "Performance Tests of Next Generation Solid Missile Propellants", NDIA 2007 Insensitive Munitions and Energetic Materials Technology Symposium, 15-18 October 2007, Miami, Florida, NDIA.

9. MD Cook, AD Wood, PR Ottley, PJ Cheese (2014); Observation of sub-detonation response in confined high density HMX based PBXs; 18th APS-SCCM and 24th AIRAPT. Journal of Physics: Conference Series, Volume 500, Issue 5, article id. 052009 (2014).
10. AAL Michalchuk, M Trestman, S Rudić, P Portius, PT Fincham, CR Pulham, CA Morrison (2019); Predicting the reactivity of energetic materials: An ab initio multi-phonon approach; Journal of Materials Chemistry A; 7, 33, 19539.
11. T Swierk, S Collignon (2019); IM Hazards Modelling & Simulation – An International Collaboration; NDIA 2019 Insensitive Munitions & Energetic Materials Technology, 21-24 October, Seville, Spain, NDIA.
12. DA Dixon, D Feller, KO Christe (2004); Enthalpies of Formation of Gas-Phase N3, N3−, N5+, and N5− from Ab Initio Molecular Orbital Theory, Stability Predictions for N5+N3− and N5+N5−, and Experimental Evidence for the Instability of N5+N3−: JACS 126, 834–843 ACS.
13. H Östmark, S Wallin, T Brinck, P Carlqvist, R Claridge, F Hedlund, L Yudina (2003): Detection of pentazolate anion (cyclo-N5−) from laser ionization and decomposition of solid p-dimethylaminophenylpentazole; Chemical Physics Letters; 379, 539–546.
14. P Coster, C Henderson, S Hunter, C Pulham et al (2014); Explosives at extreme conditions: Polymorphism of 2,4-Dinitroanisole; Acta Crystallographica Section A: Foundations and Advances; 70(a1), C896.
15. B Smit (2010); "Mitigation Techniques for Reduced Rocket Motor Vulnerability against External Thermal Stimuli", NDIA 2010 Insensitive Munitions and Energetic Materials Symposium, 11–14 October 2010, Munich, Germany.
16. P Deacon et al (2013); The Effect of Energetic Materials Ageing on System IM Response; NDIA 2013 Insensitive Munitions & Energetic Materials Technology, 7–10 October 2013, San Diego, US.
17. UK MOD (2018); Defence ordnance, munitions and explosives safety and environmental management policy (JSP 520).
18. OH Johansen et al (2008); RDX and HMX with reduced sensitivity Propellants; Explosives and Pyrotechnics; 33, 20–24.
19. R Doherty, D Watt (2008); Relationship between RDX properties and sensitivity; Propellants, Explosives and Pyrotechnics; 33:1, 4–13.
20. D Porter, P Gould, (2003); Multiscale modelling for equations of state; J. Phys. IV France; 110, 809–814.
21. R Cornish, D Porter, P Church, P Gould, T Andrews, B Proud, D Drodge, C Siviour (2007) Shock Compression of Condensed Matter, Hawaii, p. 777. AIP Publishing.
22. WG Proud (2022); Ignition and Detonation in Energetic Materials: An Introduction; Chapter 7 in Energetics Science and Technology, IOP eBooks.
23. R Mendes et al (2022); Submillimetre Spatially Resolved Observation of Detonation Phenomena; Chapter 8 in Energetics Science and Technology, IOP eBooks.
24. JB Maillet, E Bourasseau, N Desbiens, G Vallverdu, G Stoltz (2011); Mesoscopic simulations of shock-to-detonation transition in reactive liquid high explosive; Europhysics Letters 96:6. 68007. IOP.

8 Testing of Explosives and Explosive Ordnance

Mathew P. Brooks

8.1 INTRODUCTION

Test and evaluation (T&E) of explosives and explosive ordnance (EO) is crucial throughout their lifecycle, including design, production, deployment, and disposal. The primary goals are to ensure safety, performance, reliability, and compliance with regulatory standards. This involves assessing the chemical and physical properties of explosives, their behaviour under various conditions, and safety during handling, storage, and use.

The significance of T&E is paramount, especially as explosives are utilised across multiple domains and industries. Ensuring energetic materials function as intended, preventing unintended detonations or failures, is vital for the safety of personnel and the success of operations. Rigorous testing also minimises environmental impacts, ensuring compliance with regulations. EO, including bombs, grenades, and missiles, also demand evaluation. Each type requires specific test programmes to validate effectiveness and safety, encompassing stability assessments, performance testing for characteristics, and post-event analysis.

This chapter will explore various testing methods, covering chemical, physical, and environmental assessments. It will highlight key standards, equipment, and test data requirements, providing insight into the challenges and advancements in testing technologies and programmes. By examining tests, readers will gain an understanding of the complexities in evaluating energetic materials, emphasising the essential role of T&E in ensuring that explosives are safe, reliable, and effective.

T&E plays a vital role in ensuring the safe, reliable, and effective performance of explosives and EO. Through controlled tests and subsequent analysis, T&E gathers critical data to achieve a set of objectives. This chapter will explore the fundamental purposes driving explosive T&E. By fulfilling these objectives, explosive T&E provides invaluable data for the following:

- **Developing novel and improved explosives** with tailored properties for specific applications. New materials or formulations can be tested to achieve desired characteristics like increased blast wave pressure, better fragmentation patterns, or lower sensitivity.
- **Ensuring the safe and reliable performance** of existing explosives throughout their lifecycle. T&E data helps establish handling procedures, storage guidelines, and transport regulations to minimise risks.
- **Optimising the effectiveness of explosives** in various applications. By understanding explosive performance against different targets, users can develop effective detonation strategies, while industries can improve efficiency in demolition, excavation, or underwater operations.

8.2 CHARACTERISATION OF EXPLOSIVE PROPERTIES

A foundational objective of T&E revolves around understanding the inherent properties of explosives. This entails quantifying characteristics that define an explosive's behaviour, including the following:

- *Detonation Velocity*: This metric denotes the speed at which the detonation wave propagates through the explosive material. It directly influences the power and timing of the explosive effect.

 DOI: 10.1201/9781003529217-8

- *Sensitivity*: This refers to the ease with which an explosive can be initiated by stimuli such as impact, thermal energy, or friction. A thorough understanding of sensitivity is paramount for safe handling and storage practices.
- *Density*: The denser the explosive, the more of it that can be placed in a borehole of a given size. Explosive density is often expressed with the units of g/cc.

8.2.1 Vacuum Stability Test

The vacuum stability test is a critical method used to evaluate the thermal stability and decomposition characteristics of explosives and propellants, crucial for determining their safety and longevity. This test assesses how these materials behave under elevated temperatures and vacuum conditions, providing insights into their stability (STANAG 4556, 1999). The procedure involves several key steps: initially, the test sample is placed in a vacuum environment by connecting the test tube to a vacuum system, which evacuates the air to minimise atmospheric pressure influence. The test tube is then heated in a controlled environment, typically an oven, at temperatures ranging from 100 to 120°C for a predetermined duration based on the explosive type. During heating, gases evolved from the material's decomposition are collected using devices like a gas burette. Finally, the volume of gas evolved is measured over a specific time, serving as an indicator of the material's stability. The rate of gas evolution is a vital parameter in assessing the stability and safety of the material.

8.2.2 Koenen Test for Explosives

The Koenen test is an assessment tool used to evaluate the sensitivity of explosives and propellants to intense heat under confinement, determining their potential to detonate or deflagrate under high temperatures and pressure (Manual of tests and criteria, Revision 8, 2023). This test is crucial for characterising materials prone to violent reactions, thus enhancing the safety of their handling and storage. The process involves several key steps:

1. *Confinement*: A steel tube, filled with the explosive sample, is sealed with a steel plug that features a hole of a specified diameter. This hole controls the degree of confinement, and its size can vary to accommodate different levels of confinement needed for the test. A standardised test apparatus is outlined in Figure 8.1.
2. *Heating*: The confined sample is then placed either in a heating device or exposed to a regulated propane flame, where it is subjected to a uniform and controlled temperature increase. The rate of heating is carefully specified to maintain consistency across tests.
3. *Observation*: Throughout the test, the behaviour of the sample is closely monitored for any signs of reaction, such as deflagration, explosion, or the ejection of the plug.
4. *Assessment*: After the test, the reaction's severity is evaluated based on the condition of the tube and plug, including any evidence of plug ejection, tube deformation, or fragmentation. Examples of tube deformation is outlined in Figure 8.2, including condition codes in Table 8.1.

By assessing how explosive materials respond to extreme conditions, the Koenen test ensures the safe management of energetic materials.

8.2.3 Time/Pressure Test for Explosives

The Time/Pressure test is a methodology used to evaluate the ignition and burning characteristics of explosives and propellants in a confined setting. This test primarily aims to measure the pressure rise over time within a sealed vessel when an explosive material is ignited, providing essential

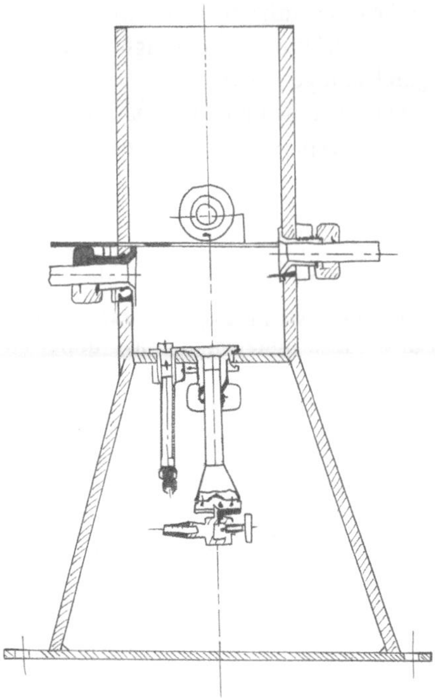

FIGURE 8.1 Standard Koenen test apparatus setup. (Adapted from Manual of tests and criteria, Revision 8, 2023.)

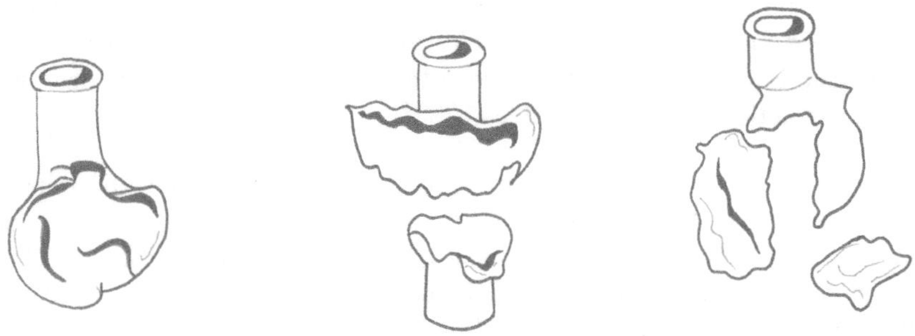

FIGURE 8.2 Tube deformation examples. (Adapted from Manual of tests and criteria, Revision 8, 2023.)

TABLE 8.1
Tube Deformation Condition Codes (Manual of tests and criteria, Revision 8, 2023)

Code	Description
O	Tube unchanged
A	Bottom of tube bulged out
B	Bottom and wall of the tube bulged out
C	Bottom of tube split
D	Wall of tube split
E	Tube split into two fragments
F	Tube fragmented into three or more mainly large pieces
G	Tube fragmented into many mainly small pieces, closing device undamaged
H	Tube fragmented into many very small pieces, closing device bulged out or fragmented

data on the material's burning rate and sensitivity to ignition, which are vital for safety assessments (Manual of tests and criteria, Revision 8, 2023). The process involves several key steps:

1. *Ignition*: The process begins by using an electric fuse or a pyrotechnic igniter to initiate the reaction within the test vessel. This is dependent on the energetic that is being tested.
2. *Confinement and Sealing*: The vessel, as outlined in Figure 8.3, is sealed to ensure that all gases produced during the reaction remain confined, allowing for precise measurement of the pressure increase.
3. *Heating*: Following ignition, the explosive material begins to burn, causing an increase in the pressure inside the vessel.

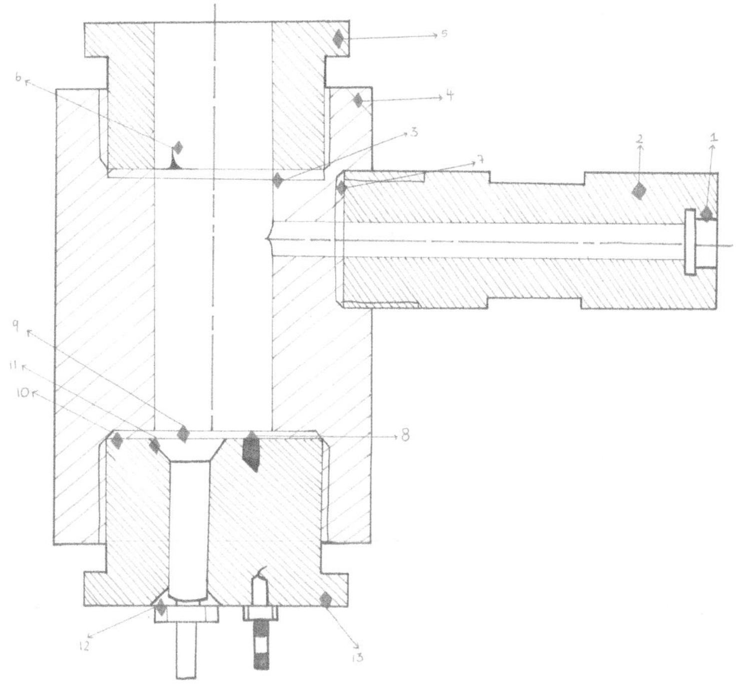

Element	Description
1	Pressure transducer thread
2	Side arm
3	Soft lead washer
4	Pressure vessel body
5	Bursting disc retaining plug
6	Bursting disc
7	Copper washer
8	Earthed electrode
9	Insulated electrode
10	Washer distorting groove
11	Insulation
12	Steel cone
13	Firing plug

FIGURE 8.3 Time/pressure test vessel. (Adapted from Manual of tests and criteria, Revision 8, 2023.)

TABLE 8.2
Time/Pressure Test Example Results (Manual of tests and criteria, Revision 8, 2023)

Energetic	Maximum Pressure (kPA)	Time for a Pressure Rise (ms)	Result
Ammonium nitrate (high-density pills)	<2070	–	–
Ammonium perchlorate (2 μm)	<2070	5	+
Nitroguanidine	>2070	400	+
Urea nitrate	>2070	400	+

4. *Pressure Measurement*: A pressure transducer or a similar device continuously records the pressure inside the vessel, capturing data until the pressure peaks or the reaction concludes.
5. *Data Analysis*: The collected pressure versus time data is analysed to determine key characteristics of the explosive material, such as the time to ignition, maximum pressure achieved, and the rate of pressure rise. A summary of example results is outlined in Table 8.2.

By assessing how explosive materials respond to controlled ignition in a confined environment, the Time/Pressure test plays an indispensable role in the comprehensive safety management of explosives, providing crucial insights that help mitigate risks associated with their use and storage.

8.2.4 Thermal Stability Test at 75°C for Explosives

The thermal stability test at 75°C is designed to assess the stability of explosive materials under elevated temperatures, crucial for evaluating their safety and reliability. The primary goal of this test is to determine how well an explosive material maintains its stability when subjected to a consistent temperature of 75°C (Manual of tests and criteria, Revision 8, 2023). This assessment is vital for identifying materials prone to decomposition or other reactions that could potentially compromise safety. The process involves several key steps:

1. *Heating*: The sample is placed in a temperature-controlled oven designed to maintain a steady temperature of 75°C. An example test setup is outlined in Figure 8.4.
2. *Duration*: The exposure time varies typically between 48 and 100 hours, tailored to meet specific test requirements.
3. *Gas Collection*: If applicable, gases emitted during the heating process are captured in a gas burette or a similar device to measure the volume and assess the extent of decomposition.
4. *Observation*: During the test, the sample is continuously monitored for any physical or chemical changes, such as colour change, melting, or gas evolution. Periodic weighing may be conducted to record any mass loss from decomposition.
5. *Post-Test Analysis*: Once the test concludes, the sample is removed from the heat source, allowed to cool, and then analysed to identify any signs of decomposition or other chemical reactions that may have occurred. A summary of example results is outlined in Table 8.3.

This test is fundamental in ensuring the safety and effectiveness of explosives by determining their thermal stability at elevated temperatures. The thermal stability test at 75°C provides crucial data, enabling manufacturers and safety personnel to identify high-risk materials and implement necessary safety measures.

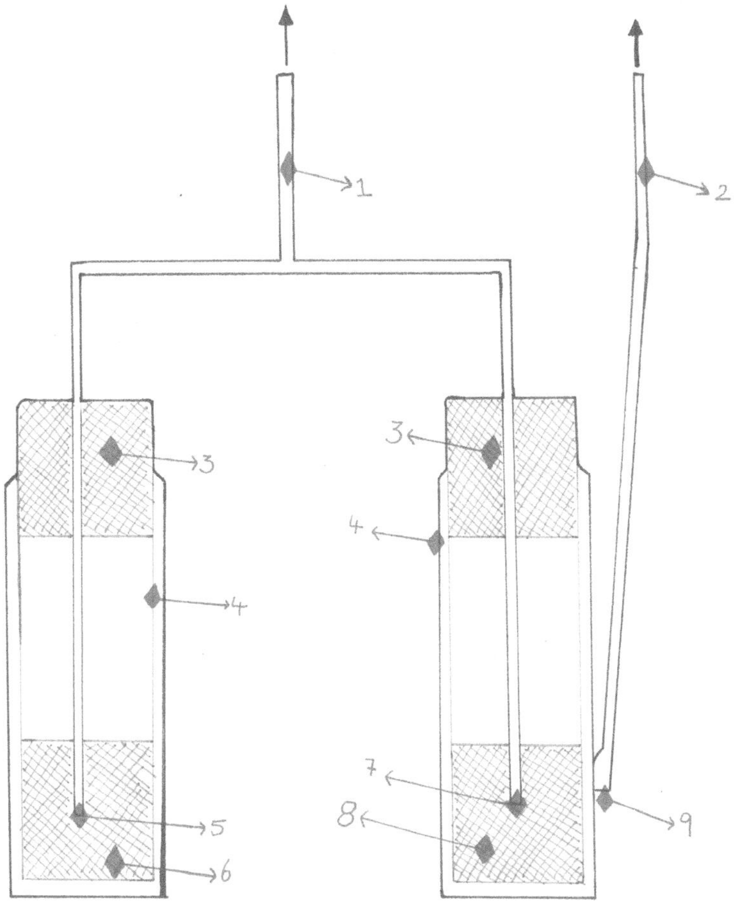

Element	Description
1	To millivoltmeter (T1–T2)
2	To millivoltmeter (T3)
3	Stoppers
4	Glass tubes
5	Thermocouple No 1 (T1)
6	Thermocouple No 2 (T2)
7	100 cm^3 of sample
8	Thermocouple No 3 (T3)
9	100 cm^3 of reference substance

FIGURE 8.4 Thermal stability test at 75°C. (Adapted from Manual of tests and criteria, Revision 8, 2023.)

TABLE 8.3
Thermal Stability at 75°C Test Example Results (Manual of tests and criteria, Revision 8, 2023)

Energetic	Observations	Result
RDX, 22% water-wetted	Mass loss <1%	–
70% AP, 16 Al, 2.5% Catocene, 11.5% binder	Oxidation reaction took place on Catocene (burning rate catalyst). Same discolouration on the surface but there is no chemical decomposition	–

8.2.5 Small-Scale Burning Test for Explosives

The small-scale burning test is a critical evaluation designed to assess the flammability and burning behaviour of explosive materials in a controlled environment. An overview of the test is outlined in Figure 8.5. The main objective of this test is to observe and document the burning characteristics of an explosive material when ignited, providing insights into potential hazards such as flame spread rate, intensity of burning, and the likelihood of producing hazardous fragments or residues (Manual of tests and criteria, Revision 8, 2023). The process involves several key steps:

1. *Ignition*: The test begins by igniting the sample using a standardised ignition source—such as a flame, spark, or hot wire—to ensure consistent and reproducible ignition conditions.
2. *Burning Observation*: The behaviour of the burning sample is observed and recorded. Key observations include the sample ignition time, characteristics of the flame (e.g., colour, height, spread rate), duration of the burning, presence of explosive reactions like pops, bangs, or flashes, and the generation of smoke or fumes.
3. *Post-Burning Analysis*: After the burning has ceased or the sample is completely consumed, the residue is carefully examined. This analysis focuses on the quantity and type of residue, as well as any fragmentation of the material.

This test provides essential data for understanding the safety risks associated with the use of explosive materials. Through detailed observation and post-burning analysis, the small-scale

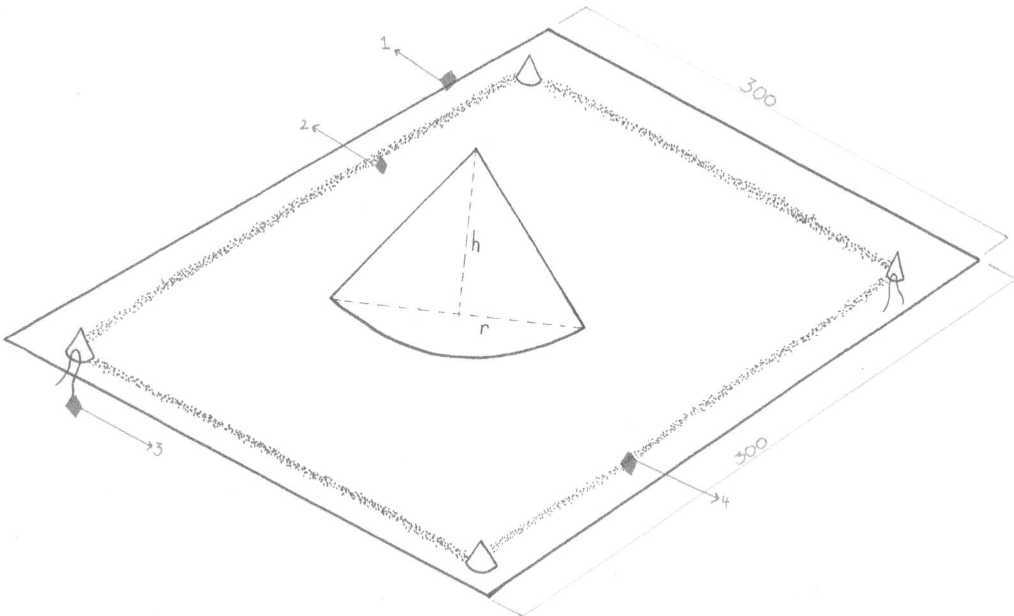

Element	Description
1	Sheet of kraft paper
2	Test substance
3	Ignition by an igniter and a few grammes of fine smokeless powder (opposite corners)
4	Trail of fine smokeless paper

FIGURE 8.5 Small-Scale burning test for explosives. (Adapted from Manual of tests and criteria, Revision 8, 2023.)

burning test helps in identifying and mitigating potential hazards, ensuring the safe handling and use of explosive substances.

8.2.6 BAM Friction Apparatus Test for Explosives

The Bundesanstalt für Materialforschung und-prüfung certification (BAM) friction apparatus test is an evaluation method used to assess the sensitivity of explosive materials to friction, aiming to measure how susceptible these materials are to ignition or explosion when subjected to mechanical friction (Manual of tests and criteria, Revision 8, 2023). This test is particularly important for determining the safety of materials during handling, processing, or transportation where frictional forces are likely to occur. The process involves several key steps:

1. *Apparatus Setup*: The sample is placed on a porcelain plate within the BAM Friction Apparatus, which includes a moving arm equipped with a steel pin or peg designed to apply a specific frictional force to the sample as outlined in Figure 8.6.

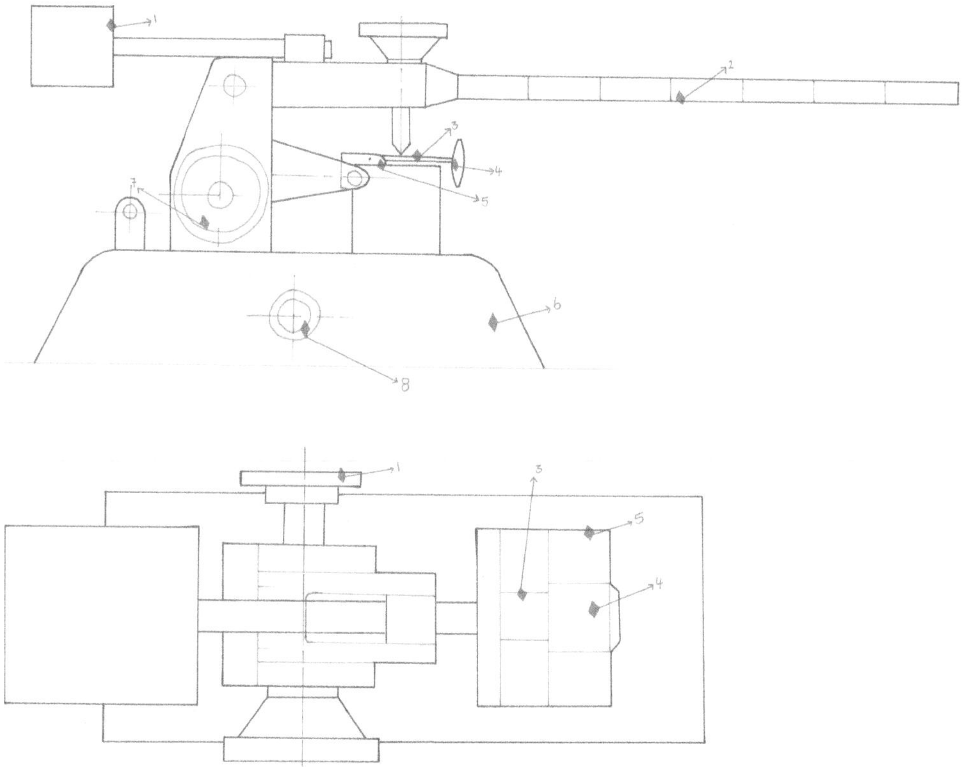

Element	Description
1	Counterweight
2	Loading arm
3	Porcelain plate held on carriage
4	Adjusting rod
5	Moveable carriage
6	Steel base
7	Setting handle
8	Directional adjustment

FIGURE 8.6 BAM friction apparatus test for explosives. (Adapted from Manual of tests and criteria, Revision 8, 2023.)

2. *Application of Force*: The steel pin or peg is pressed against the sample using a predetermined load, and then the arm is moved to generate friction. Both the force applied and the speed of the movement are carefully controlled to maintain consistent test conditions.
3. *Observation*: While the friction is applied, the sample is closely monitored for any signs of reaction such as ignition, deflagration, or explosion. The occurrence of these reactions indicates the material's friction sensitivity.
4. *Repeat Tests*: To accurately characterise the material's sensitivity, the test is conducted multiple times with various samples and different loads to determine the minimum friction force required to initiate a reaction.

The BAM Friction Apparatus test provides essential insights into the friction sensitivity of explosive materials, helping to identify risks and ensure the safety of these substances during their lifecycle.

8.2.7 UN Gap Test for Explosives

The United Nations (UN) gap test is an evaluation method used to assess the detonation sensitivity of explosives, specifically designed to determine their susceptibility to detonation under specific conditions. The primary goal of this test is to evaluate the likelihood of an explosive material initiating a detonation when subjected to a high-velocity shock wave from an adjacent charge as outlined in Figure 8.7 (Manual of tests and criteria, Revision 8, 2023). This standard test has two configurations: zero gap and 50 mm gap.

The process involves several key steps:

1. *Sample Preparation*: Two cylindrical charges, the donor (initiating charge) and the acceptor (target charge), are prepared from the explosive material. Both charges are typically made of the same material and have the same diameter to ensure consistency in testing.
2. *Setup*: The donor and acceptor charges are placed in direct contact with each other, creating a "zero gap" condition. This setup is critical as it ensures there is no spatial separation between the charges, allowing for direct transmission of the shock wave.
3. *Initiation*: The test begins with the initiation of the donor charge using a detonator or a high-explosive booster. This generates a high-velocity shock wave that propagates directly from the donor to the acceptor charge.
4. *Observation*: The behaviour of the acceptor charge upon receiving the shock wave is closely observed to determine if it undergoes detonation. The assessment typically involves examining the physical state of the acceptor charge post-test to see if it detonated.
5. *Repeat Tests*: To ensure the reliability and reproducibility of results, the test is conducted multiple times with different samples, allowing for statistically significant outcomes.

The UN gap test provides essential insights into the detonation sensitivity of explosives. By precisely evaluating how explosives respond to direct shock waves in a controlled environment, this test plays a vital role in the safety and stability assessments of explosive materials.

8.2.8 BAM Fallhammer Test for Explosives

The BAM Fallhammer test is a critical evaluation method designed to assess the impact sensitivity of explosive materials, determining their susceptibility to ignition or explosion when subjected to mechanical shocks or impacts. This test is essential for identifying explosive materials that may pose risks during handling, processing, or transportation (Manual of tests and criteria, Revision 8, 2023). The process involves several key steps:

1. *Apparatus Setup*: The test involves a BAM Fallhammer apparatus as outlined in Figure 8.8, which includes a vertically guided drop hammer of specified weight. The explosive sample is positioned on the lower anvil of the apparatus, ready for impact testing.

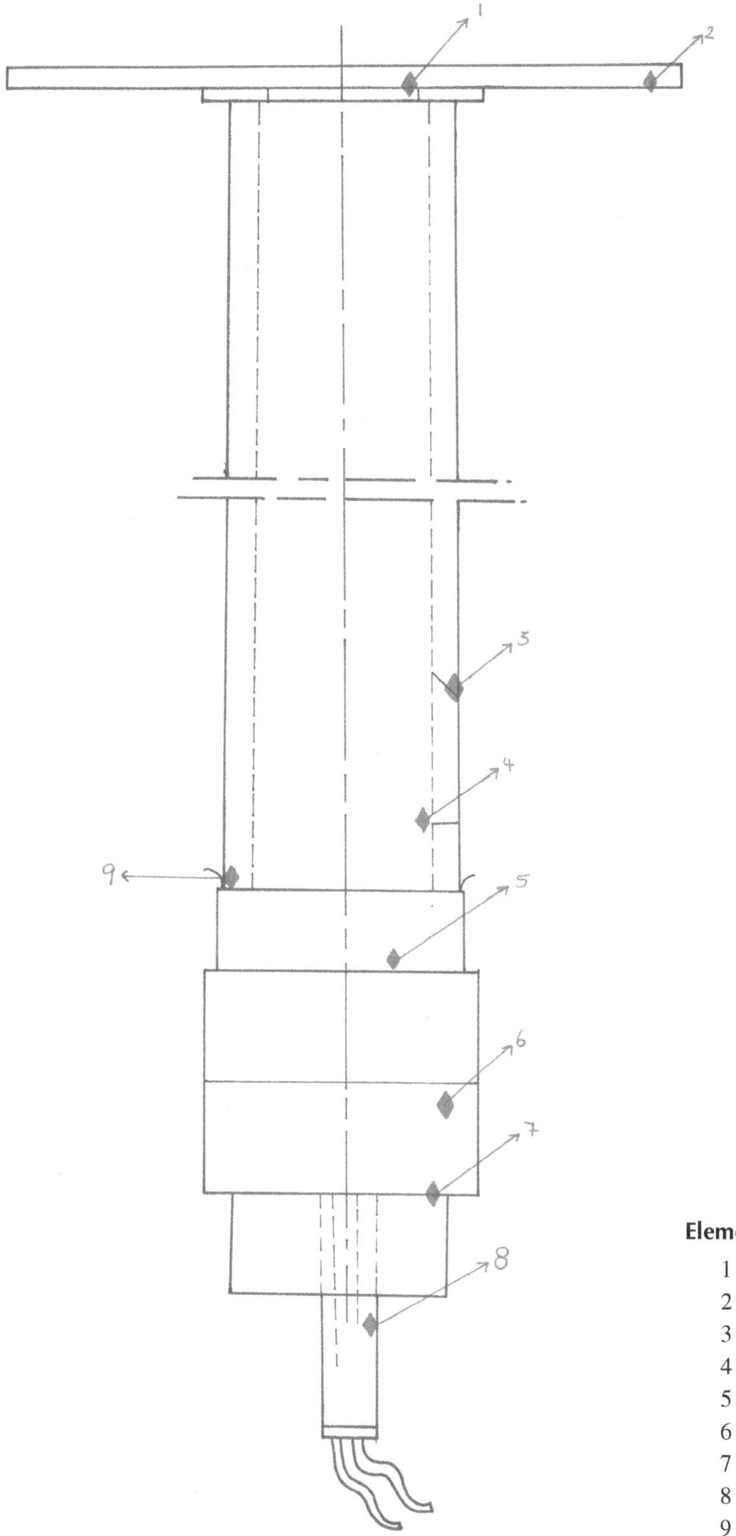

Element	Description
1	Spacers
2	Witness plate
3	Steel tube
4	Substance under test
5	PMMA spacer
6	Booster charge
7	Detonator holder
8	Detonator
9	Plastic membrane

FIGURE 8.7 UN gap test (zero gap) for explosives. (Adapted from Manual of tests and criteria, Revision 8, 2023.)

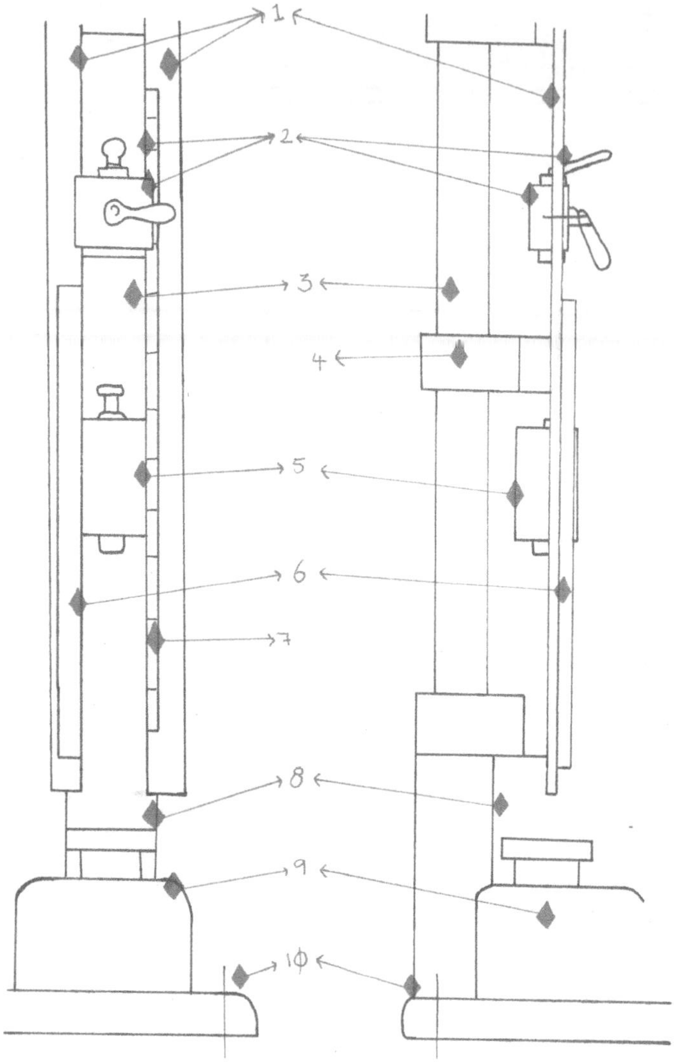

Element	Description
1	Two guides
2	Holding/releasing device
3	Column
4	Middle cross-piece
5	Drop weight
6	Toothed rack
7	Graduated scale
8	Anvil (100 mm diameter × 70 mm)
9	Steel block (230 mm × 250 mm × 200 mm)
10	Base (450 mm × 450 mm × 60 mm)

FIGURE 8.8 BAM Fallhammer test for explosives. (Adapted from Manual of tests and criteria, Revision 8, 2023.)

2. *Dropping the Hammer*: The hammer is elevated to a predetermined height and then released, allowing it to free-fall and strike the sample with a defined impact energy. The drop height, which dictates the impact energy, can be varied to test different levels of impact sensitivity.
3. *Observation*: Immediately following the impact, the behaviour of the sample is closely monitored. Observations include any signs of reaction such as ignition, deflagration, explosion, and more subtle indicators like sparks or smoke.
4. *Repeat Tests*: To accurately characterise the material's sensitivity to impact, the test is repeated multiple times with various samples and different drop heights. This repetition helps determine the minimum impact energy required to initiate a reaction.

The BAM Fallhammer test characterises the impact sensitivity of explosives, ensuring that materials prone to hazardous reactions under impact are identified and managed appropriately. This testing is fundamental to maintaining safety standards and minimising lifecycle risk.

8.3 PERFORMANCE EVALUATION

T&E extends beyond characterisation of the explosive itself, delving into its performance during application. This involves assessing how the explosive interacts with its environment and delivers its intended effects:

- *Fragmentation*: This evaluation focuses on the ability of an explosive charge to fracture a surrounding material (e.g., casing) into usable fragments acting as projectiles. Shard size, distribution, and velocity are all critical aspects.
- *Blast*: This metric measures the pressure wave generated by the detonation, employed for demolition or breaching operations. Peak pressure, impulse (pressure-time integral), and duration are all relevant factors.

8.3.1 ITOP-4.2.511 – Measurement of Unmanned Target Activated Weapon Warhead Performance

ITOP-4.2.511, *Measurement of Unmanned Target Activated Weapon Warhead Performance*, is a standard for assessing the effectiveness and performance of warheads used in unmanned target-activated weapons (UTAW). This standard ensures that these weapon systems perform reliably and safely by standardising test procedures to reflect real-world scenarios. The primary goal is to rigorously evaluate the operational effectiveness and reliability of UTAW warheads under controlled conditions, focusing on detonation characteristics, measuring lethal radius, determining fragmentation patterns, and assessing impacts on various target types (ITOP-4.2.511, 1999). The standard applies broadly to a range of UTAWs, including missiles, drones, and other remotely operated systems, with tests conducted in environments such as outdoor ranges or controlled indoor facilities based on the weapon system's intended use.

Test procedures under ITOP-4.2.511 include several key assessments, activation mechanisms of the warhead as it nears the target, timing and consistency of detonation, and the analysis of fragmentation distribution and velocity using sophisticated measurement tools like high-speed cameras and pressure gauges. Additionally, the effective lethal radius is calculated, and penetration capabilities are evaluated if applicable. After testing, the data collected is analysed to verify compliance with performance specifications, identify any deviations or inconsistencies, and suggest design improvements. Detailed reports from these tests include data and high-resolution imagery, providing a thorough analysis against performance criteria and recommending necessary modifications, thus ensuring UTAWs meet rigorous performance and safety standards.

8.3.2 ITOP 4-2-813 Static Testing of High-Explosive Munitions for Obtaining Fragment Spatial Distributions

ITOP 4-2-813, *Static Testing of High-Explosive Munitions for Obtaining Fragment Spatial Distributions*, is a standardised test focused on the evaluation of fragment dispersal characteristics of high-explosive munitions under static conditions. This standard is pivotal for assessing the lethality and effectiveness of explosive weapons, particularly regarding their fragmentation pattern and range. It is designed to mimic real-world conditions as closely as possible, providing essential data for both military applications and safety evaluations. The primary goals of ITOP 4-2-813 are to analyse fragmentation characteristics, determine the number, size, velocity, and distribution of fragments, and provide data for assessing potential battlefield effectiveness and collateral damage (ITOP-4-2-813, 1993).

This standard applies to a variety of high-explosive munitions, including grenades, bombs, and artillery shells, covering both new designs and existing stockpiles to validate specifications and improve designs. The test setup involves a detailed test range configuration, referred to as an arena (Figure 8.9) with safety barriers and fragment collection systems, alongside high-speed video equipment for capturing detonation and fragment dispersal. Munitions are prepared in a manner consistent with operational use, and advanced sensors and tracking systems, such as high-speed cameras and radar, are deployed to capture detailed data on fragment trajectories. Test procedures include controlled detonation, precise data collection, and strict safety measures to ensure the safety of personnel and equipment. Data analysis from these tests helps construct models of fragment dispersal patterns, assess munition performance consistency, and compare outcomes against design specifications. Detailed reports from these tests include fragment data, high-resolution images, and thorough analysis against performance benchmarks, ensuring that the munitions meet rigorous safety and effectiveness standards.

FIGURE 8.9 Arena trial setup (Source: author).

8.3.3 ITOP-4.2.822 Electronic Measurement of Airblast Overpressure and Impulse Noise

ITOP 4-2-822, *Electronic Measurement of Airblast Overpressure and Impulse Noise*, is a critical standardised testing protocol designed to measure the effects of airblast overpressure and impulse noise generated by explosives. This procedure is essential for evaluating the impact of explosive devices on military operations and the environment, focusing particularly on the potential damage to structures and harm to personnel (ITOP-4.2.822, 2000). It employs electronic measurement techniques, renowned for their high accuracy and reproducibility, to analyse the airblast overpressure and impulse noise from explosive detonations, ensuring explosives are used safely and effectively in both military and civilian contexts.

The primary goals of ITOP 4-2-822 are to quantify the overpressure and impulse noise characteristics of explosives, assess potential damage and auditory effects from detonations, and ensure compliance with safety standards (ITOP-4.2.822, 2000). Applicable to a diverse range of explosive materials and systems, the standard covers tests in various environments such as open fields and controlled facilities. The test setup includes specifying test site conditions, preparing explosives consistently, and setting up electronic sensors (refer to Figure 8.10) like pressure transducers and microphones for data capture. Test procedures involve safe detonation initiation, data collection on airblast parameters, and strict adherence to safety controls to protect personnel and equipment. Data from these tests is analysed to determine peak overpressure and impulse values at various distances, assess the decay rate of overpressure, and analyse noise profiles for their impact on hearing. Detailed reports from these tests provide graphical data on overpressure and noise propagation, compare measured values against safety criteria, and offer recommendations for mitigating adverse effects, thereby enhancing the operational safety and effectiveness of EO.

FIGURE 8.10 Measurement of pressure during gun system trial (Source: author).

8.4 SAFETY ASSESSMENT

Safety remains the utmost priority when handling, storing, and transporting explosives. T&E plays a pivotal role in mitigating risks by the following:

- **Evaluating the explosive's response to unintended stimuli** such as shock, heat, or electrostatic discharge. This facilitates the determination of appropriate handling procedures and safety controls.
- **Assessing the efficacy of packaging and storage methods** to ensure optimal safety during transportation and storage. T&E investigates factors like temperature control, shock absorption, and compatibility with surrounding materials.
- **Determining compatibility** between different explosives and materials to prevent accidental initiation. This includes testing combinations of explosives, boosters, detonators, and storage containers.

8.4.1 UN Series 4(b)(ii) – 12 Metre Drop Test for Unpackaged Articles, Packaged Articles and Packaged Substances

The UN Series 4(b)(ii) – 12-metre Drop Test, detailed in the United Nations Recommendations on the Transport of Dangerous Goods, is a critical standard designed to assess the safety and integrity of both unpackaged and packaged articles, including hazardous materials, when subjected to severe impacts during transportation. This test simulates an extreme scenario where a package might fall from a significant height, such as during handling or transport accidents, making it indispensable for evaluating the robustness of packaging and the stability of the contents under such conditions.

The primary objectives of this test are to evaluate the structural integrity of packaging subjected to high-impact forces, ensure that no dangerous goods leak, spill, or pose additional hazards as a result of the impact, and validate packaging designs against international safety standards. The test encompasses unpackaged articles that might be damaged or cause damage upon impact, as well as packaged articles and substances, ensuring they are safely contained during and after the impact. The testing procedure includes selecting a representative sample of the packaging and contents, dropping the package from a specified height of 12 metres onto a flat, hard surface to simulate severe impact conditions, and conducting observations and inspections post-drop to assess any breaches, deformations, leakages, or failures (refer to Figure 8.11).

Data from these tests is analysed to assess the effectiveness of the packaging design, determine necessary modifications to enhance safety, and confirm compliance with UN standards and other regulatory requirements. The results are documented in a detailed report that includes descriptions of the packaging and contents, test conditions, and observations, thereby ensuring that the test meets the stringent requirements of the UN Recommendations on the Transport of Dangerous Goods. This ensures that hazardous materials are transported safely under extreme conditions, mitigating the risks associated with their handling and transportation.

8.4.2 UN Series 6(c) – External Fire (Bonfire) Test

The UN Series 6(c) – External Fire (Bonfire) Test, detailed in the United Nations Recommendations on the Transport of Dangerous Goods, is a vital safety assessment designed to evaluate the resistance of packaging to fires during transportation. This test simulates an extreme scenario where the package is exposed to severe external fire, aiming to ensure that the packaging can protect its contents and prevent any dangerous reactions or explosions. The test is crucial for the safety controls involved in transporting hazardous materials, mimicking the thermal stresses and potential physical damage that might occur during a fire incident, especially for packaging containing flammable, explosive, or reactive substances.

FIGURE 8.11 12-metre Drop Test tower (Source: author).

The main objectives of this test are to assess the ability of packaging to withstand prolonged exposure to fire, ensure it can prevent or significantly delay the release of hazardous substances, and evaluate the structural integrity of the packaging after fire exposure. Applicable to all types of packaging such as drums, boxes, jerrycans, and composite packaging, the test setup involves selecting samples with typical contents, exposing them to an engulfing fire from an external source like a woodpile arrangement, and monitoring the fire intensity and package response with temperature sensors. The test procedure includes placing the package on a fire source, igniting the fire, and maintaining it until the package burns out or the specified test duration is completed, with

FIGURE 8.12 External fire (bonfire) test (Source: author).

continuous observations of the package behaviour, including deformation, leakage, or combustion (refer to Figure 8.12). Data from the test is analysed to assess the integrity of the package post-fire, any release of contents, and the time taken for any structural failure or content release. A report is generated detailing the test sample preparation, fire conditions, observations, and results, ensuring compliance with international standards for the transport of dangerous goods and confirming the packaging's ability to withstand and protect against external fire conditions.

8.4.3 Test 6(a) – Single Package Test

The UN Series Test 6(a) – Single Package Test is a vital safety standard from the United Nations Recommendations on the Transport of Dangerous Goods, aimed at assessing the performance of individual packaging units under transport conditions. This test evaluates the integrity and robustness of a single packaging unit to ensure it can independently protect its contents against the stresses and impacts typically encountered during shipping and handling. This is crucial for confirming that each package can withstand the rigours of transport without the support or buffering of additional packaging or palletisation, making it essential for manufacturers, shippers, and regulatory bodies to guarantee the safe transport of goods, particularly hazardous materials.

The objectives of the Test 6(a) include determining the structural integrity of the packaging under normal transport conditions, ensuring effective protection of contents from physical damage, and validating packaging design and materials for compliance with safety standards. The test is applicable to various packaging types, such as boxes, drums, and composite containers, and involves a representative sample being subjected to drops, compression, and vibration tests to simulate different transportation scenarios. The test procedures include dropping the package from a specified height to simulate handling impacts, performing a compression test to measure resistance to stacking forces, and conducting a vibration test to mimic transportation vibrations. Data from these tests is analysed to assess package integrity, evaluate design effectiveness, and identify potential weaknesses. The report documents the test conditions, results, and conclusions about the packaging's performance.

8.4.4 Test 6(d) – Unconfined Package Test

The UN Series Test 6(d) – Unconfined Package Test is an essential safety testing standard designed to evaluate the performance and integrity of packaging under unconfined conditions. This test simulates worst-case scenarios during transportation, such as drops, strikes, or impacts where the package is not mitigated by confinement or cushioned by other cargo. Its primary goal is to determine how well the packaging can protect its contents in extreme conditions, making it critical for ensuring the safety and integrity of packages, particularly those containing dangerous goods.

The objectives of the Test 6(d) include assessing the structural integrity and robustness of the packaging under severe handling and transport conditions, ensuring the package can protect its contents from damage, and verifying compliance with safety standards for transporting potentially hazardous materials. The test applies to all package types, especially those expected to endure high-risk conditions. The procedure involves selecting a representative package with standard contents or simulants, subjecting it to controlled impact tests in an open space or large testing facility, and using high-speed cameras and sensors to document the effects. Specific tests include dropping the package from various heights and orientations to simulate impacts during transport and conducting direct impact tests with objects to mimic real-world collisions. Following the tests, packages are inspected for structural failures or content damage, and the data is analysed to evaluate the impact resistance, effectiveness of the packaging design, and compliance with safety standards. The report details test conditions, results, and any recommendations for improvements or certifications if the packaging passes the tests. This ensures that the packaging adheres to the UN Recommendations on the Transport of Dangerous Goods, enhancing safety during transport and reducing the risk of accidents.

8.5 ENVIRONMENTAL ASSESSMENTS

Environmental assessments are critical in evaluating how equipment and systems perform under various environmental conditions, ensuring their reliability, durability, and functionality in real-world scenarios (STANAG 4370, 2020). This section will focus on the most widely recognised standard, MIL-STD-810H – *Department of Defense Test Method Standard: Environmental Engineering Considerations and Laboratory Tests*. This military standard provides detailed guidelines for testing the environmental effects on equipment, ranging from temperature extremes and humidity to vibration and shock. MIL-STD-810H is designed to simulate the rigorous conditions that equipment may encounter during its operational life, ensuring that it can withstand the stresses of its intended environment. By adhering to this standard, developers and testers can systematically identify potential weaknesses and improve the resilience of their products, ultimately leading to enhanced performance and longevity. The thoroughness and specificity of MIL-STD-810H make it an invaluable tool in the pursuit of robust, reliable, and mission-ready equipment across various industries, from defence to aerospace and beyond. It is not the only standard; for example, the Allied Environmental Conditions and Test Publication (AECTP) series of test standards provides an alternate, NATO-focused test programme that is relatively interchangeable with the MIL-STD-810H.

8.5.1 Low Temperature (Method 502.7)

Method 502.7: Low Temperature Test, detailed in standards like MIL-STD-810G, is designed to evaluate the performance of equipment under low-temperature conditions typical of military

FIGURE 8.13 Conditioning chamber used for low-temperature testing (Source: author).

operations and other demanding environments. This method assesses how well materials and equipment can withstand and operate in cold climates without degradation or failure, ensuring their effectiveness and survival in harsh conditions. It is particularly focused on identifying issues such as material brittleness, lubricant failures, battery performance declines, and other problems induced by cold temperatures. Compliance with Method 502.7 is critical for ensuring that equipment can function in cold environments, crucial for military and other critical applications where low temperatures are a factor (Figure 8.13).

The test encompasses a variety of equipment and materials that may be used or stored in cold environments. The procedure involves placing the equipment in a temperature-controlled test chamber that can accurately maintain the specified low temperatures, with temperature settings typically ranging from mildly cold conditions to extreme cold (below −40). The test sequence includes a preparation phase where equipment is acclimated to the test temperature, a low-temperature soak for the required duration, and a functionality check to verify operational capabilities during or after exposure. Post-test, the equipment undergoes a controlled return to room temperature to assess any recovery issues or delayed effects of cold exposure. Data collected is analysed to evaluate the immediate and long-term effects of low temperatures on equipment performance and the suitability of materials and components for cold environments, and to formulate recommendations for performance improvements. The report is then prepared, documenting test conditions, observations, and results, along with conclusions and recommendations for enhancing cold temperature performance, adhering to the MIL-STD 810 specification.

8.5.2 High Temperature (Method 501.7)

Method 501.7: High Temperature Test, detailed in MIL-STD-810G, is designed to evaluate how equipment performs under high temperature conditions, both in operational heat (induced) and

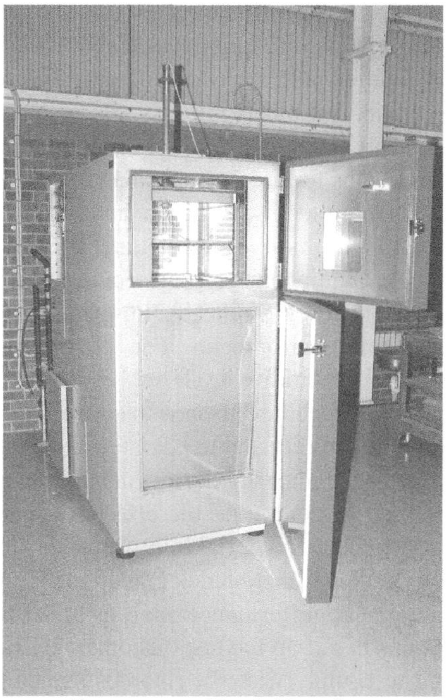

FIGURE 8.14 Conditioning chamber utilised for small-scale high-temperature tests (Source: author).

environmental heat (storage). This test ensures that equipment can withstand and function in hot climates without degradation or failure, a crucial factor for military and industrial applications in harsh environments. It assesses potential issues such as thermal expansion, cooling system performance, electronic failure, and material degradation under intense heat, ensuring that equipment adheres to military and industry standards for high-temperature performance.

The objectives of the method include assessing the robustness and operational integrity of equipment in high temperatures, identifying vulnerabilities like compromised seal integrity, lubricant failure, and electronic misoperation, and confirming the equipment's functionality in hot environments over prolonged periods. This method also applies to a broad range of equipment that are used or stored in hot climates or exposed to direct sunlight (Figure 8.14).

The test involves using a temperature-controlled chamber to simulate required high temperatures, with conditions set based on the most extreme scenarios expected during the equipment's lifecycle. The procedure includes preconditioning the equipment to the target temperature, subjecting it to heat soak under both operational and non-operational states, and conducting functionality tests to check for performance degradation or failure. Post-testing, the equipment is returned to ambient temperature to assess recovery issues or lasting effects of the exposure. Data collected helps determine the impact of high temperatures on structural integrity and operational efficiency, evaluate thermal management systems, and suggest design improvements. The report documents the test conditions and observed effects and provides conclusions and recommendations for enhancing heat tolerance, ensuring that all findings are in line with MIL-STD 810.

8.5.3 Humidity (Method 507.6)

Method 507.6: Humidity Test, outlined in MIL-STD-810G, evaluates the performance of equipment under conditions of high humidity, simulating environments with prolonged exposure to moisture.

This test is crucial for determining how well equipment can withstand and operate in humid climates without experiencing degradation, focusing on potential issues such as corrosion, condensation, electrical or mechanical failures, and material degradation caused by moisture. Adherence to Method 507.6 ensures that equipment meets military and industry standards for durability and reliability in moist environments, providing essential information for manufacturers and users about the suitability of materials and designs for use in tropical and moist climates.

The objectives of Method 507.6 are to assess the impact of high humidity on the functionality and durability of equipment, identify humidity-related vulnerabilities like corrosion, fungal growth, and short-circuiting, and ensure that equipment can perform its intended function in humid environments without premature failure. The test is applicable to various equipment types intended for use in tropical climates or high-moisture environments. The test setup involves placing equipment in a humidity-controlled chamber, with humidity levels set between 50% and 95% and temperatures varying from 20 to 60°C. The test procedure includes acclimating equipment to ensure uniform conditions, exposing it to specified humidity and temperature for extended periods, and conducting operational tests to assess performance integrity during and after exposure.

Data from the tests is analysed to determine the effects of humidity on different equipment components, identify necessary design or material improvements, and validate that the equipment can withstand humid conditions without functional degradation. The report documents the test conditions, collected data, and equipment performance analysis under humid conditions and provides conclusions and recommendations based on the test outcomes. By fulfilling the criteria of Method 507.6, equipment is confirmed to comply with the rigorous standards set by MIL-STD-810H for operation in humid environments, enhancing reliability and extending the equipment's operational lifespan.

8.5.4 Vibration (Method 514.8)

Method 514.8: Vibration Test, as outlined in MIL-STD-810, is an essential standard that sets forth procedures for testing the durability and operational integrity of military equipment under various vibration conditions. This method replicates different vibration environments to ensure that equipment can withstand and remain functional during transportation, deployment, and operational scenarios (MILSTD-810H, 2019). It is specifically designed to assess how well equipment can handle various types of vibrations that might be encountered in service, with a focus on identifying potential failures due to vibratory forces such as the loosening of fasteners, component wear, and structural integrity issues.

The primary objectives of Method 514.8 include evaluating the equipment's ability to withstand vibrations encountered during transportation (such as those from trucks, aircraft, and ships) and operational use (including machinery and engine vibrations). It also aims to identify vibration-related vulnerabilities that could lead to operational failure or degradation and to ensure that equipment maintains its functionality and structural integrity under vibratory conditions. This test is applicable to a wide range of military equipment, including electronics and weapon systems.

The test setup involves using specialised vibration tables capable of generating controlled, repeatable vibration patterns across various frequencies and amplitudes (refer to Figure 8.15). Equipment is mounted on these tables in orientations that mimic real-world installation or transport scenarios, with accelerometers and other sensors attached to monitor responses to vibrations. The test procedures include securing the equipment on the vibration table, exposing it to different types of vibration tests (such as sine vibration for simulating engine vibrations, random vibration for rough transportation environments, and mixed mode for complex conditions), and performing operational tests during and after vibration exposure to assess impacts on performance. Post-test inspections evaluate physical and functional integrity, looking for any damage or component loosening.

Data analysis from these tests assesses the equipment's structural and functional resilience, identifies resonant frequencies that cause excessive strain or damage, and evaluates the effectiveness

FIGURE 8.15 Vibration test (shaker) apparatus (Source: author).

of vibration isolation and damping measures. The report details the test setup, conditions, observations from the vibration exposure, conclusions regarding the equipment's vibration tolerance, and recommendations for enhancing vibration resistance if necessary. Adhering to Method 514.8 ensures that equipment complies with the standard for reliability, ultimately helping to prevent in-field failures and extend the service life of critical assets, aligning with the standards specified in MIL-STD-810H.

8.5.5 Shock (Method 516.8)

Method 516.8: Shock Test is a critical testing protocol for assessing the robustness of equipment against shocks or impacts that might occur during operational or transportation environments. This method is designed to simulate sudden, severe impacts or explosive conditions that equipment might face during military operations, handling, or accidental scenarios (MIL-STD-810H, 2019). By evaluating the equipment's ability to withstand high-energy impacts and shock pulses, Method 516.8 is pivotal for determining the structural integrity and operational readiness of military and industrial equipment under extreme conditions, thus ensuring reliability and safety and preventing premature failures.

The objectives of Method 516.8 include testing the equipment's structural durability and functional integrity under shock conditions, identifying potential failure points caused by high impact or explosive shocks, and confirming that the equipment can operate effectively post-exposure in accordance with its operational specifications. This test is applicable to a broad range of equipment which may experience shocks from dropping, explosive blasts, or collision impacts during their lifecycle.

The test setup involves using shock machines, drop towers, or similar devices that deliver precise shock impulses to mimic real-world impact conditions (refer to Figure 8.16). Equipment is outfitted with accelerometers and other sensors at critical points to measure shock forces and responses, with high-speed cameras capturing the events. Procedures include securely mounting the equipment

FIGURE 8.16 Admiralty Shock Test apparatus (Source: author).

on the testing platform, applying specific shock pulses like half-sine, saw-tooth, and trapezoidal shocks to reflect service conditions, and conducting operational tests during and after exposure to assess effects on functionality. Post-test, the equipment undergoes thorough inspections to check for structural and functional integrity, identifying any component shifts, breakages, or malfunctions.

Data from these tests is analysed to assess damage or changes in functionality due to shock exposure, determine the equipment's resilience to shocks without performance degradation, and evaluate the adequacy of current designs and materials in shock absorption. The report is generated, documenting the detailed test conditions, observations, results, and conclusions regarding the equipment's shock tolerance. Recommendations for design improvements are also included if necessary, ensuring that the equipment complies with MIL-STD 810 standards and is capable of enduring the harsh impacts likely to be encountered during its operational life.

8.5.6 Solar Radiation (Method 505.7)

Method 505.7: Solar Radiation Test, as delineated in MIL-STD-810G, is a crucial test that evaluates the effects of prolonged exposure to solar radiation on equipment, ensuring that devices can endure and operate effectively under the direct influence of sunlight. This test simulates the thermal and photonic effects of sunlight, focusing on potential issues such as material degradation, overheating, and photodegradation, making it indispensable for verifying that equipment intended for outdoor use can withstand the intense and varied conditions of natural sunlight (MILSTD-810H, 2019). By

Solar Radiation Chamber

FIGURE 8.17 Solar radiation test apparatus (Source: author).

replicating the natural solar spectrum, Method 505.7 helps manufacturers design equipment that is robust and durable for outdoor use, maintaining functionality and reliability even in intense solar environments.

The objectives of this method include testing the equipment's resistance to heat and light from solar exposure, evaluating the effects of ultraviolet (UV) light on material strength and colour stability, and ensuring that thermal and optical performance are not compromised under solar exposure (refer to Figure 8.17). The test is applicable to any equipment that may be exposed to direct or indirect sunlight. The test setup involves a solar radiation chamber equipped with lamps that simulate the full spectrum of solar radiation, control systems to mimic various environmental conditions, and monitor equipment to measure radiation levels and temperature. Test procedures include preparing the equipment, exposing it to various cycles of solar radiation, conducting operational tests during exposure, and inspecting the equipment post-exposure for any signs of degradation or damage. Data analysis focuses on identifying changes in material properties, assessing thermal effects on components, and evaluating thermal management systems. The report includes detailed test conditions, observational and instrumental data, conclusions about the equipment's durability and operational integrity under solar radiation, and recommendations for enhancing resistance to solar effects if necessary, adhering to MIL-STD 810.

8.5.7 Altitude (Method 500.7)

Method 500.7: Altitude Test evaluates equipment performance and reliability under low-pressure conditions akin to high altitudes. This method determines whether equipment can endure and function effectively in environments such as aircraft, mountainous regions, or other high-altitude applications, where altitude variations might impair functionality (MILSTD-810H, 2019). The test focuses on the effects of high altitude on equipment, addressing potential issues like hypobaric conditions which can cause arcing, insulation breakdown, or operational failures due to decreased air density. This testing is vital for ensuring that military and aerospace equipment performs reliably in diverse and challenging environmental conditions.

The primary objectives of Method 500.7 include testing the equipment's capability to operate at significant altitude variations where air pressure is much lower than at sea level, evaluating the effects of rapid altitude changes, and ensuring that packaging and seals maintain integrity under low-pressure conditions. This test is applicable to various equipment types, including electronics, mechanical devices, and products transported by air or used in elevated terrains. The test setup involves an altitude chamber capable of simulating high-altitude low-pressure conditions, control systems to regulate the chamber pressure, and monitoring equipment to track pressure, temperature, and humidity, as well as equipment performance (refer to Figure 8.18).

FIGURE 8.18 Pressure test vessel (Source: author).

The test procedures involve preparing the equipment within the altitude chamber and recording baseline performance metrics at normal pressure, followed by gradual reduction of chamber pressure to simulate ascent to high altitude. Equipment may also be subjected to a soak phase at high altitude for extended periods and rapid decompression to mimic sudden pressure drops typical during air transport. The final phase involves repressurisation to normal atmospheric conditions to assess any residual effects or recovery issues. Data from these tests is analysed to identify any physical deformations, leaks, performance anomalies, or operational failures, and to evaluate the equipment's recovery post-test.

The test report details test conditions, altitude profiles, observations during each test phase, and overall conclusions regarding the equipment's altitude tolerance. Recommendations for design improvements may also be included if necessary. Adhering to Method 500.7 validities equipment can withstand and function effectively under the rigorous conditions of high altitudes, enhancing reliability and safety in military and aerospace operations.

8.6 DETERMINING EFFECTIVENESS AGAINST TARGETS

Determining the effectiveness of weapons and systems against targets is a fundamental aspect of T&E in the defence and ordnance sectors. This process involves testing and rigorous analysis to ensure that weapons perform as intended in real-world scenarios, achieving desired outcomes with precision and reliability. Effectiveness assessments consider various factors, including the weapon's accuracy, destructive capability, and ability to neutralise or destroy specific targets under different conditions.

These evaluations often employ a combination of live testing, simulation, and modelling to gather data on weapon performance, enabling a thorough understanding of its strengths and limitations.

By analysing this data, researchers and engineers can make informed decisions to optimise weapon design, enhance operational strategies, and ensure that the systems meet stringent performance standards. Ultimately, determining effectiveness against targets is crucial for maintaining the technological edge, operational readiness, and mission success in defence and security operations. This section cannot be an exhaustive list of test standards and methods as all are designed around specific systems, platforms, and requirements. Careful research and community engagement is required to design a test programme to determine the effectiveness of EO against targets.

This is neither a complete nor exhaustive list of system safety or performance assessment tests. Considerable effort has been expended to ensure interchangeability and integration of a range of international and coalition weapon platforms and systems. It is imperative that a comprehensive literature scan is completed when engaging in the development and delivery of a test programme to ensure that these standards are implemented to guide risk management and decision making.

8.6.1 STANAG 4569: Procedures for Evaluating the Protection Levels of Logistic and Light Armoured Vehicles for Kinetic Energy and Artillery Threats

STANAG 4569 is a pivotal NATO standard that outlines the protection levels for logistics and light armoured vehicles against ballistic and blast threats, ensuring a common safety standard across various military platforms (STANAG 4569, 2020). This standard categorises protection into levels based on the type of munitions or blasts a vehicle should withstand, which includes kinetic energy, artillery, and IED (Improvised Explosive Device) threats. This categorisation allows for a clear and consistent understanding of vehicle survivability in combat scenarios, enhancing interoperability among NATO members and ensuring a high level of protection in military contexts.

The primary goals of STANAG 4569 are to ensure that vehicles meet a minimum required standard of protection for personnel against known threats, facilitate interoperability among NATO forces through standardised vehicle protection, and provide guidelines for vehicle manufacturers to design armour according to recognised military requirements (STANAG 4569, 2020). The standard applies to a variety of vehicles and covers a comprehensive range of threats, including small arms, heavy calibre guns, artillery shrapnel, mines, and IED blasts. Protection levels are detailed from 1 to 6, with higher levels indicating the ability to withstand more severe threats, including blast resistance, indicating the vehicle's ability to withstand mine and IED explosions.

Test setups for evaluating compliance involve using munitions or surrogates to simulate attacks under controlled conditions, preparing vehicles with necessary instrumentation like pressure sensors and dummy occupants equipped with sensors (refer to Figure 8.19). Data from these tests is analysed to assess armour penetration, spallation, survivability of dummy occupants, and the structural integrity of the vehicle post-impact. Detailed reports from these tests include test conditions, video analysis, sensor data, evaluations against the specified protection levels, and recommendations for improvements if standards are not met. By adhering to STANAG 4569, vehicles are ensured to be capable of operating in hostile environments, crucial for the safety of personnel and effectiveness of operations.

8.6.2 Nato STANAG AEP-97 MC-MOPI

The AEP-97 Multi-Calibre Manual of Proof and Inspection (MC-MOPI) serves as a critical guideline within NATO for the proof and inspection of small arms ammunition, offering standardised procedures and criteria to ensure the safety, reliability, and performance of small arms ammunition across NATO member countries (AEP-97, 2020). This framework is essential for maintaining the effectiveness of explosive materials and devices, facilitating uniformity in testing methods, and ensuring comparable data across various national and organisational boundaries.

FIGURE 8.19 Fast cook-off test (Source: author).

The primary aim of the AEP-97 MC-MOPI is to standardise proof and inspection procedures for multiple calibres of NATO small arms ammunition (AEP-97, 2020). This standardisation ensures functional interchangeability on the battlefield by providing uniform test methods, equipment, and inspection procedures that assess technical performance requirements. The scope of the AEP-97 M-CMOPI is extensive, covering everything from ammunition interchangeability principles to specific test procedures for assessing various ammunition characteristics. It includes details on NATO ammunition standardisation, test equipment, visual standards, sentencing of defects and incidents, and guidelines for various specific tests related to small arms ammunition.

The general principles outlined in the AEP-97 MC-MOPI focus on safety, consistency, and documentation. These principles ensure that all testing procedures prioritise the safety of personnel and equipment, maintain consistent testing practices across different facilities and countries, and emphasise thorough documentation and reporting of test results. This approach helps in achieving a high standard of quality control and reliability in NATO munitions. The AEP-97 MC-MOPI details procedures for a wide range of tests designed to cover the entire lifecycle of small arms ammunition. These tests are critical for ensuring that ammunition meets rigorous safety and performance standards before deployment. Specific tests are outlined in Table 8.4.

Adherence to the AEP-97 M-CMOPI ensures that NATO forces are equipped with small arms ammunition that not only meets rigorous safety and performance standards but also maintains consistency and reliability across different combat scenarios. This standardisation is vital for the effective functioning of NATO military operations, providing soldiers with the confidence that their ammunition will perform reliably in the field. By following the detailed procedures and guidelines set out in this manual, NATO continues to enhance its operational readiness and maintains a high level of safety and performance in its armament services.

8.6.3 AOP-39 Insensitive Munitions

AOP-39 is a NATO standard that outlines the tests for assessing the munitions insensitivity (STANAG 4439, 2018). It aims to ensure that munitions are safe, reliable, and perform effectively under various

TABLE 8.4
AEP-97 MC-MOPI Test Matrix

Test	Description
Combination Electronic Pressure Velocity and Action Time (EPVAT)	Verifies chamber pressure and, where relevant, gas port pressure generated by the ammunition, as well as muzzle velocity and action time (the latter defined as the time between the firing impulse (hammer/striker hitting the primer) and the projectile exiting the muzzle).
Function and casualty	The ammunition is tested in NATO Nominated Weapons to verify reliability.
Trajectory match	Verifies that the ammunition matches the ballistics of designated NATO reference ammunition, ensuring that a weapon will shoot approximately to the same point of aim with the ammunition being tested.
Precision	Verifies the precision, including accuracy and dispersion of the ammunition.
Terminal effects	Verifies the terminal ballistics, in this case hard-target penetration, of the ammunition.

conditions without unintended detonation. Insensitive munitions are designed to enhance the survivability of armed forces by reducing the likelihood of accidental explosions, thereby improving the overall safety of military operations. The standard defines the criteria and test methods for evaluating how munitions react to thermal, mechanical, and other stimuli with a summary in Table 8.5.

8.7 EXPLOSIVES AND EXPLOSIVE ORDNANCE T&E PLANNING

Planning is a critical phase in the T&E of explosives and EO. Successful T&E activities rely on detailed and structured planning to ensure all aspects of the testing process are considered, risks are mitigated, and objectives are efficiently achieved. This chapter narrates the essential components of T&E planning, underscoring its importance in obtaining reliable and valid results.

8.7.1 Objectives and Scope Definition

The first step in planning T&E activities involves setting clear objectives and defining the scope of the tests. Identifying the purpose of the T&E activities is paramount, whether it involves characterising explosive properties, evaluating performance, assessing safety, or determining effectiveness against targets. Establishing measurable success criteria that align with these objectives guides the evaluation of test results, ensuring outcomes meet the intended goals. Equally important is defining

TABLE 8.5
AOP-39 Test Matrix

Test Type	Description
Bullet impact	Evaluates munition's response to high-speed projectiles and fragments (STANAG 4241, 2018).
Slow heating	Assesses the behaviour of munitions when gradually heated to simulate fire exposure (STANAG 4382, 2020).
Fast heating	Tests munitions under rapid temperature increase conditions, mimicking fast-developing fires (STANAG 4240, 2018) (refer to Figures 8.19 and 8.20).
Sympathetic reaction	Determines if a munition will detonate when adjacent munitions explode (STANAG 4396, 2020).
Shaped charge jet	Tests the munition's resistance to penetration and detonation by a shaped charge jet (STANAG 4526, 2018).
Fragment impact	Assess the reaction of a munition impacted by a fragment (STANAG 4496, 2019).

FIGURE 8.20 Fast cook-off test setup (Source: author).

the scope of the tests. This involves determining the parameters to be tested—such as detonation velocity, sensitivity, fragmentation, and blast wave characteristics—and outlining the range and limits of these parameters to ensure coverage. Additionally, resource allocation must be carefully planned. Identifying the necessary personnel, equipment, facilities, and materials ensures that all resources are available and appropriately distributed.

8.7.2 Test Design and Methodology Development

Developing a detailed test plan is crucial for structured T&E activities. A test matrix should be created, outlining the different tests to be conducted, their sequence, and the specific conditions under which they will be performed. This matrix serves as a roadmap for the entire testing process. Selecting appropriate test methods that align with industry standards and best practices is essential. Standardised tests, such as those outlined in MIL-STD-810H, ensure consistency and comparability across different tests and evaluations. Instrumentation and data collection are key components of test design. Selecting the right instruments—such as high-speed cameras, pressure and velocity gauges, radar systems, and data acquisition systems—is critical. Ensuring these instruments are properly calibrated and validated before use guarantees accurate and reliable data. Establishing robust procedures for data collection, storage, and management ensures that data is recorded accurately and securely, making it accessible for analysis and review.

8.7.3 Execution and Monitoring

Executing the test plan involves strict adherence to established procedures. Ensuring that all personnel involved in the testing are adequately trained and understand their roles and responsibilities is essential for smooth operations. Continuous monitoring of the testing process ensures that it proceeds according to plan, and implementing controls to address any deviations or unexpected occurrences promptly is crucial for maintaining the integrity of the tests. Data analysis and interpretation follow the execution phase. Applying statistical methods to analyse the collected data helps draw meaningful conclusions. Techniques such as regression analysis, hypothesis testing, and confidence intervals are employed to ensure the results are statistically significant and practically relevant. Interpreting the results in the context of the defined objectives and success criteria ensures that conclusions are based on robust findings.

8.7.4 Reporting and Review

Documenting the entire T&E process is critical for transparency and future reference. Reports detailing the test procedures, data collected, analysis performed, and conclusions drawn should be prepared. Visual aids such as graphs, charts, and tables enhance the clarity of these reports. Conducting a thorough review of the test results and reports, and seeking feedback from stakeholders and experts, helps validate the findings and identify areas for improvement. Continuous improvement is a vital aspect of T&E. Capturing lessons learned from the activities informs future testing efforts, highlighting best practices and areas for enhancement. Implementing ongoing monitoring and surveillance programmes tracks the long-term performance and safety of EO, using these insights to refine and optimise future T&E activities. By following a structured and systematic approach to T&E planning, engineers and researchers ensure that activities are conducted efficiently, safely, and in compliance with regulatory standards. Effective planning not only enhances the reliability and validity of test results but also contributes to the overall advancement of explosive technologies.

8.8 AGEING AND SERVICE LIFE

The ageing and service life of materials are critical factors affecting explosives and EO safety, reliability, and performance. Over time, the chemical and physical properties of energetic materials can degrade due to various environmental and operational conditions.

8.8.1 Degradation of Explosives Over Time

Explosives are subject to ageing, which can significantly impact their performance and safety over time. As explosives age, they undergo various chemical and physical changes that can compromise their stability and effectiveness. Factors such as chemical decomposition, crystallisation, phase separation, mechanical stress, and degradation of containers and binders all contribute to the ageing process. These changes can alter the sensitivity, reliability, and overall performance of explosive materials, posing potential safety risks. Understanding the mechanisms of degradation and implementing proper storage, handling, and maintenance practices are crucial for ensuring the long-term stability and safety of explosives.

8.8.1.1 Chemical Decomposition

Chemical decomposition in explosives, particularly due to autocatalytic reactions in many organic compounds, involves slow decomposition that accelerates as decomposition products further propagate the reaction, compromising stability and safety. Environmental factors like heat, moisture, and oxygen exposure significantly accelerate this process. Heat increases reaction rates, moisture promotes hydrolysis, and oxygen forms unstable compounds, all of which can drastically reduce the shelf life and effectiveness of explosives. Understanding and mitigating these effects through proper storage and handling practices, including controlled temperature and humidity, are crucial for maintaining the safety and reliability of explosive materials, especially in critical applications.

8.8.1.2 Physical Changes

Physical changes in explosives, such as crystallisation and phase separation, can significantly impact their sensitivity and performance by altering their stability and potentially making them more prone to unintended detonation or less effective in their intended use. Additionally, mechanical stress from transport and handling can induce microstructural changes like cracking, which compromise the integrity of the explosive and degrade its performance, posing safety risks. Understanding and mitigating these physical changes is crucial for maintaining the safety and efficacy of explosive materials. Implementing proper handling, transport, and storage methods to minimise mechanical stress and prevent adverse physical transformations, alongside regular inspections and maintenance, is essential to preserve the explosive's stability and performance.

8.8.1.3 Material Interactions

Container and binder degradation significantly impacts the integrity and performance of explosives. Over time, the materials used in explosive casings, binders, and packaging are prone to deterioration, affecting the explosive's stability and effectiveness. For example, the migration of plasticisers in polymer-bonded explosives can lead to embrittlement, reducing the flexibility and integrity of the binder material and making it susceptible to cracking and other defects. This degradation not only compromises the protective barriers that maintain explosive stability but also heightens sensitivity to environmental factors like moisture, heat, and mechanical stress, accelerating degradation and diminishing the systems life and reliability. To counteract these effects, quality control measures during manufacturing and regular inspections during storage and handling are essential. Additionally, maintaining proper storage conditions, such as controlled temperature and humidity, can slow the degradation process. By understanding degradation mechanisms and implementing preventative measures, the safety and performance of explosives can be preserved throughout the lifecycle.

8.8.2 Methods for Assessing the Service Life of Explosive Ordnance

To ensure the safety and reliability of EO, accurately assessing their service life is essential. This involves employing a combination of methods that provide insights into how these materials degrade over time and under various conditions (STANAG 4370, 2020). By understanding the factors that influence the ageing process, it becomes possible to predict the lifespan of EO and implement necessary measures to maintain their performance and safety.

8.8.2.1 Accelerated Ageing Tests

Accelerated ageing tests assesses the longevity and reliability of explosives by simulating long-term ageing effects under condensed timelines. These tests expose explosives and EO to elevated and cyclic temperatures, high humidity levels, and other stressors, enabling the prediction service life and identify potential failure modes that could arise under operational use. This process is crucial for ensuring the safety and effectiveness of explosives throughout their intended service life. During the testing phase, the chemical and physical properties of the test sample are monitored, focusing on changes in chemical composition, crystallisation, phase separation, and mechanical integrity. Analysing this data allows for informed decision making on the long-term stability and performance, guiding the design of more reliable and durable system, optimisation of storage and handling practices, and ensuring compliance with safety standards. Accelerated ageing tests thus serve as an essential tool for enhancing explosive safety and efficacy, facilitating proactive measures to mitigate potential risks before they become critical.

8.8.2.2 Real-Time Ageing Studies

Real-time ageing studies are crucial for assessing the long-term performance and stability of explosives under actual lifecycle conditions, providing a contrast to accelerated ageing tests which simulate ageing through intensified conditions. These studies involve storing explosives and EO in environments that replicate typical storage and operational settings, periodically testing them to monitor changes in chemical and physical properties, such as stability, sensitivity, and performance. By tracking these parameters over time, researchers gain insights into the natural degradation process and identify potential lifecycle issues, offering invaluable data for manufacturers and users. This helps in understanding the longevity of explosives, enhancing planning and risk management, and validating accelerated ageing tests. Ultimately, real-time ageing studies are fundamental in ensuring the long-term safety and effectiveness of explosives, guiding the development of improved storage methods and more resilient formulations.

8.8.2.3 Non-Destructive Testing (NDT)

Non-destructive testing (NDT) is essential for ensuring the safety and reliability of explosives, providing the capability to examine the internal state of materials without causing damage. Techniques such as X-ray (refer to Figure 8.21), which analyses structures for defects, infrared (IR) spectroscopy for chemical composition and signs of degradation, and ultrasonic testing, which uses sound waves to detect internal flaws like voids or cracks, collectively offer a view of an explosive's condition. NDT's applications are crucial in preventing catastrophic failures by detecting early signs of degradation, such as crystallisation or the migration of plasticisers in polymer-bonded explosives which can lead to embrittlement. Regular NDT assessments enhance the management of an explosive's lifespan and assure performance, thereby reducing accident risks and improving safety. Incorporating NDT into regular maintenance and inspection routines allows for continuous monitoring and early detection of issues, maintaining high safety standards and extending the service life of explosive materials, which is critical for protecting personnel and equipment.

8.8.3 Implementing Surveillance Programmes

Implementing surveillance programmes is essential for managing the service life of EO and ensuring their continued safety and performance. These programmes involve the systematic monitoring

FIGURE 8.21 Example X-ray setup (Source: author).

and assessment of explosives throughout their lifecycle, from manufacturing and storage to deployment and eventual disposal. By continuously evaluating the condition of explosives, surveillance programmes help identify early signs of degradation and potential failures, allowing for timely interventions that mitigate risks and extend the service life of these critical materials.

8.8.3.1 Periodic Inspections

Periodic inspections are vital for maintaining the safety and reliability of stored explosives, conducted at predefined intervals influenced by factors like the type of explosive, storage conditions, and historical performance data. These inspections are crucial for identifying and addressing potential issues early, preventing significant degradation or failure. The scope of these inspections includes visual checks to detect physical damage such as cracks, discolouration, or crystallisation. Additionally, chemical tests are performed to detect signs of decomposition or chemical changes that might not be visible, revealing early stages of degradation. Performance tests are also essential, ensuring that the explosives meet operational requirements and function effectively under intended conditions. This proactive approach through regular and thorough inspections allows for timely interventions, such as reconditioning or disposing of degraded explosives, enhancing the safety and operational readiness of explosive materials. This robust framework of inspections and assessments ensures the continued reliability and safety of EO.

8.8.3.2 Environmental and Condition Monitoring

Environmental and condition monitoring are critical practices in managing the longevity and safety of explosives, involving the continuous or periodic measurement of key parameters such as temperature, humidity, and chemical composition that influence the ageing process of explosive materials. Employing sensors and data loggers facilitates the collection of real-time data, offering valuable insights into the storage conditions and environmental factors impacting explosives. The benefits of continuous monitoring are substantial, enabling the early detection of adverse conditions that could accelerate the degradation of explosives. For example, identification of temperature fluctuations or humidity increases allows for corrective actions, thus mitigating risks and preserving the integrity and performance of explosive materials. Additionally, this data-driven approach aids in analysing trends and patterns from the collected data, informing decisions on storage practices, inspection schedules, and maintenance activities. Understanding the environmental factors contributing to explosives' degradation enables the development of more effective strategies to extend their service life and ensure safe use, thereby playing a crucial role in the management of EO and enhancing both safety and reliability.

8.8.3.3 Data Analysis and Trend Prediction

Data analysis and trend prediction are crucial for managing the service life and safety of EO. This process begins with statistical analysis of data collected from ageing studies and surveillance programmes. Using methods such as regression analysis, users identify trends and patterns in the degradation of explosives, helping to understand how environmental factors like temperature, humidity, and mechanical stress impact the longevity and reliability of explosive materials. Predictive models, particularly those using the Arrhenius equation to describe the temperature dependence of reaction rates, play a key role in estimating the remaining service life of explosives based on observed degradation patterns. These models, rooted in empirical data, provide a framework for extrapolating future performance from current observations, allowing for accurate forecasts of when an explosive might become unsafe or ineffective. This enables timely interventions and planning. By combining statistical analysis with predictive modelling, stakeholders can proactively manage EO, implementing measures to mitigate risks, optimise storage conditions, and plan for maintenance or replacement. This data-driven approach not only enhances safety and reliability but also contributes to cost-effective inventory management, ensuring their safe and effective use over time through continuous monitoring, analysis, and modelling.

8.8.3.4 Inventory Management

Effective inventory management is crucial for ensuring EO safety and reliability. The First-In-First-Out (FIFO) method is widely used and ensures that older stock is used, minimising the use of potentially degraded materials. This systematic rotation helps maintain materials within their optimal performance period. Additionally, risk-based management prioritises inspections and use of explosives based on age, storage conditions, and criticality, allowing efficient resource allocation and focus on the most vulnerable items, enhancing safety. Combining FIFO with risk-based management boosts inventory safety and effectiveness. This approach ensures timely use of older explosives while addressing critical at-risk items, maintaining a safer inventory. Regular reviews and data-driven updates to these practices enhance safe handling of explosives. Understanding degradation mechanisms and implementing robust life-assessment methods, including surveillance programmes, helps manage the longevity and reliability of explosives effectively, ensuring operational readiness and risk mitigation.

8.9 STATISTICAL ANALYSIS AND INTERPRETATION

Statistical analysis is a cornerstone of T&E. The use of rigorous statistical methods ensures that the data collected from experiments and tests are analysed accurately and interpreted correctly, leading to meaningful and reliable conclusions. This chapter highlights the importance of statistical methods in analysing T&E data and explains how to interpret test results and draw meaningful conclusions.

8.9.1 Importance of Statistical Methods in Analysing T&E Data

Statistical methods are indispensable in the analysis of test data, providing a robust framework for interpreting complex datasets and ensuring the reliability and validity of results. For testing, where precision and accuracy are paramount, statistical techniques enable the identification of patterns, trends, and anomalies that might otherwise go unnoticed. These methods facilitate the transformation of raw data into meaningful insights, guiding decision-making processes and enhancing the overall quality of testing outcomes. By leveraging statistical tools, professionals can not only validate their hypotheses and models but also optimise testing procedures, reduce uncertainties, and ensure compliance with rigorous safety and regulatory standards. In essence, the integration of statistical methods into T&E processes is crucial for advancing the scientific understanding and technological development within the field of explosive safety and performance.

8.9.2 Interpreting Test Results and Drawing Meaningful Conclusions

Interpreting test results and drawing meaningful conclusions involves analysing the data collected during tests to ensure that the results accurately reflect the performance, safety, and reliability of the tested items. Effective interpretation goes beyond merely presenting data; it requires understanding the context, identifying significant trends, and discerning patterns that provide insights into the behaviour and characteristics of the subject under test. Drawing meaningful conclusions from these interpretations allows for informed decision making, guiding improvements in design, enhancing safety controls, and ensuring regulatory compliance. Ultimately, the ability to interpret test results effectively and derive actionable conclusions is essential for advancing technological innovations and maintaining rigorous standards in explosive safety and performance.

8.10 THE IMPORTANCE OF SAFETY

Safety is paramount in preventing accidents, injuries, and fatalities during T&E. Due to the inherently hazardous nature of explosives, stringent safety measures are necessary to protect personnel, facilities, and the environment. Ensuring safety involves rigorous methods to mitigate the high risks associated with these activities. Without comprehensive safety measures, the likelihood of catastrophic incidents increases, endangering lives and causing significant damage.

8.10.1 Risk Assessment and Management

Conducting thorough risk assessments is the first step in identifying potential hazards in T&E activities. This process involves analysing the properties of the explosives, the conditions under which they will be tested, and the environmental factors that may influence the outcome. By understanding these hazards, engineers can develop and implement strategies to mitigate risks. These strategies might include engineering controls such as blast shields, administrative controls like procedural changes, and the use of personal protective equipment (PPE) to safeguard individuals.

8.10.2 Facility Safety

The design and construction of testing facilities play a critical role in ensuring safety. Facilities must be built to withstand explosive events, incorporating features such as blast-resistant structures and proper ventilation systems. Additionally, testing sites should be located at safe distances from inhabited areas to minimise the risk of harm in the event of an explosion actively being monitored (refer to Figure 8.22). Establishing safety zones around testing areas is also crucial to restrict unauthorised access. These zones should be clearly marked and strictly enforced to prevent accidental entry by personnel or the public.

FIGURE 8.22 Test camera setup monitoring site operations (Source: author).

8.10.3 Operational Safety

Operational safety hinges on the development and adherence to Standard Operating Procedures. Procedures should cover all aspects of test activities, from the preparation and handling of explosives to their disposal and emergency response procedures. Ensuring that all personnel involved in T&E activities are adequately trained and competent is also vital. Regular training sessions and drills help maintain a high level of preparedness and reinforce the correct procedures to follow during testing.

8.10.4 Emergency Preparedness

Emergency preparedness is essential for handling accidents effectively. Developing and maintaining an emergency response plan is a critical component of this preparedness. The plan should outline procedures for dealing with explosions, fires, chemical spills, and other incidents. It should also include evacuation routes, emergency contacts, and first aid measures. Coordination with local emergency services and regulatory bodies ensures a prompt and effective response to any incidents that may occur.

8.10.5 Environmental Considerations

Environmental considerations are another important aspect of safety controls. Proper waste management practices must be implemented to handle explosive residues and by-products safely. This includes safe disposal methods and environmental monitoring to prevent contamination. Conducting environmental impact assessments for T&E activities helps identify and mitigate potential adverse effects on the environment, ensuring that the testing process does not cause harm beyond the immediate testing area.

8.11 SUPPORTING FUTURE DECISION MAKING

In the ever-evolving explosives domain, rigorous T&E is paramount to ensuring safety, reliability, and performance. This chapter has provided an overview of the fundamental principles and practices essential for the effective explosive and EO T&E. From understanding the basic properties of explosives and their ageing mechanisms to implementing advanced instrumentation and statistical analysis, each section underscores the critical role of testing.

Adherence to national and international regulations, coupled with stringent safety controls, is essential for mitigating risks and protecting personnel, facilities, and the environment. The importance of thorough risk assessments, robust facility design, well-developed operational procedures, and emergency preparedness cannot be overstated. These measures not only prevent accidents but also ensure that testing outcomes are accurate and meaningful.

As the field continues to advance, the integration of innovative technologies and methods will further enhance the effectiveness of T&E activities. By maintaining a strong commitment to safety, compliance, and continuous improvement, engineers and researchers can contribute to the development of safer and more effective EO. This chapter serves as a foundational guide, equipping professionals with the knowledge and tools necessary to excel in this critical and challenging domain.

REFERENCES

ITOP-4.2.511 (20 May 1999), *Measurement of Unmanned Target Activated Weapon (UTAW) Warhead Performance.*

ITOP-4-2-813 (01 Mar 1993), *Static Testing of High Explosive Munitions for Obtaining Fragment Spatial Distribution.*

ITOP-4.2.822 (25 Sep 2000), *Electronic Measurement of Airblast Overpressure and Impulse Noise.*

MIL-STD-810H (31 January 2019), *Department of Defense Test Method Standard: Environmental Engineering Considerations and Laboratory Tests.*

NATO Standard, AEP-97 (October 2020), *Multi-Calibre Manual of Proof and Inspection (M-C MOPI) for NATO Small Arms Ammunition*, Edition A, Version 1.

NATO Standard (STANAG) 4240 (20 November 2018), *Fast Heating Munition Test Procedures.*

NATO Standard (STANAG) 4241 (20 November 2018), *Bullet Impact Munition Test Procedures.*

NATO Standard (STANAG) 4370 (07 January 2020), *Environmental Testing.*

NATO Standard (STANAG) 4382 (09 March 2020), *Slow Heating Test Procedures for Munitions.*

NATO Standard (STANAG) 4396 (18 December 2020), *Sympathetic Reaction Munition Test Procedures.*

NATO Standard (STANAG) 4439/AOP-39 (20 November 2018), *Policy for introduction and assessment of Insensitive Munitions.*

NATO Standard (STANAG) 4496 (26 March 2019), *Fragment Impact Test Procedures for Munitions.*

NATO Standard (STANAG) 4526, (20 November 2018), *Shaped Charge Jet Munition Test Procedure.*

NATO Standard (STANAG) 4556 (22 November 1999), *Explosives: Vacuum Stability Test.*

NATO Standard (STANAG) 4569, AEP-55 (08 October 2020), *Procedures for Evaluating the Protection Levels of Logistic and Light Armoured Vehicles for KE and Artillery Threats.*

United Nations (2023), Committee of Experts on The Transport of Dangerous Goods and on the Globally Harmonised System of Classification and Labelling of Chemicals, *Manual of tests and criteria, Revision 8*, New York: United Nations.

9 Shock Tubes and Blast Simulators

Edward Chern Jinn Gan and Alex Remennikov

9.1 BACKGROUND OF SHOCK TUBES

A classical shock tube consists of a constant cross-sectional area divided into two sections by a gas-tight diaphragm as schematically shown in Figure 9.1. A shock tube is made up of the driver section which accommodates a high-pressure gas and the driven section which consists of a low-pressure region. The sudden rupture of the diaphragm leads to the initiation of a shock wave that travels into the driven section while a rarefaction wave travels in the opposite direction into the driver section.

The first laboratory studies of gasdynamic wave phenomena in tubes might be credited to Davy (1817) investigating flame speed and what is now known as deflagration in combustible gases. Similar work was continued by Bunsen (1867), Mallard and Le Chatelier (1881), and Berthelot (1881). However, the first variant of a classical compressed gas (CG) shock tube was developed and described by Vieille (1899a, 1899b). Vieille's shock tube had a diameter of 22 mm with a 6 m channel length and used materials such as cellophane, paper, and metal foil as the diaphragm. The shock tube with the cellophane diaphragm produced reproducible results when the test was repeated under the same conditions and could generate incident shock waves of $M_s \approx 2$.

However, it was not until around 30 years later that any subsequent development of the shock tube was reported. Schardin (1932) constructed a 10 m long shock tube; he had already developed the capability to visualise and capture shock waves with an optical device a few years prior (Cranz and Schardin, 1929). Around the same time, British scientists Payman and Shepherd carried out investigations using shock tubes on the ignition of gas mixtures by transmitted shock waves. This line of research was driven by concerns over coal mine safety, as methane-air explosions could be inadvertently initiated by explosives employed in mining operations (Payman, 1928; Payman and Shepherd, 1938; Payman et al., 1935; Shepherd, 1954). The use of shock tubes became especially widespread during the Second World War due to the urgent need to study blast waves. During that time, various forms of measurement techniques were developed in parallel. Some examples include using piezoelectric pressure gauges to measure blast waves by Reynolds (1943), photography of shock front positions that led to a detailed study of the principles of reflecting shock waves by Smith (1945), and the measurement of density and pressure of diffracting shock waves via the Mach–Zehnder interferometer by Bleakney et al. (1950).

Several laboratories of the US Department of Energy, Department of Defence (DoD), Army Research Laboratory, and other contractors developed some of the earliest blast simulation facilities in the 1950s (Uy et al., 2017). These initial blast simulators were essentially variants of conventional shock tubes, normally consisting of a high-pressure driver with constant area and an abruptly expanding conical test section to accommodate large targets (O'Daniel et al., 2016).

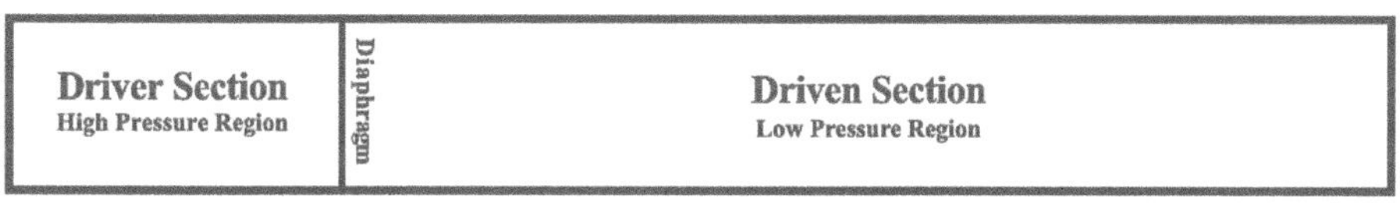

FIGURE 9.1 Schematic diagram of a shock tube.

DOI: 10.1201/9781003529217-9

While previous technological limitations meant that larger tests were confined to field testing, the ratification of the Nuclear Test Ban Treaty in 1963 fundamentally ended nuclear weapon field testing (Jacobson and Stein, 1966). This catalysed the development of new capabilities in blast loading simulation. MABS (formally Military Applications of Blast Simulators) was therefore formed in the 1960s and has been a forum for presenting such work for the past six decades. Although the scope of MABS has evolved over time and has since been renamed to Military Aspects of Blast and Shock, many lessons can be learnt particularly from some early work describing the experimental developments for blast simulation.

The emergence of supersonic flight, aerospace, and advanced missilery in the second half of the twentieth century also fuelled the popularity of the shock tube using CG drivers where they are adapted to study high particle flow velocities (Yu, 1999). An excellent historical account of the development of shock tubes around the world and in particular in the USSR and Belarus can be found in Fomin (2010).

9.2 CLASSICAL SHOCK TUBE THEORY

A Riemann problem is an initial value problem composed of a given set of conservation laws together with data consisting of two adjacent constant states (Riemann, 1860). This problem was defined by Courant and Friedrichs (1948) and analytic solutions were given.

The most common example of the Riemann problem is the shock tube, in which two gases at different pressures are separated by a diaphragm. When the diaphragm separating the high- and low-pressure sections is removed, the two gases remain separate by a contact surface which moves towards the low-pressure region (i.e., driven section). A shock wave $\vec{S_1}$ is also formed which moves into the driven section while a rarefaction wave $\overleftarrow{R}$ moves in the opposite direction. The events in an ideal, classical shock tube are schematically shown in Figure 9.2. Figure 9.2(a) presents the x–t diagram illustrating the propagation of the incident shock ($\vec{S_1}$), the rarefaction wave $\overleftarrow{R}$, and the contact surface $\vec{C}$. Figure 9.2(b)–(d) show the corresponding distributions of pressure, density, and flow velocity along the shock tube at time t_1. As observed in Figure 9.2(b) and 9.2(d), there are no changes in velocity (i.e., $u_2 = u_3$) and pressure (i.e., $P_2 = P_3$) across the contact surface $\vec{C}$ between states ② and ③.

The quantitative relationships shown in Figure 9.2 are obtained from the one-dimensional conservation equations of mass, momentum, and energy. These are presented in Equations (9.1)–(9.4). The derivation assumes ideal-gas behaviour, neglects viscous losses and frictional dissipation, and treats the diaphragm rupture as instantaneous with the rarefaction wave located precisely between the high- and low-pressure sections (Anderson, 2003; Wright, 1961).

Equation (9.1) describes the relationship of flow velocity u generated by the incident shock wave and Equation (9.2) relates the incident shock strength (P_2 / P_1) as a function of diaphragm pressure ratio (P_4 / P_1). The density ρ and temperature T ratios across the shock wave are given in Equations (9.3) and (9.4), where a is the speed of sound and γ is the ratio of specific heats.

$$u_2 = u_3 = \frac{a_1}{\gamma_1}\left[\frac{P_2}{P_1} - 1\right]\left[\frac{\frac{2\gamma_1}{\gamma_1+1}}{\frac{P_2}{P_1}+\frac{\gamma_1-1}{\gamma_1+1}}\right]^{1/2} = \frac{2a_4}{\gamma_4-1}\left[1-\left(\frac{P_2}{P_4}\right)^{(\gamma_4-1)/2\gamma_4}\right] \tag{9.1}$$

$$\frac{P_4}{P_1} = \frac{P_2}{P_1}\left[1-\frac{(\gamma_4-1)(a_1/a_4)(P_2/P_1-1)}{\sqrt{2\gamma_1\left[2\gamma_1+(\gamma_1+1)(P_2/P_1-1)\right]}}\right]^{-2\gamma_4/(\gamma_4-1)} \tag{9.2}$$

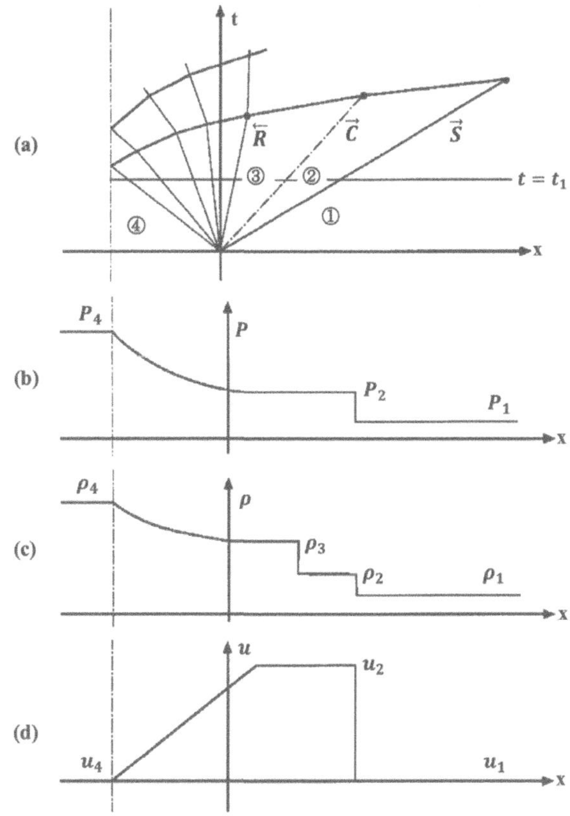

FIGURE 9.2 Events in an ideal, classical shock tube: (a) x–t diagram, (b) pressure, (c) density, and (d) flow velocity profile along the shock tube at $t = t_1$, $\vec{S_1}$ = incident shock, $\overleftarrow{R}$ = rarefaction wave, $\vec{C}$ = contact surface. (Adapted from Heilig and Igra (2001).)[1]

$$\frac{\rho_2}{\rho_1} = \frac{1 + \dfrac{\gamma+1}{\gamma-1}\left(\dfrac{P_2}{P_1}\right)}{\dfrac{\gamma+1}{\gamma-1} + \dfrac{P_2}{P_1}} \tag{9.3}$$

$$\frac{T_2}{T_1} = \frac{P_2}{P_1}\left[\frac{\dfrac{\gamma+1}{\gamma-1} + \dfrac{P_2}{P_1}}{1 + \dfrac{\gamma+1}{\gamma-1}\dfrac{P_2}{P_1}}\right] \tag{9.4}$$

9.3 CALCULATING THE PHYSICAL PROPERTIES OF SHOCK WAVES

The previous section outlined the fundamental concepts of shock tube dynamics, providing the physical picture of shock, contact, and rarefaction waves. However, to apply these ideas in practice, it is essential to quantify the physical properties of shock waves. This section therefore introduces the Rankine–Hugoniot equations, which provide convenient relationships for calculating shock properties, particularly in air, expressed as ratios relative to the ambient state. The Rankine–Hugoniot equations are utilised to calculate the physical properties of a shock wave, providing relationships that are especially convenient for shock wave studies in air. These relationships are expressed as ratios relative to the ambient properties and formulated as functions of overpressure ΔP and ambient pressure P_0. Typically, the ratio of specific heats, γ, is assumed to be 7/5, treating air as a diatomic gas.

Overpressure:

$$\Delta P = P - P_0 \tag{9.5}$$

Density:

$$\frac{\rho}{\rho_0} = \frac{2\gamma + (\gamma+1)\frac{\Delta P}{P_0}}{2\gamma + (\gamma-1)\frac{\Delta P}{P_0}} \tag{9.6}$$

Shock velocity, where C_0 is the ambient sound speed:

$$U = C_0\left(1 + \frac{[\gamma+1]\Delta P}{2\gamma P_0}\right)^{1/2} \tag{9.7}$$

Flow velocity:

$$u = \frac{\Delta P}{\gamma P_0}\frac{C_0}{\left(1 + \frac{[\gamma+1]\Delta P}{2\gamma P_0}\right)^{1/2}} \tag{9.8}$$

Absolute temperature:

$$\frac{T}{T_0} = \left(1 + \frac{\Delta P}{P_0}\right)\left(\frac{2\gamma + (\gamma-1)\frac{\Delta P}{P_0}}{2\gamma + (\gamma+1)\frac{\Delta P}{P_0}}\right) \tag{9.9}$$

Sound speed for ambient gas:

$$C_0 = \sqrt{\gamma P_0 / \rho_0} \tag{9.10}$$

Sound speed for shocked gas, where P is absolute pressure and ρ is density of the shocked gas, respectively:

$$C = \sqrt{\gamma P / \rho} \tag{9.11}$$

Dynamic overpressure:

$$q = \frac{(\Delta P)^2}{2\gamma P_0 + (\gamma-1)\Delta P} \tag{9.12}$$

Total overpressure for a subsonic case:

$$\Delta P_T = \left[P\left[1 + M^2\frac{(\gamma-1)}{2}\right]^{\frac{\gamma}{\gamma-1}}\right] - P_0 \tag{9.13}$$

Total overpressure for a supersonic case:

$$\Delta P_T = P\left[\frac{\left\{M^2\dfrac{(\gamma+1)}{2}\right\}^{\gamma}}{\left\{\dfrac{2\gamma M^2}{(\gamma+1)}\right\}-\left(\dfrac{\gamma-1}{\gamma+1}\right)}\right]^{\frac{1}{(\gamma-1)}} - P_0 \tag{9.14}$$

Reflected overpressure:

$$\Delta P_r = P_0\left[2\frac{\Delta P}{P_0}+\frac{(\gamma_0+1)\left(\dfrac{\Delta P}{P_0}\right)^2}{(\gamma_0-1)\left(\dfrac{\Delta P}{P_0}\right)+2\gamma_0}\right]-P_0 \tag{9.15}$$

9.4 VARIANTS OF SHOCK TUBES

A standard shock tube has a uniform tube and operates on a single diaphragm that separates two compartments at different pressures (Holder, 1953; Ritzel and Parks, 2015). A typical design is shown in Figure 9.3. Its biggest strength comes in the form of its simplicity in design and operation. However, for applications requiring strong shocks, much thicker diaphragms and pressure ratios are required (Martin, 1958). Moreover, the use of diaphragms means that controlling the rupture pressures can be complicated as it requires an understanding of the properties of the membrane material and/or necessitates the employment of an on-demand membrane rupture mechanism. Furthermore, the irregular rupture of diaphragms is also a known problem and has been widely reported. For example, Andreotti et al. (2015) numerically established that the most significant impact on energy loss in a shock tube is caused by the opening mechanism of the diaphragm. For example, peak pressure reduction ranges between 5% for complete rupture (i.e., fully opened) of the diaphragm and 10% reduction for a 50% opening (see Figure 9.4). It was also reported that the rupture mechanism of diaphragms changes when different rupture pressures are used (Nguyen et al., 2014).

With the abovementioned deficiencies of the conventional shock tube, innovative developments and modifications were carried out to address some of these issues and to enhance the performance

FIGURE 9.3 A 0.15 m conventional shock tube (Pennelegion et al., 1965).

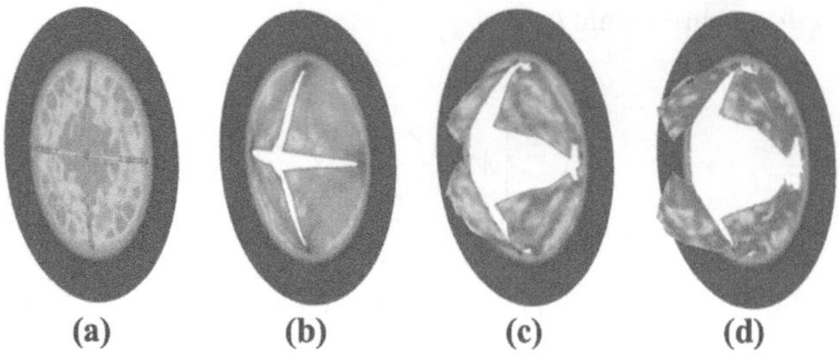

FIGURE 9.4 Diaphragm rupture process at various time instances: (a) 1.2 ms; (b) 2.4 ms; (c) 3.6 ms; (d) 7.2 ms (Andreotti et al., 2015)[2].

of the existing technology. This section presents some common variants and modifications to conventional shock tubes, with Table 9.1 providing a summary of their main features.

A double-diaphragm configuration addresses many of the shortcomings of conventional shock tubes with a single diaphragm by being able to produce a much greater final shock strength (Martin, 1958). Also, a double-diaphragm offers a greater degree of control for the timing and burst pressure

TABLE 9.1
Comparison Between Common Variants of Shock Tubes

Variant	Strengths	Weaknesses
Conventional (single diaphragm with air)	• Simple to design and operate	• Thick diaphragms and pressure ratios are required to produce strong shock waves • Requires the use of membranes which take time to set up • Difficulty in symmetrically rupturing large diaphragms • Rupture pressures are determined by the diaphragm properties, complicating the control of bursting driver pressures
Double-diaphragm	• Able to generate strong shock waves at a lower pressure ratio • Offers more control in shock wave generation • Able to distribute pressure load between diaphragms	• Requires more membranes than conventional shock tubes • More effort in setting up a test
Low-density working gas	• Allows for the generation of higher shock wave speeds at lower pressures	• Requires large amounts of helium which is expensive
Heated working gas	• Allows for the generation of higher shock wave speeds at lower pressures	• Thermal fatigue could be a problem, leading to reduced design life of shock tube
Fast-acting valves	• Able to generate very weak shock waves • Very good reproducibility of results • Eliminates the need of a diaphragm	• Shock waves generated are weaker than conventional shock tubes
Free-piston driven	• Able to generate high enthalpies	• Extremely complex to operate as it involves accelerating and decelerating large pistons
Detonation driven	• Allows the generation of high pressures and temperatures, low specific heat ratios, and high speed of sound at a relatively low pressure (high efficiency) • High repeatability • Produces key features of free-field blast waves	• Controlling the gas mix and precise ignition of driver gas are an added complexity

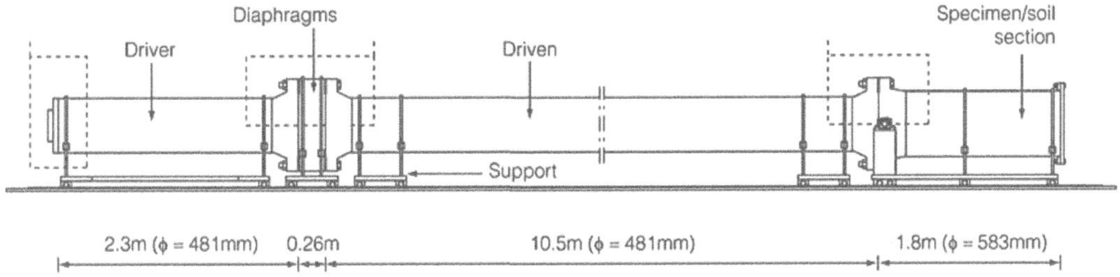

FIGURE 9.5 Schematic diagram of 583 mm shock tube with double-diaphragm configuration. (Adapted from Andreotti et al. (2015).)[3]

as compared to a typical single-diaphragm setup (Nguyen et al., 2014). In double-diaphragm configurations, the driver section is pressurised while the buffer chamber in between the driver and driven sections is pressurised to an intermediate level between the driver and driven pressures. When the desired pressure in the driver is reached, the gas in the buffer chamber is vented, causing the differential pressure between the driver and the buffer sections to reach the rupture pressure of the first diaphragm and cause it to burst, sending a pressure wave to rupture the second diaphragm and causing a shock wave to propagate along the tube (Andreotti et al., 2015; Stotz et al., 2008). Apart from that, a double-diaphragm configuration would help alleviate the pressure loads experienced by the diaphragms, allowing for the use of thinner diaphragms. Thinner diaphragms are generally desirable as they generally produce a more precise opening upon rupture, resulting in a cleaner incident shock wave with fewer disturbances (Colombo et al., 2010). Nevertheless, the benefits of the double-membrane system come with the drawbacks of needing more membranes per shot as well as requiring more effort in setting up the test (Holmberg, 2010). An example of a shock tube using the double-diaphragm configuration is given in Figure 9.5.

Another commonly adopted method to improve the performance of shock tubes is to increase the speed of sound of the working gas. A higher speed of sound allows the generation of higher shock Mach numbers at much lower driver-to-driven pressure ratios (Lu and Wilson, 2003). To increase the speed of sound of the working gas, one particular technique is to use a low-density gas such as helium in the driver section (Alay et al., 2017; Andreotti et al., 2015; Chandra et al., 2012; Colombo et al., 2010; Holder, 1953). As documented by Colombo et al. (2010), a helium driver only needs to be pressurised to 4.2 MPa rather than 56.7 MPa with the case of air when a reflected pressure of 5 MPa is desired. Alternatively, the speed of sound could be increased by raising the temperature of the working gas with a heater (Rebagay, 2017; Yu, 1999). Nevertheless, both methods come with some major drawbacks. For example, it could become prohibitively expensive as large amounts of helium would be required to pressurise the driver section, or high-pressure pumps and heaters would be required. Also, thermal fatigue could dramatically reduce the design life of the shock tube (Lu and Wilson, 2003).

In efforts to further enhance the reproducibility of results, reduce cost, as well as open up possibilities in automation, some researchers explored the idea of a diaphragmless shock tube. For example, Garen et al. (1974) introduced a new pneumatic valve to replace the diaphragm in conventional shock tubes. As diaphragmless shock tubes have the inherent ability to operate with no influence from the pressure difference between the driver and driven section, it is possible to generate extremely weak shock waves. Also, reproducibility of the results is significantly improved, as the valve can be opened at any desired pressure precisely every time. Ikui et al. (1979) implemented fast-acting valves that either operate vertically or horizontally to the shock tube axis (see Figure 9.6). These devices could generate plane shock waves as strong as conventional diaphragm shock tubes for test conditions with small initial pressure ratios. However, for test conditions with larger initial pressure ratios, conventional diaphragms were found to still produce the strongest shock waves. Hariharan et al. (2010) built a diaphragmless shock wave generator which employed a

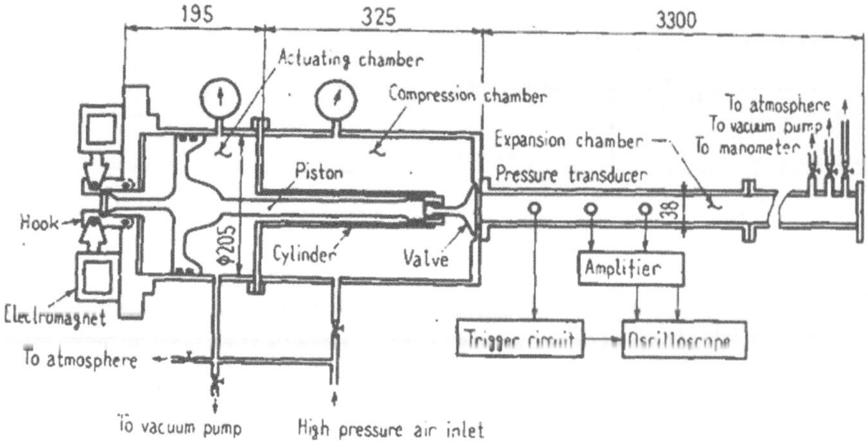

FIGURE 9.6 Fast-acting valves developed by Ikui et al. (1979).[4]

highly elastic rubber membrane as well as a fast-acting pneumatic valve. However, it was found that the rubber mechanism had to be replaced regularly and was not able to tolerate high driver pressures. In summary, while shock tubes employing fast-acting valves have some noteworthy advantages, they are generally only suited for weaker shock wave applications.

Another example for a diaphragmless shock tube is the free-piston-driven shock tube first introduced by Stalker (1967); this has been refined over the years leading to the ability to generate some of the highest enthalpies. One particular example (as shown in Figure 9.7) developed by Terao and Furushoh (1994) consisted of a detonation tube, a free-piston accelerator tube, a driver-gas tube, and a low-pressure tube, all separated with diaphragms between them. Detonation waves from the first tube drive the piston forward which causes shock waves to form in front of the piston. This causes the helium gas in the driver-gas tube to heat up and compress which causes strong shock waves to be produced in the low-pressure tube. This particular shock tube was capable of producing shock waves of $M_s \approx 15$ with the potential of even better performance with some modifications such as increasing the size and length of the tubes. Some other free-piston shock tubes are located at the University of Queensland in Brisbane, Australia (Morgan, 2000), in Göttingen, Germany (Hannemann and Beck, 2002), and the HIEST in Kakuda, Japan (Itoh, 2002). The method of using free-pistons, however, is complex as it involves rapidly accelerating and decelerating a large piston in a controlled fashion. Ojima et al. (2001) noted the complexity in safely slowing and stopping the pistons without causing severe impact damage; the authors utilised a block of alpha gel sheet to dissipate the impact of the piston when stopping it. An excellent paper comparing the performance between a free-piston and conventional shock tube can be found in Al-Falahi et al. (2008).

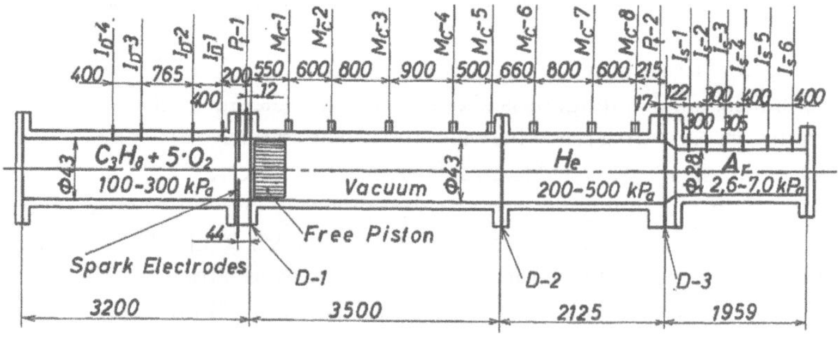

FIGURE 9.7 Free-piston driven by detonation shock tube developed by Terao and Furushoh (1994).[5]

Generally, shock tube driver designs are categorised to be either compression-driven or detonation-driven. Courtney et al. (2014) gave a good summary of their respective strengths and limitations. Typically, a compression-driven shock tube produces a shock wave formed at the leading edge of a propagating overpressure with a decay in pressure profile that mimics a blast wave to a certain extent. However, compression-driven shock tubes generate dramatic fluctuations in peak pressure between shots, pressure durations that are much longer than actual explosives, and generally do not approximate the Friedlander waveform of blast waves in an open field well (Igra and Houas, 2016). Detonation-driven shock tubes, on the other hand, produce more realistic blast waves but they are expensive and come with added risks and liabilities. The detonation-driven method is a well-established technique tracing back to early research by Hertzberg and Smith (1954), Bird (1957), Coates and Gaydon (1965), and Freiwald (1970). Oxyhydrogen and oxy-acetylene mixtures are typically used as the working gas to enable the detonation process (Lu et al., 2000). This method is highly favoured as it allows for high pressures, high temperatures, low specific heat ratios, and high speed of sound to be easily achieved in a working gas with relatively low pressure. Also, for blast simulation applications, detonation-driven drivers may generate many important characteristics of free-field blast waves, which are extremely desirable (Ritzel, 2007). Other than that, Yu (1999) asserted that the oxyhydrogen combustion driver is extremely low cost while still capable of producing strong shock waves with high enthalpies (although lower than free-piston drivers) and the required conditions for high-temperature effects, hypersonic experiments, and the physical phenomena of re-entry flows. Janardhanraj and Jagadeesha (2016) introduced a novel miniature detonation-driven shock tube that is designed to generate the required oxyhydrogen mixture with alkaline electrolysis. This device gives the advantage of added flexibility in the ability to adjust the strength of the shock wave and eliminates the requirement of storing extremely volatile, high-pressure gases. A further possible improvement is to enhance the performance of the oxyhydrogen driver by the addition of helium or argon to lower the molecular weight of the working gas while reducing the danger of the explosive mixture. All this is possible without a significant cost increase as only a small amount of helium or argon is required (Lu and Wilson, 2003). Despite the obvious advantages of detonation-driven shock tubes, they typically require precise control and metering of gas filling and mixing which is an added complexity (Martin, 1958). Excellent reviews for detonation drivers can be found in Lu et al. (2000) and Yu et al. (2006).

In addition to the different variants and modifications of the shock tube driver mechanism, the geometry of the driver and transition section as well as the choice of driver medium can significantly alter the flow environment within the facility (Ritzel and Parks, 2015). Ritzel and Parks investigated these effects through computational fluid dynamics (CFD) simulations, employing the Chinook solver to model a range of driver–driven geometries and predict the resulting waveforms. Figure 9.8 gives four examples: driver section with positive curvature and driven section with inflected curvature, driver section with linear curvature and driven section with convergent curvature, driver section with negative curvature and driven section with convergent curvature, and multiple drivers containing positive curvature and driven section with inflected curvature. These CFD-derived pressure–time profiles, adapted from their study, demonstrate how geometric shaping and driver medium directly influence shock evolution and waveform characteristics. The effect of geometry on the resulting pressure–time profiles is critical for generating tailored blast waveforms and forms the basis of the next generation of blast simulator design that will be introduced in Section 9.5.

9.5 SHOCK TUBES FOR BLAST WAVE SIMULATIONS

This section reviews shock tubes used for the purpose of blast wave simulations. These shock tubes are also known as blast simulators. Blast simulators have been used predominantly in studying the blast response of structures (Andreotti et al., 2015), blast barriers (Gan et al., 2021), blast-related injuries on organs such as the lung, ear, abdomen, head, and brain (Holmberg, 2010; Kleinschmit, 2011;

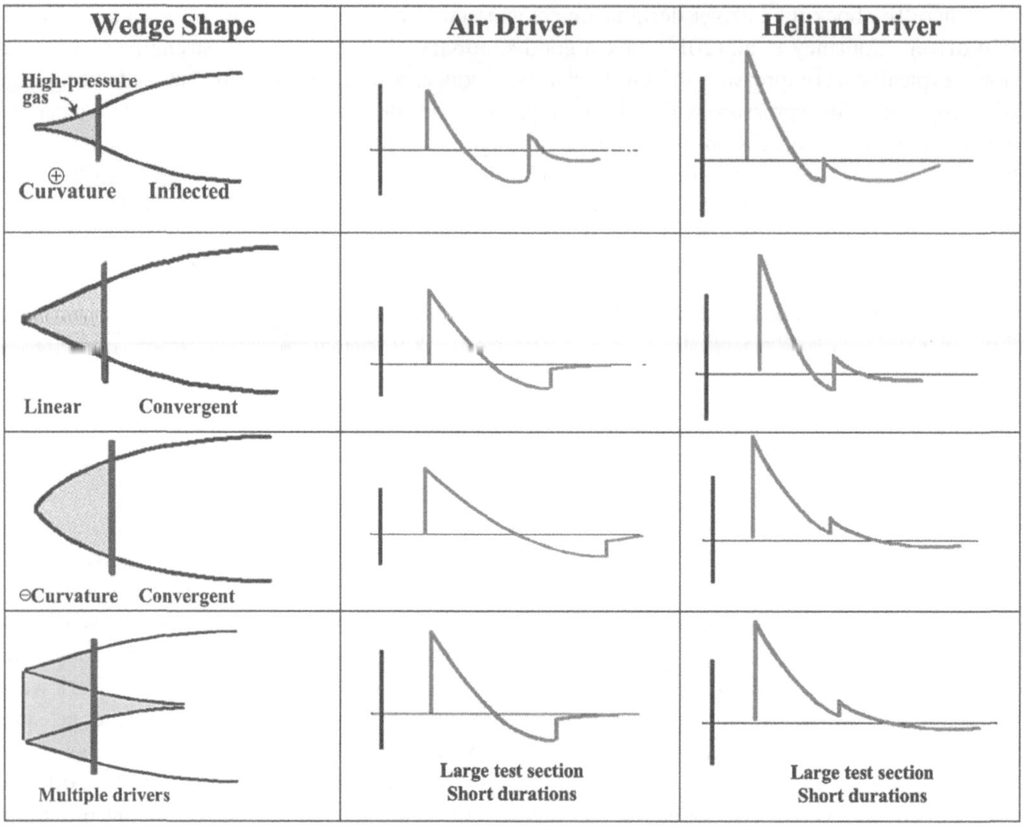

FIGURE 9.8 Effect of shock tube geometry and medium on pressure–time profile. (Adapted from Ritzel and Parks (2015).)

Maffeo et al., 2016), and indirect injuries caused by flying fragments propelled by an explosion (Ning and Zhou, 2015). Some applications of blast simulators will be further elaborated in Section 9.6.

Engineers typically investigate the effects of explosive charges with field tests or in the lab. As explained by Ritzel and Parks (2015), performing actual field tests with high explosives comes with additional complexity due to high cost, dependency on weather conditions and the availability of qualified personnel and sites, imprecise data collection resulting in a lack of reproducibility, and increased risk of accident during storage, transport, or setup of the charge. Many of these problems could be mitigated by the use of blast simulators to experimentally perform investigations. On top of that, another advantage of blast simulators is the generation of a planar wavefront, which allows for a uniform pressure distribution when testing on plane structural elements such as doors, walls, windows, etc. (Stoffel et al., 2001). However, it should be noted that planar waves are an idealisation of actual blast waves which are spherical in nature. A blast wave can be simplified to be a planar wave for large standoff distances.

While these simulators are essentially large-scale variants of the classical shock tubes discussed in the previous sections, it is important to establish and understand the difference between their intended objectives. Classical shock tubes are designed to investigate the flow properties through a shock front or flow behind a shock front that generally resembles a step function. Figure 9.9(a) illustrates this, where the generated waveform measured at the closed end of a conventional shock tube shows an initial plateau segment. The start of the waveform decay corresponds to the arrival of the reflected head of driver rarefaction (see Figure 9.9(b)).

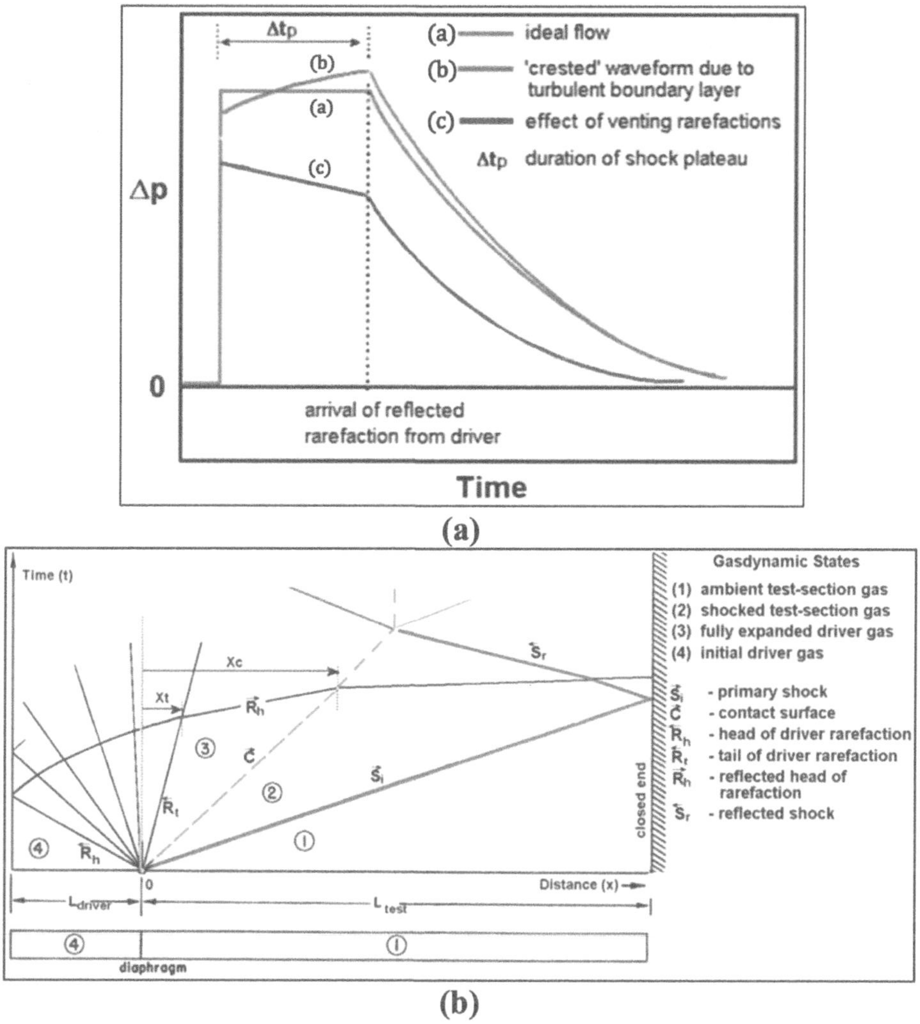

FIGURE 9.9 Closed-end shock tube: (a) typical pressure history measured at the closed end of a shock tube; (b) x–t diagram. (Adapted from O'Daniel et al. (2016).)

On the contrary, a blast simulator should ideally be able to replicate all the time-variant gasdynamic conditions of a free-field explosion (i.e., resemble the classical Friedlander waveform). However, aside from overpressure, properties such as density and velocity are crucial but are often incorrectly replicated by conventional blast simulators (Uy et al., 2017). As emphasised by Ritzel (2007), a common but serious misconception by recent research assumes that a measured static overpressure profile from a simulator would imply that all other gasdynamic conditions (e.g., density, dynamic pressure, and flow velocity) would also adhere to a free-field explosion scenario, which is not necessarily the case. Therefore, without careful consideration of the important characteristics of the generated flow, a blast simulator may potentially produce incorrect or deceptive shock wave conditions (Ritzel, 2007). Additionally, flow problems (see Figure 9.10) occurring within the common design of blast simulators often lead to various flow anomalies such as transverse reflected waves, flow separation and jetting, and anomalous longitudinal waves (Uy et al., 2017).

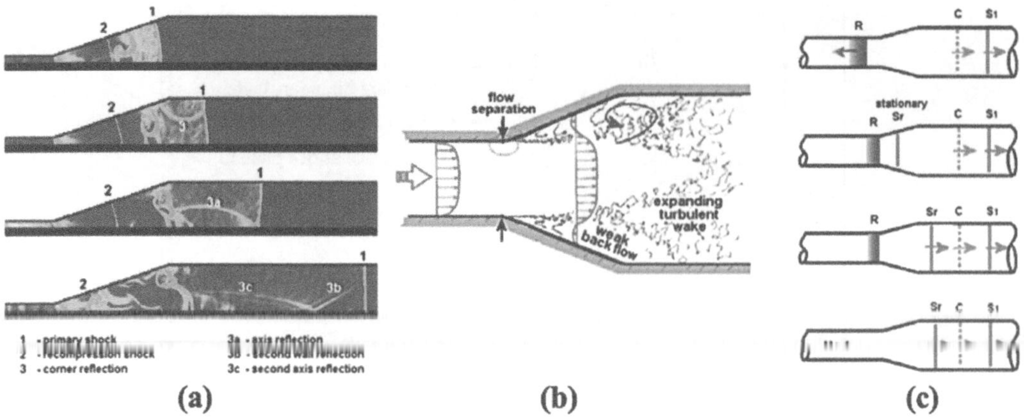

FIGURE 9.10 The typical geometry employed by conventional blast simulators causes various flow anomalies: (a) Transverse shock reflections; (b) Separation of flow and the formation of turbulent boundary layer; (c) Longitudinal wave anomalies (Uy et al., 2017).

The Lovelace blast simulator (Richmond et al., 1966) presented in Figure 9.11 illustrates numerous waveform anomalies as a result of the geometry and configuration of the blast simulator. The waveform itself does not resemble the classical Friedlander waveform and has a 'square' top and is typically seen in early blast simulators.

Conventional blast simulators often employ a constant cylindrical driver with a sudden and abrupt expansion section leading up to the test section (Alay et al., 2017; Chandra et al., 2012; Dallriva et al., 2016; Haverdings, 1981, 1983; Lloyd et al., 2009; Stolz et al., 2016). In other cases, expanding circular drivers are connected directly and abruptly to a straight test section (Ritzel, 2007) or simply in a constant cross-section for the entire length of the blast simulator (Aune et al., 2016; Colombo et al., 2010). These configurations often lead to various problems in the generated flow as illustrated in Figure 9.10 as well as being unable to develop the distinct negative phase of a free-field explosion. In addition to that, other issues specific to compressed gas (CG) drivers include the rapid cooling of the driver gas due to the sudden expansion resulting in undesirable discontinuities forming between its interface with the driven air and wave anomalies (Ritzel and Parks, 2010). Also, for shock tubes

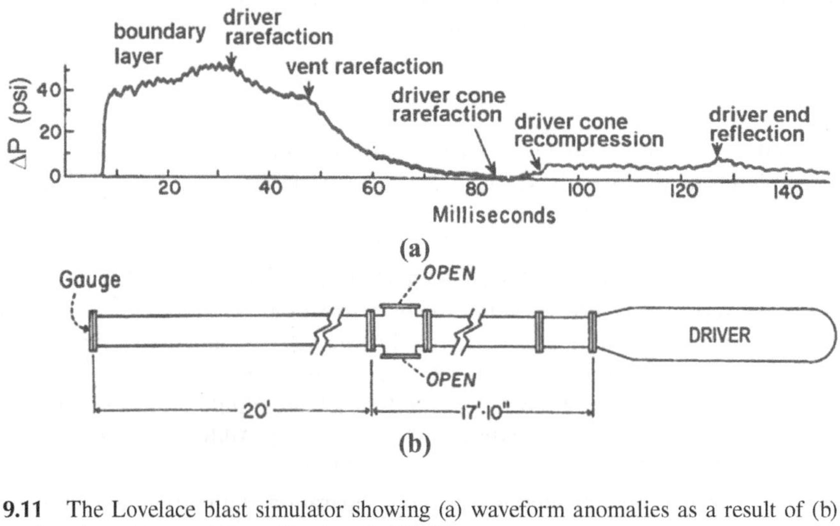

FIGURE 9.11 The Lovelace blast simulator showing (a) waveform anomalies as a result of (b) simulator geometry and settings. (Adapted from Ritzel et al. (2011).)

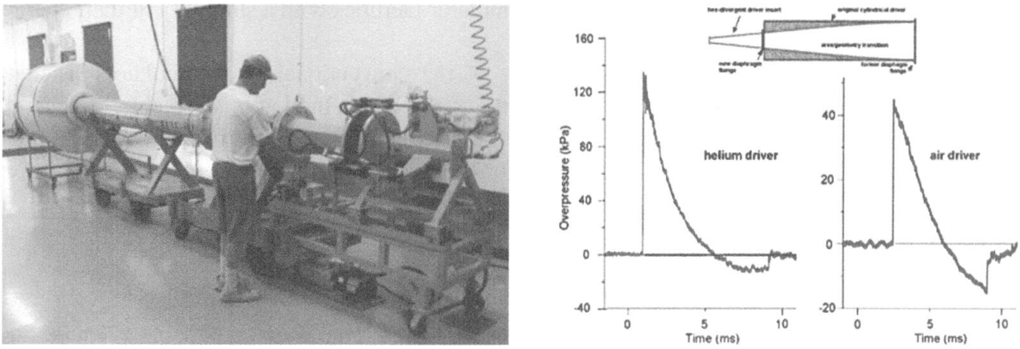

(a) (b)

FIGURE 9.12 An upgraded conventional shock tube retrofitted with an expanding conical driver as developed for the DRDC Suffield facility: (a) view of entire shock tube; (b) shock wave records recreate negative phase and secondary shock profiles well. The use of helium driver alters the shape of the waveform (Ritzel et al., 2011).

with an open end, excessive outflow tends to occur and recompression shocks moving upstream are developed, which are completely contradictory to an explosive blast scenario (Ritzel and Parks, 2010; Stewart, 2018a). It should also be noted that while static overpressure profiles resembling an explosive blast may be generated at some locations within the test section for certain configurations of a conventional blast simulator, as shown by Chandra et al. (2012), other gasdynamic conditions, particularly density and flow velocity, would often still be incorrectly simulated, which could lead to deceptive outcomes with respect to specimen loading and response (Uy et al., 2017).

Nevertheless, there are some conventional blast simulators fitted with 'hex-divergent' or expanding conical drivers operated with compressed helium or air (DRDC Suffield facility as illustrated in Figure 9.12) (Ritzel et al., 2011) or with an explosive charge (US Army Research Laboratory as illustrated in Figure 9.13) (Stewart, 2018a) attached to an expanding transition section capable of generating waveforms that resemble free-field explosions considerably well. However, the latter is successful only under very specific test section configurations and still has unresolved problems of rarefaction wave returning from the open rear evident in Figure 9.13(b) at t > 5 ms.

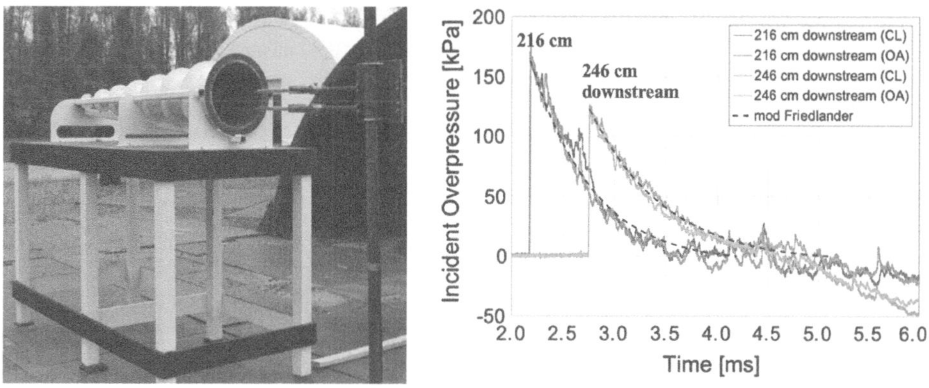

(a) (b)

FIGURE 9.13 Explosively driven conical shock tube by US Army Research Laboratory: (a) view of entire shock tube; (b) shock wave records show resemblance to Friedlander waveform but with abrupt expansion from open end rarefaction from t > 5 ms. (Adapted from Stewart (2018b).)

Select examples of recent conventional blast simulator facilities with a summary of their characteristics are given in Table 9.2.

Figure 9.14 presents the shock wave records of the conventional blast simulators listed in Table 9.2. One important takeaway is while some of the blast simulators were able to generate profiles that seem to mimic the positive phase of the Friedlander waveform reasonably well, many issues can still be observed within the records. For example, records in Figure 9.14(e)–(g) showed

TABLE 9.2
List of Examples for Conventional Blast Simulator Facilities

Conventional Blast Simulator	Size (m)	Shape	Driver	Purpose
1.8 m FAE Blast Tube at DRDC, Suffield, Canada (Ritzel, 2007)	1.8 diameter	Conical driver to tubular test section	Unconfined ethylene-air	For replicating tactical-scale nuclear blast waves
University of Ottawa Shock Tube, Ontario, Canada (Lloyd et al., 2009)	2.032 × 2.032	Cylindrical driver to square expansion cross-section	Compressed air with single or double aluminium diaphragm	To test structural components
0.23 m (9-in.) Square Shock Tube at University of Nebraska-Lincoln's Blast Wave Generation Facility, Nebraska, USA (Chandra et al., 2012)	0.23 × 0.23	Variable length cylindrical driver to square test section	Compressed nitrogen with single diaphragm	For studying blast-induced neurotrauma
0.71 m (28-in.) Shock Tube at University of Nebraska-Lincoln's Blast Wave Generation Facility, Nebraska, USA (Kleinschmit, 2011)	0.71 × 0.71	Variable length cylindrical driver to expansion section to square test section	Compressed nitrogen with Mylar diaphragm	For testing full-scale head models for researching traumatic brain injuries
Blast-STAR Shock Tube at Ernst-Mach-Institute, Baden-Württemberg, Germany (Stolz et al., 2016)	3 × 3	Cylindrical driver to expansion cross-section to square test section	Compressed air with single diaphragm	To test structural components
ERDC Blast Load Simulator, Vicksburg, Mississippi, USA (Dallriva et al., 2016)	1.22 × 1.22 or 2.44 × 2.44 (Interchangeable)	Cylindrical driver to circular expansion cross-section to square test section	Compressed air or air/helium mixture with single diaphragm	For replicating air-blast environment and studying the interaction with complex geometries
SIMLab Shock Tube Facility at Norwegian University of Science and Technology (NTNU), Trondheim, Norway (Aune et al., 2016)	0.3 × 0.3	Cylindrical driver to square test section	Compressed air with Melinex diaphragm	Simulate and generate blast loadings for studying the resulting structural behaviour
Shock Tubes at Shock Wave Testing Facility at the New Jersey Institute of Technology, New Jersey, USA (Skotak et al., 2018)	0.23 and 0.71 diameters (Two shock tubes)	Cylindrical driver to square test section	Compressed helium with Mylar diaphragm	For blast-induced injury studies

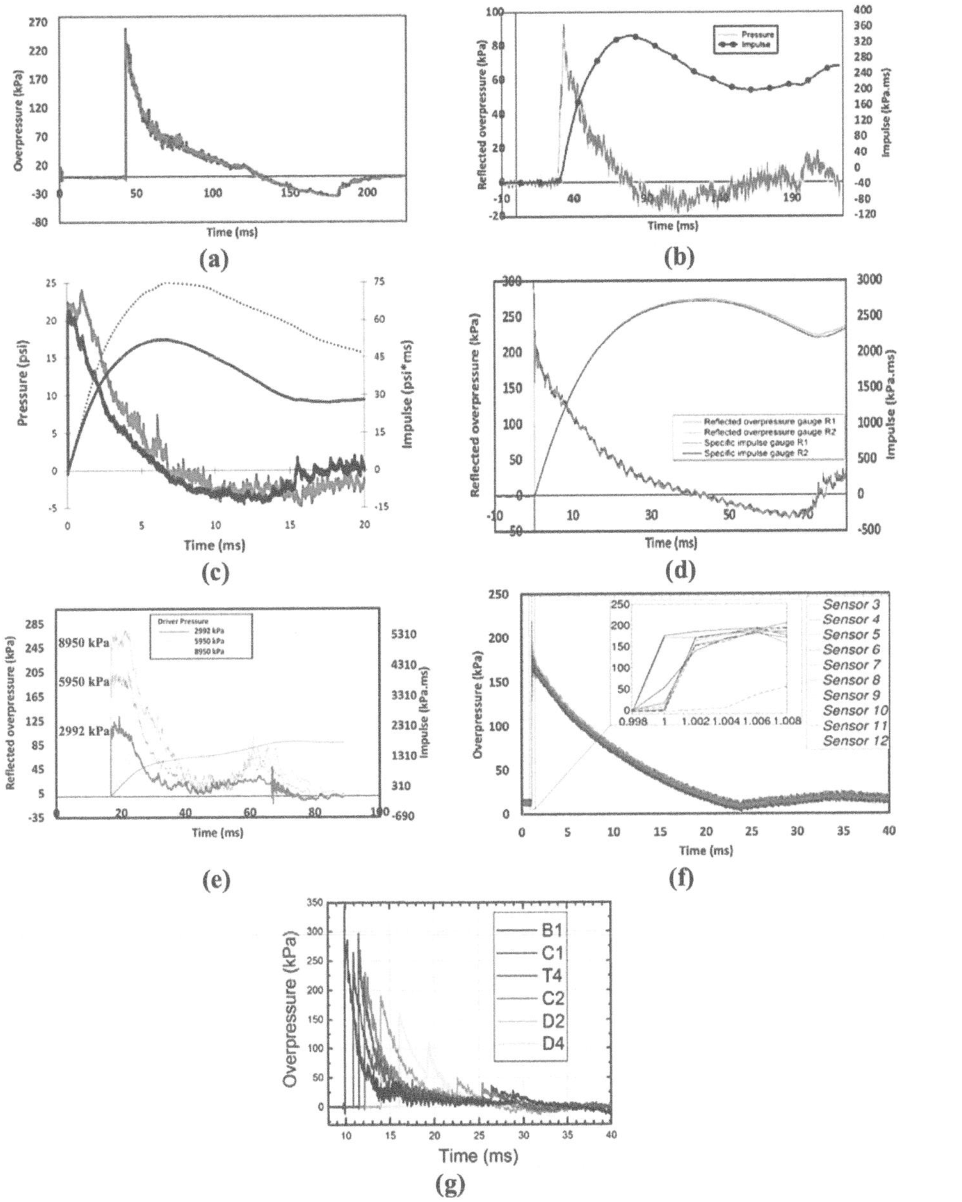

FIGURE 9.14 Example of records generated by some conventional shock simulators: (a) 1.8 m FAE blast tube at DRDC (Ritzel, 2007); (b) University of Ottawa shock tube (Lloyd et al., 2009); (c) 0.71-m (28-in.) shock tube at University of Nebraska-Lincoln's blast wave generation facility (with helium driver shown) (Kleinschmit, 2011); (d) Blast-STAR shock tube (Fraunhofer EMI, 2025); (e) ERDC blast load simulator (Dallriva et al., 2016); (f) SIMLab shock tube facility (Aune et al., 2016)[6]; (g) 0.23-m (9-in.) shock tube at the New Jersey Institute of Technology (Skotak et al., 2018).[7]

unrealistic quasi-static loading following the initial dynamic loading as a result of pressures not returning to ambient after the initial shock wave. Moreover, all records except for Figure 9.14(b) showed abrupt termination or lack of the negative phase. The negative phase in Figure 9.14(b) was achieved by venting just before the target. While the profiles of Figure 9.14 were only possible under a very specific driver length and with the use of pressurised helium in the driver (Kleinschmit, 2011). Also to be noted about the shock tube in Figure 9.14(c) is that the representations of the Friedlander waveform vary drastically between different locations of the test section. Such scenarios are contrary to actual free-field explosions that these blast simulators are trying to reproduce, where the Friedlander waveform should be represented well regardless of location or standoff distance. It is important to note that all measurements shown in Figure 9.14 are of hydrostatic pressure profiles. Other gasdynamic conditions such as dynamic pressure, density, and flow velocity are usually never measured or determined but are crucial for verifying that the generated flow conditions correctly conform to a far-field explosive environment and loading.

As previously demonstrated in Figure 9.8, the geometry of a blast simulator significantly influences the blast environment it generates. This forms the basis of the next generation design of blast simulators known as Advanced Blast Simulators (ABS), where the employed geometry complies with the characteristic of an expanding blast wave, thus addressing all problems faced with conventional blast simulators.

To address the shortcomings and limitations of conventional blast simulations described earlier, ABS were developed. The ABS concept was developed and patented by David Ritzel of Dyn-FX Consulting Ltd. (Canada) and Steven Parks of ORA, Inc. (USA) (Ritzel and Parks, 2013; Ritzel and Parks, 2015). This concept was actually first suggested several decades earlier by Cuthbertson (1967) but significant issues such as non-uniform flow in the test section could not be resolved at the time, and therefore it never went beyond the original prototype (Ritzel and Parks, 2010).

The ABS concept is governed by the principle that all blast waves which expand according to x^{β}, where x is the distance away from the point of detonation and β is between 1 and 2, would resemble a far-field blast profile. A far-field blast profile typically consists of the exponential-like decay profile, negative phase, and secondary shock (Ritzel and Parks, 2010; Uy et al., 2017).

As depicted in Figure 9.15, all the important characteristics of blast waves are a direct consequence of blast propagation in spherical geometry and can be assumed to be spherically symmetric (Ritzel and Parks, 2010; Uy et al., 2017). This means that the physics of spherical blast decay with distance can be recreated in any solid angle taken from the flow field. Therefore, a shock tube with a constantly expanding area would be ideal for simulating blast conditions from an explosive charge. As presented in Figure 9.15, the amplification of the generated blast strength is controlled by the cone angle. It is important to note that there exists an optimal and practical range for selecting the cone angles. For example, cone angles that are too small could lead to problematic boundary layer effects as well as heat transfer at the wall while angles that are too large would lead to increased requirements in the size of the facility as well as difficulty in producing a uniform blast front. Normally, a cone angle of around 8.5° is generally sufficient for most scenarios (Ritzel and Parks, 2010). Figure 9.16 gives an example of the ABS concept adapted to a blast simulator, showing the different

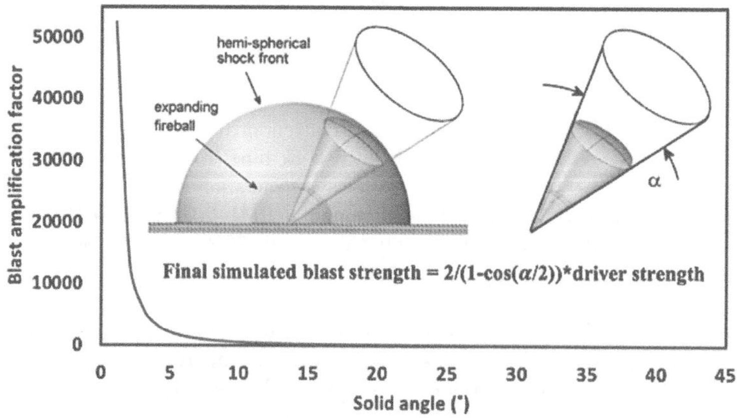

FIGURE 9.15 An idealised blast flow condition can be taken as spherically symmetric. Adjusting the solid angle will modify the final simulated blast strength. (Adapted from Ritzel and Parks (2010).)

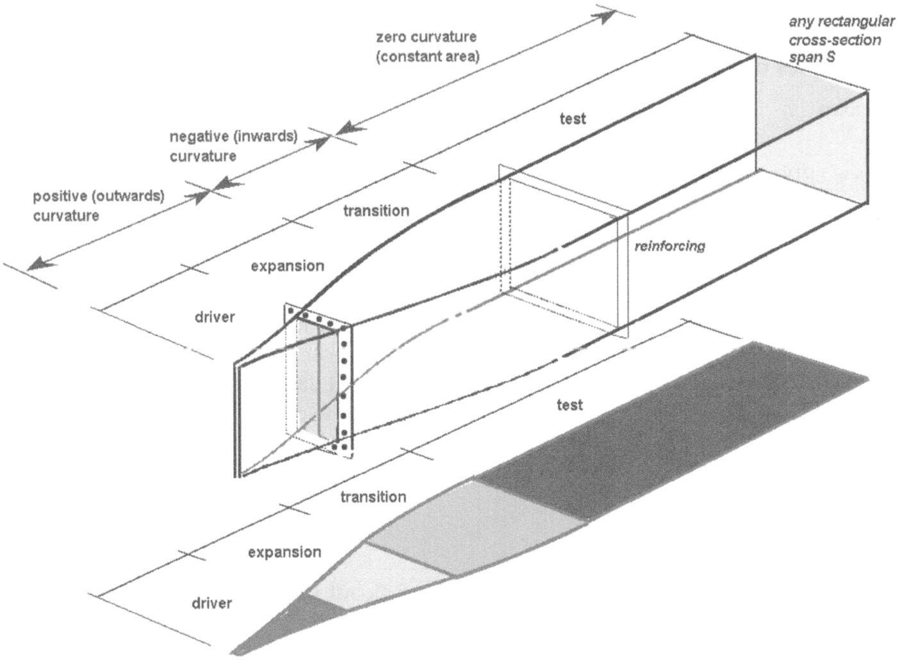

FIGURE 9.16 Illustration of the Advanced Blast Simulator (ABS) concept in the form of a blast simulator apparatus (Uy et al., 2017).

sections and the respective geometrical features, consisting of an initially expanding curvature before gradually flattening and tapering to a constant area test section. This controlled rate of area expansion allows for the wave dynamics of a free-field explosion to be accurately replicated while minimising waveform decay at the test section (Uy et al., 2017) and the occurrence of transverse waves seen in Figure 9.10.

Some examples of ABS facilities that are documented in the world today are outlined and chronologically arranged in Table 9.3, where their characteristics are summarised.

TABLE 9.3
List of Examples of ABS

Advanced Blast Simulator	Size (m)	Driver	Performance	Purpose
ABS at DRDC, Suffield, Canada (Sawyer et al., 2016)	0.0305 diameter	Compressed helium with a layered acetate diaphragm	Incident: 223 kPa and 380 kPa.ms (maximum reported in the literature)	For blast-induced injury studies
0.91 m ABS at Johns Hopkins University Applied Physics Laboratory (Maffeo et al., 2016)	0.91 × 0.91	Compressed air or helium with membrane	Incident: up to 180 kPa	To evaluate head-borne equipment: head form with helmet
Subscale 30 cm Prototype of the ABS (Parks et al., 2016; Uy et al., 2017)	0.030 × 0.012	Oxy-acetylene	Incident: 310 kPa for 6 ms (maximum reported in the literature)	As a prototype leading up to 1.2 m ABLS at ERDC

(Continued)

TABLE 9.3 (*Continued*)
List of Examples of ABS

Advanced Blast Simulator	Size (m)	Driver	Performance	Purpose
1.2 m ABLS at ERDC, Vicksburg, Mississippi, USA (O'Daniel et al., 2018)	1.2 × 1.2	Oxy-acetylene and/or ethylene-air	Incident: 310 kPa for 10 ms (maximum reported in the literature).	For concept development for the full-scale 3.7 m simulator
ABS at NFPBS at the University of Wollongong, NSW, Australia (Gan, 2021; Gan et al., 2020, 2024)	1.5 × 2.0	Oxy-acetylene or compressed gas with membrane	Incident: 200 kPa for 25 ms (current maximum) Reflected: 285 kPa for 35 ms (current maximum)	For various blast wave propagation and loading regimes, blast damage of elements of civilian and military infrastructure, blast injury protection, and others
Virginia Tech shock tube research facility (Jacques, 2020)	2.4 × 2.4	Oxy-fuel mixture	Reflected: up to 200 kPa between 5 and 20 ms	To evaluate the resilience of structural and non-structural building components subjected to extreme blast loads
(3.7 m) 12 ft ABLS at ERDC, Vicksburg, MS, USA (Parks et al., 2023)	3.7 × 3.7	Oxy-ethylene initiated by oxy-acetylene booster	Reflected: over to 600 kPa over 30 ms	Blast vulnerability assessment and protective design for army structures and facilities

All ABS facilities in the world share similar characteristics. All have geometries that resemble Figure 9.16 and usually include 'end-wave eliminator' (EWE) and venting systems (i.e., louvres) for reflected waves coming back from a reflective test target, although the designs for the EWE and louvres could be quite different between facilities as they depend on the specific requirements and circumstances.

As an example of an ABS facility, a description of the ABS at the National Facility of Physical Blast Simulation (NFPBS) in Australia is given here. The NFPBS ABS (Figure 9.17) was commissioned

FIGURE 9.17 NFPBS Advanced Blast Simulator (ABS) configuration as-built. The driver and reaction housing (RH) are rail-mounted and have sliding overlap connections with the test section such that reaction forces are not transmitted.

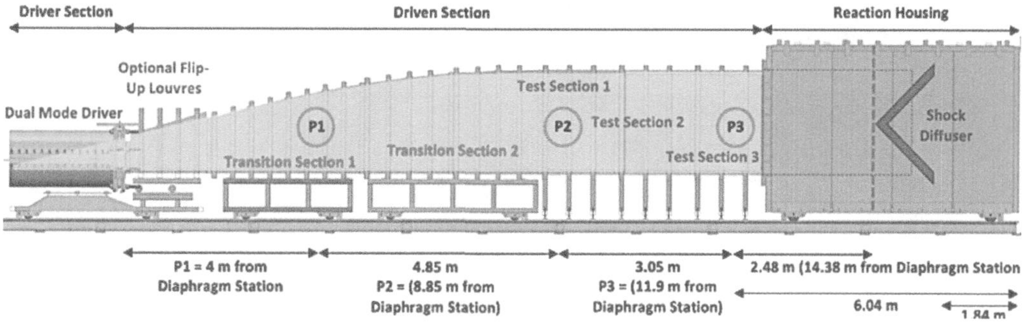

FIGURE 9.18 Schematic diagram of NFPBS ABS.

in mid-2018 (Remennikov et al., 2018) and was located within the Research and Testing Facility of the Faculty of Engineering and Information Sciences at the University of Wollongong (2018–2024). The NFPBS facility is unique within Australia and in the Southern Hemisphere as well as being one of only a few in the world, both in terms of its loading capacity and the size of specimens able to be tested, and remains the largest dual-mode ABS. An overview of the different sections of the ABS is illustrated in a schematic in Figure 9.18.

The driver has a divergent wedge-shaped profile and can operate either in compressed gas (CG) or gaseous detonation (GD) mode, depending on the requirement. Generally, the CG mode produces shock waves with a more pronounced and adjustable negative phase while GD mode produces much stronger blast simulations with weak negative phase.

With the CG mode, the ABS is operated in a similar manner to a conventional shock tube: a gas, typically air, nitrogen, or helium, is introduced into the driver, raising the pressure above ambient. A frangible diaphragm (Figure 9.19) is used to separate the high pressure contained in the driver from the ambient pressure in the driven section downstream. Upon reaching the desired pressure in the driver, the diaphragm is ruptured quickly to allow high-pressure gas to expand into the ambient

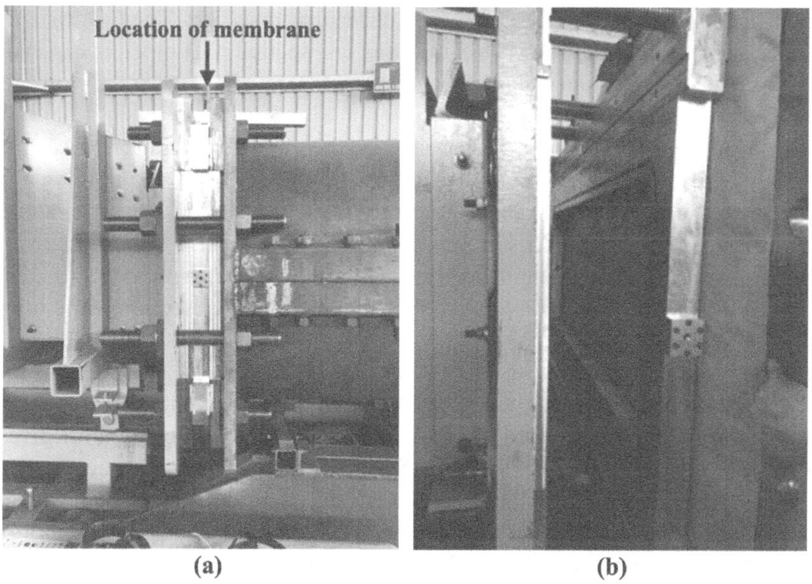

FIGURE 9.19 Compressed gas (CG) driver mode: (a) location of the diaphragm (i.e., diaphragm station); (b) clamping of a frangible diaphragm across the opening of the driver to create a gas-tight barrier.

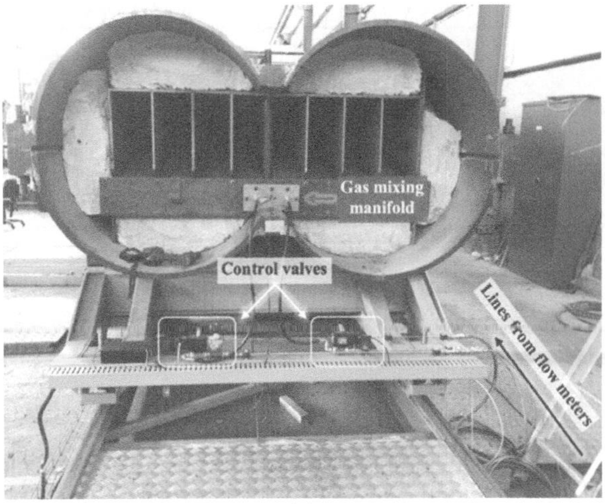

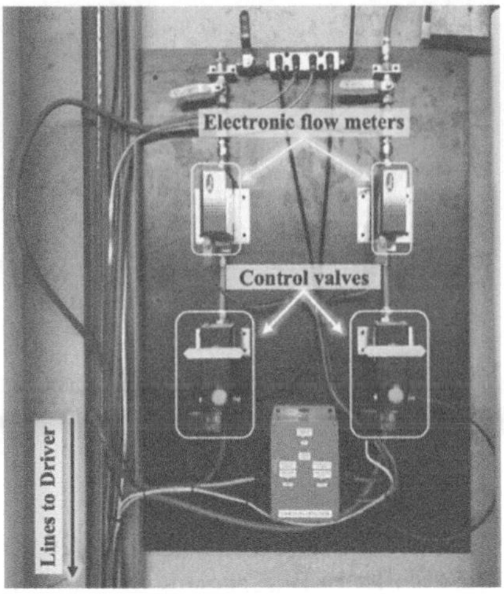

FIGURE 9.20 Gaseous detonation (GD) driver mode: (a) flammable gas delivery system; (b) electronic flow meters for precise metering of combustion gases.

pressure contained downstream. The release of the elevated pressure gas acts as a 'piston' to rapidly compress the ambient air at the interface of the high-pressure/low-pressure gas volumes to create a propagating shock wave. The characteristic 'Friedlander' blast waveform is created by the expansion of the gas out of the divergent driver and through the initial divergent transition section; once formed, the wave is smoothly re-converged into the test section.

With the GD mode (Figure 9.20), the geometry of the driver and downstream sections remains identical with the CG driver mode. However, the driver pressure is generated by the detonation of an explosive gas mixed with air and/or pure oxygen contained within light plastic bags. Typical combustible gases include acetylene (C_2H_2) and ethylene (C_2H_4). The resultant gas mixture is initiated by a powerful spark or electric match at the apex of the driver section. The GD wave transits to an air shock wave which propagates downstream to the test section. GD mode is capable of generating much higher shock levels than CG mode and has the operational advantage of not requiring the setup of a frangible diaphragm.

Downstream of the divergent-area driver section, the connecting transition section continues to geometrically expand and then smoothly and steadily re-converge the flow as a planar wave entering the constant cross-section geometry of the test section. The blast wave's rate of geometric expansion vs linear volumetric expansion is smoothly reduced to zero by the time the blast wave arrives at the test section. The length of the transition section is set to provide sufficient runout for perturbations in the initial blast to dissipate before reaching the test section while minimising the rate of change of the wall curvature.

Unlike a conventional shock tube, a fully formed blast wave is generated from the driver. Therefore, it is possible to locate targets at various distances from the first transition section as required for exposure levels similar to a free-field blast, where a closer standoff gives a stronger shock level with a shorter duration. The NFPBS ABS is designed to be modular, that is, the test section comprises three segments which can be configured as required to increase or reduce the testing standoff distance (Figure 9.21). Figure 9.22 presents the shock wave records generated with a 0.0708 m^3 (2.5 ft^3) oxy-acetylene GD driver along the length of the ABS and on the reflective wall target for the short and long ABS configurations, respectively. On each of the graphs, the waveforms which may appear as single traces are results of two identical tests superimposed, demonstrating exceptional repeatability in the results. An example of shock wave record

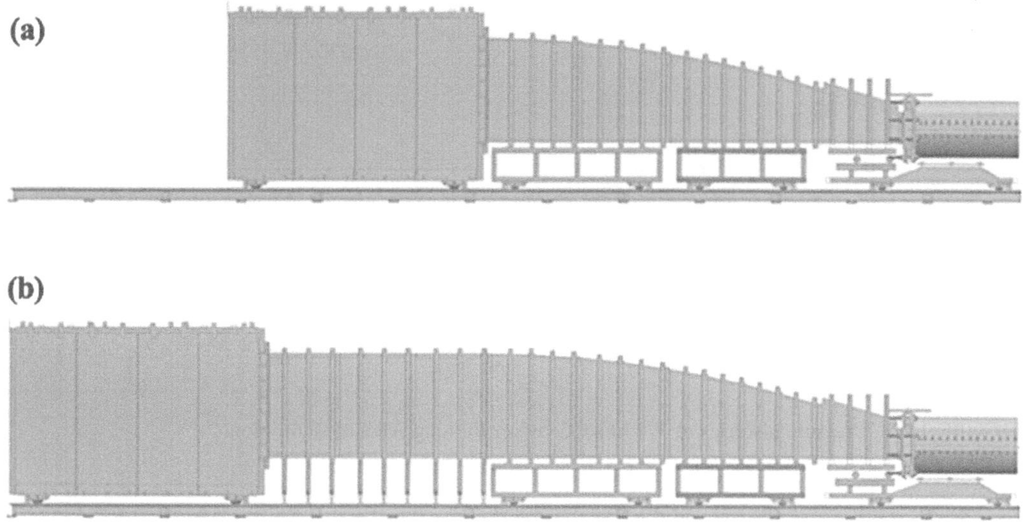

FIGURE 9.21 Modular design allows for adjustment of standoff distances: (a) reduced standoff distance (short ABS configuration); (b) full standoff distance (long ABS configuration).

generated with a CG driver pressurised to an initial static overpressure of 376 kPa with the long ABS configuration is illustrated in Figure 9.23. It can be observed that a prominent negative phase occurs after the initial blast front.

A set of louvres designed and installed at the top panel of the transition section eliminates reflected shocks propagating upstream from reflective targets that would otherwise be reflected again from the closed end of the driver to propagate downstream and interfere with experiments (Figure 9.24). The louvres can also be used to reduce the blast wave duration of the initial downstream running wave for either diffraction or reflection targets and be used for waveform tailoring (Gan et al., 2024). The louvres can be selectively fixed shut by blocking-plate assemblies or allowed to open in a controlled manner by the passing blast wave. The extent and speed of venting are controlled by the adjustable mass of the louvres as well as recoil restraints if required.

The test section provides a constant 1.5 × 2.0 m cross-sectional area within the simulator for the placement of test targets. Diffraction targets such as scale-model structures are placed within the test section, and reflective targets are mounted to the reaction frame of the reaction housing (RH) located at the end of the test section. The test section has instrumentation ports drilled through the walls, an observation window, and an access hatch. Figure 9.25 presents shock wave records

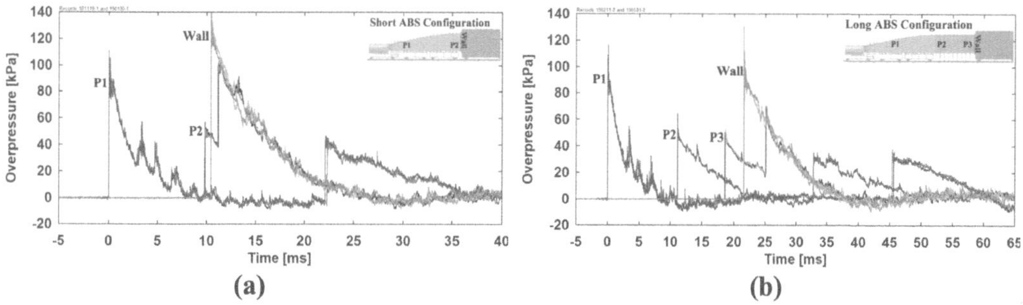

FIGURE 9.22 Pressure histories of 0.0708 m^3 (2.5 ft^3) oxy-acetylene GD driver for (a) short ABS and (b) long ABS configuration, respectively.

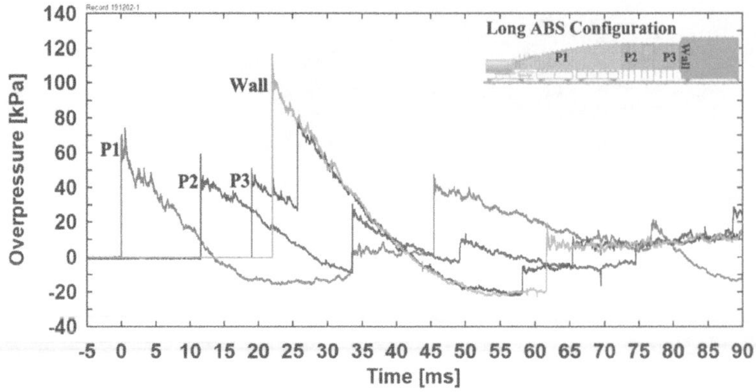

FIGURE 9.23 Pressure histories of 376 kPa CG driver for long ABS configuration.

FIGURE 9.24 Transition section louvres to allow for controlled venting of shock waves reflecting back from reflective targets: (a) louvres in a closed position prior to detonation; (b) louvres flipping open and venting reflected shocks returning from the target.

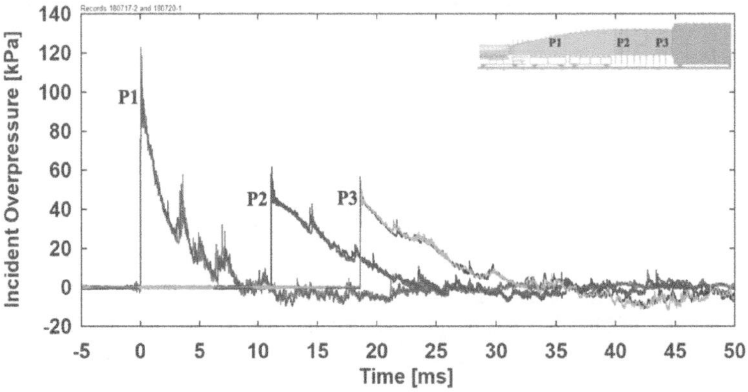

FIGURE 9.25 Pressure histories of 0.0708 m^3 (2.5 ft^3) oxy-acetylene GD driver for diffraction target

(a) (b) (c)

FIGURE 9.26 Adjustable EWE diffuser inside the RH used for diffractive tests. The porosity of EWE is adjustable. View from the upstream opening of the dump tank is shown: (a) fully opened EWE; (b) 50% closed EWE; (c) fully closed EWE.

generated with a 0.0708 m^3 (2.5 ft^3) oxy-acetylene driver along the ABS for a diffractive test scenario. As with Figure 9.22, this graph presents two identical tests superimposed.

For studies of loading and damage to diffraction targets set up in the test section, it is necessary to mitigate waves reflecting from the end of the test section once the primary blast wave has passed. Therefore, an 'end-wave eliminator' (EWE) is configured within the RH, comprising an internal shock diffuser having the form of a wedge with adjustable porosity (Figure 9.26). The porosity of the wedge is adjustable to enable optimisation of its effectiveness in dissipating waves passing into the volume of the RH, which serves as a 'dump tank'. The RH has sufficient volume to dissipate the incoming shock wave as well as mitigate noise and gas efflux into the laboratory space.

In addition, the RH also serves as the mounting surface for reflective targets such as walls or doors (Figure 9.27). The RH is constructed of heavily reinforced steel and weighs approximately

(a) (b)

FIGURE 9.27 Mounting of reflective targets: (a) test specimen mounted on reaction frame; (b) RH with a reflective target in pre-testing position.

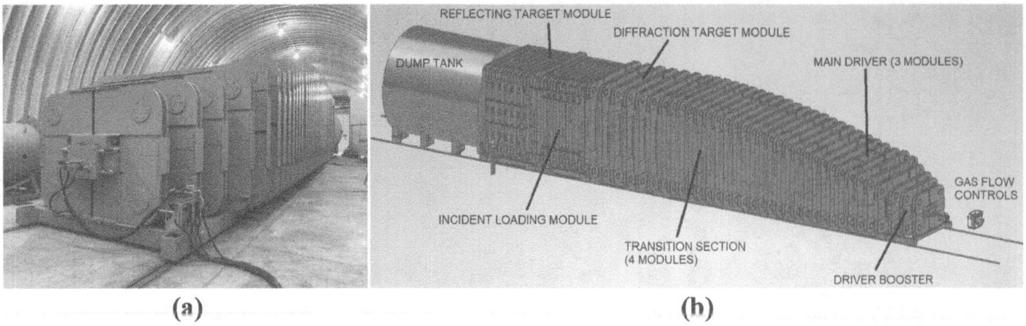

FIGURE 9.28 (a) 3.7 m ABLS at ERDC; (b) overview of the different sections and components (Parks et al., 2023).

15,500 kg; it is mounted on heavy casters and is free to roll on a track. Target reaction loads are coupled to the housing by the reaction flange and that energy is dissipated by the recoil.

Another notable ABS facility is the 3.7 m Advanced Blast Load Simulator (ABLS) at the US Army Engineer Research and Development Center (ERDC) (Figure 9.28), commissioned in late 2023 (Parks et al., 2023), which is the largest known ABS facility in the world.

Just like the NFPBS ABS, the 3.7 m ABLS at ERDC follows the principles illustrated in Figure 9.15 and the ABS characteristic shape (Figure 9.16). The ERDC ABLS utilises ethylene mixed with either air or oxygen as the reactive gas. To ensure reliable detonation, a strong shock wave is required to initiate both mixtures. This is achieved by first spark initiating the oxy-acetylene in the driver booster module, which generates the necessary shock wave to trigger the ethylene mixture in the main driver. Example records for the different driver gases are given in Figure 9.29.

In normal operation, a thin membrane is used to confine the oxy-acetylene mixture within the booster. However, if ignition timing is precisely controlled, detonation can be achieved without enclosing the oxy-acetylene cloud in a membrane or bag. This approach involves first injecting ethylene to fill both the booster and the required section of the main driver. Since ethylene has a molar mass comparable to air, it remains relatively stable within the chamber after filling. The oxy-acetylene mixture is then introduced into the booster, forming a concentrated pocket within the ethylene cloud. An electric spark ignites the oxy-acetylene, producing a strong shock wave that propagates outwards and initiates detonation of the ethylene mixture. The gas mixtures are injected through mixing nozzles at the back of the driver booster, where mechanical

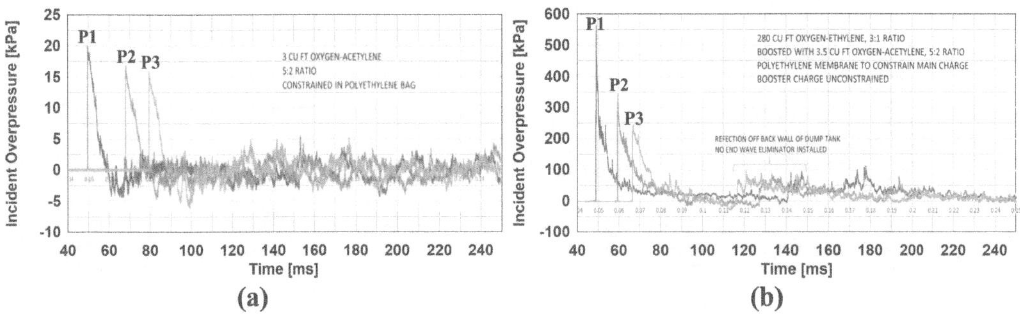

FIGURE 9.29 Pressure histories of (a) 0.085 m³ (3.0 ft³) oxy-acetylene driver and (b) 7.9 m³ (280 ft³) oxy-ethylene driver for diffraction target (Parks et al., 2023).

mixing nozzles and computer-controlled mass flow controllers regulate the gas composition before injection.

The driver also has vent louvres which are passively activated, opened by the detonation of the main charge, and provide a vent path for the reflected blast wave that propagates back towards the driver from the target. Providing the vent path minimises the secondary target loading created by the subsequent reflection of this reflected wave from the booster back to the target.

The transition section of the ERDC ABLS smoothly directs the flow from the divergent driver section into the constant-area test section. This section is made up of four modules and maintains a consistent internal width of 3.7 m along its entire length. The internal height is carefully shaped to minimise abrupt changes in cross-sectional area, ensuring a gradual and efficient transition for the flow.

The test section serves as the designated area for target placement and is composed of three modules: the Diffraction Target Module, the Incident Loading Module, and the Reflection Target Module. The test section maintains a consistent internal width and height of 3.7 m.

The Diffraction Target Module is designed for testing targets that interact with the blast wave through diffraction. These targets are positioned within the blast wave flow, and the wave diffracts around them, simulating real-world shock qualification scenarios. The module is equipped with mounting points for targets and instrumentation ports to monitor test conditions. To prevent the propagation of reflected or rarefaction waves, an EWE is installed in the dump tank. The Incident Loading Module allows for testing of targets under side-on blast loading. The side walls of this module contain removable panels that can be replaced with various test targets, such as wall or window panels. In cases where no reflective surface is present, an EWE is again required to reduce reflected waves. The removal of the side panels necessitates the installation of structural tie rods to maintain stability, ensuring the module's integrity during testing. The module is also fitted with mounting points and instrumentation ports. The Reflection Target Module is tailored for testing targets under direct, normal blast loading. It is equipped with both instrumentation and optical camera ports, allowing for comprehensive monitoring during the test.

The dump tank is a large cylindrical vessel positioned directly behind the Reflection Target Module. It functions as a reservoir for absorbing the blast wave exiting the simulator, with the energy being dissipated into the atmosphere within the tunnel enclosure. The EWE is integrated within the dump tank to further reduce wave reflections. Additionally, the dump tank provides ballast to help control the recoil response of the Reflection Target Module, with a braking mechanism installed to enhance recoil mitigation. The total mass of the dump tank is 95,000 kg.

Figure 9.30 presents some examples of the overpressure history graphs of some of the other ABSs given in Table 9.3. It is evident from the plots that all the ABS facilities were able to generate waveforms that more closely resemble a far-field blast profile than the waveforms produced with the conventional blast simulators presented in Figure 9.14. In addition to static overpressure waveforms, an ABS would also be capable of reproducing all other gasdynamic conditions of a free-field explosion (e.g., dynamic overpressure, total overpressure, temperature, density, etc.), as demonstrated in ABS characterisation studies by Gan (2021) and Gan et al. (2020).

9.6 CASE STUDIES: BLAST TESTING OF STRUCTURAL ELEMENTS WITH BLAST SIMULATORS

Blast simulators have been extensively employed in both fundamental research and applied studies to enhance blast resilience across various fields. Their applications range from evaluating structural responses and mitigating blast-induced injuries to assessing protective measures against fragmentation hazards. Beyond these primary areas, blast simulators also support the development of safety guidelines, the validation of numerical models, and the advancement of

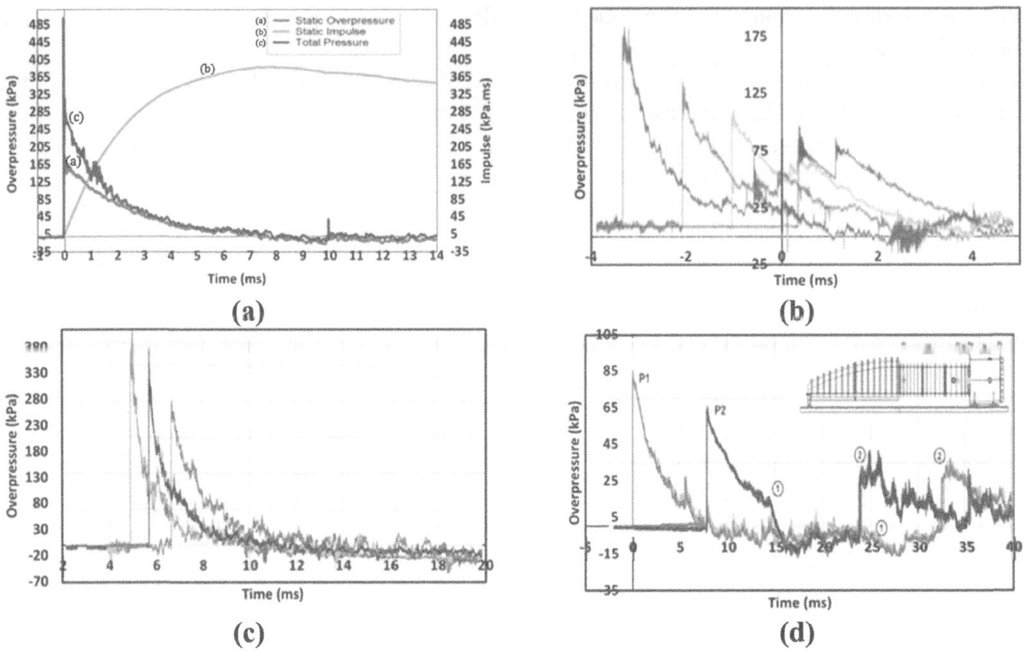

FIGURE 9.30 Example of records generated by ABS': (a) ABS at DRDC (Sawyer et al., 2016); (b) 0.91 m ABS at JHUAPL (Maffeo et al., 2016); (c) subscale 30 cm prototype of the ABS (Uy et al., 2017); (d) 1.2 m ABLS at ERDC (O'Daniel et al., 2018).

protective technologies. This section elaborates on some key applications, with specific examples based on the use of the Advanced Blast Simulator (ABS) at the National Facility of Physical Blast Simulation in Australia.

9.6.1 Explosive Resistance Testing of RC Slabs and CLT Panels

In these tests, the dynamic responses of reinforced concrete (RC) slabs and cross-laminated timber (CLT) panels when subjected to far-field blast loadings were studied.

Two specimens of 500 mm × 1950 mm were simultaneously mounted in the reaction frame in a simply supported boundary condition. The RC slabs were reinforced with 10 mm main steel bars at 100 mm intervals and 8 mm secondary bars at 200 mm intervals. The bars employed were low-ductility (L-class) grade 500 ribbed reinforcement bars. Unconfined compressive and tensile strength for the concrete were determined to be approximately 32 and 3.4 MPa, respectively, after 28 days. The CLT panels are constructed from plantation pine and several layers of lumber alternating at 90 degrees.

Figure 9.31 summarises the reflected pressures and impulses of the three tests. Additionally, equivalent TNT charge weight and standoffs are calculated for each test and are superimposed on the respective shock wave records. Figure 9.32 presents the front and rear faces of the 100-mm concrete slabs reinforced with 10-mm low-ductility reinforcing bars and CLT panels before and after the tests.

For Test 1 (Figure 9.32(a)), a dynamic load resembling a free-field explosion of 664 kg spherical TNT at 25.8 m or 328 kg hemispherical TNT at 24.2 m, respectively, was applied on the RC slabs. Maximum dynamic and permanent displacements of 57 mm and 13 mm were recorded on the RC slabs. Apart from two to three through-thickness cracks, no other damage on the specimens was observed.

For Test 2 (Figure 9.32(b)), a stronger blast load equivalent to 2002 kg spherical TNT at 33.7 or 994 kg hemispherical TNT at 31.6 m, respectively, was applied on the RC slabs. C32-S3 was found

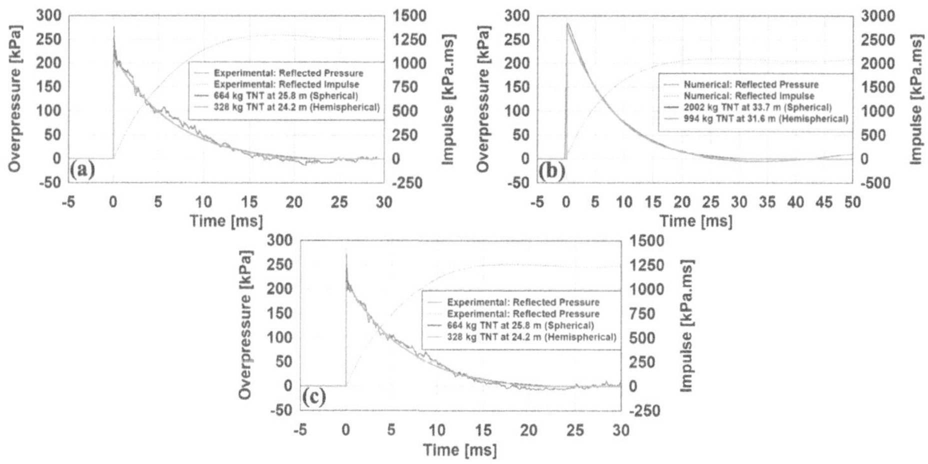

FIGURE 9.31 Blast pressure and impulse curve applied on RC slabs in (a) Test 1 (C32-S5/C32-S6), (b) Test 2 (C32-S3/C32-S4), and CLT panels in (c) Test 3 (CLT5-100/CLT3-105).

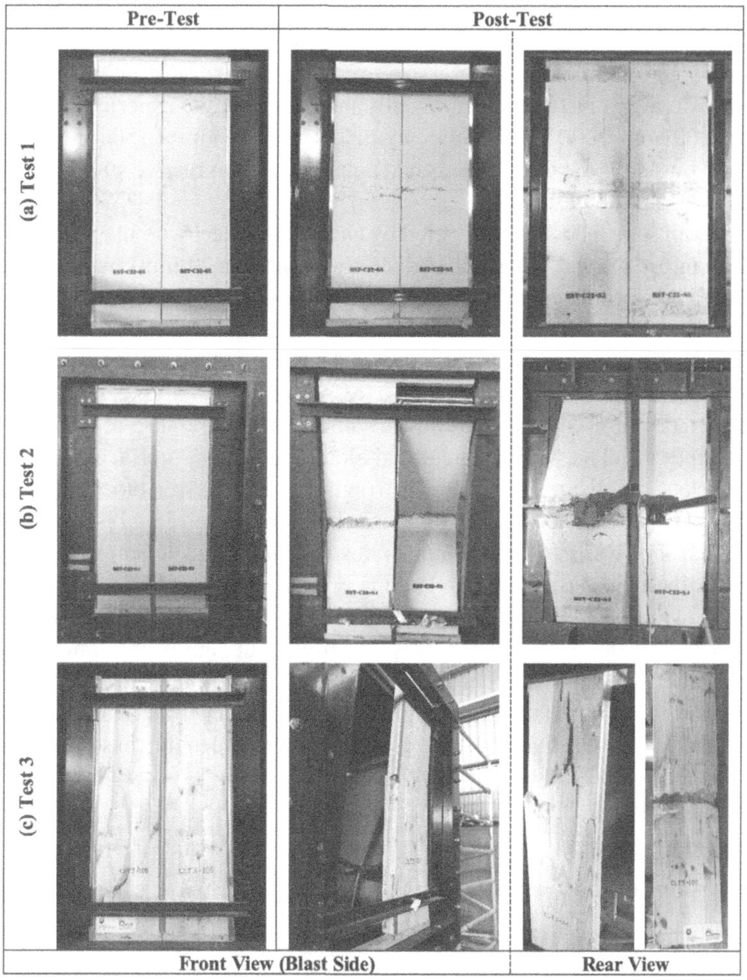

FIGURE 9.32 Front and rear sides of the mounted specimens before and after the blast tests: (a) C32-S5/C32-S6, (b) C32-S3/C32-S4, and (c) CLT5-100/CLT3-105.

to have experienced a blowout failure where it fractured into two pieces in the middle, yielding all five 10-mm steel reinforcement bars. Specimen C32-S4 only just remained intact with a 46 mm permanent displacement measured in the middle.

For Test 3 (Figure 9.32(c)), CLT panels were subjected to the same blast loading as Test 1. Specimen CLT3-105 (105 mm thick, 3 layers) was found to have experienced delamination of the rear face and developed a permanent deflection of 25 mm while CLT5-100 (100 mm thick, 5 layers) experienced a total failure, snapping into two pieces. The different damage responses shown by both CLT specimens imply that the total number of layers and the total thickness of CLT panels play an important role in altering the flexural capacity and ultimately the blast resistance of the panels.

9.6.2 Explosive Resistance Testing of Masonry Walls Coated with Recycled Rubber Concrete Protective Layers

The disposal of scrap tyres is an environmental challenge facing governments worldwide. This study investigates the performance of the new recycled rubber-based cementitious material (RCM) developed by ProtectiFlex.com to improve the resilience of masonry buildings against blast hazards. Blast simulation tests conducted with NFPBS ABS demonstrated that RCM-coated masonry walls exhibited enhanced blast resistance due to strong bonding, increased flexibility, and improved energy absorption.

The RCM is a mixture of cement, sand, rubber crumbs, and water according to the mix design developed by the authors. Uniaxial compressive strength and flexural strength tests were carried out for RCM that was sprayed onto the masonry walls and panels for the experiments using the standard size specimens (100 mm × 200 mm standard cylinders and 150 mm × 150 mm × 500 mm standard beams). The RCM used for the experiments had an average compressive strength of 10 MPa and an average modulus of rupture of 4.2 MPa.

The masonry wall specimens were prepared by anchoring a steel wire mesh (4 mm nominal diameter, 100 mm spacing) to the wall using epoxied screws, ensuring a 20-mm cover between the mesh and the surface. The RCM, formulated for high flowability, was then shotcreted onto the front and sides of one of the masonry walls in layers of 50 and 75 mm thickness and allowed to cure for 28 days to attain maximum strength before testing. Table 9.4 provides the dimensions of the masonry wall specimens.

Figure 9.33 presents snapshots of the damage response of the unprotected masonry wall specimen when subjected to a blast load with a peak reflected overpressure of 127 kPa and a positive phase duration of 19 ms. The wall disintegrated along the cement mortar layers between the concrete Besser blocks. The disintegrated mortar fragments and Besser blocks flew away with high velocity and could cause serious injuries to people behind the wall in a blast incident. The performance of bare masonry walls further demonstrates the importance of improving the resilience of masonry block walls subjected to extreme loading events.

The masonry block wall with RCM protection (50/140/75-wall) sustained higher blast loading (see Table 9.5) than the bare masonry block wall without significant damage. No cracks were observed on the front side (blast side) of the wall while several minor cracks were observed on the rear side of the slab closer to the mid-span region. The cracks penetrated through the Besser blocks up to the RCM interface on the blast side. No delamination between the masonry blocks and sprayed

TABLE 9.4
Dimensions of Masonry Wall Specimens

Test No.	Specimen	Height (mm)	Width (mm)	Thickness (mm)
1	140-wall	1810	1400	140
2	50/140/75-wall	1810	1400	265

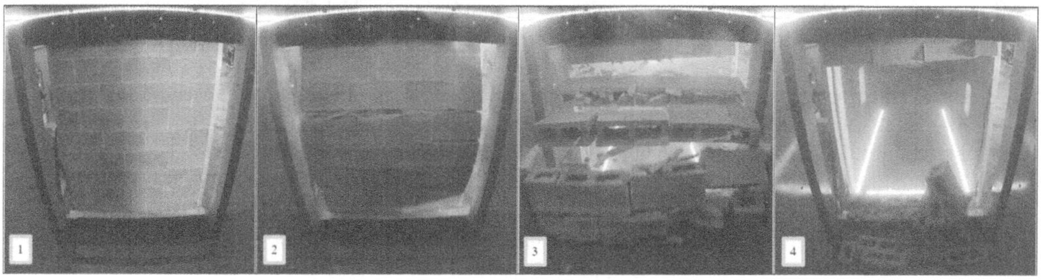

FIGURE 9.33 Damage sequence of bare masonry block wall under blast loading (140-wall).

TABLE 9.5
Summary of Test Results for Masonry Wall Testing

Test No.	Specimen	Peak Reflected Pressure (kPa)	Reflected Impulse (kPa.ms)	Duration (ms)	Peak Displacement (mm)
1	140-wall	127	945	19	Wall destroyed completely
2	50/140/75-wall	210	1540	20	11.5/−4

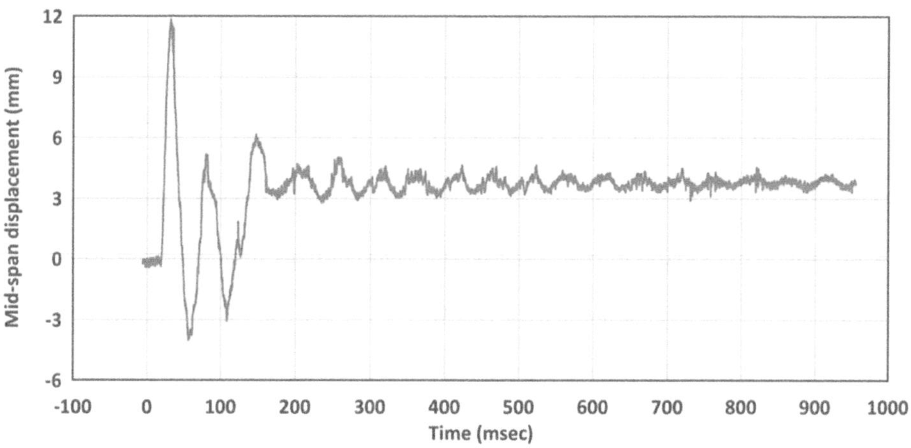

FIGURE 9.34 Mid-span displacement history of the RCM-coated 50/140/75-wall.

RCM layer was observed. The mid-span displacement history of the wall is shown in Figure 9.34. The 50/140/75-wall showed around 11.5 mm of peak mid-span displacement and a permanent displacement of 4 mm.

The blast simulation tests on the masonry block walls protected with and without RCM indicate that the RCM sprayed wall exhibits good performance with no sign of delamination and major damage or large permanent deformations under the blast loads. This also shows the superiority of the RCM to increase the blast-resistant capability of masonry block walls.

NOTES

1. Reprinted from *Handbook of Shock Waves, Volume 2: Shock Wave Interactions and Propagation*, Heilig and Igra, "Shock Waves in Channels," pp. 319–396, Copyright (2001), with permission from Elsevier.

2. Reprinted from *International Journal of Non-Linear Mechanics*, Vol. 72, Andreotti et al., "Performance of a shock tube facility for impact response of structures," pp. 53–66, Copyright (2015), with permission from Elsevier.
3. Reprinted from *International Journal of Non-Linear Mechanics*, Vol. 72, Andreotti, Colombo, Guardone, Martinelli, Riganti, and di Prisco, "Performance of a shock tube facility for impact response of structures," pp. 53–66, Copyright (2015), with permission from Elsevier.
4. Reprinted from *Bulletin of the JSME*, Vol. 22, Ikui, Matsuo, and Yamamoto, "Fast-Acting Valves for Use in Shock-Tubes (Part 2, Formation of Shock-Waves)," pp. 693–699, Copyright (1979), with permission from The Japan Society of Mechanical Engineers.
5. Reprinted from *Japanese Journal of Applied Physics*, Vol. 33, Terao and Furushoh, "Shock-Tube Using Free-Piston Driven by Detonation-Waves," pp. 2811–2816, Copyright (1994), with permission from IOP Publishing on behalf of the Japan Society of Applied Physics (JSAP).
6. Reprinted from *International Journal of Protective Structures*, Vol. 7, Aune, Fagerholt, Langseth, and Børvik, "A shock tube facility to generate blast loading on structures," pp. 340–366, Copyright (2016), with permission from Sage Publications.
7. Reprinted from *Frontiers in Neurology*, Skotak et al., "On the Accurate Determination of Shock Wave Time-Pressure Profile in the Experimental Models of Blast-Induced Neurotrauma," Vol. 9, p. 52, Copyright (2018), published under the Creative Commons Attribution (CC-BY) License.

REFERENCES

Alay, E., Skotak, M., Misistia, A. & Chandra, N. 2017. Dynamic loads on human and animal surrogates at different test locations in compressed-gas-driven shock tubes. *Shock Waves*, 28, 51–62.

Al-Falahi, A., Yusoff, M. Z. & Yusaf, T. 2008. An Experimental Evaluation of Shock Wave Strength and Peak Pressure in a Conventional Shock Tube and a Free-Piston Compressor. *ASME International Mechanical Engineering Congress and Exposition*. Massachusetts, USA, 1–10.

Anderson, J. D. 2003. *Modern Compressible Flow – With Historical Perspective*, New York, McGraw-Hill.

Andreotti, R., Colombo, M., Guardone, A., Martinelli, P., Riganti, G. & Di Prisco, M. 2015. Performance of a shock tube facility for impact response of structures. *International Journal of Non-Linear Mechanics*, 72, 53–66.

Aune, V., Fagerholt, E., Langseth, M. & Børvik, T. 2016. A shock tube facility to generate blast loading on structures. *International Journal of Protective Structures*, 7, 340–366.

Berthelot, M. 1881. Sur la Vitesse de Propagation des Phenomènes Explosifs dans le Gaz. *Comptes Rendus*, 93.

Bird, G. A. 1957. *A Note on Combustion Driven Shock. In:* Hufton, P. A. (ed.), *Hypersonic Facilities in the Aerodynamics Department, Royal Aircraft Establishment*, AGARD Report 146.

Bleakney, W., White, D. R. & Griffith, W. C. 1950. Measurements of diffraction of shock waves and resultant loading of structures. *Journal of Applied Mechanics*, 17, 493.

Bunsen, R. 1867. On the temperature of flames of carbon monoxide and hydrogen. *Ann. Phys. u. Chem*, 131, 161–179.

Chandra, N., Ganpule, S., Kleinschmit, N. N., Feng, R., Holmberg, A. D., Sundaramurthy, A., Selvan, V. & Alai, A. 2012. Evolution of blast wave profiles in simulated air blasts: Experiment and computational modeling. *Shock Waves*, 22, 403–415.

Coates, P. B. & Gaydon, A. G. 1965. A simple shock tube with detonating driver gas. *Proceedings of the Royal Society of London Series A – Mathematical and Physical Sciences*, 283, 18–32.

Colombo, M., Di Prisco, M. & Martinelli, P. 2010. A new shock tube facility for tunnel safety. *Experimental Mechanics*, 51, 1143–1154.

Courant, R. & Friedrichs, K. O. 1948. *Supersonic Flow and Shock Waves*, New York, Springer-Verlag.

Courtney, E., Courtney, A. & Courtney, M. 2014. Shock tube design for high intensity blast waves for laboratory testing of armor and combat materiel. *Defence Technology*, 10, 245–250.

Cranz, C. & Schardin, H. 1929. Kinematographie auf ruhendem Film und mit extrem hoher Bildfrequenz. *Z. Phys*, 56, 147–183.

Cuthbertson, D. W. 1967. A Description and Predicted Performance of a Conical Shock Tube Nuclear-Blast Simulator. *Military Aspects of Blast and Shock (MABS 1)*. Alberta, Canada.

Dallriva, F. D., Johnson, C. F., O'Daniel, J. L. & Dorrell, C. C. 2016. *Blast Load Simulator Experiments for Computational Model Validation – Report 1*, Vicksburg, MS, US Army Corps of Engineers (USACE) – Engineer Research and Development Center (ERDC).

Davy, H. 1817. Some researches on flame. *Philosophical Transactions*, 45.

Fomin, N. A. 2010. 110 years of experiments on shock tubes. *Journal of Engineering Physics and Thermophysics*, 83, 1118–1135.

Fraunhofer EMI. 2025. *Shock-tube facility BlastStar* [Online]. Fraunhofer Institute for High-Speed Dynamics, Ernst-Mach-Institut, EMI. Available: https://www.emi.fraunhofer.de/en/research/research-infrastructure.html#cop_17021968284 [Accessed 9 April 2025].

Freiwald, D. A. 1970. Shock wave velocity enhancement in shock tubes with solid explosive drivers by employing H_2–O_2 mixtures as the driven gas. *Journal of Applied Physics*, 41, 3557–3558.

Gan, E. C. J. 2021. *Experimental and Numerical Investigation of Shock Wave Propagation in an Advanced Blast Simulator.* PhD Thesis, University of Wollongong.

Gan, E. C. J., Remennikov, A. & Ritzel, D. 2021. Investigation of trees as natural protective barriers using simulated blast environment. *International Journal of Impact Engineering*, 158, 104004.

Gan, E. C. J., Remennikov, A. & Ritzel, D. 2024. Blast waveform tailoring using controlled venting in blast simulators and shock tubes. *Defence Technology*, 37, 14–26.

Gan, E. C. J., Remennikov, A., Ritzel, D. & Uy, B. 2020. Approximating a far-field blast environment in an advanced blast simulator for explosion resistance testing. *International Journal of Protective Structures*, 11, 468–493.

Garen, W., Synofzik, R. & Frohn, A. 1974. Shock-tube for generating weak shock-waves. *AIAA Journal*, 12, 1132–1134.

Hannemann, K. & Beck, W. H. 2002. Aerothermodynamics Research in the DLR High Enthalpy Shock Tunnel HEG. *In:* Lu, F. K. & Marren, D. E. (eds.) *Advanced Hypersonic Test Facilities.* Reston, VA: American Institute of Aeronautics and Astronautics.

Hariharan, M. S., Janardhanraj, S., Saravanan, S. & Jagadeesh, G. 2010. Diaphragmless shock wave generators for industrial applications of shock waves. *Shock Waves*, 21, 301–306.

Haverdings, W. 1981. Preliminary Design of the 2.0 m Diameter FOE-Driven Blast Simulator. *International Symposium on Military Application of Blast Simulation (MABS 7).* Medicine Hat, Alberta, Canada.

Haverdings, W. 1983. Evaluation of the 2.0 m Diameter Blast Simulator Driven by Fuel-Oxygen Explosive Charges. *International Symposium on Military Application of Blast Simulation (MABS 8).* Spiez, Switzerland.

Heilig, W. & Igra, O. 2001. Shock Waves in Channels. *In:* Ben-Dor, G., Igra, O. & Elperin, T. (eds.) *Handbook of Shock Waves, Shock Wave Interactions and Propagation.* 1st ed. San Diego, CA: Academic Press.

Hertzberg, A. & Smith, W. E. 1954. A method for generation strong shock waves. *Journal of Applied Physics*, 25, 130–131.

Holder, D. W. 1953. *A Note on Shock Tubes.* Aeronautical Research Council Current Papers, ARC/CP-110. Aeronautical Research Council.

Holmberg, A. D. 2010. *Development and Characterization of Shock Tubes for Laboratory Scale Blast Wave Simulation.* Master's Thesis, University of Nebraska, Lincoln.

Igra, O. & Houas, L. 2016. Shock Tubes. *In:* Igra, O. & Seiler, F. (eds.) *Experimental Methods of Shock Wave Research.* Switzerland: Springer International Publishing AG Switzerland.

Ikui, T., Matsuo, K. & Yamamoto, Y. 1979. Fast-acting valves for use in shock-tubes (part 2, formation of shock-waves). *Bulletin of the JSME*, 22, 693–699.

Itoh, K. 2002. Characteristics of the HIEST and Its Applicability for Hypersonic Aerothermodynamic and Scramjet Research. *In:* Lu, F. K. & Marren, D. E. (eds.) *Advanced Hypersonic Test Facilities.* Reston, VA: American Institute of Aeronautics and Astronautics.

Jacobson, H. K. & Stein, E. 1966. *Diplomats, Scientists, and Politicians: The United States and the Nuclear Test Ban Negotiations*, Ann Arbor, MI: University of Michigan Press.

Jacques, E. 2020. *Shock Tube Research Facility* [Online]. Available: http://www.ericjacques.com/courses/research/shocktube/ [Accessed 19 March 2020].

Janardhanraj, S. & Jagadeesha, G. 2016. Development of a novel miniature detonation-driven shock tube assembly that uses in situ generated oxyhydrogen mixture. *Review of Scientific Instruments*, 87.

Kleinschmit, N. N. 2011. *A Shock Tube Technique for Blast Wave Simulation and Studies of Flow Structure Interactions in Shock Tube Blast Experiments.* Master's Thesis, University of Nebraska, Lincoln.

Lloyd, A., Saatcioglu, M., Nistor, I. & Palermo, D. 2009. New Shock Tube Testing Facilities for Simulated Blast Loading of Structural and Non-Structural Components. *2nd International Workshop on Performance, Protection & Strengthening of Structures Under Extreme Loading.* Hayama, Japan.

Lu, F. K. & Wilson, D. R. 2003. Detonation driver for enhancing shock tube performance. *Shock Waves*, 12, 457–468.

Lu, F. K., Wilson, D. R., Bakos, R. J. & Erdos, J. I. 2000. Recent advances in detonation techniques for high-enthalpy facilities. *AIAA Journal*, 38, 1676–1684.

Maffeo, M., Carboni, M., Cyganik, J., Decristofano, B., Carneal, C., Zinn, D., Andrist, J., Clark, J., Alphonse, V., Kumar, R. & Nedungadi, A. 2016. Test Method Development for the Evaluation of Head Borne Equipment with a Blast Simulator. *International Symposium on Military Aspects of Blast and Shock (MABS 24).* Halifax NS, Canada.

Mallard, F.E. & Le Chatelier, H.L. 1881. Sur les Vitesses de Propagation de l'inflammation dans les Mélanges Gazeux Explosifs. *Comptes Rendus*, 93.

Martin, W. A. 1958. *A Review of Shock Tubes and Shock Tunnels*, San Diego, A Division of General Dynamics Corporation.

Morgan, R. 2000. Development of X3, a Superorbital Expansion Tube. 38th Aerospace Sciences Meeting and Exhibit, American Institute of Aeronautics and Astronautics.

Nguyen, T. T. N., Wilgeroth, J. M. & Proud, W. G. 2014. Controlling blast wave generation in a shock tube for biological applications. *Journal of Physics: Conference Series*, 500.

Ning, Y.-L. & Zhou, Y.-G. 2015. Shock tubes and blast injury modeling. *Chinese Journal of Traumatology*, 18, 187–193.

O'Daniel, J., Johnson, C., Britt, R. & Ritzel, D. 2018. Further Development of ERDC's Prototype Advanced Blast Load Simulator. *International Symposium on Military Aspects of Blast and Shock (MABS 25).* The Hague, The Netherlands.

O'Daniel, J., Johnson, C., Mcmahon, G., Steed, B., Parks, S. & Ritzel, D. 2016. Advancing ERDC's Blast Load Simulator Capabilities. *International Symposium on Military Aspects of Blast and Shock (MABS 24).* Halifax NS, Canada.

Ojima, H., Kitagawa, K., Kainuma, M., Ogawa, T., Takahashi, K., Sun, M. & Takayama, K. 2001. Characteristics of a 60mm × 150mm Vertical Diaphragmless Shock Tube. *In:* Twenty-Third International Symposium on Shock Waves, J.-., 2001, Fort Worth, Texas, USA (ed.) *International Symposium on Shock Waves (ISSW 23).* Fort Worth, Texas.

Parks, S., Johnson, C., Riegel, A., Boppe, D. & Ritzel, D. 2023. Design, Construction, and Early Commissioning Trials of the ERDC 3.7m Advanced Blast Load Simulator. *International Symposium on Military Aspects of Blast and Shock (MABS 26).* Wollongong, Australia.

Parks, S., Ritzel, D. V., O'Daniel, J. & Johnson, C. 2016. Development of a 1.2M Prototype Advanced Blast Load Simulator. *International Symposium on Military Aspects of Blast and Shock (MABS 24).* Halifax NS, Canada.

Payman, W. 1928. The detonation-wave in gaseous mixtures and the pre-detonation period. *Proceedings of the Royal Society of London. Series A, Containing Papers of a Mathematical and Physical Character*, 120, 90–109.

Payman, W. & Shepherd, W. C. F. 1938. *Sixteenth Annual Report of the Safety in Mines Research Board* (H.111.S.O., London, 1938).

Payman, W., Woodhead, D. W. & Titman, H. 1935. Explosion waves and shock waves. Part II. The shock wave and explosion products sent out by blasting detonators. *Proceedings of the Royal Society of London. Series A – Mathematical and Physical Sciences*, 148, 604–622.

Pennelegion, L., Cash, R. F. & Bedder, D. F. 1965. *Design and Operating Features of the N.P.L. 6 in. Shock Tunnel.* Reports and Memoranda No. 3449. Aeronautical Research Council.

Rebagay, R. L. 2017. *Heated Shock Tube Design and Characterization for Liquid Fuel Combustion Experiments.* Master's Thesis, Texas A&M University.

Remennikov, A., Parks, S., Ritzel, D. V., Uy, B. & Gan, E. C. J. 2018. Commissioning of the Australian National Facility for Physical Blast Simulation. *International Symposium on Military Aspects of Blast and Shock (MABS 25).* The Hague, The Netherlands.

Reynolds, G. T. 1943. *A Preliminary Study of Plane Waves Formed by Bursting Diaphragms in a Tube.*

Richmond, D. R., Gaylord, C. S., Damon, E. G. & Taborelli, R. V. 1966. *DASA-AEC-Lovelace Foundation, Blast-Simulation Facility*, Albuquerque, NM, Lovelace Foundation for Medical Education and Research.

Riemann, B. 1860. über die Fortpflanzung ebener Luftwellen von endlicher Schwingungsweite. *Abhandlungen der Königlichen Gesellschaft der Wissenschaften zu Göttingen*, 8.

Ritzel, D. V. 2007. *Upgrade of the DRDC Suffield Blast Tube Facility*, Suffield, Canada, Defence Research and Development Canada.

Ritzel, D. V. & Parks, S. 2013. *Shock Tube Apparatus for Blast Wave Simulation.* ORA, Inc., patent, Patent 2013/0042665.

Ritzel, D. V. & Parks, S. 2015. *Shock Tube Apparatus for Blast Wave Simulation.* United States patent application 13/209,545.

Ritzel, D. V. & Parks, S. A. 2010. Blast Simulation Using Shock-Tube Technology. *International Symposium on Military Aspects of Blast and Shock (MABS 21).* Jerusalem, Israel.

Ritzel, D. V., Parks, S. A., Roseveare, J., Rude, G. & Sawyer, T. W. 2011. Experimental Blast Simulation for Injury Studies. *RTO-MP-HFM-207.* Halifax, Nova Scotia, Canada.

Sawyer, T. W., Wang, Y., Josey, T., Lee, J. & Ritzel, D. V. 2016. Replicating Blast-Induced Traumatic Brain Injury in The Laboratory: A Comprehensive Approach. *International Symposium on Military Aspects of Blast and Shock (MABS 24).* Halifax NS, Canada.

Schardin, H. 1932. Bemerkungen zum Druckausgleichsvorgang in einer Rohrleitung. *Phys. Zeits.*, 33, 60–64.

Shepherd, W. C. F. 1954. Spontaneous ignition of an inflammable gaseous mixture on release from compression. *Nature*, 174, 658.

Skotak, M., Alay, E. & Chandra, N. 2018. On the accurate determination of shock wave time-pressure profile in the experimental models of blast-induced neurotrauma. *Frontiers in Neurology*, 9, 52.

Smith, L. G. 1945. *Photographic Investigation of the Reflection of Plane Shocks in Air.* NDRC Report No. A-350, OSRD Report No. 6271, Princeton University (1945).

Stalker, R. J. 1967. A study of the free-piston shock tunnel. *AIAA Journal*, 5, 2160–2165.

Stewart, J. B. 2018a. Approximating a free-field blast environment in the test section of an explosively driven conical shock tube. *Shock Waves*, 29, 355–360.

Stewart, J. B. 2018b. *Influence of Test Section Geometry on the Blast Environment in an Explosively Driven Conical Shock Tube*, ARL-TR-8335 US Army Research Laboratory.

Stoffel, M., Schmidt, R. & Weichert, D. 2001. Shock wave-loaded plates. *International Journal of Solids and Structures*, 38, 7659–7680.

Stolz, A., Millon, O. & Klomfass, A. 2016. Analysis of the resistance of structural components to explosive loading by shock-tube tests and SDOF models. *Chemical Engineering Transactions*, 48.

Stotz, I., Lamanna, G., Hettrich, H., Weigand, B. & Steelant, J. 2008. Design of a double diaphragm shock tube for fluid disintegration studies. *Review of Scientific Instruments*, 79, 125106.

Terao, K. & Furushoh, T. 1994. Shock-tube using free-piston driven by detonation-waves. *Japanese Journal of Applied Physics*, 33, 2811–2816.

Uy, B., Remennikov, A., Ritzel, D. & Parks, S. 2017. Development of the Australian National Facility for Physical Blast Simulation. 12th International Conference on Shock & Impact Loads on Structures. Singapore.

Vieille, P. 1899a. Etude sur les roles des discontinuites dans les phenomenes de propagation. *Mémoire des Poudres et Saltpêtres*, 10, 117–260.

Vieille, P. 1899b. Sur les discontinuites produites par la de′tente brusque de gas comprimes. *Comptes Rendus de l'Aca- demie des Sciences de Paris*, 129, 1228–1230.

Wright, J. K. 1961. *Shock Tubes*, London, Methuen & Co. Ltd.

Yu, H. R. 1999. Oxyhydrogen combustion and detonation driven shock tube. *Acta Mechanica Sinica*, 15, 97–107.

Yu, H. R., Chen, H. & Zhao, W. 2006. Advances in detonation driving techniques for a shock tube/tunnel. *Shock Waves*, 15, 399–405.

10 Explosive Ordnance Effects on Structures

S. Rigby

10.1 BLAST PRESSURE TERMINOLOGY

An idealised blast pressure history is shown in Figure 10.1 and is characterised by a near-instantaneous rise to peak pressure, followed by an exponential loading pulse and a subsequent period of below atmospheric pressures, after which ambient conditions are ultimately restored. These periods of above- and below-ambient pressure are termed the 'positive phase' and 'negative phase', respectively. The negative phase is caused by over-expansion of the air and a reversal of the 'blast wind' back towards the origin of the explosion. The negative phase is typically much smaller in magnitude than the positive phase but acts over a much longer duration (Rigby et al. 2014a). Blast pressures and related parameters are typically given as the magnitude above (or below) ambient pressure, P_0, which is termed 'overpressure'.

Important blast wave parameters are the following:

- t_a: Arrival time of the blast wave
- t_d^+: Positive phase duration
- t_d^-: Negative phase duration
- P^+: Peak positive phase pressure (subscripts i and r relate to incident and reflected parameters, see below)
- P^-: Peak negative phase pressure
- I_{sp}^+: Positive phase-specific impulse (integral of positive phase pressure history)
- I_{sp}^-: Negative phase-specific impulse (integral of negative phase pressure history)

A blast wave propagating through free air, or parallel/'side-on' to a reflecting surface, is known as an incident wave, and its parameters are given the subscript i. When an incident blast wave impinges onto a surface, conservation of mass/momentum/energy at the interface results in an amplification

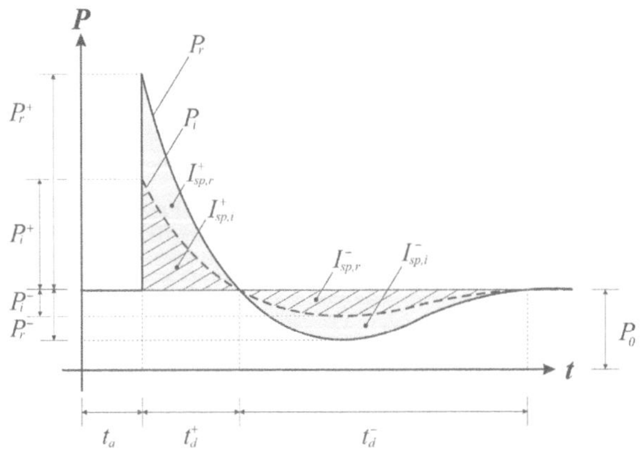

FIGURE 10.1 Idealised blast pressure history.

DOI: 10.1201/9781003529217-10

of the pressure, density, and temperature of the blast. This is termed a reflected wave, the parameters of which are given the subscript *r*.

The amount by which the blast pressure increases on reflection is termed the reflection coefficient, $C_r = P_r^+ / P_i^+$. This value depends primarily on the strength of the incident wave; C_r asymptotically approaches 2.0 (from above) with decreasing shock strength ($P_0 \gg P_i^+$) and can be as high as 20 for strong shocks ($P_i^+ \gg P_0$) when real gas effects are taken into account (US Army Materiel Command 1974). The reflection coefficient is also dependent on the rigidity of the reflecting surface and the angle of impingement of the blast wave. These two aspects will be discussed briefly in later sections.

10.2 BLAST PARAMETER PREDICTION

10.2.1 Overview

For most applications, it is reasonable to assume that the explosive energy is released instantaneously and that this energy is transferred to the shock wave in the form of internal and kinetic energy. As the blast wave expands, it spreads this energy over a much greater volume and continually does mechanical work to the air it propagates over. Because of this, a blast wave will substantially diminish in strength as it propagates, and therefore "*putting distance between the building and the bomb is extremely helpful in reducing blast effects*" (Cormie et al. 2019).

It is therefore highly important in design and consequence analysis to be able to rapidly determine how blast parameters vary with distance. Whilst experiments and high-fidelity numerical modelling can offer valuable insights, they are costly and time-consuming, and therefore unsuitable for use in situations where hundreds or thousands of individual cases need to be evaluated.

Alternatively, simple 'look-up' methods may be used, with the relevant blast parameter relationships pre-populated from previous experiments, numerical simulations, analytical work, or some combination of these. Since these methods are computationally trivial, they can satisfy the need for predictions that are rapid and accurate *enough*. More on this later.

Perhaps the most well-known and accurate method was originally developed by Kingery and Bulmash (1984). This method – hereby referred to as the 'KB method' or 'KB curves' – utilises high-degree polylogarithmic equations fit to a compilation of blast parameters from experimental tests using 1–400,000 kg charges. Technically, the KB method is a *semi*-empirical method, since close-in blast parameters – due to the associated difficulty with measuring the extremely high pressures near to the blast source – were largely derived from indirect measurement, analytical work, and rudimentary numerical analysis (Esparza 1986). This limitation is often (and in the author's opinion, unfairly) cited as reason to distrust KB predictions for the *entire range*. Again, more on this later.

Kingery and Bulmash made use of a common scaling law to allow them to condense each parameter into effectively a single relationship. This scaling law will be briefly reviewed before the KB method is discussed in detail.

10.2.2 Hopkinson–Cranz Scaling

Hopkinson–Cranz (or 'cube-root') scaling, proposed independently by Hopkinson (1915) and Cranz (1926), states that similarity exists between the blast waves produced at identical scaled distances from two explosive charges of similar geometry but different sizes (see Figure 10.2).

Consider an explosive of radius r and mass W, and another explosive with identical composition but different radius and mass, Kr and K^3W, respectively. Here, the two situations differ by the length

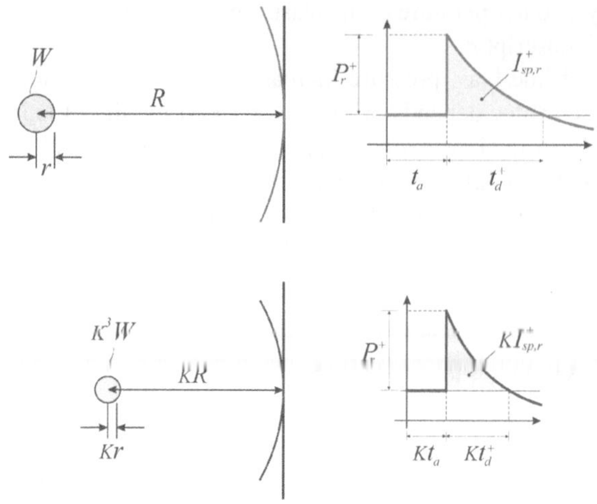

FIGURE 10.2 Hopkinson–Cranz blast wave scaling, after Baker (1973).

scale factor K, and since mass is a product of volume – three spatial dimensions – the mass scale factor is K^3. It is worth noting that in Figure 10.2, K is depicted as being less than one, but the theory holds for any reasonable value of K.

Since the total energy released by the explosive is directly proportional to charge mass, we can call the energy released by the explosives E and K^3E, respectively.

As the blast wave expands spherically, it distributes its energy in three spatial dimensions, with pressure defined as energy per unit volume. At any radial distance from the source, R, the energy will therefore be distributed over a volume proportional to R^3. Invoking geometric similitude, we can say the following:

$$\frac{E}{R^3} = \frac{K^3E}{(KR)^3}, \text{ or, since } E \propto W, \ \frac{W}{R^3} = \frac{K^3W}{(KR)^3} \tag{10.1}$$

Taking the reciprocals and cube-root of the right-hand part only:

$$\frac{R}{W^{1/3}} = \frac{KR}{(K^3W)^{1/3}} \tag{10.2}$$

This shows that the blast pressure depends only on R and $W^{1/3}$, and not K, meaning that if two situations have the same R and $W^{1/3}$, they have the same blast pressure. This introduces an important parameter in the area of blast loading, *scaled distance*, Z, defined as this very ratio:

$$Z = \frac{R}{W^{1/3}} \tag{10.3}$$

Pressure and velocity are proportional, so in order for pressure to be invariant of K – as shown previously – distance and time *must scale by the same factor*. Therefore, it is unsurprising that time *is* dependent on K (refer back to Figure 10.2).

Imagine you find yourself unlucky enough to be some distance from an explosive after it detonates, and experiencing the loading from it. If you were to take a step forward and repeat the situation with a slightly smaller explosive, you might expect the blast wave to arrive sooner (since you are closer), and the loading duration to be shorter (since the explosive is smaller and has released less total energy), despite the blast pressures probably feeling quite similar. If the amount you stepped towards the charge was proportional to the amount that the explosive radius shrunk by, then you would experience exactly that the same *profile* of loading but applied and removed sooner. The loading at 10 m from a 1000 kg charge is the same loading at 1 m from a 1 kg charge; only time (and distance) has been stretched or shrunk by a factor of ten.

More formally, if we know the *scaled* distance, then we know (or should be able to determine) peak pressure, P (positive and negative phase), *scaled* arrival time and duration, $t_a / W^{1/3}$ and $t_d / W^{1/3}$ (positive and negative phase), and *scaled* peak-specific impulse, $I_{sp} / W^{1/3}$ (positive and negative phase).

The final step is in realising that when $W = 1$, scaled and actual blast parameters are identical. This enabled Kingery and Bulmash to, effectively, relate all blast parameters back to those from a 1 kg charge and present each parameter relationship as a function of scaled distance only.

10.2.3 Kingery and Bulmash Method and UFC Curves

The KB method enables the pressure, impulse, arrival time and duration to be determined for values of scaled distance Z (distance divided by cube-root charge mass) between 0.067 and 39.67 m/kg$^{1/3}$. Predictions are provided for two main charge configurations: either a *spherical free-air burst* where a suspended charge is detonated at a height such that the entire duration of the shock wave has passed before the arrival of any waves reflected from the ground surface or a *hemispherical surface burst* where a hemisphere of explosive is detonated on a flat, rigid reflecting surface.

As trinitrotoluene (TNT) was the most commonly used and widely available explosive at the time, the W in the KB method implicitly refers to a mass of TNT explosive. Rather than generate scaled distance parameter relationships for different types of explosives, the KB method can still be used if the explosive question is converted into an equivalent mass of TNT. The *TNT equivalence* of an explosive is defined as the mass of explosive required in order to replicate the effects of a unit mass of TNT. The somewhat vague notion of 'replicate the effects of…' is partly responsible for the large number of methods available for determining TNT equivalence, and this matter remains a contentious issue. As noted by Cooper (1994), the different types of tests (plate dent, ballistic mortar, trauzl, sand crush, and air blast) "*do not necessarily measure the same output property of the sample explosive*" and therefore produce different values of TNT equivalence. It is argued that arena air blast testing produces the most reliable TNT equivalence for use in the KB method, since it is the most applicable and like-for-like scenario. TNT equivalences for modern plastic explosives, derived in this manner, tend to be around 1.2 (Farrimond et al. 2024) and are constant with respect to scaled distance (at moderate to far distances from the charge; the notion of 'near-field' and 'far-field' regions will be discussed later).

Whilst Kingery and Bulmash (1984) present their scaled distance parameter relationships both graphically and as tabulated coefficients of the up-to-25th-order polylogarithmic equations, the graphical format is the more commonly recognised. Indeed, these scaled distance charts are replicated in, amongst others, *UFC 3-340-02 Structures to Resist the Effects of Accidental Explosions* (US DoD 2008).

The KB method was derived from positive phase parameters only, whereas UFC 3-340-02 includes curves for both positive and negative phases of loading. However, as Bogosian et al. (2002) noted, "*The precise provenance of these curves is unknown at the present time*", with the suggestion that

> (n)ot all [experimental data records] were of sufficient duration and/or quality as to be able to extract negative as well as positive values, and while some had dubious peak pressure readings, others became suspect at later time and therefore could not produce reliable impulses.

The original source of the UFC negative phase parameters is unclear, although they may have been in-part derived from analytical work conducted in the 1950s by Granström (1956).

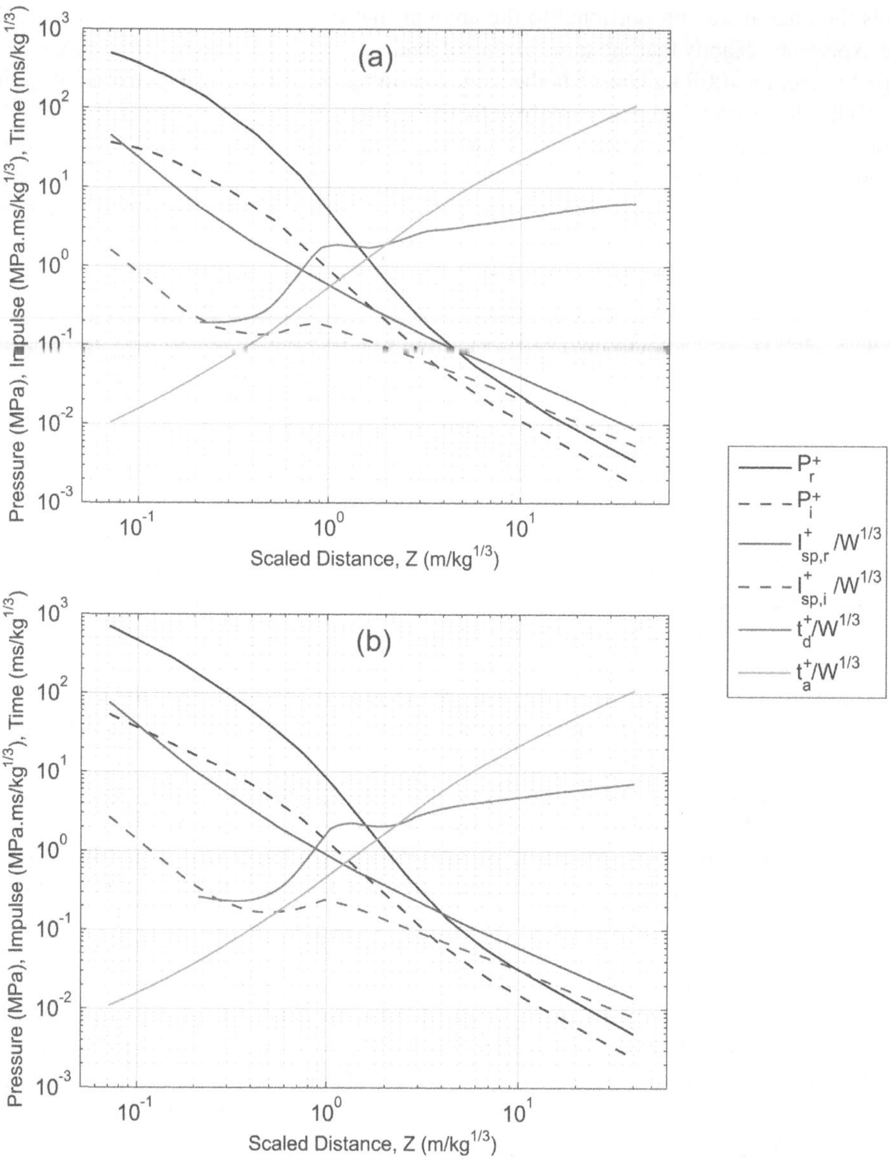

FIGURE 10.3 Incident and reflected positive phase blast wave parameters after UFC 3-340-02 (US DoD 2008): (a) spherical charges of TNT in free air and (b) hemispherical charges of TNT on the surface. Refer to Figure 10.1 for terminology.

UFC curves for positive and negative phase blast parameters are shown in Figures 10.3 and 10.4, respectively.

10.2.4 Functional Representation of the KB/UFC Curves

Shin et al. (2015) present an update to the KB/UFC curves, supplemented by high-fidelity numerical modelling. These updated results are effectively congruent with the original KB curves for $Z > 0.3$ m/kg$^{1/3}$, but with relatively significant deviations extremely close-in due to the aforementioned scarcity of directly measured data in the original Kingery and Bulmash (1984) compilation.

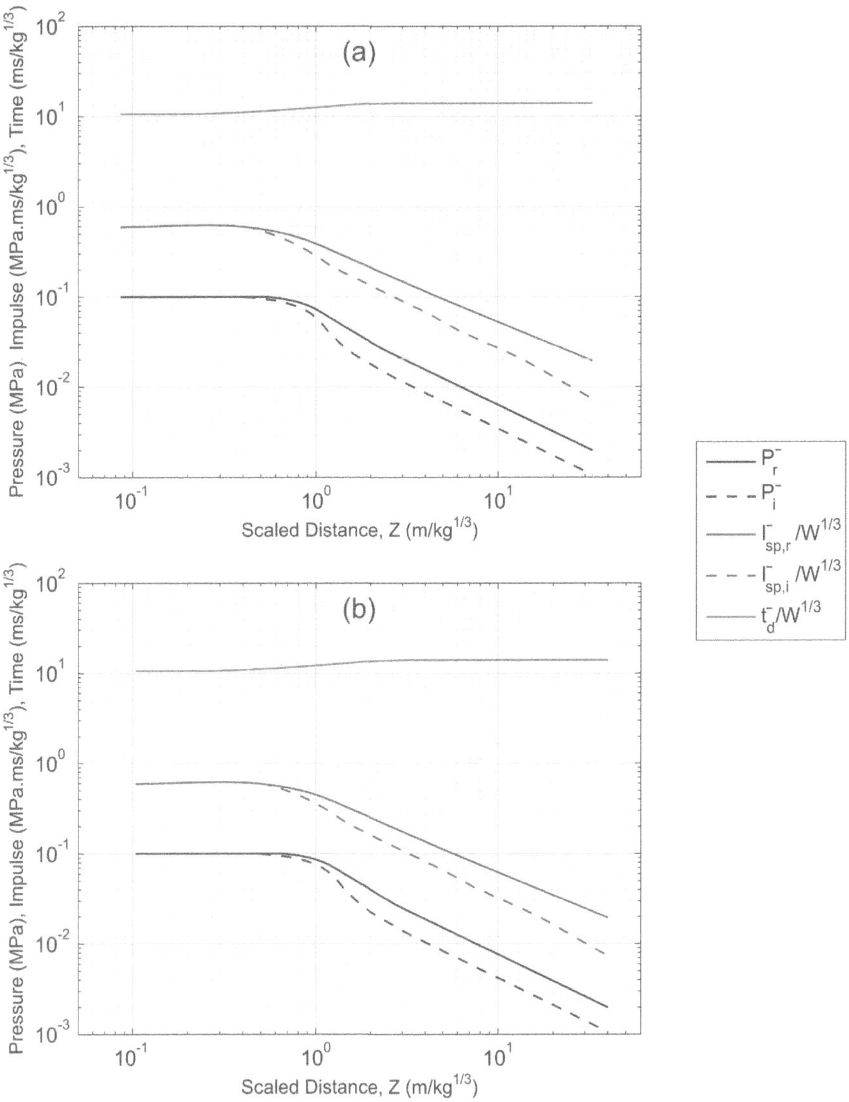

FIGURE 10.4 Incident and reflected negative phase blast wave parameters after UFC 3-340-02 (US DoD 2008): (a) spherical charges of TNT in free air and (b) hemispherical charges of TNT on the surface. Refer to Figure 10.1 for terminology.

Equations for predicting incident and normally reflected blast parameters for spherical air bursts of TNT are given in the form:

$$Y = C_0 + C_1U + \ldots C_NU^N \tag{10.4}$$

where Y is the common (base 10) logarithm of the blast parameter in question, N is the order of the polynomial, $U = K_0 + K_1 \log(Z)$, and C and K are curve-fit constants shown in Table 10.1.

Since the equations above are for spherical air bursts only, a spherical equivalence factor must be applied to hemispherical charge masses in order for the above relations to be used. This is typically 1.8 (Cormie et al. 2019) to account for 10% energy losses to the ground surface through cratering and ground shock.

TABLE 10.1
Polynomial Constants for Positive Phase Incident and Normally Reflected Peak Overpressure, Specific Impulse, and Arrival Time for $0.0553 \le Z \le 40$ m/kg$^{1/3}$, After Shin et al. (2015)

Parameter	P_i^+ (kPa)	$I_{sp,i}^+ / W^{1/3}$ (kPa.ms/kg$^{1/3}$)	P_r^+ (kPa)	$I_{sp,r}^+ / W^{1/3}$ (kPa.ms/kg$^{1/3}$)	$t_a / W^{1/3}$ (μs/kg$^{1/3}$)
K_0	−0.1239	0.05671	0.1365	0.2044	−0.1674
K_1	0.8705	0.8363	0.9839	0.6997	0.4994
C_0	2.666	2.21	4.094	3.105	3.351
C_1	−2.709	−0.3499	−2.439	−2.032	3.027
C_2	−0.06247	−2.981	−1.212	1.36	−1.24
C_3	2.676	2.746	−1.301	−1.081	−3.226
C_4	0.06411	9.522	3.953	−2.305	3.144
C_5	−4.994	−13.01	1.69	3.143	6.048
C_6	0.181	−8.244	−4.843	1.151	−3.093
C_7	5.374	12.51	0.2562	−2.901	−5.209
C_8	−0.4159	−0.5068	2.207	1.003	
C_9	−2.715	2.939	−0.7651		
C_{10}	−0.05814	1.023	−0.1251		
C_{11}	0.6429	−4.748	0.0638		
C_{12}	0.208	1.028			
C_{13}	−0.1394	−3.996			
C_{14}		1.703			
C_{15}		0.5049			
C_{16}		0.8065			
C_{17}		2.006			
C_{18}		−1.44			
C_{19}		0.6624			
C_{20}		−2.118			
C_{21}		0.08177			
C_{22}		1.678			
C_{23}		−0.6299			
C_{24}		−0.1429			
C_{25}		0.07183			

Simplified equations for negative phase parameters (hemispherical surface bursts) are provided in Rigby et al. (2014a):

$$P_r^- = \begin{cases} 101, & 0.071 < Z \le 0.668 \\ -32.9Z^2 + 13.0Z + 106, & 0.668 < Z \le 1.27 \\ -93.0Z^{-1.22}, & 1.27 < Z \le 2.78 \\ 73.0Z^{-0.978}, & 2.78 < Z \le 37.6 \end{cases} \text{(kPa)} \tag{10.5}$$

$$I_{sp,r}^- / W^{1/3} = \begin{cases} -724Z^2 + 445Z + 553, & 0.071 < Z \le 0.580 \\ 11.4Z^2 - 315Z + 752, & 0.580 < Z \le 1.19 \\ 462Z^{-0.880}, & 1.19 < Z \le 5.25 \\ 434Z^{-0.842}, & 5.25 < Z \le 37.6 \end{cases} \text{(kPa)} \tag{10.6}$$

Again, the formulae can be readily applied to spherical free-air bursts with the spherical charge mass factored by 1/1.8 to convert into an equivalent hemisphere.

10.2.5 A Brief Comment on Accuracy

The so-called near-field is characterised by extremely high magnitude ($P_r^+ \gg 100$ MPa) and very short-duration ($t_d^+ \ll 1$ ms) loading with high spatial and temporal nonuniformity (Rigby et al. 2015a). Whilst the exact bounds are difficult to rigorously quantify, the near-field is generally defined as the region where the detonation product fireball is still rapidly expanding and remains attached to the surrounding layer of compressed air (Jenkins et al. 2013), whereas in the far-field the air shock has detached from the fireball and freely propagates.

The boundary layer between the fireball and air shock is known to give rise to turbulent instabilities (Rigby et al. 2020a), which result in large localised variations in loading. Additionally, in the case of near-field reflected loads, successive reflections of the shock front of the fireball/air interface result in a highly complex loading history (Edwards et al. 1992). As such, three distinct loading regimes have been defined by Tyas (2019) as the following:

1. *Extreme near-field* ($Z < 0.5\ m/kg^{1/3}$): Surface instabilities of the expanding fireball have had little time to develop and blast load parameters are highly repeatable.
2. *Late near-field* ($0.5 \le Z < 2.0\ m/kg^{1/3}$): Surface instabilities of the expanding fireball are prominent, causing considerable surface variations and a large degree of variability in measured blast load parameters.
3. *Far-field* ($Z > 2.0\ m/kg^{1/3}$): Air shock has fully detached, local variations in the structure of air shock begin to equalise and variations in loading reduce.

Modern experimental techniques allow for *direct* measurement of blast loads in the near-field, and recent work has shown the KB/UFC curves (derived from *indirect* measurements) retain a surprising level of accuracy well into the extreme near-field (Barr et al. 2023). However, some caution should still be applied to the KB/UFC parameters in this region due to the aforementioned uncertainty in their origin. Loading in the late near-field has been observed to exhibit test-to-test variations of up to a factor of 2 (Rigby et al. 2019), whereas much smaller test-to-test variations of <10% are seen in the far-field (Rigby et al. 2014b, Farrimond et al. 2022). Whilst some studies report high degrees of variability in the far-field, it is important to note that often these feature extensive parameter variability – charge shape, composition, and height of burst (Bogosian et al. 2002), or environmental and terrain effects (Formby & Wharton 1996) – and although being reflective of true-to-life variability of explosive events, they should not be confused with findings (and generally much lower variability) from scientific testing with tightly controlled input parameters.

10.3 MODELLING BLAST PRESSURE HISTORIES

10.3.1 Positive Phase

A common function to describe the temporal variation of the positive phase of the blast load is the exponential decay of the 'modified Friedlander' equation (Friedlander 1946),

$$P(t) = \begin{cases} 0, & \hat{t} < 0 \\ P^+\left(1 - \hat{t} / t_d^+\right) \exp\left(-\alpha \hat{t} / t_d^+\right), & 0 \le \hat{t} < t_d^+ \end{cases} \tag{10.7}$$

where $\hat{t} = t - t_a$, with t being time, and α is the coefficient which controls the rate of decay of the pressure–time curve (Karlos et al. 2016). Unless stated otherwise, the equations in this section can be applied for both incident and reflected loading, with P^+ and I_{sp}^+ replaced with either P_i^+ and $I_{sp,i}^+$ or P_r^+ and $I_{sp,r}^+$.

Integrating Equation (10.7) with respect to time gives the following:

$$I_{sp}^+ = \int_0^{t_d^+} P^+\left(1-\hat{t}/t_d^+\right)\exp\left(-\alpha\hat{t}/t_d^+\right)d\hat{t} = P^+t_d^+\left(\alpha-1+\exp[-\alpha]\right)/\alpha^2 \tag{10.8}$$

Rearranging to make α the subject:

$$\alpha = P^+t_d^+\left(\alpha-1+\exp[-\alpha]\right)/\alpha I_{sp}^+ \tag{10.9}$$

which can be solved iteratively to find α for any combination of P^+, I_{sp}^+, and t_d^+, meaning that parameters determined from the KB/UFC curves can be accurately represented as a pressure load varying in time.

10.3.2 Negative Phase

In the near-field, negative phase loading parameters are very small relative to the positive phase. This is often cited as a reason to neglect the negative phase; however, it is clear from the parameter charts in Figures 10.3 and 10.4 that the negative phase impulse approaches that of the positive phase for larger scaled distances, eventually exceeding the positive phase impulse when $Z > 8$. In a free-field environment, the rising fireball creates a lower pressure near the ground surface which draws in the surrounding air. This has the effect of further lowering the air density, which has already been over-expanded after the shock wave has passed and hence results in larger negative phase impulses at larger stand-off distances (Needham 2010).

The influence of the negative phase has become an increasingly active area of research, in particular the effect of the negative phase on light cladding and glazing panels (Larcher et al. 2012), and it has been estimated that the majority of non-fatal injuries from explosions are caused by airborne glass fragments (Norville et al. 1999, Denny et al. 2023). Following the Oslo bombing on the 22 July 2011, window damage was recorded out to approximately 400 m from the centre of the explosive (Christensen & Hjort 2012), and following the ~500 t TNT-equivalent explosion (Rigby et al. 2020b) in the Port of Beirut on the 4 August 2020, glazing failure occurred several kilometres away from the epicentre (Rigby et al. 2023).

The negative phase can be represented by the Granström (1956) cubic expression:

$$P(t) = \begin{cases} -P^-\left(27\check{t}/4t_d^-\right)\left(1-\check{t}/t_d^-\right)^2, & 0 \le \check{t} < t_d^- \\ 0, & \check{t} \ge t_d^- \end{cases} \tag{10.10}$$

where $\check{t} = t-(t_a+t_d^+)$. There is no equivalent of the decay coefficient α in the negative phase expression, meaning that there is only one value of specific impulse that satisfies a pairing of peak negative phase pressure and duration. As such, it is common to adjust the duration of the negative phase, such that the specific impulse of the cubic expression matches the UFC/KB predictions. With the negative sign removed for clarity, the specific impulse is the following:

$$I_{sp}^- = \int_0^{t_d^-} P^-\left(27\check{t}/4t_d^-\right)\left(1-\check{t}/t_d^-\right)^2 d\check{t} = 9P^-t_d^-/16 \tag{10.11}$$

Rearranging gives the adjusted negative phase duration:

$$t_d^- = \frac{16 I_{sp}^-}{9P^-} \tag{10.12}$$

10.4 SIMPLE ADJUSTMENTS TO LOAD MODELS

10.4.1 Atmospheric Conditions

The Kingery and Bulmash (1984) blast wave parameters are for hemispherical and spherical blasts occurring at sea level. Sachs (1944) developed scaling laws based on a more generic form of Hopkinson–Cranz scaling (see Section 10.2.2) that can be used to account for different atmospheric conditions.

The degree of energy release over a given distance is dependent on the density, ρ, of the propagation medium, which itself is proportional to the atmospheric pressure, P_a, and inversely proportional to the temperature, T. If the *actual* value of peak overpressure from a blast event in a given medium is to be derived from a *reference* value in a different medium for the same explosive mass, the reference value must be multiplied by the pressure scaling factor, S_p,

$$S_p = \frac{P_a}{P_0} \tag{10.13}$$

where P_a is ambient pressure in the actual case, and P_0 is ambient pressure in the reference case (i.e., at sea level; 101.3 kPa). Reference arrival times and durations are scaled by the time-scaling factor, S_t,

$$S_t = \left(\frac{P_a}{P_0}\right)^{1/3} \left(\frac{T_0}{T_a}\right)^{1/2} \tag{10.14}$$

where T_a is ambient temperature in the actual case, and T_0 is ambient temperature in the reference case (15°C). Finally, the reference impulse is scaled by the impulse scaling factor, S_I,

$$S_I = \left(\frac{P_a}{P_0}\right)^{2/3} \left(\frac{T_0}{T_a}\right)^{1/2} \tag{10.15}$$

Sachs' impulse scaling was validated by Olson et al. (1960), among others.

10.4.2 Angle of Incidence

As well as assuming a semi-infinite reflecting surface, the Kingery and Bulmash's (1984) semi-empirical predictions assume that the blast wave impinges normally on a reflecting surface. That is, the angle between the direction of travel of the blast wave and the outward normal of the reflecting surface, termed the *angle of incidence*, θ, is zero. In this case, the blast is propagating perpendicular to the surface and the surface is loaded by the fully reflected wave. When the blast is propagating parallel to the surface, $\theta = 90°$, and the surface is loaded by the incident wave. Between these two extremes, $0° \leq \theta \leq 90°$, the effects of angle of incidence must be taken into account.

At angles of incidence of approximately $\theta > 40°$, a process known as 'Mach reflection' occurs. Upon reflection, the incident wave no longer 'bounces' off the reflecting surface (as with lower angles of incidence) but instead begins to 'skim' across the surface. This reflected wave, propagating with a large component of velocity parallel to the reflecting surface, coalesces with the incoming incident wave to produce a third wave, known as the *Mach Stem*. Formation of the Mach Stem is shown schematically in Figure 10.5.

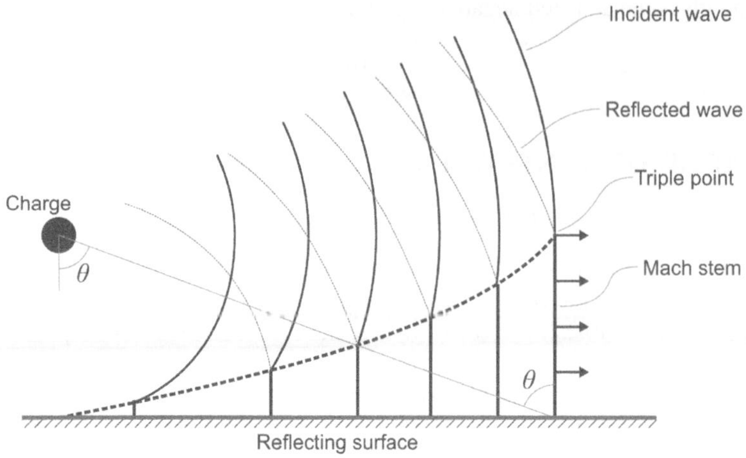

FIGURE 10.5 Formation of the Mach Stem.

The point where the three waves meet is known as the *triple point*. Between the surface and the triple point, the Mach Stem forms a vertical wave and can lead to a large amplification in the oblique pressure, often exceeding the normally reflected overpressure at that point. It is important to consider formation of the Mach Stem for situations when a conventional device detonates at some height above the ground, and also when a device is detonated inside a structure. For internal blasts, the angles of incidence between the blast wave and internal surfaces can vary over a wide range (Cormie et al. 2019).

Angle of incidence affects pressure and impulse differently (Rigby et al. 2015b). The most common way to account for angle of incidence effects on positive phase parameters is to calculate the incident pressure at the scaled slant distance (scaled distance from the explosion to the point of interest) and multiply it by the oblique reflection coefficient in Figure 10.6.

The scaled oblique-specific impulse, $I^+_{sp,\theta}$, can then be calculated from Figure 10.6, again for that particular value of incident overpressure, or by using the following expression:

$$I^+_{sp,\theta} = I^+_{sp,r}\cos^2\theta + I^+_{sp,i}\left(1+\cos^2\theta - 2\cos\theta\right) \tag{10.16}$$

Negative phase parameters have been shown to be unaffected by angle of incidence (Rigby et al. 2015b).

It is worth remarking that for close-in blast loads, any slight increase in slant distance will have a disproportionately large effect in reducing the loading. With reference to Figure 10.3(a), at the point of impingement ($\theta = 0°$) from a 1 kg spherical explosive at a stand-off distance of 0.1 m, the peak reflected pressure is almost 400 MPa. At only 0.1 m along the reflecting surface ($\theta = 45°$), the slant distance increases to 0.141 m and, for a peak incident pressure of 25 MPa (Figure 10.3(a)), the reflection coefficient (Figure 10.6(a)) is approximately 8.0, giving an oblique pressure of 200 MPa; a 50% reduction. At 0.2 m along the surface, the angle of incidence has increased to 63°, slant distance has increased to 0.224 m, incident pressure is 13 MPa, the reflection coefficient is approximately 3.0 and therefore the oblique pressure is 40 MPa; a 90% reduction.

10.4.3 Blast Wave Clearing

10.4.3.1 Mechanism

The Kingery and Bulmash's (1984) semi-empirical predictions assume that the blast wave impinges on a reflecting surface that is sufficiently large in directions normal to the direction of travel of the

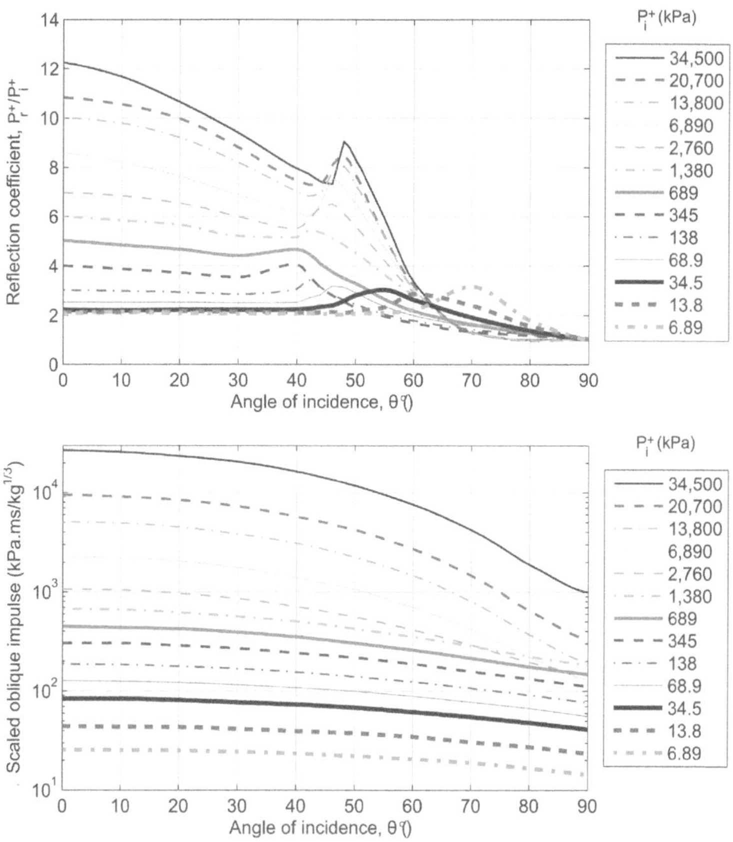

FIGURE 10.6 Reflection coefficient versus angle of incidence and oblique scaled impulse versus angle of incidence. (Adapted from Figures 2-193 and 2-194 (a) and (b) in UFC 3-340-02 (US DoD 2008).)

blast wave such that edge effects can be ignored. When this is not the case, blast wave clearing will occur.

Clearing begins the moment a blast wave reaches the free edge of a reflecting surface with finite lateral dimensions. For this illustration, we are assuming a planar incident shock front (Figure 10.7(a)). At this free edge, whilst the reflected shock front begins to reflect away from the object surface, the incident shock front continues unimpeded past the edge of the object (Figure 10.7(b)), causing diffraction around the free edge. At the same time a pressure imbalance between the lower pressure incident wave and higher pressure reflected wave initiates flow between the higher and lower pressure regions as the pressure begins to equalise. The diffraction generates a low-pressure rarefaction wave, which is driven by the flow conditions and travels along the loaded face, beginning at the boundaries and propagating in towards the centre of the loaded surface (Figure 10.7(c)). The rarefaction wave reduces the pressure acting at any point on the loaded surface it passes over and hence reduces the total positive phase impulse imparted to the surface.

10.4.3.2 UFC Simplified Clearing Corrections

The earliest observations of blast wave clearing appear to have been drawn from large-scale nuclear tests undertaken in the 1950s (Murtha 1955, Morris 1959). Accordingly, the empirical clearing corrections derived from this work assume that clearing relief begins instantaneously after arrival of the blast wavefront and is uniform across the loaded face. The cleared blast pressure *at all points on the loaded face* is assumed to decay linearly from the peak reflected pressure to the stagnation

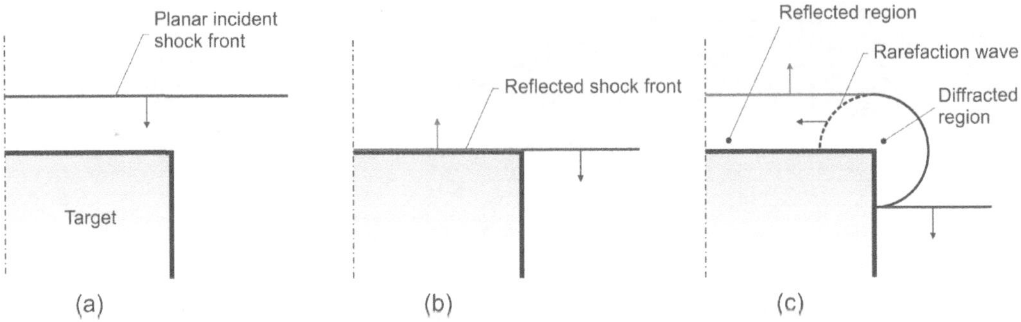

FIGURE 10.7 Diffraction of a blast wave around a finite obstacle causing the propagation of a rarefaction clearing wave. (a) Initial planar incident shock, (b) development of reflected pressures on the loaded surface, and (c) establishment of clearing rarefaction wave and diffracted region.

pressure over a characteristic clearing time, t_c, and follows the stagnation pressure thereafter, as shown in Figure 10.8.

The clearing time, t_c, is given in UFC 3-340-02 (US DoD 2008) as

$$t_c = \frac{4S}{a_0\left(1+[S/G]\right)} \tag{10.17}$$

where a_0 is the sonic velocity in the reflected region (approximately 340 m/s for weak shocks), S is height or half structure width, whichever is smallest, G is height or half structure width, whichever is largest.

The stagnation pressure is given as the sum of the incident and drag pressure,

$$p_{\text{stag}}(t) = P_i(t) + C_D q_s(t) \tag{10.18}$$

where C_D is the drag coefficient. For cubic structures loaded face on, a drag coefficient of 1.0 is considered adequate. For roof and side-wall loading, the drag coefficient varies between –0.4 for

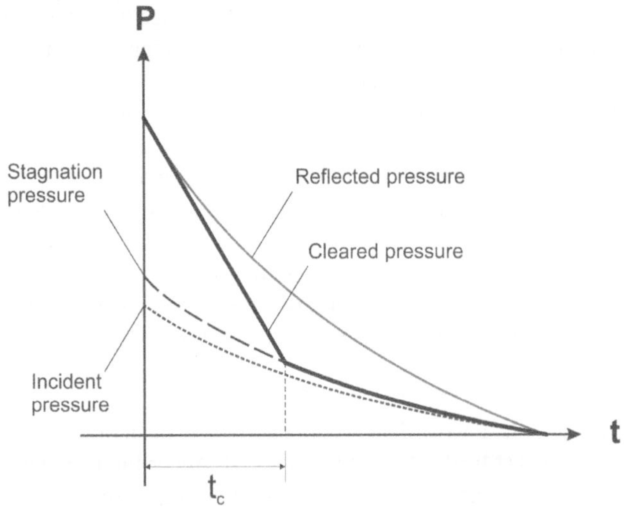

FIGURE 10.8 Clearing corrections from UFC 3-340-02 (US DoD 2008).

low pressures (< 170 kPa) and –0.2 for high pressures (> 340 kPa) (US DoD 2008). The dynamic pressure, q_s, is defined as

$$q_s = \frac{1}{2}\rho_s u_s^2 \tag{10.19}$$

and is effectively the kinetic energy of the fluid particles per unit volume, where the terms ρ_s and u_s refer to density and particle velocity immediately behind the shock front, respectively.

The above method is simple and can be used to quickly determine the level of clearing relief on a finite-sized obstacle. However, for larger reflecting surfaces subjected to smaller sized explosions, the assumption of uniform clearing relief may not hold true Rigby et al. (2014c), in which case the physics-based approach termed the Hudson (1955) method would model clearing with much greater accuracy.

10.4.3.3 The Hudson Method

Hudson (1955) calculated the spatial and temporal distribution of a clearing relief wave, caused by the impingement of a planar blast wave on a rigid surface, by approximating the rarefaction wave as an acoustic pulse. This method is based on the assumptions that the shock is weak (i.e., although the clearing wave travels through shocked air, it is acceptable to assume that it propagates at the ambient sonic velocity), the blast wave is planar (i.e., the lateral dimensions of the reflecting surface are small in comparison to the stand-off) and the depth of the obstacle is sufficiently large to ensure that no clearing waves arrive from the back face during the duration of loading. Given these assumptions, the acoustic approximation can be used to determine the pressure acting at any point on the loaded surface, giving full spatial and temporal distribution of the cleared blast pressure.

Hudson (1955) presents the contours of relief pressure (normalised against the peak incident overpressure) in terms of non-dimensional length and time parameters. Figure 10.9 shows the temporal and spatial distribution of the clearing relief wave, where Hudson's non-dimensional length scale, η, is given as

$$\eta = \frac{x}{a_0 t_d^+} \tag{10.20}$$

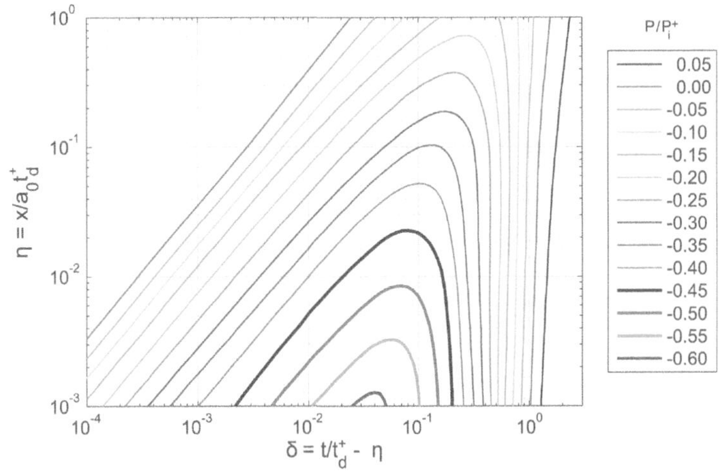

FIGURE 10.9 Spatial and temporal properties of the rarefaction relief wave (Hudson 1955).

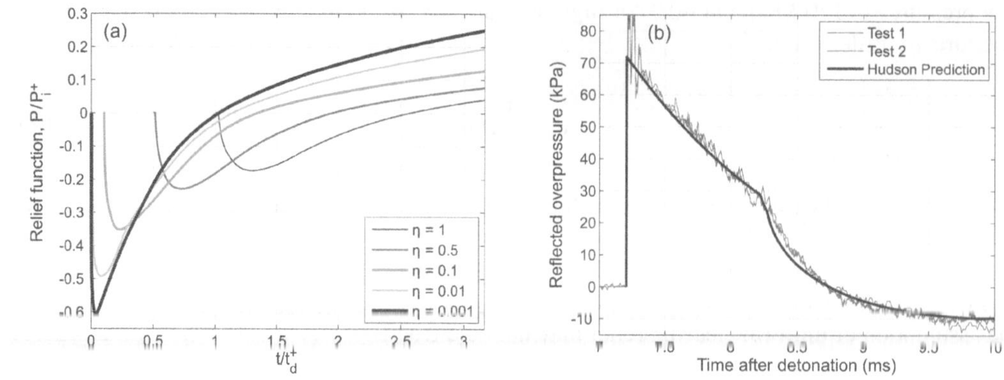

FIGURE 10.10 (a) Hudson clearing functions and (b) validation of the Hudson predictive method, adapted from the results in Tyas et al. (2011).

where x is the distance from the point of interest to the nearest free edge, a_0 is the sonic sound speed in air (assumed to be 340 m/s), and t_d^+ is the positive phase duration. Hudson's time scale, δ, is given as

$$\delta = \frac{t}{t_d^+} - \eta \tag{10.21}$$

where t is time, and η is as given in Equation (10.20).

Figure 10.10(a) shows the relief functions for select values of η. The clearing wave can be seen to become more rounded, lower in magnitude, and longer in duration as it propagates over the loaded surface. For any point on the loaded surface, $0 < \eta \le 1$. A value of $\eta = 0$ indicates a point on the free edge where clearing relief will begin immediately upon the arrival of the blast wave, whereas $\eta > 1$ indicates a point that will experience the full reflected pressure for the entire duration of the positive phase. The time-varying pressure associated with the clearing wave can be evaluated at any value of η between these bounds. The timebase can then be multiplied by the positive phase duration and the pressure multiplied by the peak incident overpressure, allowing the clearing relief function to be superimposed with the reflected pressure acting on the loaded surface to determine the cleared pressure at that point.

Tyas et al. (2011) conducted a series of experiments to validate the Hudson predictive method, with example results shown in Figure 10.10(b). In this example, the gauge location was 337.5 mm away from the vertical edge of the obstacle's loaded face and 355 mm away from the horizontal edges. The positive phase duration was 2.7 ms, giving a blast length of 0.92 m and η values of 0.37 and 0.39 to the vertical free edge and horizontal free edges, respectively. The Hudson method permits the superposition of multiple clearing waves; all three of which can be seen to arrive at roughly the same time, 8.2 ms, in Figure 10.10(b).

10.5 THE SINGLE DEGREE OF FREEDOM (SDoF) METHOD

10.5.1 Theoretical Overview

When a load is applied dynamically, the inertia of an object will contribute to the object's overall resistance to the applied forcing function. At any instant in time, therefore, the external force will be balanced by the inertia of the object *and* the resistance developed through deformation, which is linked to structural stiffness. The force balance equation varies with time:

$$F(t) = m\ddot{z}(t) + kz(t) \tag{10.22}$$

where m is the mass of the object, k is the stiffness, F is the applied load, and $\ddot{z}$ and z are acceleration and displacement, respectively. Strictly, there should be a velocity-dependent damping term in the equation of motion; however, in blast events the first few cycles of displacement only need to be considered, and therefore ignoring damping is a reasonable simplification.

For a real-life system, e.g., a simply supported beam, mass and stiffness are distributed along the length of the beam, and the displacement and acceleration vary with distance from the supports. This creates a system of time-varying equations that must be 'discretized' in both space and time in order to be solved. Finite difference (FD) and finite element (FE) schemes will spatially discretise a continuous system into a finite number of linked, discrete systems. The solution of FD/FE schemes involves the manipulation of large matrices and can be very time-consuming.

Alternatively, the continuous equation of motion of the real-life system can be converted into an equivalent, single equation. Instead of solving the complete system, a single point is chosen to represent the point of maximum displacement of a structure, and the equation is solved for that point (or degree of freedom) only. Hence why this approach is termed the 'single-degree-of-freedom' (SDoF) method.

Solving only one equation with time greatly saves on computational cost whilst still equipping the engineer with sufficient information about the response of the structure, such as the maximum displacement under an applied load, as well as reaction forces and internal stresses which can be determined from this.

10.5.2 Equation of Motion

Consider a simply supported beam with a transiently varying, spatially uniform load, as shown in Figure 10.11(a). The point of maximum displacement is at midspan and is termed z_{max}. This point of interest can be represented as an equivalent mass and equivalent spring stiffness in our equivalent model shown in Figure 10.11(b). Here, we are setting displacements, velocities, and accelerations to be equal for our real-life point of interest and our equivalent single-point system at any moment in time. Our equivalent equation of motion for our single point is therefore

$$F_e(t) = m_e\ddot{z}(t) + k_e z(t) \tag{10.23}$$

where m_e, k_e and $F_e(t)$ are the equivalent mass, equivalent stiffness, and time-varying equivalent force.

10.5.3 Transformation Factors

SDoF energy transformation is directly related to the shape that the system adopts as it is deforming. Below we will derive energy transformation factors and calculate them for a simply supported

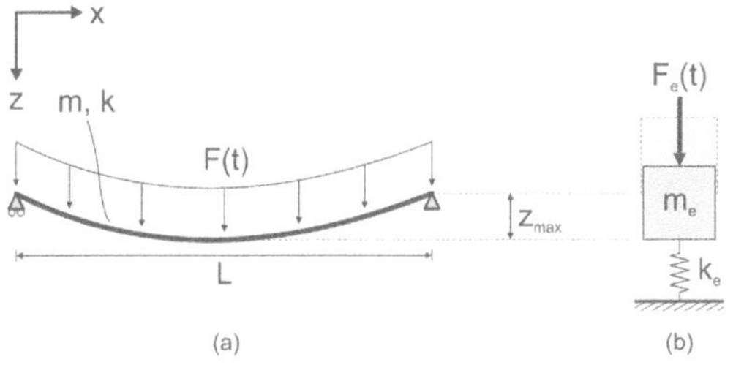

FIGURE 10.11 (a) Distributed and (b) equivalent SDOF systems.

beam as an example. We begin by transforming the stiffness and the mass of the distributed system, using the following equations:

$$F_e = K_L F \tag{10.24}$$

$$k_e = K_S k \tag{10.25}$$

$$m_e = K_M m \tag{10.26}$$

where K_L, K_S, and K_M are the load, stiffness, and mass transformation factors required to ensure that the energy states in the equivalent and real-life systems are identical at all instances in time.

It will be shown below that the stiffness and load transformation factors are identical. Our equation of motion now becomes

$$K_L F(t) = K_M m\ddot{z}(t) + K_L kz(t) \tag{10.27}$$

10.5.3.1 Conceptual Example

Consider the work done by the simply supported beam and our equivalent single point. In Figure 10.12(a), the distributed system has been simplified as nine discrete masses for illustration, with associated displacements under the applied load for each discrete mass. The work done by our SDoF system is simply the applied load multiplied by z_{max}, i.e., $K_L F \times z_{max} = K_L \times 1 \times 1 = K_L$. Here we have factored our applied load by K_L in anticipation of having to adjust the load in our SDoF model to ensure that energy is conserved.

The work done by our distributed system, in this example, is the load acting on each individual element multiplied by the displacement of that element, i.e., $2[1/9 \times 0] + 2[1/9 \times 0.5] + 2[1/9 \times 0.8] + 2[1/9 \times 0.9] + [1/9 \times 1] = 0.60$.

This means that, if we take $K_L = 1$ and set the *equivalent* load equal to the *real-life* load, we have significantly more energy in our SDoF system. This introduces the concept of a *transformation factor*, which will ensure that the energy in our SDoF system is correct. In this example, we should therefore transform our applied load in the equivalent system by multiplying it by 0.60, or setting K_L, our transformation factor, equal to 0.60.

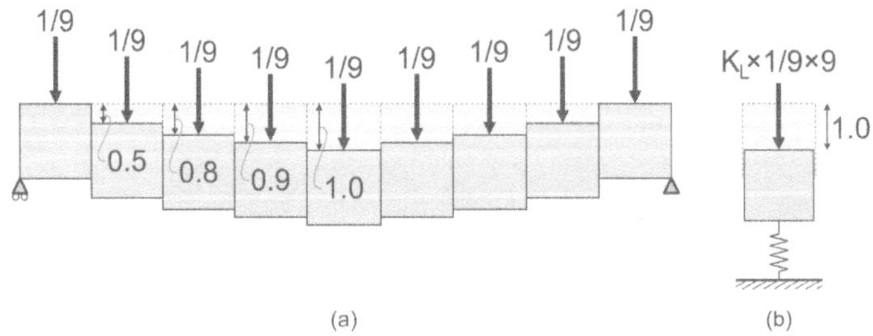

FIGURE 10.12 Work done by (a) distributed and (b) equivalent SDoF systems.

Note: this value is derived for demonstration purposes only – the deformed shape was 'guessed', and we only considered nine elements of an otherwise continuous system. A more realistic example will be shown next.

10.5.3.2 Load Transformation Factor

We saw above that the load transformation factor is calculated by equating work done between real-life and equivalent systems. The actual work done by our real-life distributed system, W_{real}, is the integral of the distributed load multiplied by the deformation at each point along the beam,

$$W_{real} = \int_0^L wz(x)\,dx \quad (10.28)$$

where w is the applied load per unit length (i.e., pressure multiplied by the loaded width), therefore $F = wL$. The work done by our equivalent SDoF system, W_{SDoF}, is simply the total load multiplied by the displacement of our SDoF, which itself is the maximum displacement of the beam,

$$W_{SDoF} = K_L wL z_{max} \quad (10.29)$$

Equating the two and rearranging to make K_L the subject:

$$K_L = \int_0^L \frac{wz(x)\,dx}{wLz_{max}} \quad (10.30)$$

If we define the normalised deformed shape profile as $\phi(x) = z(x)/z_{max}$, Equation (10.30) reduces to

$$K_L = \int_0^L \frac{\phi(x)\,dx}{L} \quad (10.31)$$

10.5.3.3 Stiffness Transformation Factor

The internal strain energy of our real-life system is given as

$$SE_{real} = \int_0^L \frac{k(x)}{2L} z(x)^2\,dx \quad (10.32)$$

As Hooke's law gives $k(x) = wL/z(x)$, we can replace $k(x)/L$ in Equation (10.32) with $wL/Lz(x) = w/z(x)$. We are then left with a single $z(x)$ as a numerator and, when equating this to strain energy in the SDoF system, our 'stiffness' transformation factor will be identical to the load transformation factor:

$$K_S = K_L \quad (10.33)$$

Hence, the term 'stiffness transformation factor' is rarely used, and instead 'load transformation factor' is used to describe both the load and stiffness transformation. This can be easily justified given the fact that stiffness is force per unit displacement, and since displacement is equal between real-life and equivalent systems, stiffness must be factored in the exact same way as force is.

10.5.3.4 Mass Transformation Factor

The kinetic energy of our real-life system is given as

$$KE_{real} = \int_0^L \frac{m}{2L} \dot{z}(x)^2\,dx \quad (10.34)$$

and the kinetic energy of our SDoF system is given as

$$KE_{SDoF} = \frac{K_M m}{2} \dot{z}_{max}^2 \tag{10.35}$$

Given that, at any point on the beam, $z \propto \dot{z}$, therefore $\dot{z}(x)^2 / \dot{z}_{max}^2 = \phi(x)^2$. Equating the kinetic energy between the two systems and rearranging gives

$$K_M = \int_0^L \frac{\phi(x)^2 \, dx}{L} \tag{10.36}$$

10.5.3.5 Derivation of K_L and K_M for a Simply Supported Beam

The deformed shape of a simply supported beam of length L, under a uniformly distributed load, w, is:

$$z(x) = \frac{w}{24EI}\left(x^4 - 2Lx^3 + L^3x\right) \tag{10.37}$$

Given that the peak midspan displacement is

$$z_{maz} = \frac{5wL^4}{384EI} \tag{10.38}$$

We can divide Equation (10.37) by Equation (10.38) to determine the normalised deflected shape, $\phi(x) = z(x) / z_{max}$:

$$\phi(x) = \frac{16}{5L^4}\left(x^4 - 2Lx^3 + L^3x\right) \tag{10.39}$$

Inserting Equation (10.39) into Equation (10.31):

$$K_L = \int_0^L \frac{16}{5L^5}\left(x^4 - 2Lx^3 + L^3x\right) dx \tag{10.40}$$

$$K_L = \frac{16}{5L^5}\left[\frac{x^5}{5} - \frac{Lx^4}{2} + \frac{L^3x^2}{2}\right]_0^L \tag{10.41}$$

$$K_L = \frac{16}{5}\left[\frac{1}{5} - \frac{1}{2} + \frac{1}{2}\right] = \frac{16}{25} = 0.64 \tag{10.42}$$

Inserting Equation (10.39) into Equation (10.36):

$$K_M = \int_0^L \frac{256}{25L^9}\left(x^4 - 2Lx^3 + L^3x\right)\left(x^4 - 2Lx^3 + L^3x\right) dx \tag{10.43}$$

$$K_M = \int_0^L \frac{256}{25L^9}\left(x^8 - 4Lx^7 + 4L^2x^6 + 2L^3x^5 - 4L^4x^4 + L^6x^2\right) dx \tag{10.44}$$

$$K_M = \frac{256}{25L^9}\left[\frac{x^9}{9} - \frac{Lx^8}{2} + \frac{4L^2x^7}{7} + \frac{L^3x^6}{3} - \frac{4L^4x^5}{5} + \frac{L^6x^3}{3}\right]_0^L \tag{10.45}$$

$$K_L = \frac{256}{25}\left[\frac{1}{9} - \frac{1}{2} + \frac{4}{7} + \frac{1}{3} - \frac{4}{5} + \frac{1}{3}\right] = \frac{256}{25} \times \frac{31}{630} = 0.50 \tag{10.46}$$

The load and mass transformation factors can be calculated in a similar manner for any given loading distribution and support conditions, provided the assumed deformation profile can be calculated from static structural mechanics. Transformation factors have been calculated for common loading and support arrangements and are shown in Table 10.2.

At this point, it is worth noting that so far only time-invariant transformation factors have been considered. This inherently assumes that

- The loading distribution remains constant at all instants in time
- The beam deforms in the same single mode throughout its response
- The mode itself is the same as that produced if the load was applied statically

Whilst there are methods available for calculating temporally evolving load transformation factors, this begins to get more complex and moves away from the elegance and simplicity of the SDoF method.

The maximum beam resistance, R_u, is also given in Table 10.2 in terms of the moment capacity at midspan, M_m, and moment capacity at the supports, M_s. The spring constant, k, is also shown in Table 10.2, in terms of EI, the bending stiffness of the element.

SDoF resistance-deflection *functions* will be discussed later.

TABLE 10.2
Transformation Factors for Beams and One-Way Slabs (Biggs 1964)

Loading Diagram	Strain Range	Load Factor, K_L	Mass Factor, K_M	Maximum Resistance, R_u	Spring Constant, k
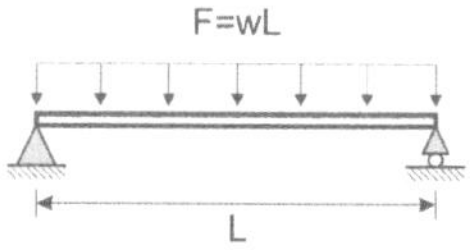	Elastic	0.64	0.50	$\frac{8M_m}{L}$	$\frac{384EI}{5L^3}$
	Plastic	0.50	0.33	$\frac{8M_m}{L}$	0
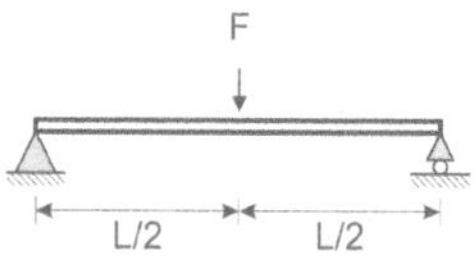	Elastic	1.00	0.49	$\frac{4M_m}{L}$	$\frac{48EI}{L^3}$
	Plastic	1.00	0.33	$\frac{4M_m}{L}$	0
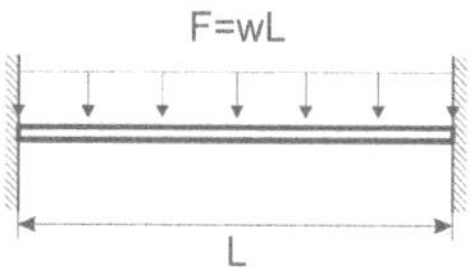	Elastic	0.53	0.41	$\frac{12M_s}{L}$	$\frac{384EI}{5L^3}$
	Elastic/plastic	0.64	0.50	$\frac{8(M_s+M_m)}{L}$	$\frac{384EI}{5L^3}$
	Plastic	0.50	0.33	$\frac{8(M_s+M_m)}{L}$	0
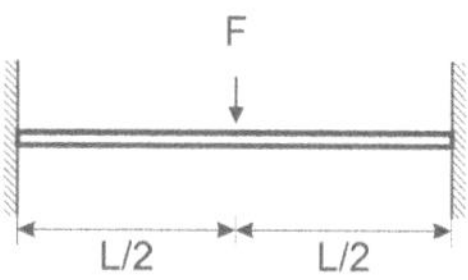	Elastic	1.00	0.37	$\frac{4(M_s+M_m)}{L}$	$\frac{192EI}{L^3}$
	Plastic	1.00	0.33	$\frac{4(M_s+M_m)}{L}$	0

10.6 LINEAR ELASTIC SDOF SYSTEMS

10.6.1 Closed-form Solution

The angular frequency of the SDoF system, ω, in radians per unit time, is given as

$$\omega = \sqrt{k_e / m_e} \tag{10.47}$$

Since there are 2π radians in one angular revolution, the time period of each oscillation, T (in seconds), also known as the 'natural period', is given as

$$T = \frac{2\pi}{\omega} = 2\pi\sqrt{\frac{m_e}{k_e}} \tag{10.48}$$

and therefore the natural frequency (in Hz) of the system is

$$f = \frac{1}{2\pi}\sqrt{\frac{m_e}{k_e}} \tag{10.49}$$

The equation of motion of a structure subjected to a finite duration dynamic load can be solved directly, given the following conditions:

1. The response remains elastic throughout (i.e., the resistance-deflection function is linearly increasing)
2. The loading is suddenly applied and its magnitude decreases linearly over time

If the loading is defined to decay linearly, rather than exponentially over the entire positive phase duration, then the specific impulse of this pressure history will exceed that predicted by the UFC/KB method. Hence, a 'linear positive phase duration' is instead used here:

$$t_{d,lin} = \frac{2I^+_{sp,r}}{P^+_r} \tag{10.50}$$

such that the impulse given by integrating the linear expression now exactly matches the semi-empirical predicted impulse. The forcing function is therefore given as

$$F_e(t) = \begin{cases} F_e^+\left(1 - \frac{t}{t_{d,lin}}\right), & t \le t_{d,lin} \\ 0, & t > t_{d,lin} \end{cases} \tag{10.51}$$

where F_e^+ is peak force, i.e., pressure multiplied by the loaded area of the structural component, multiplied by the load transformation factor. The time-varying displacement of the system (Biggs 1964) can be evaluated purely as a function of the equivalent force, stiffness, and angular frequency and is computed in two stages: response during load application ($t \le t_{d,lin}$) and response after load application ($t > t_{d,lin}$),

$$z(t) = \begin{cases} \frac{F_e^+}{k_e}\left(1 - \cos[\omega t]\right) + \frac{F_e^+}{k_e t_{d,lin}}\left(\frac{\sin[\omega t]}{\omega} - t\right), & t \le t_{d,lin} \\ \frac{F_e^+}{\omega k_e t_{d,lin}}\left(\sin[\omega t] - \sin\left[\omega\left(t - t_{d,lin}\right)\right]\right) - \frac{F_e^+}{k_e}\cos[\omega t], & t > t_{d,lin} \end{cases} \tag{10.52}$$

10.6.2 Reaction Forces

In addition to calculating structural deflections, the determination of reaction forces is required when designing structures to resist dynamic loading. Such expressions for dynamic support reactions can be used to analyse the direct shear response of a structural element under consideration, in order to determine the dynamic shear force and hence *required shear strength* of the element. In contrast to static loading, where the reaction forces are only a function of the applied load and its location/distribution, the reactions generated by dynamic loading are also affected by the inertia of the structural element.

Hence the dynamic reactions are derived based on a force equilibrium that includes inertial forces, which is dependent on the deforming shape of the structure. Because of this, reaction forces must be calculated from consideration of the *real* structure, as opposed to the single-point equivalent system. Figure 10.13 shows the loading arrangement and dynamic free-body diagram for a simply supported beam, length L with midspan point load, F.

Since the shear force, Q, at midspan is zero, dynamic equilibrium can be considered in half-symmetry with the maximum moment, M, acting at the centre. As the centre of the beam is deflected by z, the inertial resistance to motion is distributed along the span of the beam proportional to the deflected shape, ϕ, which for this arrangement is given as

$$\phi(x) = \left(\frac{3x}{L}\right) - 4\left(\frac{x}{L}\right)^3 \tag{10.53}$$

The resultant inertia force, I, acts at a lever arm, x, from the end of the beam, which can be calculated from the deflected shape in Equation (10.53):

$$x = \frac{16L}{50} \tag{10.54}$$

To determine the magnitude of the dynamic reaction force, V, we must consider moment equilibrium about the resultant inertial force:

$$\frac{16VL}{50} - M + \frac{F}{2}\left(\frac{1}{2} - \frac{16}{50}\right)L = 0 \tag{10.55}$$

For convenience, the dynamic bending moment, M, can be expressed in terms of a resisting force, R, via

$$M = \frac{RL}{4} \tag{10.56}$$

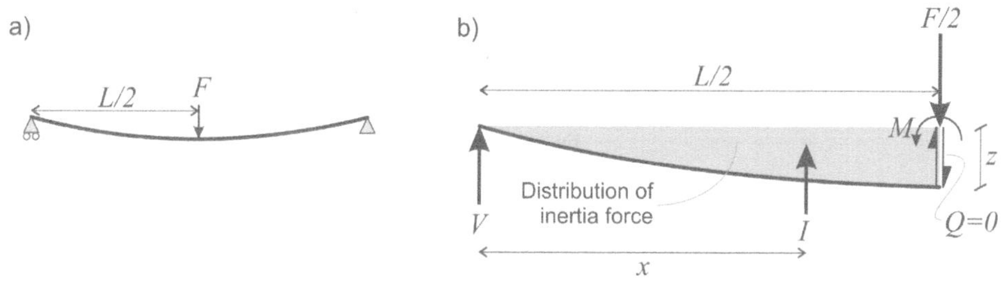

FIGURE 10.13 Determination of dynamic reactions: (a) loading arrangement and (b) free-body diagram.

TABLE 10.3
Dynamic Reaction Forces for Beams and One-Way Slabs (Biggs 1964)

Loading Diagram	Strain Range	Dynamic Reaction, $V(t)$
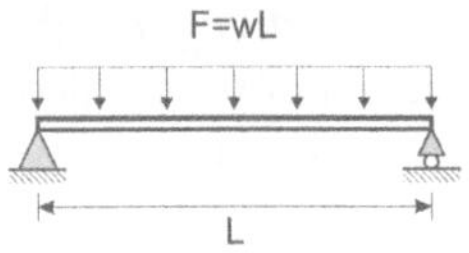	Elastic	$0.39R(t)+0.11F(t)$
	Plastic	$0.38R_m+0.12F(t)$
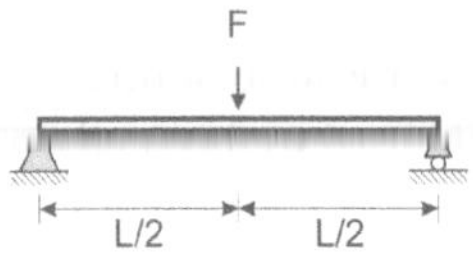	Elastic	$0.78R(t)-0.28F(t)$
	Plastic	$0.75R_m-0.25F(t)$
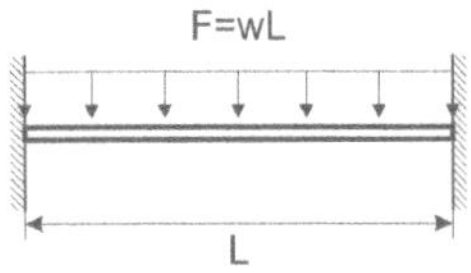	Elastic	$0.36R(t)+0.14F(t)$
	Elastic/plastic	$0.39R(t)+0.11F(t)$
	Plastic	$0.38R_m+0.12F(t)$
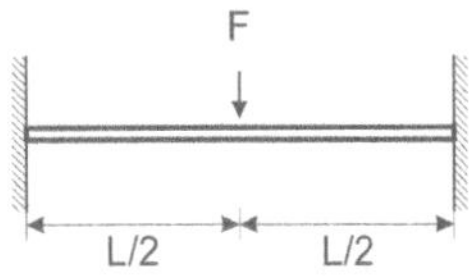	Elastic	$0.71R(t)-0.21F(t)$
	Plastic	$0.75R_m-0.25F(t)$

where R is the static load that would produce the same deflection that the structure is currently experiencing. Substituting Equation (10.56) into Equation (10.55) gives:

$$V = 0.78R - 0.28F \tag{10.57}$$

Thus the dynamic reaction is a function of load and resistance, both of which are functions of time. Since this equation must hold true for static loading, and since in that case $R = F$, the sum of the two coefficients must equal 0.50 for all load/support conditions (Biggs 1964).

Table 10.3 shows dynamic reaction forces for common loading and support arrangements. When a structure is deforming plastically, its resistance function no longer varies with time and is hence replaced with R_m, the maximum resistance. SDoF resistance functions will be covered in more detail in later sections.

10.6.3 Dynamic Load Factor

An important and easy-to-calculate parameter is the *dynamic load factor*, which relates the dynamic deflection to the fictitious deflection that would be caused had the peak force been applied as a constant static load:

$$\mathrm{DLF}(t) = \frac{z(t)}{z_{st}} \tag{10.58}$$

and

$$\text{peak DLF} = \frac{z_{max}}{z_{st}} \tag{10.59}$$

where the static deflection, z_{st}, is simply the load divided by stiffness,

$$z_{st} = \frac{F_e^+}{k_e} \tag{10.60}$$

Dividing Equation (10.52) by Equation (10.60) allows DLF-time histories to be evaluated for two phases of response, during load application and after removal of load:

$$\mathrm{DLF}(t) = \begin{cases} \left(1-\cos[\omega t]\right) + \dfrac{\sin[\omega t]}{\omega t_{d,lin}} - \dfrac{t}{t_{d,lin}}, & t \le t_{d,lin} \\ \dfrac{1}{\omega t_{d,lin}}\left(\sin[\omega t] - \sin\left[\omega\left(t - t_{d,lin}\right)\right]\right) - \cos[\omega t], & t > t_{d,lin} \end{cases} \tag{10.61}$$

As can be seen from Equation (10.61), the variation of dynamic load factor with time is dependent on the value of $\omega t_{d,lin}$. Since ω is inversely proportional to the natural period, T, we can say that the response of a structure to a suddenly applied load is directly proportional to the ratio of the loading duration to the natural period:

$$\mathrm{DLF} \propto \frac{t_{d,lin}}{T} \tag{10.62}$$

This is a key concept in the analysis of structural response under blast loads and forms the basis of the approximate relations in the following section. The concept of DLF varying with $t_{d,lin}/T$ is illustrated in Figure 10.14, which shows DLF-time histories for elastic structures loaded with a long-duration load ($t_{d,lin}/T = 5.00$) and a short-duration load ($t_{d,lin}/T = 0.05$). It is worth noting that whilst short-duration loads induce a relatively low DLF, these are typically associated with much larger pressures and hence larger values of z_{st}, and therefore the total displacement may be similar to, or even exceed, a lower magnitude load with much larger duration.

Figure 10.15 shows the variation of peak DLF with $t_{d,lin}/T$ through solution of Equation (10.61) for a wide range of loading durations. The peak values from Figure 10.14 ($t_{d,lin}/T = 5.00$, peak DLF = 1.90, and $t_{d,lin}/T = 0.05$, peak DLF = 0.16) are indicated by the blue and red circles,

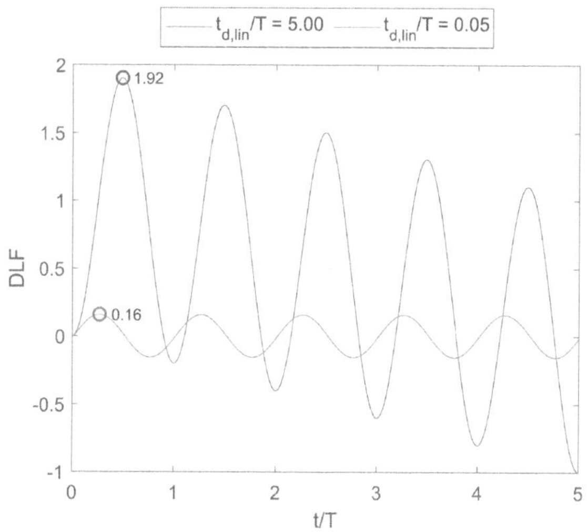

FIGURE 10.14 Comparison of DLF-time histories for a structure loaded with a long-duration load ($t_{d,lin}/T = 5.00$) and a structure loaded with a short-duration load ($t_{d,lin}/T = 0.05$).

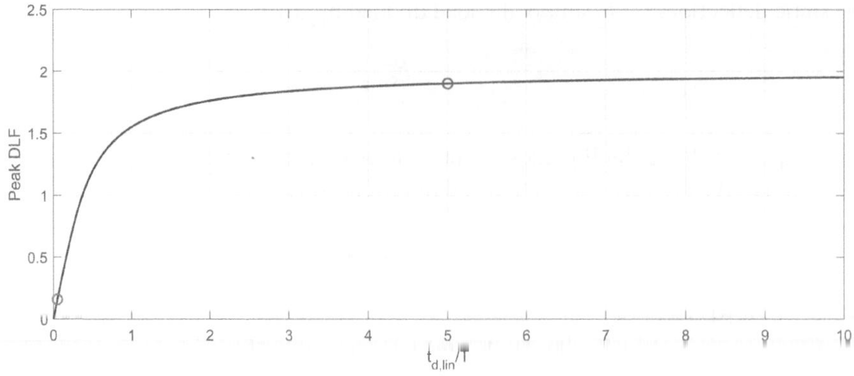

FIGURE 10.15 Variation of peak DLF with $t_{d,lin} / T$.

respectively. It can be seen that for shorter loading durations, the relationship between DLF and $t_{d,lin} / T$ is linear, and that for longer loading durations the DLF asymptotically approaches a value of 2.0. These loading regimes are discussed in the next section.

10.7 DYNAMIC RESPONSE LIMITS

10.7.1 Relationship Between Load Duration and Response

With reference to Figure 10.16, it is clear that the response of elastic structures to blast loading falls into three distinct regimes: (a) impulsive; (b) dynamic, and; (c) quasi-static loading.

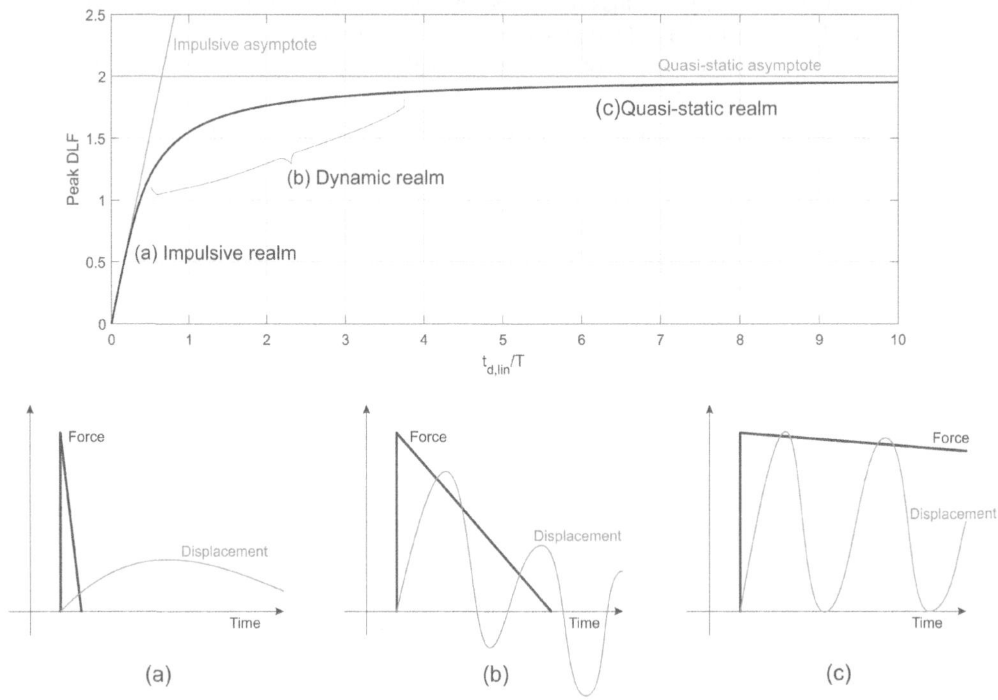

FIGURE 10.16 Approximate regimes for: (a) impulsive; (b) dynamic, and; (c) quasi-static loading. (Adapted from Cormie et al. (2019).)

For impulsive loading (Figure 10.16(a)), the load duration, $t_{d,lin}$, is much shorter than T. In this case, the load has finished acting before the structure has had chance to deform to any significant degree. Thus, we can say that displacement is a function of stiffness, mass, and *impulse*. Impulsive loading is associated with low values of DLF.

In the reverse situation, i.e., when $t_{d,lin}$ is much longer than T (Figure 10.16(c)), as the duration of loading approaches infinity, our situation becomes more and more like that of a suddenly applied, constant load. Thus, we can say that displacement is a function of stiffness and *peak load*. As the situation approaches that of a suddenly loaded structure, our DLF asymptotically approaches a value of 2.0 as will be formally derived shortly.

In between these two regimes, where $t_{d,lin}$ and T are similar (Figure 10.16(b)), the response of the system is more complex and dependent on the time-varying values of load and impulse. However, for impulsive and quasi-static loading conditions, we can estimate the peak structural displacement using simple energy balances.

10.7.2 Impulsive Energy Balance

For low values of $t_{d,lin} / T$, the structure displaces a negligible amount during the loading duration, hence the work done by the load is effectively zero. Accordingly, the energy initially imparted to the system is only in the form of kinetic energy. As the structure deforms, this kinetic energy converts into elastic strain energy until, at the moment of peak deflection, the kinetic energy is zero and has been converted entirely into strain energy (ignoring damping), hence

$$\frac{m_e \dot{z}_0^2}{2} = \frac{k_e z_{max}^2}{2} \tag{10.63}$$

Impulse is the product of mass and velocity (temporal integral of $F = ma$), so

$$\dot{z}_0 = \frac{I_e^+}{m_e} \tag{10.64}$$

where I_e^+ is the equivalent total impulse (i.e., specific impulse multiplied by the loaded area of the structural component, multiplied by the load transformation factor). Substituting Equation (10.64) into Equation (10.63) yields

$$\frac{I_e^{+2}}{2m_e} = \frac{k_e z_{max}^2}{2} \tag{10.65}$$

So that, for small values of $t_{d,lin} / T$, peak displacement is approximately given as:

$$z_{max} = \sqrt{\frac{I_e^{+2}}{k_e m_e}} \tag{10.66}$$

10.7.3 Derivation of Impulsive Asymptote

Since I_e^+ is equal to the integral of the equivalent force-time history, which we have assumed to be in the form of a triangular reverse ramp load,

$$I_e^+ = \frac{F_e^+ t_{d,lin}}{2} \tag{10.67}$$

Substituting this expression into Equation (10.63) and simplifying

$$\frac{\left(F_e^+ t_{d,lin}\right)^2}{4m_e} = k_e z_{max}^2 \tag{10.68}$$

Rearranging

$$\frac{k_e z_{max}}{F_e^+} = \frac{1}{2}\sqrt{\frac{k_e}{m_e}}\, t_{d,lin} \tag{10.69}$$

Since F_e^+ / k_e is the static deflection, the left-hand side equals the dynamic load factor, DLF. Given this, and also given the formula for natural period (in Equation 10.48), Equation (10.69) can be reduced to

$$\text{DLF} = \pi \frac{t_{d,lin}}{T} \tag{10.70}$$

Therefore, the gradient of the impulsive asymptote in Figures 10.15 and 10.16 is, rather pleasingly, π.

10.7.4 Quasi-Static Energy Balance

For large values of $t_{d,lin} / T$, the load is effectively held constant whilst the structure is deforming. Hence, at the moment of peak deflection, the kinetic energy is zero and the work done by the equivalent load is directly balanced by the elastic strain energy:

$$F_e^+ z_{max} = \frac{k_e z_{max}^2}{2} \tag{10.71}$$

Rearranging gives the expression for peak displacement at large values of $t_{d,lin} / T$.

$$z_{max} = \frac{2F_e^+}{k_e} \tag{10.72}$$

Again, since F_e^+ / k_e is the static deflection, it can be seen that the DLF for true quasi-static loading conditions ($t_{d,lin} / T = \infty$) exactly equals 2.0, and this value is approached asymptotically with increasing $t_{d,lin} / T$.

The peak deflection under quasi-static conditions is independent of inertial resistance of the structure (i.e., is not a function of mass). The analysis is therefore greatly simplified as, rather than having to solve any time-varying equations of motion, we only need to multiply the static deflection by a factor of 2.

For a more formal derivation, we will consider the case of an elastic system loaded suddenly by a load that is then held constant. Therefore, the equivalent equation of motion, at all instances of time, is balanced by the time-invariant, externally applied force, F_e^+. Dividing the equation of motion (10.23) through by m_e gives

$$\ddot{z}(t) + \frac{k_e}{m_e} z(t) = \frac{F_e^+}{m_e} \tag{10.73}$$

The Solution to this Equation takes a Harmonic Form

$$z(t) = A\sin(\omega t) + B\cos(\omega t) + \frac{F_e^+}{k_e} \tag{10.74}$$

Since the structure is initially loaded from rest, the initial displacement, z_0, is zero. Hence, at $t = 0$:

$$0 = A\sin(0) + B\cos(0) + \frac{F_e^+}{k_e} \tag{10.75}$$

so that

$$B = -\frac{F_e^+}{k_e}$$

Differentiating Equation (10.74) with respect to time yields

$$\dot{z}(t) = A\omega\cos(\omega t) - B\omega\sin(\omega t) \tag{10.76}$$

At $t = 0$, the initial velocity is also zero, hence

$$0 = A\omega\cos(0) - B\omega\sin(0) \tag{10.77}$$

so that

$$A = 0$$

Substituting the values of A and B into Equation (10.74) gives the closed-form equation of motion of an elastic structure under a suddenly applied constant load:

$$z(t) = \frac{F_e^+}{k_e}\left[1 - \cos(\omega t)\right] \tag{10.78}$$

Since we have already defined the static deflection, $z_{st} = F_e^+ / k_e$, Equation (10.78) can be simplified as

$$z(t) = z_{st}\left[1 - \cos(\omega t)\right] \tag{10.79}$$

Since $T = 2\pi / \omega$ (Equation 10.48), it can be readily seen that when $t = T/2$, $\omega t = \pi$, and $\cos(\omega t) = -1$. Therefore, the displacement in Equation (10.79) reaches a maximum value of 2.0, as seen previously.

10.7.5 Summary

In summary, the three loading regimes can be considered to approximately lie within the following limits:

a. Impulsive, $t_{d,lin} / T \leq 0.4$
b. Dynamic, $0.4 < t_{d,lin} / T \leq 4.0$
c. Quasi-static, $t_{d,lin} / T > 4.0$

10.8 NONLINEAR SDOF SYSTEMS

10.8.1 Resistance-Deflection Function

The closed-form solution derived previously assumes linear resistance functions and linearly decaying loading as, in this situation, the equation of motion has a closed-form solution as the relationship between displacement, force, and resistance can be readily determined. Whilst closed-form solutions allow rapid appraisal of structural response under dynamic loading scenarios, there are severe limitations on the complexity of the problems that can be analysed. When concerned with loads that do not vary linearly with time (Figure 10.17(a)), or structural systems with nonlinear resistance functions (Figure 10.17(b)), the equation of motion cannot be solved directly and instead we must rely on numerical approximations.

Specifically, the most appropriate numerical solution is one of 'explicit time-stepping', where the response of the structure is divided into a series of very small time-steps. Dynamic equilibrium is only solved at each time-step rather than as a continuous function at *any and all* time-steps, as with the closed-form solutions. The solution at each time is used as input values to calculate equilibrium at the next step in time, and so on, until a solution time is reached and the analysis terminates.

A simpler problem, although one that still necessitates the use of numerical solutions to the SDoF equation of motion, is the consideration of yielding and the accumulation of plastic work. Under static loading, formation of a plastic hinge would result in a mechanism and would be deemed a structural failure. Under dynamic loading, however, a yielding structure may possess sufficient inertia to resist the applied load. In this case, plastic work is actually beneficial, since it dissipates energy.

The resistance-deflection function of a structural element is defined simply by the force required to cause a particular level of deflection at the point of maximum displacement of the element (Cormie et al. 2019). When a member undergoes plastic deformation, it requires little or no additional load to cause larger amounts of displacement, and therefore the resistance-deflection function flattens out and has a shallower gradient than the elastic part of the curve. In blast resilient design, it is common to assume that a member offers no additional resistance during plasticity, and therefore the resistance-deflection curve is usually simplified as a bilinear 'perfectly plastic' function as shown in Figure 10.18 (Biggs 1964).

As the displacement increases, the resistance increases linearly with a slope, k_e, with an elastic resistance of $k_e z$. This linearity increases until the elastic limit, z_E, is reached, at which point the maximum resistance, $R_{u,e}$, has been attained and the structure enters plasticity. As the displacement increases further, the spring resistance remains constant at $R_{u,e}$. This resistance is maintained until the structure reaches its ductility limit and fails, or the loading event ceases.

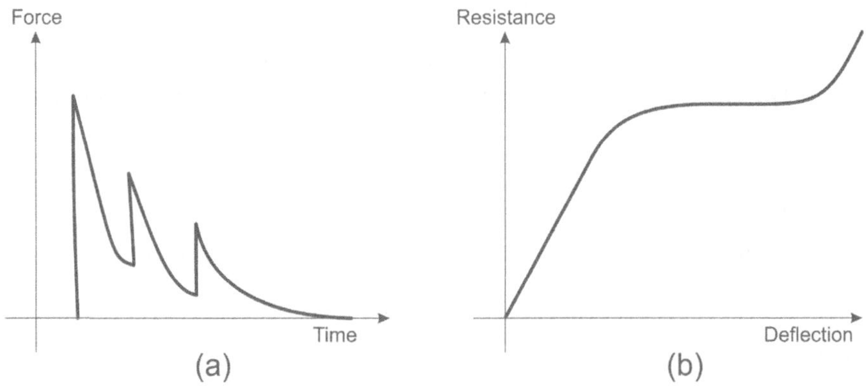

FIGURE 10.17 (a) Nonlinear loading history; (b) nonlinear resistance-deflection function.

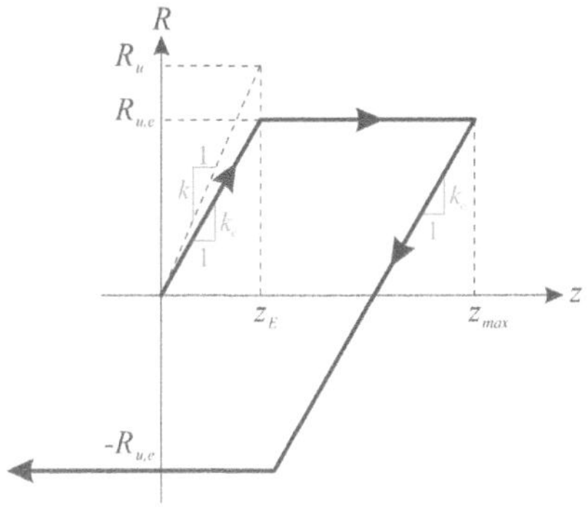

FIGURE 10.18 Resistance-deflection function of an elastic–plastic SDoF system (Biggs 1964).

Mathematically, the maximum or ultimate resistance can therefore be defined as

$$R_u = kz_{max} \tag{10.80a}$$

or

$$R_{u,e} = k_e z_{max} \tag{10.80b}$$

If the displacement reaches a maximum value, z_{max}, that is less than the ductility limit of the structure, it will begin to 'rebound'. Here, the system again behaves elastically as the resistance decreases along a line parallel to the initial elastic slope. The resistance will continue to decrease with decreasing displacement until a spring force of $-R_{u,e}$ is attained, whereby the system returns to plasticity. After loading is complete, the system will continue to oscillate elastically about the point of maximum deflection until some time later when its velocity has damped out and the system returns to rest.

10.8.2 Numerical Solution

The SDoF equation of motion (10.23) can be solved numerically using the linear acceleration method (Biggs 1964) depicted in Figure 10.19. Assuming the acceleration varies linearly across the time-step, Δt, the acceleration at any time, t_j, between times t_i and $t_{i+\Delta t}$, is given as

$$\ddot{z}_j = \ddot{z}_i + \frac{\ddot{z}_{i+\Delta t} - \ddot{z}_i}{\Delta t}\left(t_j - t_i\right) \tag{10.81}$$

The velocity is given as

$$\dot{z}_j = \dot{z}_i + \int_{t_i}^{t_j} \ddot{z}\,\mathrm{d}t \tag{10.82a}$$

or

$$\dot{z}_j = \dot{z}_i + \ddot{z}_i\left(t_j - t_i\right) + \frac{\ddot{z}_{i+\Delta t} - \ddot{z}_i}{2\Delta t}\left(t_j - t_i\right)^2 \tag{10.82b}$$

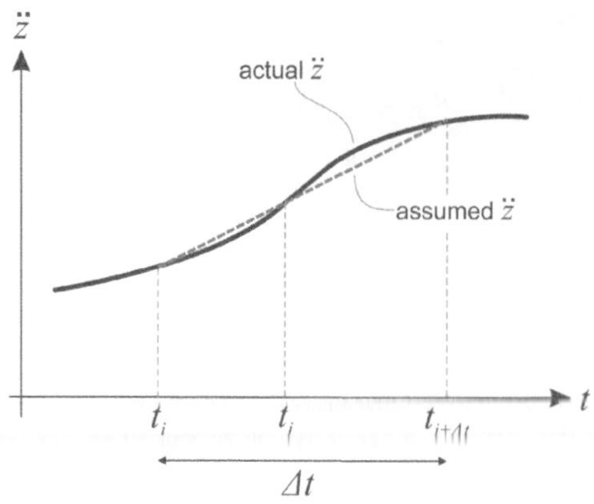

FIGURE 10.19 Linear acceleration method.

which, at $t_j = t_{i+\Delta t}$, becomes

$$\dot{z}_{i+\Delta t} = \dot{z}_i + \frac{\Delta t}{2}\left(\ddot{z}_i + \ddot{z}_{i+\Delta t}\right) \tag{10.83}$$

The displacement at $t_{i+\Delta t}$, is given as

$$z_{i+\Delta t} = z_i + \int_{t_i}^{t_{i+\Delta t}} \dot{z}\, \mathrm{d}t \tag{10.84a}$$

or

$$z_{i+\Delta t} = z_i + \dot{z}_i \Delta t + \frac{(\Delta t)^2}{6}\left(2\ddot{z}_i + \ddot{z}_{i+\Delta t}\right) \tag{10.84b}$$

In order to determine the displacement at $t_{i+\Delta t}$, the acceleration at $t_{i+\Delta t}$ must be known. The acceleration at $t_{i+\Delta t}$ (from Equation 10.23) is

$$\ddot{z}_{i+\Delta t} = \frac{F_e\left(t_{i+\Delta t}\right)}{m_e} - \frac{k_e z_{i+\Delta t}}{m_e} \tag{10.85}$$

which can be substituted into Equation (10.84) to yield

$$z_{i+\Delta t} = \frac{z_i + \dot{z}_i \Delta t + \dfrac{(\Delta t)^2}{3}\ddot{z}_i + \dfrac{(\Delta t)^2}{6}\dfrac{F_e\left(t_{i+\Delta t}\right)}{m_e}}{1 + \dfrac{(\Delta t)^2}{6}\dfrac{k_e}{m_e}} \tag{10.86}$$

In order to initiate the analysis, the initial acceleration must be known. Assuming zero initial velocity and displacement, the initial acceleration under a zero-rise-time load, from Equation (10.85), is given as

$$\ddot{z}_0 = \frac{F_e^+}{m_e} \tag{10.87}$$

The displacement at the first time-step, $t = \Delta t$, can be calculated by substituting the value of $\ddot{z}_0$ into Equation (10.86) (again, assuming $z_0 = 0$ and $\dot{z}_0 = 0$). The acceleration at $t = \Delta t$ can then be found using Equation (10.85), and finally the velocity at $t = \Delta t$ can be found using Equation (10.83). The steps are then repeated and the equation of motion for the SDoF system is solved at each time-step until the solution time is reached.

The linear acceleration method does not require the complete mathematical description of force and resistance functions and instead can use any arbitrary time-varying load function and/or displacement-varying resistance function. One drawback (when compared to the closed-form solution) is that the accuracy of the solution is dependent on time-step size. A greater accuracy requires a greater number of calculations (smaller Δt), but given the simplicity of the calculations at each time-step, convergence is rarely difficult to achieve.

10.9 GRAPHICAL AND TABULATED METHODS FOR STRUCTURAL RESPONSE

10.9.1 Biggs Response Charts

Biggs (1964) presents the results from rigorous analysis of elastic–plastic SDoF systems subjected to linearly decaying blast loads with peak force and load duration denoted by F_e^+ and $t_{d,lin}$, respectively.

The SDoF equation of motion was solved using explicit numerical time-stepping for various combinations of load duration and peak force. The load duration was varied between 0.1 and 20 times the natural period, T (i.e., the 'time ratio', $t_{d,lin} / T$, varied between 0.1 and 20), and the peak force was varied between 0.5 and 10 times the maximum spring resistance, $R_{u,e}$ (i.e., the 'resistance ratio' of the equivalent SDoF system: $R_{u,e} / F_e^+$, varied between 0.1 and 2.0). Note that the resistance ratio of the real-life system, R_u / F^+, is identical to that of the equivalent system since in the equivalent system the load transformation factors cancel.

Biggs presented the results graphically in non-dimensional form in what have become known as 'response charts'. These charts are repeated here in Figures 10.20 and 10.21. From knowledge of the equivalent properties of an SDoF system, the peak deflection and time taken to reach peak deflection can readily be determined from these two charts.

In Figure 10.20, the peak dynamic deflection, z_{max}, is divided by the elastic limit, z_E, and plotted against the time ratio. Each curve represents a different value of $R_{u,e} / F_e^+$, with lower values of $t_{d,lin} / T$ indicating a shorter load duration and lower values of $R_{u,e} / F_e^+$ indicating a higher applied load relative to the resistance of the SDoF system. Any value of above unity on the y-axis indicates that the system has undergone some plastic deformation. Clearly, longer duration, higher magnitude loads result in the largest levels of plastic deformation. When $R_{u,e} / F_e^+ = 2.00$, the applied load is exactly half the resistance, and the peak dynamic deflection tends towards the elastic limit with increasing $t_{d,lin} / T$. This is consistent with linear elastic behaviour at the quasi-static asymptote, as derived in Section 10.7.4.

In Figure 10.21, the time taken to reach peak deflection, t_{max}, is divided by the load duration and plotted against the time ratio. Any value above unity on the y-axis indicates that peak displacement is reached sometime after the end of the equivalent linear positive phase.

10.9.2 Dimensionless P–I Diagram for Elastic Structures

Whilst either analytical or numerical solution of the SDoF equations of motion allow the analyst to determine the overall transient behaviour of a structural system under a blast load, it is often only the peak response which is of interest. Or, more specifically, whether this peak respond exceeds a certain specified limit.

Consider the response chart in Figure 10.20, as an example. If we are analysing a structural member with a peak allowable deflection, z_{max}, exactly equal to the elastic limit, z_E, then any combination of peak force and load duration that results in a normalised displacement *above* the horizontal line

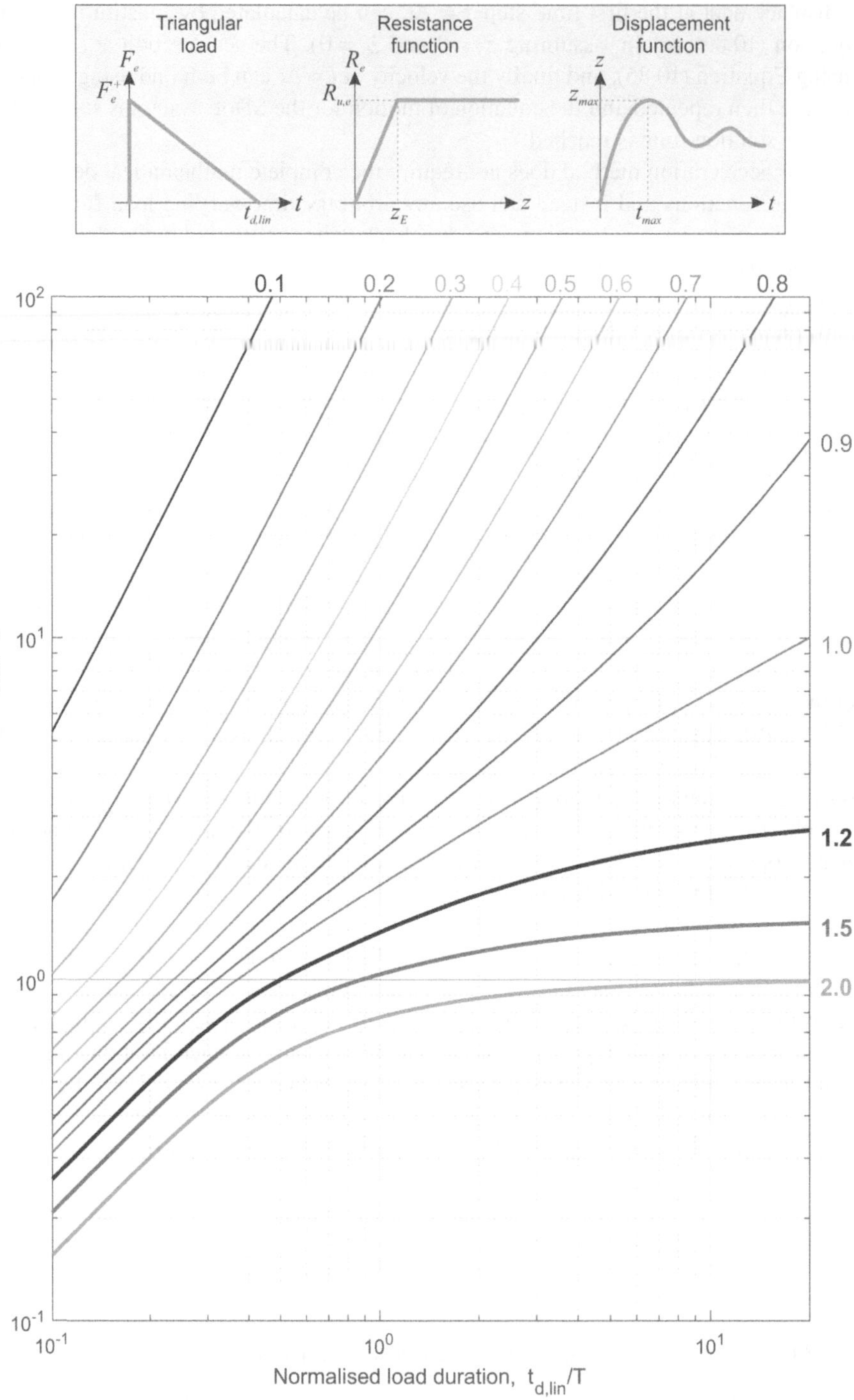

FIGURE 10.20 Peak deflection of an elastic–plastic SDoF system under a triangular load. Numbers next to curves are resistance ratio, $R_{u,e} / F_e^+$ (Biggs 1964).

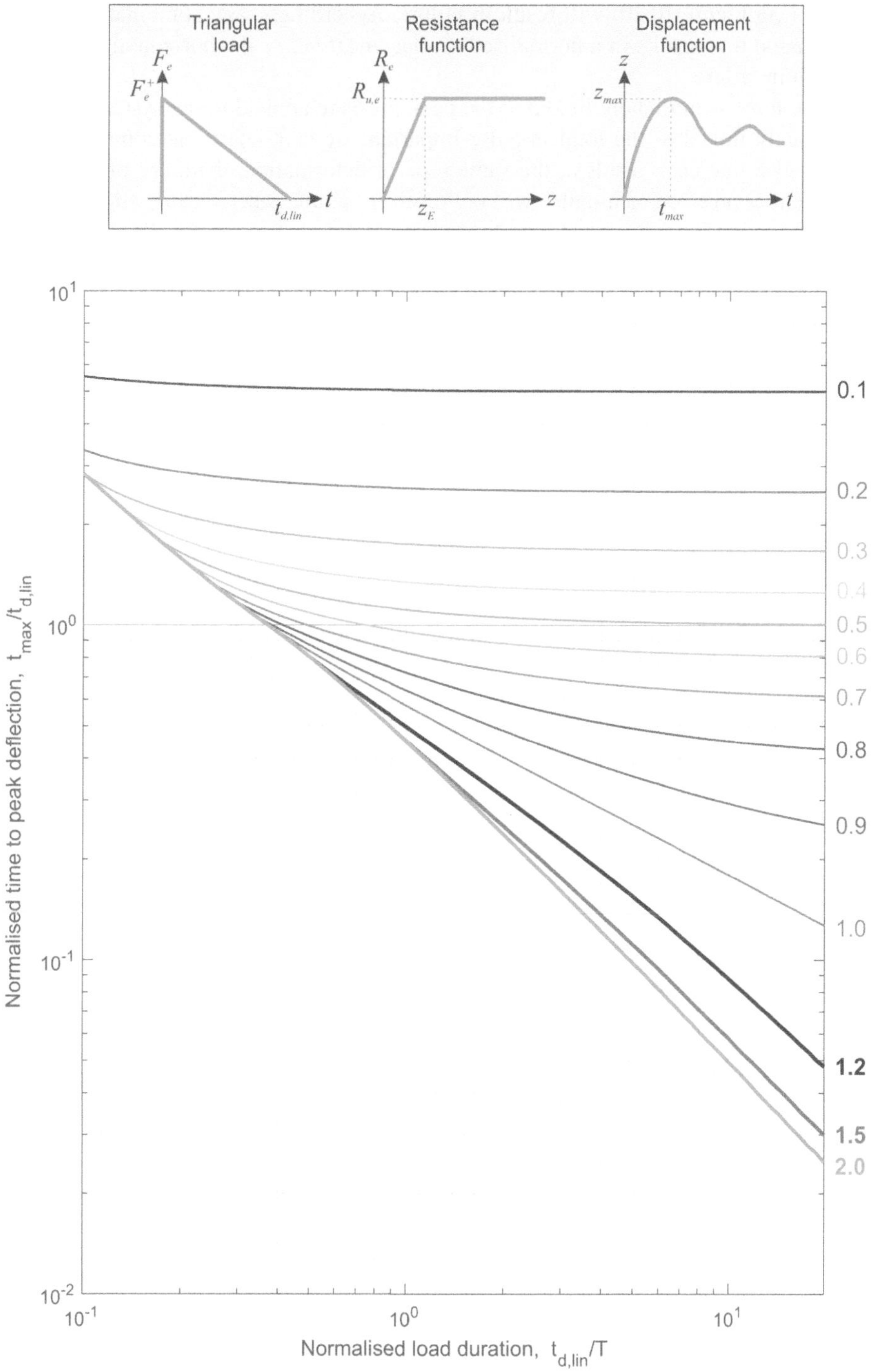

FIGURE 10.21 Time to peak deflection of an elastic–plastic SDoF system under a triangular load. Numbers next to curves are resistance ratio, $R_{u,e}$ / F_e^+ (Biggs 1964).

at $z_{max} / z_E = 1$ (in Figure 10.20) will result in failure. Accordingly, any combination of peak force and load duration that results in a normalised displacement *below* the horizontal line at $z_{max} / z_E =$ will not result in failure.

Since peak force is intimately linked to the peak pressure applied to the system, and load duration is intimately linked to the total impulse imparted; we can determine combinations of pressure and impulse that each result in the same type of deformation or failure of the system. Any reduction in either pressure or impulse will not result in failure, whereas any (slight) increase will. Charts of these pairs of pressure and impulse are known as a 'pressure–impulse (P–I) diagram', or 'iso-damage curves'. The pressure–impulse (P–I) diagram is a useful design tool that permits easy assessment of response to a specified load. With a maximum displacement or damage level defined, the diagram indicates the combinations of load and impulse that will cause failure or a specific damage level (Krauthammer 2008).

For elastic systems subjected to linearly decaying blast loads, it is possible to convert the response spectrum in Figure 10.15 into a P–I diagram by manipulating the axes in the following manner.

The quasi-static asymptote given in Equation (10.72) can be rewritten as

$$\frac{2F_e^+}{k_e z_{max}} = 1 \tag{10.88}$$

where the left-hand side is proportional to the maximum load divided by the maximum resistance. Therefore, the quasi-static asymptote will become a straight horizontal line when plotted on a graph with vertical axis $2F_e^+ / k_e z_{max}$.

Similarly, the horizontal axis of a P–I diagram can be specified such that the impulse asymptote (Equation 10.70) is plotted as a straight line. Given that the dynamic load factor is the ratio of dynamic deflection to static deflection, DLF = $z_{max} / \left(F_e^+ / k_e\right)$, the impulsive asymptote can be expressed as

$$\frac{k_e z_{max}}{F_e^+} = \pi \frac{t_{d,lin}}{T} \tag{10.89}$$

Recalling that $T = 2\pi\sqrt{m_e / k_e}$ (Equation 10.48), after some manipulation the above equation can be rearranged as

$$\frac{F_e^+ t_{d,lin}}{2z_{max}\sqrt{k_e m_e}} = 1 \tag{10.90}$$

Since $F_e^+ t_{d,lin} / 2$ is by definition equal to the total area-integrated impulse, I_e^+, the above equation reduces to

$$\frac{I_e^+}{z_{max}\sqrt{k_e m_e}} = 1 \tag{10.91}$$

Therefore, the impulsive-static asymptote will become a straight vertical line when plotted on a graph with horizontal axis $I_e^+ / z_{max}\sqrt{k_e m_e}$.

Once an allowable peak deflection (or damage level), z_{max}, is defined, this P–I diagram can be used to determine the combinations of load and impulse that will result in a greater deflection than this allowable valuable and hence cause failure (or damage). A dimensionless P–I diagram for an elastic system subjected to a linearly decaying blast load is shown in Figure 10.22. Combinations of pressure and impulse that fall to the left of and below the curve will not induce

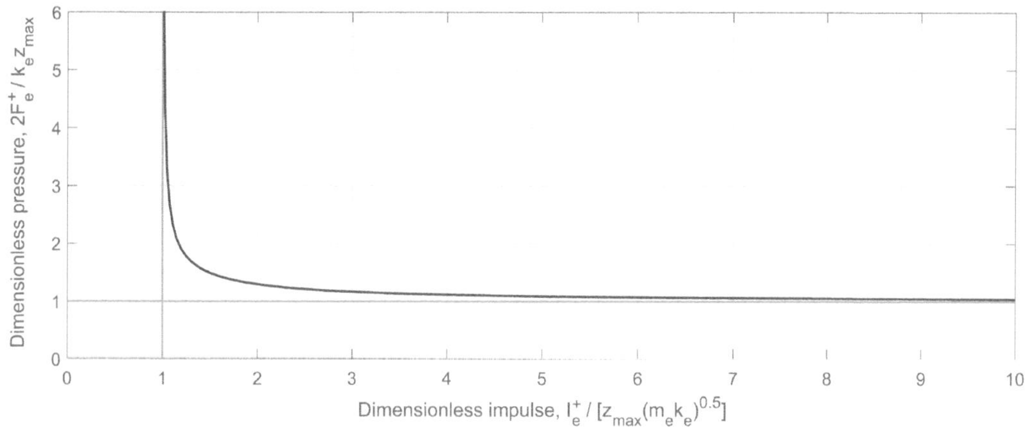

FIGURE 10.22 Dimensionless pressure–impulse diagram for an elastic system subjected to a linearly decaying blast load.

failure while those to the right and above the graph will produce damage in excess of the allowable limit.

10.9.3 P–I Diagrams for Brick-Built Houses

P–I diagrams do not necessarily need to be derived analytically, nor presented in dimensionless form. For example, Jarrett (1968) compiled a database of damage to typical homes and factory buildings from various sources, including records of bombs dropped on the UK during the Second World War. It was found that the distance at which various classes of damage were sustained by the 'average British dwelling house' could be approximated by the following relation:

$$R = \frac{KW^{1/3}}{\left(1 + \left[\frac{3175}{W}\right]^2\right)^{1/6}} \tag{10.92}$$

where R is the threshold safe stand-off in metres, W is the charge mass in kg TNT, and K is an empirical constant which differs depending on the specified damage level. The damage was divided into the classes in Table 10.4.

TABLE 10.4
Damage Classes for Brick-Built Houses (Jarrett 1968)

Class	Damage Constant, K	Description
A	3.77	Almost complete demolition
B	5.55	50–75% external brickwork destroyed or rendered unsafe and requiring demolition
C_b	9.52	*Houses uninhabitable*: partial or total collapse of roof, partial demolition of one to two external walls, severe damage to loadbearing partitions requiring replacement
C_a	27.8	*Not exceeding minor structural damage*: partitions and joinery wrenched from fixings
D	55.5	*Remaining inhabitable after repair*: some damage to ceilings and tiling, more than 10% window glass broken

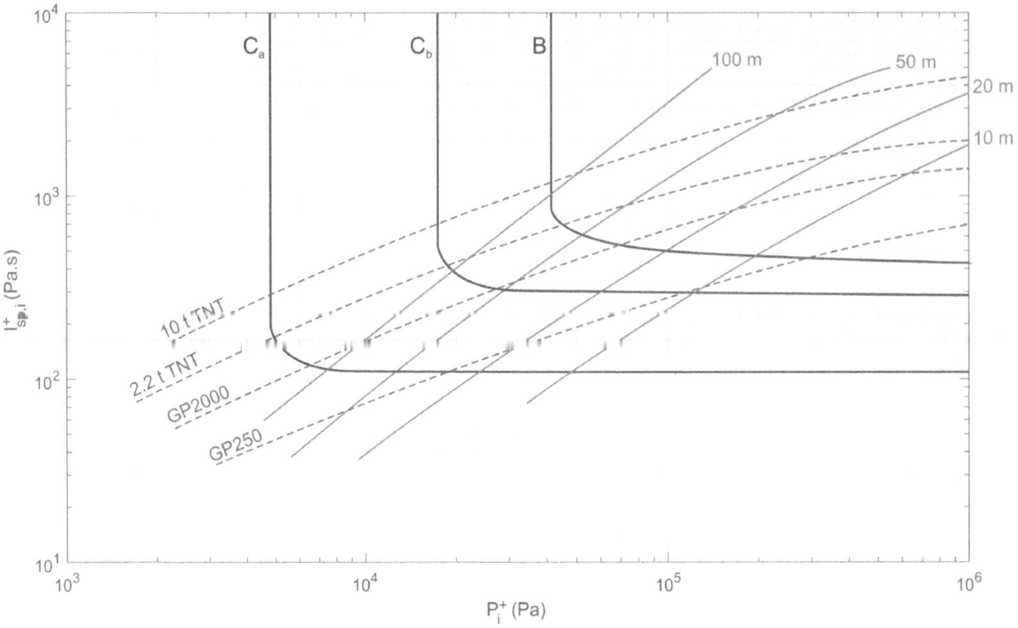

FIGURE 10.23 Iso-damage curves (classes B–C_a) for brick-built houses, with stand-off and charge mass overlaid (Smith & Hetherington 1994). GP2000 and GP250 denote 'general-purpose' 2000 and 250 lb bombs (907 and 113 kg, respectively).

Smith and Hetherington (1994) presented Jarrett's equation in P–I space, with peak incident pressure, P_i^+, and peak incident specific impulse, $I_{sp,i}^+$, on the horizontal and vertical axes, respectively (see Figure 10.23). These iso-damage curves are currently used in the evaluation of safe stand-off distances for explosive storage in the UK and can also be used with reasonable confidence to predict the damage to other structures such as small office buildings and light-framed factories (Cormie et al. 2019).

10.9.4 Tabulated Response Limits and Damage Levels

Tabulated response limits are shown in Table 10.5 according to the following protection levels (Cormie et al. 2019), and Table 10.6 shows a summary of damage levels associated with different levels of static incident pressure.

- *Category 1*: for the protection of personnel and equipment from the effects of primary and secondary fragments and falling portions of the structure under the action of blast loading
- *Category 2*: for the protection of structural elements themselves from collapse under the action of blast loading

10.10 MATERIAL AND STRUCTURAL RESPONSE TO DYNAMIC LOADING

10.10.1 Spalling

Sudden application of a pressure load can bring about modes of structural response which are not observed under static (or gradual) loading. When a load is suddenly applied to the surface of a structure, a global distribution of internal stress is not established instantaneously. Instead, stress is transmitted from the point of loading in the form of a propagating stress wave. When a compressive

TABLE 10.5
Rotation and Deformation Limits for Common Structural Components for Categories 1 and 2 Protection Levels

Element		Protection Category	Support Rotation	z_{max} / z_E
Structural steel beams and plates		1	2°	10
		2	12°	20
Cold-formed steel floor and wall panels		1[a]	1.25°	1.75
		1[b]	4°	6
Steel–concrete–steel composite		1	2°	–
		2	5°	–
Reinforced concrete beams and slabs		1	2°	–
		2	4°	–
Glazing frames:	Aluminium	2	3°	5
	Stainless steel	2	5°	10
	Timber	2	–	2

Source: Adapted from Cormie et al. (2019).
[a] Without Tensile Membrane Action
[b] With Tensile Membrane Action

TABLE 10.6
Summary of Static Overpressure Values for General Structural Damage (Center for Chemical Process Safety 1994)

Incident Overpressure (kPa)	Description of Damage
0.15	Annoying noise
0.2	Occasional breaking of large window panes already under strain
0.3	Loud noise; sonic boom glass failure
0.7	Breakage of small windows under strain
1	Threshold for glass breakage
2	'Safe distance', probability of 0.95 of no serious damage beyond this value; some damage to house ceilings; 10% window glass broken
3	Limited minor structural damage
3.5–7	Large and small windows usually shattered; occasional damage to window frames
5	Minor damage to houses and similar structures
8	Partial demolition of houses (made uninhabitable)
7–15	Corrugated asbestos shattered. Corrugated steel or aluminium panels fastenings fail, followed by buckling; wood panel fastenings fail; panels blown in
10	Steel frame of clad building slightly distorted
15	Partial collapse of walls and roofs of houses
15–20	Concrete or cinderblock walls, not reinforced, shattered
18	Lower limit of serious structural damage. 50% destruction of brickwork or houses
20	Heavy machines in industrial buildings suffered little damage; steel frame building distorted and pulled away from foundations
20–28	Frameless, self-framing steel panel building demolished; rupture of oil storage tanks
30	Cladding of light industrial buildings ruptured
35	Wooden utility poles snapped
35–50	Nearly complete destruction of houses
50–55	Unreinforced brick panels, 250–350 mm thick, fail by shearing or flexure
70	Probable total destruction of buildings

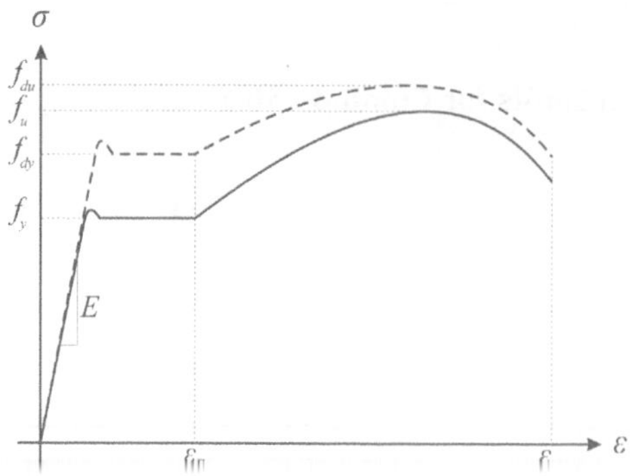

FIGURE 10.24 Typical static and dynamic stress–strain curves for steel.

stress wave reaches a free boundary, it will reflect as a tensile wave. In materials such as concrete and masonry, which possess a significantly lower tensile strength than compressive, a propagating stress pulse (which itself may be below the compressive strength of the material) may exceed the tensile strength of the material when it reflects. Practically, this means that the rear surface of a concrete or masonry structure subjected to blast loading is particularly prone to this type of failure, known as *spalling*.

Spalling generally results in the formation of small concrete fragments as the failure zone propagates through the thickness of the element (Cormie et al. 2019), with the fragments ejected at up to 60 m/s, although the depth of the different spall layers and the velocities with which spall is ejected is greatly dependent on the shape of the propagating stress wave (McVay 1988). When the so-called spall depth extends to half the thickness of the element, then breaching failure is expected to occur.

General equations for predicting the required section thickness, h (in m), to avoid breaching/spalling failure from a spherical TNT charge are given below as a function of scaled distance (Walley 1978, as cited in Cormie et al., 2019):

$$\text{if } h / W^{1/3} > 0.03Z^{-0.62}, \text{ then no breaching should occur} \tag{10.93a}$$

$$\text{if } h / W^{1/3} > 0.07Z^{-0.62}, \text{ then no spalling should occur} \tag{10.93b}$$

where W and Z are charge mass and scaled distance as defined previously.

10.10.2 Strain Rate Sensitivity

10.10.2.1 Steel

Engineering properties of steel relevant for protective design are the following: strength, expressed in terms of yield stress and ultimate tensile strength; ductility, expressed in terms of percent elongation at rupture; and rigidity, expressed in terms of modulus of elasticity (US DoD 2008).

When loaded statically, structural steel generally exhibits a linear stress–strain relationship until the static yield stress, f_y, is reached. Beyond the yield point, steel has significant ductility and can elongate substantially beyond the elastic limit, ϵ_E, with little increase in stress, known as the 'yield plateau'. After this, the steel begins to exhibit strain hardening until it reaches ultimate strength, f_u, with a substantial drop in stress thereafter. Eventually, the steel will reach a maximum value of strain, ϵ_u, and fracture. Typical static and dynamic stress–strain curves for steel are shown in Figure 10.24.

Representative values for steel are the following: $\epsilon_E = 0.01–0.02$, $\epsilon_u = 0.07–0.30$, $E = 200–210$ GPa, $f_y = 275–355$ MPa, and $f_u = 410–470$ MPa.

When loaded dynamically, the mechanical properties of structural steel are altered in the following manner (US DoD 2008):

1. The yield stress increases substantially to the dynamic yield stress value, f_{dy}. The ratio of f_{dy} / f_y is known as the *dynamic increase factor* (*DIF*).
2. The modulus of elasticity, E, will remain insensitive to the rate of loading.
3. The ultimate strength increases slightly to the dynamic ultimate strength, f_{du}. However, the percentage increase is less than that for the yield stress.
4. The strain at failure, ϵ_u, either remains unchanged or is slightly reduced due to increased strain rate.

DIFs for steel are shown in Table 10.7.

10.10.2.2 Reinforced Concrete

A concrete matrix contains aggregate particles of varying size and strength. When a crack forms under low loading rates, the crack path is influenced by the relative stiffness and strength of the constituent particles: if a crack is propagating through the concrete mortar and encounters a stiffer aggregate, it will be able to change direction and follow a path offering less resistance. For geometric compatibility to be satisfied under rapid loading, however, the crack will be forced to follow a shorter path and hence may propagate through the stiffer particles. Fracture of the stiffer particles requires a greater amount of energy and hence the concrete will appear to have a greater strength.

As with steel, the ultimate strength of concrete increases with loading rate, whilst the modulus of elasticity remains relatively unchanged. With reinforced concrete subjected to dynamic loading, the DIF for concrete and steel reinforcement may differ and each should be applied separately. DIFs for reinforced concrete are shown in Table 10.7.

TABLE 10.7
Dynamic Increase Factors (DIFs) for Steel and Reinforced Concrete, After US DoD (2008)

		Yield Stress		
Type	**Loading Range**	**Bending**	**Tension/Compression**	**Ultimate Stress**
S235/275	'High pressure'	1.36	1.24	1.10
S355		1.24	1.15	1.05
S700		1.12	1.07	1.00
S235/275	'Low pressure'	1.29	1.24	1.10
S355		1.19	1.15	1.05
S700		1.09	1.07	1.00
	Typical[a]	1.20	1.10	1.05
Type	**Loading range**	**Bending**	**Compression**	**Shear**
Concrete	'Close-in'	1.25	1.16	1.10
Reinforcing bars, yield		1.23	1.13	1.10
Reinforcing bars, ultimate		1.05	–	1.00
Concrete	'Far'	1.19	1.12	1.10
Reinforcing bars, yield		1.17	1.10	1.10
Reinforcing bars, ultimate		1.05	–	1.00

[a] Representative values taken from Cormie et al. (2019)

10.10.2.3 Tabulated Values

DIFs for steel and concrete, under a range of stress states and loading intensities, are shown in Table 10.7 (US DoD 2008).

The assumed strain rates for steel members in bending are 0.1 *ε/s* for the 'low pressure' design range and 0.3 *ε/s* for the 'high pressure' design range (US DoD 2008). For steel members in tension and/or compression, the assumed strain rates are 0.02 and 0.05 *ε/s*, respectively. Lower strain rates are selected for tension/compression members as these are likely to be secondary members subjected to reaction forces from primary members, which have a significantly longer rise time then applied blast forces.

The assumed strain rates for reinforced concrete structures in bending are $0.1e^{-3}$ *ε/s* for the 'far' design range and $0.3e^{-3}$ *ε/s* for the 'close-in' design range (US DoD 2008). For reinforced concrete members in compression, these values are $0.2e^{-3}$ and $0.5e^{-3}$ *ε/s*, respectively. Assumed strain rates for shear elements are not specified, although it is acknowledged that they are 'more conservative' than the DIFs for bending or compression, due to "the need to prevent brittle shear and bond failure and to account for uncertainties in the design process for shear and bond".

10.10.3 Deformation Modes

10.10.3.1 Far-Field Versus Near-Field Loading

As described above, the sudden nature of blast loading can cause spalling and dynamic strength enhancement effects. Additionally, the *spatial distribution* of blast loading on the surface of the structure can greatly influence the mode of response. Far-field loading, Figure 10.25(a), is produced by charges at long stand-off distances relative to the height or span of the structure, resulting in a low interaction curvature and a time-varying load with a high degree of spatial uniformity.

Near-field loading, Figure 10.25(b), is produced by charges at short stand-off distances relative to the height or span of the structure. Here, the resulting high interaction curvature produces a spatially varying transient load. The pressure and specific impulse imparted to the epicentral regions of the structure are much greater than those generated at the periphery, and the varying arrival time of the shock front also means that loading may have ceased directly under the explosion before loading of the rest of the structure has even started.

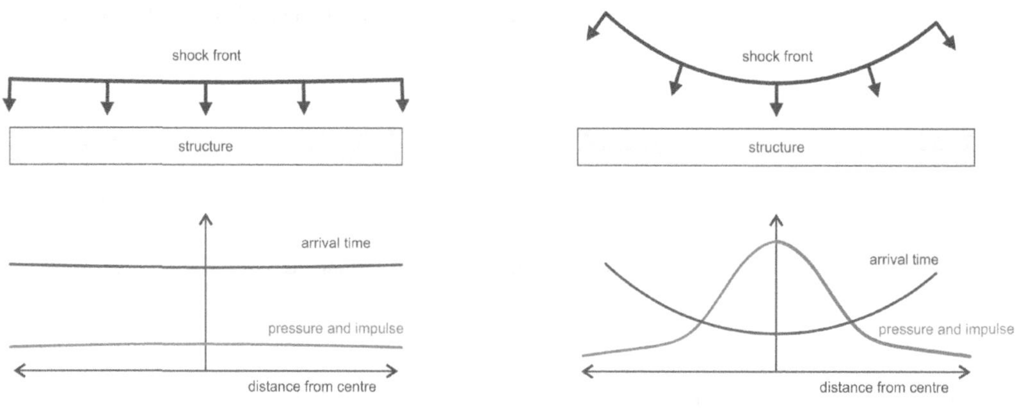

(a) Far-field (b) Near-field

FIGURE 10.25 Blast interaction diagram and loading parameters associated with far-field (a) and near-field (b) blast loading.

10.10.3.2 Deformation Under Uniform Loading

Figure 10.26(a) shows the evolving deformation of a ductile structure when exposed to a spatially uniform dynamic load, with vertical displacements exaggerated for clarity. If the structure were loaded statically, there would be sufficient time available for the stresses induced by the load to be transferred to the supports through flexure. The structure would therefore deform in a single, global deformation mode.

As the rate of loading increases, the time available for the development of global equilibrium reduces, and instead a transient deformation mechanism will evolve. Under spatially uniform dynamic loading, all locations of the structure are initially mobilised with the same velocity. As motion is locally restricted in the region of the supports, a plastic hinge develops. As the centre of the structure continues to deflect, the hinge migrates inward until, eventually, the familiar triangular deflected shape associated with uniform static loading is formed. Although the residual shape of the structure is similar to that produced by static loading, the mechanism for its development is very different.

Figure 10.26(b) shows typical deformation modes of a brittle structure subjected to a spatially uniform dynamic load. As well as global bending, a global shear failure mode exists. If the stresses developed at the supports exceed the shear strength of the structure during deformation, the main body will be 'punched through'.

10.10.3.3 Deformation Under Concentrated Loading

Figure 10.27(a) shows the evolving deformation of a ductile structure when exposed to a spatially concentrated dynamic load. Due to the high interaction curvature of the loading, Figure 10.25(b), the vast majority of the momentum is imparted in the area nearest the blast, causing an epicentral

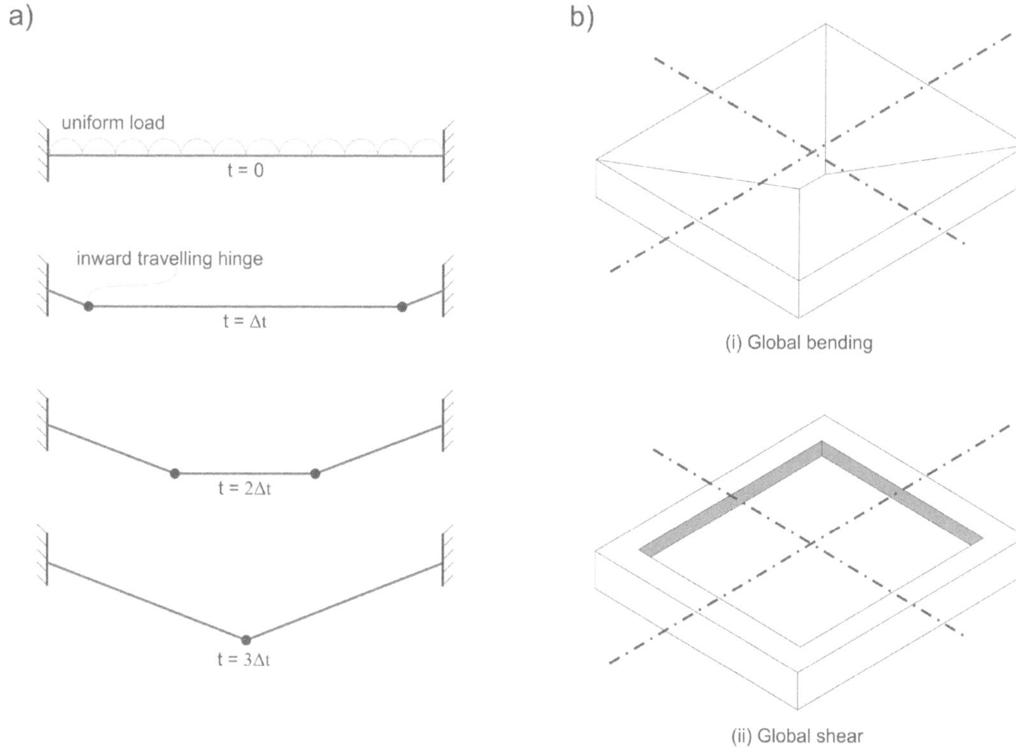

FIGURE 10.26 Deformation modes under spatially uniform dynamic loads: (a) evolving profile of a ductile structure and (b) residual profiles of a brittle structure.

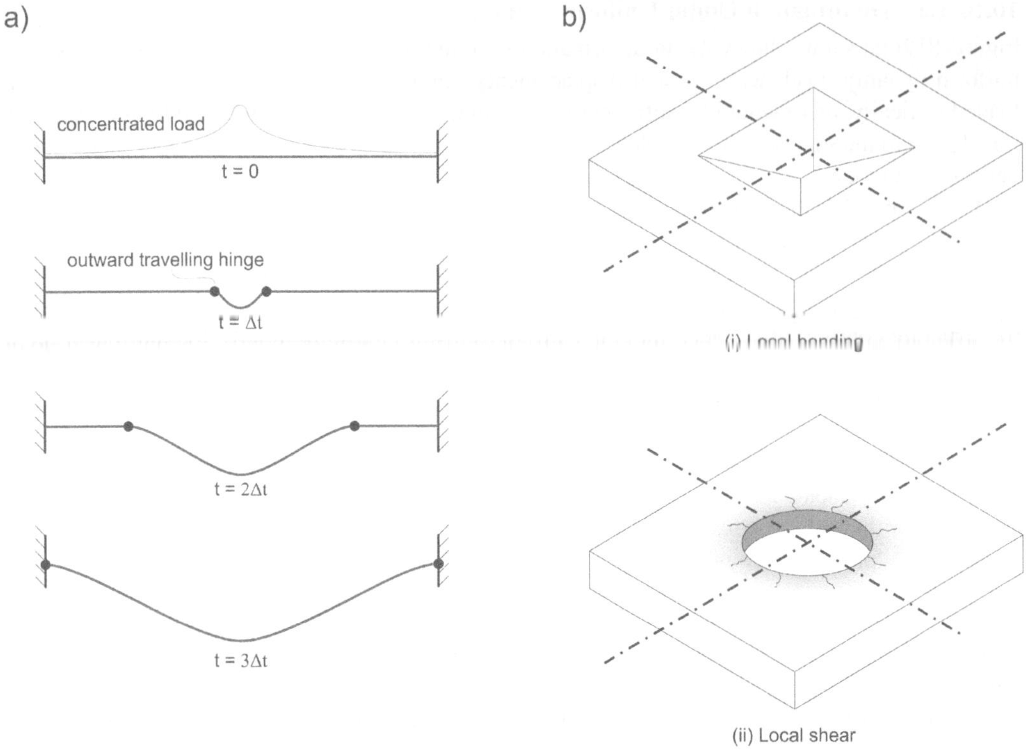

FIGURE 10.27 Deformation modes under spatially concentrated dynamic loads: (a) evolving profile of a ductile structure and (b) residual profiles of a brittle structure.

region of the structure to be mobilised with significantly higher velocity than nearby regions. Initially, there is insufficient time for the stress caused by this movement to be transferred to the supports and so a local 'dish' forms, bounded by inertial support or plastic hinges.

If the structure possesses sufficient ductility, the plastic hinge may migrate outwards towards the supports as the epicentral region transfers its momentum outwards to regions which are deforming with a lower velocity. The structure may eventually adopt a peak 'triangular' deformation shape, similar to that caused by a spatially uniform load, but again the manner by which the deformed shape evolves is very different.

In more brittle structures such as reinforced concrete panels, the in-plane velocity gradients generated by the localised momentum may cause local shear cracks through the thickness of the material. These effectively isolate the central part of the structure from the rest so that the final (or residual) deformation mode is either a local dish or shear plug failure, as illustrated in Figure 10.27(b).

10.11 SUMMARY

10.11.1 Key Parameters

Key parameters which influence blast loading are as follows:

- *Scaled distance,* $Z = R / W^{1/3}$ *(m/kg$^{1/3}$)*: Generally, peak pressure and specific impulse decrease substantially with increasing scaled distance, whereas loading duration increases modestly.

- *Charge mass*, W *(kg TNT)*: For a given scaled distance, time and impulse parameters are scaled by $W^{1/3}$: a larger explosive mass will result in longer loading durations for the same value of peak pressure.
- Atmospheric conditions such as increased pressure or ambient temperature, although the effects on loading parameters are relatively minor.
- *Structural span,* L *(m)*: Large differences in slant distance between the point of impingement (angle of incidence $\theta = 0°$) and the edge of the structure may result in substantial variations in loading parameters along the span. Substantial temporal variations in loading may also be seen; in the extreme case the loading event may be completed at the point of impingement before remote parts of the structure have even begun to be loaded. Spatial and temporal variations may be particularly amplified in the case of near-field loading with short stand-off distances to the point of impingement, as a small increase in distance along the loaded surface leads to larger relative increases in slant distance, and substantial reduction in loading parameters.
- *Size of reflecting surface*: In the extremes, a surface of infinite height and width will experience the full reflected pressure history, whereas a surface of zero height and width will offer no resistance to the flow and experience only the incident pressure history. In between, blast wave clearing may occur, a physical process caused by diffraction of the blast wave around the loaded edge which results in successive reduction of pressure along the loaded surface as a 'relief wave' propagates inwards. For relatively small structures, simple corrections exist, whereas for structures whose dimensions are comparable to the physical length of the blast, complex spatial and temporal distributions of pressure are experienced and are best evaluated using more sophisticated physics-based approaches such as the Hudson (1955) method.

Key parameters which influence structural response are as follows:

- Structural properties such as distributed mass, m (kg), and stiffness, k (N/m), and the resulting natural period, T (s). Damping is typically ignored as its influence on the first few cycles of displacement is negligible.
- Peak force, F^+, (N) and loading duration, t_d^+ (s) for exponential loading, or more typically $t_{d,lin}$ for a simplified reverse ramp load with preserved impulse for exponential loading, I^+ (N.s).
- Support conditions and loading arrangement, since these ultimately influence the normalised deflected shape, ϕ (–), used to evaluate mass and load transformation factors (which themselves govern the energy balance).
- The ratio of loading duration to the natural period: $t_{d,lin} / T$ (–). Low values indicate impulsive loading whereby structural response is near-zero during load application and work done is zero throughout. Large values indicate quasi-static loading where, at the point of peak displacement, work done is balanced by internal strain energy. These simple energy relations enable peak displacement to be calculated with simplified two- or three-parameter equations. For intermediate values of $t_{d,lin} / T$ (–), the energy balance is complex and time-varying and therefore the equations of motion should be solved directly, either analytically (elastic response, linear load only) or numerically (arbitrary loading and resistance functions permitted).
- Elastic deflection limit, z_E (m), since this influences the peak resistance, R_u (N) and therefore the ratio of resistance to applied loading: R_u / F^+(–).
- Ductility, or strain at failure, as this dictates how much deformation can be achieved in plasticity/post-yield. Ductility also influences the mode of deformation which itself is dependent on loading distribution
- Strain rate, since these influences increase in apparent strength under dynamic loading, are termed the DIF.

10.11.2 Limitations

The methods discussed in previous sections have the following limitations:

- Kingery and Bulmash/UFC loading parameters in the near-field were compiled from very few, if any, direct measurements, and as such their validity is disputed.
- The above loading predictions necessitate that the explosive is formed of either a sphere of TNT (with no obstructions between the explosive and the structure) or a hemisphere of TNT (on a flat, rigid reflecting surface, again with no obstacles between the explosive and the structure). It is commonly accepted that a different explosive composition can be expressed as an equivalent mass of TNT, using the concept of 'TNT equivalence', however, exact values of this metric for different explosives are still disputed, owing to the numerous different techniques with which this can be quantified. The explosives are also assumed to be centrally detonated.
- Whilst load parameters in the far-field have been shown to be highly repeatable, and modern techniques allow for extreme near-field loads to be quantified with commensurate accuracy, there exists a region between these two (roughly $0.5 \leq Z < 2.0$ m/kg$^{1/3}$), where surface instabilities of the expanding fireball result in large variations in loading parameters.
- The UFC (US DoD 2008) clearing corrections implicitly assume a small reflecting surface relative to the physical length of the blast wave.
- The Hudson (1955) method assumes a planar, weak shock ($Z > 2.0$ m/kg$^{1/3}$) impinging normally on a rigid surface with no existing flow conditions in the direction of travel of the clearing wave and sufficient depth such that no diffraction waves arrive from the rear of the structure.
- Load and mass transformation factors in the SDoF method are assumed to remain constant throughout the deformation. This means that, if the system is expected to enter plasticity, then a weighted average should be taken between the elastic and plastic factors (Biggs 1964).
- The normalised deflected shape, ϕ (–), is assumed to be that of the same structure (e.g., span and support conditions) if it were loaded with the same distributed load in a static manner. This means that evolving shape profiles discussed in Section 11.10.3 are not considered.
- Concrete spalling and shear failure is not normally considered in SDoF models.
- DIFs listed in Table 11.7 are representative 'design' values for two selected strain rates. In reality, DIF varies continuously with $\dot{\epsilon}$, however, the tabulated values "are satisfactory for design and the determination of the actual DIF values is unwarranted" (US DoD 2008).

ACKNOWLEDGEMENTS

This chapter is based on course notes from *MAC470 – Blast and Impact Effects on Structures*, taught at the University of Sheffield. Although the material has been substantially re-worked and re-structured in recent years, SE Rigby very gratefully acknowledges the combined efforts of DJ Pope and A Tyas in developing the first version of these notes.

REFERENCES

Baker, W. E. (1973), *Explosions in air*, University of Texas Press, Austin, TX.

Barr, A. D., Rigby, S. E., Clarke, S. D., Farrimond, D. & Tyas, A. (2023), 'Temporally and spatially resolved reflected overpressure measurements in the extreme near field', *Sensors* **23**(2), 964.

Biggs, J. M. (1964), *Introduction to Structural Dynamics*, McGraw-Hill, New York, NY.

Bogosian, D., Ferritto, J. & Shi, Y. (2002), Measuring uncertainty and conservatism in simplified blast models, *in 30th Explosives Safety Seminar*, Atlanta, GA, pp. 1–26.

Center for Chemical Process Safety (1994), *Guidelines for Evaluating the Characteristics of Vapor Cloud Explosions, Flash F, and BLEVEs. Appendix B: Effects of Explosions on Structures*, American Institute of Chemical Engineers, New York, NY.

Christensen, S. O. & Hjort, Ø. J. S. (2012), Back calculation of the Oslo bombing on July 22nd, using window breakage, *in 22nd Symposium on the Military Aspects of Blast and Shock (MABS22)*, Bourges.

Cooper, P. (1994, 24–29 July), Comments on TNT equivalence, *in 20th International Pyrotechnics Seminar*, Colorado Springs, CO, pp. 1–12.

Cormie, D., Mays, G. & Smith, P. (2019), *Blast effects on buildings, 3rd ed.*, Thomas Telford, London.

Cranz, C. (1926), *Lehrbuch der basllistik*, Springer, Berlin.

Denny, J., Batchelor, J. & Al-Hajj, S. (2023, 3–8 December), Estimating blast exposures from the 2020 beirut explosion and examining correlation with blast injuries, *in 26th International Symposium on Military Aspects of Blast and Shock (MABS26)*, Wollongong.

Edwards, D. H., Thomas, G. O., Milne, A., Hooper, G. & Tasker, D. (1992), 'Blast wave measurements close to explosive charges', *Shock Waves* **2**, 237–243.

Esparza, E. (1986), 'Blast measurements and equivalency for spherical charges at small scaled distances', *International Journal of Impact Engineering* **4**(1), 23–40.

Farrimond, D., Rigby, S., Clarke, S. & Tyas, A. (2022), 'Time of arrival as a diagnostic for far-field high explosive blast waves', *International Journal of Protective Structures* **13**(2), 379–402.

Farrimond, D., Woolford, S., Tyas, A., Rigby, S., Clarke, S., Barr, A., Whittaker, M. & Pope, D. (2024), 'Far-field positive phase blast parameter characterisation of RDX and PETN based explosives', *International Journal of Protective Structures* **15**(1), 141–165.

Formby, S. & Wharton, R. (1996), 'Blast characteristics and TNT equivalence values for some commercial explosives detonated at ground level', *Journal of Hazardous Materials* **50**(2–3), 183–198.

Friedlander, F. G. (1946), 'The diffraction of sound pulses. I. Diffraction by a semi-infinite plane', *Proceedings of the Royal Society A: Mathematical, Physical and Engineering Sciences* **186**(1006), 322–344.

Granström, S. A. (1956), Loading characteristics of air blasts from detonating charges, Technical Report 100, Transactions of the Royal Institute of Technology, Stockholm.

Hopkinson, B. (1915), British ordnance board minutes, 13565.

Hudson, C. C. (1955), Sound pulse approximations to blast loading (with comments on transient drag), Technical Report SC-TM-191-55-51, Sandia Corporation, NM, USA.

Jarrett, D. E. (1968), 'Derivation of British explosives safety distances', *Annals of the New York Academy of Sciences* **152**(1), 18–35.

Jenkins, C. M., Ripley, R. C., Wu, C., Horie, Y., Powers, K. & Wilson, W. H. (2013), 'Explosively driven particle fields imaged using a high speed framing camera and particle image velocimetry', *International Journal of Multiphase Flow* **51**, 73–86.

Karlos, V., Solomos, G. & Larcher, M. (2016), 'Analysis of the blast wave decay coefficient using the Kingery–Bulmash data', *International Journal of Protective Structures* **7**(3), 409–429.

Kingery, C. N. & Bulmash, G. (1984), Airblast parameters from TNT spherical air burst and hemispherical surface burst, Technical Report ARBRL-TR-02555, US Army BRL, Aberdeen Proving Ground, MD, USA.

Krauthammer, T. (2008), *Modern Protective Structures*, CRC Press: Taylor & Francis Group, Boca Raton, FL.

Larcher, M., Solomos, G., Casadei, F. & Gebbeken, N. (2012), 'Experimental and numerical investigations of laminated glass subjected to blast loadings', *International Journal of Impact Engineering* **39**(1), 42–50.

McVay, M. (1988), Spall damage of concrete structures, Technical Report SL-88-22, US Army Corps of Engineers Waterways Experiment Station, Vicksburg, MS, USA.

Morris, W. E. (1959), Shock diffraction in the vicinity of a structure, Technical Report Operation UpshotKnothole, WT-786, Sandia Corporation, MD, USA.

Murtha, J. P. (1955), Blast loading of structures in the regular reflection and low mach-stem regions, Technical Report SC-TM-5112, Sandia Corporation, MD, USA.

Needham, C. E. (2010), *Blast Waves*, Springer, Berlin.

Norville, H. S., Harvill, N., Conrath, E. J., Shariat, S. & Mallonee, S. (1999), 'Glass-related injuries in Oklahoma City bombing', *Journal of Performance of Constructed Facilities* **13**(2), 50–56.

Olson, W. C., Patterson, J. D. I. & Williams, J. S. (1960), The affect of atmospheric pressure on the reflected impulse from air blast waves, Technical Report BRL-1241, Aberdeen Proving Ground, Maryland, USA.

Rigby, S. E., Knighton, R., Clarke, S. D. & Tyas, A. (2020a), 'Reflected near-field blast pressure measurements using high speed video', *Experimental Mechanics* **60**(7), 875–888.

Rigby, S. E., Ratcliff, A., Clarke, S., Farrimond, D., Fay, S. & Wholey, W. (2023), 'Consequences of the 2020 Beirut explosion', *Fire and Blast Information Group (FABIG) Technical Newsletter* **88**, 6–10.

Rigby, S. E., Tyas, A., Curry, R. J. & Langdon, G. S. (2019), 'Experimental measurement of specific impulse distribution and transient deformation of plates subjected to near-field explosive blasts', *Experimental Mechanics* **59**(2), 163–178.

Rigby, S., Fay, S., Tyas, A., Warren, J. & Clarke, S. (2015b), 'Angle of incidence effects on far-field positive and negative phase blast parameters', *International Journal of Protective Structures* **6**(1), 23–42.

Rigby, S., Lodge, T., Alotaibi, S., Barr, A., Clarke, S., Langdon, G. & Tyas, A. (2020b), 'Preliminary yield estimation of the 2020 Beirut explosion using video footage from social media', *Shock Waves* **30**(6), 671–675.

Rigby, S., Tyas, A., Bennett, T., Clarke, S. & Fay, S. (2014a), 'The negative phase of the blast load', *International Journal of Protective Structures* **5**(1), 1–20.

Rigby, S., Tyas, A., Bennett, T., Fay, S., Clarke, S. & Warren, J. (2014c), 'A numerical investigation of blast loading and clearing on small targets', *International Journal of Protective Structures* **5**(3), 523–574.

Rigby, S., Tyas, A., Clarke, S., Fay, S., Reay, J., Warren, J., Gant, M. & Elgy, I. (2015a), 'Observations from preliminary experiments on spatial and temporal pressure measurements from near-field free air explosions', *International Journal of Protective Structures* **6**(2), 175–190.

Rigby, S., Tyas, A., Fay, S., Clarke, S. & Warren, J. (2014b, 16–17 October), Validation of semi-empirical blast pressure predictions for far field explosions – Is there inherent variability in blast wave parameters?, *in 6th International Conference on Protection of Structures Against Hazards (PSH14)*, Tianjin.

Sachs, R. G. (1944), The dependence of blast and air blast on ambient pressure and temperature, Technical Report BRL-466, Aberdeen Proving Ground, Maryland, USA.

Shin, J., Whittaker, A. S. & Cormie, D. (2015), 'Incident and normally reflected overpressure and impulse for detonations of spherical high explosives in free air', *Journal of Structural Engineering* **141**(12), 04015057.

Smith, P. & Hetherington, J. (1994), *Blast and Ballistic Loading of Structures*, Butterworth-Heinemann, Oxford.

Tyas, A. (2019, 14–15 December), Blast loading from high explosive detonation: What we know and what we don't know, *in 13th International Conference on Shock and Impact Loads on Structures*, Guangzhou.

Tyas, A., Warren, J., Bennett, T. & Fay, S. (2011), 'Prediction of clearing effects in far-field blast loading of finite targets', *Shock Waves* **21**(2), 111–119.

US Army Materiel Command (1974), *Explosions in Air Part One*, Department of the Army, Alexandria, VA, AMCP-706-181.

US DoD (2008), *Structures to Resist the Effects of Accidental Explosions*, US DoD, Washington, DC, UFC 3-340-02.

Walley, F. (1978), *Directorate of Civil Engineering Services*, Property Services Agency. UK Department of the Environment, London, UK.

11 Modelling of Explosive Effects

Zongjun Li

11.1 INTRODUCTION

11.1.1 NECESSITY OF MODELLING EXPLOSIVE EFFECTS

Explosive events involving highly dynamic and complex phenomena that cross multiple physical scales and disciplines are of great importance in military and civilian applications but also in industrial contexts. Many consequential effects, including high temperatures, extreme pressures, shock waves, and fragmentation, are typically generated by these events. These destructive effects subsequently pose a significant threat to both society and public safety. To effectively mitigate the risk and to optimise the design of explosive ordnance systems, precise predictions of these explosive effects are indeed essential for the advancement of key engineering applications. These include, but are not limited to, the development of munitions, the dynamic response of target materials, as well as the engineering of blast-resistant structures.

Experimental observation has long been the foundation of explosive-effects research. Physical testing provides essential data, but its scope is constrained by safety protocols and scale limits. Laboratory experiments cannot fully capture the multi-scale interactions that occur during an explosion – between shock waves, fragmentation, and structural response – so complete replication of real-world events is often impossible.

Over the last few decades, computational modelling has matured into an indispensable complement to experiment. Modern simulation tools can represent fluid–structure interaction and material behaviour at high strain rates, enabling controlled, reproducible studies across a wide range of loading conditions. Numerical methods such as finite element analysis (FEA) and computational fluid dynamics (CFD) now allow detailed parametric investigations that overcome many experimental limitations.

11.1.2 PHYSICAL PRINCIPLES OF MODELLING

The fundamental basis of explosive effect modelling through computational methods depends on the three conservation laws of mass, momentum, and energy. The partial differential equations derived from these principles enable the representation of explosive phenomenon evolution and their media interactions through state parameter changes of pressure and density and internal energy and temperature. The equation of state (EOS) serves as the main relationship which links thermodynamic properties to predict material response under extreme stress conditions. Accurate EOS formulations are therefore essential for determining reliable modelling outcomes. In detonation modelling, particularly for the gas products, the Jones–Wilkins–Lee (JWL) EOS is widely adopted to characterise the expansion behaviour of detonation products (Lee et al., 1968), whilst the Mie–Grüneisen EOS is commonly employed to describe the response of solids under high-pressure and high-temperature conditions (Mie, 1903). Together, these formulations establish the pressure–volume–energy relationships necessary for describing the variations of the thermodynamic properties under explosive loading.

Beyond the thermodynamic descriptions, the mechanisms of the interactions and the dynamic responses of materials also play a key role in the modelling of explosive effects, such as the elastic–plastic deformation and rate-dependent failure mechanisms of targets to the extreme loadings. To present a better modelling outcome, many complex governing equations are required to be employed

DOI: 10.1201/9781003529217-11

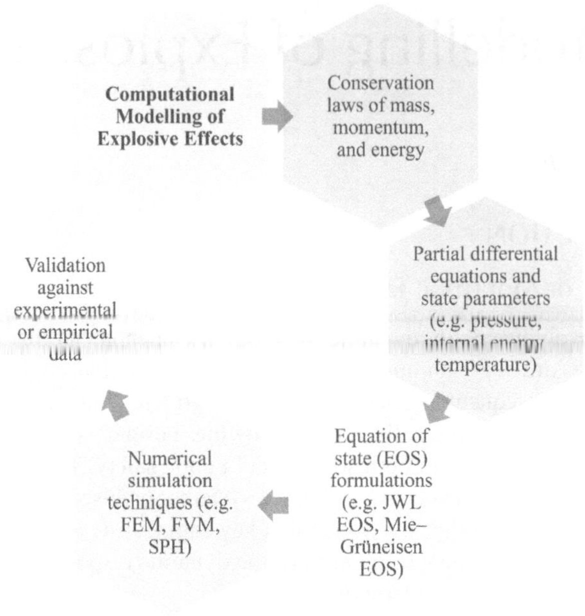

FIGURE 11.1 Flowchart of the computational modelling process of explosive effects.

together. Whereby advanced numerical techniques, such as finite element method (FEM), finite volume method (FVM), and mesh-free approaches like smoothed particle hydrodynamics (SPH), are necessary to govern the sophisticated equations and complex interactions. These techniques are established based on multi-physics coupling to capture the complex nature of explosive events, where fluid–structure interaction, shock-induced heating, and thermal–mechanical effects play significant roles. To determine the accuracy and reliability of the numerical methods, the predicted results need to be compared with experimental measurements or empirical data, typically referred to as the validation process. The validation process serves to confirm that the selected constitutive models and EOS formulations and failure criteria match their intended applications. The flowchart of a typical modelling process of explosive effects is shown in Figure 11.1.

11.2 OVERVIEW OF COMPUTATIONAL MODELLING APPROACHES

The field of computational modelling of explosive effects has shown substantial growth throughout the past few decades. The advancement of explosive research has resulted from uniting advanced numerical methods with multi-scale modelling techniques that improve the simulation of complex physical processes. The following section outlines the historical progression of computational methods in explosive research together with essential numerical techniques and modelling strategies for improving simulation-based analysis predictions.

11.2.1 Evolution of Modelling in Explosive Research

The evolution of modelling of explosive effects has progressed significantly over the past century. In the early twentieth century, studies relied primarily on the experimental data and semi-empirical techniques. Models like the Hopkinson–Cranz scaling law were established during this period to provide a basic understanding of blast wave attenuation and to predict the structural and material responses under the blast overpressures (Cranz, 1926; Hopkinson, 1915). However, the reliability and applicability of these models remained limited, as they could only provide legitimate results under idealised and simplified conditions (Wei and Hargather, 2021). The modelling of explosive effects under more complex conditions became possible with the advancement of computing technologies shown in the late twentieth century. Numerical methods, like FEM and finite difference method (FDM), were developed and adapted to solve the governing conservation equations of mass, momentum, and energy by discretising variables within defined boundaries. Established upon these concepts, FVM and mesh-free techniques, such as SPH, further expanded modelling capabilities, enabling simulations to handle more complex scenarios where large deformations and highly nonlinear material behaviour were involved (Gharehdash et al., 2020). In recent years, there has been a growing trend towards multi-scale and multi-physics modelling, aimed at achieving a more comprehensive understanding of explosive effects (Hao et al., 2024). A more direct link between micro-scale material behaviour, such as fracture, damage, and reaction kinetics, and the macro-scale structural response under extreme loading, could be formed through the multi-scale approaches. In addition, more complicated scenarios, such as fluid–structure interactions, chemical energy release, and thermodynamic effects, could be integrated into the advanced multi-physics coupling techniques to further capture the interdependent mechanisms associated with the propagation of the shock waves (Liu et al., 2024; Ma et al., 2024). These developments have transformed computational research on explosive effects, enabling modern simulations not only to complement experimental studies but, in many cases, to reproduce observed behaviours with high accuracy.

11.2.2 Common Computational Methods

The modelling of explosive effects relies on FEM and SPH methods which provide different advantages and limitations based on their computational efficiency and accuracy and their applicability to various scenarios. The four most common approaches, FEM, FVM, SPH, and multi-scaled modelling, are discussed in detail below.

11.2.2.1 Finite Element Methods

The numerical technique FEM stands as one of the most frequently used methods in computational mechanics. The method serves structural response analysis in explosive research because it effectively models material deformation and failure when subjected to blast or impact loads through Lagrangian, Eulerian, or Arbitrary Lagrangian–Eulerian formulations (Li et al., 2017). The method divides a continuum domain into finite elements which connect at nodes to enable the variable approximation of stress, strain, and displacement. There are several commercial software packages such as ABAQUS, ANSYS AUTODYN, and LS-DYNA employing FEM to conduct numerical simulations. Most common behaviour and material response associated with explosive effects, such as structural deformation and failure under high strain rates, interaction between blast waves and solid boundaries, fracture and fragmentation of materials, can be simulated within the software packages with reasonable and accurate solutions provided. Yet, the method shows limited effectiveness when dealing with extreme deformations and fluid–dynamics interactions (Narváez-Muñoz et al., 2025).

11.2.2.2 Finite Volume Methods

The main difference between FVM and FEM exists because the former utilises local conservation through discrete volume integration instead of using variational principles. FVM finds its extensive application in CFD because it effectively simulates turbulence and shock waves and high-speed flow phenomena in each control volume. OpenFOAM software is commonly used with FVM to simulate blast wave propagation along with shock wave interactions with different surfaces and obstacles and high-speed gas dynamics of detonation products. The implementation of FVM for compressible flow and shock dynamics requires high-resolution meshes near shock fronts to achieve accurate predictions (Myers et al., 2023). The method faces restrictions when used for solid materials and structural response because it struggles to represent stress distribution and fracture mechanics properties (Matos et al., 2024). In addition, this method also uses control volume discretisation instead of solid continuum representation which limits its ability to model stress distribution and fracture mechanics (Hawez et al., 2021). Therefore, other approaches such as FEM or SPH (mesh-free method) are needed to be combined into a hybrid approach to simulate structural or material responses when FVM is used.

11.2.2.3 Smoothed Particle Hydrodynamics

SPH is one of the widely used mesh-free methods that represents a continuum domain as a collection of discrete particles. Similar to FEM and FVM, field variables, including density, velocity, and pressure, are incorporated within the particles. These particles interact with their neighbouring particles through smoothing functions, which distribute their influence within a predefined kernel radius function. Unlike FEM and FVM, the SPH method operates independently from fixed meshes as it does not require mesh discretisation of the target area, which makes it particularly effective for explosive loading where large deformations, fragmentations, and free-surface flows are typically involved (Lin et al., 2014). The mesh-free nature of SPH also prevents the potential convergence issues which are typically shown in FEM or FVM for the complex interface cases. Such a fact makes it exceptionally suitable for modelling the structures involving highly deformed boundaries, such as multi-media wave propagations, underwater explosions, and fluid–structure interactions. Moreover, SPH can be incorporated with other techniques like FEM, to deliver stable and accurate results in the events that are associated with high-velocity impacts, such as blast-induced fragmentation, cavitation phenomena, and material failure analysis under high strain rates (Wang et al., 2023). However, the efficiency of SPH depends on two main factors which include selecting appropriate kernel functions and determining suitable smoothing parameters and particle resolutions. Whilst increasing particle density improves predictive capability, it normally also leads to higher computational cost and may significantly reduce the overall efficiency of the modelling.

11.2.3 Multi-Scale Modelling

The main objective of multi-scaled modelling is to improve the accuracy of the computational models by combing micro-scale material responses with macro-scale effects. This concept has not been properly adapted because of its complexity. Currently, different scaled models are developed with a focus on the initiation and detonation of the explosives only (Saurel et al., 2018).

11.2.3.1 Micro-Scale Quantum Mechanical and Molecular Dynamics Models

Quantum mechanics (QM) and molecular dynamics (MD) techniques have been proposed to examine the initiation and detonation behaviour of various explosives. This model proposes that large molecules are subjected to shock or impact loading according to the relative strengths of chemical and molecular bonds. The initial shock kinetic energy transforms into low-frequency lattice vibrations that then propagate to high-frequency chemical bond vibrations. As the vibrational modes initiated, the chemical bonds start to break when sufficient kinetic energy is absorbed, which consequently causes thermal decomposition. In theory, this process requires certain low-frequency vibrations to overcome energy barriers, which can facilitate the up-pumping process through energy transformation. However, the connection between the up-pumping and large-scale thermal combustion is still unclear, which limits this model from being used in other studies rather than the simulation of the initiation of explosives (Gu et al., 2024). In addition, if the type of ignition is primarily thermal and requires hot spots on the micrometre scale, performing QM calculations with a significant number of atoms is beyond current technological capabilities (Christensen et al., 2016).

11.2.3.2 Meso-Scale Modelling

Materials with complex and heterogeneous structures resist proper modelling through both atomic-scale and continuum macro-scale approaches at this particular scale. It is, therefore, required to perform the analysis at the meso-scale grain. Several approaches have been developed to simulate the initiation and detonation of explosives under this category. The modified burn models simplify microscopic events to enable large-scale engineering calculations by using shear banding as their primary localisation mechanism. The hot spot regions in statistical models receive probabilistic treatment whilst direct grain-scale simulations exist in multiple forms (Kim et al., 2016). This approach uses continuum mixture theory with FEM to simulate burn-to-detonation modelling at the described scale. The approach uses macroscopic continuum models to incorporate constituent elements within meso-scaled level by managing large deformations as well as modelling heterogeneity (Horie, 2014). The stochastic characteristics of explosive microstructures demand analysis of external stimulus responses through both time- and space-based multi-scale interactions. The meso-scale of explosive behaviour remains poorly understood despite recent modelling progress because of limited experimental validation. General comparisons of time and length scales among the multi-scaled modelling methods are illustrated in Figure 11.2.

11.2.3.3 Continuum Macro-Scale Modelling

The macro-scale modelling of the explosive effects is normally conducted through numerical methods, such as FEM, as shown in Figure 11.2. Currently, macro-scaled models are developed with the focus on the initiation and progression of the burning effects of explosives. The initiation of explosive models at large scales begins with burn model activation to perform numerical analyses. The burn models typically consist of two fundamental elements which include the internal energy function (EOS) and the rate law that controls reaction progression. The internal energy function (EOS) displays the distribution of reacting explosive mass and heat-release amounts whilst the rate law determines how the reaction advances through a progress variable. To accurately simulate the continuous phenomenon of the modelling, growth models are usually incorporated into the burn or initiation model. The ignition and growth (I&G) model functions as a widely used method which combines pressure-dependent growth models with reaction rate laws to model explosive initiation and burning processes. The approach operates based on the empirical parameter adjustments to match reactive flows across different explosives and environmental conditions and material characteristics (Hobbs and Kaneshige, 2014). On the other hand, other models are also available to represent the modelling at macro-scale level. The Johnson–Tang–Forest model functions as a macro-scale model which is incorporated with reaction modelling throughout different stages (Johnson et al., 1985). The model follows the reaction sequence through a four-stage process which begins with shock-induced hot spot formation and then intermediate decomposition and explosive mass heating before ending with final decomposition. The hot spot reactions were described through a first-order rate equation since the reaction speeds of hot spots were assumed to be much faster than the surrounding material. The model simplifies the initiation process by uniting mass fraction and heat transfer into one parameter

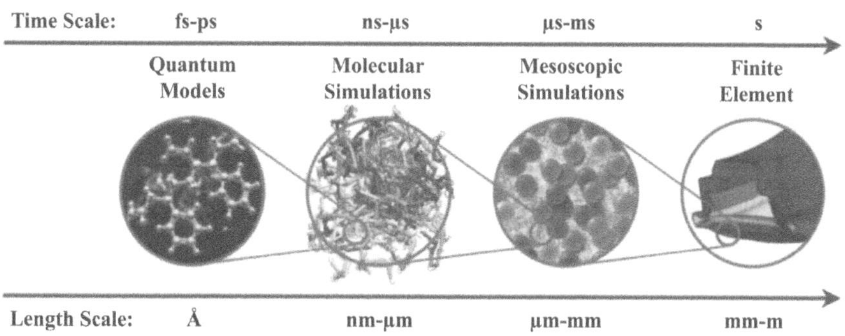

FIGURE 11.2 Representative length and time scales associated with the modelling approaches from the quantum to macroscopic modelling. (Adapted from Dequidt et al. (2020).)

which establishes heat transfer as the dominant factor (Horie, 2014). These macro-scaled models improve large-scale ignition and detonation simulation capabilities but they do not accurately show mechanistic processes at detailed spatial levels (Lizundia et al., 2021).

11.3 BLAST EFFECTS

The major destructive effects of an explosion can be generally classified into two categories: blast effects and fragmentation effects. A typical damage effective area of an explosion on the ground is illustrated in Figure 11.3. In the initial stage of an explosion, a massive amount of energy is released in an extremely short duration, producing high-pressure and -temperature gases within a much-confined space. This energy performs work on the surroundings through the expansion and dynamic motion of the generated gases, resulting in the formation of a shock wave. A large portion of the energy propagates outwards as a shock wave, causing damage to the environment nearby, which is commonly referred to as the blast effect. This effect typically weakens rapidly due to the expansion of the gases and the dissipation of energy within the transmission medium, typically air, during propagation. Consequently, the effective range of blast effects is generally limited to a relatively small distance from the point of detonation. On the other hand, fragments, either generated from the weapon casing or pre-existing within the ordnance, are propelled outwards as the high pressure from the gas expansion is converted into the initial kinetic energy of the fragments. These fragments disperse and travel away from the detonation point. The kinetic energy of the fragments decreases primarily due to aerodynamic resistance, such as air drag, rather than diminishing with distance in the manner of the shock wave. Therefore, the destructive radius of fragmentation effects is typically greater than that of blast effects, as shown in the Figure 11.3.

During the energy release process of an explosion, the generated gases and their rapid expansion and motion represent one of the primary mechanisms by which work is done on the surrounding environment. The typically high temperature and pressure gases produced from the detonation dissipate the energy to the surrounding medium nearby, driving the formation of shock waves and accelerating fragments. To accurately analyse and predict these effects, it is essential to understand the fundamental properties of the gases involved. Specifically, the relationship between pressure, volume, temperature, and density, which are generally defined by the EOS, forms the basis for describing the dynamic behaviour of the gases.

11.3.1 Gaseous Equations of State

The research field of gas dynamics treats the targeted gases as compressive continuums whilst ignoring their viscosity and thermal conductivity (Wang et al., 2022). This simplification is justified because, during explosions, pressure forces and inertial effects dominate the flow behaviour,

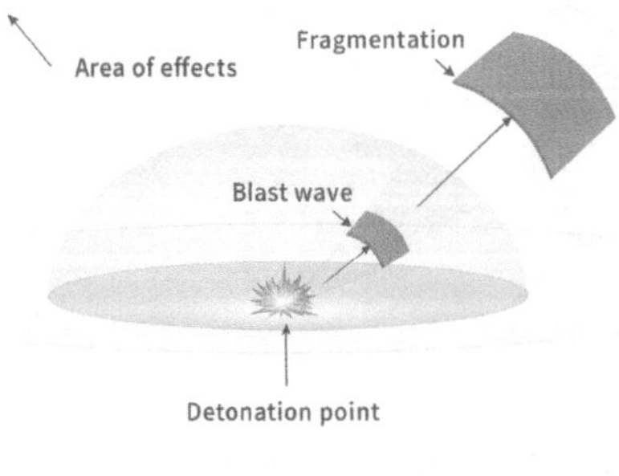

FIGURE 11.3 Effects area for detonation of a typical explosive ordnance. (Adapted from Hazell (2022).)

making the contribution of viscosity negligible. Furthermore, the thermal conductivity of gases is generally too low to produce a significant influence, particularly under the short timescales and high velocities associated with detonations (Liu et al., 2018). The behaviour of gases under these conditions is described by the EOS, which defines the relationship between pressure, density, and temperature required for modelling.

An ideal gas, also known as a perfect gas, can be described by the following EOS:

$$PV = RT \tag{11.1}$$

where P is the pressure, V denotes the specific volume, T stands for the absolute temperature, and R is the specific gas constant, the value of which depends on the type of gas. For any perfect gas, R is related to the universal gas constant $\bar{R}$ by

$$R = \frac{\bar{R}}{\bar{M}} \tag{11.2}$$

where $\bar{R}$ is the universal gas constant $\left[8.314\ J/(mol \cdot K)\right]$ and $\bar{M}$ is the molecular mass of the gas. For instance, in the case of hydrogen, with a molecular mass of $\bar{M} = 2.016\ g/mol$, the specific gas constant is calculated as

$$R = {}^{8.314\ J/(mol \cdot K)}\!\big/_{2.016\ g/mol} = 4.124\ J/(g \cdot K)$$

By definition, a perfect gas is the one in which intermolecular forces are ignored and the gas molecules are assumed to occupy no volume. As a result, the gas volume can theoretically approach zero as the pressure increases without bound. Whereby the specific internal energy e is considered a function of temperature alone:

$$e = e(T) \tag{11.3}$$

In accordance with the definition of specific heat at constant volume $\left[C_V = (\partial e/\partial T)_V\right]$, the change in specific internal energy is given by

$$de = C_V dT \tag{11.4}$$

The change in specific enthalpy, h, for a perfect gas can be written as

$$dh = de + d(PV) = de + d(P/\rho) = de + Rd(T) \tag{11.5}$$

where ρ is the density. This indicates that the specific enthalpy of a perfect gas is also a function solely of temperature. According to the definition of specific heat at constant pressure $\left[C_P = (\partial h/\partial T)_P\right]$,

$$dh = C_P dT \tag{11.6}$$

Substituting Equation (11.6) into Equation (11.5),

$$C_P dT = C_V dT + RdT \tag{11.7}$$

Which leads to the well-known relationship:

$$C_P - C_V = R \tag{11.8}$$

It should be noted that both C_P and C_V in the above equation are generally functions of temperature and are not constant over a wide range of temperatures. If a perfect gas undergoes a thermodynamic process between two equilibrium states, the changes in specific internal energy and specific enthalpy are given by

$$\begin{cases} e_2 - e_1 = \int_1^2 C_V\, dT \\ h_2 - h_1 = \int_1^2 C_P\, dT \end{cases} \tag{11.9}$$

An expression for the change in specific entropy S of a perfect gas can be derived from the thermodynamic relation:

$$TdS = dh - VdP \tag{11.10}$$

which leads to:

$$S_2 - S_1 = \int_1^2 dS = \int_1^2 \frac{C_P}{T} dT - R \ln \frac{P_2}{P_1} \tag{11.11}$$

For an isentropic process, where $dS = 0$, this reduces to

$$\int_1^2 \frac{C_P}{T} dT = R \ln \frac{P_2}{P_1} \tag{11.12}$$

In the case where the specific heat C_P is assumed to be constant, the above integral can be simplified to

$$\frac{C_P}{R} \ln \frac{T_2}{T_1} = \ln \frac{P_2}{P_1} \tag{11.13}$$

or equivalently,

$$\left(\frac{T_2}{T_1}\right)^{C_P/R} = \frac{P_2}{P_1} \tag{11.14}$$

Introducing the ratio of specific heats:

$$\gamma = \frac{C_P}{C_V} = \frac{C_V + R}{C_V} = 1 + \frac{R}{C_V} \tag{11.15}$$

and substituting this into Equation (11.13), then

$$\frac{T_2}{T_1} = \left(\frac{P_2}{P_1}\right)^{\gamma-1/\gamma} = \left(\frac{\rho_2}{\rho_1}\right)^{\gamma-1} = \left(\frac{V_1}{V_2}\right)^{\gamma-1} \tag{11.16}$$

And the relationship between pressure and density for an isentropic process can be written as

$$\begin{cases} \dfrac{P_2}{P_1} = \left(\dfrac{\rho_2}{\rho_1}\right)^{\gamma} \\ P = B\rho^{\gamma} \end{cases} \tag{11.17}$$

Equation (11.17) represents the isentropic EOS for an ideal gas. It is well known that the molecules of any real gas occupy a finite volume. Specifically, when a gas is subjected to elevated pressures, its density increases accordingly. Under such conditions, the volume of the gas molecules can no longer be ignored. Furthermore, the average distance between molecules decreases as the density increases. As a result, the intermolecular forces become increasingly significant, whereby the internal energy of the gas is not solely a function of temperature. It links to the density (or the specific volume) as well. Consequently, the EOS for an ideal gas, which assume negligible molecular volume and intermolecular forces, are not effective for describing the behaviour of a real gas under high pressure conditions. To accurately characterise the properties of gases at elevated pressures, it is therefore of importance to utilise more advanced EOS that consider molecular volume and intermolecular interactions.

11.3.1.1 Van der Waals Equations of State

One of the earliest and most well-known EOS that considers the finite volume of gas molecules and the intermolecular forces is proposed by Van der Waals. The Van der Waals equation can be expressed as

$$nRT = \left[P + a\left(\frac{n}{V}\right)^2\right](V - nb) \tag{11.18}$$

where n is the number of moles and can be defined as the inverse of the average molecular weight ($n = 1/M$), P is the total pressure, a is a constant that accounts for the attractive forces between gas molecules, and b is a constant that accounts for the volume occupied by the gas molecules. The equation can also be written in terms of the representation of gas density:

$$P + a\rho^2 = \frac{n\rho}{1 - b\rho} RT \tag{11.19}$$

In the above equation, $a\rho^2$ represents the contribution of intermolecular attractive forces to the pressure, normally referred as the cold pressure. The term on the right-hand side corresponds to the heat pressure, associated with the thermal motion of molecules. In either form of the Van der Waals equations, only the attractive forces between molecules are considered, whilst the covolume factor is treated as a constant specific to the type of gases. Hence, these equations are suitable for describing the behaviour of gases under conditions that do not involve extremely high pressures.

11.3.1.2 Equations of State Under Extremely High Pressures

At extremely high pressures, like the gases of detonation products with densities ranging from 1.0 to 2.3 kg/m^3, the gas molecules are basically in permanent contact condition (Liu et al., 2018). The pressure is mainly generated by the repulsive forces between molecules. Under such conditions, both the internal energy and the pressure of the gas can be considered as comprising two distinct components:

$$\begin{cases} E = E_K(V) + E_T(V,T) \\ P = P_K(V) + P_T(V,T) \end{cases} \tag{11.20}$$

where the subscript K denotes the cold (or elastic) contribution arising from intermolecular interactions, and the subscript T stands for the thermal (or heat) contribution associated with molecular motion. The cold pressure component $P_K(V)$ consists of two terms, a repulsive term and an attractive term, represented as

$$P_K(V) = AV^{-\gamma} - BV^{-m}, \text{ where } \gamma \neq C_P / C_V \tag{11.21}$$

where A and B are material constants, V is the specific volume, γ describes the repulsive exponent, and m describes the attractive exponent. Based on the theory of solids, the exponents satisfy the relations $\gamma > m$, and $m > 2$. At large volumes (i.e., low density cases), the intermolecular distances are substantial, causing $P_K(V)$ to become negligibly small. Under such condition, the attractive forces are the dominating forces, and the cold pressure mainly reflects the effects of intermolecular attraction. However, as the density increases and the intermolecular distance decreases, repulsive forces become increasingly significant. As adequately high densities, the cold pressure is generated almost completely by the repulsive interactions, with the contribution of attractive forces becoming negligible, whereby the pressure of the gas can be simplified to

$$P = AV^{-\gamma} + P_T(V,T) = AV^{-\gamma} + f(v) \cdot T \tag{11.22}$$

where $f(v)$ is a function of volume describing the thermal contribution. In the case of extremely high pressures, the cold pressure becomes dominant relative to the heat pressure. Consequently, the thermal effects can be neglected, and Equation (11.22) further simplifies to

$$P = AV^{-\gamma} = A\rho^{\gamma} \tag{11.23}$$

It should be noted, however, that whilst this simplified form may describe the general pressure–density relationship at extreme pressures, it may not be suitable for precise quantitative estimation in certain applications, where more comprehensive models are required.

In summary, the EOS for gases provide basic frameworks for describing gas behaviour under various conditions. The ideal gas law offers a simplified model at low pressures where molecular volume and interactions are negligible. The Van der Waals equations incorporate molecular volume and attractive forces, which are suitable for moderate pressures. At very high pressures, where repulsive forces dominate due to molecular contact, advanced forms of EOS consider both cold and heat pressure components. Each formulation reflects the requirement to describe the properties of gases under different pressure conditions.

11.3.2 Explosive Material Models

The computational analysis of blast effects depends on gas-form products produced by explosive detonation because these products create the expansion work that generates shock waves and causes fragmentation and structural loading. The EOS for detonation products provides an accurate way to model the pressure–volume–internal energy relationship during expansion. The development of computational models for explosive effects relies on these models because accurate blast wave propagation and environmental interaction require precise thermodynamic descriptions of detonation gases. Different explosive compositions and their specific applications have led to the development of various EOS formulations which range from empirical to semi-empirical models throughout the past decades.

11.3.2.1 Jones–Wilkins–Lee Equation of State

The JWL EOS (Lee et al., 1968) is a widely used empirical model designed to characterise the pressure–volume–energy relationship of detonation products during the expansion stage of explosives. Unlike equations that consider the chemical reactions, the JWL EOS focuses more on the thermodynamic work performed by the detonation gases. This model is capable of describing the expansion work and has become the standard for evaluating explosive work capability in both research and engineering applications.

The parameters of the JWL EOS are typically determined using experimental methods, most notably from the cylinder tests, which serves as the standardised procedure for quantifying the detonation product characteristics and validating the work potential of explosives (Ning and Ma, 2015). The JWL EOS has been well-recognised in various engineering applications and studies, particularly in computational modelling of explosive effects. Its general formulation expresses the pressure of detonation products as

$$P = A \cdot \exp(-R_1 V) + B \cdot \exp(-R_2 V) + \frac{\omega E}{V} \tag{11.24}$$

where V is the relative specific volume (volume divided by initial volume), E is the specific internal energy, A, B, R_1, R_2, and ω are empirical constants specific to the explosive. The first two terms $A\exp(-R_1 V)$ and $B\exp(-R_2 V)$ represent the higher-pressure contributions of the detonation products, whilst the last term $\omega E / V$ states the contribution of gas expansion at lower pressures. The specific internal energy can be expressed as the similar form:

$$E = \frac{A}{R_1}\exp(-R_1 V) + \frac{B}{R_2} \cdot \exp(-R_2 V) + \frac{C}{\omega V^{\omega}} \tag{11.25}$$

In isentropic expansion, the pressure–volume relationship satisfies:

$$-\left(\frac{\partial P}{\partial V}\right)_S = AR_1 \exp(-R_1 V) + BR_2 \exp(-R_2 V) + \frac{C(1+\omega)}{V^{\omega+2}} = \rho_0 D^2 \tag{11.26}$$

where ρ_0 is the initial density and D is the detonation wave velocity. For large-specific volumes, the exponential terms of the JWL equation can be ignored, and the new EOS in this situation turns to

$$P = \frac{\omega E}{V} \tag{11.27}$$

which represents the gas-like behaviour at very low pressures and large explosions. The JWL EOS requires six constants (A, B, R_1, R_2, ω, and C) but three constants remain independent because of the constraints used during parameter fitting. These parameters can be determined for different types of explosives through the cylinder tests, which records the velocity and deformation of a metal tube subjected to internal blast loading. Table 11.1 lists the tabulated parameters of JWL EOS for several typical explosives.

TABLE 11.1
JWL EOS Parameters for Explosives

	ρ_0 (g/cm^3)	P (GPa)	D (m/s)	A (GPa)	B (GPa)	R_1	R_2	ω	E_0 (GPa)
TNT	1.630	21.0	6930	373.77	3.7471	4.15	0.9	0.35	6.0
Comp B	1.717	29.5	7980	524.23	7.6780	4.20	1.1	0.34	8.5
HMX	1.891	42.0	9110	778.28	7.0714	4.20	1.0	0.30	10.5
PETN	1.770	33.5	8300	617.05	16.926	4.40	1.2	0.25	10.1
OCTOL	1.821	34.2	8480	748.60	13.380	4.50	1.2	0.38	9.6
PBX-9501	1.840	37.0	8800	852.40	18.020	4.55	1.3	0.38	10.2

Source: Data adapted from Ning and Ma (2015).

11.3.2.2 Becker–Kistiakowsky–Wilson Equation of State

The Becker–Kistiakowsky–Wilson (BKW) EOS is one of the earliest models formulated to describe the thermodynamic behaviour of explosive detonation products (Cowan and Fickett, 1956). Like ideal gas model and JWL model, the BKW EOS, a semi-empirical model, is also designed to represent the pressure–volume–energy relationship of detonation gases as they expand and cool. The model integrates concepts of chemical equilibrium to represent the complex gas mixtures produced by high explosives. Unlike the purely empirical expressions, the BKW EOS combines theoretical thermochemistry with empirical calibration. It expresses the pressure as

$$P = \frac{RT}{V}\left(1 + Xe^{\beta X}\right) \tag{11.28}$$

where P is the pressure, V is the molar gas volume, X is a function of covolumes expressed by $X = K / VT^{\alpha}$ (where $\alpha = 0.25$) in which K could be considered as an additive function of covolumes of constituent of the gases, and β is a material constant. Cowan and Fickett contributed a significant development of BKW EOS by introducing the constant θ to the temperature term of X (Cowan and Fickett, 1956). In addition, a modified function of K was redefined as $\kappa \sum x_i k_i$ in which κ is a constant, x_i is the mole fraction of component i, and k_i is the individual geometrical covolumes (covolume factors). Then, the comprehensive expressions of BKW EOS can be represented as

$$\begin{cases} \dfrac{PV}{RT} = 1 + Xe^{\beta X} \\ X = \dfrac{K}{V(T+\theta)^{\alpha}} \\ K = \kappa \sum x_i k_i \\ x_i = n_i / \sum n_i \end{cases} \tag{11.29}$$

Equation (11.29) consists of four variable material constants (α, β, κ, and θ) as well as covolumes of gaseous detonation products (k_i). The values of these constants and covolumes are required to be adjusted to fit measured detonation properties for different types of explosives. The calibration of such values has been proposed over the past several decades. In this study, two of the typical explosives, TNT and RDX, are chosen to be presented as examples shown in Table 11.2. The values of covolumes of main detonation products have been proposed distinctly in different studies, detailed examples are shown in Table 11.2 (Suceska et al., 2023).

11.3.2.3 Kihara–Hikita–Tanaka Equation of State

The Kihera–Hikita–Tanaka (KHT) EOS is another semi-empirical model developed to describe the thermodynamic behaviour of detonation products from condensed explosives (Tanaka, 1986). It is

TABLE 11.2
BKW EOS Parameters for Explosives

	$\rho_0\,(g/cm^3)$	P (GPa)	$D\,(m/s)$	α	β	$\kappa\,(cm^3K^{\alpha})$	θ (K)	E_0 (GPa)
TNT	1.63	21.0	6930	0.50	0.0959	12.685	400	6.0
RDX	1.80	34.0	8750	0.54	0.1810	14.150	400	9.8

Source: Data adapted from Suceska et al. (2023).

mainly designed to address limitations in existing models, like JWL model, by introducing a more flexible mathematical structure capable of fitting experimental data with better accuracy across Chapman–Jouguet (Jouguet, 1901) and far-field conditions. Similar to BKW EOS, KHT model also utilises the repulsive potential but with modified parameters to improve the accuracy of prediction. For the virial expansion process, it employs a power series representation of the pressure as a function of density. The KHT EOS is typically expressed as

$$\begin{cases} \dfrac{PV}{RT} = \dfrac{1 + aX + bX^2 + cX^3 + dX^4 + eX^5}{1 - \alpha X} \\ X = \left(\dfrac{\lambda}{PV}\right)^{3/n} V^{-1} \\ \lambda^{3/n} = \displaystyle\sum_{i=1}^{N} x_i \lambda_i^{3/n} \end{cases} \tag{11.30}$$

where $a, b, c, d, e, n, \lambda_i$, and α are material constants, x_i is the mole fraction of component i, similar to the BKW model. The performance of computational stability has improved since KHT EOS has employed a power series approximation and more empirical parameters (Yang et al., 2023).

11.3.3 Case Study: Influence of Apex Angle on Liner Velocity in Shaped Charge

To demonstrate the application of explosive models in computational modelling, this section presents a case study of shaped charge liner velocity as influenced by variations in apex angle. To focus on the effect of apex angle and to distinguish the study from military applications, only the charge and liner are modelled; the outer casing and confinement effect are excluded. The shaped charge is cylindrical with dimensions shown in Figure 11.4.

A nonlinear FEM is developed using ABAQUS/Explicit, employing the JWL EOS to examine the relationship between liner apex angle and tip velocity during the detonation process of shaped charge. The liner material is determined as copper and composition B is selected as the explosive charge. Both the liner and explosive volumes are modelled with four-node bilinear axisymmetric quadrilateral elements with reduced integration and hourglass control (CAX4R). Surface-to-surface contact interactions are applied, with separation permitted after shock loading (Li et al., 2024). Due to the symmetry of loading and geometry, a symmetric model is developed to optimise

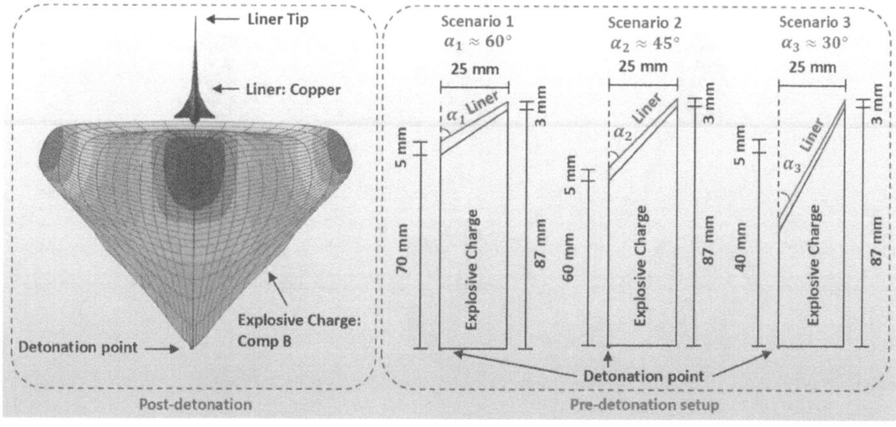

FIGURE 11.4 Pre-detonation and post-detonation setup of shaped charge diagram.

TABLE 11.3
Input Parameters of Material Models in ABAQUS

Copper	ρ (g/cm^3)	G (GPa)	*Rate* C	*Rate* ε	*EOS* $(U_S - u_P)$	C_0 (LT^{-1})	s	γ_0
	8.960	53	0.025	1		3940	1.489	2.02
	J-C model		A (MPa)	B (MPa)	n	m	T_{melt} (°C)	T_{tran} (°C)
			86	291	0.31	1.09	1083	27
Comp B (JWL)	ρ (g/cm^3)	D (m/s)	A (GPa)	B (MPa)	$E_{detonation}$ (kJ/m^3)	R_1	R_2	ω
	1.717	7980	524	7678	4950	4.2	1.1	0.34

computational efficiency. Mesh sizes of 3 × 3 mm for the explosive and 1 × 1 mm for the liner balance accuracy and computational cost. Symmetric boundary conditions restrict horizontal displacement for both the charge and liner.

11.3.3.1 Material Models

Two material models, of a copper liner and composition B explosive, are employed in this computational analysis. The elastic behaviour of the liner is defined by its shear response, whilst its plasticity is represented using a modified Johnson–Cook (JC) model coupled with a solid material pressure–volume EOS. The explosive volume is modelled with the JWL EOS mentioned in Section 11.3.2.1. The input parameters used in this study are listed in Table 11.3.

11.3.3.2 Numerical Results

A representation of the formation of the copper-shaped charge jet with an apex angle of 45° is shown in Figure 11.5. Upon initiation of detonation, a shock wave is formed and expands from the detonation point upwards to the liner within the cylinder of explosive charge. Eventually, the shock wave front engulfs the liner and locally deforms the copper liner to very high strains over an extremely short period. After the formation of the jet tip (>20 μs), a residual slug follows the jet at a reduced velocity.

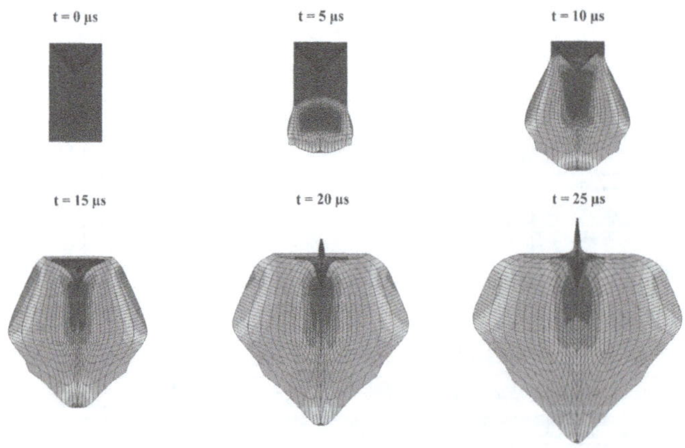

FIGURE 11.5 Formation of a copper-shaped charge jet with apex angle of 45°.

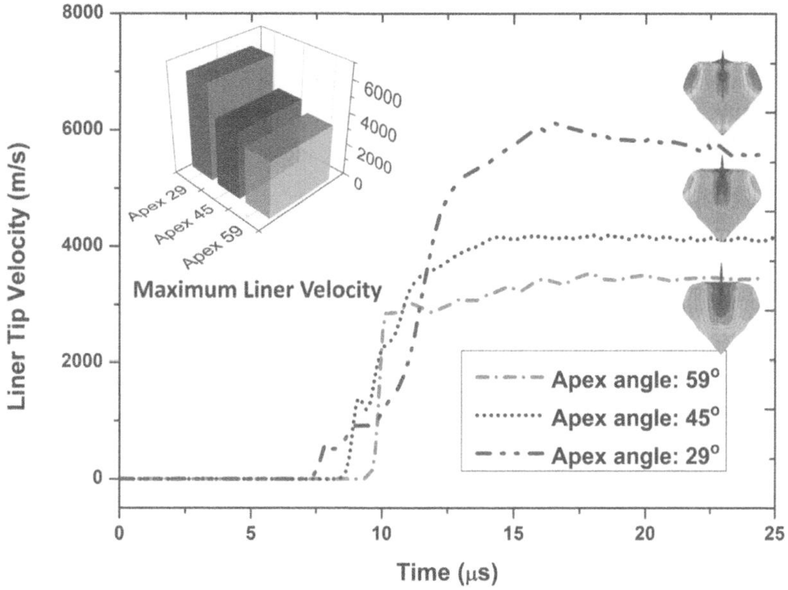

FIGURE 11.6 Liner velocity history diagrams of different apex angles.

11.3.3.3 Influence of Apex Angle on Liner Velocity

The variation of liner tip velocity histories for three shaped charge simulations with differing apex angles of 29°, 45°, and 59° is shown in Figure 11.6. An initial delay around 9–11 μs before the onset of acceleration is represented due to the variation of the distance between the liner tip and the detonation point. The liner with an apex angle of 29° achieves the highest peak velocity of 6122 m/s, whilst the liner with a 59° apex angle reaches a significantly lower velocity of 3528 m/s. The intermediate case of 45° follows a moderate trajectory. The results indicate that the liner velocity increases with the decrease of the apex angle.

11.3.4 Modelling of Blast-Induced Damage

The explosive force generates blast waves through rapid energy releases which create short periods of intense pressure on surrounding targets. The reliable predictions of the blast-induced damage are highly dependent on the accuracy of the shock wave parameters estimations and blast effects on both soft and hard targets (Ameri et al., 2021). Many analytical and empirical models have been proposed over the last several decades to assess the blast overpressure and its damage effects on the nearby structures. The assessment of this matter needs the creation of scaling laws together with fundamental expressions for shock propagation modelling.

11.3.4.1 Scaling Laws in Blast Analysis

The prediction of blast-induced effects involves the computational generalisation through the normalisation of explosive mass and standoff distance effects. The Hopkinson–Cranz scaling laws, also known as the cube-root scaling law, is the most widely used method to predict blast wave overpressure from large-scale explosions based on small-scale testing results (Cranz, 1926; Hopkinson, 1915). Details of the description and explanations of the scaling laws are covered in the previous chapter. As noted previously, the scaled distance for an explosive event is given by

$$Z = \frac{R}{W^{1/3}} \tag{11.31}$$

where Z is the scaled distance ($m / kg^{1/3}$), R is the standoff distance (m), and W is the TNT equivalent charge mass (kg). This relationship enables normalised representation of blast effects and facilitating predictive analysis across different charge sizes. To calculate the mass for the non-TNT explosives, a conversion factor is normally recommended for blast analysis. The conversion to TNT equivalent to non-TNT explosives is expressed as

$$W_{eq} = W_{exp} \times TNT\ conversion\ factor \tag{11.32}$$

where W_{eq} is the TNT equivalent mass, W_{exp} is mass of the non-TNT explosive, TNT conversion factor (Q_{exp} / Q_{TNT}) is the ratio of the mass specific energy of the particular explosive to the mass specific energy of TNT. Conversion factors for some high explosives taken from Hazell (2022) and Ullah et al. (2017) are shown in Table 11.4. Therefore, for example, a 10-kg charge of HMX converts to 12.56 kg of TNT since the conversion factor of HMX to TNT equivalence is 1.256.

11.3.4.2 Analytical Models for Predicting Blast Overpressure

Many analytical and empirical expressions (Huang et al., 2021; Ullah et al., 2017) have been proposed to estimate peak overpressure P_s as a function of scaled distance Z (Equation (11.31)), accounting for different detonation scenarios and environments.

For spherical airburst conditions, key models (units: $P_s - MPa$, $Z - m / kg^{1/3}$) include the following:

Sadovsky (1952) model:

$$P_s = \frac{0.085}{Z} + \frac{0.3}{Z^2} + \frac{0.82}{Z^3} \tag{11.33}$$

TABLE 11.4
Conversion Factors for Some High Explosives

Explosive	Mass Specific Energy Q_{exp} (kJ / kg)	TNT Equivalence Q_{exp} / Q_{TNT}
Lead azide	1540	0.340
Lead styphnate	1910	0.423
Amatol 80/20 (80% ammonium nitrate 20% TNT)	2650	0.586
60% Nitroglycerin dynamite	2710	0.600
Octol 70/30 (70% HMX 30% TNT)	4500	0.994
Tetryl	4520	1.000
TNT	4520	1.000
C-4 (91% RDX 9% plasticiser)	4870	1.078
Pentolite 50/50 (50% PETN 50% TNT)	5110	1.129
Comp B (60% RDX 40% TNT)	5190	1.148
RDX (cyclonite)	5360	1.185
Semtex	5660	1.250
HMX	5680	1.256
PBX 9404	5770	1.277
PETN	5800	1.282
Nitroglycerin (liquid)	6700	1.481

Source: Adapted from Hazell (2022).

Brode (1955) model:

$$P_s = \begin{cases} \dfrac{0.67}{Z^3} + 0.1, \text{ when } P_s > 1 \\ \dfrac{0.0975}{Z} + \dfrac{0.1455}{Z^2} + \dfrac{0.585}{Z^3} - 0.0019, \text{ when } 0.01 \le P_s \le 1 \end{cases} \tag{11.34}$$

Henrych and Major (1979) model:

$$P_s = \begin{cases} \dfrac{1.38}{Z} + \dfrac{0.543}{Z^2} - \dfrac{0.035}{Z^3} + \dfrac{0.000613}{Z^4}, \text{ when } 0.05 \le Z \le 0.3 \\ \dfrac{0.607}{Z} - \dfrac{0.032}{Z^2} + \dfrac{0.209}{Z^3}, \text{ when } 0.3 < Z \le 1 \\ \dfrac{0.0649}{Z} + \dfrac{0.397}{Z^2} + \dfrac{0.322}{Z^3}, \text{ when } 1 < Z \le 10 \end{cases} \tag{11.35}$$

Mills (1987) model:

$$P_s = \frac{1.772}{Z^3} - \frac{0.114}{Z^2} + \frac{0.108}{Z} \tag{11.36}$$

The Hopkins–Brown and Bailey (1998) model:

$$P_s = \begin{cases} -1.245 + \dfrac{1.935}{Z} + \dfrac{0.2353}{Z^2} - \dfrac{0.01065}{Z^3}, \text{ when } 0.05 \le Z \le 1.15 \\ \dfrac{0.0707}{Z} + \dfrac{0.3602}{Z^2} + \dfrac{0.4891}{Z^3}, \text{ when } 1.15 \le Z \le 40 \end{cases} \tag{11.37}$$

Huang et al. (2021) model:

$$P_s = \frac{0.082}{Z} + \frac{0.265}{Z^2} + \frac{0.686}{Z^3}, \text{ when } 1 \le Z \le 15 \tag{11.38}$$

For hemispherical surface bursts, different models have been summarised below for various environmental conditions, all the expressions have been modified to have the unit of megapascals (MPa) for the overpressure estimation.

Newmark and Hansen (1961) model – Normal ground:

$$P_s = 0.6784\frac{W}{R^3} + 0.294\left(\frac{W}{R^3}\right)^{1/2} \tag{11.39}$$

Wu and Hao (2005) model – Normal ground:

$$P_s = \begin{cases} 1.059 \times Z^{-2.56} - 0.051, \text{ when } 0.1 \le Z \le 1 \\ 1.008 \times Z^{-2.01}, \text{ when } 1 < Z \le 10 \end{cases} \tag{11.40}$$

Huang et al. (2021) model – Normal ground:

$$P_s = \frac{0.102}{Z} + \frac{0.399}{Z^2} + \frac{1.26}{Z^3}, \text{ when } 1 \le Z \le 15 \tag{11.41}$$

Huang et al. (2021) model – Rigid ground (e.g. reinforced concrete floor):

$$P_s = \frac{0.106}{Z} + \frac{0.43}{Z^2} + \frac{1.4}{Z^3}, \ when\ 1 \leq Z \leq 15 \tag{11.42}$$

Huang et al. (2021) model – Soil (soft) ground:

$$P_s = \frac{0.41}{Z} + \frac{0.69}{Z^2} + \frac{0.668}{Z^3}, \ when\ 1 \leq Z \leq 15 \tag{11.43}$$

Huang et al. (2021) model – Underwater:

$$P_s = \frac{52.4}{Z^{1.13}} \tag{11.44}$$

These models (Equations (11.33)–(11.44)) serve as the basic evaluation tools for assessing the blast damage to a target. One of the key parameters used in damage assessment is the peak overpressure: the higher the overpressure experienced by a target, the greater the likelihood and severity of structural or functional damage. The above models provide recommendations for peak overpressure as functions of scaled distance for various conditions. Once the scaled distance is calculated or determined, the corresponding correlation should be utilised to estimate the overpressure for the specific scenario. For example, for a surface detonation occurring on a rigid or reinforced concrete surface ground, the overpressure should be estimated by Equation (11.42) to present the threshold value.

11.3.4.3 Blast-Induced Damage on Targets

The extent of blast damage depends on both overpressure intensity and target vulnerability characteristics. The process of accurate damage assessment needs two crucial elements: the peak overpressure should be estimated correctly based on the specified environment and explosion conditions from the previous section, and established reference data should be used for evaluating the severity of the structural or functional damage of the targets. The data sets bridge the connections between observed damage thresholds and overpressure levels across different target types. Tables 11.5–11.7 summarise the expected degrees of damage for various categories, including personnel, civilian infrastructure, and military targets such as aircraft, naval vessels, and armoured vehicles, under different overpressure conditions.

Table 11.7 lists the reference criteria for blast-induced damage on military targets. The lower limit shown in this table represents the threshold for damage initiation; overpressures below this value are generally associated with no damage or only negligible effects on the target. Whilst on

TABLE 11.5
Vulnerability Characteristics of Personnel to Overpressure

Degrees of Personnel Damage	Overpressure (MPa)
No damage	< 0.02
Minor injury or tinnitus	0.02 – 0.03
Moderate injury (internal injury or eardrum rupture)	0.03 – 0.05
Severe wound (fracture or internal bleeding)	0.05 – 0.1
Death or fatal injury	> 0.1

Source: Adapted from Huang et al. (2021).

TABLE 11.6
Blast-Induced Damage on Civilian Infrastructure

	Overpressure (kPa)				
Infrastructure	**Slight Scratch**	**Minor Damage**	**Moderate Damage**	**Severe Damage**	**Complete Damage**
Wooden buildings	0.05–3	3–9	9–12	12–20	20–30
Brick buildings (single level)	0.05–7	7–15	15–25	25–35	35–45
Brick buildings (multistorey)	2–5	5–10	10–20	20–30	30–40
Reinforced concrete structures	3–10	10–30	30–50	50–80	80–100
Steel frame structures	3–5	5–20	20–30	30–50	50–80
Railway trains	0.01–3	3–4	4–10	10–20	
Railway	0.1–10	10–15	15–30	30–50	

Source: Adapted from Huang et al. (2021).

the other hand, the upper limit represents the threshold for the onset of severe or complete damage, any value exceeding this threshold is considered to cause the failure of the target. The damage severity depends on the overpressure levels between the two limits. When the calculated values are approaching the lower limit, the damage mode is expected to be the light damage whereas the values closer to the upper limit suggest a moderate to heavy damage mode.

The gradation listed in Tables 11.5–11.7 offers a useful reference framework for understanding blast effects as a function of overpressure intensity. However, it might not be applicable to use the data from the tables directly for the specific case in the real world. The real case should be evaluated in a case-by-case basis due to the variation of the loading factors, such as loading angle and location of the damage. The reference values mentioned above can only serve as guidance for the general assessment. Practical damage assessments need to examine multiple variables which include but are not limited to the target position accuracy and production flaws as well as surface orientation and blast loading circumstances.

TABLE 11.7
Blast-Induced Damage on Military Targets

Target	Degrees of Damage	Overpressure (MPa)
Armoured vehicles	Light-armoured vehicles and self-propelled artillery, various degrees of damage	0.035–0.3
	Tanks and other types of heavy armoured vehicles, various degrees of damage	0.045–1.5
Naval vessels	Minor to moderate damage, pressing mode or third-degree damage	0.028–0.043
	Severe damage, heavy damage mode, or second-degree damage	0.07–0.085
Aircraft	Fighter and bomber: minor damage, pressing mode or third-degree damage, transport aircraft: moderate to severe damage, heavy damage mode, or second-degree damage	0.02–0.05
	Piston engine aircraft are completely destroyed, destruction mode, jet aircraft have moderate to severe damage, heavy damage mode, or second-degree damage	0.05–0.1
	All types of aircraft were completely destroyed, corresponding to destruction mode or first order/degree damage	> 0.1

Source: Adapted from Huang et al. (2021).

11.3.4.4 Examples of Blast-Induced Damage

Example 11.1

A heavily armoured vehicle (HAV) was subjected to an air blast resulting from the detonation of an artillery shell at a distance of 10 m from its turret. The shell contains 39 kg of Pentaerythritol tetranitrate (PETN). Calculate the overpressure and estimate the damage that is experienced by this vehicle.

SOLUTION:

The explosive here is PETN, we need to calculate the TNT equivalent mass based on Table 11.4:

$$W = 39 \times 1.282 = 50\ kg$$

Then the scaled distance can be calculated by

$$Z = \frac{R}{W^{1/3}} = \frac{10}{50^{1/3}} = 2.714\ m / kg^{1/3}$$

The blast overpressure can be estimated based on the analytical/empirical models [taking Mills model (Equation (11.36)) and Huang et al.'s model (Equation (11.38)) as examples]:

Mills' model: $P_s = \frac{1.772}{Z^3} - \frac{0.114}{Z^2} + \frac{0.108}{Z} = 0.113\ MPa$

Huang et al.'s model: $P_s = \frac{0.082}{Z} + \frac{0.265}{Z^2} + \frac{0.686}{Z^3}\ (1 \le Z \le 15) = 0.101\ MPa$

According to Table 11.7, various degrees of damage are expected for a HAV subjected to an overpressure ranging from 0.045 to 1.5 MPa. As mentioned earlier, no or little damage is expected when the overpressure is below this range, whilst a heavy damage or complete destruction mode is expected when exceeding the upper threshold. The calculated values (approximately 0.1 MPa) from the above equations all fall close to the minimum damage range (less than 10% of the upper threshold). Therefore, light damage (or low to moderate damage) is expected for the HAV.

Example 11.2

An artillery shell that contains 30 kg of Pentolite 50/50 was detonated 10 m away from a two-story building made of reinforced concrete. Assume the detonation location was a normal ground condition, what would be the peak overpressure experienced by the building? Theoretically, is a single artillery shell sufficient to completely destroy the building? (By considering the blast effect only)

SOLUTION:

The explosive here is Pentolite 50/50; we need to calculate the TNT equivalent mass based on Table 11.4:

$$W = 30 \times 1.129 = 33.87\ kg$$

Then the scaled distance can be calculated by (the distance between the detonation point and the building)

$$Z = \frac{R}{W^{1/3}} = \frac{10}{33.87^{1/3}} = 3.091\ m / kg^{1/3}$$

The blast overpressure can be estimated based on the analytical/empirical models of normal ground condition (Equation (11.41)):

$$P_s = \frac{0.102}{Z} + \frac{0.399}{Z^2} + \frac{1.26}{Z^3} \left(1 \le Z \le 15\right) = 0.1175\ MPa > 0.1MPa\ \left(100\ kPa\right)$$

The peak overpressure experienced by the building is estimated as 117.5 kPa, which exceeds 100 kPa of the upper limit of the damage criteria for reinforced concrete structures. As indicated in Table 11.6, such pressure level is expected to cause complete damage to the concrete building. Therefore, in theory, a single artillery shell is sufficient to destroy the reinforced concrete building.

11.4 FRAGMENTATION EFFECTS

When an explosive detonates, a portion of the released energy is consumed to rupture the casing and accelerate the resulting fragments. The remaining energy contributes to the formation of a shock front and the associated blast wave. Once ejected, the fragments attain very high velocities and can quickly overtake the shock front, continuing their trajectory beyond it (Driels, 2021). This occurs because, as explained in the last section, the shock front rapidly loses velocity over distance, whereas fragment deceleration due to air resistance is more gradual. Hence, the effective range of fragment-induced damage, whilst dependent on the characteristics, typically extends beyond the radius of blast-induced damage. A further consideration involves fragments generated by blast-driven devices such as shells, pipe bombs, or improvised explosive devices (Hazell, 2022). These sources produce fragments with diverse shapes and velocities. Accurately assessing the damage potential of such fragments requires the understanding of their initial motion and the mechanics of their acceleration during the explosive events.

11.4.1 Initiation and Propagation of Fragments

The Gurney model serves as the standard method to determine the speed of casing fragments that result from detonation. The semi-empirical model serves as a practical tool to calculate fragment velocities through explosive mass and ordnance shape measurements. The model makes an ideal assumption about energy distribution between fragments and detonation products to simplify the calculation of initial fragment velocities.

11.4.1.1 Gurney Model

During the Second World War, Gurney developed a simplified approach in estimating the terminal velocity of casing accelerated by explosive charges (Gurney, 1943). The experimental results revealed that the key parameters governing fragment velocity were the mass of the casing material (M) and the mass of the explosive charge (C). Gurney assumed that the chemical energy released during detonation was converted into kinetic energy, shared between the explosive products and the accelerating metal fragments. Based on this assumption, a set of expressions were established to correlate the fragment velocity with the mass-to-charge ratio (M / C).

11.4.1.1.1 Cylindrical Geometry

The detonation gases are assumed to exhibit a linear velocity distribution from the centre to the outside of the explosive cylinder. The quantity of interest is the initial velocity of the surrounding fragments, denoted as v_0. In this formulation, some of the factors are neglected, for instance, the propagation of stress waves through the material, the energy dissipated in plastic deformation and fracture, and the finite duration of the detonation front as it travels along the length of the cylinder. The velocity of the gas v can be expressed as

$$v = v_0 \left(\frac{r}{a} \right) \tag{11.45}$$

where a is the internal radius of the cylinder and r is the general radius by which the state of the v is represented. The total kinetic energy (KE) is equivalent to the sum of the kinetic energy of the external shell and the kinetic energy of the gases:

$$KE = \frac{1}{2}\sum m_i v_0^2 + \frac{1}{2}\int v^2\, dm_g \tag{11.46}$$

where m_i is the mass of fragments and m_g is the mass of the gases. For a cylindrical shape function, the element of the mass of gases dm_g can be expressed as

$$dm_g = \rho dv = 2\pi\rho dr \tag{11.47}$$

where ρ is the density of the explosive. If all the fragments are assumed to have the same velocity and we substitute Equation (11.45) into Equation (11.46), then

$$KE = \frac{1}{2}Mv_0^2 + \frac{1}{2}\int_0^a v_0^2 \frac{r^2}{a^2} 2\pi r\rho dr \tag{11.48}$$

Simplified to

$$KE = \frac{1}{2}Mv_0^2 + \frac{1}{4}\pi v_0^2 a^2 \rho \tag{11.49}$$

And the density of the explosive can be expressed as

$$\rho = \frac{C}{\pi a^2} \tag{11.50}$$

Therefore, Equation (11.49) turns to

$$KE = \frac{1}{2}Mv_0^2 + \frac{1}{4}Cv_0^2 \tag{11.51}$$

Based on the assumption that no energy is dissipated in plastic deformation and fracture, the kinetic energy is calculated therefore equal to the chemical energy of the explosive. Hence,

$$EC = \frac{1}{2}Mv_0^2 + \frac{1}{4}Cv_0^2 \tag{11.52}$$

Rearranging Equation (11.52) into the expression of the initial velocity, then

$$v_0^2 = \frac{2EC}{M + \frac{1}{2}C} \tag{11.53}$$

or

$$v_0 = \sqrt{2E}\left(\frac{M}{C} + \frac{1}{2}\right)^{-\frac{1}{2}} \tag{11.54}$$

The Gurney equations are normally presented in the above form, where the initial velocity is shown as a function of mass-to-charge ratios. The term $\sqrt{2E}$ always appears in Gurney equations and has the units of velocity.

11.4.1.1.2 Spherical Geometry

The assumption of the kinetic energy of fragments and gases equalling the chemical energy of the explosive remains the same. Hence,

$$KE = \frac{1}{2}Mv_0^2 + \frac{1}{2}\int v^2\, dm_g \tag{11.55}$$

For a spherical shape function, the element of the mass of gases dm_g can be expressed as

$$dm_g = 4\pi r^2 \rho dr \tag{11.56}$$

Like cylindrical geometry, all the fragments are assumed to have the same velocity, then

$$KE = \frac{1}{2}Mv_0^2 + \frac{1}{2}\int_0^a v_0^2 \frac{r^2}{a^2} 4\pi r^2 \rho dr \tag{11.57}$$

Simplified to

$$KE = \frac{1}{2}Mv_0^2 + \frac{2}{5}\pi v_0^2 a^3 \rho \tag{11.58}$$

For a spherical shape,

$$v = \frac{4}{3}\pi a^3 \text{and } 4\pi a^3 \rho = 3v\rho = 3C \tag{11.59}$$

Substituting Equation (11.59) into Equation (11.58), then

$$v_0 = \sqrt{2E}\left(\frac{M}{C} + \frac{3}{5}\right)^{-\frac{1}{2}} \tag{11.60}$$

The derivations for other geometries, such as symmetric flat sandwich shape, open-faced sandwich shape, and asymmetric sandwich shape, can also follow the similar procedures shown in the above two geometries. One thing to note is that the zero velocity is not assigned to the centre of the explosive charge for the asymmetric sandwich condition. Instead, the velocity distribution is expressed as a linear variation fashion with the zero velocity assigned by considering the mass of the case distribution on both sides. The Gurney equations for various geometries are shown in Figure 11.7.

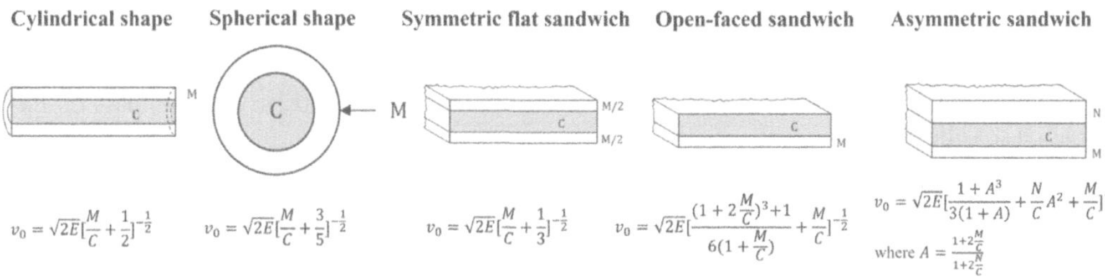

FIGURE 11.7 Gurney equations for different geometries. (Adapted from Hazell (2022).)

TABLE 11.8
Gurney Constants for Certain Explosives

Explosive	Density (g / cm^3)	E (kcal / g)	$\sqrt{2E}$ (mm / μs)
RDX	1.77	1.03	2.93
		0.96	2.83
Composition C-3	1.60	0.86	2.68
TNT	1.63	0.67	2.37
		0.71	2.44
Composition B (RDX/TNT)	1.72	0.87	2.71
		0.91	2.77
		0.87	2.70
		0.86	2.68
HMX	1.89	1.06	2.97
PBX-9404	1.84	1.01	2.90
Tetryl	1.62	0.75	2.50
TACOT	1.61	0.54	2.12
Nitro methane	1.14	0.69	2.41
PETN	1.76	1.03	2.93
EL506D	1.46	0.62	2.28
		0.75	2.50
EL506L	1.56	0.58	2.20

Source: Data adapted from Liu et al. (2018).

11.4.1.1.3 Gurney Constant

In the derivation of initial velocities for various geometries shown in Figure 11.7, a parameter keeps cropping up, namely $\sqrt{2E}$. This term is commonly referred to as the Gurney constant or Gurney energy, which actually has the units of velocity. The Gurney constant for a given explosive can be determined experimentally by measuring the terminal velocities of metal plates or shells accelerated by detonation (Kennedy, 1970). It is important to know that Gurney model only estimates the terminal velocity attained by the flyer plate, not the full acceleration history. In practical applications, the flyer initiates motion from rest and gains velocity progressively. Hence, a sufficient standoff distance has to be maintained between the flyer plate and the target to allow for proper acceleration. Representative values of Gurney constants for different types of explosives are listed in Table 11.8.

Example 11.3

Calculate the initial velocity of the metal from a symmetric flat sandwich bomb that is made from steel ($\rho_{steel} = 7800\ kg / m^3$) and is 400 mm in length, 100 mm in height, and 200 mm in width with the thickness of 5 mm for each plate. Assume the internal section is fully filled with HMX.

SOLUTION:

To calculate the initial velocity from Gurney model, the mass of explosive and the mass of the casing shall be calculated first.

The mass of the explosive charge (C):

$$C = \rho_{HMX} \times V_{HMX} = 1890 \times 0.4 \times 0.2 \times (0.1 - 0.005 \times 2) \approx 13.61\ kg$$

The mass of the steel case (M):

$$M = \rho_{steel} \times V_s = 7800 \times 0.4 \times 0.2 \times 0.005 \times 2 = 6.24\ kg$$

Based on Table 11.8, Gurney constant of HMX is 2.97 mm / μs.
According to the shape function for a symmetric flat sandwich (Figure 11.7),

$$V_0 = \sqrt{2E}\left[\frac{M}{C}+\frac{1}{3}\right]^{-1/2} = 2.97\times\left(\frac{6.24}{13.61}+\frac{1}{3}\right)^{-0.5} \approx 3.338\ mm/\mu s = 3338\ m/s$$

Example 11.4

Calculate the initial velocity of the metal casing from a spherical bomb that is made from steel ($\rho_{steel} = 7800\ kg/m^3$) with an external diameter of 300 mm and an internal diameter of 280 mm. Assume that the internal part is fully filled with TNT.

SOLUTION:

To calculate the initial velocity from Gurney model, the mass of explosive and the mass of the casing shall be calculated first.

The mass of the explosive charge (*C*):

$$C = \rho_{TNT}\times V_{TNT} = 1630\times 0.0115 \approx 18.75\ kg$$

The mass of the steel case (*M*):

$$M = \rho_{steel}\times V_s = 7800\times(0.0141-0.0115) = 20.28\ kg$$

Based on Table 11.8, Gurney constant of TNT is 2.37 mm / μs.
According to the shape function for a symmetric flat sandwich (Figure 11.7),

$$V_0 = \sqrt{2E}\left[\frac{M}{C}+\frac{3}{5}\right]^{-1/2} = 2.37\times\left(\frac{20.28}{18.75}+\frac{3}{5}\right)^{-0.5} \approx 1.828\ mm/\mu s = 1828\ m/s$$

11.4.1.2 Fragment Size Prediction

For an explosive weapon in the real world, fragment distribution around the casing is inherently non-uniform. The lateral regions normally produce the highest fragment density, whilst the nose and base produce fewer but significantly larger fragments. Although it is challenging to predict the exact fragment size and quantity resulting from a detonation, simplifying assumption of uniform distribution is often adapted to facilitate analytical solution. Mott's fragmentation theory remains a fundamental model in describing natural fragmentation process. Fragmentation in general should be caused by material microstructure rather than by pre-formed notches or manufactured defects. Mott claimed that in an ideal situation, when a homogeneous cylinder casing is subjected to internal blast loading, it would undergo uniform expansion until fragmentation occurs, leading to an infinite number of fragments (Mott, 1947). However, in reality, once a crack initiates, the associated free surface causes a localised stress release. This generates a release wave that propagates through the material, impeding further crack growth. The eventual number and size of fragments are therefore governed by the interplay between the rate of hoop strain development and the velocity of this release wave (Hazell, 2022). Mott claimed that the average fragment length is proportional to a number of properties, which can be expressed as

$$l_f = \sigma_f\sqrt{\frac{1+\varepsilon_f}{\rho_0 B}} \tag{11.61}$$

where l_f is the average fragment length, σ_f is the true stress at fracture of the shell casing, ε_f is the plastic strain at facture, ρ_0 is the density of the casing, and B is the strain hardening characteristic of the casing from the true stress-strain curve of a material:

$$\sigma = Y_0 + B\ln(1+\varepsilon) \tag{11.62}$$

where σ is the true stress, ε is the true strain, and Y_0 is the yield strength. It is important to note that this standard model does not account for the anomalously large fragments often originating from the nose or base sections of a warhead (Driels, 2021). On the other hand, the number of fragments with a specific known mass can be estimated by the two-dimensional Mott distribution:

$$N = \frac{Mexp\left[-\left(2m_{min} / m_0\right)^{0.5}\right]}{m_{min}} \tag{11.63}$$

where N is the number of fragments, M is the mass of the casing, m_{min} is the mass of the smallest fragment, and m_0 is the average mass of fragments, which can be calculated by

$$m_0 = \frac{CD_0^2}{v_0^2} \tag{11.64}$$

where C is a constant of 60.10^6, D_0 is the outside diameter of shell, and v_0 is the initial fragment velocity.

Another fragment size prediction method has been carried out by Grady (1982), by which the model has shown a very good fit to the scenario where armour is attacked by an IED. In this study, the average diameter of the fragment could be linked to the strain rate by

$$d = \left(\frac{\sqrt{20}K_{Ic}}{\rho_0 c_0 \dot{\varepsilon}}\right)^{2/3} \tag{11.65}$$

where d is the fragment diameter, K_{Ic} is the fracture toughness, ρ_0 is the density of the casing, c_0 is the speed of sound in the casing material, and $\dot{\varepsilon}$ is the strain rate.

Huang et al. (2021) have proposed a method for predicting the fragment size of an ultra-ductile metallic casing:

$$m_{fi} = 2\tilde{A}^2 \frac{\breve{\vartheta}_i^2 \left(r_i + \vartheta_i\right)^3}{r_i^2}\left(1 + 0.5\alpha_i\right) \tag{11.66}$$

where r_i is the average internal radius, ϑ_i is the average thickness of the shell, $\tilde{A}$ is the constant related to the type of the explosive [e.g. $\tilde{A} = 0.53\left(g \cdot cm^{-3}\right)^{1/2}$ for comp B, $0.42\left(g \cdot cm^{-3}\right)^{1/2}$ for TNT], and α_i is the ratio of C/M. The total number of fragments can be simply estimated by $N = M / m_{fi}$. One thing to note is that this particular estimation method should only be considered in the ultra-ductile material case.

11.4.1.3 Drag on Fragments

All objectives in motion through the atmosphere, including explosive fragments, are subjected to aerodynamic drag. This resistive force acts opposite to the direction of motion and significantly influences the trajectory and velocity decay of the fragments. A full derivation of the velocity from the drag force is given by (Hazell, 2022). The decelerated velocity on a freely travelling fragment, such as a projectile or ballistic body, for a certain travelling distance can be described by the following expression (Hazell, 2022):

$$v = v_0 \exp\left[-C_D \rho_{air} x \left(\frac{A}{2m}\right)\right] \tag{11.67}$$

where v is the velocity of the fragment after travelling distance x, v_0 is the initial velocity at detonation and can be estimated by the Gurney model, m is the mass of the fragment, ρ_{air} is the air density, A is the presented area of the fragment, and C_D is the coefficient of drag on the fragment. The term A and C_D

in the above equation can be estimated by following the material characterisation of the fragment. The term A is the projected area of the fragment during the travelling in air, if the fragment is a bullet from a gunshot (where the fragment does not rotate during the travelling), then the area A can be calculated by

$$A = \pi \frac{d^2}{4} \tag{11.68}$$

where d is the diameter of the bullet. If the fragment is generated from the blast loading of a detonation of explosive, the area A can be estimated through a mathematical expectation:

$$A = \hat{F} m^{2/3} \tag{11.69}$$

and,

$$\hat{F} = \begin{cases} 3.07 \times 10^{-3}, \textit{ sphere shaped fragment} \\ 3.09 \times 10^{-3}, \textit{ cubic shaped fragment} \\ 1.03 \times 10^{-3} \cdot \dfrac{1.446 + 1.844(l/d)}{(l/d)^{2/3}}, \textit{ cylindrical fragment} \\ 1.03 \times 10^{-3} \cdot \dfrac{(l_1/l_3)(l_2/l_3) + (l_1/l_3) + (l_2/l_3)}{(l_1 l_2 l_3^2)^{2/3}}, \textit{ rectangular (cuboid) shape} \end{cases} \tag{11.70}$$

The drag coefficient C_D is a dimensionless parameter that varies with both the geometry of the fragment and its velocity. A higher value of C_D corresponds to increased aerodynamic resistance, resulting in a more rapid reduction in fragment velocity. Particularly, C_D exhibits significant variation in the transonic regime (approximately Mach 0.8 to Mach 1.2), introducing additional complexity to accurate predictions, as shown in Figure 11.8(a). Nevertheless, for simplified estimations, the velocity decay of a fragment due to drag can be reasonably approximated by assuming a constant value of C_D based on the shape of the fragment only, as shown in Figure 11.8(b).

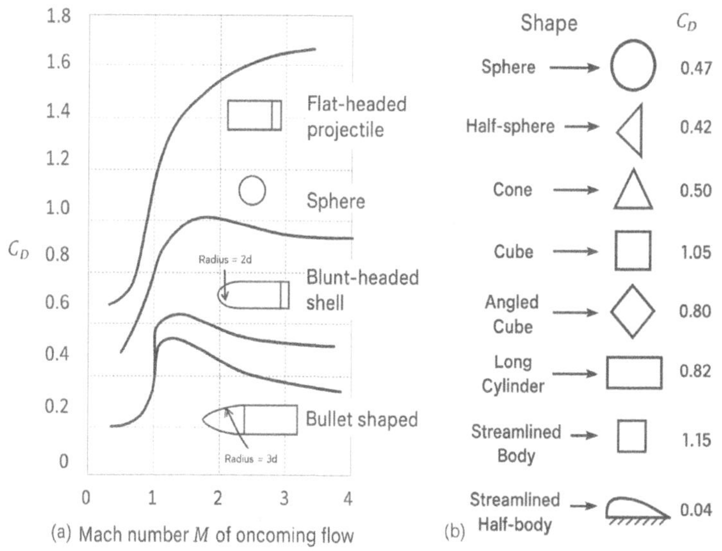

FIGURE 11.8 Drag coefficient of various shapes of fragments: (a) Drag coefficient as a function of Mach number; (b) constant values of C_D based on the geomoties of gragments (conventional method). (Adapted from Hazell (2022) and Massey and Ward-Smith (1998).)

Example 11.5

A cubic-shaped fragment with a mass of 20 g is flying from an explosion with an initial velocity of 500 m/s. Assuming a $C_D = 1.05$, calculate the velocity drop-off (residual velocity) in the air at 1000 m.

SOLUTION:

To calculate the velocity drop-off, the presented area A should be calculated first.

$$A = \hat{F}m^{2/3} = 3.09 \times 10^{-3} \times 0.02^{2/3} = 2.28 \times 10^{-4} m^2$$

Based on Equation (11.67),

$$v = v_0 \exp\left[-C_D \rho_{air} x \left(\frac{A}{2m}\right)\right]$$

$$= 500 \times \exp\left[-1.05 \times 1.225 \times 1000 \times \left(\frac{2.28 \times 10^{-4}}{40 \times 10^{-3}}\right)\right] = 0.327 \ m/s$$

Example 11.6

A sphere-shaped fragment with a mass of 10 g is flying from an explosion with an initial velocity of 800 m/s. Calculate the velocity drop-off in the air at 100 m.

SOLUTION:

To calculate the velocity drop-off, the presented area A should be calculated first.

$$A = \hat{F}m^{2/3} = 3.07 \times 10^{-3} \times 0.01^{2/3} = 1.42 \times 10^{-4} m^2$$

The drag coefficient C_D is not provided, the value of 0.47 can be obtained from Figure 11.8. Based on Equation (11.67),

$$v = v_0 \exp\left[-C_D \rho_{air} x \left(\frac{A}{2m}\right)\right]$$

$$= 800 \times \exp\left[-0.47 \times 1.225 \times 100 \times \left(\frac{1.42 \times 10^{-4}}{20 \times 10^{-3}}\right)\right] = 530 \ m/s$$

11.4.2 Modelling of Fragment Penetration

The computational analysis of explosively formed fragment penetration serves as a fundamental tool to study material and armour responses under high-velocity impact conditions. The simulation of projectile–target interactions enables researchers to conduct effective studies about penetration depth and damage evolution and residual fragment behaviour. The method provides an efficient way to assess protection systems and structural resilience through simulations instead of experimental testing limitations. A thorough analysis demands precise representations of fragments and target materials which requires suitable solid EOS and dependable constitutive models.

11.4.2.1 Structural Material Models

In the computational simulations of fragment–target interactions under blast and impact loading conditions, materials often exhibit large deformation and nonlinear behaviour due to the intensity of dynamic loads. The use of an appropriate dynamic constitutive model is therefore of importance in accurately capturing the response of materials under such extreme conditions. This section presents a selection of commonly used and well-established dynamic constitutive models for different types of materials.

11.4.2.1.1 Johnson–Cook Model

The JC strength model was developed in the early 1980s to describe the plastic flow behaviour of metallic materials under extreme conditions, such as high strain rates, large deformation, and elevated temperatures (Johnson and Cook, 1983). It is still commonly used nowadays in computational modelling of high-velocity impact, penetration, and explosion events. The general form of the JC model is expressed as

$$\sigma = \left[A + B\varepsilon_p^n\right]\left[1 + C\ln\left(\frac{\dot{\varepsilon}_p}{\dot{\varepsilon}_0}\right)\right]\left[1 - \left(\frac{T - T_0}{T_m - T_0}\right)^m\right] \tag{11.71}$$

where σ is the equivalent stress, A is the yield stress, B is the strain hardening modulus, C is a constant of strain rate sensitivity coefficient, n is a constant of strain hardening exponent, m is a constant of thermal softening exponent, T is the current temperature, T_0 is the original (room) temperature, T_m is the melting temperature of the metal, ε_p is the equivalent plastic strain, $\dot{\varepsilon}_p$ is the plastic strain rate, and $\dot{\varepsilon}_0$ is the reference strain rate. The parameters A, B, and n can be obtained from quasi-static tests at room temperature. The parameter C is calibrated using high strain rate tests, such as split Hopkinson pressure bar tests. The parameter m is typically obtained from elevated temperature tests at constant strain rate. Table 11.9 lists these parameters for different metals.

11.4.2.1.2 Zerilli–Armstrong Model

The Zerilli–Armstrong (ZA) model was developed to address the limitations of empirical models in capturing the influence of microstructure on the dynamic plastic behaviour of metals (Zerilli and Armstrong, 1994). The model was conducted based on the dislocation mechanics in the microstructural level for both body-centred cubic (BCC) and face-centred cubic (FCC) metals. The mechanisms of thermal activation, strain rate sensitivity, and dislocation interactions were incorporated into this semi-empirical model to describe the plastic flow stress (Zerilli and Armstrong, 1987). The general form of the ZA model can be expressed as

$$\sigma = C_1\varepsilon_p^n \exp\left(-C_2T + C_3T\ln\dot{\varepsilon}_p\right) + C_4 + C_5d^{-1/2} + C_6\varepsilon_p^m \tag{11.72}$$

TABLE 11.9
JC Model Parameters for Different Metal Alloys

Material	A (MPa)	B (MPa)	n	C	m
AISI-1045 medium carbon steel	50	176	0.517	0.095	0.662
Copper	90	292	0.310	0.025	1.090
Al 5083-H116	167	596	0.551	0.001	0.859
AZ31 Mg alloy	175	2332	0.350	0.063	1.000
Al 6061-T6	324	114	0.420	0.002	1.340
Inconel 601	424	1496	0.837	0.010	1.200
AA 7075-T651	527	575	0.720	0.017	1.610
Ti–6Al–4V	988	762	0.414	0.015	1.516
ARMOX 440T steel	1210	1543	0.584	0.014	1.000
AISI 4140 Chrome-Moly steel	1450	910	0.450	0.034	0.328

Source: Data adapted from Sirigiri et al. (2022).

The general form of the expression can be further categorised into two different expressions for BCC and FCC metals separately as

For BCC structure metals:

$$\sigma = C_1 \exp\left(-C_2 T + C_3 T \ln \dot{\varepsilon}_p\right) + C_4 + C_5 d^{-1/2} + C_6 \varepsilon_p^m \tag{11.73}$$

For FCC structure metals:

$$\sigma = C_1 \varepsilon_p^n \exp\left(-C_2 T + C_3 T \ln \dot{\varepsilon}_p\right) + C_4 + C_5 d^{-1/2} \tag{11.74}$$

where σ is the flow stress, T is the temperature, ε_p is the equivalent plastic strain, $\dot{\varepsilon}_p$ is the plastic strain rate, C_1, C_2, C_3, C_4, C_5, C_6, n, and m are material-dependent constants. Table 11.10 shows the specific values of these material constants for several common metals.

11.4.2.1.3 Johnson–Holmquist Model

The Johnson–Holmquist (JH-2) model was specifically developed to describe the dynamic response of brittle materials, such as ceramics, under extreme loading conditions. Unlike the above two models, the JH-2 model is aimed at capturing the features of brittle failure, including pressure-dependent strength, progressive damage evolution, and strain rate sensitivity (Johnson and Holmquist, 1994). Its mathematical expression enables the numerical simulation of the material behaviour from initial intact state through damaged response to the ultimate failure state. The general form of the JH-2 model can be expressed as

$$\sigma^* = \sigma_i^* - D\left(\sigma_i^* - \sigma_f^*\right) \tag{11.75}$$

where σ^* is the normalised equivalent stress, σ_i^* is the intact strength, σ_f^* is the damaged strength, and D is the damage factor ($0 \le D \le 1$). The normalised equivalent stress can be caused by

$$\sigma^* = \frac{\sigma}{\sigma_{HEL}} \tag{11.76}$$

where σ_{HEL} is the Hugoniot elastic limit stress and σ is the stress that can be calculated from:

$$\sigma = \sqrt{1/2\left[\left(\sigma_x - \sigma_y\right)^2 + \left(\sigma_x - \sigma_z\right)^2 + \left(\sigma_y - \sigma_z\right)^2 + 6\left(\tau_{xy}^2 + \tau_{yz}^2 + \tau_{xz}^2\right)\right]} \tag{11.77}$$

TABLE 11.10
ZA Model Parameters for Different Metal Alloys

Material	C_1 (MPa)	C_2 (K^{-1})	C_3 (K^{-1})	C_4 (MPa)	C_5 (MPa / mm)	C_6 (MPa)	n	m
Copper	890	0.0028	0.000115	46.5	5.0	0	0.5	0
Tungsten	2490	0.0666	0.000307	20	25.6	860	0	0.443
Molybdenum	937	0.0036	0.000107	0	22.7	647	0	0.401
Zirconium	600	0.0024	0.000132	21	7.9	76	0	0.510
Titanium	1100	0.0023	0.000170	54	14.9	300	0	0.500
Iron	1033	0.0069	0.000415	0	22.0	266	0	0.289

Source: Adapted from Ning and Ma (2015).

and the intact and damaged strength from the general form can be defined as

$$\begin{cases} \sigma_i^* = A(P^* + T^*)^N (1 + C \ln \dot{\varepsilon}^*) \\ \sigma_f^* = B(P^*)^M (1 + C \ln \dot{\varepsilon}^*) \end{cases} \tag{11.78}$$

where A, B, C, M, and N are material-dependent constants. P^* is the normalised pressure (P / P_{HEL}), T^* is the normalised tensile hydrostatic pressure (T / P_{HEL}), T is the maximum tensile hydrostatic pressure, and $\dot{\varepsilon}^*$ is the normalised strain rate.

The damage factor D presents the extent of damage for the brittle material, where $D = 0$ represents the intact state and $D = 1$ indicates the complete damage state. The factor D can be calculated by

$$D = \sum \Delta\varepsilon_p / \varepsilon_p^f \tag{11.79}$$

where ε_p is the plastic strain and ε_p^f is a function of pressure, normally expressed as

$$\varepsilon_p^f = D_1 (P^* + T^*)^{D_2} \tag{11.80}$$

in which D_1 and D_2 are the material-dependent constants. Table 11.11 shows the JH-2 model parameters for AD-90 Alumina ceramics.

11.4.2.1.4 Drucker–Prager Model

The Drucker–Prager model was mainly developed to describe the yielding behaviour of pressure-sensitive and frictional materials, such as soils and concrete. It represents a pressure-dependent extension of the von Mises criterion for material whose shear strength increases with the confining pressure (Drucker and Prager, 1952). The general form of this criterion model can be expressed as

$$K = \alpha I_1 + \sqrt{J_2} \tag{11.81}$$

where I_1 is the first stress invariant, J_2 is the second invariant of the deviatoric stress tensor, α and K are the material constants related to the internal friction angle and cohesion by

$$\begin{cases} \alpha = \dfrac{2 \sin\phi}{\sqrt{3}(3 - \sin^2\phi)} \\ K = \dfrac{6c \cos\phi}{\sqrt{3}(3 - \sin^2\phi)} \end{cases} \tag{11.82}$$

TABLE 11.11
JH-2 Model Parameters for AD-90 Alumina

AD-90	$\rho(g/cm^3)$	$G(GPa)$	$T(GPa)$	A	B	C
	3.625	109.7	0.262	0.88	0.28	0.007
	D_1	D_2	$HEL(GPa)$	β	M	N
	0.02	0.83	5.3	1.0	0.60	0.64

Source: Adapted from Ning and Ma (2015).

where c is the cohesion and ϕ is the friction angle for the material. The criterion can be implemented with either associated or non-associated flow rules in the solver, like ABAQUS and LS-DYNA, with the utilisation of a plastic potential function.

11.4.2.1.5 Holmquist–Johnson–Cook Model

The Holmquist–Johnson–Cook (HJC) model was established to describe the dynamic behaviour of brittle-alike materials such as concrete and ceramics. It incorporates the JC and JH-2 model mentioned above to present a unified framework with consideration of strain rate dependence, pressure hardening, and damage evolution for impact events of concrete and other types of brittle materials (Holmquist et al., 1993). Its general form for concrete materials is expressed as

$$\sigma^* = \left[A(1-D)+BP^{*N}\right](1+C\ln\dot{\varepsilon}^*) \tag{11.83}$$

where σ^* is the normalised equivalent stress ($\sigma^* = \sigma / f_c^{'}$), $f_c^{'}$ is the quasi-static compressive strength of concrete, p^* is the normalised pressure ($P^* = P / f_c^{'}$), $\dot{\varepsilon}^*$ is the normalised dimensionless strain rate ($\dot{\varepsilon}^* = \dot{\varepsilon} / \dot{\varepsilon}_0$), A is the normalised cohesive strength, B is the normalised pressure hardening factor, C is the strain rate sensitivity coefficient, N is the pressure hardening exponent, and D is the accumulated damage factor. It is noted that the HJC model is often paired with a linear or polynomial EOS for the $P-\mu$ description. To better present the $P-\mu$ relationship of concrete under compression, the EOS of concrete can normally be categorised into three regions: linear elastic region, transition region, and dentification region. The EOS of concrete under compression can be expressed as

$$P = \begin{cases} \mu K_{elastic},\ when\ P < P_1\ (Linear\ elastic\ limit) \\ P_1 + \dfrac{P_2 - P_1}{\mu_2 - \mu_1}(\mu - \mu_1),\ when\ P_1 \le P \le P_2\,(Transition\ limit) \\ k_1\bar{\mu} + k_2\bar{\mu}^2 + k_3\bar{\mu}^3,\ when\ P > P_2\,(Dentification\ region) \end{cases} \tag{11.84}$$

and the EOS of concrete under tension can be expressed as

$$P = \begin{cases} \mu K_{elastic}\ (Linear\ elastic\ tensile\ region) \\ \left[(1-\bar{\bar{\mu}})K_{elastic} + \bar{\bar{\mu}}k_1\right]\mu\ (Tensile\ transition\ region) \\ k_1\mu\ (Dentification\ region) \end{cases} \tag{11.85}$$

where μ, $\bar{\mu}$, and $\bar{\bar{\mu}}$ in the compression and tension expressions are represented as

$$\mu = \frac{\rho}{\rho_0} - 1;\ \bar{\mu} = \frac{\mu - \mu_2}{1+\mu_2};\ \bar{\bar{\mu}} = \frac{\mu_{max} - \mu_1}{\mu_2 - \mu_1}$$

In addition, P_1 and μ_1 are the pressure and volumetric strain when the internal gaps are closed, P_2 and μ_2 is the upper limit of the true elastic pressure, $K_{elastic}$ is the bulk modulus of elasticity, k_1, k_2, and k_3 are the bulk modulus terms.

11.4.2.1.6 Blatz–Ko Model

The Blatz–Ko model was developed to describe the large deformation behaviour of hyperelastic compressible materials, such as rubber. The model is suitable for rubber-like materials that exhibit both significant volumetric changes and nonlinear stress-strain responses, which can be used in various scenarios like impact and vibration events (Blatz and Ko, 1962). The Blatz–Ko model utilises the deformation tensor and deformation gradient to define the strain energy, which can be expressed as

$$W = \frac{G}{2}\left(I_1 J^{-2/3} - 3\right) + \frac{K}{2}\left(J^2 - 1 - 2\ln J\right) \tag{11.86}$$

Which can also be simplified into the expression of stress tensor:

$$\sigma_1 = G\left(\frac{1}{V} I_1 - V^{\frac{1}{1-\upsilon}} \delta_1\right) \tag{11.87}$$

where W is the strain energy per unit volume, G is the shear modulus, K is the bulk modulus, I_1 is the right Cauchy–Green deformation tensor, J is the determinant of the deformation gradient, σ_1 is the Piola–Kirchhoff stress tensor, V is the specific volume, and υ is the Poisson's ratio.

11.4.2.2 Mie–Grüneisen Equation of State

Under shock or high strain-rate impact loading conditions, the material response is dominated by compressibility and irreversible heating, hence the constitutive strength models alone mentioned in the previous section are not able to fully predict the pressure field or the propagation of compressive and release waves without an EOS. Both fragment and target materials respond differently from their original states to the change of the conditions, including the change in pressure, temperature, and strain rate. An accurate EOS enables the relation needed to convert mechanical work and plastic dissipation into pressure and temperature changes, to compute post-shock states from Hugoniot data, and to model unloading and rarefaction behaviour. It also enables coupling between thermal and mechanical fields, supports realistic treatment of phase change or pore collapse where applicable, and is required for accurate prediction of wave speeds and transmitted loads. Many EOS have been proposed over the past several decades, for instance, Bridgman EOS (Bridgman, 1946) are used to describe the pressure–volume relationship of a material during an isothermal process whilst Murnaghan EOS (Murnaghan, 1944) are employed to describe the pressure–volume relationship during an isentropic process (Ning and Ma, 2015). However, for a high-velocity impact event, neither temperature nor entropy should be prioritised for analysing the material responses. Instead, a comprehensive EOS that can be used to link pressure, volume, and energy is more appropriate for the dynamic loading scenarios, such as the Mie–Grüneisen EOS (Mie, 1903), in which a new thermodynamic parameter Γ (Grüneisen coefficient) that represents the change in pressure with respect to internal energy at constant volume is introduced:

$$\Gamma = V\left(\frac{\partial P}{\partial e}\right)_v \tag{11.88}$$

According to the first and second law of thermodynamics,

$$de = Tds - Pdv \tag{11.89}$$

For an isentropic process ($ds = 0$),

$$\left(\frac{\partial e}{\partial T}\right)_V = T\left(\frac{\partial S}{\partial T}\right)_V \equiv C_V \tag{11.90}$$

The differential of volume can be expressed as

$$dV = \left(\frac{\partial V}{\partial P}\right)_T dP + \left(\frac{\partial V}{\partial T}\right)_P dT \tag{11.91}$$

From the above equations,

$$\left(\frac{\partial P}{\partial T}\right)_V = -\left(\frac{\partial P}{\partial V}\right)_T\left(\frac{\partial V}{\partial T}\right)_P = -V\left(\frac{\partial P}{\partial V}\right)_T \cdot \frac{1}{V}\left(\frac{\partial V}{\partial T}\right)_P = k_T\alpha \tag{11.92}$$

Thus, the Grüneisen coefficient becomes:

$$\Gamma = V\frac{\left(\frac{\partial P}{\partial T}\right)_V}{\left(\frac{\partial e}{\partial T}\right)_V} = \frac{k_T\alpha V}{C_V} \tag{11.93}$$

Assuming the Grüneisen coefficient Γ is only a function of volume, Equation (11.88) can be integrated into

$$P = \frac{\gamma}{V}e + \Pi(V) \tag{11.94}$$

or:

$$P = P_k(V) + \frac{\Gamma}{V}\left[\mathrm{e} - e_k(\mathrm{V})\right] \tag{11.95}$$

This is the Mie–Grüneisen EOS, which can be combined with reference state data from the Hugoniot curve:

$$P - P_H = \Gamma\rho\left(e - e_H\right) \tag{11.96}$$

The Rankine–Hugoniot energy relation defines

$$e_H - e_0 = \frac{1}{2}P_H\left(\frac{1}{\rho_0} - \frac{1}{\rho}\right) \tag{11.97}$$

where e_0 is the specific internal energy, and ρ_0 is the original density. Substituting Equation (11.97) into Equation (11.96),

$$P = P_H\left(1 - \frac{\Gamma\mu}{2}\right) + \Gamma\rho\left(e - e_0\right) \tag{11.98}$$

where μ is the compression, referred to as the ratio of the change in volume to the initial volume or expressed as $\mu = \rho / \rho_0 - 1$.

Based on the propagation of shock wave, the shock front can be expressed as

$$U_S = c_0 + \mathrm{S} \cdot u_P \tag{11.99}$$

The relationship between the shock front velocity U_S (sometimes referred to as D) and particle velocity u_P can be used in the mass conservation equation:

$$\rho\left(U_S - u_P\right) = \rho U_S \tag{11.100}$$

From the above equations, the shock front velocity and particle velocity can be expressed in terms of local sound of speed c_0 and compression μ:

$$U_S = \frac{c_0(1+\mu)}{1-(S-1)\mu} \tag{11.101}$$

and

$$u_P = \frac{c_0\mu}{1-(S-1)\mu} \tag{11.102}$$

Substituting Equations (11.101) and (11.102) into the momentum conservation equation gives the Hugoniot pressure,

$$P_H = \frac{\rho_0 c_0^2 \mu(1+\mu)}{[1-(S-1)\mu]^2} \tag{11.103}$$

Now let's combine Equations (11.103) and (11.98):

$$P = \frac{\rho_0 c_0^2 \mu(1+\mu)}{[1-(S-1)\mu]^2}\left(1-\frac{\Gamma\mu}{2}\right)+\Gamma\rho(e-e_0) \tag{11.104}$$

Equation (11.104) is the widely adapted form of Mie–Grüneisen EOS in computational software such as AUTODYN for describing the shock response of materials. The EOS input parameters for several common metallic materials are shown in Table 11.12.

11.4.2.3 Case Study: Numerical Modelling of Penetration Mechanisms

The penetration mechanics of explosive-formed fragments can be represented with the employment of the material models mentioned in the above section. Here in this section, a computational modelling of explosive formed fragments striking a bi-layer ceramic/metal target is performed using

TABLE 11.12
Mie–Grüneisen EOS Parameters for Common Metal Alloys

Material	$\rho_0\,(g/cm^3)$	$c_0\,(mm/\mu s)$	S	Γ
Silver	10.49	3.23	1.60	2.5
Gold	19.24	3.06	1.57	3.1
Copper	2.93	3.94	1.49	2.0
Steel	7.85	3.57	1.92	1.8
1006 Steel	7.89	4.57	1.49	2.2
Tin	7.29	2.61	1.49	2.3
Tungsten	19.22	4.03	1.24	1.8
Zinc	7.14	3.01	1.58	2.1
Al 2024	2.79	5.33	1.34	2.0
Al 6061	2.70	5.35	1.34	2.0
Al 7075-T6	2.80	5.20	1.36	2.2
Brass	8.45	3.73	1.43	2.0

Source: Adapted from Ning and Ma (2015).

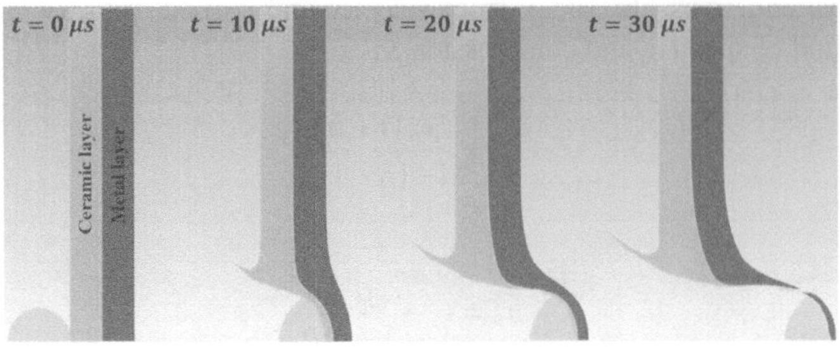

FIGURE 11.9 Schematic diagram of the fragment penetration progress.

ANSYS AUTODYN. The performance of two different shapes of fragments, of cylindrical one and spherical one, was evaluated and compared in this analysis with a diameter of 10 mm for steel sphere and cylinder (height of 6.5 mm). Both impactors have been assigned with the same mass (4 g) and velocity (1500 m/s), and the target is designed to be 100 mm in diameter for a circular-face bi-layered sample with the ceramic face to be the initial contact face. Mesh sizes of 1 × 1 mm for both impactor and target have been selected to balance the computational cost and accuracy. Symmetric boundary conditions have been assigned accordingly. The schematic diagram of the penetration progress in AUTODYN is illustrated in Figure 11.9.

Three different material models, including steel 4030, Al 6061-T6, and Al_2O_3-99.5, have been employed in this analysis, detailed input parameters for the corresponding constitutive models are shown in Table 11.13.

TABLE 11.13
Input Parameters of Material Models in AUTODYN

Steel 4340	EOS: Linear	$K(\text{GPa})$	$T_0(\text{K})$	$C_V(\text{J/kg}\cdot\text{K})$	$\rho(\text{g/cm}^3)$	$G(\text{GPa})$	Θ_ε
		159	300	477	7.83	77	2.0
	JC model	$A(\text{MPa})$	$B(\text{MPa})$	$\dot{\varepsilon}_0(\text{s}^{-1})$	n	C	m
		792	510	1.0	0.26	0.014	1.03
	JC damage	$T_{melt}(\text{K})$	D_1	D_2	D_3	D_4	D_5
		1793	0.05	3.44	−2.12	0.002	0.61
Al 6061-T6	EOS: Shock	$C_1(\text{m/s})$	$T_0(\text{K})$	$C_V(\text{J/kg}\cdot\text{K})$	$\rho(\text{g/cm}^3)$	Γ	S_1
		5240	300	885	2.70	1.97	1.4
	Steinberg Guinan model		$G(\text{GPa})$	$Y_{max}(\text{MPa})$	$Y_0(\text{MPa})$	$T_{melt}(\text{K})$	β
			27.6	680	290	1220	125
			Θ_ε	$G'_T(\text{Pa/K})$	G'_P	G'_Y	n
			0.5–2.0	−17000	1.8	0.0189	0.1
Al_2O_3-99.5	EOS: Polynomial	$K(\text{GPa})$	$A_2(\text{MPa})$	$A_3(\text{MPa})$	$T_1(\text{MPa})$	$\rho(\text{g/cm}^3)$	Θ_ε
		231	−160	2774	231	3.89	2.0
	JH-2 model	$G(\text{GPa})$	$HEL(\text{MPa})$	A	N	C	B
		152	6.57	0.88	0.64	0.007	0.28
	JH-2 damage	M	F_{max}	$HTL(\text{kPa})$	D_1	D_2	β
		0.6	1.0	−262	0.01	0.7	1.0

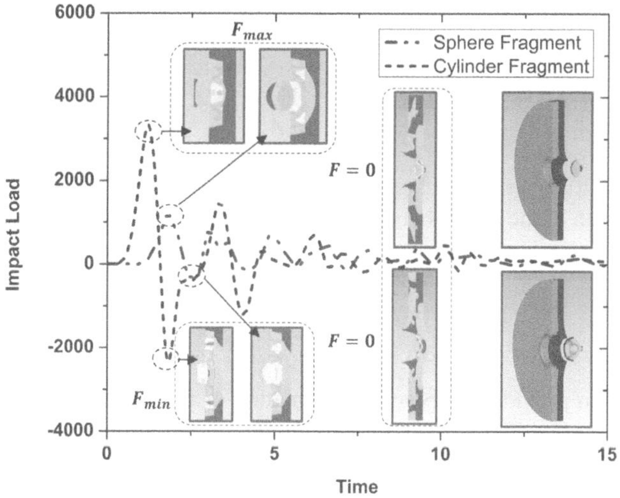

FIGURE 11.10 Schematic diagram of the loading histories.

The loading histories associated with the fragment penetration are illustrated in Figure 11.10. The maximum impact force occurs in the initial phase of penetration, when the stress wave has not yet reached the back layer. At this stage, both fragments are mainly resisted by front ceramic layer. The magnitude of the peak load is significantly affected by the geometry of the fragment, whereby the maximum load from the cylindrical fragment is approximately four times greater than that of the spherical fragment. The minimum load is observed when the compressive stress wave reaches the free end surface, and a release wave is reflected back towards the impactor. The impact force drops to zero at approximately 8 μs, and complete perforation of the bi-layered target is observed at around 12 μs for both types of fragments.

11.4.3 Modelling of Fragmentation-Induced Damage

High-velocity fragments or debris from explosions create serious threats to structures and personnel through fragmentation-induced damage. The prediction of such damage through computation depends on the calculated distribution of fragments and their velocities and masses. The risk assessment depends on hit probability analysis to determine fragment impact probability on targets and vulnerability evaluation to measure damage severity after impact. These components create a functional framework for studying and forecasting fragmentation effects.

11.4.3.1 Terminal Effects: Hit Probability of Fragmentation

In assessing the terminal effects of fragmentation, one of the key factors is to determine the probability that fragments will strike a target. If it is assumed that fragments are dispersed isotopically in all directions, then the proportion of the fragments generated likely to hit a given target can be estimated based on a geometric consideration.

The surface area over which the fragments are distributed is the surface area of a sphere centred at the point of detonation, given by

$$\textit{Total Surface Area} = 4\pi r^2 \tag{11.105}$$

where r is the distance between the detonation point and the target. If a target is presented with a projected area A, for example, the frontal cross-section of a vehicle or structure, the fraction of fragments expected to strike the target:

$$P_{hit} = \frac{A}{4\pi r^2} \tag{11.106}$$

where P_{hit} is the fraction of the hit probability. To determine the actual number of fragments impacting the target, this fraction is multiplied by the total number of fragments produced by the warhead:

$$N_{impact} = N_{total} \cdot \frac{A}{4\pi r^2} \tag{11.107}$$

This simple relation provides a first-order estimation of the interaction between fragments and target based solely on geometry and range, forming the basis for more advanced models involving lethality thresholds and vulnerability analysis.

11.4.3.2 Vulnerability Assessment for Fragmentation Effects

Once the number of impacting fragments has been estimated, the next step in assessing the fragmentation effect involves determining the extent of damage these fragments inflict upon various types of targets. Central to this evaluation is the kinetic energy of each fragment at the point of impact. The kinetic energy E_k of a fragment can be calculated by using the kinetic expression:

$$E_k = \frac{1}{2}mv^2 \tag{11.108}$$

where m is the mass of a fragment, and v is the velocity upon reaching the target after a certain distance, and such value can be obtained from the drag on fragments section. This kinetic energy value represents the energy delivered to the target and forms the basis for quantifying the likelihood and severity of resulting damage.

Different types of targets, such as unprotected personnel, armoured vehicles, or reinforced concrete structures, correspond to different thresholds of the vulnerability. Tables 11.14 and 11.15 summarise the expected extent of damage for various categories, including personnel, aircraft, and armoured vehicles, for their respective vulnerability assessments. These thresholds are typically represented in the tabulated forms, indicating certain kinetic energy levels required to achieve specific damage outcomes. One thing to note is that these tables only provide a relatively coarse estimation and can only be used as a reference table. Exact fragmentation effect analysis should be conducted on a case-by-case basis. For example, consider a scenario in which a person is struck by a 5-mm

TABLE 11.14
Fragmentation Damage Effects for Different Types of Targets

	Kinetic Energy of Fragment (kJ)		
Target	**Light Damage (P_{hit} = 0.1)**	**Moderate Damage (P_{hit} = 0.5)**	**Heavy Damage (P_{hit} = 0.9)**
Personnel	0.1	1	4
Aircraft	4	10	20
Armoured vehicle	10	500	1000

Source: Adapted from Huang et al. (2021).

TABLE 11.15
Fragmentation Initiation Damage for Various Types of Targets

Target	Minimum Mass of Fragment (g)	Kinetic Energy (J)
Personnel	1	8–10
Horse	4	19
Aircraft	–	25–40
Aircraft (metal case)	2–11.6	100–250
7 mm plate armour	–	200–220
10 mm plate armour	–	350
13 mm plate armour	–	590
16 mm plate armour	–	1040
Aircraft engine	25–50	90–135
Aircraft fuel tank	20–30	20–30

Source: Adapted from Huang et al. (2021).

diameter projectile weighing approximately 0.5 g, travelling at a velocity of 1170 m/s. Whilst the calculated energy of approximately 0.3 kJ falls between the threshold for light and moderate damage, it is important to note that such an impact is not likely to render the person a continuous threat or result in an immediate elimination, which highly depends on the location of the hitting locations.

11.5 CASE STUDIES OF EXPLOSIVE EFFECTS

Example 11.7

A hypothetical mortar round contains 2 kg of RDX explosive material inside a cylindrical body with a 2 kg steel casing shell which detonates 5 m away from a 10 m^2 projected area of an armoured fighting vehicle (AFV). The detonation takes place directly above a reinforced concrete surface that remains rigid. It is assumed that the steel casing can generate 500 evenly distributed spherical fragments. Evaluate and compare the blast effects and the cumulative fragmentation-induced damage on this AFV (use the sum/accumulated value of the fragment damage).

SOLUTION:

1. *Blast effect analysis*:
   Mass of charge should be calculated into TNT equivalence:

$$TNT\ equivalence: 2 \times 1.185 = 2.37\ kg$$

Scaled distance analysis:

$$Z = \frac{5}{2.37^{\frac{1}{3}}} = 3.75\ m / kg^{1/3}$$

Blast overpressure evaluation (for rigid ground condition, Equation (11.42)):

$$P_s = \frac{0.106}{Z} + \frac{0.43}{Z^2} + \frac{1.4}{Z^3}\ (1 \le Z \le 15) = 0.085\ \ MPa$$

0.085 *MPa* is close to the lower limit of the heavily armoured vehicle damage range (0.045 – 1.5 *MPa*); therefore, light damage is expected for the AFV to the blast effects.

2. *Fragmentation effect analysis*:
   The initial velocity of the fragments can be estimated by the Gurney model:

$$v_0 = \sqrt{2E}\left[\frac{M}{C}+\frac{1}{2}\right]^{-\frac{1}{2}} = 2.93\times\left[\frac{2}{2}+\frac{1}{2}\right]^{-\frac{1}{2}} = 2.392\ mm/\mu s = 2392\ m/s$$

   Note that the mass of the charge in the Gurney equation does not need to be converted to TNT equivalence.
   Characterisation of a single fragment:

$$C_D = 0.47; m = \frac{2000}{500} = 4\ g = 0.004\ kg$$

$$A = \hat{F}m^{2/3} = 3.07\times10^{-3}\times0.004^{2/3} = 7.736\times10^{-5}m^2$$

   Impact velocity of the fragments after travelling certain distance:

$$v = v_0 \exp\left[-C_D\rho_{air}x\left(\frac{A}{2m}\right)\right]$$

$$= 2392\times\exp\left[-0.47\times1.225\times5\times\left(\frac{7.736\times10^{-5}}{8\times10^{-3}}\right)\right] = 2326m/s$$

   Fragmentation damage evaluation:

$$\text{Total surface area}: 4\pi r^2 = 314\ m^2$$

$$N = \frac{10}{314}\times500\approx16$$

   Kinetic energy of a single fragment can be calculated as

$$KE = \frac{1}{2}mv^2 = 0.5\times0.004\times2326^2 = 10.8\ kJ$$

   Hence, the accumulative kinetic energy:

$$10.8\times16 = 172.8\ kJ;\ 10 < 172.8 < 500$$

According to Table 11.14, the value falls into the range of light damage criterion. Therefore, light damage is expected for the AFV to the fragmentation effects. In summary, light damage is expected for both blast and fragmentation effects.

Example 11.8

By considering the blast and fragmentation effects of a hypothetical 115 mm cylindrical tank gun projectile equipped with an impact fuze, predict a lethal range of this ordnance using the given information. The total mass of this projectile is 17.82 kg, which includes a steel case and explosive charge of 3.13 kg A-IX-2 (assuming the TNT equivalence conversion factor of A-IX-2 is 1.54 and the Gurney constant of A-IX-2 is 2.8 mm/μs), assuming 500 spherical fragments are all evenly generated from the case. Predict the theoretically effective lethality range (detonation not directly in contact with the target) of this projectile for a lightweight armoured vehicle with a projected area of 4 m^2.

SOLUTION:

1. *Basic information of this weapon*:
The total mass is 17.82 kg, and the mass of charge is 3.13 kg of A-IX-2 (C). Therefore, the mass of the outer casing (M) is 17.82 – 3.13 = 14.69 kg. The TNT equivalence of the charge: C = 1.54 × 3.13 = 4.82 kg. The shape of the fragments is spherical, and the total number of fragments is 500.
2. *Blast effect analysis*:
The type of fuze is impact, and the ground condition is assumed to be normal ground condition, based on Equation (11.41):

$$P_s = \frac{0.102}{Z} + \frac{0.399}{Z^2} + \frac{1.26}{Z^3} \quad (1 \le Z \le 15)$$

According to Table 11.5, 300 kPa, or 0.3 MPa is the upper limit of the overpressure range for destroying the vehicle. Hence,

$$0.3 = \frac{0.102}{Z} + \frac{0.399}{Z^2} + \frac{1.26}{Z^3} \quad (1 \le Z \le 15)$$

Using the mathematic solver function, such as goal seek function,

$$Z = 2.02\ m / kg^{1/3};\ Z = \frac{R}{W^{1/3}}$$

Therefore, the distance R can be calculated from the above equation:

$$R = 3.41\ m \approx 3.4\ m$$

The theoretically effective destroying range of 3.4 m is achieved for a blast overpressure of 300 kPa.
3. *Fragmentation effect analysis*:
The frontal (projected) area of the target (light-armoured vehicle) is 4 m^2 ($A_{frontal} = 4\ m^2$). According to Equations (11.106) and (11.107),

$$P_{hit} = \frac{A_{frontal}}{4\pi r^2} \times N_{frag}$$

For 100% hit probability,

$$r = \sqrt{\frac{N_{frag} \times A_{frontal}}{4\pi P_{hit}}} = 12.62\ m \approx 12.6\ m$$

Validation of the effective range in the fragmentation effect analysis:

Characteristics of fragment:

$$m = 14.69 / 500 = 0.02938\ kg$$

Size prediction of a single fragment:

$$Volume = \frac{m}{\rho} = \frac{0.2938}{7800} = 3.77 \times 10^{-6}\ m^3$$

The corresponding diameter of a single fragment can be calculated as 19.3 mm, which is considered to be a very large piece of projectile.

The projected area of one fragment:

$$A = \hat{F} m^{2/3} = 3.07 \times 10^{-3} \times 0.2938^{2/3} = 0.000292\ m^2$$

Initial velocity of fragments:

$$v_0 = \sqrt{2E}\left[\frac{M}{C}+\frac{1}{2}\right]^{-\frac{1}{2}} = 1229\ m/s$$

The impact velocity @ 12.6 m:

$$v = v_0 \exp\left[-C_D \rho_{air} x \left(\frac{A}{2m}\right)\right] = 1185\ m/s$$

The kinetic energy of a single piece of fragment:

$$E_k = \frac{1}{2}mv^2 = 21.1\ kJ$$

Since the reference data for light-armoured vehicles is not available, the data sourced from Table 11.14 is used in this study. Accordingly, such energy is not sufficient to cause complete destroying damage for the vehicle. The value of the calculated kinetic energy falls between the light and moderate damage mode. Hence, light to moderate damage is expected for this vehicle. If the upper damage value of 1000 kJ was selected for the destroy evaluation, the number of the hitting fragments would be approximately 47. According to Equations (11.106) and (11.107), the distance would be approximately 1.8 m, which is smaller than the blast range. Therefore, for this particular hypothetical weapon, the effective range for destroying the light-armoured vehicle is 3.4 m.

Example 11.9

A hypothetical air-to-air missile, equipped with a proximity fuze, detonated at a distance of 10 m from a fighter aircraft at an altitude of 2000 m. The temperature was 0°C. The missile explosive charge section is assumed to be located within a cylindrical 25 kg Octanitrocubane (ONC) body with 2.38 TNT equivalent conversion factor and 3.08 Gurney constant along with a 15 kg steel casing. Assume the explosive device produced 2000 evenly distributed spherical fragments all from the casing. The target aircraft exposed profile area reached 10 m². Assess and compare the blast-induced damage and fragmentation-induced damage on the aircraft. Use the cumulative (summed) fragmentation energy to evaluate the fragmentation effect. The air pressure can be estimated by

$$P_{air} = 100 \times \left(\frac{44331.514 - z}{11880.516}\right)^{1/0.1902632}$$

where P_{air} is the pressure in Pascals and z is the altitude in metres.

SOLUTION:

1. *Basic information analysis*:

$$T_0 = 0°C = 273K$$

$$P_0 = P_{air} = 100 \times \left(\frac{44331.514 - z}{11880.516}\right)^{1/0.1902632} = 79495.53\ Pa$$

According to Equation (11.1),

$$Pv = RT,\ hence\ \frac{1}{v_0} = \rho_0 = \frac{P_0}{RT_0} = \frac{79495.53}{287.1 \times 273} = 1.014\ kg/m^3$$

2. *Blast effect analysis*:
   Mass of charge should be calculated into TNT equivalence:

$$TNT\ equivalence: 25 \times 2.38 = 59.5\ kg$$

   Scaled distance:

$$Z = \frac{10}{59.5^{\frac{1}{3}}} = 2.56\ m/kg^{1/3}$$

   Overpressure estimation:

$$P_s = \frac{0.082}{Z} + \frac{0.265}{Z^2} + \frac{0.686}{Z^3}\ (1 \le Z \le 15) = 0.113 MPa > 0.1\ MPa$$

   According to Table 11.7, such a calculated value exceeds the limit of the aircraft damage, which indicates that *"all types of aircraft completely destroyed, corresponding to destruction mode or first order/degree damage"*; therefore, a heavy damage model or even complete destroy damage mode was expected for the blast effect.
3. *Fragmentation effect analysis*:

   Initial velocity calculation:

$$v_0 = \sqrt{2E}\left[\frac{M}{C} + \frac{1}{2}\right]^{-\frac{1}{2}} = 3.08 \times \left[\frac{15}{25} + \frac{1}{2}\right]^{-\frac{1}{2}} = 2.937\ mm/\mu s = 2937\ m/s$$

   Characterisation of a single fragment:

$$C_D = 0.47; m = \frac{15}{2000} = 0.0075\ kg$$

$$A = \hat{F}m^{2/3} = 3.07 \times 10^{-3} \times 0.0075^{2/3} = 1.176 \times 10^{-4} m^2$$

   Impact velocity @ 10 m,

$$v = v_0 \exp\left[-C_D \rho_{air} x\left(\frac{A}{2m}\right)\right]$$

$$= 2937 \times \exp\left[-0.47 \times 1.014 \times 10 \times \left(\frac{1.014 \times 10^{-4}}{0.0075 \times 2}\right)\right] = 2829\ m/s$$

   Fragmentation damage evaluation:

$$N = \frac{10}{4\pi r^2} \times 2000 \approx 16$$

$$KE = \frac{1}{2}mv^2 \times 16 = 478\ kJ > 20\ kJ$$

According to Table 11.14, such a calculated value exceeds the heavy damage limit of an aircraft. An overpressure of 0.113 MPa reached the aircraft which typically causes either heavy damage or complete destruction. The cumulative fragmentation energy reaches 478 kJ which indicates a heavy damage or complete destroy mode as well. Under these conditions, both the blast and fragmentation effects support a conclusion of heavy damage or destruction of the aircraft.

Example 11.10

A hypothetically general-purpose bomb is considered for an aerial training scenario. The ordnance is assumed to have a total mass of 200 kg comprising 80 kg TNT charge housed within a 120 kg steel casing. On detonation, the steel casing is assumed to shatter into 15,000 uniformly distributed, cube-shaped fragments.

a. Theoretically, what is the minimum safe distance for a soft target of a project area of 2 m^2 from this weapon, considering the fragmentation effect?
b. If the steel casing is replaced with a carbon fibre reinforced polymer (CFRP) casing, assuming the casing would disintegrate or melt upon detonation and not contribute to fragmentation, what would the revised minimum safe distance be under the same condition (by considering blast effect only)?

SOLUTION:

a. The frontal area of a soft target is 2 m^2, so the hit probability and range determination from the fragmentation effect:

$$P_{hit} = \frac{A_{frontal}}{4\pi r^2} \times N_{frag}$$

According to Table 11.14, the threshold limit for the light damage is 10%, then,

$$r = \sqrt{\frac{N_{frag} \times A_{frontal}}{4\pi P_{hit}}} = 154.5\ m$$

Validation of this range:
Characteristics of the fragments:

$$m = \frac{120}{15000} = 0.008\ kg$$

Size prediction of a single fragment:

$$Volume = \frac{m}{\rho} = \frac{0.008}{7800} = 1.026 \times 10^{-6} m^3$$

By which the size of a single fragment equivalents to a cubic fragment with an edge length of 10 mm, which is considered to be a reasonably large piece of projectile (it is still considered to be a lethal size of fragment for a soft target when the impact velocity is high).

The projected area of one fragment:

$$A = \hat{F} m^{2/3} = 3.09 \times 10^{-3} \times 0.008^{2/3} = 1.236 \times 10^{-4} m^2$$

The initial velocity of fragments:

$$v_0 = \sqrt{2E}\left[\frac{M}{C} + \frac{1}{2}\right]^{-\frac{1}{2}} = 1676\ m/s$$

Velocity @ 154.5 m:

$$v = v_0 \exp\left[-C_D \rho_{air} x\left(\frac{A}{2m}\right)\right] = 361\ m/s$$

Such impact velocity of a cubic fragment with an edge length of 10 mm is still to be considered as dangerous. Therefore, other criteria should be considered as well.

According to Table 11.14, the threshold limit of kinetic energy for the light damage of a soft target is 0.1 kJ, hence,

$$KE = \frac{1}{2}mv^2 = 0.5 \times 0.008 \times velocity^2 = 0.1\ kJ$$

$$v = 158\ m/s$$

Taking this velocity into Equation (11.67),

$$v = v_0 \exp\left[-C_D \rho_{air} x \left(\frac{A}{2m}\right)\right] = 158\ m/s$$

$$x = 238\ m$$

Therefore, a minimum safe distance of 238 m is theoretically considered through the fragmentation effect analysis.

In addition, the initiation damage caused by fragmentation stated in Table 11.15 can be employed as a reference as well, accordingly:

$$KE = \frac{1}{2}mv^2 = 0.5 \times 0.008 \times velocity^2 = 0.01\ kJ$$

$$v = 50\ m/s$$

Taking this velocity into Equation (11.67),

$$v = v_0 \exp\left[-C_D \rho_{air} x \left(\frac{A}{2m}\right)\right] = 50\ m/s$$

$$x = 353\ m$$

Therefore, a minimum distance of 353 m is required to initiate the damage of a soft target by considering the fragmentation effect.

b. Since the CFRP casing will disintegrate upon detonation, only the blast effect is required in this situation. Assuming the ordnance is equipped with an impact fuze, the minimum overpressure of 0.02 MPa is set as the reference based on Table 11.5. Therefore,

$$0.02 = \frac{0.102}{Z} + \frac{0.399}{Z^2} + \frac{1.26}{Z^3}\ (1 \le Z \le 15)$$

$$Z = 8.336\ m/kg^{1/3}$$

$$Z = \frac{R}{W^{1/3}}$$

$$R = 36\ m$$

Therefore, a minimum safe distance of 36 m is theoretically considered through the blast effect analysis.

11.6 SUMMARY

Explosive effects can be evaluated by using the aforementioned numerical and analytical approaches. The accuracy of the results, however, is dependent on many factors such as the selection of modelling techniques and material models, as well as the comprehensive consideration of the representative circumstances. Either numerical simulations or analytical approaches have their own set of compromises. One should be mindful that, in most cases, these models can only deliver the approximations of reality, not the exact replications of the real explosive events or effects.

REFERENCES

Ameri, A., Wang, H., Li, Z., Quadir, Z., Rickard, W. D., Gonzalez, M., Hazell, P. J. & Escobedo-Diaz, J. P. 2021. Ferrite phase transformation in dual-phase steel under shock loading. *Materials Science Engineering: A*, 802, 140690.

Blatz, P. J. & Ko, W. L. 1962. Application of finite elasticity to the deformation of rubbery materials. *Transactions of the Society of Rheology*, 6, 223–251.

Bridgman, P. W. 1946. Recent work in the field of high pressures. *Reviews of Modern Physics*, 18, 1.

Brode, H. L. 1955. Numerical solutions of spherical blast waves. *Journal of Applied Physics*, 26(6), 766–775.

Christensen, A. S., Kubar, T., Cui, Q. & Elstner, M. 2016. Semiempirical quantum mechanical methods for noncovalent interactions for chemical and biochemical applications. *Chemical Reviews*, 116, 5301–5337.

Cowan, R. & Fickett, W. 1956. Calculation of the detonation properties of solid explosives with the Kistiakowsky-Wilson equation of state. *The Journal of Chemical Physics*, 24, 932–939.

Cranz, C. 1926. *Lehrbuch der Ballistik*, Berlin, Springer-VCH.

Dequidt, A., Garruchet, S., Latour, B., Martzel, N., Blaak, R., Munch, E., Seeboth, N. & Malfreyt, P. 2020. A multiscale modelling approach for the design of new polymer materials. *Computer Aided Chemical Engineering*, 48, 25–30.

Driels, M. R. 2021. *Weaponeering for the Warfighter*, Virginia, American Institute of Aeronautics and Astronautics.

Drucker, D. C. & Prager, W. 1952. Soil mechanics and plastic analysis or limit design. *Quarterly of Applied Mathematics*, 10, 157–165.

Gharehdash, S., Barzegar, M., Palymskiy, I. B. & Fomin, P. A. 2020. Blast induced fracture modelling using smoothed particle hydrodynamics. *International Journal of Impact Engineering*, 135.

Grady, D. Fragment size prediction in dynamic fragmentation. AIP Conference Proceedings, 1982. American Institute of Physics, 456–459.

Gu, F., Li, Q. L. & Xiao, J. J. 2024. Analytical techniques in molecular simulation and its application in energetic materials. *Langmuir*, 40, 27137–27164.

Gurney, R. W. 1943. *The Initial Velocities of Fragments from Bombs, Shell, Grenades*, Aberdeen Proving Ground, MD: Army Ballistic Research Laboratories.

Hao, Y. M., Wei, X. D., Li, Q. & Zhao, G. F. 2024. A feasible approach for engineering-scale 3D blasting numerical modelling incorporating explosive charges and layout design. *Computers and Geotechnics*, 170.

Hawez, H. K., Sanaee, R. & Faisal, N. H. 2021. A critical review on coupled geomechanics and fluid flow in naturally fractured reservoirs. *Journal of Natural Gas Science and Engineering*, 95.

Hazell, P. J. 2022. *Armour: Materials, Theory, and Design*, Boca Raton, FL: CRC Press.

Henrych, J. & Major, R. 1979. *The Dynamics of Explosion and Its Use*, Amsterdam: Elsevier Scientific Publishing Company.

Hobbs, M. L. & Kaneshige, M. J. 2014. Ignition experiments and models of a plastic bonded explosive (PBX 9502). *Journal of Chemical Physics*, 140.

Holmquist, T., Johnson, G. & Cook, W. A computational constitutive model for concrete subjected to large strains. High Strain Rates and High Pressures, 14th International Symposium on Ballistics 1993. Quebec City, Canada, 591–600.

Hopkins-Brown, M. A. & Bailey, A. 1998. *Explosion Effects*, Cranfield: Cranfield University.

Hopkinson, B. 1915. *British Ordnance Board Minutes No. 13565*. Kew, UK: The National Archives.

Horie, Y. 2014. Hot spots, high explosives ignition, and material microstructure. *Explosion, Shock Wave and High-Energy Reaction Phenomena II*, 767, 3–12.

Huang, G., Feng, S., Liu, H. & Wang, T. 2021. *Explosion Technology and Its Application*, Beijing: Beijing Institute of Technology Press.

Johnson, G. R. & Cook, W. H. A constitutive model and data for metals subjected to large strains, high strain rates and high temperatures. Proceedings of the 7th International Symposium on Ballistics, 1983. The Hague, The Netherlands.

Johnson, G. R. & Holmquist, T. J. An improved computational constitutive model for brittle materials. AIP Conference Proceedings, 1994. American Institute of Physics, 981–984.

Johnson, J., Tang, P. & Forest, C. J. J. O. A. P. 1985. Shock-wave initiation of heterogeneous reactive solids. *Journal of Applied Physics*, 57, 4323–4334.

Jouguet, E. 1901. Sur la propagation des discontinuités dans les fluides. *Comptes Rendus de l'Académie des Sciences*, 132, 673–676.

Kennedy, J. E. 1970. *Gurney Energy of Explosives: Estimation of the Velocity and Impulse Imparted to Driven Metal*, Albuquerque, NM: Sandia Laboratories.

Kim, S., Miller, C., Horie, Y., Molek, C., Welle, E. & Zhou, M. 2016. Computational prediction of probabilistic ignition threshold of pressed granular Octahydro-1,3,5,7-tetranitro-1,3,5,7-tetrazocine (HMX) under shock loading (vol 120, 115902, 2016). *Journal of Applied Physics*, 120.

Lee, E., Hornig, H. & Kury, J. 1968. *Adiabatic Expansion of High Explosive Detonation Products*, Livermore, CA: Univ. of California Radiation Lab. at Livermore.

Li, Z., Khennane, A., Hazell, P. J. & Brown, A. D. 2017. Impact behaviour of pultruded GFRP composites under low-velocity impact loading. *Composite Structures*, 168, 360–371.

Li, Z., Zhou, Y., Cai, F., Zhu, Z. & Sui, L. 2024. Experimental and numerical analyses of flexural behaviour of hybrid LRS FRP-concrete-steel double-skin tubular beams with eccentric encased steel tube. *Construction and Building Materials*, 448, 138177.

Lin, J., Naceur, H., Coutellier, D. & Laksimi, A. 2014. Efficient meshless SPH method for the numerical modeling of thick shell structures undergoing large deformations. *International Journal of Non-Linear Mechanics*, 65, 1–13.

Liu, N., Zeng, F., Wang, J., Wang, J. T., Zhang, P., Li, K. & Yu, Y. Q. 2024. Large scale massively parallel numerical simulation for blast effects on structures. *Physica Scripta*, 99.

Liu, Y., Wu, Y., Huang, F. & Wang, X. 2018. *Fundamentals of Explosion Physics*, Beijing: Beijing Institute of Technology Press.

Lizundia, E., Sipponen, M. H., Greca, L. G., Balakshin, M., Tardy, B. L., Rojas, O. J. & Puglia, D. J. G. C. 2021. Multifunctional lignin-based nanocomposites and nanohybrids. *Green Chemistry*, 23, 6698–6760.

Ma, J. J., Li, R., Li, C. H., Chen, J. J., Lin, Y. X. & Huang, L. C. 2024. Further development of GDEM for the modeling of multi-scale dynamic response of rock subjected to blasting and impact. *International Journal for Multiscale Computational Engineering*, 22, 49–65.

Massey, B. S. & Ward-Smith, J. 1998. *Mechanics of Fluids*, Boca Raton, FL: CRC Press.

Matos, H., Galuska, M., Javier, C., Kishore, S., Leblanc, J. & Shukla, A. J. F. 2024. A review of underwater shock and fluid–structure interactions. *Flow*, 4, E10.

Mie, G. 1903. Zur kinetischen Theorie der einatomigen Körper. *Annalen der Physik*, 316, 657–697.

Mills, C. A. The design of concrete structures to resist explosions and weapon effects. Proceedings of the 1st International Conference on Concrete for Hazard Protections, 1987. Ediburgh, UK.

Mott, N. F. 1947. Fragmentation of shell cases. *Proceedings of the Royal Society of London. Series A. Mathematical Physical Sciences*, 189, 300–308.

Murnaghan, F. D. 1944. The compressibility of media under extreme pressures. *Proceedings of the National Academy of Sciences*, 30, 244–247.

Myers, C., Palmer, T. & Palmer, C. 2023. A hybrid finite volume-smoothed particle hydrodynamics approach for shock capturing applications. *Computer Methods in Applied Mechanics and Engineering*, 417.

Narváez-Muñoz, C., Hashemi, A. R., Hashemi, M. R., Segura, L. J. & Ryzhakov, P. B. 2025. Computational ElectroHydroDynamics in microsystems: A review of challenges and applications. *Archives of Computational Methods in Engineering*, 32, 535–569.

Newmark, N. M. & Hansen, R. J. 1961. *Design of Blast Resistant Structures*. In Harris, C. M. & Crede, C. E. *Shock and Vibration Handbook* (Vol. 3), New York: McGraw-Hill.

Ning, J. & Ma, T. 2015. *Computational Explosion Mechanics*, Beijing: National Defense Industry Press.

Sadovsky, M. A. 1952. *Mechanical Action of Air Shock Waves from Explosions according to Experimental Studies*, Moscow, Academy of Sciences of the USSR: Physics of Explosion.

Saurel, R., Fraysse, F., Furfaro, D. & Lapebie, E. 2018. Multiscale multiphase modeling of detonations in condensed energetic materials (Reprinted from An International Journal Computers and Fluids). *Computers & Fluids*, 169, 213–229.

Sirigiri, V. K. R., Gudiga, V. Y., Gattu, U. S., Suneesh, G. & Buddaraju, K. M. 2022. A review on Johnson Cook material model. *Materials Today: Proceedings*, 62, 3450–3456.

Suceska, M., Chan, H. Y. S., Stimac, B. & Dobrilovic, M. 2023. BKW EOS: History of modifications and further improvement of accuracy with temperature-dependent covolumes of polar molecules. *Propellants, Explosives, Pyrotechnics*, 48, e202100278.

Tanaka, K. Detonation properties of high explosives calculated by revised Kihara-Hikita equation of state. *Proceedings of the Eighth Symposium (International) on Detonation: July 15-19, 1985, Albuquerque Convention Center, Albuquerque, New Mexico* (NSWC MP 86-194, pp. 548–557). Naval Surface Weapons Center.

Ullah, A., Ahmad, F., Jang, H.-W., Kim, S.-W. & Hong, J.-W. 2017. Review of analytical and empirical estimations for incident blast pressure. *KSCE Journal of Civil Engineering*, 21, 2211–2225.

Wang, B., Li, H., Xing, H. & Li, X. 2022. Modelling of gas-driven fracturing and fragmentation in liquid CO_2 blasting using finite-discrete element method. *Engineering Analysis with Boundary Elements*, 144, 409–421.

Wang, Z., Chen, W., Hao, H., Dong, Y. & Huang, Z. 2023. Numerical prediction of blast fragmentation of reinforced concrete slab using ALE-FEM-SPH coupling method. *Finite Elements in Analysis and Design*, 220, 103948.

Wei, T. & Hargather, M. J. 2021. A new blast wave scaling. *Shock Waves*, 31, 231–238.

We, C. & Hao, H. 2005. Modling of simultaneous ground shock and airblast pressure on nearby structures from surface explosions. *International Journal of Impact Engineering*, 31(6), 699–717.

Yang, Z.-H., Rong, J.-L. & Zhao, Z.-T. 2023. Study on the prediction and inverse prediction of detonation properties based on deep learning. *Defence Technology*, 24, 18–30.

Zerilli, F. J. & Armstrong, R. W. 1987. Dislocation-mechanics-based constitutive relations for material dynamics calculations. *Journal of Applied Physics*, 61, 1816–1825.

Zerilli, F. J. & Armstrong, R. W. Constitutive relations for the plastic deformation of metals. AIP Conference Proceedings, 1994. American Institute of Physics, 989–992.

12 Quantitative Explosive Risk Assessment of Blast and Fragmentation Hazards

Mark G. Stewart

12.1 INTRODUCTION

An issue often facing asset owners and regulatory agencies is how safe is safe enough, and to ensure that expenditure on proposed protective measures is invested in a manner that optimises public safety in a cost-effective manner. This includes an assessment of safety and damage hazards from blast wave propagation and the trajectory of explosively induced fragments.

Personnel and infrastructure safety risks from explosions in gas, oil, chemical and other process industries have often been estimated from probabilistic risk assessment (PRA) or quantified risk assessment (QRA) for decades (e.g., Huang et al. 2017; Stewart and Melchers 1997). These methods have been applied to reliability-based load factors for explosive blast loading and reliability-based design of structural components (e.g., Campidelli et al. 2013; Olmati et al. 2017; Stewart 2018, 2019). These methods, with some modification, are also well suited to explosive ordnance (EO), including fragmentation (e.g., Grisaro and Dancygier 2019; Haring et al. 2009; Qin and Stewart 2021; Sielicki et al. 2021; Van der Voort and Weerheijm 2013). Hence, the chapter explores how probabilistic safety and hazard modelling of blast and fragmentation can better inform a quantitative explosive risk assessment (QERA). A QERA may be used to (i) define an explosive safety distance based on the safety or damage risk of explosive hazards being less than a societal acceptable risk, or (ii) assess the cost-effectiveness of a protective measure by comparing the benefits in terms of lives saved and/or losses averted with the cost of the risk mitigating measure. This chapter will describe the QERA procedure used by the Australian Defence Force to assess the safety and damage to people and infrastructure from the accidental detonation of munitions and other explosives (eDEOP-101 2023). A QERA is an accepted, robust and reasonable way of calculating safety and hazard risks, and the calculated fatality risks may then be compared with existing Worker Health and Safety (WHS) and hazardous facilities planning quantitative risk acceptance criteria. While the terminology differs, this is the same philosophy and method as a QRA or PRA applied to many hazardous systems. Moreover, eDEOP-101 (2023) notes that nations including the UK and the US are also currently using quantitative risk assessment for authorising explosive activities.

12.2 DEFINITION OF RISK

The standard definition of risk adopted for many applications (ISO 31000, 2018; Stewart and Melchers 1997) is

$$(\text{Risk}) = (\text{Hazard}) \times (\text{Consequences}) \tag{12.1}$$

The nomenclature can vary from discipline to discipline, but in the context of safety risks for EO, these terms are defined as follows:

- *Hazard* – The likelihood and location of a fire or accident or malevolent act that causes EO to detonate. The magnitude of the hazard may be measured by blast overpressure, primary fragmentation of the EO, debris (secondary fragmentation) of a surrounding structure and thermal effects.

DOI: 10.1201/9781003529217-12

- *Consequences* – How will people and infrastructure be affected by the hazard? What are the life-safety, economic and social costs if people are harmed or infrastructure is damaged? The criticality of the consequences will depend on the exposure – e.g., the time of day, the location of damage and crowd density.

A risk assessment will combine these two measures in a way to estimate the overall risk to people, operations and infrastructure. Risk may be measured as an expected loss such as annual fatality risk, or expected monetary loss per year.

An area of particular difficulty for risk analysis is where the potential consequences are extremely large or severe yet the probability of these consequences actually occurring is estimated to be extremely low. These are termed "low probability – high consequence" events or hazards. The likelihood and location of a fire or accident or malevolent act that causes EO to detonate is a low-probability event, and blast overpressure, fragmentation and thermal effects will result in high (or catastrophic) consequences. Hence, a risk assessment for EO constitutes "low probability – high consequence" events or hazards.

A risk assessment needs to be tailored to the needs of the duty holder and other decision-makers. There are many tools and methods available for conducting a risk assessment (e.g., Stewart and Melchers 1997). Analysis methods can be qualitative, quantitative or a combination of the two.

12.2.1 Qualitative Approaches

The simplest approach to risk analysis is to use subjective assessments of risks and to rank these in a subjective manner. The top-ranking risk items are then those considered to require attention, such as through amelioration techniques, risk management approaches or a more thorough risk analysis to determine whether risk acceptance criteria are likely to be exceeded.

Expert opinions are critical in analyses of this type. The experts must understand both the consequences and the likelihood of occurrence of the consequences and then rank each according to the estimated effect and the estimated magnitude. These estimates must then be combined to obtain an estimate for the risk associated with each identified consequence. Where a number of experts are consulted, consensus may be difficult to obtain. In this case, meetings at which the experts can discuss the issues may be appropriate in an attempt to develop a consensus opinion, or implementation of the Delphi (or similar consensus-seeking) technique may be necessary to obtain a degree of consistency among the expert opinions (e.g., Ayyub 2001).

A refinement of this process is to keep consequences and probability of occurrence of the consequences separate. Each is assessed using a qualitative descriptor, such as "high", "medium" and "low" and the results are entered into a risk matrix – e.g., Figure 12.1 shows risk matrices used by Australian Department of Defence Explosives Regulations (eDEOP-101 2023) and the Royal Australian Navy (RAN). A further subjective judgement is then desirable to assess each of the matrix entries. ISO 13824 (2020) surmises that

> risk matrixes do not pretend to express the exact level of risk, but are useful to highlight the different contributions to risk from the underlying hazards. Risk matrixes and similar tables can be feasible tools during risk meetings aimed at identifying and evaluating hazards in a direct manner.

A key problem is the amount of subjectivity required in selecting a likelihood and a consequence, and then whether the obtained risk is "low", "high" etc. For example, the harmonised RAN risk matrix given in Figure 12.1b shows that the risk is "low" if likelihood is rare but consequences are catastrophic. Whereas a similar risk matrix in the Australian Department of Defence Explosives Regulations (eDEOP-101 2023) given in Figure 12.1a shows that for a rare likelihood and a catastrophic consequence the obtained risk is "high". So two different risk matrices used by Defence

(a)

Consequence Rating					
Likelihood Rating	**Catastrophic (1)**	**Major (2)**	**Moderate (3)**	**Minor (4)**	**Insignificant (5)**
Almost Certain (A)	Extreme 1	Extreme 3	Extreme 8	High 11	High 16
Likely (B)	Extreme 2	Extreme 5	High 9	High 15	Medium 20
Possible (C)	Extreme 4	Extreme 7	High 12	Medium 18	Low 23
Unlikely (D)	Extreme 6	High 10	Medium 17	Low 21	Low 24
Rare (E)	High 13	High 14	Medium 19	Low 22	Low 25

(b)

Likelihood		Consequence				
		Minor (A)	Moderate (B)	Major (C)	Critical (D)	Catastrophic (E)
5	Almost Certain	(A5) LOW	(B5) MEDIUM	(C5) HIGH	(D5) VERY HIGH	(E5) VERY HIGH
4	Probable	(A4) LOW	(B4) MEDIUM	(C4) HIGH	(D4) HIGH	(E4) VERY HIGH
3	Occasional	(A3) VERY-LOW	(B3) LOW	(C3) MEDIUM	(D3) HIGH	(E3) HIGH
2	Improbable	(A2) VERY LOW	(B2) VERY LOW	(C2) LOW	(D2) MEDIUM	(E2) MEDIUM
1	Rare	(A1) VERY LOW	(B1) VERY LOW	(C1) VERY-LOW	(D1) LOW	(E1) LOW

FIGURE 12.1 Risk matrices: Australian Department of Defence Explosives Regulations (eDEOP-101 2023) (a), and Royal Australian Navy (ANP3411-0010) (b).

give two very different risk levels – one is "low" and one is "high". These are two contradictory outcomes and do little to inform a duty holder about the actual level of risk.

Another problem with qualitative assessments is that there is little indication on an absolute scale how serious the risk might be, particularly for comparison to other risk sources. It may not, therefore, be very useful for making rational decisions, although it might be of use in preliminary analyses or risk screening. For example, if there is consensus that risks are "low" or "medium", then no further treatment is needed. However, if qualitative risks are "high" or "very high", it is difficult to choose which risks need treatment, and to what extent. According to WorkSafe Victoria, a risk matrix is "often too coarse a tool to distinguish between options" (WorkSafe 2011). Moreover, ISO 13824 (2020) comments that

> In practice, qualitative estimation is often used, as a preliminary risk estimation, to obtain a general indication of the level of risk and to reveal the risks that shall be considered. Later, it can be necessary to undertake more specific or quantitative estimation on the revealed risk.

A better way of analysing risks is to turn to a quantitative method.

12.2.2 Quantitative Approaches

In quantitative estimation, numerical values (rather than descriptive scales used in qualitative estimation) are used for both consequences and probability of occurrence based on data from a variety of sources. Such an assessment is termed a QRA or a PRA. In the context of EO, the appropriate method is a QERA. This is termed a probabilistic safety analysis in the UK.

According to the Australian Department of Defence Explosives Regulations (eDEOP-101 2023), a QERA

> calculates the risk from a potential explosion site (PES) to personnel situated at an adjacent exposed site (ES). A QERA is based on experimental and historical data and can be considered a detailed investigation of a potentially hazardous situation. Conducting a QERA is a complex task that must be undertaken and reviewed by qualified and experienced personnel.

A flowchart of the QERA process is shown in Figure 12.2 which is consistent with the international standard Risk Management ISO 31000 (2018). Quantifying probabilities of events, exposure and consequences can be challenging. However, databases of incidents and high-level computational modelling of the consequences of blast overpressure, fragmentation and thermal effects are available – these will be reviewed and discussed in the following sections.

A key feature of a QERA (or any other quantitative risk assessment) is that it provides robust and evidence-based advice on the actual level of risk, and this can be compared directly with acceptable risk criteria that is well established in society. The process of quantification is fully transparent, as are assumptions and data gaps during this process. Moreover, sensitivity analyses can be conducted to test the robustness of decisions – this is particularly important if there is uncertainty about event frequencies, exposure or consequences.

The utility of a QERA can best be illustrated by typical risk criteria reported in eDEOP-101 (2023):

- Public fatality risk criteria – fatality risk level of 1×10^{-6} per year as the limit for risk acceptability for residential area exposure

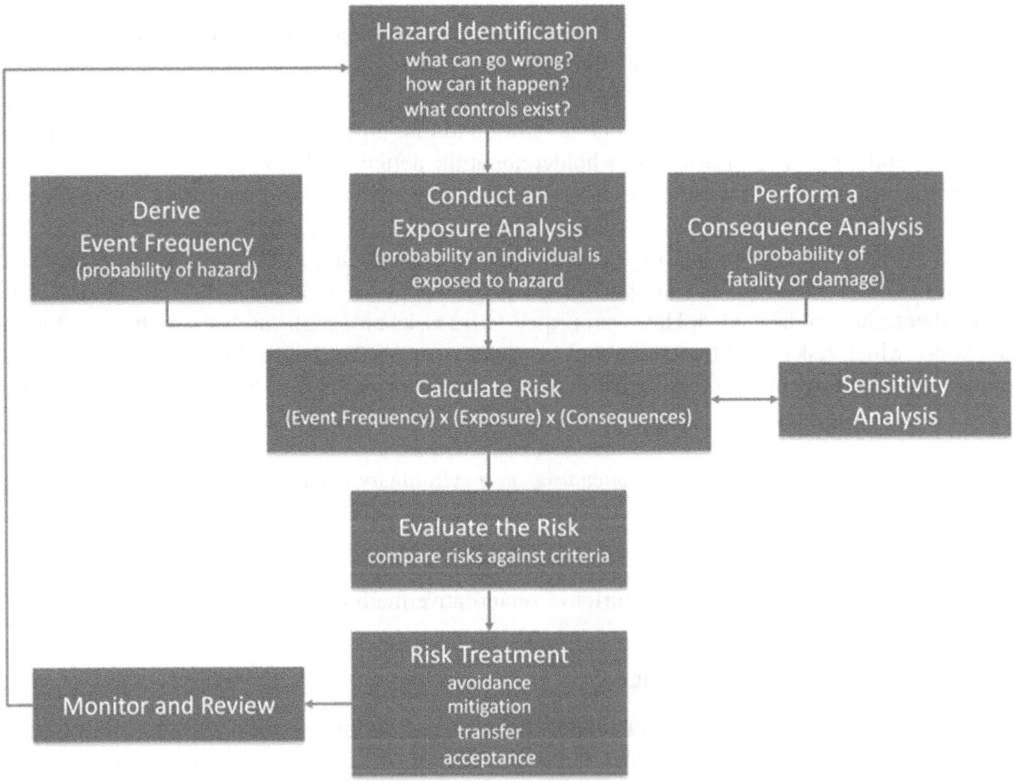

FIGURE 12.2 Flowchart of QERA process.

- Worker fatality risk criteria – fatality risk level of 50×10^{-6} per year as the limit for risk acceptability for workers
- Injury risk criteria – An incident explosion overpressure at residential areas should not exceed 7 kPa at frequencies of more than 50×10^{-6} per year.

Another risk criterion is the use of cost–benefit analysis (CBA) based on As Low As Reasonably Practicable (ALARP) or So Far As Is Reasonably Practicable (SFARP) principles to assess if mitigating measures are necessary.

The QERA methodology meets key benchmarks against legislation, ADF regulations and WHS guidelines and practices for high hazard industries. In other words, it represents best practice in Australia and internationally.

12.3 EXPLOSIVE ORDNANCE LICENCING REQUIREMENTS IN AUSTRALIA

The work herein considers current ADF licencing requirements for the storage and transport of explosives, as listed within eDEOP-101 (2023) *Department of Defence Explosive Regulations, Issue 4* published by the Australian Department of Defence. A key feature of eDEOP-101 (2023) is that it specifies quantity (or safety) distances (QDs) for a large range of scenarios suited to military EO. For example, if an exposed person or building (i.e., ES or exposed site) is within a QD, then this is deemed as "unsafe" by the regulator.

There are many aspects to QDs, where a variety of QD types are used within many different sections of eDEOP-101. The basis for QDs is explained in eDEOP-101 as "The QD are based on an extensive series of trials and a careful analysis of all available data on accidental explosions in different countries. However, QD are subject to uncertainty owing to the variability of explosions". The source of some of the QDs described in eDEOP-101 is the UK and the US, some dating back to the 1960s (e.g., Jarrett 1968; Kinney et al. 1981; Lyman 1979). However, the degree of safety and the conservatism (or not) of QD values is unknown.

The QD functions are all empirical and deterministic (eDEOP-101 2023). For example, the quantity distance for an inhabited building is

$$D = 5.5Q^{1/2} \quad Q < 4500 \text{ kg} \tag{12.2}$$

where Q in the NEQ in kg (i.e., net equivalent quantity in terms of TNT), and D is in m. The deterministic nature of QDs assumes that the outcome for any explosive event is always the same. Such an assumption is rarely valid. Observations of repeated explosives trials clearly show variability and uncertainty, both in the input parameters and the resultant blast pressures (see Netherton and Stewart 2010; Stewart 2023; Stewart et al. 2019, 2024). The calculation of most likely blast loads and fragmentation effects is only possible using probabilistic methods, such as through QERA techniques. Added to this is that a QERA can assess likely casualty and damage losses, which allows the costs and safety and economic benefits for different risk mitigation measures to be compared.

It would appear that eDEOP-101 (2023) deterministic calculations are based on what is thought to be a reasonable, and perhaps deliberately conservative approach (for more details see Stewart and Netherton 2019); however, these safety distance calculations do not appear to have any basis in terms of expected or calculated risk. However, they are of little use to any decision-maker attempting to determine what effectiveness a risk mitigation measure may have for their particular circumstances. A degree of conservatism may be appropriate for the design and management of new EO facilities where the incremental cost of safety is minor. However, if an existing facility is assessed as unsafe, when in reality it is safe, the cost associated with relocating the existing facilities will be considerable. Further, overly conservative QDs may result in less-than-optimal use of (limited) real estate.

12.4 QUANTIFIED EXPLOSIVE RISK ASSESSMENT (QERA)

eDEOP-101 (2023) uses the following notation to calculate the individual annual fatality risk at any potential explosive site (PES):

$$\text{Individual Fatality Risk} = \text{IR} = F_e \times P_f \times (E_{\text{individual}} \times E_{\text{PES}}) \tag{12.3}$$

where

- F_e is event frequency or the probability that an event will occur at the PES per year,
- P_f is the probability of fatality given an explosive event and the presence of a person,
- $E_{\text{individual}}$ is the exposure period for the particular individual at an ES expressed as a fraction of a year, and
- E_{PES} is the period time the potential hazard is present at the PES expressed as a fraction of a year.

The US *Ammunition And Explosives Safety Standards* Department of Defense 6055.09-Std (DoD 2024) and the *Approved Methods and Algorithms for DoD Risk-Based Explosives Siting* (DDESB 2009) define their siting acceptance criteria based on individual fatality risk (IR). However, risk criteria for land use safety planning in Australia, such as the NSW Government's Hazardous Industry Advisory Plan, conservatively define individual fatality risk for 24-hour-per-day exposure (DoP 2011a, b). In this case, $E_{\text{individual}} \times E_{\text{PES}} = 100\%$. This aligns with eDEOP-101 (2023) where

$$\text{Potential Risk} = \text{PR} = F_e \times P_f \tag{12.4}$$

where this defines a worst-case scenario for any single individual.

The probability of fatality (P_f) given an explosive event is (eDEOP-101 2023), assuming mutually exclusive events is

$$P_f = P_{\text{Blast}} + P_{\text{Fragmentation}} + P_{\text{Debris}} + P_{\text{Thermal}} \tag{12.5}$$

where

- P_{Blast} is the fatality probability resulting from blast effects,
- $P_{\text{Fragmentation}}$ is the fatality probability resulting from primary fragment effects,
- P_{Debris} is the fatality probability resulting from debris (or secondary fragment effects, including ejecta debris), and
- P_{Thermal} is the fatality probability resulting from thermal radiation effects.

Primary fragments originate from the explosive item, secondary fragments (debris) originate from the PES, and ejecta are debris originating from the foundation of the PES.

Note that DDESB (2009) has a different formulation than Eqn. (12.5) as P_{Blast}, $P_{\text{Fragmentation}}$, P_{Debris} and P_{Thermal} are modelled (less conservatively) as non-mutually exclusive events, leading to

$$\begin{aligned} P_f = {} & P_{\text{Blast}} + (1 - P_{\text{Blast}})P_{\text{Fragmentation}} + (1 - P_{\text{Blast}})(1 - P_{\text{Fragmentation}})P_{\text{Debris}} \\ & + (1 - P_{\text{Blast}})(1 - P_{\text{Fragmentation}})(1 - P_{\text{Debris}})P_{\text{Thermal}} \end{aligned} \tag{12.6}$$

The US Department of Defense Explosives Safety Board (DDESB) specifies a number of approved methods and algorithms, which, collectively, are the basis of the Safety Assessment for Explosives Risk (SAFER©) software model used to perform:

> ... quantitative risk analysis to determine risk to people from a potential explosion site (PES) by conducting a sequential calculation of probability of event, explosives effects, and personnel exposure. SAFER can be used to assess the risks to people in inhabited building distance (IBD), public traffic route (PTR), and intraline distance scenarios.

This method of calculating explosive safety distances is a significant improvement in terms of providing risk-based advice to decision-makers. Moreover, the DDSEB allows for uncertainty modelling, as is now listed as a "major design goal" for SAFER (DDESB 2009). The uncertainty model within SAFER includes many factors that contribute to uncertainty in the expected risk estimate. This includes probabilistic uncertainties in the model and its inputs, as well as "real world variation in explosive effects and vulnerabilities" (DDESB 2009).

However, there remain some fundamental issues with respect to the DDESB (2009) determination of blast loads. For example, their blast-load effects fatality analysis (P_{blast}) assumes that the explosion produces a blast wave described by pressure and impulse defined deterministically using Kingery–Bulmash's simplified equations (Kingery and Bulmash 1984). Thus, whilst the DDESB has presented a robust quantitative risk assessment procedure, the problem is that the source of their blast-load calculations remains deterministic and does not account for the effect that post-detonation airblast variability (i.e., uncertainty of NEQ, range and airblast model error) may have on fatality and damage risks.

12.4.1 Event Frequency – F_E

The probability of a fire or accident or malevolent act that causes EO to detonate or cause another hazard is defined as the event frequency. Intentional or malevolent acts may include insider threats or terrorism.

Accident or incident databases in Australia and internationally are available for many initiating events – these generally contain information on failure events, their cause, consequences and time of occurrence. In some cases, and at best, they will provide indicative probabilities only as accidents that lead to the explosion of EO are very low-probability events – i.e., likely to be lower than 1 in a thousand per PES per year. For example, according to DDESB (2009), the likelihood of missile detonation during loading/unloading is 1.9×10^{-5} or one chance in 50,000 per year.

Most databases, however, are a compilation of accidents and their initiating events. In these cases, the probability of the event can be inferred, for example, by dividing the number of accidents by the estimated total number of PES during that time period. Estimating the latter is subject to considerable uncertainty.

There may well be some identified event scenarios for which there are no incident databases or probabilistic information available. In these instances, the event frequency may be estimated from comparisons with other events where probabilities are known. Or by estimating the likelihoods of each component of a sequence of initiating events needed to result in the detonation of munitions – e.g., the probability of fire in a compartment multiplied by probability that fire/heat reaches a magazine multiplied by likelihood that munitions within the magazine detonate.

Qualitative and quantitative information obtained from Defence personnel (or others), if of sufficient quality, may also provide useful event frequency data. Such data relies, of course, heavily on the experience and knowledge of the people consulted and their ability to make judgements and to convey opinions. Typically, they may be asked to estimate the average frequencies and/or to estimate the range of frequencies. From this information it may be possible to derive point estimates or a probability distribution for event frequencies.

Finally, a "break-even" analysis may be useful where the maximum value of the probability of an explosive event is back-calculated to ensure that the risk does not exceed the risk criteria. For instance, if the potential risk criteria are 1×10^{-6} fatalities per year (one chance in a million), and the probability of a fatality given an explosive event is P_f = 10%, then Eqn. (12.4) shows that the break-even event frequency (F_e) is PR/P_f or 1×10^{-5} (or one chance in a hundred thousand) events per year. If Defence are satisfied that the actual probability of accidental detonation is less than 1×10^{-5} per year, then the QERA shows that fatality risks are at an acceptable level.

12.4.2 Exposure – $E_{individual}$, E_{PES}

The exposure period of Defence personnel, workers and the public is more tractable than determining event frequency. The period of time the potential hazard is present at the PES (E_{PES}) may be estimated from EO records. For example, if EO is stored in a magazine for 300 days per year, then E_{PES} = 300/365= 82%. The exposure period for a particular individual at an ES, expressed as a fraction of a year ($E_{individual}$), may also be obtained from an examination of worker and personnel rosters. For example, dockside workers and ADF personnel may be working or on duty for an average of 10 hours per working day for 200 days per year, leading to $E_{individual}$=(10/24) × (200/365) = 23%.

As noted above, risk criteria for residential areas in Australia conservatively define individual fatality risk for 24-hour-per-day exposure where $E_{individual} \times E_{PES}$ = 100%.

12.4.3 Consequences

The consequences arise from the following EO hazards: (i) blast overpressure, (ii) primary fragmentation of the weapon, debris (secondary fragmentation) of PES and ES (such as buildings and magazines) and (iii) thermal effects. This can lead to life-safety, economic and social costs if people are harmed or infrastructure is damaged.

12.4.3.1 EO Hazards

12.4.3.1.1 Blast Overpressure

The modelling of blast waves within complex blast environments is not feasible via the idealised free-field blast-load model proposed by Kingery and Bulmash (1984) that is used in ConWep (1991) and UFC 3-340-02 (2014). In complex geometric environments, the best available models are those that include constitutive relationships of materials within the blast-load environment – these are computational fluid dynamics (CFD) models (e.g., Air3D 2001; ProSAir 2009; Viper::Blast 2019). CFD models can account for explosive shape, orientation, method of detonation/deflagration and other material and geometric realities; hence, they are often viewed as accurate in most circumstances, and increases in computational power mean that CFD modelling can now move from the deterministic to the probabilistic domain which bodes well for uncertainty modelling of post-detonation blast loads (Viper::Blast 2019).

12.4.3.1.2 Fragmentation and Thermal Effects

The main EO hazard to humans is often fragmentation, and it is the risk from fragmentation that is often used to derive QD distances. Modelling of primary and secondary (debris) fragmentation hazards is difficult due to the many parameters that must be taken into account – e.g., initial fragment velocity and direction, mass and fragment distribution, drag coefficients, fragment kinetic energy (KE), throw range, and fragment density probabilities. The initial velocity of fragments is based on the Gurney equation for different types of high explosives (Gurney 1943). Gurney equations are provided for cylindrical- and spherical-shaped charges, which are applicable to primary fragmentation for most military ordnance. The dispersion and density of fragments are based on maximum throw distance and are modelled probabilistically to reflect uncertainty in fragment velocity and direction – for example, see the US *Approved Methods and Algorithms for DoD Risk-Based Explosives Siting* (DDESB 2009) for more details, including thermal hazards. The literature has many descriptions of primary fragments from individual items of EO (e.g., Gold et al. 2006), and how such fragments disperse when unimpeded (e.g., Haring et al. 2009; Qin and Stewart 2021, 2024).

12.4.3.2 Human Vulnerability

The two organs most susceptible to injury from blast overpressure are the inner ear and the lungs. The probability of fatality from blast overpressure (P_{blast}) may be estimated based on the probit equation as a function of peak pressure, impulse, and mass of the human body for lung rupture, whole body displacement, and skull fracture due to translation of the body as it impacts hard surfaces (DDESB 2009; Mannan 2012; TNO 1992). These are empirical methods widely used by the US Department of Defense and the process industries. An alternative is the Axelsson model that represents the human thorax as a mass-spring system (Axelsson and Yelverton 1996) to predict lung injury. The probability of eardrum rupture can also be estimated from available models (e.g., Mannan 2012).

Casualties from fragmentation are influenced by the shape, mass, velocity and density of fragments. Fragments with KE over 79 J are deemed hazardous by the United Nations (2015) and NATO (2015). The probabilities of fatality, major and minor injuries as a function of the KE of the fragment hitting an average standing person (i.e., human vulnerability) may be modelled by a lognormal function based on the weighted average (Qin and Stewart 2021) of vulnerabilities for different body parts (i.e., head, thorax, abdomen and limbs) from relevant physiological studies (RCC 2000). Similarly, the lognormal parameters for major and minor injuries can also be obtained (see Figure 12.3). The probability that an individual will be hit by a fragment is based on fragment trajectory, density and the projected area of the exposed person. The probability of a fatality from fragmentation P_{frag} is the product of the probability that an individual will be hit by a fragment and the probability of fatality if hit by a fragment. Building collapse/damage and glazing hazards may increase the probability of fatal and major injuries if an individual is inside a vulnerable building or structure – see DDESB (2009) for more details.

12.4.3.3 Building Damage, Value of Life and Other Consequences

In addition to casualty risks, the duty holder may be concerned about damage risks and economic and social consequences. An accurate prediction of structural damage to building and other infrastructure first requires that information about structural details (window type and thickness, material properties of load-bearing members etc.) be known. If this information is known, then detailed structural modelling can be used to infer specific damage levels such as window breakage and column or wall collapse. An indication of the vulnerability of buildings is possible using some simple-to-use damage models, as will be discussed below.

An accurate prediction of structural damage to building can be obtained from detailed structural engineering models to infer specific damage levels such as window breakage and column or wall collapse. Extreme blast loads induce very high rates of loading and cause structural elements to

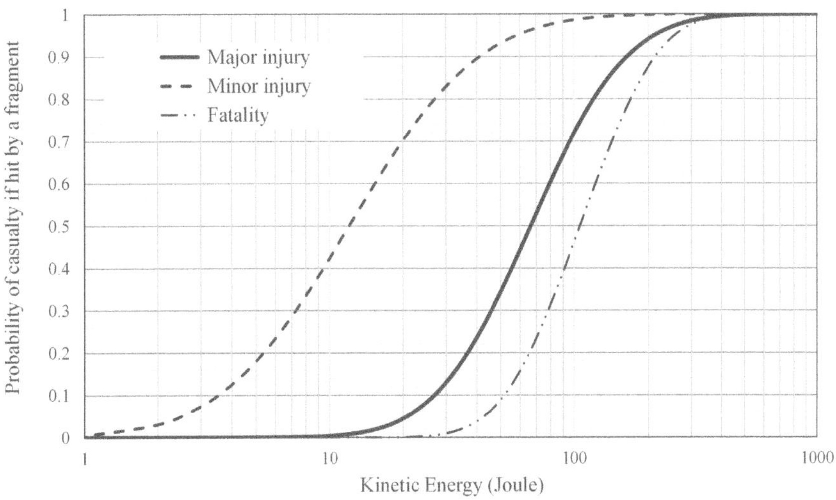

FIGURE 12.3 Vulnerability curves for an average standing person if hit by a fragment (e.g., Qin and Stewart 2021).

undergo large inelastic deformations. Very high strain rates can increase steel and concrete stiffness and strength by a factor of six. Increasing the rate of loading will thus increase the load-carrying capacity of structural elements. Strain rate-dependent constitutive models thus need to be included when modelling structural response and such models for concrete and steel are readily available in the literature.

Experience suggests that the non-linear finite element analysis (FEA) models developed for concrete, steel and masonry buildings are accurate for blast-resistant design. Strain rate-dependent constitutive models for steel, concrete and masonry are also readily available in the literature. LS-DYNA, AUTODYN or other software may be used as the computational tool as they are widely used FEA programmes that use explicit time integration to analyse the non-linear dynamic response of three-dimensional structures. These and similar packages can include material and geometry non-linearity with a non-linear time history analysis to obtain the dynamic load-deflection response.

Damage may also be predicted from pressure–impulse (P–I) curves. For example, P–I curves for three damage levels for reinforced concrete (RC) columns with or without fibre-reinforced polymer (FRP) strengthening subject to explosive blast loading are shown in Figure 12.4 (Hao et al. 2016). The damage levels considered were D = 0.2, 0.5 and 0.8 representing low damage, medium damage and high damage (collapse), respectively. P–I curves are widely available for many structural elements, including glazing.

In addition to FEA and P–I diagrams, other methods to design or assess structural damage and the retrofit of existing buildings include the following:

- Single-degree-of-freedom
- Multi-degree-of-freedom
- Response spectra

Finally, an indication of the vulnerability of buildings is possible using some simple-to-use damage models. For example, the US Federal Emergency Management Agency *Reference Manual to Mitigate Potential Terrorist Attacks Against Buildings* (FEMA 2011) and the US Department of Defense *DoD Minimum Antiterrorism Standards For Buildings* (UFC 3-340-02 2014) provide damage prediction for unstrengthened (i.e., conventional construction, not designed to be blast-resistant)

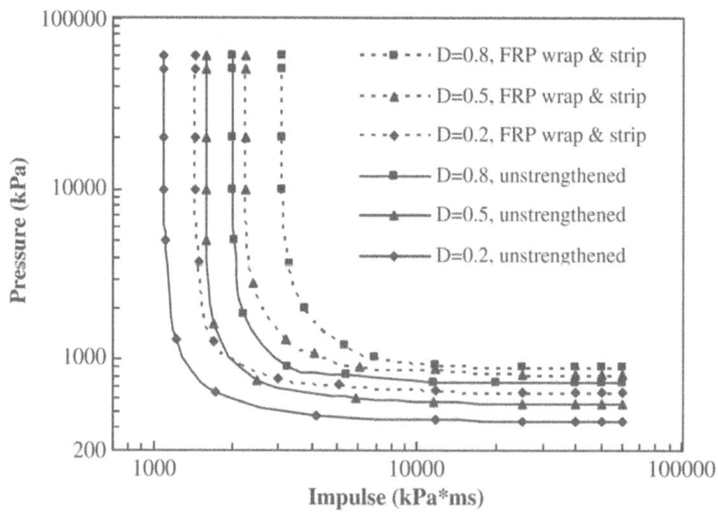

FIGURE 12.4 Comparison of P–I curves of unstrengthened and FRP strengthened RC columns (Hao et al. 2016).

TABLE 12.1
Damage Predictions (FEMA 2011)

Incident Pressure (kPa)	Damage Level	Potential Structural Damage	Potential Door and Glazing Hazards	Potential Injury
7.6	Low	Damaged – repairable. Minor deformations of non-structural elements and secondary structural members and no permanent deformation in primary structural members.	Glazing will break but will remain in the window frame. Doors will stay in frames but will not be reusable.	Some minor injuries, but fatalities are unlikely.
12.4	Medium	Damaged – unrepairable. Major deformation of non-structural elements and secondary structural members, and minor deformation of primary structural members, but progressive collapse is unlikely.	Glazing will break but fall within 1 m of the wall or otherwise not present a significant fragment hazard. Doors may fail, but they will rebound out of their frames, presenting minimal hazards.	Majority of personnel suffer significant injuries. There may be a few (<10%) fatalities.
15.9	High	Heavily damaged – onset of structural collapse. Major deformation of primary and secondary structural members, but progressive collapse is unlikely. Collapse of non-structural elements.	Glazing will break and is likely to be propelled into the building, resulting in serious glazing fragment injuries, but fragments will be reduced. Doors may be propelled into rooms presenting serious hazards.	Majority of personnel suffer serious injuries. There are likely to be a limited number (10%–25%) of fatalities.

buildings subject to above-ground explosions where blast load is measured as incident pressure (see Table 12.1). Other damage criteria are available from TNO (1992).

Consequences of loss comprise (i) direct costs, including loss of life and physical damage, (ii) indirect costs such as business losses, and (iii) social losses, which is the effect of the level of anxiety within society about safety following a catastrophic event. A CBA tends to measure benefits and consequences in monetary units. For instance, the value of preventing a fatality is often referred to as the value of a statistical life or VSL. The VSL is A$5.5 million when assessing Australian Government regulations (OBPR 2010), adjusted for inflation to 2024. The OECD (2012) recommends a VOSL for the European Union in the range of USD 3 million to 9 million, with a base value of USD 6 million (adjusted for inflation to 2024). This is not the value of a life, but a measure of the statistically estimated amount of money worth spending to avert a fatality. The cost of physical damage is relatively straightforward if the extent of damage is known. The estimation of indirect and social consequences of extreme events such as terrorism or natural hazards is well studied, and guidelines exist for their estimation (e.g., EU 2022). However, aggregate losses may err on the conservative side by placing a high premium on indirect and social losses (e.g., Mueller and Stewart 2011).

12.4.4 Risk Evaluation and Risk Treatment

Important phases of a QERA are risk evaluation (appraisal) and risk treatment. Evaluating the risk calculated in a QERA against risk criteria is a method for determining a risk management (or treatment) action. This is one of the key advantages of a quantitative risk assessment – it can ensure that actual safety and damage risks to workers and the public are at a level acceptable to duty holders, regulators and society. It is an evidence-based assessment of safety and damage risks.

Often the concepts of ALARP or SFARP are enshrined within WHS legislation. For example, Section 17 of the Australian 2011 Work Health and Safety Act (WHS 2017) states that

A duty imposed on a person to ensure health and safety requires the person:

a. *to eliminate risks to health and safety, so far as is reasonably practicable; and*
b. *if it is not reasonably practicable to eliminate risks to health and safety, to minimise those risks so far as is reasonably practicable.*

This is followed by Section 18 of the Act:

In this Act, reasonably practicable, in relation to a duty to ensure health and safety, means that which is, or was at a particular time, reasonably able to be done in relation to ensuring health and safety, taking into account and weighing up all relevant matters including:

a. *the likelihood of the hazard or the risk concerned occurring; and*
b. *the degree of harm that might result from the hazard or the risk; and*
c. *what the person concerned knows, or ought reasonably to know, about:*
   i. *the hazard or the risk; and*
   ii. *ways of eliminating or minimising the risk; and*
d. *the availability and suitability of ways to eliminate or minimise the risk; and*
e. *after assessing the extent of the risk and the available ways of eliminating or minimising the risk, the cost associated with available ways of eliminating or minimising the risk, including whether the cost is grossly disproportionate to the risk.*

The concepts of ALARP and SFARP are largely interchangeable and describe the level to which risks are expected to be controlled, noting that ALARP allows the application of "broadly acceptable" for risks, whereas SFARP does not (see Figure 12.5). The practical implementation of these concepts involves considering a risk in terms of the trouble, time and money needed to control it. The risk criteria described below are based on these considerations, and particularly the concept of minimising the risk while ensuring that the cost is not grossly disproportionate to the risk.

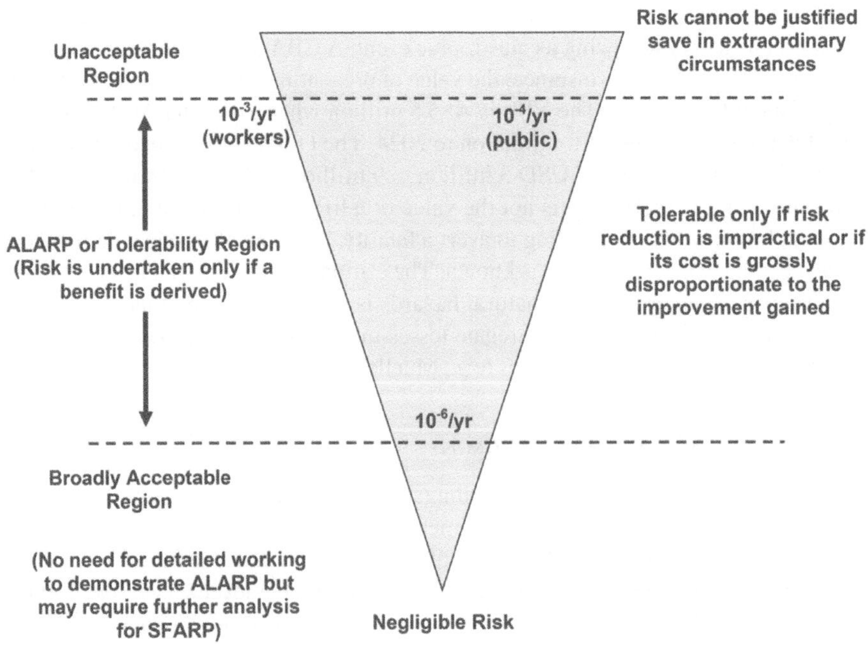

FIGURE 12.5 Fatality risks (based on HSE 2013a).

It is beyond the scope of the present chapter to stipulate exact risk criteria as this is an issue that requires a discussion of existing worker and public risks, WHS regulations, comparison of risk criteria adopted internationally and trade-offs between public safety and costs. The final decision is the prerogative of the duty holder or government regulators.

If risks fail to be acceptable or SFARP, then risk treatment may be needed to mitigate the risk by reducing the threat, hazard or damage likelihood, exposure and/or consequences. Risk treatment might include the following:

- reactive controls (reduce the likelihood of the event), and/or
- proactive controls (mitigate the consequences).

Finally, eDEOP-101 (2023) states that

> The decision to approve or not approve an EO activity must not be based purely on the numeric outcome alone because the purpose of a QERA is to provide a sensitivity analysis of specific areas in terms of event frequency, consequences, and exposure that have the potential to affect assessed risk levels.

This statement is applicable to any quantitative risk assessment where assessed risks are compared with risk criteria and is echoed by other regulators such as the NSW Government Department of Planning and Environment (DoP 2011a), as well as in texts on the subject (Stewart and Melchers 1997).

12.4.4.1 Fatality Risks

Figure 12.5 shows the consensus risk acceptance criteria for fatality risks. These quantitative safety goals are used by the US Nuclear Regulatory Commission, UK Health and Safety Executive, Australian and European hazardous industrial development regulators, US environmental carcinogenic exposure regulators and others (e.g., Mueller and Stewart 2011; Stewart and Melchers 1997). These safety goals or targets have been developed to provide overall guidance to duty holders and regulators, but any specific decision will necessarily involve other criteria as well, much of it non-quantifiable and/or emotional, that may also be important in balancing the preferences of interested parties.

The willingness to accept risk is influenced by a variety of psychological, social, cultural, and institutional processes depending on such qualities as the uncontrollability of the risks; the dread (or fear) they inspire; their involuntary nature or catastrophic potential; whether they can be preventively controlled, are certain to be fatal, can easily be reduced, result in an inequitable distribution of risk, threaten future generations or affect one personally; whether they are increasing or not observable, unknown to those exposed, new or unfamiliar, and unknown to science; and whether they have immediate effect or affect a large number of people. It follows that the definition of an acceptable risk is not hard-and-fast but should be interpreted with some flexibility. These safety goals or targets have been developed to provide overall guidance to operators and regulators, but any specific decision will necessarily involve other criteria as well, much of it non-quantifiable and/or emotional, that may also be important in balancing the preferences of interested parties.

Note that eDEOP-101 (2023) does not define or recommend risk criteria; however, reference is made to the NSW Government 2011 Hazardous Industry Planning Advisory (DoP 2011a). Risk criteria for land use safety planning for NSW suggest that risks to individual members of the public in residential areas (including hotels) should not exceed 1×10^{-6} or one in a million fatalities per year (see Table 12.2). However, individual industrial worker fatality risks should not exceed 50×10^{-6} fatalities per year. This risk criterion is also widely adopted in other states in Australia (AICE 2009).

TABLE 12.2
Risk Criteria for NSW Government Department of Planning (DoP 2011a)

Land Use	Suggested Risk Criteria (Risk in a Million Per Year)
Hospitals, schools, child-care facilities, old-age housing	0.5
Residential, hotels, motels, tourist resorts	1
Commercial developments, including retail centres, offices and entertainment centres	5
Sporting complexes and active open space	10
Industrial	50

Note, individual fatality risk is defined by the NSW Department of Planning as *"the risk of death to a person at a particular point"*. The individual fatality risk criteria for land use safety planning are the peak individual risk – i.e., a conservative measure where risk contours are plotted on the "basis that the individual is exposed for the full duration of the hazardous incident and no account is taken of evasive actions or protection by clothing, buildings, etc." (DoP 2011b). In other words, this is equivalent to the potential risk (PR), described by eDEOP-101 (2023), and as given by Eqn. (12.4).

Alternatively, the US Department of Defense mandates the following siting acceptance criteria (DoD 2024), which are based on individual fatality risks (IR), as expressed by Eqn (12.3):

• Any one related individual (worker)	$IR \leq 1\times10^{-4}$ fatalities per year
• All related individuals (all workers)	$IR \leq 1\times10^{-3}$ fatalities per year
• Any one unrelated individual	$IR \leq 1\times10^{-6}$ fatalities per year
• All unrelated individuals	$IR \leq 1\times10^{-5}$ fatalities per year

The DoD criteria are similar to that adopted by Singapore, Switzerland and other countries. (e.g., Linmei et al. 2010).

It is beyond the scope of the chapter to stipulate exact risk criteria for ordnance safety as this is an issue that requires a discussion of existing worker risks, WHS and Defence regulations, comparison of risk criteria adopted internationally and trade-offs between worker safety and costs.

It may be instructive to compare the risk criteria with public and worker risks from other hazards (see Table 12.3). The annual fatality risks for work and traffic accidents, cancer, homicides, natural disasters and many other hazards are well above the NSW Government public risk criteria of one in a million.

A risk criterion for societal risk may also be appropriate; however, there is less agreement on quantitative safety goals for this risk metric, and any guidance is indicative only (see Figure 12.6). Individual risks are expressed in terms of fatalities per year, fatalities per year of exposure etc. Societal risks are typically represented in terms of an F–N curve which is a plot of cumulative frequency (F) of n or more fatalities versus number of fatalities (N). This preference is reflected in a typical F–N curve, where for example an individual annual fatality risk of 10^{-4} has the same preference as an event killing ten people with a frequency of 10^{-6} per year. This shows an increasingly risk-averse behaviour as the consequences increases; however, from a purely rational viewpoint, this may be viewed as a somewhat illogical approach to increasing life-safety, and it is severely doubtful if such an approach can lead to efficient and rational decisions (e.g., Faber and Stewart 2003). Despite the long tradition for using F–N diagrams in risk analysis, it should be noted that these do not provide a consistent means for comparing the risks between different activities. Finally, while individuals are often risk-averse, the role of government and

TABLE 12.3
Annual Fatality Risks (Stewart and Mueller 2018).

Hazard	Territory	Period	Annual Fatality Risk
World War II	Worldwide	1939–1945	1 in 221
Cancers	US	2015	1 in 545
All accidents	US	2014	1 in 2350
Traffic accidents	Australia	2015	1 in 20,000
Homicide	US	2014	1 in 20,000
Traffic accidents	UK	2015	1 in 36,000
Homicide	Australia	2014	1 in 56,000
Industrial accidents	US	2014	1 in 66,000
Natural disasters	US	1999–2008	1 in 480,000
Terrorism	Western Europe	1970–2016	1 in 3,500,000
Terrorism	US	1970–2016	1 in 4,000,000
Terrorism	Australia (incl. Bali attacks)	1970–2015	1 in 8,000,000
Lightning	US	2006–2015	1 in 9,800,000
Commercial aviation accidents	Worldwide	2002–2015	1 in 13,000,000

regulators is to be risk-neutral and focus on mean estimates of risk and ignore the temptation to be risk-averse (e.g., Sunstein 2002).

12.4.4.2 Injury Risks

Because society is concerned with the risk of injury as well as the risk of death, eDEOP-101 (2023) and the NSW Department of Planning and Environment also describe an injury risk criterion where an incident explosion overpressure at residential areas should not exceed 7 kPa at frequencies of more than 50×10^{-6} per year. The 7 kPa threshold is the level above "which significant effects to people and property may occur" (DoP 2011a).

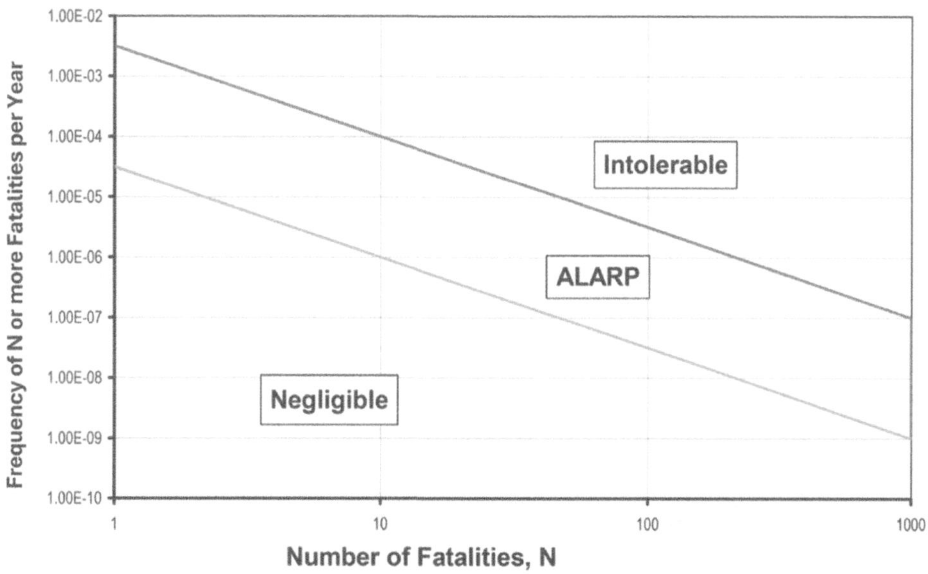

FIGURE 12.6 F–N curve showing societal risk tolerability (DoP 2011a).

12.4.4.3 Damage Risks

The NSW Department of Planning and Environment also describes a building damage risk criterion where an incident explosion overpressure should not exceed 14 kPa at frequencies of more than 50×10^{-6} per year. The 14 kPa threshold is the level above which "significant damage to buildings may occur" (DoP 2011a). An incident pressure level of 14 kPa may also damage nearby piping and equipment which may facilitate accident propagation (DoP 2011a).

12.4.4.4 SFARP and Cost–Benefit Analysis

As discussed above, a key concept of SFARP is to minimise the risk while ensuring that the cost is not grossly disproportionate to the risk. For example, when considering *Reasonably Practicable* judgements, the UK Health and Safety Executive (HSE) and the Work Health and Safety Act 2011 advise that the appropriate rule is that the measure must be adopted unless the sacrifice is *grossly disproportionate* to the risk.

A CBA can be performed to inform choices between risk mitigation options. However, under SFARP these conditions, the usual rules of CBA – that the measure should only be adopted if its benefit exceeds the costs – do not apply. Hence, even if the costs outweigh benefits, a mitigating measure could still be reasonably practicable to introduce – i.e., when balancing costs against safety, a CBA must bias the calculation in favour of safety. How much costs can outweigh benefits before being judged "grossly disproportionate" depends on the factors surrounding the risk. For example, the larger the risk, the greater can be the disproportion between the cost and the benefit. The UK HSE provides some guidelines for disproportionate factors (DF, or sometimes referred to as the gross disproportionate factor, GDF), noting that DFs of 3 are common for workplace environments but may extend as high as 10 under some circumstances (HSE 2013b). Hence, the cost of a mitigation measure is acceptable if the benefit-to-cost ratio (BCR) exceeds 1/DF – for example, if DF=10, then the BCR for a mitigating measure has to exceed 1/10 or 0.1 for a cost not to be grossly disproportionate to the risk. The value of DF may depend on the level of risk and whether risks are broadly acceptable or tolerable. The DF is also only applied to life-safety benefits, and not to benefits arising from reduced building damage or other direct or indirect economic or social losses.

The BCR for a mitigation measure is

$$\text{BCR} = \frac{\Delta\text{R} \times (\text{Existing Risk})}{\text{Cost of Mitigating Measure}} = \frac{\Delta\text{R} \times (\text{Hazard Likelihood}) \times \left(\text{DF} \times \text{L}_{\text{life-safety}} + \text{L}_{\text{economic}}\right)}{\text{Cost of Mitigating Measure}} \tag{12.7}$$

where ΔR is the percentage risk reduction that may arise from a combination of reduced hazard, vulnerability and/or consequences or exposure, DF is the disproportionate factor, $\text{L}_{\text{life-safety}}$ is the life-saving consequences (or losses), and $\text{L}_{\text{economic}}$ is the direct and indirect economic consequences. The existing risk represents the "business as usual" or "do nothing" scenario – that is, the risk before risk mitigation measures are implemented.

12.4.5 Uncertainty Modelling and Sensitivity Analyses

A conventional QERA should use point estimates (or "best-estimates") for event frequencies, exposure and consequences. Such numbers reflect the fact that (i) they are the best that can be obtained given the available information, and (ii) they are estimates rather than precisely determined quantities.

In many cases the probability of occurrence of an event, exposure or consequence cannot be described adequately by a point estimate value. An example is when there is a large amount of uncertainty about an event frequency. In this case it is appropriate to retain the uncertainties associated with the event frequency by modelling event frequency as a probability distribution and allow the risk analysis to propagate these uncertainties throughout the analysis. This may be termed a

PRA (e.g., Stewart and Melchers 1997) or more simply "uncertainty modelling". Uncertainty modelling can consider aleatory (random variation inherent in real-life events) and epistemic (knowledge uncertainty inherent in the model of the world and associated scientific algorithms) uncertainties. The US *Approved Methods and Algorithms for DoD Risk-Based Explosives Siting* (DDESB 2009) allows the use of uncertainty modelling in risk estimation.

Sources of uncertainty and variability in a QERA might include the following:

- event frequency
- exposure
- NEQ of hazardous material contributing to the event
- location (and range) of PES and ES
- hazard and consequence model uncertainty
- structural capacity of PES (e.g., steel strength and thickness)
- response of buildings and other infrastructure to blast overpressure and fragments
- human vulnerability to blast overpressure and fragments

Monte-Carlo computer simulation is a convenient tool to include uncertainty modelling in a risk analysis (e.g., Stewart and Melchers 1997). The outcome of uncertainty modelling is a probability distribution of hazards, consequences and/or fatality risks. This will provide the decision-maker with additional advice regarding the likelihood that a specific risk criterion is exceeded. For example, a decision to accept a risk would be easier if uncertainty modelling found there is a 99% chance that the fatality risk will be less than a risk criterion. On the other hand, if the likelihood of non-exceedance reduces to 50%, then the decision-maker may be less inclined to accept that level of risk.

An illustrative example of uncertainty modelling and its effect on fatality risk is shown in Figure 12.7. In this example, the probability of fatality is treated as a probabilistic variable as the human vulnerability to blast overpressure, and fragments is highly uncertain (e.g., DDESB 2009). If the probability of fatality (P_f) is uncertain, then the fatality risk is also uncertain and can be represented by a probability distribution, as shown in Figure 12.7.

Figure 12.7 shows that the best estimate (mean) potential risk is 6×10^{-6} fatalities per year, which is nearly ten times lower than the risk criteria for workers of 50×10^{-6} fatalities per year and, hence, easily satisfies the NSW Government's risk criteria for land use safety planning. However, due to the high uncertainties, the 90th percentile of potential risk (i.e., only a 10% chance of being exceeded) increases to 24×10^{-6} fatalities per year – this risk is still lower than the NSW Department of Planning and Environment risk criteria of 50×10^{-6} fatalities per year. The probability of exceeding

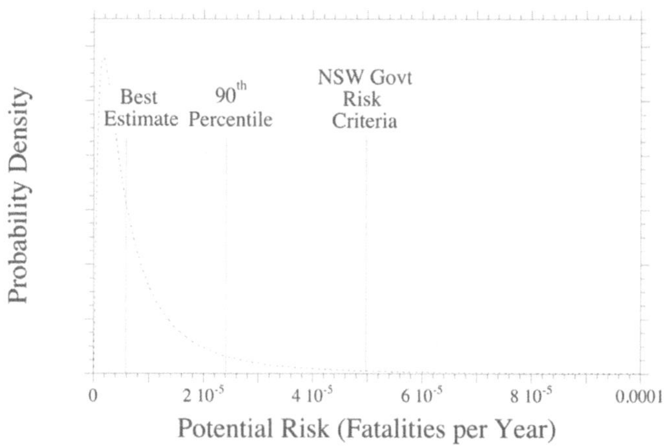

FIGURE 12.7 Probability distribution of potential risks considering uncertainty modelling.

the risk criteria is a very low 1% – meaning that there is a 99% chance that actual risk is less than the risk criteria. Hence, a risk-averse decision-maker might also deem that the risks posed by this EO scenario are acceptable. For more details see Stewart and Netherton (2019).

Uncertainty modelling is essentially one form of a sensitivity analysis. An alternate approach is to undertake a detailed sensitivity analysis of input parameters to assess their influence on the final result. These approaches are able to test the robustness of the results, and more importantly, the robustness of decisions. It can also help to reveal safety-critical processes or tasks, as well as interdependencies within and between systems. Finally, a QERA is in itself not a decision-making tool, but a risk-informed tool – i.e., it provides additional information to the decision-maker to help them make a more well-informed decision.

The robustness of a decision is also maximised if the risk assessment discusses and lists all assumptions. A key advantage of a QERA is that the risk assessment process is fully transparent. All assumptions and quantifications are explicitly stated and justified. The veracity of the evidence justifying key assumptions and quantifications can be fully tested. This allows for stakeholders to better understand the inputs in the analysis and how these inputs affect the final results and decisions.

12.4.6 Quality Assurance and Peer Review

Since the outcomes of a risk assessment can affect significantly the operations and management of EO facilities, as well as regulatory and licensing decisions, it is important that some form of control be exercised. Usually this consists of quality assurance measures and peer review. Quality assurance procedures tend to be focussed on the review of internal procedures and practices. Peer review consists of an independent and critical review, conducted by recognised experts in the area.

12.5 CONCLUSIONS

This chapter described the principles of a quantified explosive risk assessment (QERA) as specified in the Australian Department of Defence Explosives Regulations (eDEOP-101). The UK, the US and many other countries are also currently using quantitative risk assessment for authorising explosive activities. While the terminology differs, this is the same philosophy and method as a QRA or PRA applied to many hazardous systems. A QERA provides a risk-based decision tool to assess How Safe is Safe Enough? for the determination of safety distance and risk reduction measures to prevent fatality and injury from blast pressure, fragmentation and thermal hazards. An advantage of QERA is that uncertainty and variability are explicitly considered within a probabilistic framework. These variabilities may be considerable, and it is important to recognise that the world is not deterministic. A QERA is an accepted, robust and reasonable way of calculating safety and hazard risks, and the calculated fatality risks will then be compared with existing WHS and hazardous facilities planning quantitative risk acceptance criteria. This helps inform risk-based advice for decision-makers.

ACKNOWLEDGEMENTS

The support of the Australian Research Council grant DP210101487 is gratefully acknowledged. This chapter draws on some content from Stewart, M.G. and Netherton, M.D. (2019), A Probabilistic Risk-Acceptance Model for Assessing Blast and Fragmentation Safety Hazards, *Reliability Engineering and System Safety*, 191: 106492.

REFERENCES

AICE (2009), *Guidelines for Developing Quantitative Safety Risk Criteria*, Center for Chemical Process Safety, American Institute of Chemical Engineers, Wiley, New Jersey.

Air3D (2001), A Computational Tool for Airblast Calculations, developed by Rose, T.A., in: An Approach to the Evaluation of Blast Loads on Finite and Semi Infinite Structures, PhD Thesis, Cranfield University, Royal Military College of Science, Shrivenham, Oxfordshire (UK).

Axelsson, H. and Yelverton, J.T. (1996), Chest wall velocity as a predictor of non-auditory blast injury in a complex wave environment. *The Journal of Trauma: Injury, Infection, and Critical Care*, 40: 31–37.

Ayyub, B.M. (2001), *Elicitation of Expert Opinions for Uncertainties and Risks*, CRC Press, Boca Raton.

Campidelli, M., Razaqpur, A.G. and Foo, S. (2013), Reliability-based load factors for blast design, *Canadian Journal of Civil Engineering*, 40(5): 461–474.

ConWep (1991), *Conventional Weapons Effects Program*, Prepared by D.W. Hyde, US Waterways Experimental Station, Vicksburg, MS.

DDESB (2009), *Approved Methods and Algorithms for DoD Risk-Based Explosives Siting*, Department of Defense Explosives Safety Board (DDESB), Alexandria, VA, 21 July 2009.

DoD (2024), *DoD Ammunition and Explosives Safety Standards, DESR 6055.09-Std, February 23, 2024.* United States Department of Defense (DoD), Office of The Deputy Under Secretary of Defense, Washington, DC.

DoP (2011a), *Risk Criteria for Land Use Safety Planning, Hazardous Industry Planning Advisory No 4*, NSW Government Planning, Sydney, January 2011.

DoP (2011b), *Hazard Analysis, Hazardous Industry Planning Advisory No 6*, NSW Government Planning, Sydney, January 2011.

eDEOP-101 (2023), *Australian Department of Defence Explosives Regulations, Issue 4*, Australian Government, Canberra, ACT.

EU (2022), *Security by Design: Protection of Public Spaces from Terrorist Attack*, Joint Research Centre, European Union, Luxemburg.

Faber, M. and Stewart, M.G. (2003), Risk assessment for civil engineering facilities: Critical overview and discussion, *Reliability Engineering and System Safety*, 80(2): 173–184.

FEMA (2011), *Reference Manual to Mitigate Potential Terrorist Attacks Against Buildings*, FEMA 426, Federal Emergency Management Agency, Washington, DC, October 2011.

Gold, V.M., Baker, E.L., Poulos, W.J. and Fuchs, B.E. (2006), PAFRAG modeling of explosive munitions performance. *International Journal of Impact Engineering*. 33: 294–304.

Grisaro, H.Y. and Dancygier, A.N. (2019), Model of fragmentation distribution over a protective wall. *International Journal of Impact Engineering*, 128: 1–10.

Gurney, R.W. (1943). *Initial Velocities of Fragments from Bombs, Shell, Grenades*. Report Ballistic Research Laboratories, Report No. 405.

Hao, H., Li, Z.-H. and Shi, Y. (2016), Reliability analysis of RC columns and frame with FRP strengthening subjected to explosive loads, *Journal of Performance of Constructed Facilities*, 30(2): 04015017.

Haring, I., Schonherr, M. and Richter, C. (2009) Quantitative hazard and risk analysis for fragments of high explosive shells in air. *Reliability Engineering and System Safety*, 94(9): 1461–1470.

HSE (2013a), *Risk Analyses or 'predictive' Aspects of COMAH Safety Reports Guidance for Explosives Sites*, Health & Safety Executive, United Kingdom, https://www.hse.gov.uk/comah/assessexplosives/

HSE (2013b), *HSE Principles for Cost Benefit Analysis (CBA) in Support of ALARP Decisions*, Health & Safety Executive, United Kingdom, https://www.hse.gov.uk/

Huang, Y., Ma, G. and Li, J. (2017), Multi-level explosion risk analysis (MLERA) for accidental gas explosion events in super-large FLNG facilities, *Journal of Loss Prevention in the Process Industries*, 45: 242–254.

ISO 31000 (2018), *Risk Management: Guidelines*, International Organization for Standardization, Geneva, Switzerland.

ISO 13824 (2020), *General Principles on Risk Assessment of Systems Involving Structures*, International Organization for Standardization, Geneva.

Jarrett, D.E. (1968), Derivation of the British explosives safety distances, *Annals of the New York Academy of Sciences*.

Kingery, C.N. and Bulmash, G. (1984), *Airblast Parameters from TNT Spherical Air Burst and Hemispherical Surface Burst, Technical Report ARBRL-TR-02555*, US Army Armament Research and Development Centre, Ballistic Research Laboratory, Aberdeen Proving Ground, MD.

Kinney, G.F., Sewell, R.G.S. and Graham, K.J. (1981), *Definitive Quantity-Distance Tables for Stores and Explosives*, Naval Weapons Center, China Lake, CA.

Linmei, A.L., Hao, L.Y. and Lian, Y.C. (2010), *Risk Benchmarks for the Siting of Military Explosive facilities, DSTA Horizons 2010*, Defence Science and Technology Agency, Singapore, 136–147.

Lyman, O.R. (1979), The history of the quantity distance tables for explosive safety, *Memorandum Report ARBBRL-MR-02925*, US ARMY Armament Research & Development Command, Ballistic Research Laboratory, Aberdeen Proving Ground, MD.

Mannan, S. (2012), *Lees' Loss Prevention in the Process Industries*, Butterworth-Heinemann, Amsterdam.

Mueller, J. and Stewart, M.G. (2011), *Terror, Security, and Money: Balancing the Risks, Benefits, and Costs of Homeland Security*, Oxford University Press, Oxford and New York.

NATO (2015). *AASTP-1 NATO Guidelines for the Storage of Military Ammunition and Explosives.* The North Atlantic Treaty Organization (NATO), Bruxelles.

Netherton, M.D. and Stewart, M.G. (2010), Blast load variability and accuracy of blast load prediction models, *International Journal of Protective Structures*, 1(4): 543–570.

OBPR (2010), *Best Practice Regulation Handbook*, Office of Best Practice Regulation, Australian Government, Canberra, June 2010.

Olmati, P., Vamvatsikos, D. and Stewart, M.G. (2017), Safety factor for structural elements subjected to impulsive blast loads, *International Journal of Impact Engineering*, 106(8): 249–258.

ProSAir (2009), *Propagation of Shocks in Air, A Computational Fluids Dynamics Code Developed by the Department of Engineering Systems and Management*, Cranfield University, Defence Academy of the United Kingdom, Shrivenham, Oxfordshire (UK).

Qin, H. and Stewart, M.G. (2021), Casualty risks induced by primary fragmentation hazards from high-explosive munitions, *Reliability Engineering and System Safety*, 213: 107874.

Qin, H. and Stewart, M.G. (2024), Mitigating casualty risks from primary fragmentation hazards, *International Journal of Protective Structures*, 15(4):703–721.

RCC (2000). *Common Risk Criteria for National Test Ranges; Inert Debris, Supplement to Standard 321-00.* Range Commanders Council, U.S. Army White Sands Missile Range, NM.

Sielicki, P.W., Stewart, M.G., Gajewski, T., Malendowski, M., Peksa, P., Al-Rifaie, H., Studziński, R. and Wojciech, S. (2021), Field test and probabilistic analysis of irregular steel debris casualty risks from a person-borne improvised explosive device, *Defence Technology*, 17(6): 1862–1863.

Stewart, M.G. (2018), Reliability-based load factor design model for explosive blast loading, *Structural Safety*, 71:13–23.

Stewart, M.G. (2019), reliability-based load factors for airblast and structural reliability of RC columns for protective structures, *Structure and Infrastructure Engineering*, 15(5): 634–646.

Stewart, M.G. (2023), Spatial variability of explosive blast loading and its effect on damage risks to reinforced concrete buildings, *Engineering Structures*, 285: 115650.

Stewart, M.G., Baldacchino, H. and Netherton, M.D. (2019), Observed airblast variability and model error from repeatable explosive field trials, *International Journal of Protective Structures*, 10(3): 306–329.

Stewart, M.G. and Melchers, R.E. (1997), *Probabilistic Risk Assessment of Engineering Systems*, Chapman & Hall, London.

Stewart, M.G. and Mueller, J. (2018), *Are We Safe Enough? Measuring and Assessing Aviation Security*, Elsevier, New York.

Stewart, M.G. and Netherton, M.D. (2019), A probabilistic risk-acceptance model for assessing blast and fragmentation safety hazards, *Reliability Engineering and System Safety*, 191: 106492.

Stewart, M.G., Netherton, M.D., Qin, H. and Li, J. (2024), Explosive field trial for repetitive testing of VBIEDs to probabilistically measure blast and fragmentation hazards, *International Journal of Protective Structures*, 15(4): 937–963.

Sunstein, C.R. (2002), *The Cost-Benefit State: The Future of Regulatory Protection*, ABA Publishing, American Bar Association, Chicago, IL.

TNO (1992), Methods for the Determination of Possible Damage to People and Objects Resulting from Releases of Hazardous Materials (TNO Green Book), CPR 16E, Directorate-General of Labour of the Ministry of Social Affairs and Employment, The Hague, The Netherlands.

UFC 3-340-02 (2014), Structures to resist the effects of accidental explosions, Unified Facilities Criteria (UFC) Manual Number: 3-340-02. US Department of Defense (DoD), Washington, DC.

United Nations (2015), *Formulae for ammunition management*, International Ammunition Technical Guideline, United Nations Office for Disarmament Affairs, New York.

Van der Voort, M. and Weerheijm, J. (2013), A statistical description of explosion produced debris dispersion, *International Journal of Impact Engineering*, 59, 29–37.

Viper::Blast (2019), *Modelling and simulation of explosive detonation, blast loads on structures, weapon effects and injury.* www.viper-blast.com.

WHS (2017), *Work Health and Safety Act 2011*, No. 137, 2011, Compilation No. 8, Commonwealth of Australia, 21 September 2017.

WorkSafe (2011), *Guidance Note: Requirements for Demonstration*, WorkSafe Victoria, April 2011.

13 Human Vulnerability Subjected to Blast and Fragments

Mansoor Ali Khan, Zain Ali Khan, and William Graham Proud

13.1 FRAMING INJURIES

At the time of writing this chapter, late 2024, there are two widely reported conflicts involving significant explosive munition use: Russia–Ukraine (February 2022 to Present) and Gaza–Israel (October 2023 to present). In both cases, there has been a significant displacement of people. The number of civilian casualties in Ukraine in the first 18 months is around 12,000 killed and 24,000 wounded. While in Gaza, the total number of civilian deaths in the first six months (including Israeli civilians and hostages of the Hamas attack in October 2023) stands at around 40,000 killed, 80,000 wounded (BBC News, accessed 11 September 2024). The largest demographic within the casualties of Gaza are children, while in Ukraine, the largest demographic is male adults.

Without entering into the political dimensions or justifications, these numbers starkly show the difference between a country where the civilian population can move to 'rear' areas away from the immediate conflict (Ukraine) and those areas where the civilian population cannot extract itself from the main conflict zone (Gaza).

It is well-attested that from the 19th century onwards, civilians are increasingly directly targeted in conflict. Similarly, there is a continuum between 'dissent', 'terrorism', 'internal policing', 'counter-insurgency', 'low level conflict', 'civil war', and 'extended war'. Such terms often define the terms of description. This is an area where medicine, social sciences, engineering, natural sciences, and economics interact and is deeply fascinating.

In this chapter we will discuss the issues, research, and outcomes of injuries caused by munitions, bombs, unexploded ordnance, and improvised explosive devices. This will be framed through data taken from the Global Terrorism Database (LaFree and Dugan, 2007), which currently contains data from 1970 to 2020. Figure 13.1(a) shows the number of attacks recorded in the Global Terrorism Database, while Figure 13.1(b) gives some detail on the type of attack.

If we consider the UK population of ~70 million, in the period 2015–2020, on average, there are 150–200 significant domestic gas explosions per year, many linked to decaying infrastructure. However, when we consider the resulting number of people killed by those gas explosions, it is ~2–3 per year, with ~30–40 injured per year. In comparison, those poisoned by carbon monoxide are ~40–50 per year. These values are low compared to the 500–600 people murdered per year in the UK or the 1400–1600 killed in road accidents. It is clear, outside of a widely defined scenario of civil conflict, terrorism, or war, that blast and fragment injuries are a small factor in overall mortality in many countries. However, this should not induce complacency in the storage and movement of explosives and munitions.

13.2 BLAST WAVES AND FRAGMENTS

The specific details of blast and fragment formation are dealt with elsewhere in this book, see Chapter 11, and the index.

The fundamental process is the rapid transformation, by chemical reaction, from a solid or liquid to a hot, high-pressure gas. This gas violently expands, pushing against the surrounding air, driving a blast wave outwards. This wave, the blast wave, will move at supersonic velocity, decaying in intensity with distance while reflecting off barriers, moving around obstacles. These reflections

DOI: 10.1201/9781003529217-13

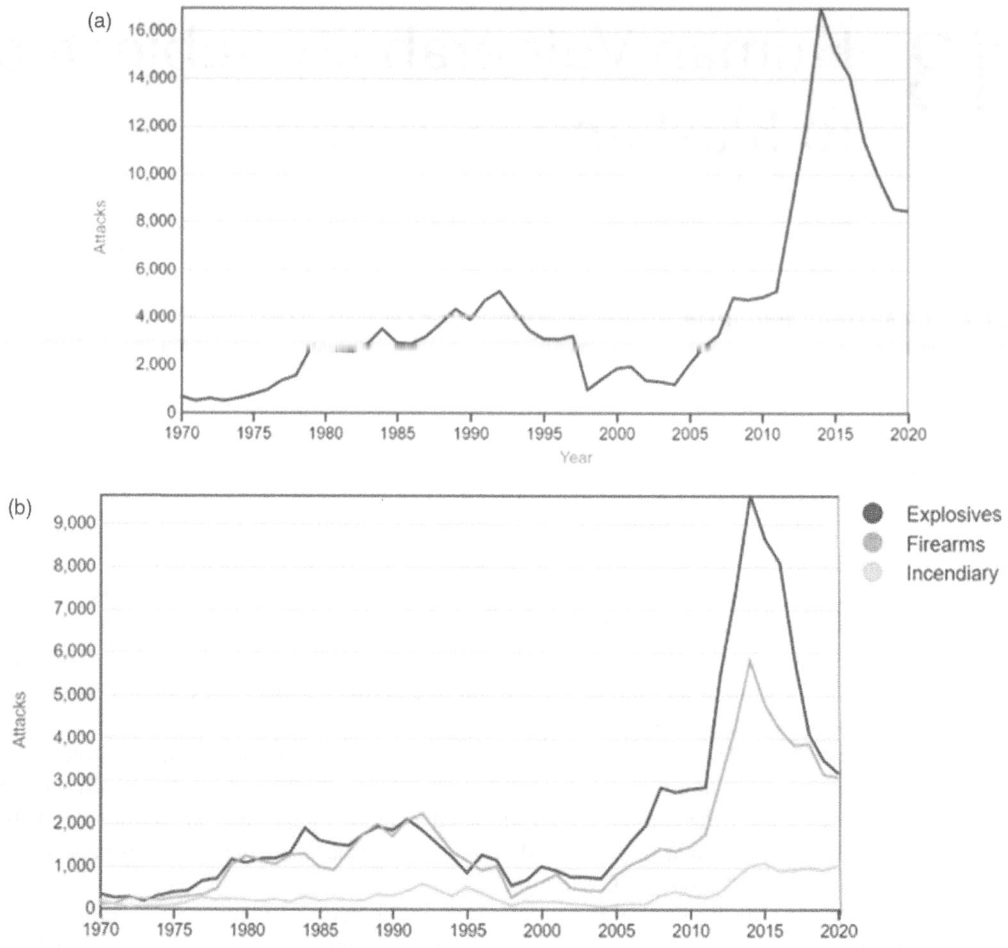

FIGURE 13.1 (a) The number of attacks per year from 1970 to 2020. Approximately half of these attacks resulted in the death of one person or more. The peak years of 2014 and 2015 include over 3000 attacks in one country, Iraq. (b) The principal weapons used, explosives, indicate explosives alone; firearms may indicate the use of firearms with grenades and other light munitions. The increased incidence of incendiaries is a notable feature of recent years.

may cause local strengthening of the blast. Initially, the gas atoms and molecules in the explosive reaction products move at several kilometres per second.

If there is a casing around the explosive charge, this will be accelerated and fragmented. The two basic theories associated with this are the Mott Fragmentation Criteria, which broadly describes the range of fragment sizes and are highly dependent on the casing material and its thickness. The velocity of the fragments is predicted via the Gurney energy, which is strongly dependent on the energy content of the energetic material. Fragments can initially be moving at 2–3 kilometres per second. Close to the explosive charge, there will be hot, high-temperature products at several thousand Kelvin, which can heat and burn objects through radiative heating or hot gas washing over surfaces.

Energetic materials themselves are divided into high-explosives, which detonate, release energies of megajoules per kilogram over microseconds of time. This class contains materials such as trinitrotoluene (TNT). Low-explosives will deflagrate (a technical term meaning 'burn') often with a degree of violence, releasing megajoules of energy but over milliseconds. This group contains many gun-powders, propellants. Pyrotechnics are the final class and represent systems such as flares, smokes/obscurants.

13.3 SURVIVABILITY

In the 1960s, there was a concerted effort to estimate and understand the level of blast, blast duration, and orientation of the person affected. Often these studies were conducted on a range of animals of different sizes, weights, and lung capacity. The style and type of test arrangement would not be granted ethical approval in the current research environment, so these reports are difficult to read. The test animals were placed in a field, restrained, and subject to blast. The number of immediate deaths was counted, and the test animals were then checked regularly over the next few days and longer-term survivability assessed. These studies are not for the squeamish, and it may be argued that extending rodent and ruminant physiology to the primate physiology of humans is not a completely coherent mapping. However, these do provide a solid, first-order approximation of survival; in the first-order, larger body mass helps survival. The types of results are as displayed in Figure 13.2.

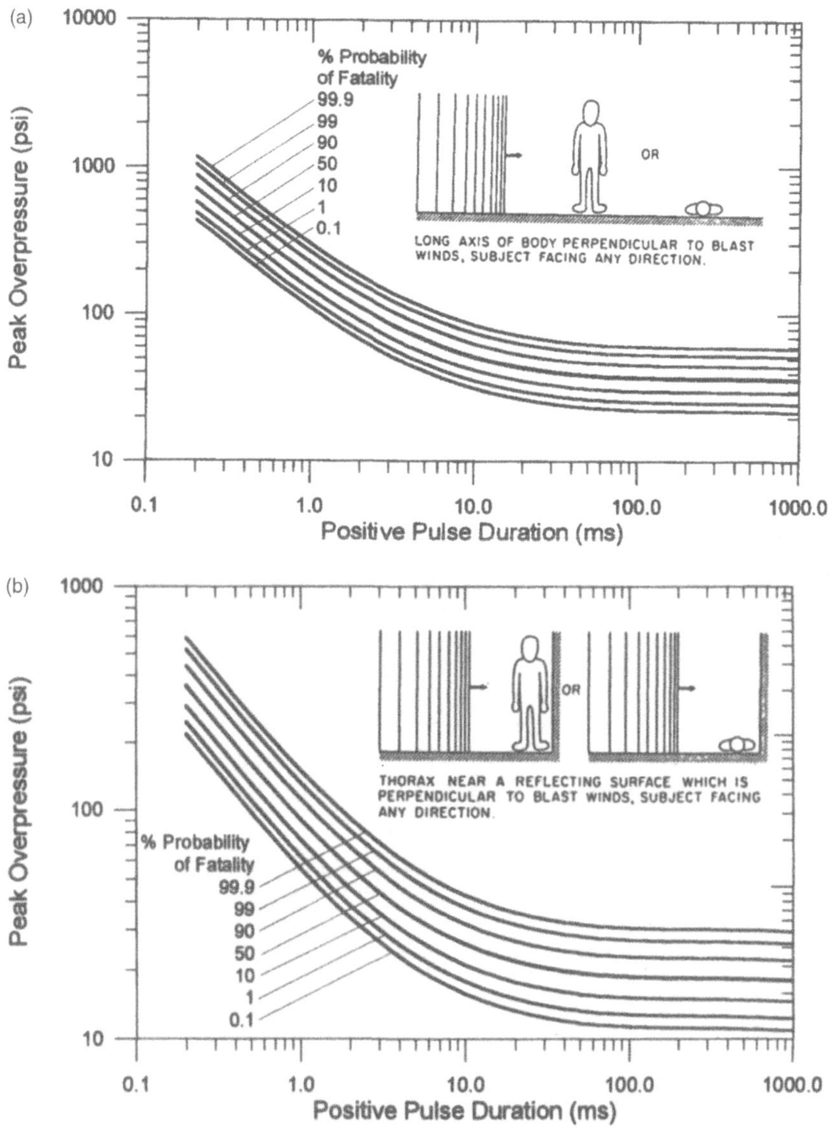

FIGURE 13.2 Example of estimated blast-survival curves from the 1960s, for (a) person far from a boundary, (b) being close to a boundary. The axes are in pounds per square inch, as this data originates from the USA. The main scaling parameter between the animal test subjects and humans is based on the average body mass. (Extracted from (data from Bowen et al. (1968), graphs redrawn from Cooper (1996)).)

These studies are key in showing the difference between protection against bullets and fragments. Bullets and fragments need solid, strong layers against penetration and high-speed sound to spread the incoming energy over a large area. Blast requires an 'easily' crushable layer to absorb the energy. Blast protection should not overly increase the cross-sectional area of the target to the blast waves, to prevent increased energy transfer to the target; this is not a significant consideration for ballistic armour.

13.4 OVERVIEW OF EXPLOSIVE EVENTS AND MECHANISMS

Explosive events occur when energy is rapidly released, generating a high-pressure wave that can cause significant damage to structures, objects, and living beings. These events can result from a variety of explosive materials, including chemical, nuclear, or mechanical sources, each with its unique mechanisms of action. Understanding the different types of explosive events and their effects on the human body is crucial for mitigating injury and developing protective strategies (Khan and Nott, 2021; Magnus et al., 2018).

13.4.1 Types of Explosions

Explosions are broadly classified into two main categories based on the speed of the chemical reaction and the energy release: high-order and low-order explosions.

High-order explosions involve rapid combustion and the near-instantaneous release of energy, generating a shock wave that travels faster than the speed of sound. High-order explosives, such as TNT (Krzystała et al., 2016) and RDX (research department explosive), produce intense overpressure and fragmentation (Banas et al., 2023).

Low-order explosions involve slower combustion processes and do not generate a shock wave. Instead, they produce a blast of lower intensity, often resulting in fire and structural damage. Low-order explosives include flammable gas leaks or fireworks, which cause injuries primarily through thermal burns, projectiles, and localised overpressure (Banas et al., 2023; Khan and Nott, 2018).

13.4.2 Blast Wave Propagation

When an explosion occurs, the rapid release of energy results in the formation of a blast wave, a high-pressure wave that expands outward from the explosion's epicentre. This wave is characterised by a sharp rise in pressure, followed by a rapid decrease in pressure as it moves away from the explosion (D'yachenko and Manyuhina, 2006; Khan and Nott, 2018, 2021; Loubeau et al., 2006; Magnus et al., 2018, 2020). The blast wave can be divided into two primary phases:

- *Positive Pressure Phase*: This phase occurs immediately after the explosion and is characterised by a sudden increase in pressure, known as overpressure. The high-intensity shock wave exerts significant force on objects and individuals in its path, compressing air and causing primary blast injuries.
- *Negative Pressure Phase*: After the initial shock wave passes, there is a negative pressure phase, where a vacuum effect is created as the air rushes back towards the explosion site. This phase can exacerbate injuries by pulling debris and fragments towards the point of explosion, leading to secondary injuries. See Figure 13.3.

13.4.3 Fragmentation and Debris Dispersal

In explosive events, fragmentation is a major cause of injury. Fragments can come from the explosive device itself (such as metal casings), surrounding materials (concrete, glass, etc.), or improvised projectiles. The speed, shape, and size of fragments influence the severity of injuries.

- Primary fragments are intentionally designed to cause harm, such as shrapnel from grenades or bombs.

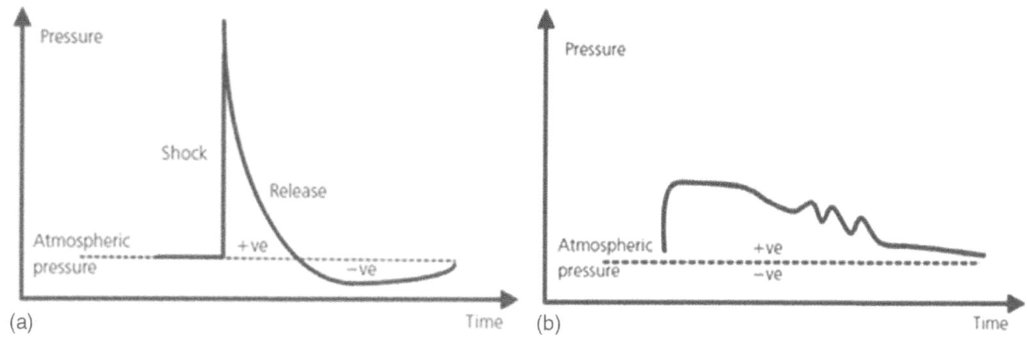

FIGURE 13.3 A schematic overview of blast wave shapes, (a) in an open field, the classic 'Freidlander' form, and (b) a longer, lower pulse seen within a vehicle. (Taken from Magnus et al. (2021).)

- Secondary fragments include environmental debris, which becomes lethal when propelled by the blast wave. Secondary fragments can vary in size and shape, leading to a wide range of injuries, from superficial lacerations to deep penetrative wounds.

The dispersal of fragments is unpredictable, as their trajectory depends on factors such as the explosive material, the type of casing, and the surrounding environment. The kinetic energy transferred by these fragments increases their ability to penetrate the human body, damaging vital organs and tissues.

13.4.4 Environmental Factors and Blast Impact

The impact of an explosion can vary significantly depending on the environment in which it occurs. Open spaces allow the blast wave to disperse more quickly, reducing the pressure felt by those at a distance. In contrast, enclosed environments (such as buildings, vehicles, or tunnels) trap the blast wave, amplifying its effects and increasing the likelihood of severe injuries. Structural collapse and fire may also compound injuries in such settings.

In an urban environment, blast reflection off buildings and walls can intensify the overpressure in certain areas, causing more severe injuries to individuals in these zones. Additionally, enclosed spaces increase the potential for secondary fragments, as materials from walls, windows, and equipment are shattered and thrown by the blast wave.

13.5 BLAST WAVE DYNAMICS

When an explosive device detonates, it releases energy rapidly, creating a shock wave that propagates outward at supersonic speeds (Gao et al., 2022). This blast wave is a sharp rise in atmospheric pressure, followed by a rapid decline. The pressure wave generated is known as overpressure—the amount of pressure exerted above normal atmospheric levels.

A blast wave consists of two phases (Geng et al., 2015; Magnus et al., 2021):

- Positive pressure phase occurs immediately after detonation and is characterised by a sudden spike in pressure. The intensity of this phase diminishes as the wave travels further from the source. The positive pressure phase causes most of the damage in an explosion, compressing air and creating a powerful force that can crush objects and human bodies.

- Negative pressure phase occurs after the initial shock wave passes, a vacuum effect is created as air is pushed outwards. This is followed by the air rushing back towards the point of explosion. This negative pressure phase, or under pressure, can pull debris and shrapnel inward, contributing to the overall destruction and injury potential. While less intense than the positive phase, it is longer-lasting and exacerbates the damage caused by the initial blast.

13.5.1 Differences Between Blast Waves in Open and Enclosed Spaces

The environment in which an explosion occurs plays a critical role in how the blast wave behaves and the severity of injuries (Magnus et al., 2021; Roh et al., 2020). The effects of a blast wave differ significantly between open and enclosed spaces:

- *Open Spaces*: In an open environment, the blast wave can expand freely, dissipating more quickly as it moves away from the source. Although individuals closer to the explosion face greater risks due to the higher overpressure, the rapid dissipation of energy generally reduces the blast's impact on those farther away. Injuries are usually confined to direct exposure or from fragments and debris propelled by the blast (Isaac et al., 2023).
- *Enclosed Spaces*: In confined environments (e.g., buildings, tunnels, vehicles—see Figure 13.3b), the blast wave can reflect off surfaces, amplifying the pressure. This leads to a phenomenon called blast reflection, where shock waves bounce off walls and other surfaces, creating higher pressure zones and reinforcing the destructive force of the explosion. The contained nature of the space prevents the wave from dispersing, often resulting in more severe injuries for anyone within the enclosure. Additionally, debris and fragments from the surrounding environment are more likely to be propelled at high speeds, increasing the risk of secondary injuries (Cullis, 2001).

13.5.2 Fragment Trajectories and Kinetics

Fragments are a major source of injury in explosive events. These fragments are dispersed by the explosive force, and their trajectory and behaviour are influenced by several factors, including the explosive force, the shape of the object, and any interactions with environmental obstacles (Zhang et al., 2001).

The initial energy from the explosion imparts velocity to fragments, sending them flying in all directions. The distribution of energy influences how far and fast fragments travel. Fragment shape plays a significant role in its trajectory. Irregularly shaped objects, such as shattered building materials, may tumble or deviate from a predictable path, while more streamlined fragments like metal shrapnel can travel farther and more accurately. Obstacles such as walls, vehicles, or terrain can alter the path of fragments by causing deflections or absorbing part of the energy, which can influence the direction and speed of the fragments, further complicating injury patterns.

13.5.3 Speed, Mass, and Shape of Fragments and Their Effect on Injury Severity

The lethality of fragments in an explosion depends heavily on their speed, mass, and shape. These factors determine how much kinetic energy is transferred upon impact, which in turn influences the severity of injury.

Fragments from an explosion can travel at speeds ranging from hundreds to thousands of metres per second. Higher-speed fragments carry greater kinetic energy, making them more likely to penetrate human tissue, cause deep wounds, or damage vital organs. The velocity of fragments decreases as they travel farther from the explosion, but even at reduced speeds, they can still inflict severe harm.

Heavier fragments carry more momentum and are capable of penetrating deeper into the body. Small, lightweight fragments, while still dangerous, may cause more superficial injuries or be stopped by barriers such as clothing or body armour. Larger, denser objects such as concrete blocks or metal debris pose a greater threat to structural integrity and human safety.

Fragment shape influences how easily an object can penetrate the body. Sharp, jagged fragments (e.g., broken glass or shrapnel) are more likely to cause lacerations, puncture wounds, and deep tissue damage. Rounded or blunt fragments may not penetrate as easily, but they can cause significant blunt force trauma, breaking bones or rupturing internal organs.

13.6 HUMAN VULNERABILITY TO PRESSURE AND FORCE

Different parts of the human body have varying tolerance levels to overpressure. The severity of injury depends on the amount of overpressure applied and the duration of exposure.

13.6.1 Primary, Secondary, Tertiary, and Quaternary Blast Effects

Medically, blast injuries are categorised into four distinct types (Khan and Nott, 2018; Loftis et al., 2019; Magnus et al., 2021; Patterson and Hamernik, 1997), based on the mechanism of injury (Figure 13.4):

- *Primary Blast Injuries*: These are caused by the direct effects of the blast wave (overpressure) on the body, particularly affecting gas-filled organs such as the lungs, ears, and gastrointestinal system. These injuries can lead to pulmonary barotrauma (blast lung), tympanic membrane rupture, and bowel perforations (Yeh and Schecter, 2012).
- *Secondary Blast Injuries*: Resulting from fragmentation, these injuries occur when shrapnel, debris, or other objects are propelled by the blast wave, striking the body at high

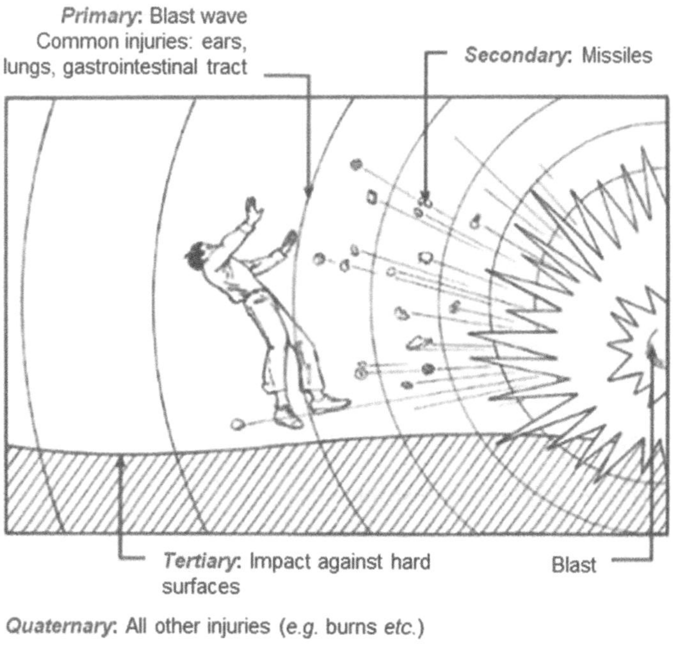

FIGURE 13.4 An overview of the differing classification of blast injuries. (Taken from Magnus et al. (2018).)

velocity. These fragments can cause lacerations, punctures, and blunt trauma to tissues, muscles, and organs (Delannoy et al., 2020; Korkmaz et al., 2023).

- *Tertiary Blast Injuries*: These are caused by the displacement of the body due to the blast wave, leading to blunt force trauma. Victims may be thrown into walls, ground surfaces, or other objects, resulting in fractures, head injuries, and internal trauma (Pasha et al., 2015; Poopitaya and Kanchanaroek, 2009; Rezaei et al., 2014).
- *Quaternary Blast Injuries*: These encompass all other effects of the explosion, such as burns, inhalation of toxic gases, and exposure to radiation or chemical agents. These injuries are often less immediate but can result in long-term health issues, including respiratory damage, infections, and chronic pain (Moga et al., 2021). See Figure 13.4.

The ears are highly sensitive to pressure changes. Overpressure as low as 5–15 psi can rupture the tympanic membrane (eardrum), leading to hearing loss and disorientation. Ear injuries are often the first detectable injury in blast victims (Hirsch, 1966; Liang et al., 2019).

The lungs are especially vulnerable to blast waves. Overpressure of around 30–40 psi can cause significant pulmonary barotrauma, leading to lung contusions, haemorrhaging, and in severe cases, air embolisms. The sudden compression and rapid expansion of air-filled spaces within the lungs make them prone to blast injuries, often referred to as blast lung (Cooper, 1996; Goatly et al., 2019; Stottmeister et al., 2020).

Hollow organs, such as the intestines, are also susceptible to blast-induced damage. Overpressure can cause bowel perforations and internal bleeding, particularly when victims are close to the explosion. Air trapped in the digestive system can expand rapidly, leading to tearing or rupturing of tissues (Clifford et al., 1984; Harmon and Haluszka, 1983).

13.6.2 Effects on Soft Tissue, Bones, and Internal Organs

Soft tissues, such as skin, muscles, and blood vessels, are highly vulnerable to both blast waves and fragments. Blast waves can cause lacerations, bruising, and tissue tears, while high-velocity fragments may cut through skin and muscle, leading to severe bleeding and contamination of wounds (Shuker, 2010).

Bones, though stronger than soft tissue, can still be fractured or shattered by the blunt force of a blast or by impact from flying debris. The severity of fractures varies depending on the proximity to the explosion and the size and mass of the impacting fragment. Long bones (e.g., femur, tibia) and the skull are particularly susceptible to fractures from tertiary blast effects, where victims are thrown against surfaces (Harmon and Haluszka, 1983).

Internal organs, particularly those with high fluid content (such as the liver, kidneys, and spleen), are prone to injury due to the rapid pressure changes associated with blast waves. Organs can be ruptured or torn, resulting in internal bleeding and life-threatening complications. The heart is vulnerable to blunt trauma caused by tertiary blast effects, which may lead to cardiac contusions or even rupture (Kokkinakis and Drikakis, 2023; Yang et al., 1996).

13.6.3 Primary Blast Injuries

13.6.3.1 Mechanisms of Primary Blast Injuries

Primary blast injuries result from the direct impact of a blast wave on the human body, particularly affecting air-filled structures. These injuries are caused by the sharp rise in pressure, known as overpressure, which can cause significant damage to the pulmonary, auditory, and gastrointestinal systems. Unlike secondary blast injuries, which involve projectiles, primary blast injuries occur from the blast wave itself and are most severe for those closest to the explosion (Magnus et al., 2021; Phillips, 1986).

13.6.3.2 Air Pressure Effects on Lungs, Ears, and Gastrointestinal Systems

The human body, particularly air-filled structures, is vulnerable to the rapid compression and decompression forces of a blast wave. The following systems are most susceptible to primary blast injuries:

- The lungs are highly sensitive to pressure changes due to their large surface area and the air-filled alveoli. The rapid increase in air pressure compresses the lungs, causing barotrauma and severe internal damage. Blast lung, characterised by pulmonary contusions and haemorrhage, is a common and life-threatening consequence of primary blast exposure (Kirkman and Watts, 2011; Stuhmiller et al., 1991).
- The ear is the most pressure-sensitive organ in the body, making it particularly vulnerable to blast waves. Even relatively low levels of overpressure can rupture the tympanic membrane (eardrum), leading to hearing loss, disorientation, and balance problems (Edwards and Clasper, 2016).
- The gastrointestinal system, especially hollow organs such as the intestines, is susceptible to pressure changes. The rapid compression and expansion of air within these organs can result in intestinal rupture, perforation, and internal bleeding, which may go undetected initially but lead to severe complications if untreated (Turégano-Fuentes et al., 2014).

13.6.3.3 Risk of Blast Lung, Perforated Eardrums, and Intestinal Rupture

The risk of primary blast injuries depends on the proximity to the blast and the magnitude of the overpressure:

- Blast Lung is one of the most severe primary blast injuries. It results from alveolar rupture and bleeding into lung tissues, leading to hypoxia, pneumothorax (air leakage into the chest cavity), and respiratory distress. Victims may have trouble breathing, coughing up blood, and diminished oxygen exchange. Blast lung is often fatal without immediate medical intervention (Smith, 2011).
- Perforated eardrums are a relatively common injury in blast events. Perforated eardrums can occur at overpressure levels as low as 5 psi. Victims may experience immediate hearing loss, tinnitus (ringing in the ears), and dizziness. Although eardrum rupture is not typically life-threatening, it can lead to long-term hearing impairment and increase the risk of infections (Cave et al., 2007; Cho et al., 2013).
- The intestines, being air-filled structures, can rupture under high pressure, leading to internal bleeding, peritonitis (inflammation of the abdominal lining), and infection. This type of injury may not present immediate symptoms, making diagnosis and treatment more challenging (Turégano-Fuentes et al., 2014).

13.6.3.4 Detailed Analysis of Vulnerable Organs

13.6.3.4.1 Pulmonary System

The pulmonary system is particularly prone to primary blast injury due to the fragility of the alveoli and their susceptibility to rapid pressure changes. The following conditions are common in victims of blast lung:

- Barotrauma is caused by an overpressure in the airways, causing alveolar rupture, allowing air to escape into areas where it should not be, such as the pleural space (pneumothorax) or the subcutaneous tissues (subcutaneous emphysema). This can lead to difficulty breathing, chest pain, and, in severe cases, collapse of the lung (Horrocks, 2001; Phillips, 1986).
- In a pneumothorax, air enters the pleural space, causing the lung to collapse. This can result in diminished oxygen exchange and severe respiratory distress. Blast waves can also

lead to tension pneumothorax, a life-threatening condition in which air accumulates in the chest cavity and compresses the lungs and heart (Guy et al., 1998; Horrocks, 2001).
- The high-energy blast wave can cause vascular damage within the lungs, leading to pulmonary haemorrhage. Blood leaking into the alveoli further reduces oxygen exchange and can lead to respiratory failure (Kirkman and Watts, 2011).

13.6.3.4.2 Auditory System

The auditory system is often the first part of the body to be injured in a blast due to the sensitivity of the tympanic membrane to changes in air pressure:

- The tympanic membrane (eardrum) is a thin, delicate structure that can rupture when subjected to sudden overpressure. A perforated eardrum typically results in immediate hearing loss, pain, and tinnitus. Although the membrane can heal over time, permanent hearing damage may occur, especially with repeated exposure to blasts (Cho et al., 2013).
- In addition to eardrum perforation, blast waves can damage the inner ear structures, leading to temporary or permanent hearing loss. Sensorineural hearing loss, which affects the cochlea and auditory nerve, is common in severe blast cases and can result in lasting impairment (Ritenour et al., 2008).

13.6.3.4.3 Gastrointestinal System

The gastrointestinal system can sustain significant injuries from blast waves, particularly the hollow organs like the intestines and stomach:

- The rapid compression and expansion of air within the intestines can lead to the rupture of the intestinal wall, resulting in perforations. Also, the rapid expansion of air can cause venous infarction of the bowel, which can result in delayed perforations. This allows the contents of the intestines to leak into the abdominal cavity, potentially causing peritonitis and sepsis if not treated promptly (Cripps and Cooper, 1997; Khan, 2018; Paran et al., 1996).
- Blast-induced tears in the gastrointestinal tract can cause internal bleeding, which may be difficult to detect initially. Blood loss from such injuries can lead to shock and other life-threatening complications if left untreated (Khan, 2018; Mathews and Koyfman, 2015).

13.6.4 Secondary Blast Injuries (Fragmentation Effects)

13.6.4.1 Injuries Caused by High-Velocity Fragments

Secondary blast injuries are caused by high-velocity fragments propelled by the explosion. These fragments, which can include shrapnel from the explosive device, debris from the surrounding environment, or parts of structures, cause a wide range of injuries, often with severe consequences. The nature of these injuries depends on the size, shape, and speed of the fragments, as well as the part of the body affected.

13.6.4.2 Lacerations, Punctures, and Blunt Trauma

Secondary blast injuries typically result in a variety of trauma types, primarily lacerations, punctures, and blunt trauma:

- Shrapnel and debris from an explosion often have sharp edges, leading to deep cuts and lacerations. These wounds can range from superficial to life-threatening, depending on the depth and location of the injury. Large, jagged fragments can tear through muscle and skin, often requiring extensive surgical intervention.

- Small, high-velocity fragments can penetrate the body, creating deep puncture wounds. These wounds can be deceptive, appearing small on the surface while causing significant internal damage. Puncture wounds are particularly dangerous as they can damage vital organs or blood vessels and are often accompanied by severe internal bleeding or infection.
- While most secondary blast injuries involve sharp or high-velocity fragments, some debris can cause blunt trauma. Large fragments or objects propelled by the blast may strike victims with great force, leading to fractures, bruising, and organ damage without penetrating the skin. This type of trauma can result in broken bones, concussions, or internal injuries.

13.6.4.3 Damage to Skin, Muscles, Bones, and Blood Vessels

The human body is vulnerable to a range of secondary blast injuries affecting multiple tissues:

- The skin is often the first line of defence against fragments, and blast debris frequently causes lacerations, abrasions, and burns. Deep lacerations can damage underlying tissues and increase the risk of infection if not treated promptly.
- Fragments can tear through muscle tissue, causing pain, loss of function, and muscle damage. Extensive muscle injuries may lead to loss of mobility or require reconstructive surgery, particularly in areas with dense muscle tissue such as the legs or torso.
- Secondary blast fragments can fracture or shatter bones. Injuries to long bones, such as the femur or tibia, may result in complicated fractures that require surgical fixation. Shattered bone fragments may also exacerbate the damage by acting as secondary projectiles within the body.
- Damage to blood vessels can lead to severe bleeding and haemorrhage. Punctures or lacerations to major arteries or veins are particularly life-threatening and require immediate medical attention. Uncontrolled bleeding can result in shock or death if not managed quickly (Cannon et al., 2017; Griffiths, 2011).

13.6.4.4 Ballistics of Fragment Penetration Through Various Tissues

The behaviour of fragments when penetrating human tissue is influenced by ballistic dynamics. Key factors such as velocity, size, shape, and material of the fragments determine the extent of injury.

The speed at which a fragment is propelled directly impacts the depth and severity of tissue penetration. High-velocity fragments are more likely to cause deep puncture wounds, penetrate bone, and damage internal organs. Lower-velocity fragments may cause more superficial injuries but can still lead to significant trauma if they strike vital areas.

The ability of a fragment to penetrate the body also depends on the type of tissue it encounters. For example, fragments easily pass through softer tissues like skin and muscle, while denser tissues such as bone can reduce the velocity of the fragment or cause it to shatter.

13.6.4.5 Fragment Size, Shape, and Material

The size, shape, and material composition of fragments play a critical role in the resulting injury:

- Larger fragments typically cause more extensive damage, as they can sever muscles, bones, and blood vessels more effectively than smaller fragments. However, smaller fragments may travel deeper into the body and lodge in vital organs, complicating treatment.
- The shape of a fragment affects how it interacts with tissue. Sharp, jagged fragments create more severe lacerations and are more likely to penetrate deeply. In contrast, blunt or rounded fragments may cause less penetration but can result in significant blunt force trauma and internal injuries.
- The material of the fragment, whether it's metal, concrete, or glass, also influences injury outcomes. Metal fragments, especially those from the explosive casing, are often more lethal due to their sharpness and density, while glass and concrete may shatter upon impact, creating secondary projectiles within the body.

13.6.4.6 Medical Response to Fragmentation Injuries

13.6.4.6.1 Immediate Interventions

Prompt and effective medical intervention is crucial in treating fragmentation injuries to prevent further complications such as haemorrhage, infection, or shock:

- Uncontrolled bleeding is the primary concern in fragmentation injuries. The immediate application of tourniquets or pressure dressings is critical in controlling external bleeding, particularly when major arteries or veins are affected. In cases of internal bleeding, emergency surgical intervention may be required to locate and repair damaged blood vessels (Cannon et al., 2017; Clough and Khun, 2019; Benítez et al., 2021).
- Cleaning and dressing wounds to prevent infection is a vital step in managing fragmentation injuries. Debris embedded in wounds must be removed carefully to avoid further tissue damage, and deep lacerations or puncture wounds often require sutures or staples. In more severe cases, surgical debridement may be necessary to remove damaged tissue and foreign material (Duchesne et al., 2018).

13.6.4.6.2 Long-Term Care

Once the immediate threat is managed, long-term care focuses on restoring function and preventing complications:

- Many victims of fragmentation injuries require multiple surgeries to repair damaged tissue, remove embedded fragments, and address complications such as fractures or organ damage. For example, orthopaedic surgeries may be necessary to set broken bones, while vascular surgery might be needed to repair damaged arteries or veins (Duchesne et al., 2018).
- Physical therapy and rehabilitation are critical components of recovery (Smits et al., 2024), particularly for those with muscle or bone injuries. Patients may need extended rehabilitation to regain strength and mobility, especially if limb function has been impaired or amputations were required. Psychological support may also be necessary, as blast survivors often suffer from post-traumatic stress disorder (PTSD) (Chen and Huang, 2011; Rosenfeld and Ford, 2010) or other mental health challenges due to their experiences.

13.6.5 Tertiary Blast Injuries

13.6.5.1 Impact and Acceleration Forces

In blast events, the range of injuries extends beyond the direct effects of the explosion and fragments. Tertiary and quaternary blast injuries occur because of the body being displaced by the force of the blast or due to environmental factors such as heat, toxic products, as well as chemicals and pathogens present in the environment. These injuries can complicate medical management and lead to long-term health consequences (Kirkman et al., 2011).

Tertiary injuries occur when the blast wave physically displaces the body, propelling it into solid objects or throwing individuals to the ground. These injuries are caused by both impact and acceleration forces that act on the body during this displacement.

13.6.5.2 Blunt Trauma from Body Displacement

The force of the blast can cause individuals to be thrown through the air or slammed into nearby walls, ground, or other objects. These violent displacements result in blunt trauma, which is characterised by the following:

- High-impact collisions with hard surfaces often lead to fractures in the extremities, ribs, and pelvis. The nature of these fractures depends on the angle and force of the impact. For instance, falls from a height or horizontal displacement into rigid objects can cause complex fractures that require surgical repair (De Candole, 1967).

- Blunt force trauma can also cause contusions (bruises) and damage to internal organs without penetrating the skin. The force from the blast may compress organs such as the liver, spleen, or kidneys, leading to internal bleeding and organ damage that requires immediate medical intervention.

13.6.5.3 Common Injuries

Tertiary blast injuries frequently result in traumatic brain injuries (TBIs), which occur when the head strikes a surface or is subjected to rapid acceleration-deceleration forces (Kocsis and Tessler, 2009).

TBIs are a major concern in tertiary blast injuries, especially when the head is impacted. Depending on the severity, TBIs can range from mild concussions to severe brain damage that results in cognitive, sensory, or motor deficits. Symptoms of TBIs include loss of consciousness, confusion, memory loss, and behavioural changes. Immediate medical attention is necessary to assess and manage brain injuries (Bochicchio et al., 2008; Mathews and Koyfman, 2015).

High-energy trauma can lead to broken bones, particularly in the arms, legs, spine, and ribs. Complex fractures may require surgical fixation, and in some cases, victims may suffer permanent disability or require long-term rehabilitation.

13.6.6 Quaternary Injuries

Quaternary injuries are those caused by environmental factors associated with the explosion, such as burns, toxic inhalation, and exposure to hazardous chemicals or radiation. These injuries are not directly caused by the blast wave but are still life-threatening and require specialised treatment (DuBose et al., 2016).

13.6.6.1 Thermal Burns and Inhalation Injuries from Toxic Gases

The intense heat generated by explosions often results in thermal burns, while the combustion process can release toxic gases and particles, leading to inhalation injuries.

The heat from the explosion can cause severe burns, particularly if the blast ignites clothing or nearby materials. Burns may range from first-degree (superficial) to third-degree (deep tissue), depending on proximity to the heat source. Victims with extensive burns may experience fluid loss, infection, and require skin grafts or reconstructive surgery (Beaven and Parker, 2015; Bumbaširevic et al., 2006).

Inhalation of hot gases, smoke, or toxic chemicals can damage the respiratory system. Blast events often release gases such as carbon monoxide, ammonia, or other combustion byproducts that irritate the lungs and airways. Inhalation injuries can lead to inflammation of the airways, pulmonary oedema (fluid in the lungs), and, in severe cases, respiratory failure. Victims may require supplemental oxygen, airway management, or mechanical ventilation (Luks, 2010).

13.6.6.2 Special Considerations for Chemical and Radiological Exposure

Explosions in industrial or military settings can involve chemical or radiological hazards in addition to the blast itself.

Certain blast events, particularly those involving industrial accidents or terrorist attacks, may release hazardous chemicals into the air. Victims exposed to these substances may suffer from chemical burns, respiratory distress, or systemic toxicity. Decontamination is a crucial first step in managing these injuries, followed by specific treatments to counteract the effects of the chemical agent.

In rare instances, explosions may involve radioactive materials, resulting in radiological exposure. Victims exposed to radiation may suffer from acute radiation sickness, characterised by nausea, vomiting, and damage to bone marrow and other tissues. Long-term exposure increases the risk of cancer and other serious health issues. Immediate containment and specialised medical care are essential for those affected by radiological incidents (Khan and Frankel, 2015a, 2015b).

13.6.7 Cumulative Effects of Multiple Injury Types

In blast events, individuals often suffer from multiple types of injuries simultaneously, leading to more complex medical management and worse outcomes.

Tertiary and quaternary injuries frequently occur in conjunction with primary and secondary blast injuries, meaning victims may suffer from both blunt trauma and lacerations, or burns, along with fractures. For instance, a victim thrown into a wall by the blast may also suffer shrapnel wounds and thermal burns. The cumulative effects of these injuries increase the risk of complications such as infection, multi-organ failure, and death.

The combination of blast-related injuries can make recovery more difficult. For example, a patient with both lung damage from inhalation injury and multiple fractures faces challenges in respiratory therapy and mobility. The burden on the healthcare system is also greater when dealing with multiple traumas, as victims may require prolonged hospitalisation, multiple surgeries, and extensive rehabilitation.

The cumulative effect of blast injuries can lead to long-term physical and psychological disabilities. Survivors may experience chronic pain, impaired mobility, or permanent damage to vital organs. Psychological effects such as PTSD and depression are also common, particularly when blast injuries result in permanent disfigurement or loss of function.

13.6.8 Biomechanics of Blast and Fragmentation Injuries

13.6.8.1 Modelling Human Response to Blast and Fragments

Understanding the biomechanics of blast and fragmentation injuries is critical for improving injury prevention, protective measures, and medical response. The study of human response to blasts and high-velocity fragments involves computational modelling, experimentation, and biomechanical analysis to identify thresholds for injury and the factors that influence the severity of trauma (Price et al., 2017; Solomos et al., 2020).

Computational models play a key role in simulating how the human body responds to blasts and fragmentation events. These models help researchers and engineers predict injury patterns, assess the effectiveness of protective gear, and improve understanding of injury mechanisms(Chanda and Callaway, 2018).

Finite Element Analysis (FEA) is one of the most used computational methods in studying blast injuries (Chanda and Callaway, 2018; Gupta and Przekwas, 2015). This technique divides the human body into thousands of tiny elements, allowing researchers to simulate how different body tissues (e.g., bones, muscles, organs) deform under the pressure of a blast or the impact of a fragment. FEA provides valuable insight into the specific locations and types of injuries that may occur under various conditions.

Computational Fluid Dynamics (CFD) is another modelling technique used to study how blast waves travel through air and interact with the human body (Chanda and Callaway, 2018; Sarvghad-Moghaddam, 2015). By analysing how pressure waves propagate in different environments (e.g., open fields versus enclosed spaces), CFD models help assess the risk of injury based on variables such as blast intensity, distance from the explosion, and environmental factors like walls or barriers.

Human Body Simulations, such as crash-test dummies modified for blast exposure, allow researchers to test different protective systems and assess their performance under blast conditions (Gupta and Przekwas, 2015). These simulations often integrate both FEA and CFD data to provide comprehensive insights into the complex interactions between blast forces and the human body.

13.6.8.2 Experimentation in Controlled Environments

In addition to computational models, researchers use controlled experiments to validate their findings and refine injury thresholds.

Animal testing, particularly using species with similar anatomical and physiological characteristics to humans (e.g., pigs), helps researchers study the biological response to blast waves and fragment impacts. These experiments provide valuable data on tissue damage, haemorrhaging, and organ trauma, which are then extrapolated to human injuries. Although ethically sensitive, animal models continue to offer key insights into the real-world effects of blasts and fragmentation that cannot be captured by models alone (Risling and Davidsson, 2012).

Crash-test dummies modified to simulate blast scenarios are frequently used to measure the mechanical forces exerted on the human body during a blast event. Dummies equipped with pressure sensors and accelerometers help track the exact forces experienced by different body regions, providing critical information on how blunt trauma, overpressure, and fragmentation forces distribute across the body. These tests are often performed in specialised facilities using controlled explosions (Hayda et al., 2004; McKay, 2010).

13.6.9 Biomechanical Thresholds for Injury

13.6.9.1 Understanding the Injury Thresholds for Overpressure and Fragment Impact

Understanding the thresholds at which blast waves and fragments cause injury is vital for developing protective systems and medical interventions. These thresholds vary depending on the specific injury type, the body part affected, and the intensity of the blast.

The human body can tolerate low levels of overpressure, but as blast pressure increases, tissues begin to fail. Pulmonary barotrauma (lung injury) is one of the most common effects of primary blast waves, typically occurring at overpressure levels above 40–80 kPa. Other sensitive structures include the eardrums, which rupture at pressures around 35 kPa (Greer, 2006), and the gastrointestinal system, which may suffer perforation at higher pressure levels. The threshold for fatal overpressure injuries is generally above 100 kPa (Greer, 2006; Hubbard, 2016), depending on the duration of exposure and environmental factors.

Fragmentation injuries result from shrapnel or debris travelling at high speeds. Injury thresholds for fragments are influenced by velocity, mass, and surface area of the projectile. Larger, blunt fragments require even higher velocities to penetrate deep tissues. The presence of personal protective equipment (PPE) like body armour alters these thresholds by absorbing and redistributing the force of the fragments.

13.6.9.2 Factors Influencing Injury Severity

Several factors influence the biomechanical response of the human body to blasts and fragmentation, and thus the severity of injury.

The orientation and positioning of the body in relation to the blast can significantly affect injury outcomes. For example, a person facing the explosion may experience more severe abdominal and chest injuries, while someone positioned sideways might be more vulnerable to limb injuries. Furthermore, body parts closer to the ground may be shielded from direct blast exposure but are more likely to be injured by ground shock or secondary fragments (e.g., debris) (Mathews and Koyfman, 2015).

The use of body armour, helmets, and eye protection can drastically reduce injury severity by absorbing and dissipating the energy from blast waves and fragments. Modern military body armour, designed to withstand shrapnel and bullet impacts, helps protect vital organs from both primary and secondary blast effects. However, armour is not always effective in preventing injuries to exposed areas such as the limbs, face, and neck, or against tertiary forces like being thrown into objects (Yang, 2023).

The presence of walls, vehicles, or other barriers between an individual and the explosion significantly influences the outcome. In open environments, the blast wave disperses rapidly, reducing overpressure and fragment velocities as they spread. However, in enclosed spaces such as buildings or vehicles, the blast wave reflects off surfaces, amplifying pressure and potentially causing more severe injuries. Barriers can also block or deflect fragments, but in some cases, they may shatter and create additional projectiles.

13.7 PROTECTION AND MITIGATION STRATEGIES

13.7.1 Personal Protective Equipment (PPE)

In the context of blast and fragmentation injuries, developing effective protection and mitigation strategies is essential to minimising injury and enhancing survivability. These strategies encompass a combination of PPE, structural protections, and technological innovations aimed at reducing the impact of blast waves and fragment penetration.

PPE plays a crucial role in protecting individuals from both the direct and indirect effects of blast and fragmentation injuries. Modern military and law enforcement personnel, as well as civilians in high-risk environments, rely on various forms of PPE to shield vital body regions from trauma.

13.7.1.1 Helmets, Body Armour, and Blast-Resistant Clothing

Helmets are designed to protect the head from both blunt force and penetrating injuries. Helmets are a key component of PPE. Military helmets, such as the Advanced Combat Helmet (ACH), are constructed using Kevlar or composite materials to absorb shock and resist fragmentation (Kulkarni et al., 2013). Helmets are particularly important for preventing TBIs resulting from blasts and projectiles. However, they do not provide complete protection against severe blast waves or the concussive force of large explosions.

Modern body armour consists of ballistic-resistant materials designed to shield the torso and other vital areas from fragment penetration and bullet impacts (Hada and Kumar, 2019). There are various types of body armour, ranging from soft armour, which is made of flexible materials like Kevlar and offers protection against low-velocity fragments, to hard armour with ceramic or metal plates capable of stopping high-velocity projectiles and shrapnel. While body armour significantly reduces the risk of fatal injury, it does not protect against blast overpressure, and exposed areas like the neck and limbs remain vulnerable (Ciesielska-Wróbel, 2017).

Specialised clothing designed for blast protection can offer limited protection against burns and shrapnel. This includes flame-resistant fabrics to prevent burns from explosions and blast shorts that protect the lower body from injury. While this clothing adds an additional layer of protection, its effectiveness is limited by the wearer's mobility and the intensity of the blast (Zhou et al., 2005).

13.7.1.2 Limitations and Performance in Real-World Scenarios

While PPE is critical in reducing injury severity, there are several limitations in real-world scenarios:

- PPE typically covers only the head, chest, and torso, leaving other body parts like the arms, legs, and face vulnerable to injury. Blast waves and fragments can still cause severe damage to unprotected areas.
- High-level protection often comes with the drawback of increased weight, which can limit mobility. Soldiers wearing full body armour may experience fatigue, reduced agility, and slower reaction times, which can be detrimental in fast-paced, dangerous environments.
- PPE is primarily designed to protect against fragments and bullets. However, blast overpressure—the sudden increase in atmospheric pressure caused by a blast wave—can still cause internal injuries such as lung damage or brain trauma, even if the individual is wearing full armour.

13.7.2 Structural and Environmental Protection

In addition to personal protective gear, structural protections such as barriers, fortified buildings, and armoured vehicles are critical in reducing the impact of blasts and fragments.

13.7.2.1 Blast Barriers, Bunkers, and Vehicle Armour

Blast barriers are walls or structures specifically designed to absorb and deflect the energy of an explosion. These barriers are used in both military and civilian environments to protect personnel from blast effects. Reinforced concrete, steel, and sandbagged barriers are commonly used to block or redirect the force of an explosion away from sensitive areas (Goel and Matsagar, 2014).

Underground bunkers or heavily reinforced above-ground shelters provide substantial protection against both direct blasts and fragmentation. Built with thick walls and ceilings, often using steel-reinforced concrete, bunkers can withstand significant overpressure and prevent fragments from penetrating. Bunkers are especially effective in military settings, where they offer safe havens for soldiers during attacks (Skrabacz and Wołoch, 2024).

Armoured vehicles used by military and security forces provide mobility and protection in conflict zones. These vehicles are equipped with heavy armour plating to resist blasts and high-velocity fragments. Innovations in vehicle armour, such as the MRAP (Mine-Resistant Ambush Protected) vehicles (Dikici, 2014), are designed to protect against both roadside bombs and larger explosive devices. The 'V-shaped' hull design of these vehicles helps deflect the blast force away from the occupants.

13.7.2.2 Importance of Distance, Shielding, and Positioning

The distance from the blast source is one of the most critical factors in reducing injury risk. The inverse square law explains that the farther a person is from the explosion, the lower the intensity of the blast wave and the fewer fragments reach them. Strategic positioning to increase distance from potential blast sites is essential in high-risk environments (Khan and Nott, 2018).

Using physical objects such as walls, barriers, or natural terrain features (e.g., hills or ditches) can provide effective shielding against blasts. Shielding helps reduce exposure to both overpressure and fragmentation.

Proper positioning is crucial for maximising the protective benefits of structural barriers. For instance, lying flat on the ground with feet pointed towards the blast source reduces the body's surface area exposed to the blast wave, minimising the risk of injury.

13.7.3 Technological Innovations in Protection

Ongoing research into materials science and protective design is driving technological advancements that offer enhanced protection against blast and fragmentation injuries. These innovations aim to improve both the effectiveness and comfort of PPE and structural protection.

13.7.3.1 New Materials and Designs for Reducing Blast Impact and Fragment Penetration

Advances in composite materials, including ultra-high-molecular-weight polyethylene (UHMWPE), carbon fibre, and nanomaterials, are making PPE and armour both lighter and more effective. These materials offer improved ballistic resistance while maintaining flexibility and mobility, reducing the physical strain on users (Padhy and Kandasubramanian, 2024).

Research into energy-absorbing foams (Toboła et al., 2023) is being used to design helmets and body armour with better protection against blast overpressure. These foams can dissipate the energy from shock waves, helping reduce the risk of TBIs and internal organ damage.

Some new armour designs incorporate adaptive technologies that change their protective properties in response to the environment. For example, armour that becomes more rigid upon impact can protect against both blunt force and penetration from high-velocity fragments.

13.7.3.2 Research into Wearable Technologies and Smart Armour

Wearable sensors embedded in body armour or helmets are being developed to monitor vital signs and detect blast overpressure levels. These sensors provide real-time data to medical personnel, helping them quickly assess the risk of internal injuries, such as lung damage or concussions, even when external injuries are not immediately visible.

Smart armour technologies use embedded electronics and advanced materials to enhance protective capabilities. These systems can monitor impact forces, detect fragment penetration, and adjust the armour's properties in real time. For instance, smart fabrics that harden upon detecting an imminent threat can increase protection against shrapnel and blast debris.

In military applications, powered exoskeletons are being researched to provide both mobility assistance and enhanced protection. These wearable machines can help soldiers carry heavier loads, reduce fatigue, and potentially integrate armour that provides additional protection against blast forces (Loftis et al., 2019).

13.8 MEDICAL MANAGEMENT AND TREATMENT OF BLAST INJURIES

13.8.1 Emergency Response to Blast Trauma

Effective medical management of blast injuries requires a multi-disciplinary approach that addresses the unique and complex nature of trauma caused by explosions. The treatment of blast injuries begins with emergency response and triage, followed by surgical interventions to manage severe wounds, and concludes with long-term rehabilitation to help survivors recover both physically and psychologically.

Immediate response to blast trauma is critical to increasing the chances of survival and minimising long-term damage. The chaotic nature of blast events requires quick and organised triage, stabilisation, and transport to advanced medical facilities.

13.8.2 Triage and Field Stabilisation of Victims

In the aftermath of a blast, triage prioritises patients based on the severity of their injuries. Triage protocols are designed to rapidly assess patients' vital signs, airway, and circulation to determine who requires immediate intervention. Victims with life-threatening injuries, such as severe bleeding or compromised airways, are treated first. Due to the unpredictability of blast injuries, medical responders must be prepared for a range of trauma types, including polytrauma (injury to multiple systems) (Khan and Nott, 2021).

Immediate stabilisation in the field is vital for preventing further injury and maintaining critical functions. Common interventions include the following:

- Ensuring that victims can breathe is a top priority, often requiring the use of emergency airway devices (e.g., oropharyngeal airways) or even emergency tracheostomies in cases of airway obstruction.
- Tourniquets, pressure dressings, and haemostatic agents are essential for controlling severe bleeding from blast-induced wounds. Rapid blood loss is a common cause of death in blast incidents, making early intervention crucial.
- Victims suffering from shock due to blood loss may require intravenous (IV) administration of blood to stabilise their blood pressure and circulation.

13.8.3 Critical Care for Pulmonary, Abdominal, and Head Injuries

Blast lung (pulmonary barotrauma) is one of the most common primary blast injuries, characterised by alveolar rupture, pneumothorax, and haemorrhage. Victims may require immediate oxygen support, and in severe cases, mechanical ventilation to manage hypoxia (Khan and Frankel, 2013; Matsushima et al., 2014). Advanced trauma centres may use positive pressure ventilation or extracorporeal membrane oxygenation (ECMO) to assist in oxygenating blood.

Primary blast waves and secondary fragments can cause significant damage to the gastrointestinal system, leading to internal bleeding, perforations, and organ rupture. Field stabilisation may include managing external bleeding and preparing patients for emergency surgery to repair internal damage.

Blast-induced TBIs are common and often difficult to detect in the field. Early signs of TBI, such as altered mental status or loss of consciousness, require rapid intervention. Patients with severe TBI may need immediate neurosurgical intervention to relieve intracranial pressure or control brain haemorrhage.

13.8.4 Surgical Interventions

Surgery is often necessary to manage the complex injuries caused by blasts and fragments. Surgeons must be prepared to address not only the visible trauma but also internal injuries that may be life-threatening (Griffiths, 2011).

13.8.4.1 Managing Fragmentation Wounds, Amputations, and Complex Fractures

Shrapnel and debris from explosions can cause deep penetration wounds. Surgical intervention typically involves debridement to remove foreign objects and damaged tissue, followed by repair of muscles, nerves, and blood vessels. Laparotomy may be required to manage internal bleeding or organ damage caused by fragments.

Severe blast injuries, particularly those affecting the limbs, may necessitate amputation. This can occur immediately due to the blast or later because of infection or vascular compromise. Surgeons aim to preserve as much tissue as possible while ensuring that the wound is clean and well-vascularised to support healing and future prosthetic use.

The intense forces of a blast can cause compound fractures, where bones are shattered or exposed through the skin. These fractures often require external fixation to stabilise the bones and facilitate healing. In cases where bone fragments are missing or severely damaged, reconstructive surgery, including bone grafting, may be necessary.

13.8.4.2 Treatment of Secondary Complications

Due to the contamination often present in blast injuries, patients are at high risk for developing infections. Prophylactic antibiotics are typically administered to prevent infections in open wounds, while surgical debridement removes infected or necrotic tissue. In severe cases, patients may develop sepsis, requiring aggressive antibiotic therapy and supportive care.

Injuries to blood vessels can lead to haemorrhage or ischemia (lack of blood flow to tissues). Vascular repair may involve direct suturing of blood vessels, the use of vascular grafts, or, in some cases, endovascular techniques such as stent placement to restore circulation. Without prompt treatment, vascular injuries can result in limb loss or life-threatening complications.

13.8.4.2.1 Rehabilitation and Long-Term Recovery

After the initial medical and surgical interventions, patients face a long road to recovery. Blast injuries often result in permanent disabilities, and comprehensive rehabilitation is necessary to restore function and improve quality of life.

13.8.4.3 Physical and Psychological Recovery

Patients recovering from blast injuries typically require physical therapy to regain strength, mobility, and coordination. Those with amputations undergo prosthetic fitting and training to adapt to their new limitations. Occupational therapy helps patients relearn daily activities and regain independence.

Survivors of blast trauma often suffer from psychological effects, including PTSD, anxiety, and depression. Cognitive-behavioural therapy (CBT), eye movement desensitisation and reprocessing (EMDR), and other psychological interventions are critical in addressing the emotional and mental health challenges faced by survivors. Early psychological support can reduce the long-term impact of trauma.

13.8.5 Addressing PTSD and Long-Term Disability in Survivors

PTSD is a common consequence of surviving a blast event, with symptoms including flashbacks, nightmares, hypervigilance, and emotional detachment. Therapeutic interventions aimed at managing PTSD focus on trauma processing and coping strategies. In addition to therapy, some patients may benefit from medication to manage anxiety and depression.

Many blast survivors face permanent disabilities, such as limb loss, vision impairment, or hearing loss. Assistive devices (e.g., wheelchairs, hearing aids) and long-term physical rehabilitation programs are essential to improving their quality of life. For military personnel and civilians alike, social reintegration and support systems are crucial in helping them navigate their new physical limitations and return to normalcy.

13.8.5.1 Psychological Impact of Blast and Fragmentation Trauma

13.8.5.1.1 Psychological and Emotional Consequences

Blast and fragmentation trauma not only cause physical injuries but also have profound psychological and emotional effects on survivors. The nature of these traumatic events, which often involve sudden and violent explosions, can lead to lasting mental health challenges. Addressing the psychological impact is a critical component of comprehensive care, as the emotional consequences can be just as debilitating as physical injuries.

Survivors of blast and fragmentation trauma often experience a wide range of psychological reactions, from acute stress to chronic mental health conditions. Understanding these responses is essential for providing appropriate care and support.

13.8.5.1.2 Post-Traumatic Stress Disorder (PTSD) and Its Prevalence in Blast Survivors

PTSD is one of the most common psychological outcomes of exposure to life-threatening events like blasts. It can manifest through intrusive thoughts, flashbacks, nightmares, and hypervigilance. Survivors may re-experience the traumatic event through these recurring memories, which can severely disrupt their daily lives and relationships (Fares et al., 2017).

The prevalence of PTSD among blast survivors is particularly high, especially among military personnel and first responders who are frequently exposed to explosive events. Studies indicate that combat-related PTSD is more likely to develop in individuals exposed to blasts than in those injured through other mechanisms, such as gunfire. Civilian survivors of terrorist attacks or industrial accidents involving explosions are also vulnerable to developing PTSD (Fares et al., 2017; Jehel et al., 2003).

Symptoms of PTSD may not appear immediately after the event and can take weeks, months, or even years to manifest. Survivors may experience emotional numbness, avoidance of trauma-related triggers, and difficulty connecting with others.

13.8.5.1.3 Coping Mechanisms and Psychological Support

Coping mechanisms among survivors of blast trauma vary widely. Some individuals may rely on social support systems, such as family and friends, while others turn to avoidance behaviours or substance abuse as maladaptive coping strategies. Developing healthy coping mechanisms, such as engaging in hobbies, mindfulness practices, or physical activity, is crucial for recovery.

Psychological support in the aftermath of a blast injury is essential. Early interventions, such as crisis counselling and psychological first aid, can help stabilise survivors emotionally and provide them with initial coping tools. CBT and exposure therapy are commonly used to help individuals process their trauma and reduce PTSD symptoms (Shubina, 2015).

Peer support programs, especially among veterans or those who have experienced similar trauma, can offer invaluable emotional relief. Sharing experiences in a structured environment helps survivors feel less isolated and more understood (Drebing et al., 2018; Hamblen et al., 2019).

13.8.5.1.4 The Role of Mental Health in Recovery

Mental health plays a central role in the overall recovery process for blast and fragmentation trauma survivors. Emotional resilience and psychological well-being are intertwined with physical healing, and a holistic approach that integrates mental health care into the medical treatment plan is crucial for comprehensive recovery.

13.8.5.1.5 Integrating Psychological Care into Medical Treatment

Early psychological intervention is key to addressing the mental health challenges that arise from blast trauma. Mental health care should begin as soon as the patient is physically stabilised. Psychiatric evaluations and mental health screenings can identify individuals at high risk of developing PTSD, depression, or anxiety, allowing for early intervention.

Trauma-informed care should be incorporated into every stage of treatment. Medical professionals, including surgeons and nurses, need to be aware of the potential psychological effects of blast injuries and provide care that considers the patient's emotional state. This may include offering counselling services during hospitalisation and involving mental health professionals in the treatment team from the outset.

Family involvement is also essential. The psychological impact of blast trauma extends beyond the survivor to their family and loved ones. Providing families with counselling and education about the psychological challenges their loved one may face can help create a supportive environment that fosters recovery.

13.8.5.1.6 Rehabilitation Programs Addressing Emotional Resilience

Rehabilitation programs designed for blast survivors should integrate mental health care alongside physical therapy and medical rehabilitation. These programs help patients rebuild their lives by addressing both the physical and psychological consequences of trauma.

Emotional resilience is a key focus in these rehabilitation programs. Survivors are taught how to cope with emotional triggers, manage stress, and develop strategies for dealing with future challenges. Group therapy, individual counselling, and mindfulness techniques are commonly used to enhance emotional strength and promote mental well-being.

Addressing co-morbid conditions such as depression, anxiety, and substance abuse is also critical to ensuring a full recovery. Many blast injury survivors struggle with these co-morbid conditions, which can exacerbate physical healing and increase the risk of long-term disability if left untreated. Comprehensive care programs often provide a multidisciplinary approach, involving psychologists, psychiatrists, and social workers to address these challenges.

13.9 ADVANCES IN RESEARCH AND FUTURE DIRECTIONS

13.9.1 Current Research on Blast Injury Prevention

Research on blast injuries and fragmentation trauma has advanced significantly in recent years, driven by the need to better understand injury mechanisms and develop effective prevention and mitigation strategies. As technology progresses, so too do innovations in protective gear and injury prevention. However, many challenges remain, particularly in addressing the diverse impacts of blast injuries on vulnerable populations and improving interdisciplinary collaboration (Brokaw et al., 2023).

Ongoing research efforts focus on understanding the mechanics of blast injuries and improving prevention strategies through advancements in material science, biomechanics, and protective equipment.

13.9.2 Ongoing Studies in Material Science, Biomechanics, and Protective Gear

Researchers are continuously exploring new materials and composites for use in protective gear, such as body armour and helmets. The goal is to develop materials that can better absorb and disperse the energy from blast waves and shrapnel, thereby reducing the severity of injuries. Studies on nanomaterials, ultra-lightweight composites, and smart fabrics that can adapt to changing conditions are showing promise.

Understanding the biomechanical response of the human body to blasts and fragmentation is critical for designing better protective gear and medical interventions. These studies help identify key vulnerabilities in the human body, such as the head, thorax, and limbs, and inform the design of more effective armour systems and safety equipment (Chanda and Callaway, 2018).

The development of advanced PPE is a key focus of research. Current studies aim to improve the performance of helmets, body armour, and blast-resistant clothing under real-world conditions. Researchers are investigating ways to make protective gear lighter, more flexible, and more comfortable, while enhancing its ability to protect against blast overpressure and high-velocity fragments.

13.9.3 Emerging Technologies in Injury Mitigation

Technological innovation plays a critical role in advancing injury mitigation strategies. New tools and systems designed to detect, measure, and respond to blasts and impacts are being developed, while cutting-edge materials for protection are also on the horizon.

13.9.3.1 Innovations in Blast Sensors, Wearable Impact Detection Systems

One area of innovation is the development of blast sensors that can measure the intensity of a blast in real time. These sensors, which can be embedded in helmets, body armour, and vehicles, provide immediate data on the strength and nature of a blast. This information is valuable for both medical personnel and engineers, as it helps assess the likelihood of injuries and the performance of protective gear during the explosion.

Wearable technology a focus for injury mitigation. Wearable sensors can monitor the forces exerted on the body during a blast or impact, providing data that can be used to assess injury risk and trigger immediate medical interventions. These systems are being integrated into military uniforms and sports equipment, offering potential applications beyond the battlefield, such as in law enforcement, construction, and other high-risk industries.

13.9.4 Future Materials for Body Armour and Structural Reinforcement

Researchers are working on developing next-generation materials that can provide better protection against blast forces and shrapnel while remaining lightweight and easy to wear. Graphene-based materials, meta-materials, and liquid armour are among the emerging technologies that promise to revolutionise the field of body armour. These materials are designed to harden upon impact or dynamically respond to changes in the environment, offering superior protection with minimal weight or bulk.

Beyond individual protection, advances in materials for structural reinforcement are also being explored. New materials for blast-resistant walls, bunkers, and vehicle armour are being developed to improve the ability of buildings and transportation systems to withstand explosive forces. This is particularly important for civilian infrastructure in areas prone to terrorist attacks or industrial accidents involving explosions.

13.9.5 Future Challenges and Opportunities

While significant progress has been made in understanding and mitigating blast injuries, several challenges remain. Addressing these challenges will require a multi-disciplinary approach and continued collaboration across research fields.

13.9.6 Addressing Gaps in Understanding Blast Effects on Vulnerable Populations

One of the key areas where research is still lacking is the effect of blasts on vulnerable populations, including the elderly, children, and individuals with pre-existing medical conditions. These groups may respond differently to blast forces due to factors such as weaker musculoskeletal systems, thinner skin, or age-related physiological changes. There is a growing need for targeted studies that explore how these populations experience blast trauma and how protective measures can be adapted to meet their unique needs.

Children may suffer more severe injuries from blast impacts due to their smaller size and developing bodies.

13.9.7 Enhancing Collaboration Between Military, Medical, and Engineering Fields

Another critical challenge lies in enhancing collaboration between different fields involved in blast injury prevention and treatment. Historically, the military, medical, and engineering communities have each worked on separate aspects of blast injury prevention and care. However, integrated research efforts are increasingly recognised as essential for achieving comprehensive solutions.

Military research on blast injuries and protective technologies often generates valuable data that can inform civilian applications, such as in disaster response or urban planning. Medical professionals bring insight into injury mechanisms and patient outcomes, while engineers provide the technical expertise needed to design new protective materials and systems. By fostering closer collaboration between these disciplines, more effective solutions can be developed to address the multifaceted nature of blast trauma.

13.10 CONCLUSION

13.10.1 Summary of Human Vulnerability to Blast and Fragments

The study of human vulnerability to blast and fragmentation injuries reveals a complex interplay of physical and psychological factors that demand a comprehensive approach to prevention and treatment. This section summarises key findings, reviews current protection strategies, and looks ahead to future research and global efforts aimed at mitigating these injuries.

Understanding the impact of blast and fragmentation injuries involves analysing how these traumatic events affect the human body and the effectiveness of various protection strategies.

Blast and fragmentation injuries can be categorised into primary, secondary, tertiary, and quaternary effects, each with distinct mechanisms and implications for human health. Primary injuries result from the direct impact of overpressure on internal organs, leading to conditions such as blast lung and gastrointestinal perforations. Secondary injuries arise from high-velocity fragments causing lacerations, punctures, and blunt trauma. Tertiary injuries occur due to the displacement of the body, resulting in fractures and TBIs. Quaternary injuries include burns, inhalation injuries, and exposure to toxic substances.

Advances in material science, biomechanics, and protective gear have led to significant improvements in injury prevention. Current research focuses on developing more effective PPE, including helmets, body armour, and blast-resistant clothing. Innovations in blast sensors and wearable impact detection systems are enhancing our ability to measure and respond to blasts in real time. Future materials, such as graphene-based composites and smart fabrics, hold promise for even greater protection.

The psychological consequences of blast injuries, including PTSD and other mental health issues, are as significant as the physical injuries. Effective psychological support and integration of mental health care into the rehabilitation process are crucial for the overall recovery of survivors.

13.10.2 Outlook

The future of blast injury research and prevention is marked by ongoing advancements and global initiatives aimed at reducing human vulnerability.

The field of blast injury research is continuously evolving, with ongoing studies addressing gaps in understanding and improving protective technologies. Emerging research in materials science and biomechanics promises to yield new insights into injury mechanisms and prevention strategies. The integration of cutting-edge technologies, such as wearable sensors and smart armour, is expected to enhance both individual and structural protection against blasts.

Efforts to mitigate blast injuries are not confined to any single sector or region. Global initiatives are increasingly focused on reducing human vulnerability in high-risk environments, including military, civilian, and industrial settings. International collaborations between military, medical, and engineering fields are fostering innovation and sharing best practices. Furthermore, research and development are expanding to address the needs of vulnerable populations, such as children and the elderly, who are disproportionately affected by blast trauma.

A key factor in advancing blast injury prevention and treatment is the enhancement of interdisciplinary collaboration. Combining expertise from military research, medical care, and engineering disciplines enables a more comprehensive approach to addressing the multifaceted challenges of blast injuries. Future research will benefit from these integrated efforts, leading to more effective solutions and improved outcomes for blast survivors.

In conclusion, while significant progress has been made in understanding and mitigating blast injuries, ongoing research and technological advancements are essential for further reducing human vulnerability. By continuing to address both the physical and psychological impacts of blast trauma and fostering global collaboration, we can enhance protection strategies and improve the quality of life for individuals affected by such devastating events.

REFERENCES

Ang, B.W.K., Yeow, C.H. and Lim, J.H., 2023. A critical review on factors affecting the user adoption of wearable and soft robotics. Sensors, 23(6), p.3263.

Banas, A.M., Banas, K. and Breese, M.B., 2023. Classification of the residues after high and low order explosions using machine learning techniques on Fourier transform infrared (FTIR) spectra. Molecules, 28(5), p.2233.

BBC News, https://www.bbc.co.uk/news accessed 11 September 2024.

Beaven, A. and Parker, P., 2015. Treatment principles of blast injuries. Surgery (Oxford), 33(9), pp.424–429.

BenÍtez, C.Y., Ottolino, P., Pereira, B.M., Lima, D.S., Guemes, A., Khan, M. and Ribeiro, M.A.F., 2021. Tourniquet use for civilian extremity hemorrhage: Systematic review of the literature. Revista do Colegio Brasileiro de Cirurgioes, 48, p.e20202783.

Bochicchio, G.V., Lumpkins, K., O'connor, J., Simard, M., Schaub, S., Conway, A., Bochicchio, K. and Scalea, T.M., 2008. Blast injury in a civilian trauma setting is associated with a delay in diagnosis of traumatic brain injury. The American Surgeon, 74(3), pp.267–270.

Bowen, I.G., Fletcher, E.R. and Richmond, D.R., 1968. Estimate of man's tolerance to the direct effects of air blast (No. DASA2113).

Brokaw, E.B., S. Brungart, D., M. Byrne, R., A. Flamme, G., Gupta, R., Jokel, C.R., Kujawa, S.G., Lalis, L., L. McKinley, R., Murphy, W.J. and W. Spencer, R., 2023. Recommendations for a military health system auditory blast injury prevention standard. Military Medicine, 188(Supplement_6), pp.176–184.

Bumbaširevic, M., Lesic, A., Mitkovic, M. and Bumbaširevic, V., 2006. Treatment of blast injuries of the extremity. JAAOS-Journal of the American Academy of Orthopaedic Surgeons, 14(10), pp.S77–S81.
Cannon, J.W., Khan, M.A., Raja, A.S., Cohen, M.J., Como, J.J., Cotton, B.A., Dubose, J.J., Fox, E.E., Inaba, K., Rodriguez, C.J. and Holcomb, J.B., 2017. Damage control resuscitation in patients with severe traumatic hemorrhage: A practice management guideline from the Eastern Association for the Surgery of Trauma. Journal of Trauma and Acute Care Surgery, 82(3), pp.605–617.
Cave, K.M., Cornish, E.M. and Chandler, D.W., 2007. Blast injury of the ear: Clinical update from the global war on terror. Military Medicine, 172(7), pp.726–730.
Chanda, A. and Callaway, C., 2018. Computational modeling of blast induced whole-body injury: A review. Journal of Medical Engineering & Technology, 42(2), pp.88–104.
Chen, Y. and Huang, W., 2011. Non-impact, blast-induced mild TBI and PTSD: Concepts and caveats. Brain Injury, 25(7–8), pp.641–650.
Cho, S.I., Gao, S.S., Xia, A., Wang, R., Salles, F.T., Raphael, P.D., Abaya, H., Wachtel, J., Baek, J., Jacobs, D. and Rasband, M.N., 2013. Mechanisms of hearing loss after blast injury to the ear. PLoS One, 8(7), p.e67618.
Ciesielska-Wróbel, I.L., 2017. Contemporary personal ballistic protection (PBP). Textiles for Advanced Applications, 20, p.213.
Clifford, C.B., Moe, J.B., Jaeger, J.J. and Hess, J.L., 1984. Gastrointestinal lesions in lambs due to multiple low-level blast overpressure exposure. Military Medicine, 149(9), pp.491–495.
Clough, R.A. and Khan, M., 2019. Initial CABC: Advances that have led to increased survival in military casualties. Trauma, 21(4), pp.247–251.
Cooper, G.J., 1996. Protection of the lung from blast overpressure by thoracic stress wave decouplers. Journal of Trauma and Acute Care Surgery, 40(3S), pp.105S–110S.
Cooper, P.W., 1996. Explosives Engineering. New York: Wiley-VCH.
Cripps, N.P.J. and Cooper, G.J., 1997. Risk of late perforation in intestinal contusions caused by explosive blast. British Journal of Surgery, 84(9), pp.1298–1303.
Cullis, I.G., 2001. Blast waves and how they interact with structures. BMJ Military Health, 147(1), pp.16–26.
De Candole, C.A., 1967. Blast injury. Canadian Medical Association Journal, 96(4), p.207.
Delannoy, Y., Plu, I., Sec, I., Delabarde, T., Taccoen, M., Tracqui, A. and Ludes, B., 2020. Terrorist attacks: Cutaneous patterns of gunshot and secondary blast injuries. Forensic Sciences Research, 5(3), pp.208–213.
Dikici, E., 2014. Mine resistant ambush protected vehicle (MRAP) program and the DoD AT&L reform policy. Journal of Management and Information Science, 2(2), pp.22–26.
Drebing, C.E., Reilly, E., Henze, K.T., Kelly, M., Russo, A., Smolinsky, J., Gorman, J. and Penk, W.E., 2018. Using peer support groups to enhance community integration of veterans in transition. Psychological Services, 15(2), p.135.
DuBose, J., Plurad, D.S. and Rhee, P.M., 2016. Blast injuries. In Penetrating Trauma: A Practical Guide on Operative Technique and Peri-Operative Management (pp. 541–547). Berlin, Heidelberg: Springer Berlin Heidelberg.
Duchesne, J., Inaba, K. and Khan, M.A. eds., 2018. Damage Control in Trauma Care: An Evolving Comprehensive Team Approach. Switzerland: Springer.
D'yachenko, A.I. and Manyuhina, O.V., 2006. Modeling of weak blast wave propagation in the lung. Journal of Biomechanics, 39(11), pp.2113–2122.
Edwards, D.S. and Clasper, J., 2016. Blast injury mechanism. In Blast Injury Science and Engineering: A Guide for Clinicians and Researchers (pp. 87–104). Cham: Springer International Publishing.
Fares, J., Gebeily, S., Saad, M., Harati, H., Nabha, S., Said, N., Kanso, M., Rassoul, R.A. and Fares, Y., 2017. Post-traumatic stress disorder in adult victims of cluster munitions in Lebanon: A 10-year longitudinal study. BMJ Open, 7(8), p.e017214.
Geng, J., Mander, T. and Baker, Q., 2015. Blast wave clearing behavior for positive and negative phases. Journal of Loss Prevention in the Process Industries, 37, pp.143–151.
Gao, Y., Wang, S., Zhang, J., Jia, X., Liang, C. and Ma, F., 2022. Effects of underwater explosion depth on shock wave overpressure and energy. Physics of Fluids, 34(3).
Goatly, G., Guidozzi, N. and Khan, M., 2019. Optimal ventilator strategies for trauma-related ARDS. BMJ Military Health, 165(3), pp.193–197.
Goel, M.D. and Matsagar, V.A., 2014. Blast-resistant design of structures. Practice Periodical on Structural Design and Construction, 19(2), p.04014007.
Greer, A., 2006. Numerical modeling for the prediction of primary blast injury to the lung. MASc Thesis.
Griffiths, R., 2011. Oxford desk reference' major trauma. British Journal of Anaesthesia, 107(2), pp.282–283.

Gupta, R.K. and Przekwas, A.J., 2015, July. A framework for multiscale modeling of warfighter blast injury protection. In The 6th International Conference on Computational Methods (ICCM2015, 14–17 July 2015), (Vol. 7). Auckland: ScienTech Publisher.

Guy, R.J., Cripps, N.P.J. and Glover, M.A., 1998. The pathophysiology of primary blast injury and its implications for treatment. Part I: The thorax. Journal of the Royal Naval Medical Service, 84(2).

Hada, J.S. and Kumar, I., 2019. Body armours-The Indian review. Man-Made Textiles in India, 47(1).

Hamblen, J.L., Grubaugh, A.L., Davidson, T.M., Borkman, A.L., Bunnell, B.E. and Ruggiero, K.J., 2019. An online peer educational campaign to reduce stigma and improve help seeking in veterans with posttraumatic stress disorder. Telemedicine and e-Health, 25(1), pp.41–47.

Harmon, J.W. and Haluszka, M., 1983. Care of blast-injured casualties with gastrointestinal injuries. Military Medicine, 148(7), pp.586–588.

Hayda, R., Harris, R.M. and Bass, C.D., 2004. Blast injury research: Modelling injury effects of landmines, bullets, and bombs. Clinical Orthopaedics and Related Research (1976–2007), 422, pp.97–108.

Hirsch, F.G., 1966. Effects of overpressure on the ear: A review.

Horrocks, C.L., 2001. Blast injuries: Biophysics, pathophysiology and management principles. BMJ Military Health, 147(1), pp.28–40.

Hubbard, W.B., 2016. Investigating injury pathology of blast-induced polytrauma and assessing the therapeutic role of hemostatic nanoparticles after blast exposure.

Isaac, O.S., Alshammari, O.G., Pickering, E.G., Clarke, S.D. and Rigby, S.E., 2023. Blast wave interaction with structures – An overview. International Journal of Protective Structures, 14(4), pp.584–630.

Jehel, L., Paterniti, S., Brunet, A., Duchet, C. and Guelfi, J.D., 2003. Prediction of the occurrence and intensity of post-traumatic stress disorder in victims 32 months after bomb attack. European Psychiatry, 18(4), pp.172–176.

Khan, M.A. and Frankel, H.L., 2013. Adjuncts to ventilation part I: Nitric oxide, surfactants, prostacyclin, steroids, sedation, and neuromuscular blockade. Current Problems in Surgery, 50(10), pp.424–433.

Khan, M.A. and Frankel, H.L., 2015a. Trauma in austere environments: Cold injuries and hypothermia. In Trauma Team Dynamics: A Trauma Crisis Resource Management Manual (pp. 237–241). Cham: Springer International Publishing.

Khan, M.A. and Frankel, H.L., 2015b. Nuclear injuries. In Trauma Team Dynamics: A Trauma Crisis Resource Management Manual (pp. 249–252). Cham: Springer International Publishing.

Khan, M. and Nott, D.M., 2018. Ballistics and blast injury … Morgan McMonagle. In Trauma: Code Red (pp. 37–62). Boca Raton, FL: CRC Press.

Khan, M. and Nott, D., 2021. Fundamentals of Frontline Surgery (p. 260). Boca Raton, FL: CRC Press.

Khan, M., 2018. Abdominal trauma. In Trauma: Code Red (pp. 189–232). Boca Raton, FL: CRC Press.

Kirkman, E. and Watts, S., 2011. Characterization of the response to primary blast injury. Philosophical Transactions of the Royal Society B: Biological Sciences, 366(1562), pp.286–290.

Kirkman, E., Watts, S. and Cooper, G., 2011. Blast injury research models. Philosophical Transactions of the Royal Society B: Biological Sciences, 366(1562), pp.144–159.

Kocsis, J.D. and Tessler, A., 2009. Pathology of blast-related brain injury. Journal of Rehabilitation Research and Development, 46(6), pp.667–672.

Kokkinakis, I.W. and Drikakis, D., 2023. Internal explosions and their effects on humans. Physics of Fluids, 35(4).

Korkmaz, İ., Çelikkaya, M.E. and Atıcı, A., 2023. Secondary blast injury: Radiological characteristics of shrapnel injuries in children. Emergency Radiology, 30(3), pp.307–313.

Krzystała, E., Mężyk, A. and Kciuk, S., 2016. Minimisation of the explosion shock wave load onto the occupants inside the vehicle during trinitrotoluene charge blast. International Journal of Injury Control and Safety Promotion, 23(2), pp.170–178.

Kulkarni, S.G., Gao, X.L., Horner, S.E., Zheng, J.Q. and David, N.V., 2013. Ballistic helmets–their design, materials, and performance against traumatic brain injury. Composite Structures, 101, pp.313–331.

LaFree, G. and Dugan, L., 2007. Introducing the global terrorism database. Terrorism and Political Violence, 19(2), pp.181–204.

Liang, J., Smith, K.D., Gan, R.Z. and Lu, H., 2019. The effect of blast overpressure on the mechanical properties of the human tympanic membrane. Journal of the Mechanical Behavior of Biomedical Materials, 100, p.103368.

Loftis, K.L., Mazuchowski, E.L., Clouser, M.C. and Gillich, P.J., 2019. Prominent injury types in vehicle underbody blast. Military Medicine, 184(Supplement_1), pp.261–264.

Loubeau, A., Sparrow, V.W., Pater, L.L. and Wright, W.M., 2006. High-frequency measurements of blast wave propagation. The Journal of the Acoustical Society of America, 120(3), pp.EL29–EL35.

Luks, F.I., 2010. Blast injuries-and the pivotal role of trauma surgeons. Acta Chirurgica Belgica, 110(5), pp.517–520.

Magnus, D., Khan, M.A. and Proud, W.G., 2018. Epidemiology of civilian blast injuries inflicted by terrorist bombings from 1970–2016. Defence Technology, 14(5).

Magnus, D., Sory, D.R., Lee, J., Khan, M.A. and Proud, W.G., 2020. November. Study of soft material blast mitigation effects using a shock tube. In AIP Conference Proceedings (Vol. 2272, No. 1, p. 040009). AIP Publishing LLC.

Magnus, D., Brown, K.A., Khan, M. and Proud, W.G., 2021. Patterns of injury. In Fundamentals of Frontline Surgery (pp. 7–23). Boca Raton, FL: CRC Press.

Mathews, Z.R. and Koyfman, A., 2015. Blast injuries. The Journal of Emergency Medicine, 49(4), pp.573–587.

Matsushima, K., Khan, M. and Frankel, H.L., 2014. Trauma-related critical care. Scandinavian Journal of Surgery, 103(2), pp.138–142.

Mckay, B.J., 2010. Development of Lower Extremity Injury Criteria and Biomechanical Surrogate to Evaluate Military Vehicle Occupant Injury during an Explosive Blast Event. Michigan: Wayne State University.

Moga, H.K., Parmar, A.P. and Gohil, N.D., 2021. An autopsy case of death due to AC compressor blast – A rare case illustrating primary, secondary, tertiary & quaternary blast injuries. Journal of Forensic and Legal Medicine, 80, p.102173.

Padhy, V. and Kandasubramanian, B., 2024. Polymer nanocomposites of ultra-high molecular weight polyethylene. Polymer Bulletin, 81(17), pp.15259–15292.

Paran, H., Neufeld, D., Shwartz, I., Kidron, D., Susmallian, S., Mayo, A., Dayan, K., Vider, I., Sivak, G. and Freund, U., 1996. Perforation of the terminal ileum induced by blast injury: delayed diagnosis or delayed perforation? Journal of Trauma and Acute Care Surgery, 40(3), pp.472–475.

Pasha, N.A., Khan, R.S. and Noordin, S., 2015. Bomb blast injuries: Tertiary care hospital in-patient experience over the last 20 years. JPMA: Journal of Pakistan Medical Association, 65(11), pp.132–135.

Patterson Jr, J.H. and Hamernik, R.P., 1997. Blast overpressure induced structural and functional changes in the auditory system. Toxicology, 121(1), pp.29–40.

Phillips, Y.Y., 1986. Primary blast injuries. Annals of Emergency Medicine, 15(12), pp.1446–1450.

Poopitaya, S. and Kanchanaroek, K., 2009. Injuries of the thoracolumbar spine from tertiary blast injury in Thai military personnel during conflict in southern Thailand. Journal of the Medical Association of Thailand, 92(Suppl 1), pp.S129–S134.

Price, M.A., Nguyen, V.T., Hassan, O. and Morgan, K., 2017. An approach to modeling blast and fragment risks from improvised explosive devices. Applied Mathematical Modelling, 50, pp.715–731.

Rezaei, A., Salimi Jazi, M., Karami, G. and Ziejewski, M., 2014. A computational study on brain tissue under blast: primary and tertiary blast injuries. International Journal for Numerical Methods in Biomedical Engineering, 30(8), pp.781–795.

Risling, M. and Davidsson, J., 2012. Experimental animal models for studies on the mechanisms of blast-induced neurotrauma. Frontiers in Neurology, 3, p.30.

Ritenour, A.E., Wickley, A., Ritenour, J.S., Kriete, B.R., Blackbourne, L.H., Holcomb, J.B. and Wade, C.E., 2008. Tympanic membrane perforation and hearing loss from blast overpressure in Operation Enduring Freedom and Operation Iraqi Freedom wounded. Journal of Trauma and Acute Care Surgery, 64(2), pp.S174–S178.

Roh, T., Lee, Y., Lee, W. and Yoh, J.J., 2020. Understanding the effects of blast loads on open spaces and enclosed structures in simulations and experiments. Shock Waves, 30(7), pp.843–854.

Rosenfeld, J.V. and Ford, N.L., 2010. Bomb blast, mild traumatic brain injury and psychiatric morbidity: A review. Injury, 41(5), pp.437–443.

Sarvghad-Moghaddam, H., 2015. Computational biomechanics of blast-induced traumatic brain injury: Role of loading directionality, head protection, and blast flow mechanics. Doctoral Dissertation, North Dakota State University.

Shubina, I., 2015. Cognitive-behavioral therapy of patients with Ptsd: literature review. Procedia-Social and Behavioral Sciences, 165, pp.208–216.

Shuker, S.T., 2010. Facial skin-mucosal biodynamic blast injuries and management. Journal of Oral and Maxillofacial Surgery, 68(8), pp.1818–1825.

Smith, J.E., 2011. Blast lung injury. Journal of the Royal Naval Medical Service, 97(3).

Skrabacz, A. and Wołoch, F., 2024. Protective structures as a device of ensuring the safety of victims in a war situation. Inżynieria Bezpieczeństwa Obiektów Antropogenicznych, (1), pp.1–12.

Smits, I., Heil, S., van de Krol, E., Edwards, M., Priesterbach, A., Stirler, V. and Koenders, N., 2024. Rehabilitation outcome of a severe combat blast injury: A case report. European Journal of Physiotherapy, 26(6), pp.359–364.

Solomos, G., Larcher, M., Valsamos, G., Karlos, V. and Casadei, F., 2020. A survey of computational models for blast induced human injuries for security and defence applications. JRC (Joint Research Centre) Technical Reports, Luxembourg: Publications Office of the European Union, 2020.

Stottmeister, A., von Ramin, M. and Schneider, J.M., 2020. On models of blast overpressure effects to the thorax. SN Applied Sciences, 2(12), p.2076.

Stuhmiller, J.H., Phillips, Y.Y. and Richmond, D.R., 1991. The physics and mechanisms of primary blast injury. Conventional Warfare: Ballistic, Blast, and Burn Injuries, 5, pp.241–270.

Toboła, W., Papis, M., Jastrzębski, D. and Perz, R., 2023. Experimental research of energy absorbing structures within helmet samples made with the additive manufacturing method-preliminary study. Acta of Bioengineering & Biomechanics, 25(1).

Turégano-Fuentes, F., Pérez-Diaz, D., Sanz-Sánchez, M., Alfici, R. and Ashkenazi, I., 2014. Abdominal blast injuries: Different patterns, severity, management, and prognosis according to the main mechanism of injury. European Journal of Trauma and Emergency Surgery, 40(1), pp.151–160.

Yang, C., 2023. Defense against blast injury. In Explosive Blast Injuries: Principles and Practices (pp. 117–141). Singapore: Springer Nature Singapore.

Yang, Z., Wang, Z., Tang, C. and Ying, Y., 1996. Biological effects of weak blast waves and safety limits for internal organ injury in the human body. Journal of Trauma and Acute Care Surgery, 40(3S), pp.81S–84S.

Yeh, D.D. and Schecter, W.P., 2012. Primary blast injuries—An updated concise review. World Journal of Surgery, 36(5), pp.966–972.

Zhang, F., Frost, D.L., Thibault, P.A. and Murray, S.B., 2001. Explosive dispersal of solid particles. Shock Waves, 10(6), pp.431–443.

Zhou, W., Reddy, N. and Yang, Y., 2005. Overview of protective clothing. In Textiles for Protection (pp. 3–30). Cambridge: Woodhead Publishing.

14 Sacrificial Claddings for Blast Protection

Ola Wattad and Hezi Y. Grisaro

14.1 BACKGROUND AND MOTIVATION

In recent years, the increasing frequency of incidents generating blast waves, whether from accidental explosions in industrial settings or intentional attacks, has emphasized the importance of blast-resistant design in modern construction. The standard structural designs often focus on resisting natural forces such as wind and earthquakes, but they may fall short when exposed to the intense pressures and rapid energy release associated with blast (USACE 2008; ASCE 2011; Krauthammer 2008; Federal Emergency Management Agency (FEMA) 2003; United States Department of Defense 2003; Department of the Army 1986). Consequently, there is a growing need to incorporate blast resistance into the design criteria for critical infrastructure and high-risk facilities. This proactive approach aims not only to protect human life but also to minimize structural damage and ensure the continued functionality of essential buildings in the aftermath of such events. By integrating blast mitigation strategies early in the design process, engineers can create more resilient structures capable of withstanding both conventional and unconventional threats (Hao et al. 2016; Ngo et al. 2007; Hetherington and Smith 2014; Federal Emergency Management Agency (FEMA) 2003; United States Department of Defense 2003; Department of the Army 1986; Jones, Rajendran, and Brebbia 2002; CSA-Group 2012).

To ensure the resilience of structures under blast threats, various assessment and design methods have been developed with the objective of minimizing occupant injuries and reducing structural damage. These methods fall broadly into two categories: traditional design approaches and novel protective technologies. Traditional methods for blast-resistant design typically focus on enhancing the inherent strength and rigidity of a structure to resist blast forces (USACE 2008; ASCE 2011; Krauthammer 2008; Mays, Smith, and Cormie 2019; Genç et al. 2024). Conventional designs often employ reinforced concrete, steel reinforcement, incorporating steel fibers in concrete (USACE 2008; ASCE 2011; Krauthammer 2008; Mays, Smith, and Cormie 2019; Genç et al. 2024; Jones, Rajendran, and Brebbia 2002), unidirectional passive dampers in steel structures, creating composite structures such as using Fiber Reinforced Polymer (FRP) retrofit techniques for masonry walls (Buchan and Chen 2007), implementation of stiffeners with spacing in steel I-sections (Grisaro et al. 2021; Grisaro, Packer, and Seica 2019), and other high-strength materials to increase structural capacity against extreme loads (Maalej et al. 2012; Zhou et al. 2023; Rebentrost and Wight 2013). These approaches rely heavily on advanced engineering calculations and simulations to model the expected loads and structural response, allowing engineers to anticipate and mitigate potential failures. However, while traditional methods are effective in many scenarios, they often result in heavier, costlier structures and may not fully address the complex dynamics of blast waves. Furthermore, this retrofitting approach does not add a dedicated protective shield to the structure; instead, it involves significantly increasing structural stiffness and capacity, often resulting in a massive overhaul for what is ultimately a rare and extreme scenario. Nevertheless, traditional design approaches define the damage to be directed toward and absorbed by the primary structural components themselves; thus, damage is inevitable. In such cases, the primary structural components would likely become unserviceable, requiring significant repair, retrofitting, or even complete replacement to restore functionality and safety.

DOI: 10.1201/9781003529217-14

Considering the limitations of the traditional approach of blast-resistant design, the second approach introduces the implementation of sacrificial cladding systems as an innovative solution for structural protection. Those claddings are specifically designed for blast resilience, and do not affect the structural shield in terms of design consideration, i.e., non-structural elements. Unlike conventional methods, sacrificial claddings are designed to absorb and redirect blast energy away from primary structural elements, thereby minimizing the risk of critical damage. These non-structural layers are often composed of materials with unique energy-dissipating properties, such as metallic foams, honeycomb structures, and even unconventional materials like repurposed beverage cans. These claddings serve as a buffer between the blast wave and the main structure. By sacrificing themselves during a blast event, these layers protect the essential components of the structure, significantly enhancing resilience while simplifying post-event repairs and replacements.

This approach of using sacrificial claddings offers a promising shift in blast protection design, focusing on absorbing and dissipating energy through replaceable, non-structural layers. As research progresses, a wide variety of materials, configurations, and design principles are being explored to optimize the effectiveness of these protective layers under different blast conditions. In the following sections, a deeper delve into the fundamental aspects of sacrificial cladding systems is presented, examining their materials, mechanisms, and practical considerations that are essential for their successful application in blast-resistant design.

14.2 INTRODUCTION TO SACRIFICIAL CLADDING

Sacrificial layers, or claddings, are specially manufactured materials to absorb and dissipate energy during extreme events, such as explosions, thereby protecting main structural components from severe damage. The concept of sacrificial claddings dates back to ancient fortification practices and has evolved significantly with advancements in military and civil engineering (Mays, Smith, and Cormie 2019; American Society of Civil Engineers 2023).

Historically, while the concept of intentional sacrificial layers did not exist in its modern form, early civilizations used layered stone or mud-brick walls in fortifications to protect critical structures (Maduabum 2024; BibleWalks 2024; Derat 2024; Stephenson 2005). These layered defenses, though not designed specifically as sacrificial elements, dissipated energy from impacts and siege weaponry, serving as the earliest examples of protective claddings. During World Wars I and II, military engineers began incorporating reinforced concrete layered with various materials to shield bunkers, ammunition stations, and other strategic installations. Basic sacrificial elements, such as sandbags and concrete panels, were employed to absorb energy from bomb blasts, offering a primitive yet effective form of blast protection (Toy 1965; IHBC NewsBlog 2024; Warfare History Network 2024; Engines of Our Ingenuity 2024).

The Cold War era, with its heightened nuclear threat, marked a pivotal moment in blast-resistant construction. Engineers developed sacrificial concrete layers, shock-absorbing mats, and layered steel barriers specifically designed to absorb impact forces and shield underground bunkers and essential infrastructure. This period also saw early research into materials like foams and lightweight concrete composites, with initial findings suggesting their potential for energy absorption in blast scenarios (American Society of Civil Engineers 1997; National Research Council 1995).

In the 1970s and 1980s, modern developments introduced foams and composite materials, leading to an increase in research on cellular structures like metal foams, which could absorb energy through densification and compaction. Aluminum foams, in particular, emerged as promising materials for blast energy absorption. By the late 20th century, sacrificial claddings began to be investigated for potential use in civilian infrastructure, particularly in high-risk industrial settings such as chemical plants where explosion risks are prominent. While practical implementation remains limited, research has demonstrated their potential for mitigating blast-induced damage in these environments (Ashby et al. 2000; Hao and Li 2016; Hanssen, Enstock, and Langseth 2002; Reid and Peng 1997). The 21st century has witnessed significant advancements in material science and

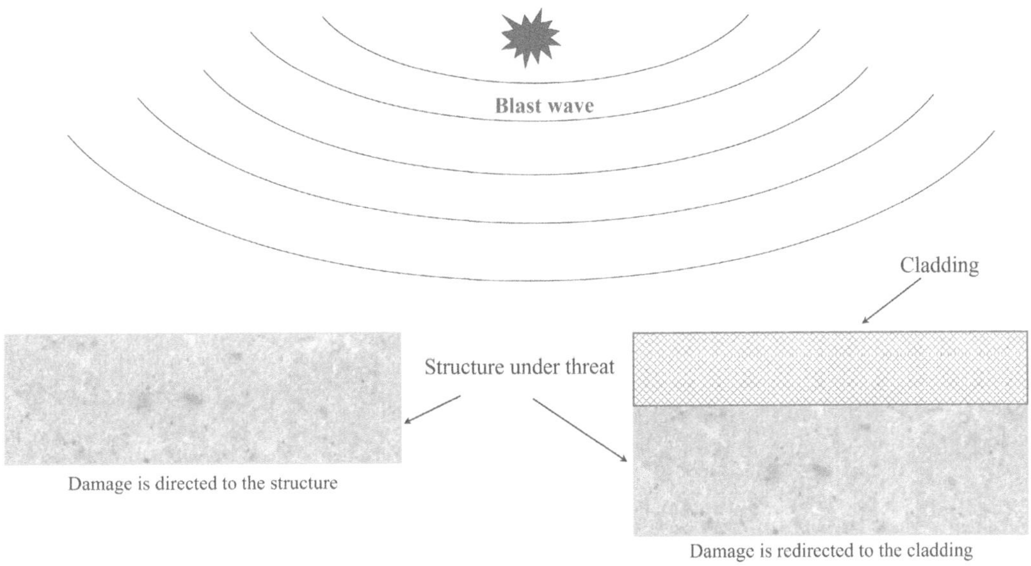

FIGURE 14.1 The concept of using sacrificial cladding.

computational modeling, allowing engineers to simulate blast loads and optimize sacrificial layers for specific conditions. Novel materials, such as advanced metal foams, polymers, and hybrid composites, have been engineered to compress and deform, effectively absorbing energy and protecting the primary structure. Today, research focuses on optimizing these sacrificial materials through experimental testing and advanced modeling. Engineers now use tools like pressure–impulse (P–I) diagrams to design claddings tailored to withstand various blast intensities, further enhancing the protective capabilities of sacrificial claddings (Wattad and Grisaro 2024a, 2024b; Zhou et al. 2015; Su, Liu, and Hou 2024; Xia, Wu, and Bennett 2017).

Through the historical evolution and modern development of sacrificial claddings, the underlying principle remains consistent: to strategically absorb and manage the energy from blast or impact loads, thereby reducing the risk of catastrophic failure in the protected structure (see Figure 14.1). This concept forms the foundation of the subsequent sections, which delve into the materials, mechanical behavior, modeling strategies, and evaluation methods that define the design and implementation of sacrificial claddings.

14.3 MATERIALS AND CONFIGURATIONS

Over the years, various materials have been developed and implemented as sacrificial claddings. The key concept in selecting these materials is to ensure that the cladding can undergo significant plastic deformation, absorbing and dissipating energy to protect the primary structure. By enduring large plastic deformations, sacrificial claddings mitigate the intensity of the stress transmitted to the underlying structure, thereby reducing structural damage and enhancing resilience under extreme loads.

A class of material that has been widely researched as a promising candidate for this purpose is metallic foam. Metallic foams are a class of materials characterized by low densities and unique physical, mechanical, thermal, electrical, and acoustic properties. From an environmental perspective, these foams are recyclable and, importantly, non-toxic. Due to these properties, metal foams offer the potential for lightweight energy absorption and thermal insulation at a relatively reasonable cost.

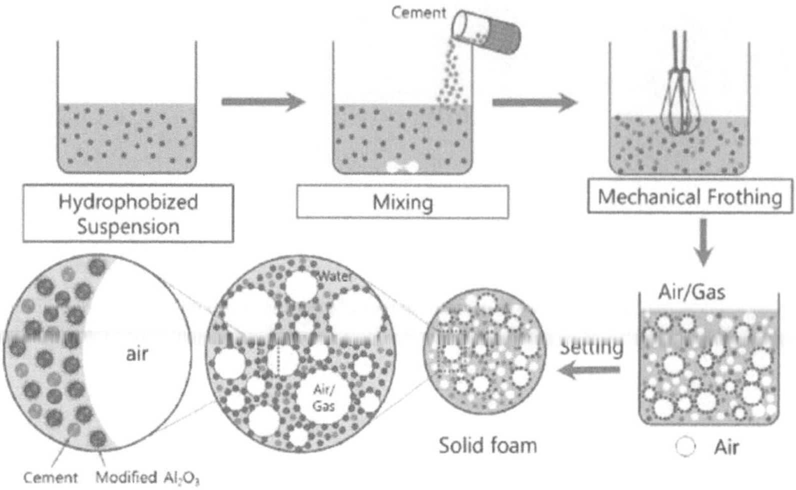

FIGURE 14.2 Production process of metallic foams. (Adapted from Pokhrel et al. (2013).)

The production of metallic foams is still evolving, with new processing techniques continually being developed to enhance their properties and manufacturability. These foams feature low density and weight due to their cellular structure, which can be achieved through various processing methods. Two primary approaches are air-filling and casting processes. Air-filling processes involve injecting air bubbles into liquid metal or adding a foaming agent that releases gas when heated, creating a closed-cell structure. Casting processes, on the other hand, involve casting molten metal into wax or polymer templates to produce an open-cell structure. The properties of the resulting foam depend on the metal used and the specific production process (refer to Figure 14.2), determining the relative density, cell size, and cell topology (open or closed cells) (Ashby et al. 2000).

Extensive research has been conducted on metallic foams (refer to Figure 14.3), particularly those made of aluminum, due to their favorable weight-to-strength ratio and energy-absorbing capabilities. Studies vary widely in focus, from developing models to predict foam behavior under dynamic loads to optimizing parameters like foam thickness, weight, and air-cell distribution. Key areas of research include optimizing the foam thickness and density to maximize energy absorption while minimizing weight, fine-tuning cell size and distribution to improve impact resistance, and enhancing manufacturing processes to ensure consistency in foam properties. Researchers have also investigated how varying the cell geometry and distribution influences stress distribution across the foam, aiming to tailor foam properties for specific applications (Wattad and Grisaro 2024a, 2024b;

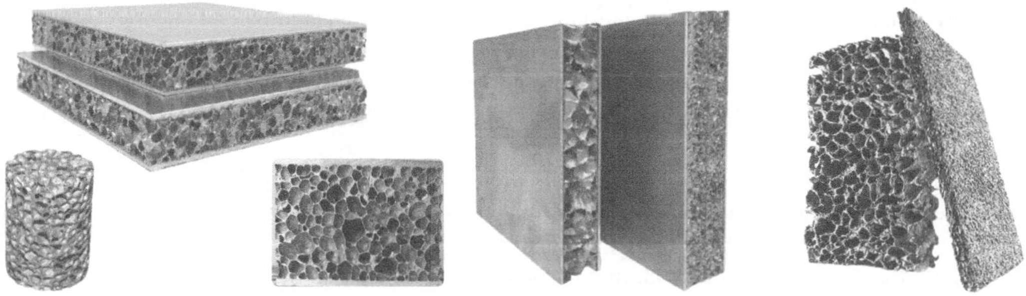

FIGURE 14.3 Commercial metallic foams.

Xia et al. 2014; Xia, Wu, and Bennett 2017; Ma and Ye 2005, 2007; Karagiozova 2011; Hanssen, Enstock, and Langseth 2002). Altogether, metallic foams continue to show great promise as an effective, lightweight, and adaptable solution for sacrificial cladding in blast and impact mitigation.

Another type of sacrificial cladding is a layer composed of miscellaneous geometric recurring shapes. These claddings typically consist of a series of identical or similar geometric elements arranged in a structured pattern between one or two solid plates (see Figure 14.4). The individual shapes, which may include tubes, honeycomb cells, square columns, or even beverage cans, are designed to absorb and dissipate energy upon compression. These elements can be arranged in various configurations, such as stacked or interlocked patterns, depending on the intended load distribution and energy dissipation requirements.

The shapes in these claddings can be either hollow or filled with a secondary material to enhance their energy-absorbing capacity. For example, tubes may be left empty for lightweight applications or filled with materials such as foam concrete to increase their density and strength under loading. The material choice and internal structure of these elements contribute significantly to the overall performance of the cladding, allowing for customization based on the anticipated blast conditions. This structural arrangement has been shown to produce the large plastic deformations essential for the overall concept of sacrificial claddings, dissipating energy while protecting the main structure. Due to their geometry and deformation behavior, these claddings can also be considered a form of cellular material, as they share similar mechanisms of energy absorption and stress distribution.

This design approach enables a high degree of adaptability; by altering the geometry, arrangement, or filler material, engineers can fine-tune the cladding response to specific types of loads. Studies have shown that such claddings are effective in distributing the load across the structure while reducing peak stresses, making them suitable for applications requiring high resilience against dynamic impacts. The modular nature of these layers also allows for easy replacement of individual components after an impact event, further enhancing the practicality of using geometric shapes in sacrificial claddings.

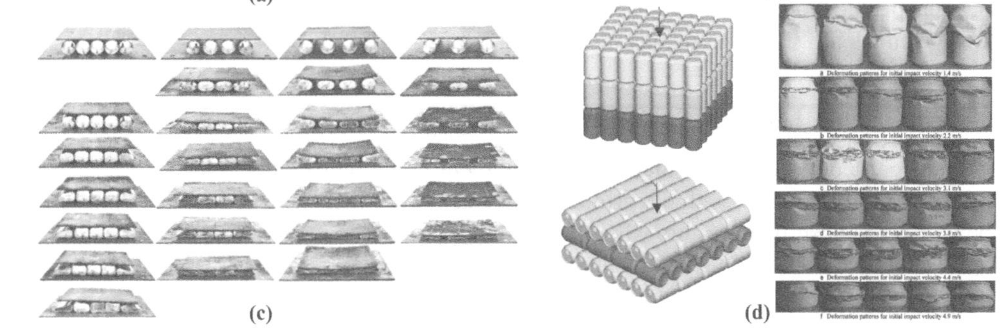

FIGURE 14.4 Sacrificial cladding miscellaneous geometric recurring shapes: (a) foam concrete-filled square tubes, (b) tubular cores, (c) foam-filled tubular cores, and (d) recyclable cans. (Adapted from Gu, Wang, and Zhai (2023); Yuen, Cunliffe, and du Plessis (2017); Wang, Xu, and Yuen (2019); Palanivelu et al. (2011), respectively.)

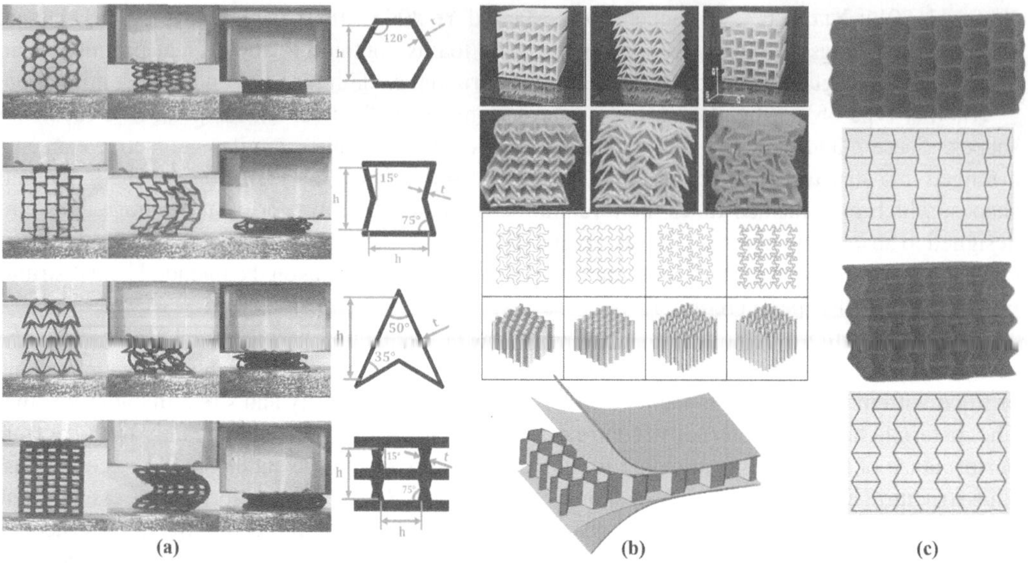

FIGURE 14.5 Sacrificial cladding made of 3D-printed auxetic structures. (Adapted from (a) Blake et al. (2024); (b) Junio et al. (2023); (c) Critchley et al. (2022), respectively.)

The last type of sacrificial cladding to be discussed involves employing 3D-printed auxetic structures (see Figure 14.5). Auxetic materials are characterized by a negative Poisson's ratio, meaning they expanded perpendicular to the direction of compression. This unique property allows them to absorb and dissipate energy more effectively than conventional materials. Advances in additive manufacturing have enabled the fabrication of complex auxetic geometries, such as re-entrant honeycombs and chiral structures, which were previously difficult to produce using traditional manufacturing methods (Critchley et al. 2022; Joseph, Mahesh, and Harursampath 2021; Bohara et al. 2023; Yan et al. 2022; Junio et al. 2023; Tunay and Cetin 2023). Moreover, the lightweight nature of these cellular structures makes them advantageous for applications where weight is a critical factor.

These 3D-printed auxetic claddings are composed of intricate lattice structures that can deform under blast and impact loads, undergoing large plastic deformations that enhance energy absorption. Studies have demonstrated that auxetic honeycomb-cored sandwich panels exhibit superior blast resistance compared to traditional honeycomb designs. The ability to tailor the geometry and material properties through additive manufacturing allows engineers to optimize the cladding performance for specific threat levels and loading conditions. The structural behavior of auxetic claddings under dynamic loads shows promising results in mitigating blast effects.

14.4 MATERIAL FEATURES AND MECHANICAL BEHAVIOR

The unique structure of cellular materials makes them an attractive choice for use as sacrificial claddings under blast loads, with the primary advantage being their replaceability after sustaining damage. Energy absorption occurs when the cladding undergoes uniaxial compression, resulting in significant deformation. During this process, inner buckling mechanisms dissipate a substantial amount of energy, effectively reducing the load transmitted to the primary structure.

The global stress–strain response of these materials is characterized by three main regimes (refer to Figure 14.6 for graphical visualization). Initially, a transient elastic regime is observed, followed by an extended plastic regime during which large strains accumulate, while the stress stabilizes at the plateau stress of the cladding material (σ_{pl}). The third regime begins at the full compaction of the material, marked by the compaction strain ε_D, where stress rises rapidly, indicating stress hardening.

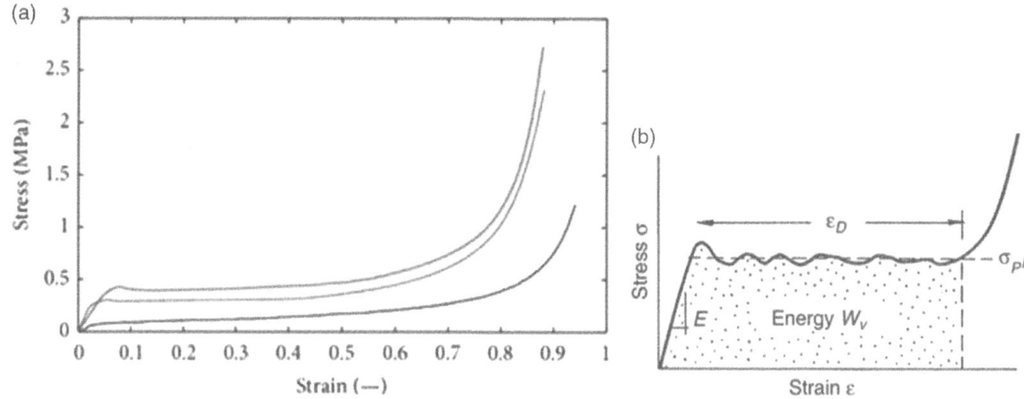

FIGURE 14.6 Stress–strain relationship describing the general shape of the sacrificial claddings. (a) Experimental curve. (b) Schematic curve. (Adapted from Ousji et al. (2016); Ashby et al. (2000), respectively.)

This fundamental stress–strain behavior is key to the effectiveness of materials used as sacrificial cladding. Essentially, any material with a microstructure capable of producing this type of global response can be considered for cladding applications and will adhere to the same constitutive laws and modeling approaches applicable to cellular materials, which will be discussed in the following text. Accordingly, in terms of the types of materials that were discussed previously, the stress–strain behavior derived from experimental results is depicted in Figure 14.7, where the described three-stage behavior can be observed in the graphs.

Several constitutive laws are available to describe the cellular material stress–strain relationship. The most common laws are outlined as follows:

- *Rigid-Perfectly-Plastic-Locking (RPPL)*: or Rigid-Perfectly-Plastic-Rigid (RPPR), was initially developed for modeling wood crushing (Reid and Peng 1997) and later adapted for cellular materials. This model neglects the elastic phase and assumes an immediate transition to the plastic regime at the plateau stress. Upon reaching full compaction, the stress rises sharply, translating into the dynamic response to the dynamic compaction stress (refer to Figure 14.8(a)). This model is relatively computationally efficient and is commonly used for high-intensity, short-duration events like blasts, where the cladding is expected to compress fully without significant elastic recovery. However, its simplicity can limit accuracy in scenarios where elastic behavior or gradual compaction is significant (Harrigan et al. 2005; Pattofatto et al. 2007a, 2007b).
- *Elastic-Perfectly-Plastic-Locking (EPPL)*: This model incorporates an initial elastic regime, providing a more realistic representation of the material response before reaching the plastic plateau.

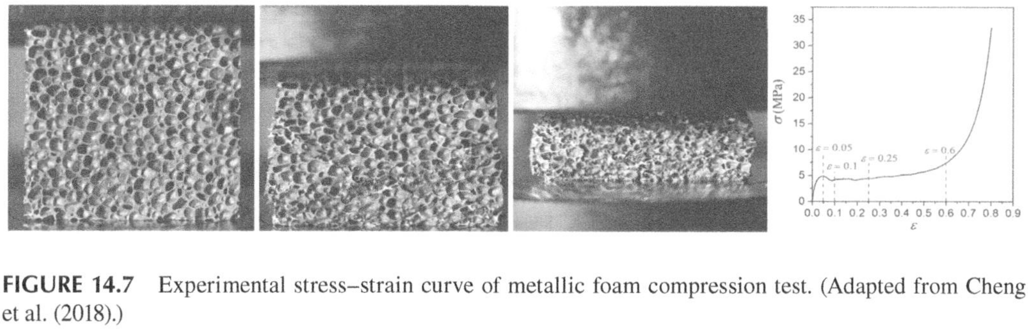

FIGURE 14.7 Experimental stress–strain curve of metallic foam compression test. (Adapted from Cheng et al. (2018).)

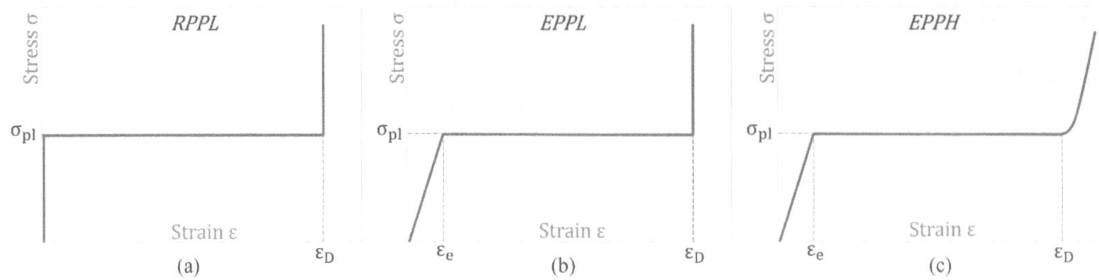

FIGURE 14.8 Metallic foam under compressive stress. (a) The RPPL constitutive law. (b) The EPPL constitutive law. (c) The EPPH constitutive law.

Once the elastic limit is surpassed, the material follows a response similar to the RPPL model, with a constant plateau stress until compaction strain is reached (refer to Figure 14.8(b)). This model is useful in applications where initial elastic deformation plays a role in energy absorption, such as in moderate blast loads or impacts where the foam does not compress fully and may undergo unloading and reloading cycles (Lopatnikov et al. 2003, 2004).

- *Elastic-Perfectly-Plastic-Hardening (EPPH)*: The model builds on the EPPL model by introducing strain hardening after the compaction strain, where stress increases progressively as the compaction wave travels through the foam. This gradual increase in stress reflects a more complex compaction behavior, where the foam structure provides increasing resistance under higher strain levels. The EPPH model is valuable for dynamic analyses where the cladding is expected to undergo significant deformation with varying stress levels, such as in high-intensity impacts or blasts, where a prolonged compaction phase influences the overall protective response (Li and Meng 2002) (refer to Figure 14.8(c)).

In summary, the key points regarding the presented constitutive laws are as follows: RPPL: suitable for high-speed simulations and simplified models, that describe rapid compaction after locking; computationally simple but less precise for partial compaction. EPPL: Adds realism with the elastic regime, making it useful for applications with moderate load intensities or reloading, making it less applicable for simplified models. EPPH: Most detailed, capturing progressive compaction with strain hardening, ideal for complex dynamic loading cases; making it suitable for high-ranked models, and inapplicable for simplified models (Zhu, Chou, and Yang 2011; Gupta and Kidane 2024).

The constitutive approaches introduced above are correlated to material properties of the cladding, i.e., the elastic and the compaction strains ε_e and ε_D, respectively, and the plateau stress σ_{pl}. The material parameters vary within the cladding thickness when considering cellular materials with a general density distribution ($\rho(z)$), where z is the location coordinate along the thickness. The mechanical properties are also a function of the location along the cladding thickness (Daxner 2010; Dornheim et al. 2024; Karagiozova and Langdon 2018; Niu and Yan 2016). Their general expressions for the mentioned parameters read:

$$\varepsilon_e = \frac{C_1}{\alpha_2}\varepsilon_{ys}\left(\frac{\rho(z)}{\rho_s}\right)^{1-n} \tag{14.1}$$

$$\varepsilon_D(z) = \left(\alpha - \lambda\frac{\rho(z)}{\rho_s}\right) \tag{14.2}$$

$$\sigma_{pl}(z) = C_1\sigma_{ys}\left(\frac{\rho(z)}{\rho_s}\right)^{m} \tag{14.3}$$

TABLE 14.1
Scaling Laws for the Mechanical Properties of Metallic Foams

Mechanical Properties	Open-Cell Foam	Closed-Cell Foam
Young's modulus E	$E=(0.1-4)E_s\frac{\rho^2}{\rho_s}$	$E=(0.1-1.0)E_s\times 0.5\frac{\rho^2}{\rho_s}+0.3\frac{\rho}{\rho_s}$
Shear modulus G	$G\approx\frac{3}{8}E$	$G=\frac{3}{8}E$
Compressive strength, σ_c	$\sigma_c=(0.1-1.0)\sigma_{cs}\frac{\rho^{3/2}}{\rho s}$	$\sigma_c=(0.1-1.0)\sigma_{cs}\times 0.5\frac{\rho^{2/3}}{\rho_s}+0.3\frac{\rho}{\rho_s}$
Tensile strength σ_t	$\sigma_t\approx(1.1-1.4)\sigma_c$	$\sigma_t\approx(1.1-1.4)\sigma_c$
Densification strain ε_D	$\varepsilon_D=(0.9-1.0)\times 1-\frac{4\rho}{\rho_s}+0.4\frac{\rho^3}{\rho_s}$	$\varepsilon_D=(0.9-1.0)\times 1-\frac{4\rho}{\rho_s}+0.4\frac{\rho^3}{\rho_s}$

Source: Adapted from Ashby et al. (2000).

where z is the reference location along the foam, ρ_s and σ_{ys} are the density and yield stress of the metal used to produce the cellular material, respectively, and α, λ, C_1, n, and m are empirical parameters obtained from experiments and vary among the different metals or materials used in cladding production (Ashby et al. 2000). In Table 14.1, an overview of the empirical parameters range assigned to different types of metallic foams can be found, which are used in the expressions for the mechanical properties.

14.5 MODELING STRATEGIES

Several modeling approaches have been developed to capture the complex behavior of sacrificial claddings under dynamic loading, each with unique assumptions, computational demands, and degrees of precision. These strategies range from simplified one-dimensional models to highly detailed hydrodynamic finite element simulations. The choice of model often depends on the desired balance between computational efficiency and the level of detail required to capture phenomena such as shock front propagation, local stress attenuation, and material deformation.

The simplest approaches, such as the shock front theory (SFT), provide a computationally efficient means of simulating cladding compaction with a minimal degree of freedoms (DoFs) and assumptions that simplify wave propagation dynamics. More complex strategies, such as the spring mass theory (SMT), add granularity by discretizing the cladding into multiple masses connected by non-linear springs, allowing for detailed insights into localized stress variations within the material. Two-dimensional modeling and finite element methods offer even greater precision by incorporating both the thickness and span of the cladding, as well as complex material behaviors and interactions. These advanced methods, however, require substantial computational resources and expertise in numerical modeling. The following sections present each of these modeling strategies, discussing their foundational assumptions, applications, and the trade-offs involved in applying each approach to sacrificial claddings under blast and impact loads.

14.5.1 The Shock Front Theory

The SFT is one of the simplest and most computationally efficient approaches for modeling the compaction of sacrificial claddings under dynamic loads. Initially developed to simulate wood crushing (Reid and Peng 1997), it was later adapted to simulate the compaction of cellular materials

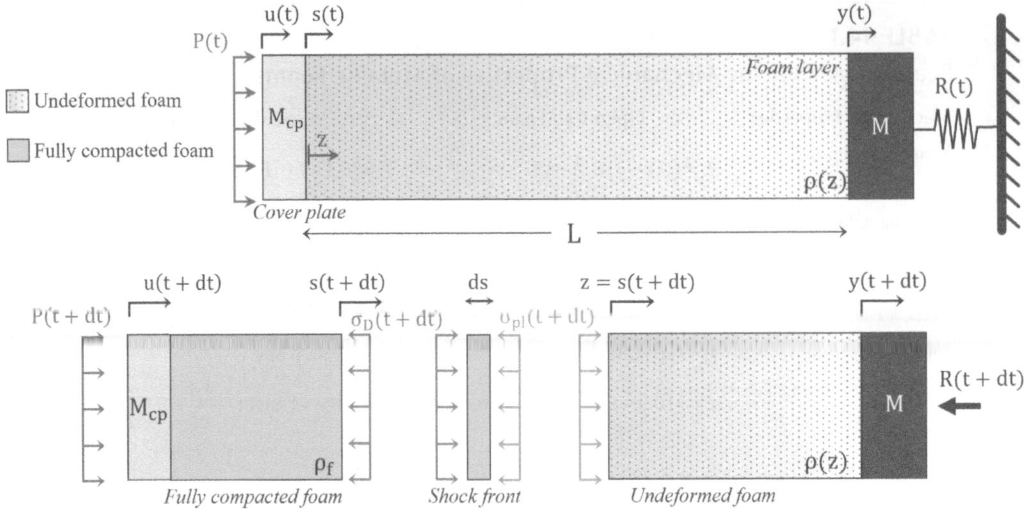

FIGURE 14.9 Shock front propagation along the cladding thickness. (Adapted from Wattad and Grisaro (2024b).)

(Hanssen, Enstock, and Langseth 2002), (as demonstrated in Figure 14.9). The SFT approach assumes that, under dynamic loading, a plastic wave propagates through the thickness of the cladding, resulting in distinct compacted and non-compacted regimes. In this framework, the compacted regime is subjected to dynamic compaction stress, while the non-compacted regime endures the plateau stress of the cladding material. The discontinuity between these two regimes defines the shock front, which travels along the cladding thickness as the material undergoes compaction. The SFT is attractive for its low computational cost and simplicity, as it requires only a few DoFs, typically one DoF defining the deflection of the loaded end and another for the shock front location, which are both mathematically correlated. This simplicity makes it feasible to model large-scale problems or conduct parametric studies with minimal computational resources.

Over the years, numerous variations and assumptions have been incorporated into SFT-based models to explore a range of conditions. Conditions in these studies assume: a rigid boundary for the protected structure, neglecting the structural response during cladding compaction (Hanssen, Enstock, and Langseth 2002; Karagiozova, Langdon, and Nurick 2012; Shen, Yu, and Lu 2013; Ma and Ye 2007; Liang et al. 2017; Karagiozova 2011; Reid and Peng 1997; Zhou et al. 2015), structural response that could not catch the structural resistance (Ma and Ye 2005; Ye and Ma 2007), neglecting some of the model limitations (Xia, Wu, and Li 2015; Wu and Zhou 2011), and others that made various assumptions to create realistic conditions (Wattad and Grisaro 2024a, 2024b). The theory has also been used to investigate the effects of uniform versus non-uniform density distributions in claddings, with some studies considering linear (Wattad and Grisaro 2024b; Xia, Wu, and Bennett 2017) or exponential density gradients across the thickness (Zhou et al. 2015). For example, Zhou et al. (2015) developed an analytical model for cladding with an exponential density distribution along its thickness and validated it against finite element simulations, with the condition of a fixed end resembling a later-on structural response. They found that the density configuration of the cladding could be fine-tuned to minimize displacement, with higher densities near the loaded surface enhancing energy absorption under specific blast loads. Similarly, in the work of Wattad and Grisaro (2024b), a SFT-based model was developed for a general density function, and using a parametric study, they evaluated the linear and exponential density function and corroborated Zhou et al.'s findings.

Despite its utility, the SFT has certain limitations when applied to cellular materials under dynamic loads. In the study of Li and Reid (2006), concerns were raised regarding the potential for stress enhancements during shock propagation through a cellular medium, particularly under impact conditions. Their findings suggested that the SFT may not fully capture the complexity of stress transmission in cellular materials, as shock enhancements observed experimentally were not adequately reflected by the model. Additionally, studies indicated that while SFT-based models are suitable for blast loads, they might not always accurately predict stress distributions for direct impact loads or cases where the cladding is subjected to sudden deceleration.

In specific applications, some adaptations of the SFT consider the presence of dual shock fronts within the cladding, which can occur when high-intensity loads generate multiple compaction waves through the material (Ma and Ye 2007). These dual-front models account for additional shock interactions within the cladding, especially in non-uniform density claddings or those with complex density profiles. However, such configurations require additional modeling assumptions and parameters, making them more complex than the single-shock model traditionally used in the SFT.

When accounting for the coupled response of the cladding and the protected structure, most studies that used the SFT have traditionally assumed a decoupled approach, whereby the structural response only occurs after the cladding compaction has ended. In other words, the structure experienced an attenuated stress, generated from the stress–strain curve (i.e., the plateau stress), which lasts over a longer period of time compared to the initial blast pressure. This assumption has provided a simplified framework for assessing energy dissipation and structural protection; however, it limits the understanding of the simulated work between the two components of the dynamic system. Only a limited number of studies have investigated the direct coupling between cladding deformation and structural response. In scenarios that incorporate structural response, some models assume linear-elastic behavior for the structure, while others consider non-linear structural resistance, allowing for plastic deformation. These variations in assumptions have led to diverse applications of the SFT, with studies investigating both air-blast loads and mass impacts.

In summary, while the SFT provides a computationally efficient framework for analyzing sacrificial cladding under dynamic loading, it operates under certain simplifying assumptions that may limit its applicability to certain blast and impact scenarios. The absence of structural coupling, assumptions of uniform or non-uniform density distributions, and variations in load types (e.g., blast versus impact) each introduce limitations to the model fidelity. Nonetheless, SFT remains a foundational approach for initial analyses of sacrificial claddings, offering a balance between simplicity and practical utility for assessing energy dissipation in protective layers.

14.5.2 The Spring Mass Theory

The SMT, also known as the multi-DoFs approach, is a more advanced strategy than the SFT to model the behavior of sacrificial claddings under blast loads. This approach was discussed in Li and Meng (2002) and Qi, Yang, and Gao (2013) and provides a one-dimensional detailed discretization of the cladding by dividing it into multiple masses connected by non-linear springs, each with a DoF (refer to Figure 14.10). Unlike the SFT, which assumes a simplified, single-shock propagation through the cladding, the SMT accounts for complex local interactions and "meso-scopic" stress oscillations within the cladding thickness. This approach enables the study of stress variations and wave propagation across the cladding, which could not be captured by the simpler SFT.

In the SMT, the cladding is modeled as a series of masses connected by springs, governed by non-linear constitutive laws, such as the EPPH model. This model captures the non-linear behavior of the cladding material, accounting for loading, unloading, and reloading phases during compaction. The ability to model these phases allows the SMT to represent the energy dissipation and resistance properties of the cladding more precisely over time, while also increasing the computational complexity. By discretizing the thickness of the cladding in this way, the SMT provides a detailed deflection field throughout the layer and captures the progression of stress waves and the rebound phenomenon as they propagate through the cladding.

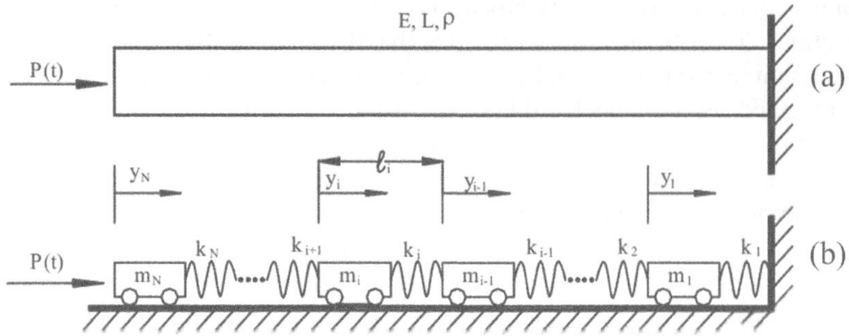

FIGURE 14.10 (a) Cellular material subjected to a pulse loading. (b) The simplified one-dimensional model of 10(a) with N discrete lump masses. (Adapted from Li and Meng (2002).)

One of the key advantages of the SMT is its ability to reveal localized stress attenuation within the cladding. Studies have shown that, due to local and mesoscopic behavior, stress may attenuate within the transmitted stress–time history to the rigid support (Li and Meng 2002; Qi, Yang, and Gao 2013). This phenomenon is not observed in SFT-based models, which lack the resolution to capture such detailed wave interactions. Nevertheless, as SMT enables a more nuanced view of global phenomena, the transition between compacted and uncompacted regimes within the cladding is vague, mainly because of the constitutive models adopted in this strategy. Accordingly, defining these regimes requires careful calibration, as the definition of a fully compacted spring may need to be specified manually.

Nonetheless, the SMT has faced some criticism in the literature. Harrigan, Reid, and Seyed Yaghoubi (2010) argued that the SMT may inaccurately predict stress enhancements during the dynamic compaction of cellular materials like foam claddings. They compared the SMT predictions with those of an extended shock theory, which incorporates non-linear post-locking behavior, and found significant differences between the two approaches. According to Harrigan and Reid, the SFT predicts a constant stress in the compacted regime during the initial phase of compaction, followed by a reduction in stress as compaction progresses, while the SMT predicts higher stress peaks during wave propagation. They noted that, due to these divergences, the SMT should be applied cautiously, especially for cases where compaction waves play a dominant role in the material response. To address the highlighted points of concern in the work by Li and Meng (2002); Xi, Dai, and Li (2023) introduced a contact spring between masses to mitigate issues observed during numerical simulations; however, this modification did not eliminate the observed stress enhancements. The persistence of stress enhancements suggests that the phenomenon identified by Li and Meng may reflect some inherent aspect of the material behavior.

Despite its limitations, the SMT offers valuable insights into the dynamic behavior of sacrificial claddings under blast loading. It provides a more refined analysis of wave propagation, localized stress attenuation, and the detailed response of the cladding, making it a useful tool for applications where a high level of precision is required. However, due to its computational demands, the SMT should be applied selectively and often relies on specific assumptions about the cladding and cladded structure. Comprehensive studies that integrate the SMT with a responsive cladded structure remain limited, indicating an area for future research to fully understand the interaction between sacrificial claddings and the primary structure.

14.5.3 Two-Dimensional Discretization – Advanced Multi-DOFs Approach

The two-dimensional discretization approach considers discretization in two directions: the loaded span and the cladding thickness. This approach adds an additional spatial dimension to the modeling of the cladding compared to previously discussed approaches. By incorporating both directions, it captures the spatial variations in cladding behavior more comprehensively. The approach assumes uniaxial compaction of the foam at each spatial coordinate but allows for variations in compaction along the loaded span. As a result, there is no single uniaxial compaction value for the entire structure, necessitating a high number of DoFs to represent the cladding accurately (see Figure 14.11). This strategy enables the model to consider non-uniform loading distributions, varying compaction states, and differential effective resistance across the cladding surface area, providing a more detailed analysis of structural response under complex loading conditions. Additionally, as the approach considers uniaxial compaction for a certain location of the loaded span, this approach allows for the flexibility to use any of the previously discussed compaction models at these locations, carrying over their respective advantages and limitations.

An example of this modeling strategy is demonstrated in the work of (Wu and Sheikh 2013), who applied a two-dimensional model to a protected structural member clad with aluminum foam. They divided the structure into small regions, categorizing these regions into hinged and non-hinged regimes. The non-hinged regimes were modeled as quadratic beam elements, using Timoshenko beam theory (Timoshenko and Goodier 1951) to capture bending and shear deformation. The sacrificial cladding was discretized with multiple uniaxial springs governed by high-order stress–strain models, transferring pressure through connecting springs to the protected member (in this case, none of the compaction models presented above was adopted). The model accounted for spatial

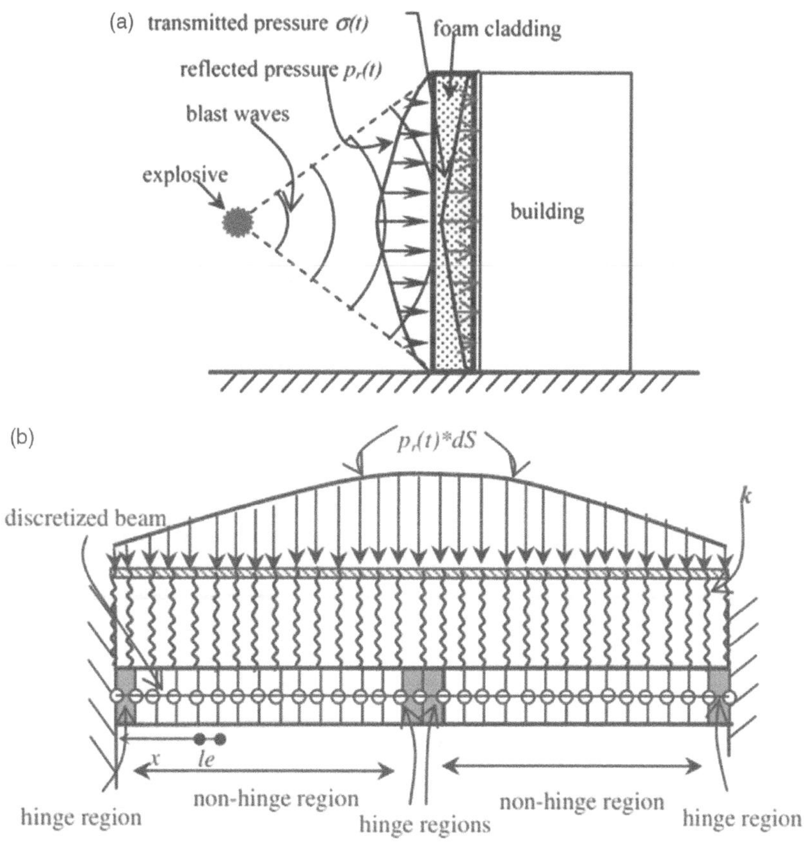

FIGURE 14.11 (a) Mitigation mechanism of metallic foam as sacrificial cladding. (b) Load cladding structure interaction model based on the two-dimensional discretization approach. (Adapted from Wu and Sheikh (2013).)

variations in structural response across the foam surface and throughout the cladding thickness by employing a large set of springs with non-linear constitutive laws. This approach enabled a detailed deflection-time history for various sections of the structure, and the model was validated against field test data, demonstrating its accuracy in capturing the dynamic response.

However, the increase in DoFs in this two-dimensional approach significantly complicates the formulation and solution of governing equations, requiring substantial computational resources. In some cases, this complexity may not yield a proportional benefit in accuracy or insight, making simpler models more appealing for certain applications.

14.5.4 Detailed Finite Element Methods

The Finite Elements Model (FEM), particularly in its detailed form, is one of the most advanced strategies for modeling the dynamic response of sacrificial claddings under blast loads. This high-resolution approach allows for an intricate representation of the material behavior of the cladding, which is especially relevant for cellular and complex materials such as metallic foams and other porous structures. Compared with simpler modeling approaches, detailed FEM simulations provide a more precise and comprehensive representation of the energy absorption and dissipation characteristics of the cladding, although at the cost of increased computational demand (refer to Figure 14.12).

For sacrificial claddings, FEM modeling is commonly implemented using commercial software packages such as ABAQUS (Smith 2009), Ansys-Autodyn (DeSalvo and Swanson 1985), and LS-DYNA (Livermore Software Technology Corporation (LSTC) 2023). These software tools offer

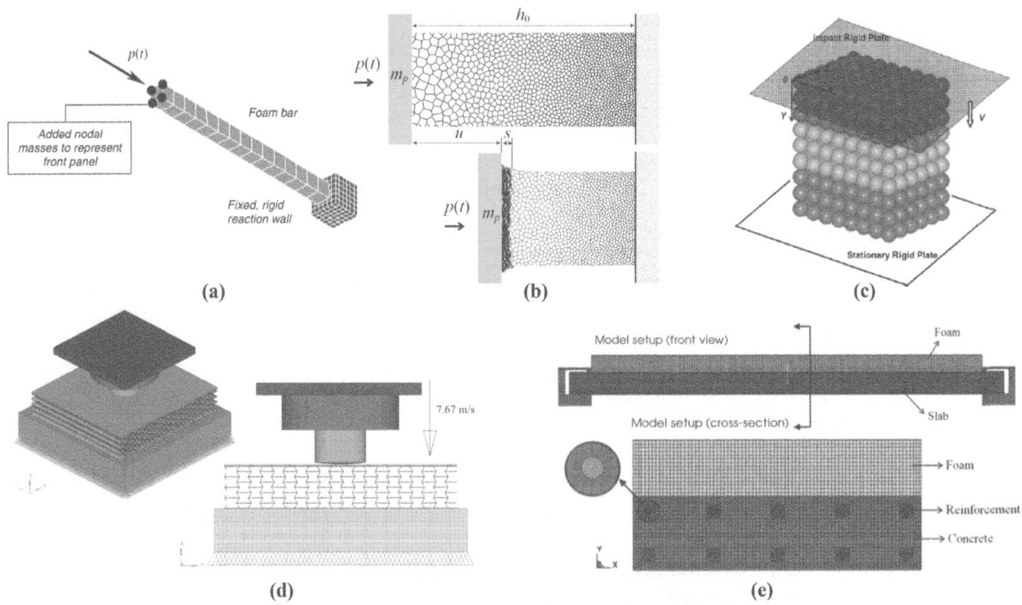

FIGURE 14.12 FEM generation of: (a) Aluminum foam cladding using LS-DYNA (Livermore Software Technology Corporation (LSTC) 2023). (b) Continuous-density graded cellular materials, using ABAQUS/Explicit (Smith 2009). (c) MHS system with stepwise density gradient, using LS-DYNA and ABAQUS/Explicit (Livermore Software Technology Corporation (LSTC) 2023; Smith 2009). (d) Auxetic honeycomb-cored sandwich cladding, using ABAQUS/Explicit (Smith 2009). (e) FEM Concrete slab cladded with aluminum foam sacrificial layer, using LS-DYNA (Livermore Software Technology Corporation (LSTC) 2023). (Adapted from Hanssen, Enstock, and Langseth (2002); Liang et al. (2017); Zhou et al. (2015); Qi et al. (2017); Xia et al. (2014), respectively.)

specialized features and advanced material models specifically designed to capture the complex responses of sacrificial claddings, including hydrodynamic and cellular material behaviors. In FEM simulations of sacrificial claddings, some studies have implemented simplified conditions, such as assuming uniaxial compaction across the cladding material. Under this assumption, the cladding undergoes compression along a single axis, simplifying the model while still capturing essential deformation and energy absorption characteristics. Additionally, certain models assume a fixed boundary condition for the cladded structure, where the structure does not dynamically respond during the compaction of the cladding (Hanssen, Enstock, and Langseth 2002; Karagiozova, Langdon, and Nurick 2012; Shen, Yu, and Lu 2013; Ma and Ye 2007; Liang et al. 2017, 2019; Liu et al. 2015; Ding et al. 2020; Zhou et al. 2015; Langdon et al. 2010; Liu, Zhang, and Yang 2016). This fixed-wall assumption can simplify the simulation, as it removes the need to model the coupled response of the cladded structure, focusing instead on the cladding behavior under loading.

One of the primary advantages of using detailed FEM for sacrificial claddings is its ability to account for the intricate deformation mechanisms and stress–strain relationships that characterize cellular and porous materials. With high-fidelity FEM, it can incorporate complex constitutive models for the cladding, allowing for the accurate representation of non-linearity in material behavior, such as complex material density distributions, rebound waves, elastic waves, etc. This level of detail enables engineers to predict the performance of the sacrificial cladding with a high degree of precision under various loading conditions, making it invaluable for designs requiring optimized blast mitigation.

However, the detailed FEM approach comes with several challenges and limitations. First, the computational resources required for these simulations are substantial, particularly for hydrodynamic simulations and cellular materials, which often require a fine mesh and an extensive set of material parameters.

Additionally, the complexity of setting up such models necessitates specialized expertise in FEM analysis and material modeling. Moreover, accurately defining the material properties and boundary conditions requires considerable experience and skill, as small variations in parameters can significantly impact simulation results. Furthermore, while FEM provides a highly detailed view of cladding behavior, the increased precision may not always justify the computational cost, especially when simpler models may yield sufficiently accurate insights for preliminary design purposes.

In summary, detailed FEM is a powerful tool for modeling sacrificial claddings in blast protection applications, offering unmatched accuracy and capabilities. Nonetheless, its resource-intensive nature and the expertise required for effective implementation make it a technique best reserved for applications where high precision is essential and computational resources are available.

14.6 TESTING, EVALUATION METHODS, AND GAPS

14.6.1 Testing

Most experimental work on sacrificial claddings for blast protection has focused on testing the stress–strain response of the cladding materials under controlled conditions. These tests often involve laboratory-based setups designed to characterize the material capacity for energy absorption and plastic deformation. However, only a limited number of studies have conducted full-scale field tests, which provide insights into the realistic effectiveness of these claddings. Field tests are crucial as they simulate actual blast environments and allow for a more comprehensive understanding of how the cladding interacts with the primary structure under blast loads.

Field tests on auxetic re-entrant honeycomb sandwich panels demonstrated that this type of cladding effectively absorbs blast energy, providing significant protection to steel plate structures under near-field blast conditions. However, the study also highlighted that there is a threshold of blast intensity beyond which the cladding effectiveness diminishes (Kalubadanage et al. 2021). This

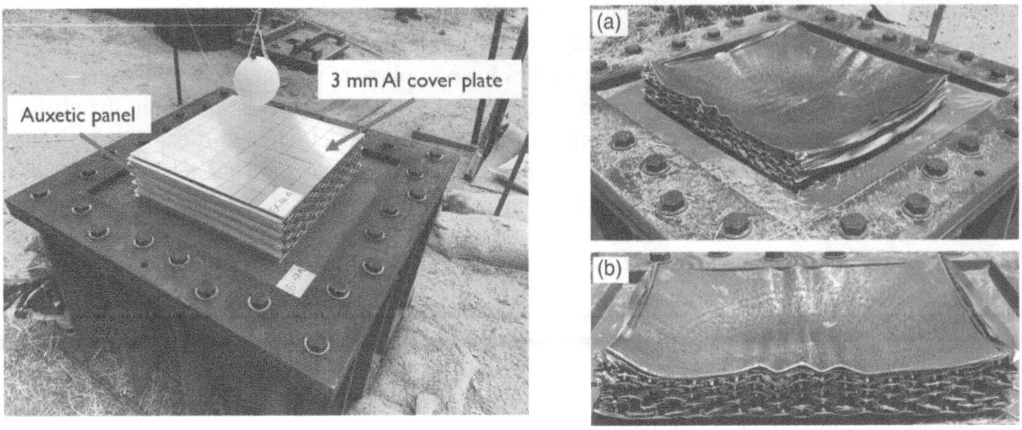

FIGURE 14.13 Experimental setup for blast testing, alongside (a) and (b) the post-blast deformations of 3D-printed auxetic claddings. (Adapted from Kalubadanage et al. (2021).)

finding suggests that while auxetic structures offer promising blast mitigation, their application may be limited to specific load intensities (see Figure 14.13).

In another study, reinforced resin panels with insulation layers, applied to reinforced concrete structures, showed substantial reductions in structural damage and displacement when compared to both unprotected structures and those with traditional steel jacketing. The reinforced resin cladding was particularly effective in absorbing blast energy, making it a viable alternative to more conventional cladding systems (Codina, Ambrosini, and de Borbón 2017).

Studies on calcium silicate–based mineral foam cladding indicated a significant reduction in maximum deflection, with certain foam configurations decreasing deflection by up to 70%. This result demonstrates the potential of mineral foams in blast mitigation and highlights the importance of foam structure and material choice in optimizing cladding effectiveness (Aminou et al. 2024). Similarly, hybrid multi-cell tubes used as cladding for reinforced concrete panels were found to significantly reduce mid-span deflection, further indicating their suitability for blast load mitigation (Abada, Ibrahim, and Jung 2021).

Research on polymeric foams as crushable cores demonstrated that open-cell polyurethane foam was the most effective in blast energy absorption among the foams tested. This study underscored the importance of selecting appropriate foam types and front plate masses for optimizing energy dissipation in cladding systems, as the performance varied significantly across different configurations (Ousji et al. 2016) (see Figure 14.14).

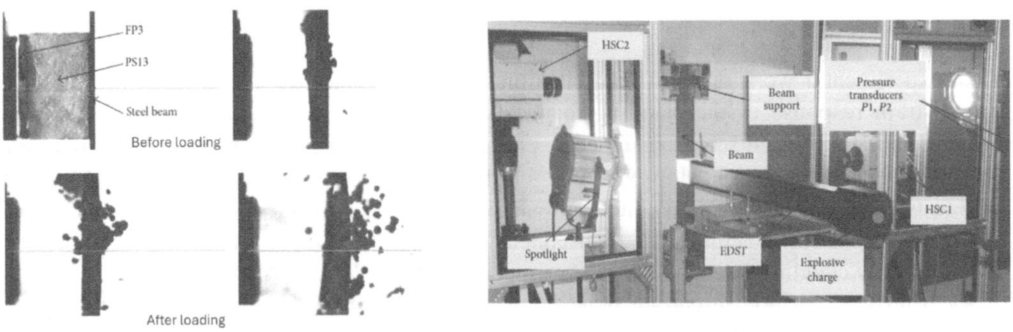

FIGURE 14.14 Experimental setup and results of sacrificial cladding with expanded polystyrene cores for blast mitigation. (Adapted from Ousji et al. (2016).)

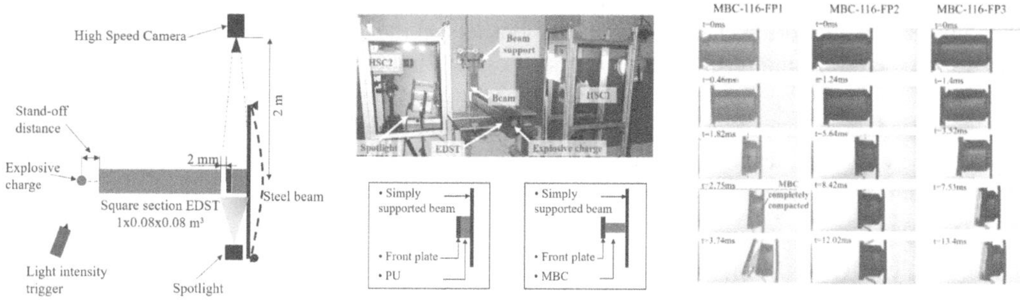

FIGURE 14.15 Experimental setup of impact testing of sacrificial cladding made of metallic beverage cans alongside the chronological crushing event of cladding. (Adapted from Ousji et al. (2024).)

Additionally, research on low-density crushable core systems made of metallic beverage cans provided insights into the importance of cladding configuration. The study suggested that these cores could transform hazardous blast loads into lower-pressure loads over an extended time span, potentially enhancing the protective performance of sacrificial claddings (Ousji et al. 2024) (see Figure 14.15).

Further emphasizing the potential of aluminum foam cladding, in 2009 a series of experimental and field blast tests were conducted (Zhu et al. 2009) and adapted in the work of Wu and Sheikh (2013) to investigate its effectiveness in protecting reinforced concrete slabs. Their study demonstrated that metallic foam cladding significantly mitigated blast effects, primarily by controlling the transmitted pressure to the slabs. Field tests were conducted with varying standoff distances and foam configurations. These tests validated that the foam layer effectively reduced the maximum deflection of the slab while absorbing high energy at nearly constant stress levels during the plateau phase of the foam compression behavior. The authors highlighted the importance of maintaining the structural integrity of the foam under extreme loads by employing a metal cover plate to prevent disintegration.

Field testing on aluminum foams further highlights the capabilities of metallic foams as sacrificial claddings. In the work of Xia et al. (2016), claddings composed of four layers of aluminum foams, each with a distinct density, on reinforced concrete slabs. These configurations were initially characterized through compressive testing to determine the stress–strain response of each layer, providing essential material properties for further evaluation. The study then conducted field tests under various blast scenarios, demonstrating the ability of layered aluminum foam claddings to absorb significant energy and reduce transmitted pressures (refer to Figure 14.16). The findings

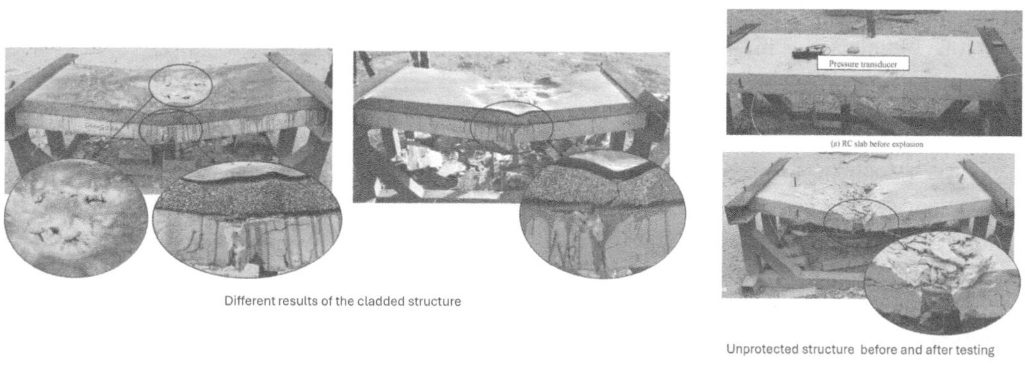

FIGURE 14.16 Experimental results adapted from Xia et al. (2016), illustrating the post-test condition of bare and clad structures, using graded-density aluminum foam sacrificial cladding under blast loading.

highlight the role of density distribution in optimizing blast mitigation performance and underscore the importance of tailoring cladding configurations to specific threat levels.

Despite the promising results from these studies, it is important to note that field tests are relatively rare, and most experimental studies rely on controlled laboratory conditions. Moreover, studies examining the effects of varying cladding densities under field conditions are limited. This lack of comprehensive field testing, particularly with density variations, restricts a full understanding of how different configurations perform under realistic blast scenarios.

14.6.2 Evaluation Methods

The performance of sacrificial cladding is commonly evaluated by evaluating the peak displacement of the cladded structure and comparing it to a bare (unprotected) structure under similar loading conditions. This approach serves as a practical indicator of the ability of the cladding to mitigate damage to the primary structure. However, Wattad and Grisaro (2024a, 2024b) emphasized that it is essential to consider the effect of the additional mass due to the cladding itself, which may affect the dynamic response in addition to the ability of the cladding to absorb energy. They propose comparing the performance of the cladded structure to that of a structure with an equivalent additional mass, but without the energy-absorbing capabilities of the cladding. In this manner, the effect of the alternating mass of the system can be isolated, and the capabilities of the cladding material are isolated in terms of energy dissipation and blast mitigation, which are adequately evaluated.

Despite the focus on energy absorption as a key benefit of sacrificial cladding, there is a notable lack of studies that directly quantify and evaluate the actual energy absorbed and dissipated by the cladding during a blast event. Most research infers energy absorption based on displacement measurements, yet this approach does not fully capture the cladding performance in terms of energy dissipation. Direct calculations or experimental evaluations of energy absorption would provide a more accurate assessment of the effectiveness of the cladding; however, limited studies in this area are conducted, while neglecting the response of the cladded structure (Zhou et al. 2015; Karagiozova 2011; Tan et al. 2005; Ma and Ye 2007), or preliminarily evaluating the energy of the sacrificial for specific models, (Su, Liu, and Hou 2024). The previous highlights the need for methodologies that can quantitatively measure the energy absorbed by the cladding materials, offering a more comprehensive understanding of their role in blast mitigation.

Nevertheless, in the limited studies that directly examined and calculated the energy absorption capabilities of sacrificial claddings, the dynamic response of the cladded structure itself, during the cladding response, was often neglected due to simplifying assumptions. This approach, while useful for events where the structure response is actually neglectable during the cladding response, overlooks the coupling between the two components of the system, which plays a crucial role in real-world scenarios. As a result, there remains a gap in understanding how the energy absorption of the cladding integrates with and affects the overall response of the protected structure, limiting the scope of knowledge in this field.

14.6.3 Gaps

Several gaps remain in the evaluation of sacrificial claddings for blast protection. Firstly, energy absorption, a key concept in assessing the effectiveness of cladding materials, is often indirectly evaluated through peak displacements rather than direct energy measurements. This omission can lead to an incomplete understanding of how much energy is truly absorbed by the cladding versus transmitted to the primary structure.

Moreover, there is a lack of comparative studies that evaluate different modeling strategies against each other. Each modeling approach, from simplified theories to detailed finite element simulations, has unique assumptions and computational demands, yet no comprehensive analysis has been conducted to assess which models provide the best balance of accuracy and efficiency in predicting sacrificial cladding performance. Additionally, optimization studies focused on cladding density, a parameter that strongly influences energy absorption and blast mitigation, are also scarce, leaving a gap in the development of guidelines for selecting optimal density configurations.

14.7 CONCLUSIONS

The chapter presented an overviewed of the use of sacrificial cladding as a novel tool for structural protection. It discussed the rationale behind using this type of technology when compared to widely spread alternatives. It initially reviewed the evolution of using this type of technology and focused on the recent developments in this field. This chapter has explored the potential of sacrificial claddings as an effective approach for mitigating blast-induced damage to structures. The section included an overview of the materials commonly used in sacrificial layers, highlighting the unique properties of metal foams, geometric configurations, and advanced 3D-printed auxetic structures. Each of these materials offers distinct advantages in terms of energy absorption, customization, and adaptability for various blast scenarios, making them suitable choices for sacrificial cladding applications.

The chapter also examined the mechanical behavior of these materials under dynamic loading, particularly focusing on their stress–strain responses. Understanding these properties is crucial for predicting how these materials will perform under high-stress conditions. Various constitutive models, including the RPPL, EPPL, and EPPH models, were discussed to capture this behavior accurately.

In addressing modeling strategies, a review of several approaches was conducted, from simplified methods like the SFT to more advanced methods, such as detailed FEM simulations. These models each offer unique insights but vary in complexity, computational demands, and applicability. The choice of model depends on the desired balance between accuracy and efficiency, with more complex models providing detailed predictions but at a higher computational cost.

Testing and evaluation methods were also discussed, with most studies focusing on measuring peak displacement or stress–strain curves to assess the cladding protective performance. Field testing, while limited, has provided valuable insights into the real-world effectiveness of these claddings, although the absence of studies on the impact of varying densities and direct energy absorption measurements highlights areas for further research. Additionally, the importance of considering the cladding's added mass was emphasized, as this factor significantly influences the dynamic response of the overall system.

Despite significant advancements in understanding sacrificial claddings, several gaps remain in current research. Energy absorption, a critical metric of cladding performance, is often indirectly measured, with limited studies directly calculating the energy absorbed by the cladding. Moreover, there is a need for comparative evaluations of different modeling strategies and for optimization studies that consider the influence of density configurations on blast mitigation.

In summary, sacrificial claddings represent a promising solution for enhancing structural resilience in extreme loading scenarios. Future research should aim to address the limitations identified in this chapter, particularly through comprehensive field tests, direct energy absorption measurements, and the development of optimized cladding configurations. Such advancements will enable more precise, effective, and adaptable designs for protecting structures against increasingly severe blast and impact threats.

REFERENCES

Abada, Mahmoud, Ahmed Ibrahim, and Sung Ju Jung (Mar. 2021). "Improving Blast Performance of Reinforced Concrete Panels Using Sacrificial Cladding with Hybrid-Multi Cell Tubes". *Modelling* 2.1, pp. 149–165. ISSN: 2673-3951. DOI: 10.3390/MODELLING2010008.

American Society of Civil Engineers (1997). *Design of Blast-Resistant Buildings in Petrochemical Facilities.* Reston, VA: ASCE Publications.

American Society of Civil Engineers (2023). *Blast Protection of Buildings*. 2nd ed. Reston, VA: American Society of Civil Engineers. ISBN: 9780784415719.

Aminou, Aldjabar et al. (Oct. 2024). "Experimental and Numerical Evaluation of Calcium-Silicate-Based Mineral Foam for Blast Mitigation". *Applied Sciences* 14.21, p. 9656. ISSN: 2076-3417. DOI: 10.3390/APP14219656.

ASCE (2011). Standard ASCE/SSEI 59-11: Blast Protection of Buildings. Reston, VA: American Society of Civil Engineers (ASCE). ISBN: 9780784476505.

Ashby, Michael F et al. (2000). *Metal Foams: A Design Guide*. Oxford: Elsevier.

BibleWalks (2024). *Mud Bricks*. URL: https://www.biblewalks.com/mudbricks/.

Blake, Peter et al. (2024). "3D-Auxetic Elastomeric Cellular Structures for Impact Protection". *International Journal of Protective Structures* 15.4. DOI: 10.1177/20414196241284296.

Bohara, Rajendra P et al. (2023). "Anti-Blast and -Impact Performances of Auxetic Structures: A Review of Structures, Materials, Methods, and Fabrications". *Engineering Structures* 276, p. 115377. DOI: 10.1016/j.engstruct.2022.115377.

Buchan, Paul A. and Jian-Fei Chen (July 2007). "Blast Resistance of FRP Composites and Polymer Strengthened Concrete and Masonry Structures – A State-of-the-Art Review". *Composites Part B: Engineering* 38.5–6, pp. 509–522. ISSN: 1359-8368. DOI: 10.1016/J.COMPOSITESB.2006.07.009.

Cheng, Ying et al. (2018). "Compressive Properties and Energy Absorption of Aluminum Foams with a Wide Range of Relative Densities". *Journal of Materials Engineering and Performance* 27.8, pp. 4016–4024. ISSN: 1544-1024. DOI: 10.1007/s11665-018-3514-4.

Codina, Ramón, Daniel Ambrosini, and Fernanda de Borbón (June 2017). "New Sacrificial Cladding System for the Reduction of Blast Damage in Reinforced Concrete Structures". *International Journal of Protective Structures* 8.2, pp. 221–236. ISSN: 2041420X. DOI: 10.1177/2041419617701571.

Critchley, Richard et al. (2022). "Blast Mitigation Using Polymeric 3D Printed Auxetic Re-Entrant Honeycomb Structures: A Preliminary Study". *International Journal of Protective Structures* 13.3, pp. 469–486. DOI: 10.1177/20414196211052062.

CSA-Group (2012). *S850-12 Design and Assessment of Buildings Subjected to Blast Loads*. CSA Group (Canadian Standards Association), Mississauga, ON, Canada.

Daxner, Thomas (2010). "Finite Element Modeling of Cellular Materials". In: *Cellular and Porous Materials in Structures and Processes*. Springer, pp. 47–106. https://doi.org/10.1007/978-3-7091-0297-8_2

Department of the Army (1986). *Army Technical Manual 5-855-1*. Washington, DC: Department of the Army.

Derat, Marie-Laure (2024). *Educating the Nation in Ethiopia*. Centre français des études éthiopiennes. URL: https://books.openedition.org/cfee/819?lang=en.

DeSalvo, Gabriel J and John A Swanson (1985). *ANSYS Engineering Analysis System User Manual*. Houston, PA: Swanson Analysis Systems.

Ding, Yuanyuan et al. (2020). "Blast Alleviation of Sacrificial Cladding with Graded and Uniform Cellular Materials". *Materials* 13.24. DOI: 10.3390/ma13245616.

Dornheim, Johannes et al. (2024). "Neural Networks for Constitutive Modeling: From Universal Function Approximators to Advanced Models and the Integration of Physics". *Archives of Computational Methods in Engineering* 31, pp. 1097–1127. DOI: 10.1007/s11831-023-10009-y.

Engines of Our Ingenuity (2024). *Engines of Our Ingenuity: Episode 3130*. URL: https://engines.egr.uh.edu/episode/3130.

Federal Emergency Management Agency (FEMA) (Dec. 2003). *EMA 427: Primer for Design of Commercial Buildings to Mitigate Effects of Terrorist Attacks*. Washington, D.C.: Federal Emergency Management Agency (FEMA).

Genç, Og˘uz Kag˘an et al. (Oct. 2024). "Blast-Resistant Design of Reinforced Concrete Slabs with Auxetic-Shaped Reinforcement Layout". *Buildings* 14.11, p. 3392. ISSN: 2075-5309. DOI: 10.3390/BUILDINGS14113392.

Grisaro, Hezi Y, Jeffrey A Packer, and Michael V Seica (2019). "Weak Axis Response of Steel I-Sections Subjected to Close-in Detonations". *Journal of Constructional Steel Research* 160, pp. 189–206. ISSN: 0143-974X. DOI: 10.1016/j.jcsr.2019.05.025.

Grisaro, Hezi Y, Michael V Seica et al. (2021). "Strong-Axis Response of Steel I-Sections Subjected to Close-in Detonations". *International Journal of Protective Structures* 12.3, pp. 347–376. DOI: 10.1177/2041419620984487.

Gu, Liquan, Yonghui Wang, and Ximei Zhai (July 2023). "Behaviour of a Sacrificial Cladding with Foam Concrete-Filled Square Tubes under Impact Loads". *Buildings* 13.7, p. 1774. ISSN: 2075-5309. DOI: 10.3390/BUILDINGS13071774.

Gupta, Vijendra and Addis Kidane (Dec. 2024). "Designing Density-Graded Cellular Materials for Tailored Constitutive Response". *Composites Part B: Engineering* 287, p. 111793. ISSN: 1359-8368. DOI: 10.1016/J.COMPOSITESB.2024.111793.

Hanssen, Arve G., Lars Enstock, and Magnus Langseth (July 2002). "Close-Range Blast Loading of Aluminium Foam Panels". *International Journal of Impact Engineering* 27.6, pp. 593–618. ISSN: 0734-743X. DOI: 10.1016/S0734-743X(01)00155-5.

Hao, Hong, Yifei Hao et al. (Aug. 2016). "Review of the Current Practices in Blast-Resistant Analysis and Design of Concrete Structures". *Advances in Structural Engineering* 19.8, pp. 1193–1223. ISSN: 20484011. DOI: 10.1177/1369433216656430.

Hao, Hong and Zhong-Xian Li, eds. (2016). *Advances in Protective Structures Research*. Boca Raton, FL: CRC Press.

Harrigan, John J., Steve R. Reid, and Ali Seyed Yaghoubi (Aug. 2010). "The Correct Analysis of Shocks in a Cellular Material". *International Journal of Impact Engineering* 37.8, pp. 918–927. ISSN: 0734-743X. DOI: 10.1016/J.IJIMPENG.2009.03.011.

Harrigan, John J et al. (2005). "High Rate Crushing of Wood Along the Grain". *International Journal of Mechanical Sciences* 47.4, pp. 521–544. ISSN: 0020-7403. DOI: 10.1016/j.ijmecsci.2004.12.013.

Hetherington, J. and P. Smith (2014). *Blast and Ballistic Loading of Structures*. CRC Press. URL: https://books.google.co.il/books?id=eURZDwAAQBAJ.

Jones, Norman, Annamalai M. Rajendran, and Carlos A. Brebbia (Jan. 2002). "Protective Design of Concrete Buildings Under Blast Loading". *Sustainability* 2, p. 612.

Joseph, Athul, Vinyas Mahesh, and Dineshkumar Harursampath (June 2021). "On the Application of Additive Manufacturing Methods for Auxetic Structures: A Review". *Advances in Manufacturing* 9.3, pp. 342–368. ISSN: 2195-3597. DOI: 10.1007/S40436-021-00357-Y.

Junio, Raí Felipe et al. (2023). "Development and Applications of 3D Printing-Processed Auxetic Structures for High-Velocity Impact Protection: A Review". *Engineering* 4.1, pp. 54–70. DOI: 10.3390/eng4010054.

Kalubadanage, Dulara et al. (Nov. 2021). "Close-in Blast Resistance of Large-Scale Auxetic Re-Entrant Honeycomb Sandwich Panels". *Journal of Sandwich Structures and Materials* 23.8, pp. 4016–4053. ISSN: 15307972. DOI: 10.1177/1099636220975450.

Karagiozova, Dora, Genevieve S. Langdon, and Gerald N. Nurick (Oct. 2012). "Propagation of Compaction Waves in Metal Foams Exhibiting Strain Hardening". *International Journal of Solids and Structures* 49.19–20, pp. 2763–2777. ISSN: 0020-7683. DOI: 10.1016/J.IJSOLSTR.2012.03.012.

Karagiozova, Dora (Dec. 2011). "Velocity Attenuation and Force Transfer by a Single- and Double-Layer Claddings Made of Foam Materials". *International Journal of Protective Structures* 2.4, pp. 417–438. ISSN: 20414196. DOI: 10.1260/2041-4196.2.4.417.

Karagiozova, Dora and Genevieve Sarah Langdon (2018). "Impact of Cellular Materials". In: *Encyclopedia of Continuum Mechanics*. Springer, pp. 1–22. DOI: 10.1007/978-3-7091-0297-8_6.

Kindred, Bob (2024). *A Brief History of Reinforced Concrete*. IHBC NewsBlog. https://newsblogs.ihbc.org.uk/? p = 34938.

Krauthammer, Theodor (2008). *Modern Protective Structures*. Boca Raton. ISBN: 9780429118890. DOI: 10.1201/9781420015423.

Langdon, Gevevieve S et al. (2010). "Fracture of Aluminium Foam Core Sacrificial Cladding Subjected to Air-Blast Loading". *International Journal of Impact Engineering* 37.6, pp. 638–651. ISSN: 0734-743X. DOI: 10.1016/j.ijimpeng.2009.07.006.

Li, Qing Ming and Haowei Meng (2002). "Attenuation or Enhancement—A One-Dimensional Analysis on Shock Transmission in the Solid Phase of a Cellular Material". *International Journal of Impact Engineering* 27.10, pp. 1049–1065. DOI: 10.1016/S0734-743X(02)00016-7.

Li, Qing Ming and Stephen R. Reid (Nov. 2006). "About One-Dimensional Shock Propagation in a Cellular Material". *International Journal of Impact Engineering* 32.11, pp. 1898–1906. ISSN: 0734-743X. DOI: 10.1016/J.IJIMPENG.2005.04.006.

Liang, Minzu, Xiangyu Li et al. (2019). "Compaction Wave Propagation in Layered Cellular Materials Under Air-Blast". *International Journal of Applied Mechanics* 11.01, p. 1950003. DOI: 10.1142/S1758825119500030.

Liang, Minzu, Zhibin Li et al. (2017). "Theoretical and Numerical Investigation of Blast Responses of Continuous-Density Graded Cellular Materials". *Composite Structures* 164, pp. 170–179. ISSN: 0263-8223. DOI: 10.1016/j.compstruct.2016.12.065.

Liu, Chengjun, Ying Xiang Zhang, and Chunhui Yang (Nov. 2016). "Numerical Modelling of Mechanical Behaviour of Aluminium Foam Using a Representative Volume Element Method". *International Journal of Mechanical Sciences* 118, pp. 155–165. ISSN: 0020-7403. DOI: 10.1016/J.IJMECSCI.2016.08.021.

Liu, Jiagui et al. (2015). "A Theoretical Study of Shock Front Propagation in the Density Graded Cellular Rods". *International Journal of Impact Engineering* 80, pp. 133–142. ISSN: 0734-743X. DOI: 10.1016/j.ijimpeng.2015.02.001.

Livermore Software Technology Corporation (LSTC) (2023). *LS-DYNA Keyword User's Manual*. Livermore, CA. URL: https://www.lstc.com/.

Lopatnikov, Sergey L, Bazle A Gama, Md. Jahirul Haque et al. (2003). "Dynamics of Metal Foam Deformation during Taylor Cylinder–Hopkinson Bar Impact Experiment". *Composite Structures* 61.1, pp. 61–71. ISSN: 0263-8223. DOI: 10.1016/S0263-8223(03)00039-4. URL: https://www.sciencedirect.com/science/article/pii/S0263822303000394.

Lopatnikov, Sergey L, Bazle A Gama, Md. Jahirul Haque et al. (2004). "High-Velocity Plate Impact of Metal Foams". *International Journal of Impact Engineering* 30.4, pp. 421–445. ISSN: 0734-743X. DOI: 10.1016/S0734-743X(03)00066-6.

Ma, Guowei W and Zhiqiang Q Ye (2005). "Analysis of Foam Claddings for Blast Alleviation". *International Journal of Impact Engineering* 34, pp. 60–70. ISSN: 0734-743X. DOI: 10.1016/j.ijimpeng.2005.10.005.

Ma, G W and Z Q Ye (2007). "Energy Absorption of Double-Layer Foam Cladding for Blast Alleviation". *International Journal of Impact Engineering* 34.2, pp. 329–347. ISSN: 0734-743X. DOI: 10.1016/j.ijimpeng.2005.07.012.

Maalej, M. et al. (Nov. 2012). "Review of Potential Structural Applications of Hybrid Fiber Engineered Cementitious Composites". *Construction and Building Materials* 36, pp. 216–227. ISSN: 0950-0618. DOI: 10.1016/J.CONBUILDMAT.2012.04.010.

Maduabum, Anthony (2024). "An Insight into the History of Earthen Architecture". *International Research Journal of Modernization in Engineering Technology and Science* 6.7, pp. 2369–2375. URL: https://www.irjmets.com/uploadedfiles/paper/issue_7_july_2024/60295/final/fin_irjmets1721382590.pdf.

Mays, Geoff C., Peter D. Smith, and David Cormie (2019). *Blast Effects on Buildings*. 3rd ed. Bingley, UK: Emerald Publishing Limited. ISBN: 9780727761477.

National Research Council (1995). *Protecting Buildings from Bomb Damage: Transfer of Blast-Effects Mitigation Technologies from Military to Civilian Applications*. Washington, DC: National Academies Press.

Ngo, Tuan et al. (2007). "Blast Loading and Blast Effects on Structures – An Overview". *Electronic Journal of Structural Engineering* 1.1, pp. 76–91. DOI: 10.56748/ejse.671.

Niu, B and J Yan (2016). "A New Micromechanical Approach of Micropolar Continuum Modeling for 2-D Periodic Cellular Material". *Acta Mechanica Sinica* 32, pp. 456–468. DOI: 10.1007/s10409-015-0492-8.

Ousji, Hamza, Bachir Belkassem, Aldjabar Aminou et al. (Aug. 2024). "Sacrificial Cladding Design for Blast Mitigation Using Low Density Crushable Core Systems". *International Journal of Protective Structures*. ISSN: 2041420X. DOI: 10.1177/20414196241268039.

Ousji, Hamza, Bachir Belkassem, Mohamed Abderaouf Louar et al. (2016). "Experimental Study of the Effectiveness of Sacrificial Cladding Using Polymeric Foams as Crushable Core with a Simply Supported Steel Beam". *Advances in Civil Engineering* 2016. DOI: 10.1155/2016/8301517.

Palanivelu, Sivakumar et al. (July 2011). "Low Velocity Axial Impact Crushing Performance of Empty Recyclable Metal Beverage Cans". *International Journal of Impact Engineering* 38.7, pp. 622–636. ISSN: 0734-743X. DOI: 10.1016/J.IJIMPENG.2011.02.008.

Pattofatto, S et al. (2007a). "Shock Enhancement of Cellular Structures under Impact Loading: Part I Experiments". *Journal of the Mechanics and Physics of Solids* 55.12, pp. 2652–2670. DOI: 10.1016/j.jmps.2007.04.005.

Pattofatto, S et al. (2007b). "Shock Enhancement of Cellular Structures under Impact Loading: Part II Analysis". *Journal of the Mechanics and Physics of Solids* 55.12, pp. 2671–2686. DOI: 10.1016/j.jmps.2007.04.004.

Pokhrel Ashish, Seo Dong Nam, Lee Seung Taek, and Kim Ik Jin (2013). "Processing of Porous Ceramics by Direct Foaming: A Review". *Journal of the Korean Ceramic Society* 50.2, p. 93. DOI: 10.4191/kcers.2013.50.2.093.

Qi, Chang, Alex Remennikov et al. (Nov. 2017). "Impact and Close-in Blast Response of Auxetic Honeycomb-Cored Sandwich Panels: Experimental Tests and Numerical Simulations". *Composite Structures* 180, pp. 161–178. ISSN: 0263-8223. DOI: 10.1016/J.COMPSTRUCT.2017.08.020.

Qi, Chang, Shu Yang, and Renjing Gao (2013). "Numerical Simulation of Force Enhancement by Cellular Material under Blast Load". *Advances in Mechanical Engineering* 5, p. 328651. DOI: 10.1155/2013/328651.

Rebentrost, Markus and Gordon Wight (2013). "Ch 24: Investigation of UHPFRC Slabs under Blast Loads". In: *Designing and Building with UHPFRC*. Ed. by François Toutlemonde and Jacques Resplendino. Wiley, pp. 363–376. DOI: 10.1002/9781118557839.ch24.

Reid, Stephen R and Choi Peng (1997). "Dynamic Uniaxial Crushing of Wood". *International Journal of Impact Engineering* 19.5, pp. 531–570. ISSN: 0734-743X. DOI: 10.1016/S0734-743X(97)00016-X.

Shen, Cheng-Jun, Tong-Xi Yu, and Guoxing Lu (2013). "Double Shock Mode in Graded Cellular Rod under Impact". *International Journal of Solids and Structures* 50.1, pp. 217–233. ISSN: 0020-7683. DOI: 10.1016/j.ijsolstr.2012.09.021.

Smith, Michael (2009). *ABAQUS/Standard User's Manual, Version 6.9*. Providence, RI: Dassault Systèmes Simulia Corp..

Stephenson, Charles (2005). *Fortifications and Siegecraft: Defense and Attack through the Ages*. Oxford, UK: Osprey Publishing.

Su, Juan, Kunyu Liu, and Xiaomeng Hou (2024). "Methodology for Predicting Dynamic Response of Multi-Layer Aluminum Foam Protected RC Slabs under Blast Loading". *Journal of Building Engineering* 82, p. 108145. ISSN: 2352-7102. DOI: 10.1016/j.jobe.2023.108145.

Tan, Paul J. et al. (Oct. 2005). "Dynamic Compressive Strength Properties of Aluminium Foams. Part II—'Shock' Theory and Comparison with Experimental Data and Numerical Models". *Journal of the Mechanics and Physics of Solids* 53.10, pp. 2206–2230. ISSN: 0022-5096. DOI: 10.1016/J.JMPS.2005.05.003.

Timoshenko, Stephen and James Norman Goodier (1951). *Theory of Elasticity*. New York: McGraw-Hill.

Toy, Sidney (1965). *The Evolution of Fortification: An Illustrated History of Ancient, Medieval & Modern Fortifications*. Princeton, NJ: Princeton University Press.

Tunay, Merve and Erhan Cetin (2023). "Energy Absorption of 2D Auxetic Structures Fabricated by Fused Deposition Modeling". *Journal of the Brazilian Society of Mechanical Sciences and Engineering* 45.9, p. 395. DOI: 10.1007/s40430-023-04423-3.

United States Department of Defense (2003). *Department of Defense (DoD) Guidelines*. Washington, DC: United States Department of Defense.

USACE (2008). *Unified Facilities Criteria (UFC) – Structures to Resist the Effects of Accidental Explosions*. Washington, D.C.: U.S. Army Corps of Engineers.

Wang, Cheng, Bin Xu, and S Chung Kim Yuen (2019). "Numerical Analysis of Cladding Sandwich Panels with Tubular Cores Subjected to Uniform Blast Load". *International Journal of Impact Engineering* 133, p. 103345. 0734-743X. DOI: 10.1016/j.ijimpeng.2019.103345.

Warfare History Network (2024). *The Concrete Fleet of WWII*. URL: https://warfarehistorynetwork.com/article/the-concrete-fleet-of-wwii/.

Wattad, Ola and Hezi Y Grisaro (2024a). "On the Efficiency of Uniform Aluminum Foam as Energy-Absorbing Sacrificial Cladding for Structural Blast Mitigation". *Engineering Structures* 308, p. 118047. ISSN: 0141-0296. DOI: 10.1016/j.engstruct.2024.118047.

Wattad, Ola and Hezi Y Grisaro (May 2024b). "Modeling Non-Uniform Aluminum Foams as Sacrificial Layers for Blast Mitigation of Protected Structures—A Theoretical Approach". *Journal of Building Engineering*, p. 109527. ISSN: 2352-7102. DOI: 10.1016/J.JOBE.2024.109527.

Wu, Chengqing and Hamid Sheikh (May 2013). "A Finite Element Modelling to Investigate the Mitigation of Blast Effects on Reinforced Concrete Panel Using Foam Cladding". *International Journal of Impact Engineering* 55, pp. 24–33. ISSN: 0734-743X. DOI: 10.1016/J.IJIMPENG.2012.11.006.

Wu, Chengqing and Yun Zhou (Sept. 2011). "Simplified Analysis of Foam Cladding Protected Reinforced Concrete Slabs against Blast Loadings". *International Journal of Protective Structures* 2.3, pp. 351–366. ISSN: 20414196. DOI: 10.1260/2041-4196.2.3.351.

Xi, Cheng-Quan, Zheng-Jun Dai, and Qing-Ming Li (2023). "Further Study on the Dynamic Loading Transmission in Cellular Solids Based on One-Dimensional Mass-Spring Model". *International Journal of Impact Engineering* 171, p. 104389. ISSN: 0734-743X. DOI: 10.1016/j.ijimpeng.2022.104389.

Xia, Ye, Chengqing Wu, and Terry Bennett (2017). "An Analytical Model of Linear Density Foam–Protected Structure under Blast Loading". *International Journal of Protective Structures* 8.3, pp. 454–472. DOI: 10.1177/2041419617721543.

Xia, Ye, Chengqing Wu, and Zhong-Xian Li (Sept. 2015). "Optimized Design of Foam Cladding for Protection of Reinforced Concrete Members under Blast Loading". *Journal of Structural Engineering* 141.9, p. 06014010. ISSN: 0733-9445. DOI: 10.1061/(ASCE)ST.1943-541X.0001190.

Xia, Ye, Chengqing Wu, Zhong-Xian Liu et al. (2016). "Protective Effect of Graded Density Aluminium Foam on RC Slab Under Blast Loading – An Experimental Study". *Construction and Building Materials* 111, pp. 209–222. DOI: 10.1016/j.conbuildmat.2016.02.092.

Xia, Ye, Chengqing Wu, Fangrui Zhang et al. (Dec. 2014). "Numerical Analysis of Foam-Protected RC Members under Blast Loads". *International Journal of Protective Structures* 5.4, pp. 367–390. ISSN: 2041420X. DOI: 10.1260/2041-4196.5.4.367.

Yan, Zichen et al. (2022). "Anti-Blast Performance of 3D-Printed Sandwich Panels with Auxetic Hexagonal and Regular Hexagonal Honeycomb Cores". *Engineering Structures* 272, p. 114996. DOI: 10.1016/j.engstruct.2022.114996.

Ye, Z. Q. and G. W. Ma (Jan. 2007). "Effects of Foam Claddings for Structure Protection against Blast Loads". *Journal of Engineering Mechanics* 133.1, pp. 41–47. ISSN: 0733-9399. DOI: 10.1061/(ASCE)0733-9399(2007)133:1(41).

Yuen, Chung Kim, Steeve Gareth Cunliffe, and Mark C. du Plessis (Dec. 2017). "Blast Response of Cladding Sandwich Panels with Tubular Cores". *International Journal of Impact Engineering* 110, pp. 266–278. ISSN: 0734-743X. DOI: 10.1016/J.IJIMPENG.2017.04.016.

Zhou, Hongyuan et al. (2015). "Energy Absorption of Graded Foam Subjected to Blast: A Theoretical Approach". *Materials & Design* 84, pp. 351–358. DOI: 10.1016/j.matdes.2015.06.124.

Zhou, Ying et al. (Nov. 2023). "Flexible Materials and Structures for Mitigating Combined Blast and Fragment Loadings – A Review". *International Journal of Impact Engineering* 181, p. 104759. ISSN: 0734-743X. DOI: 10.1016/J.IJIMPENG.2023.104759.

Zhu, C et al. (2009). "Protection of Reinforced Concrete Structures against Blast Loading". School of Civil and Environmental Engineering, Univ. of Adelaide, Adelaide, Australia.

Zhu, Feng, Clifford C Chou, and King H Yang (2011). "Shock Enhancement Effect of Lightweight Composite Structures and Materials". *Composites Part B: Engineering* 42.5, pp. 1202–1211. ISSN: 1359-8368. DOI: 10.1016/j.compositesb.2011.02.014.

15 Drone Delivery Systems
Perspectives from the War in Ukraine

Oleksandra Molloy

15.1 INTRODUCTION

Throughout history, people have converted everyday devices and available technology into instruments of war. Through necessity, technological advancements were used to protect and ensure their survival by developing mechanisms and countermeasures to defend against acts of aggression. With numerous advancements in aviation, chemistry, aerodynamics, nuclear science, to name a few, more efficient weapons, and associated delivery systems have been developed. Any weapon or weapon systems' effectiveness depends on a viable delivery platform (Nichols et al., 2022); hence the main focus of this chapter.

Uncrewed aerial systems, commonly known as drones, have shaped the ways modern wars are being fought, offering enhanced operational efficiency in various combat scenarios (Molloy, 2024a). The war in Ukraine is the first war that has seen drone warfare on an unprecedented scale at all levels of operations, from individual soldiers buying commercial drones online and then adapting them for military use on the frontlines, which provides greater situational awareness and an information effect on adversary. Ukraine has demonstrated the proliferation of drone technology in the air, on the ground, and at sea. Uncrewed systems operations range from tracking enemy forces, identifying targets, guiding artillery, or dropping munitions; through to unmanned combat aerial vehicles (UCAVs) that are being launched to launch long-range strikes deep into Russia; uncrewed ground vehicles (UGVs) used to deliver explosives, driven as a kamikaze drone, or save the wounded soldiers; and unmanned surface vehicles (USV) launched attacks on Russian naval forces in the Black Sea.

Three years since the start of the full-scale invasion, tactical UAVs account for 60–70% of damaged and destroyed Russian systems (Reynolds, 2025). In addition, these systems account for 80% of casualties of daily combat losses (Thomas, 2025). In other words, they are twice as effective as every other weapon in the Ukrainian arsenal put together (Hambling, 2025b). This is a remarkable turnaround for weapons which did not officially exist in the Ukrainian military at the start of the war.

As drone systems are the most widely used weapon on the battlefield, the pace of development and production of innovative weapon models by Ukrainian manufacturers is high. This is evident from the statistics of the number of weapons and military equipment models approved for operation since the beginning of 2025 – over 600, as reported by the Ministry of Defence of Ukraine. Most of them are domestically produced. For comparison, in 2024, over 300 domestic new weapons and military equipment models were codified in the first half of the year. In the first half of 2025, the UAV arsenal was replenished with over 270 new models. This is almost three times more than last year for the same period (Rozgon, 2025b).

Drones can carry ammunition of different types and weights and fly into trenches and windows of buildings, hatches of tanks, enemy dugouts, armoured vehicles, and cars (see Figure 15.1). Drones are considered a powerful military capability, and their weapons payloads are useful not only in the air but also on the ground and at sea. These systems have a range of roles, including conducting

DOI: 10.1201/9781003529217-15

FIGURE 15.1 Soldier UAV operator launches an army drone with a bomb to drop into enemy fortifications and trenches. Concept using military robots in modern warfare. (Photo credit: Parilov. Obtained from Adobe Stock, Weapons And Drones Ukraine Images – Browse 2466 Stock Photos, Vectors, and Video | Adobe Stock.)

aerial intelligence, surveillance, and reconnaissance (ISR), striking targets (including deep strike), serving as a 'drone carrier' increasing the range of FPV drones, supporting other legacy systems (i.e., adjusting artillery strikes), delivering cargo, as well as conducting search and rescue, mining and de-mining, and providing humanitarian support (Molloy, 2024a).

Since 2022, more rapid experimentations have been conducted with drones striking targets (see Figure 15.2). In fact, the Ukrainian Territorial Defence Forces crafted Molotov cocktails (costing approximately AUD $5) filled in beer bottles fastened via a thin plastic band to the DJI Mavic drone to strike Russian targets. A grenade or Molotov cocktail was attached to deliver 'cargo' to adversary's positions and targets. In the absence of advanced weapons and artillery shells across the frontline, this has become the only solution at some places that stopped adversary's invasion (Crumley, 2022). Hence, initial experimentation with attaching a munition to a drone was basic and unreliable, posed a risk to an engineer who attached explosives, and sometimes dropped prior to reaching to its designated area (Pleasance, 2024). For example, since 2024, the FPV drones have been equipped with an RPG-18 grenade launcher that is capable of firing a single cumulative grenade weighing 1.4 kg at a distance of up to 200 m. It is believed that if mass-produced, it could reduce the number of injuries to Ukrainian military personnel when equipping initiation boards and improvised explosive devices (IEDs). The risk of injuries to engineers while equipping kamikaze drones is quite high and, while not commonly discussed, should not be underestimated (Struck, 2024).

Overall, UAVs are widely used by both Ukrainian and Russian armed forces for reconnaissance, surveillance, and information gathering, as well as for direct targeting on the battlefield. UAVs are also used to deliver various anti-personnel and anti-tank mines to targets. These actions are often carried out at night and can lead to the unexpected presence of mines in areas that were previously cleared (de-mined). Great caution should be exercised around all downed UAVs due to the hazards posed by munitions and the potential presence of counter-defusal elements. Depending on the type

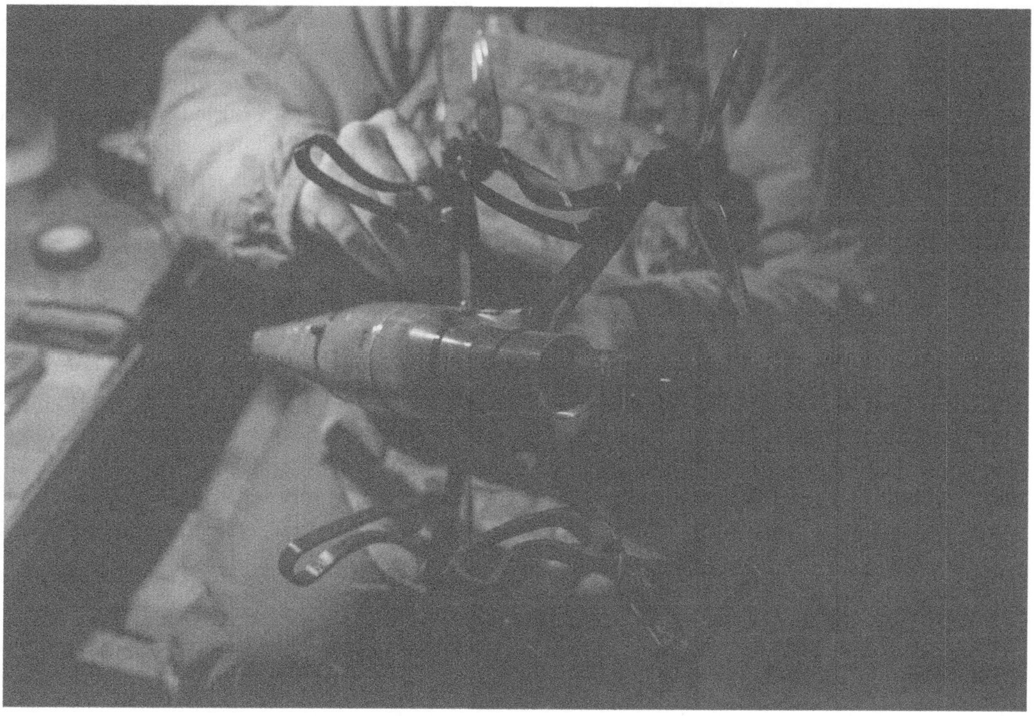

FIGURE 15.2 A Magyar birds' drone operator attaching a warhead to an FPV drone. (Photo credit: Magyar Birds. Reproduced with permission from the owner.)

of UAV, various munitions can be used accordingly. A new direction in the ammunition industry has been created, the development of drone munitions (Seddon, 2025).

15.2 CATEGORIES OF DRONES USED FOR DELIVERY OF EXPLOSIVES

Traditionally, military-graded drones have become more technologically advanced, from the MQ-1 Predator and MQ-9 Reaper used for targeted strikes to the RQ-4 Global Hawk and MQ-4C Triton used for high-altitude surveillance. For example, MQ-9 Reapers used by the U.S. Central Intelligence Agency were equipped with advanced sensors and precision-guided munitions, like the AGM-114 Hellfire missile, and capable of conducting targeted strikes with pinpoint accuracy. As the global war on terror pivoted coalition states away from conventional strategies focused on occupying territory and towards a strategy that dismantled enemy networks using intelligence-led strike systems, drone technology enabled high-value individuals to be targeted by ground forces or directly with weapons carried on the drones themselves. These systems moved beyond the control of intelligence agencies into widespread military use (Singer, 2009).

A characteristic of drone technology is that the use of these platforms is sufficiently inexpensive for state and non-state actors alike. The U.S. military first encountered the use of commercially available drones by a terrorist group when they were employed by ISIS to carry out aerial attacks (dropping grenades) in Mosul, Iraq, in 2016 (Larter, 2017).

While previously benign, the auditory cue of a drone overhead now signalled imminent danger as it meant that a grenade or other explosive device could be dropped at any moment. Since that time, almost all American soldiers deployed to Iraq have been trained on drone recognition, most have received basic training on drone defeat, and some have been armed with commercially available counter-drone systems. This training has significantly reduced experiences of 'a shock' from seeing a drone in action (Schmidt and Schmitt, 2016). However, modern warfare, as in Ukraine, has seen the first large-scale deployment of smaller commercial-off-the-shelf (COTS) drones – by both Ukrainian and Russian forces – for real-time intelligence gathering and direct combat engagements, illustrating the tactical versatility of these systems. Today, drones are regarded by Ukraine as a powerful military capability, and their weapons payloads are useful not only in the air but also on the sea and land. Drones are now being used to adjust artillery fire, conduct aerial reconnaissance, maintain situational awareness, strike enemy targets, mine and de-mine some areas, deliver cargo, and evacuate the wounded (Molloy, 2024).

Drones are generally classified based on distinct characteristics such as function, size, payload, geographical range, flight endurance, and altitude. According to the NATO classification, the drone dynamics in Ukraine has seen the deployment of Class I and Class III drones, where Class I drones are less than 150 kg, and Class III drones are greater than 600 kg (NATO, 2014). In the war in Ukraine, small military drones – both fixed-wing and rotary, integrated with the ground units – have been commonly used for surveillance, target acquisition, battle damage assessment, and information warfare (Pettyjohn, 2024). Class III medium-altitude long-endurance (MALE) drones are large surveillance and strike drones. They are useful for gathering information over extended periods of time and can execute remote strikes. The focus of drone use in Ukraine has been on Class I to III drones that can be used for precise payload delivery, surveillance, nuisance/loitering, and cyber/hacking (Molloy, 2024).

15.2.1 Class III Drones

The Turkish-built Bayraktar TB2 is a vivid example of the large (Class III) armed military drones that captured Ukrainian headlines and the public imagination in the first days and weeks of the Russian full-scale invasion of Ukraine. It was used for ISR and tactical strike missions (Yackley, 2022). The acquisition of Bayraktar TB2 drones was presented as a deterrent against Russia's aggression and provided just as important as the supply of Javelin Anti-Tank Guided Missile. It can carry four laser-guided bombs or two guided anti-tank missiles and could effectively destroy, for example, anti-aircraft missile systems, electronic warfare stations, and command posts. The Bayraktar TB2's low speed (i.e., 130 km/h) and size (i.e., 12 m) made it vulnerable on a battlefield with dense anti-air defence systems and electronic warfare at high density (Peria-Peigne, 2023). As such, the TB2 disappeared over the skies of Ukraine from September 2023 due to the evolving dynamic of the war (Molloy, 2024b).

15.2.2 Class II Drones

Ukraine and Russia have also employed long-range platforms, some capable of striking up to 3000 km (Kesteloo, 2025). For example, Ukraine has developed its own drones to strike targets deep inside Russia such as military factories and oil refineries. Payload for long-range drones can carry 20 kg (Sutton, 2025) or loaded with a 250 kg bomb (Denisova, 2025). Russia employed Shahed long-range attack drones to fly into Ukrainian targets, including civilian infrastructure and energy facilities. The Shahed drone, designed in Iran but now built in Russia, has delivered several different warheads over the past three years:

- In 2024, these drones were equipped with the TBBCh-50 warhead (50 kg) with thermobaric and fragmentation effects. It's known for its characteristic body coated with premade fragmentation in the form of 9-mm metal balls.
- The Shahed drones were reported to use the BST-52 warhead, incorporating additional shaped charge and incendiary effect during the modernisation process (Synaivska, 2024).
- A new KOFZBCh warhead with a combined anti-tank, fragmentation, high-explosive, and incendiary effect. This warhead has an increased warhead of 90 kg (see Figure 15.3).

FIGURE 15.3 An engineer is setting up an FPV drone with munitions. (Photo credit: Magyar birds. Reproduced with permission from the owner.)

The KOFZBCh is an acronym for 'shaped-charge-fragmentation-high-explosive-incendiary warhead'. The explosive remained unchanged – TGF-35P2 which is TNT mixed with phlegmatised RDX, same as in the standard 50-kg OFZBCh-50 warhead that lacked the anti-armour effect. The incendiary action is achieved by using a powder mixture of metal hydrides, giving a high combustion temperature of up to 3500°C.T

- The Iranian-made new warhead also weighs 90 kg and combines shaped-charge, high-explosive, and fragmentation action. The device belongs to the OLA-series explosives, based on octogen and aluminium with an intermediate charge made of RDX (Synaivska, 2024).
- Recent images show Shahed warheads packed with small ball bearings and other shrapnel. These fragmentation warheads cause casualties over a much wider radius than explosives alone (Hambling, 2025a).

15.2.3 Class I Drones

While large drones with missiles can be destructive under conditions of air superiority, small drones are proving crucial for battlespace awareness of infantry and manoeuvring units. Additionally, low-cost, one-way kamikaze attack drones offer another way to deliver explosives by flying into the target. The success of UAS-enabled targeting networks currently depends on their effective integration into a battle management architecture that leverages multi-modal information across command and control (C2) nodes, and on their ability to remain resilient to degraded communications environments (Borsari and Davis, 2023).

To date, both Ukrainian and Russian military forces have used a range of UAS. One of the emblematic UAVs has been a first-person view (FPV), a low-cost reusable attack drone, such as the Chinese-made DJI Mavic quadcopters. Russia has used DJI Mavic and the Orlan-10 (and later Orlan-30) multi-use UAS, which can be equipped with various payloads, usually up to 1.5–2 kg, and electronic warfare (EW) capabilities. Another type of drones, bombers, used by Armed Forces of Ukraine is usually a large hexacopter (or octocopter) drone capable of lifting a payload of up to 15 kg to a height of up to 400 m (Babaiev, 2025). The Ukrainian-made Vampire drone is a hexacopter that has six arms and can carry much heavier payload. It is fitted with a thermal camera because it carries out missions mainly at night. It is very loud so it's easy to spot and shoot down during the day, but it can attack a wider variety of targets. It is powerful to carry and drop repurposed mortar shells, even anti-tank mines. Hence, it can be used to strike troops and hardened targets like bunkers, trenches, and tanks (Pleasance, 2024).

The smaller drones tend to be quadcopters that were repurposed from their original commercial use. These drones are fitted with a carrying cradle underneath which are made via 3D printing. The bombs tend to be grenades which have been specifically adapted with 3D-printed parts like tail fins and nose fuses so that they fall straight downwards. Because the grenades are small, their main targets are Russian troops on foot, but at times they can be used to attack an armoured vehicle or a tank by throwing them through an open roof hatch and setting light to the fuel or ammunition inside. Another example is a hand grenade that was adapted with an x-shaped piece of plastic which replaces the pin, and it snaps when it hits the ground and activates the grenade.

Furthermore, both sides have employed, in a single platform, UAS that are fitted with loitering munitions (LMs) for a broad range of missions, including the suppression and destruction of enemy air defences, ISR, communication nodes, data transfer, and strike. LMs are autonomous missiles designed to stay airborne for some time, identify a target, and then attack. These systems blend features of both drones and precision-guided weapons. The term 'loiter' indicates the amount of time between launch and detonation. LM UAS represents a bridge between precision-guided weapons and future autonomous weapon systems (Goldstein and Waechter, 2023). The warhead of an LM drone ranges from 2 to 15 kg. Even a 15 kg warhead is sufficient to disable critical radar assets. However, the total destruction would require heavier warheads or more powerful munitions, like ATACMS (Defense Express, 2025c).

15.3 TYPES OF EXPLOSIVE ORDNANCE DROPPED BY UAVs: AN OVERVIEW

In the war in Ukraine, the use of UAVs has become dominant in the sky and a threat with various types of payloads, which can strike various targets (i.e., enemy vehicles, personnel, strategic targets). In broader terms, drone munitions used in the war in Ukraine can be broadly divided into two categories – non-specialised and specialised.

15.3.1 Non-Specialised Munitions (Improvised)

Non-specialised munitions for the UAVs are understood as standard explosives that are overall not designed to be used with drones, but which are used on the battlefield. The AFU realised that they could use these explosives for precise strikes on the enemies' locations (see Figure 15.3). Non-specialised drone munitions can consist of modified military-grade ammunition, such as grenades, mortar shells, submunitions (extracted from cluster munitions and aerial cluster bombs), and mines. Drone munitions based on IEDs typically consist of a small military-grade explosive charge placed in a fragmentation casing and equipped with a contact detonator, which can be electric or mechanical and is designed to activate upon impact with the ground or a target. Sometimes, the detonation systems of military-grade munitions are modified so that these munitions can be used by UAVs. These modifications often involve the intentional removal of safety features and priming the detonator, which usually increases the sensitivity of the munition and the associated danger (Seddon, 2025).

FIGURE 15.4 A Ukrainian drone operator is preparing to launch a drone with a grenade. (Photo credit: Oleksandr Volosyanskyi. Reproduced with permission from the owner.)

Typically, these include hand grenades, VOG, or mines. Overall, non-specialised ammunition is usually used with commercial drones, which were originally not intended for delivering strikes from the sky. For example, a commercial drone, DJI Mavic 3, has a payload capacity of about 0.5 kg. Such 'lifting' capabilities limit the range of ammunition that can be used on such a drone. Most often, this includes up to two hand grenades RG-D5 or grenades for under-barrel grenade launchers (VOG), each weighing about 250–300 g (see Figure 15.4). With a slight advantage, one could try to carry a hand grenade F-1 weighing about 600 g (Synytsin, 2023).

The capabilities of hand grenades and VOGs differ, but their principle of operation is generally the same – an explosion and the scattering of shrapnel over a range of approximately 3–12 m (for VOGs) and up to 25–30 m for hand grenades. The explosive power and the injury from shrapnel from hand grenades and VOGs are often sufficient to sustain an injury to enemy personnel. Although grenades are mainly used against infantry, there are occasional reports of their use against vehicles, though such cases are still rare. For instance, the Ukrainian thermobaric grenade RGT-27S2 was used against an enemy BMP, which destroyed the vehicle. Its weight is equal to that of the F-1 and is about 600 g. Commercial drones with greater payload capacities allow for the lifting of more serious munitions. For example, the DJI Matrice 300 RTK drone can carry up to 3–4 kg and can lift an 82 mm mortar shell weighing around 3–3.5 kg into the sky. The Australian PPDS drone made of cardboard can also do this with its payload capacity of 3–5 kg. However, 60 mm mines, which weigh significantly less, can also be used (Synytsin, 2023).

Indeed, such munitions are generally not often used for equipping drones, as a payload capacity of 3–5 kg is still quite rare for civilian drones, and they cost three to four times more than the DJI Mavic 3. Finally, the specifics of the munitions themselves imply that the fuse is set upon firing from a mortar, so to trigger it, one has to hang it on the drone with the detonator set, which is not very safe.

FPV drones, or 'kamikaze drones' as they are sometimes called, are disposable weapons designed for attacking targets on the battlefield. They can be used against armoured combat vehicles, supply transport vehicles, trenches, and field fortifications, as well as groups of personnel in the open. FPV drones are often equipped with suspended cumulative warheads, repurposed from RPG grenades, or malfunctioning anti-tank missiles. The initiation device is usually an improvised electric switch, mounted on the nose of the warhead. The warheads of FPV drones that have failed to detonate may be in a very sensitive state, and consideration should be given to their selective detonation using means of explosive ordnance disposal (Seddon, 2025).

15.3.2 Specialised Explosives

As the name suggests, specialised munitions are those that were originally manufactured for use with UAVs. One type of such munitions is the so-called LM, which for brevity is simply called a kamikaze drone. The advantage of such munitions is that these are specifically designed for UAV use and include capabilities for mounting, targeting, and other features that soldiers usually have to do themselves for other systems. Importantly, these bombs can be modified to fit specific armed drones (i.e., Bayraktar TB2), which have improved capabilities compared to their civilian counterparts.

Since 2021, Ukrainian defence manufacturers have begun producing special ammunition for drones against enemy equipment. For example, the Kyiv plant 'Mayak' launched the production of the cumulative projectile RKG-1600 (weighing 1.6 kg) based on the Soviet cumulative grenade RKG-3, and the cumulative bomb KZ-4800 (weighing 4.8 kg) for dropping from drones. In wartime, the 'Mayak' plant began the production of special drone munitions in the form of fragmentation bombs (weighing 0.5 and 1 kg) to strike the enemy's infantry. For each drone, a specific munition was made, using 3D-printed parts. This enabled the production of powerful aerial bombs for drones capable of destroying the enemy's military equipment. Importantly, the wind often deviates dropped munitions from the planned trajectory, so stabilisers resembling fins or arrow feathers were attached to them. While previously they were mostly made from bottle caps, pieces of plastic, or shuttlecocks for badminton, most recently, they were printed on 3D printers, calculating the aerodynamic properties of bombs on computers.

Since 2022, mechanisms for payload release have been bought online, although they were originally designed for carrying balloons, souvenirs, and other entertainment items. The mechanism comprises a light sensor and 'releases' the projectile as soon as the flashlight on the drone is turned on by the operator. A lightweight 30 mm ammunition for under-barrel grenade launchers, VOG-17 (350 g), which explodes on impact, has been most commonly attached to drones for bombing. The explosive filling may vary depending on the types of grenades, but due to their low power, they are most often used to destroy personnel or damage light equipment (Zafra, 2024).

15.4 TYPES OF WARHEADS AND EXPLOSIVES FOR UAVs

According to the Defence Ministry of Ukraine, within the six months since the beginning of 2024, there were 60 samples of weapons codified and used by the Armed Forces of Ukraine. Ninety per cent of these weapons were ammunition for UAVs. Specifically, these included high-explosive fragmentation, cumulative fragmentation, combined action, and multi-purpose. They also vary in weight: from several hundred grammes to more than several kilogrammes, depending on the size and weight of a specific type of UAV (Ukraine, 2024).

Tables 15.1 and 15.2 show selected examples of warheads and explosives for UAVs. Overall, the types of explosives attached to drones can be thermobaric, cumulative, and high-explosive fragmentation (Babaiev, 2025).

15.4.1 Thermobaric Warhead

Thermobaric warhead is designed for targeting inside buildings and lightly armoured vehicles. The lethal power of the thermobaric charge spreads within a radius of 15 m in open space, being most effective when used against enemy's fortifications, and often used to hit dugouts. This weapon burns up oxygen in the surrounding air, creating a destructive vacuum. These are used in enclosed spaces and are ideal payloads for small drones. Thermobarics are about blast rather than a solid explosive

TABLE 15.1
Selected Examples of Warheads and Explosives for UAVs (Ukraine, 2023)

Thermobaric Warhead	Cumulative Warhead	High-Explosive Fragmentation Warhead
Grenade RSHG-2. Explosives: fuel mixture LP-30T (1160 g, equivalent to 3 kg of TNT) and a charge made of OKFOL (octogen phlegmatised with organic lacquer).	PG-7L. Explosive: OKFOL	OG-7V. Explosive: A-IX-1 (composition 1 based on RDX).
TBG-7V. Explosives: fuel mixture LP-30T (1800 g) and a block made of OKFOL.	PG-7S. Explosive: OKFOL	OG-9M. Explosive: DT (dinitro naphthalene – TNT).

like RDX; they typically disperse a cloud of liquid and ignite it in an explosive fireball (Wildegger-Gaissmaier, 2004). In general, a thermobaric explosive (TBX) consists of a certain central charge (called the core), which is usually a high-explosive, and an external secondary charge (fuel-rich formulation). Therefore, the detonation of TBX consists of a dual action: (1) firstly, anaerobic action (without air oxygen) inside the conventional high-explosive core occurs; (2) then aerobic delayed burning action of the fuel mixture of the outer charge happens which depends mainly on the consumption of the surrounding air (Yen and Wang, 2012).

TABLE 15.2
Main Explosive Ordnance Used in Specific Munitions for UAVs (Ukraine, 2023)

Explosive Substance	Munitions
Lead azide	Detonator caps
TNRC	Detonator caps
TNT (trinitrotoluene)	TM-62 (except TM-62B), OZM-72, PMN, PDM-6, POMZ-2, MON-100, MON-200, BPM, SRM, YARM, PDM-1, PDM-2, PDM-3, MS-3, SZ-1, SZ-3, SZ-3a, SZ-6, KZ-3, MZU, POM-2
Tetryl	PMN-2, PMN-4, KZ-1, KZ-2, MVCh-62, MVP-62, VPZ-1, MVD-62, MVN-72, MVN-62, VPDM-1, VPDM-1M, VPDM-2, MVZ-62, M.ZU-2, VZRP, VOZ-1, ML-7, OZM-72 (usually in auxiliary detonators)
TEN (tetryl-based explosive)	MVN-80, PFM-1, PFM-1S, PTM-3
TG-40	TM-83, PMN-2, KZ-5, KZ-6, KZ-7, KZK, KZU-2, POM-1, PTM-3
TG-50	MPM, SZ-6, SZ-3A, KZ-4, KZU, VZZ-2
MS	TM-62 (except TM-62B), SPM, UPM
TGA-16	BPM, TM-62 (except TM-62B)
PVV-4	SZ-6M
PVV-5A	MON-90, MON-50, SZ-1P, SZ-4P, ML-7
PVV-7	DKRP-4
PVV-12C	PTM-1S
EVV-11	SZ-1E
A-1X-1	MON-50, MON-90, SPM, KZK, KZU-2, ZRP-2, 03-1(KZ), OG-7V, VOG-17, VZZ-2, PG-9
A-1X-2	03-1(FZ), OFSP
Hexfol-5 (A-1X-10)	KZ-5, KZ-6, KZ-7
VS-6D	PFM-1, PFM-1S
OKFOL	PG-7

Thermobaric warheads deliver more energy than standard explosives and produce a blast wave with lower peak overpressure but greater duration (i.e., a matter of milliseconds), which is far more destructive to both buildings and human lungs (Hambling, 2024a). While a thermobaric warhead will not injure people hundreds of metres away like a fragmentation weapon, it does more damage in a narrow zone.

For example, the Ukrainian RGT27-S2 thermobaric hand grenade can be used as a drone bomb to demolish abandoned Russian vehicles. The blast and high temperature of the fireball ensure that vehicles are left burned out and unusable. The peculiarity of the thermobaric munitions is that after the initiator is detonated, the shock wave causes the movement of a thermobaric mixture that interacts with air and with itself, forming a fire cloud. The temperature of this cloud ranges from 2000 to 2600°C and spreads over the entire area (Kushnikov, 2024).

15.4.2 Cumulative Warhead

The cumulative ammunition is simply a shaped charge warhead and can penetrate armour up to 400 mm in thickness and can be used against lightweight armoured vehicles (Galka, 2022).

This type of warhead can weigh 2 kg and is capable of penetrating up to 400 mm of steel armour. This is sufficient to penetrate a tank when striking a weakened area, and to damage other armoured vehicles on the battlefield (Shumlyanskyi, 2024). Small commercial drones that carry these warheads can be LM or kamikaze drones. Russia's Shahed-136 and Geran-2 carry multi-purpose warheads, including the cumulative one, and ZALA Lancet is equipped with a cumulative warhead, weighing 3 kg. The U.S. Switchblade 600 is equipped with a cumulative anti-tank warhead with 2 kg of explosives, which is similar to the multi-purpose warhead.

15.4.3 High-Explosive Fragmentation Warhead

High-explosive fragmentation for drones is designed to inflict significant damage upon impact, making it effective against a variety of ground targets. It is designed to target personnel in open spaces. It destroys personnel within a 15 × 3 m area using 150 damaging fragments. It is designed to scatter small metal balls or splinters upon detonation (Defense Express, 2024a). This type of warhead has been used in long-range kamikaze UAVs like Russia's Shahed-136, Geran-2 type, or Gerbera UAVs. The munition for a Shahed-136 drone included the BSF-50, a 50 kg warhead that combines high-explosive fragmentation, and incendiary effects, as well as TBBCh-50M, which contains a thermobaric explosive composition and a fragmentation effect. However, one of the largest documented warheads for Shahed-136 is the BCh-90, a 90 kg shaped charge warhead with an additional fragmentation effect, which is constructed with a composite fragmentation liner of steel and zirconium, a metal used to generate both fragmentation and incendiary effects. Integrating the larger warhead of 90 kg requires switching to a smaller tank on the munition – leading to a significantly decreased range.

Russia's Gerbera drones – consisting of a plywood and foam fuselage, engine, basic onboard computer, and a soft polyethylene fuel tank – were first spotted in July 2024 during an attack on Kyiv Oblast. These UAVs are designed to imitate Shahed-136/Geran-2 OWA drones and are used en masse to overwhelm Ukrainian air defence systems. The Gerbera drone, from a foam decoy to a drone that can carry 5 kg high-explosive fragmentation warheads, poses a greater threat than before (Zoria, 2025). This is why the Armed Forces of Ukraine could not ignore the Gerbera drones because it was unclear whether these were decoys or real ones with explosives, until they reached their targets; hence, they needed to down each of them (Burgess, 2024). Recent evidence suggests that the Gerbera drone was equipped with a serial Russian BCh-2.5 high-explosive fragmentation warhead filled with OKFOL, an explosive mixture typically used in shaped charges (see Figure 15.5).

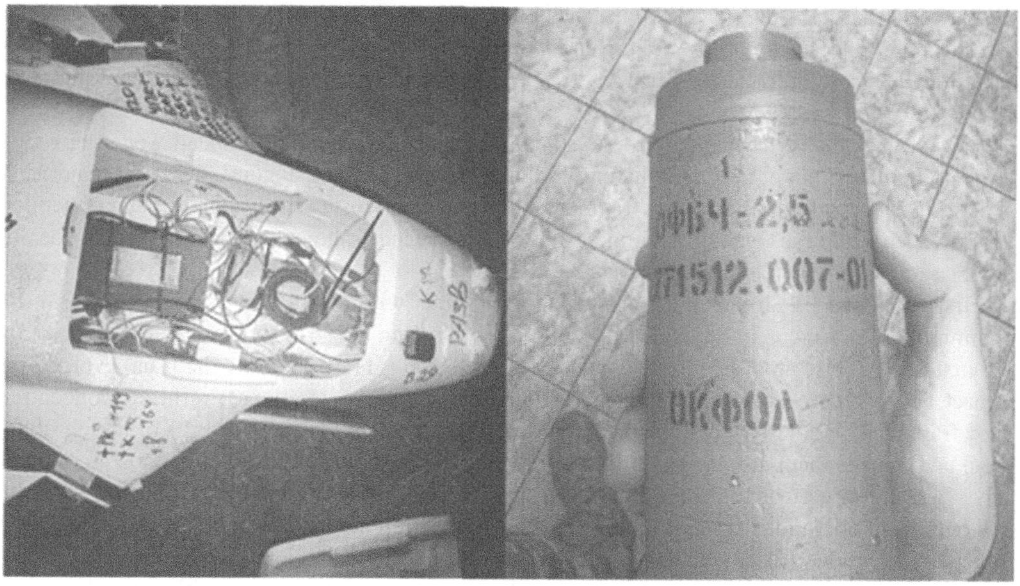

FIGURE 15.5 Wreckage of Russia's Gerbera decoy drone (L) and a warhead found in it. (Photo credit: Telegram channel of Serhii Flesh t.me/serhii_flash. Reproduced with permission from the owner.)

15.5 MUNITION FOR UAVs: SELECTED EXAMPLES FROM THE WAR IN UKRAINE

15.5.1 Munition for FPV Drones

Civilian-grade FPV drones, with their origin from racing drones and a relatively low price between US $500 and $1000, have become one of the most emblematic types of drones in the war in Ukraine. These drones were converted to carry warheads and are able to precisely deliver a designated payload to an adversary's locations. With no existing ready-made warheads for FPV drones, soldiers attached everything from mortar bombs to grenades to rocket warheads, and even 'Molotov cocktails'. Since 2022, this has meant strapping warheads (i.e., RPG-7 or RKG-3 anti-tank hand grenade) to drones and flying them into targets (Hambling, 2024b). Although it is called a grenade, and it does have some fragmentation effects, it is in fact a shaped-charge system (Hazell, 2022). An RPG munition (with the cost of between US $100 and $500) can neutralise a T-90M Russian main battle tank (MBT) which is one of the most advanced tanks in the Russian arsenal. Further, it is equipped with protection specifically designed to counter such a threat (Recognition, 2023). The RPG-7, an anti-tank grenade launcher dating back to the Soviet era, is known for its simplicity and reliability, but not necessarily for its ability to penetrate sophisticated modern armour like that of the T-90M. The fact that such ammunition succeeded in disabling the tank highlights two critical points. Firstly, even the most advanced military technologies can be vulnerable to ingenious tactics and seemingly obsolete weapons when used effectively. Secondly, there are legitimate questions about the reliability and real capabilities of Russian explosive reactive armour (ERA), which has failed to stop the munition.

The hefty weapon, like the RKG-3, is supposed to be thrown over an enemy tank, in a long arc, so that it would strike the thinner top armour. Given that it could only be thrown approximately 18 metres, dangerously close to a tank, the RKG-3 represents more or less a suicide weapon. As an FPV warhead, though, it can strike from longer range with no risk to a drone operator. In addition, some FPVs were equipped with anti-personnel fragmentation warheads and others with thermobaric charges to destroy buildings and bunkers (Hambling, 2024b). One of the notable advantages of FPV drones is their cost-effectiveness in engaging a wide range of targets, including tanks, armoured vehicles, aircraft, and naval assets, some of which are valued at tens of millions of dollars (Molloy, 2024b).

FPV drones are light weapons by the standards of artillery, with a warhead weighing only a few kilogrammes, but enable demolishing buildings and destroying dugouts. The vast majority of general-purpose munitions are blast/fragmentation weapons. Most of the weight is metal rather than explosive, either in the form of a metal casing or pre-formed fragments, which generally does far more damage over a wider area than blast.

First cases of FPV drones packing grenades or shells with airburst fuses were notable in the autumn of 2023. Remotely detonated in mid-air and blasting tiny fragments over a wide area, the airburst drones pose an ever greater danger to unprotected infantry than do the usual explode-on-impact FPV drones. Remotely detonated airburst drones with directional warheads concentrate their fragments in a narrow cone. However, operating such drones requires skills of merely flying them and aiming them at a target, in the chaos of actual combat and when drone operators themselves are the targets of the enemy's own drones. Skills come with experience through training and battlefield operations (Axe, 2024). FPV crews use different types of munitions to hit different targets. A selection is discussed briefly below.

Armour-piercing – Upon impact, a hot jet of metal shoots out of the munition at velocities of between 10 and 12 km/s, piercing the armour of the vehicle (Mike Collett-White, 2024). Ukraine has developed EFP-S armour-piercing ammunition warhead for FPV drones, capable of penetrating armoured vehicles (see Figure 15.6). It is aimed at use against lightly armoured vehicles, including

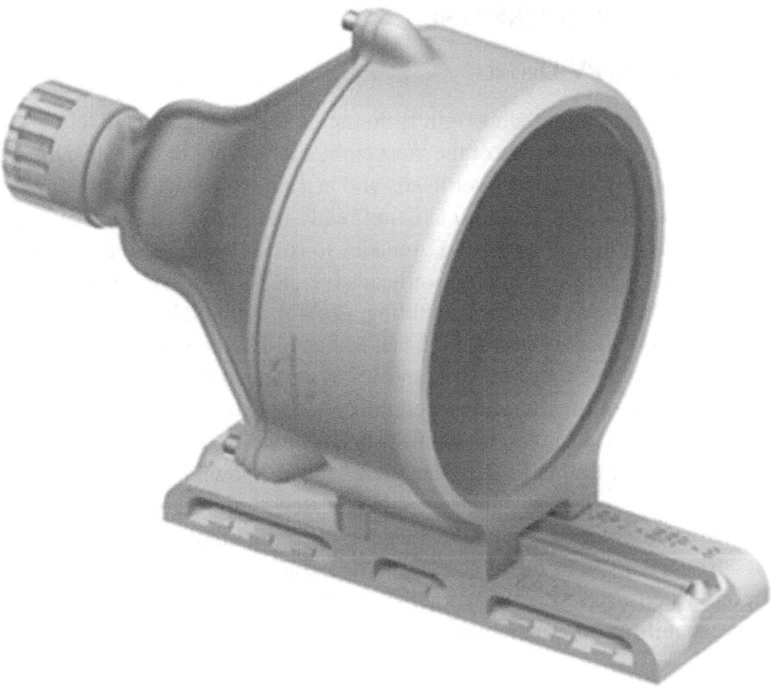

FIGURE 15.6 Ukrainian EFP-S armour-piercing ammunition of the shock wave dynamics (SWD) family for drones. (Photo credit: Kushnikov, 2024; Ukraine developed new EFP-S armour-piercing ammunition for FPV – Militarnyi; Reproduced with permission from the owner.)

self-propelled guns and armoured personnel carriers (APCs), which are protected by slat armour and anti-drone equipment. The EFP-S ammunition's armour penetration capacity is 25 mm, with a damage range spanning from 1.5 to 50 m. The striking element weighs 165 g, attaining velocity of 1800 m/s. The ammunition design is attached to the FPV drone using nylon clamps (Defense Express, 2024b).

Thermobaric is often used to hit dugouts. This weapon burns up oxygen in the surrounding air, creating a destructive vacuum. The Armed Forces of Ukraine have demonstrated successful use of FPVs with thermobaric warheads against Russian-held buildings and fortifications. They are most effective when the explosion occurs within the targeted structure. This is challenging with other weapons, but a skilled pilot can fly through an open doorway or window (Hambling, 2024a). One of the tactics of the use of several FPV drones demonstrates that a leading FPV blows a hole in the roof of a building, and the follow-up FPV drone flies through the hole with a thermobaric warhead. Thermobaric warheads allow FPV pilots to effectively target individual soldiers even if they are sheltering in buildings or bunkers. Even well-dug-in troops are now in the danger zone, and the experienced operators can achieve these precise missions.

In the early stage of the use of FPV drones, small payloads like VOG-17 grenades for anti-personnel roles or RKG-3 anti-tank grenades to take on vehicles were used. As the war progressed, more potent RPG-7 rounds were strapped to the drones (see Figure 15.7), including custom explosively formed penetrator (EFP) charges, which significantly boosted their anti-armour capabilities. This upgrade made them a formidable threat to tanks on both sides. Combat engineers have become highly resourceful in crafting detonators for the payload as well. Some rigs use two coiled wires hung in front of the drone, separated by an air gap. When the drone hits its target, the wires cross, completing the circuit and triggering the explosive. Recent iterations can be remotely detonated mid-air to create an airburst effect (a rain of lethal shrapnel) down on infantry.

Ukrainian drones use regular ammunition, adapted for dropping from reconnaissance UAVs/ suspended under an FPV drone. In addition, the Ukrainian Defence Forces use FPV drones with a warhead that strikes targets by air bursting a high-explosive or shrapnel warhead.

FIGURE 15.7 Magyar birds attach a warhead to an FPV drone. (Photo credit: Magyar Birds, reproduced with permission from the owner.)

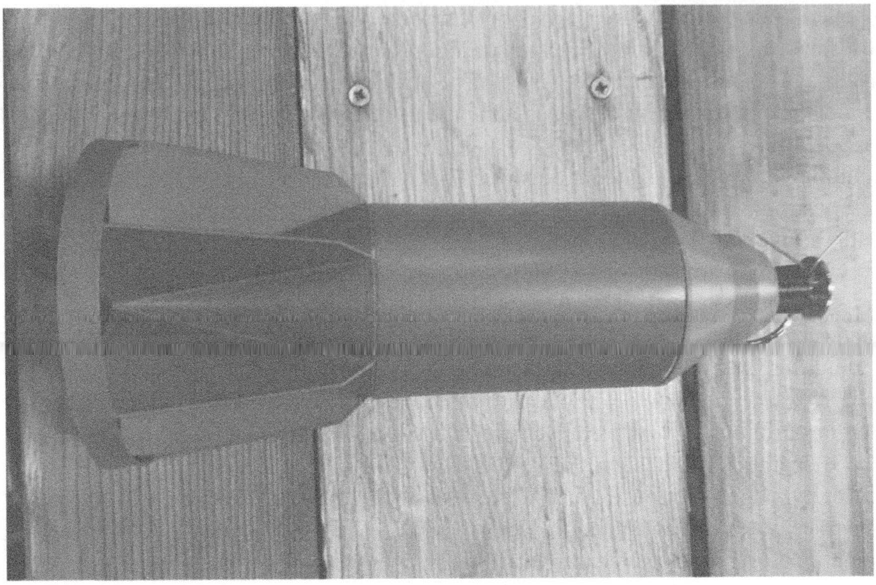

FIGURE 15.8 Ukrainian shrapnel ammunition for dropping from FPV drones. (April 2023. Photo credit: Steel Hornets. Obtained from Ukrainian volunteers develop new ammunition for FPV drones – Militarnyi; reproduced with permission from the owner.)

The shrapnel munition has a mechanical nose pin and metal balls as striking elements, used to defeat enemy manpower (see Figures 15.8 and 15.9). The diameter of the case is 63 mm, and depending on the size and payload, it ranges from 800 to 2800 g of total combat weight. The case is filled with 4.5 mm ball bearings in each round. For airstrikes with this ammunition, UAV operators

FIGURE 15.9 Striking elements for aviation ammunition. (Photo credit: Steel Hornets Obtained from Ukrainian volunteers develop new ammunition for FPV drones – Militarnyi; reproduced with permission from the owner.)

FIGURE 15.10 Cutting up an artillery shell to remove explosives. (Photo credit: United 24 obtained from shell to Javelin: what drone ammunition is made of – Militarnyi reproduced with the permission of the owner.)

use the dive method or detonate an FPV drone in the air directly above the Russians. This approach enables using FPV drones several times in a row (Kozatskyi, 2023).

The ammunition for FPV drones and bombers is also turned from defective artillery shells, which remain in dozens in the Ukrainian military workshops. The commander of the third Regiment of the Special Operations Forces reported that there is a range of ammunition that they receive to fill the warheads of Ukrainian drones, including unusable artillery shells, the damaged Javelin, and NLAW anti-tank missiles that still contain explosives (see Figure 15.10). 'Everything you know, and even things you don't, can be turned into a weapon. You just have to be creative' – he said (Kyshnikov, 2025).

Approximately 50 military personnel work in the workshop conveyor line to produce drone warheads (Figures 15.11 and 15.12). For example, one group is tasked with cutting metal rods into monotonous elements that will become fragments. Another group loads shrapnel into the 3D-printed bodies of combat units for FPV drones. The third group fills these cases with explosives. PG-7 grenades for the RPG-7 grenade launcher are also repurposed, via adding counterweights and feathers to them so that they can be dropped from bombing drones. Some grenades have an additional rim of fragmentation jacket so that when detonated, the anti-tank projectile can also hit enemy infantry.

In October 2024, the Cabinet of Ministers of Ukraine allowed the use of elements from unexploded ordnance for the use of the Armed Forces of Ukraine. According to the Defence Minister of Ukraine, Rustam Umerov,

> Approximate data shows that up to 20% of explosives in the line of fight do not explode. Hence, we have developed a mechanism which will help not only utilise dangerous objects, but change this process for the use of our forces.
>
> **(Kushnikov, 2024)**

A directional airburst warhead can affect enemy.

FIGURE 15.11 Filling an explosive device with ready-made fragments. (Photo credit: United 24 obtained from shell to Javelin: what drone ammunition is made of – Militarnyi; reproduced with permission from the owner.)

FIGURE 15.12 Processing of PG HEAT grenades into free-fall bombs. (Photo credit: United 24. Obtained from shell to Javelin: what drone ammunition is made of – Militarnyi; reproduced with permission from the owner.)

In February 2025, it was reported that the Russian forces were using 'Molniya' fixed-wing FPV-type drone with TM-62 anti-tank mines with 10 kg of explosives and plastic explosive or a double charge from the RPO-A Shmel rocket-assisted thermobaric weapon (flamethrower). Previously, this drone was seen with improvised warheads, including KZ-6 demolition munition or anti-armour grenade from an RPG-7 (Defense Express, 2025a). In May 2025, a Molniya-type drone was packed with a bag of nails. Upon explosion, the nails scatter as lethal shrapnel, inflicting maximum injuries on people within the blast radius of up to 40 m. The danger from this is that if to neutralise this drone in the air, the impact on the ground remains the same as downing a standard UAV carrying a high-explosive warhead (Vialko, 2025).

15.5.1.1 Incendiary Payloads

In early September 2024, the Ukrainian military introduced the 'Dragon Fire' drone, designed to release thermite – a highly incendiary mix of powdered aluminium and ferrous oxide (rust) that burns at temperatures up to 3000°C. The drone creates a flamethrower-like effect, igniting tree lines, foliage, and any flammable materials, like camouflage netting used by enemy troops. It has been used in warfare since World War I, primarily in artillery shells and hand grenades, and has expanded its use to UAVs (Kesteloo, 2025). Soon after its introduction by Ukrainian military, Russians attempted to copy it, but less successfully.

Incendiary warheads – These warheads can be of different types – based on burning mixtures (i.e., napalm), based on thermite compounds or based on white phosphorus (prohibited by the UN convention (Mitrofanov, 2025)). For example, a napalm-based warhead can be used to set fire to shelters for enemy personnel made using flammable elements – wood, plastic, tyres, and buildings and structures whose structural elements or interior elements can support combustion. In addition, FPV drones with incendiary warheads can be used to strike at identified enemy ammunition storage sites in order to increase the likelihood of their detonation (Kesteloo, 2025).

15.6 EXPLOSIVE PAYLOAD INTEGRATION AND RELEASE SYSTEMS

The Geneva International Centre for Humanitarian Demining reported that some of the munitions that can be attached to UAVs are shown below (GICHD, 2022).

15.6.1 RGD-5

RGD-5 is a common offensive grenade (see Figure 15.13) that has been in use since the 1950s. The reduced fragmentation means a reduced effective range of 15–20 m. The M-10 fuse thread accepts the standard UZRGM-2 fuse and can also accept compatible switches such as an MUV series fuse with an MD-5 detonator. As with other grenades the UZRGM-2 has a 3.2–4.2 second delay. The grenade is usually marked with a black capital T to indicate a TNT fill. However, the markings differ depending on where the grenade was manufactured.

15.6.2 VOG-17/VOG-17M

The VOG-17 is a projected HE/FRAG grenade. It is typically fired by AGS-17 grenade launcher and can be used as a form of light local indirect fire support. The VMG is a point detonating and graze fuse that arms by setback and centrifugal force. A VMG-M version, used on the VOG-17M, includes a pyrotechnic delay self-destruct feature that is initiated by setback on launch. The setback process is based on the principle of the use of intense inertial force during launch to initiate fuse functions. In the VMG-M fuse, setback ignites the pyrotechnic self-destruct delay, ensuring that every launched VOG-17M grenade either detonates on impact or safely self-destructs after a set time. In fact, when fired, setback force drives a small firing pin or inertia

FIGURE 15.13 RGD – 5, hand grenade. (Photo credit: Ryzhkov Oleksandr. Obtained from Adobe Stock. RGD-5 stock photos, royalty-free images, vectors, video).

element against a primer. The primer lights a delay composition, and if the grenade doesn't hit a target within the set delay time (i.e., 14–17 seconds), the fuse self-destructs automatically. It should be noted that the VMG-M contains a small rotor that maintains the primer out of line until the holding devices are removed during the arming sequence. The VMG has no such rotor and thus contains an inline primer and therefore is not bore safe. VOG-17s with VMG fuses, even when found as AXO, should be handled accordingly. The grenade is projected by means of a single-based nitrocellulose propellant charge initiated on indentation of the primer by a firing pin in the AGS-17.

15.6.3 VOG-25

The VOG-25 is a projected HE/FRAG grenade largely used with the GP-25 UGL. Often used as an indirect fire weapon, it is claimed to have an effective range up to 400 m. As with other projected grenades, the muzzle velocity is relatively low (77 m/s). The forces required to arm the fuse are relatively low. The VMG-K is a point-initiating fuse with a pyrotechnic self-destruct initiated on setback. Factory markings are usually visible on the nose of the VMG-K fuse. The windshield of the grenade will typically show indentations on impact. Such impact damage and an indentation on the primer, along with propellant scorch marks, are the easiest way to identify a blind.

The Russian industry, similar to Ukraine, has started production of drone ammunition. The gravity bombs were developed based on the VOG-25 40 mm HE-Frag grenades for hand grenade launchers (see Figure 15.14). The mass of the original ammunition is 250 g, with a mass of 48 g being the explosive fill. This ammunition is reported to have been used with the Orlan-10 reconnaissance drone.

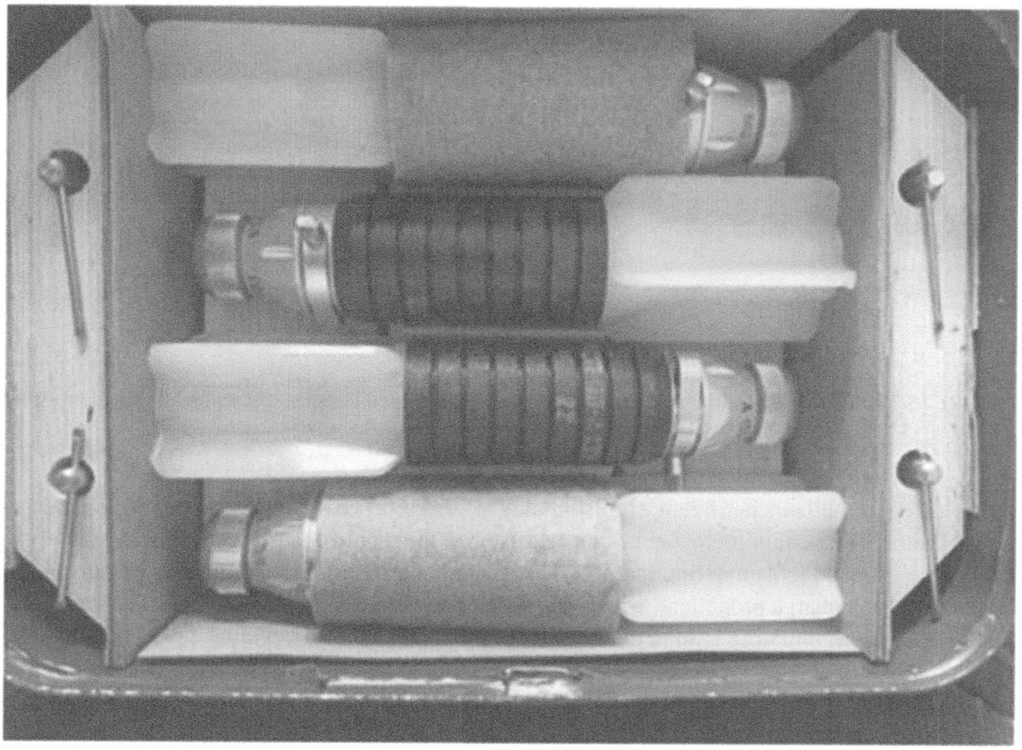

FIGURE 15.14 Russian bomblets for UAVs developed on the basis of the VOG-25 grenade. (Photo from open sources; obtained from Militarnyi. A separate category of ammunition for UAVs was created in Ukraine – Militarnyi; with permission from the owner.)

15.6.4 RKG-3

The RKG-3, once a hand-thrown anti-tank grenade that was thrown onto tanks and armoured vehicles, has been repurposed as a drone-dropped anti-armour weapon. Its built-in parachute tail is used for accurate top-attack strikes from quadcopters, especially in urban combat where drones can drop it onto tanks, or vehicles in confined spaces. The RKG-3 is a high-explosive anti-tank (HEAT) grenade that uses a shaped charge (cumulative charge) to penetrate armour (see Figure 15.15). An RKG-3M version also exists, which employs a copper rather than steel-shaped charge liner. A longer RKG-3EM version with an enhanced standoff configuration is also in circulation. The easiest way to tell if the item has been armed is the deployment of a small white drogue chute. Like the RGO/RGN the RKG-3 has an impact fuse. The cocked striker may function when an impact force moves the last holding devices (sleeve and locking balls). The cocked striker fuse is known to be highly dangerous. EOD operators are advised to be highly cautious. There are indications this item has been adapted for use as a small anti-armour bomb labelled the RKG-1600 to be deployed by UAV. In this configuration the fusing in the handle is replaced by what is assumed to be an in-line impact inertia fuse.

15.6.5 Recoilless Ammunition

15.6.5.1 PG-7M

The PG-7M is a HEAT rocket-propelled grenade which is fired from the RPG-7 range of 40 mm calibre launchers. The PG-7M functions in a similar way to other PG-7 type rounds. The round is

FIGURE 15.15 RKG-3 anti-tank hand grenade isolated. (Photo credit: Evgeny. Obtained from Adobe Stock, RKG-3 stock photos, royalty-free images, vectors, video.)

propelled from the launcher by an expulsion charge which is screwed onto the rear of the sustain rocket motor. The sustain motor ignites a safe distance in front of the firer and provides the thrust to take the warhead to its maximum range. The round is fin stabilised, and the fins are surrounded by the expulsion charge propellant, which is protected by a waxed cardboard tube. The warhead is fitted with a point-initiating base detonating (PIBD) fuse which also contains a pyrotechnic-based self-destruct mechanism which causes warhead to function at a maximum range of approximately 950 metres. The PG-7M differs from the PG-7V not only in shape but also in the plastic liner used in the windshield of the warhead in order to minimise short circuit blinds due to the round glancing off an object before the piezo fuse can impact the target. The PG-7M is still in production. It can easily be mistaken for the PG-7VS, which looks externally almost identical but has a 72 mm warhead diameter and different markings: PG-7VS (ПГ-7ВС). Notably the PG-7VS uses OKFOL rather than A-IX-1. Some sources suggest that more recent versions use a VP-22 fuse. Indicators of a potentially armed fuse include the seals covering the sustain motor venturis being blown and an indentation on the percussion primer at the motor end of the device. Fired PG-7M rounds which have failed to function should be destroyed in situ, and consideration should be given to reducing the hazard posed by the shaped charge warhead.

15.6.6 Miscellaneous

15.6.6.1 RKG-1600 Aerial Bomb

This appears to be a repurposed RKG-EM grenade – deployed by a UAV. Some sources state that this device was developed by a conventional manufacturer in Ukraine and is designated the RKG-1600. The warhead targets the thinner armour on top of an armoured vehicle. The fuse is inevitably different from the standard RKG-EM. It is possible that there are limited safety features with the fuse and that any item found should be considered in line and armed.

15.6.7 Anti-Vehicle Munition

15.6.7.1 TM-62M

The TM-62M is a metal-bodied, AV, blast mine, which was developed from the earlier TM-57. The mine has a single, large, Bakelite, central fuse well, with a metal-cased booster screwed onto the bottom of the fuse well. Like all other mines in the TM-62 series, the TM-62M will accept all fuses that were developed for the TM-62 series, the TM-72 series, and the TM-80 series of mines. Therefore, potentially this mine can be fitted with a range of magnetic-influence fuses. The TM-62M is typically fitted with the pressure-actuated, MVCh-62 pressure fuse,

FIGURE 15.16 Russian anti-tank mine TM-62M with a TNT block installed on it, necessary for the clearance of mines during detonation. The inscription in Russian on the photo: 'TNT block 400 grams with M10x1 thread' (Photo credit: Akexandr. Obtained from Adobe Stock. TM-62M stock photos, royalty-free images, vectors, video.)

which contains a cocked striker retained by ball bearings. It is armed by removing the safety clip from the arming button. This begins a clockwork arming delay where a spring-loaded striker moves from the horizontal to the vertical and in line with the detonator. Once armed, a weight greater than 150 kg will initiate the fuse. Figure 15.16 shows the mine with an MVP minimum metal fuse. If deployed in the field in this way, it would be a waste of a fuse and would indicate a potential lack of training or lack of alternative fuses for those who emplaced the mine. The TM-62M is confirmed as being widely used in Ukraine since 2014. It is known to be used at roadblocks among other locations (GICHD, 2014).

The types of warheads used in UAVs have been evolving. For example, in 2024, kamikaze drones, like Molniya and Privet-82 launched by Russia, mostly carried RPG-7 shaped-charge anti-tank grenade or similar explosive, like KZ-6 munitions used in Lancet's warheads. However, since January 2025, Russian forces switched to much larger and more powerful explosive payloads (i.e., 10-kg anti-tank TM-62 mines with at least 7 kg of explosive filling). However, a heavier payload means shorter flight range and hindered controllability. In addition, the UAVs with this payload are required to be deployed from high-rise buildings to compensate for the initial loss of altitude due to heavy payload.

15.7 ANTI-DRONE AMMUNITION

In June 2025, Ukraine's Bravel researchers developed NATO 5.56 mm anti-drone rifle ammo to strengthen infantry drone defence. Newly designed anti-drone rifle rounds can significantly increase the chances of infantry neutralising fast-moving aerial threats (i.e., Mavic-type quadcopters or FPV drones). This is achieved through a specialised projectile design for 5.56×45mm cartridges that

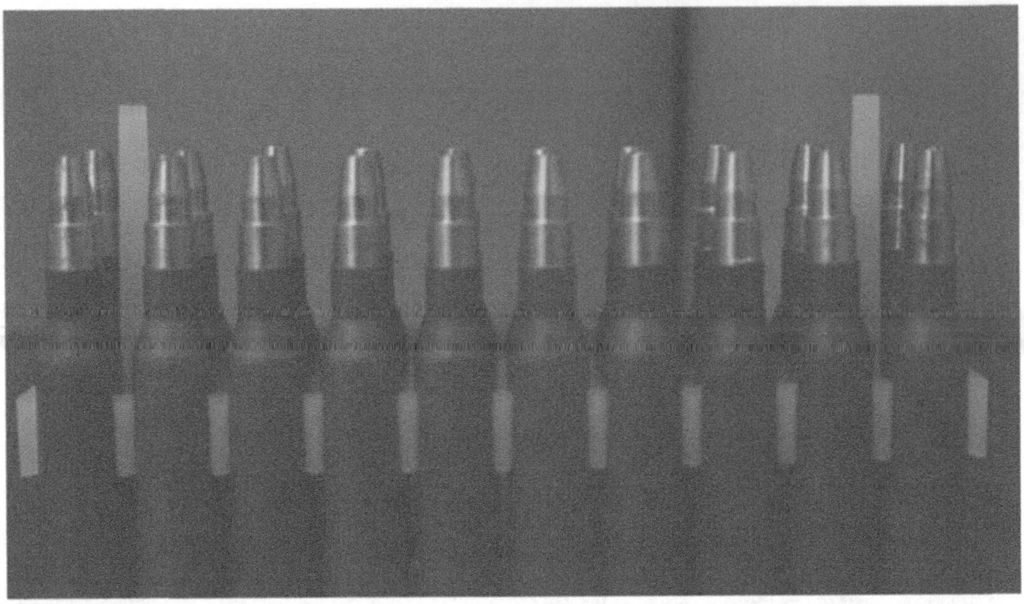

FIGURE 15.17 A pre-production batch of the Russian 5.45 × 39 mm multi-round cartridges. (Photo credit: Militarnyi; Ukraine Codifies Anti-Drone Ammunition for Assault Rifles – Militarnyi with permission from the owner.)

can be used in standard rifles (Defense Express, 2025b). Standard assault rifles, such as AR-15, are commonly used in Ukraine to strike the threat of drones. The round resembles a multi-projectile cartridge – upon firing, the bullet fragments in the air into several parts. When fired in rapid bursts, the rounds mimic a shotgun blast – dramatically improving the chances of hitting fast, low-flying drones. This significantly increases the density of fire and, consequently, boosts the chances of hitting the target by two to three times. It is expected that each infantry soldier will carry a dedicated magazine loaded with these anti-drone rounds, which can be quickly attached to a rifle and used to open fire upon detecting enemy UAVs. Bravel stated that production is ramping up, with the goal of supplying each soldier with one magazine of the counter-drone rounds. The design lets soldiers quickly swap magazines without switching weapons, a critical advantage under fire (Martyniuk, 2025). While specific characteristics and development details remain undisclosed, it is possible that the new ammunition will be produced in several variants to fit other commonly used rifle calibres within the military.

Similar ammunition in rifle calibres is also being developed in Russia (see Figure 15.17). In March, the developers of such a round claimed they had achieved acceptable accuracy, with round fragments spreading within a 25-cm diameter when fired from a distance of 100 m. Currently, neither side has serially produced anti-drone ammunition in widespread use. As a result, soldiers on both sides of the front often resort to crafting improvised versions of such rounds themselves. However, the effectiveness of these improvised rounds is often questionable – even against lightweight plastic UAVs – and they tend to deliver inconsistent results or, in some cases, may even damage the weapon.

For now, Ukraine's anti-drone rifle ammo looks like a promising tool – but whether it will turn into a real game changer or remain just a filler depends on scaling up production, battlefield feedback, and – most importantly – timely and sufficient supplies. Without consistent delivery and proven effectiveness in real combat, even the best lab results may not be enough to make a lasting difference.

15.8 CONCLUSION

Today, drones are regarded by Ukraine as a powerful military capability, and their weapons payloads are useful not only in the air but also on the sea and land. Drones now have a range of roles, including adjusting artillery fire, conducting aerial reconnaissance, maintaining situational awareness, striking enemy targets, mining and de-mining some areas, delivering cargo, and evacuating the wounded (Molloy, 2024). The use of various types of weapons on remote-controlled multi-copters has opened a new chapter in battlefield tactics. While it started with commercial drones to drop hand grenades, the role of drones as delivery systems has evolved to allow more precise strikes with heavier ammunition, night bombing, the use of kamikaze drones as anti-tank, anti-aircraft systems, interceptors of aerial targets, and thermite mixture carriers.

Ukraine created a special category of ammunition for drones, while Russia has also started the production of drone ammunition. With an increase in both specialised and improvised munitions, there is a lack of a standardised approach. This approach may reduce the development costs by using effective munition designs, minimising training, and ensuring compatibility with the drones of different types and existing stockpiles. In addition, 3D printing is promising and can be cost-effective compared to traditional manufacturing methods.

In 2025, Ukraine is expanding its drone production, and it is planned to supply more than 4 million drones to the front. Hence, there is a clear demand for munition for UAVs (Ukrinform, 2025). In July 2025, the Minister of Defence of Ukraine announced that new Government grant programme would scale up the production of explosives. According to the Defence Minister of Ukraine, the grants will enable companies to establish or expand production facilities, and to procure components, systems, and other resources required for the manufacture of explosives (Ukraine, 2025). This will enable domestic ammunition production for the Defence Forces of Ukraine and provide ways of working with defence industry as key drivers of innovation.

Drone weaponisation is a complex and contentious issue that raises a growing concern about their use in combat, in light of an increasing number of countries that have armed drones. Despite their use for surveillance and reconnaissance purposes in the previous conflicts, they began to be armed with various types of weapons. The use of drones in the war in Ukraine underscores their rapid development and evolution of a separate category as drone ammunition. Drones are capable of carrying a wide range of weapons, including bombs, missiles, rockets, and machine guns. The type of weapon carried by a drone depends on its size, design, and intended use. Smaller drones may carry lighter weapons (i.e., grenades), while larger drones can carry heavier payloads (i.e., missiles and bombs) or precision-guided munitions, allowing them to strike targets with high accuracy.

Overall, the weaponised drones offer insurgents and non-state actors a cheap and accessible way to inflict damage and casualties. These types of attacks pose significant threats to military personnel and equipment, as well as civilian infrastructure and strategic support where collateral damage from explosives can be devastating. As drones have shaped the way modern wars are being fought and will be present in future wars, military institutions, which understand the 'drone threat', will not only invest in various types of drones for their contexts of operation but also in a distinct category of munitions designed specifically for drones.

REFERENCES

Axe, D. 2024. A tiny drone with a directional airburst warhead could mow down enemy infantry in a 60-degree arc. Available: https://www.forbes.com/sites/davidaxe/2024/03/21/a-tiny-drone-with-a-directional-airburst-warhead-could-mow-down-enemy-infantry-in-a-60-degree-arc/ [Accessed August 3, 2025].

Babaiev, B. 2025. 'Vampire postman' in sky: Military shows how drones assist soldiers in Ukraine's Vovchansk. Available: https://newsukraine.rbc.ua/news/vampire-postman-in-sky-military-shows-how-1744477939.html [Accessed August 3, 2025].

Borsari, F. & Davis, G. 2023. An Urgent Matter of Drones. *Center for European Policy Analysis*, 27.

Burgess, A. 2024. Russia is unleashing record drone attacks on Ukraine with foam decoys and new Shahed warheads. Available: https://www.abc.net.au/news/2024-11-29/russia-largest-drone-strike-foam-gerbera-/104652372 [Accessed July 20, 2025].

Crumley, B. 2022. Ukraine reportedly adapts small drones to drop Molotov cocktails in war with Russians. Available: https://dronedj.com/2022/03/11/ukraine-reportedly-adapts-small-drones-to-drop-molotov-cocktails-in-war-with-russians/ [Accessed August 2, 2025].

Denisova, K. 2025. Ukraine deploys long-range drone capable of 2,000 km strike, military says. Available: https://kyivindependent.com/ukraine-deploys-long-range-drone-capable-of-2-000-km-strike-military-says/ [Accessed August 2, 2025].

Defense Express. 2024a. Shahed with a fragmentation warhead strikes Ukrainian city: What makes it so dangerous. Available: https://en.defence-ua.com/weapon_and_tech/shahed_with_a_fragmentation_warhead_strikes_ukrainian_city_what_makes_it_so_dangerous-12615.html [Accessed August 4, 2025].

Defense Express. 2024b. Ukraine developed advanced EFP-3 Armor-piercing ammunition for FPV drones. Available: https://en.defence-ua.com/news/

Defense Express. 2025a. Russian Molniya-2 Kamikaze drones now equipped with thermobaric warheads. Available: https://en.defence-ua.com/weapon_and_tech/russian_molniya_2_kamikaze_drones_now_equipped_with_thermobaric_warheads-13563.html [Accessed August 2, 2025].

Defense Express. 2025b. Ukraine develops anti-drone ammo: Gamechanger or filler? Available: https://en.defence-ua.com/weapon_and_tech/ukraine_develops_anti_drone_ammo_gamechanger_or_filler-15068.html [Accessed August 2, 2025].

Defense Express. 2025c. What Ukrainian drones with 15 kg warheads can do and why that's enough to cripple Russian air defenses in Crimea. Available: https://en.defence-ua.com/news/what_ukrainian_drones_with_15_kg_warheads_can_do_and_why_thats_enough_to_cripple_russian_air_defenses_in_crimea-14361.html [Accessed August 1, 2025].

Galka, N. 2022. It will be difficult for the Russians to fight against it. Interview with the co-developer of the Ukrainian drone RAM II. *Building TECH* [Online]. Available: https://suspilne.media/332296-rosianam-bude-vazko-z-nim-borotisa-intervu-zi-spivrozrobnikom-ukrainskogo-drona-ram-ii/[Accessed August 2, 2025].

GICHD 2022. Explosive ordnance for Ukraine. First Edition. The Geneva International Centre for Humanitarian Demining.

Goldstein, L. & Waechter, N. 2023. Chinese Strategists Evaluate the Use of 'Kamikaze' Drones in the Russia-Ukraine War. *Rand, Blog.*

Hambling, D. 2024a. Thermobaric drones blast Russian positions. Available: https://www.forbes.com/sites/davidhambling/2024/01/23/thermobaric-drones-blast-russian-positions/ [Accessed August 3, 2025].

Hambling, D. 2024b. Why Ukraine is cannibalizing brand new weapons for drone warheads. Available: https://www.forbes.com/sites/davidhambling/2024/11/27/why-ukraine-is-cannibalizing-brand-new-weapons-for-drone-warheads/ [Accessed August 2, 2025].

Hambling, D. 2025a. Horror as Russian drones target Kyiv with mystery cluster bombs. Available: https://www.forbes.com/sites/davidhambling/2025/05/02/confusion-as-russian-drones-target-kyiv-with-mystery-cluster-bombs/ [Accessed August 2, 2025].

Hambling, D. 2025b. New RUSI report: Drones now inflicting two thirds of Russian losses. Available: https://www.forbes.com/sites/davidhambling/2025/02/18/new-report-drones-now-destroying-two-thirds-of-russian-targets/.

Hazell, P. J. 2022. *Armour: Materials, theory, and design*, 2nd Ed. CRC Press.

Kesteloo, H. 2025. Ukrainian Long-Range Drone Completes 1,864-Mile Test, Signals Escalation in Aerial Capabilities. *Dronexl* [Online]. Available: https://dronexl.co/2025/03/19/ukraine-long-range-drone-1864-mile-test/# [Accessed August 2, 2025].

Kozatskyi, S. 2023. FPV drones: Weapons that changed the modern war. Available: https://militarnyi.com/en/articles/fpv-drones-weapons-that-changed-the-modern-war/ [Accessed August 2, 2025].

Kushnikov, K. 2024. New thermobaric warhead discovered in Shahed. *Militarnyi* [Online]. Available: https://militarnyi.com/en/news/new-thermobaric-warhead-discovered-in-shahed/ [Accessed August 3, 2025].

Kyshnikov, V. 2025. From shell to Javelin: What drone ammunition is made of. Available: https://militarnyi.com/en/news/from-shell-to-javelin-what-drone-ammunition-is-made-of/ [Accessed August 3, 2025].

Larter, D. B. 2017. SOCOM commander: Armed ISIS drones were 2016's 'most daunting problem'. Available: https://www.defensenews.com/digital-show-dailies/sofic/2017/05/16/socom-commander-armed-isis-drones-were-2016s-most-daunting-problem/.

Martyniuk, Y. 2025. Forbes: Ukraine's new rifle ammo explodes mid-air to kill drones—And might blow up Russia's whole strategy. Available: https://euromaidanpress.com/2025/07/16/ukraine-drone-killing-ammo/ [Accessed August 10, 2025].

Collett-White, M., Scarr, S., & Char, P. 2024. How are drone combat in Ukraine is changing warfare. Available: https://www.reuters.com/graphics/UKRAINE-CRISIS/DRONES/dwpkeyjwkpm/ [Accessed August 2, 2025].

Mitrofanov, A. 2025. Incendiary warheads for kamikaze UAVs of all types: an objective need of the SVO. *Top War* [Online]. Available: https://en.topwar.ru/263544-zazhigatelnye-boevye-chasti-dlja-bpla-kamikadze-vseh-tipov-obektivnaja-potrebnost-svo.html [Accessed August 1, 2025].

Molloy, O. 2024a. Drones in modern warfare: lessons learnt from the war in Ukraine. *Australian Army Occasional Paper*, 29, 1–116.

Molloy, O. 2024b. How are drones changing the modern warfare? Available: https://researchcentre.army.gov.au/library/land-power-forum/how-are-drones-changing-modern-warfare [Accessed August 1, 2025].

NATO. 2014. NATO STANAG 4670 – ATP-3.3.7. Guidance for the Training of Unmanned Aircraft Systems (UAS) Operators. 3rd ed.

Nichols, R. K., Sincavage, S., Mumm, H., Lonstein, W., Carter, C., Hood, J. P., Mai, R., Jackson, M., Monnik, M. & Mccreight, R. 2022. *DRONE DELIVERY OF CBNRECy–DEW WEAPONS Emerging Threats of Mini-Weapons of Mass Destruction and Disruption (WMDD)*. New Prairie Press.

Peria-Peigne, L. 2023. TB2 Bayraktar: Big strategy for a little drone. *Briefing DE L'IFRI*. Security Studies Centre.

Pettyjohn, S. 2024. *Evolution not revolution: Drone warfare in Russia's 2022 invasion of Ukraine*. Center for a New American Security.

Pleasance, C. 2024. *How Ukraine's grenade-dropping drones changed war*. Daily Mail World.

Recognition, A. 2023. Ukrainian soldiers use FPV drone armed with RPG rocket warhead to target Russian T-90M tank. Available: https://www.armyrecognition.com/focus-analysis-conflicts/army/conflicts-in-the-world/russia-ukraine-war-2022/ukrainian-soldiers-use-fpv-drone-armed-with-rpg-rocket-warhead-to-target-russian-t-90m-tank.

Reynolds, J. W. A. N. 2025. *Tactical developments during the third year of the Russo–Ukrainian war*. The Royal United Services Institute for Defence and Security Studies.

Rozgon, O. 2025b. The Ministry of Defense has approved over 120 new types of weapons for use by the Armed Forces of Ukraine. *Ukrainian National News*.

Schmidt, M. S. & Schmitt, E. 2016. Pentagon confronts a new threat from ISIS: Exploding drones. *New York Times*, 11.

Seddon, B. 2025. Explosive ordnance: Guide for Ukraine. GIACHD - The Geneva International Centre for Humanitarian Demining, GIACHD.

Shumlyanskyi, D. 2024. A new warhead was produced for the Warmate UAV. Available: https://militarnyi.com/uk/news/dlya-bpla-warmate-vyrobuvaly-novu-bojovu-chastynu/ [Accessed August 4, 2025].

Singer, P. W. 2009. *Wired for war: The robotics revolution and conflict in the 21st century*. Penguin.

Struck, J. 2024. Grenade launcher meets drone, Ukrainians test FPV drone armed with RPG-18. Available: https://www.kyivpost.com/post/38865 [Accessed August 3, 2025].

Sutton, H. I. 2025. Guide to Ukraine's long range attack drones. Available: https://southfront.press/overview-of-ukrainian-long-range-drones-capable-to-attack-russian-rear-areas-covert-shores/#:~:text=The%20type%20was%20introduced%20in%202023%20and%20has,to%20attack%20Moscow%20and%20other%20targets%20in%20Russia. [Accessed August 2, 2025].

Synaivska, S. 2024. Russia modifies Shahed-136 drone warhead for the third time, adding shaped charge and incendiary effect and releasing photos. Available: https://global.uaobozrevatel.org/events/russia-modifies-shahed-136-drone-warhead-for-the-third-time-adding-shaped-charge-and-incendiary-effect-and-releasing-photos.html [Accessed August 3, 2025].

Synytsin, M. 2023. Revenge from the skies: What Ukrainian drones drop on the heads of the occupiers. Available: https://war.telegraf.com.ua/ukr/tehnologii/2023-04-06/5786050-mest-s-nebes-chto-ukrainskie-drony-sbrasyvayut-na-golovy-okkupantov [Accessed August 3, 2025].

Thomas, R. 2025. Drones now account for 80% of casualties in Ukraine-Russia war. Available: https://www.army-technology.com/news/drones-now-account-for-80-of-casualties-in-ukraine-russia-war/ [Accessed August 2, 2025].

Ukraine, M. O. D. O. 2024. *Ministry of Defence of Ukraine in partnership with Victory Drones are launching a project to recruit UAV operators*. Ministry of Defence of Ukraine.

Ukraine, M. O. D. O. 2025. *Denys Shmyhal: The Government launches a grant program to scale up the production of explosives*. Ministry of Defence of Ukraine.

Ukraine, T. D. F. O. 2023. *The use of missiles, grenades, and engineering munitions as a combat component for various types of UAVs.* Territorial Defence Forces of Ukraine.

UKRINFORM. 2025. Minister Umerov: Ukraine to supply over 4M drones to front by year end. *UKRINFORM*, June 28, 2025.

Vialko, M. K. A. D. 2025. Cluster munitions and nails: How dangerous could Russia's drones become? Available: https://newsukraine.rbc.ua/analytics/cluster-munitions-and-nails-how-dangerous-1746202359.html [Accessed August 2, 2025].

Wildegger-Gaissmaier, A. E. 2004. Aspects of thermobaric weaponry. *Military Technology*, 28, 125–126.

Yackley, A. J. 2022. Ukraine army hails Turkish drones but Ankara plays down weapons sales. Available: https://www.ft.com/content/7529a862-7928-4b4e-a4c5-b2f50747d1f6 [Accessed July 22, 2025].

Yen, N. H. & Wang, L. Y. 2012. Reactive metals in explosives. *Propellants, Explosives, Pyrotechnics*, 37, 143–155.

Zafra, M., Hunder, M., Rao, A. & Kiyada, S. 2024. 'How Drone Combat In Ukraine Is Changing Warfare.' Available: https://www.reuters.com/graphics/UKRAINE-CRISIS/DRONES/dwpkeyjwkpm/ [Accessed 26 March 2024].

Zoria, Y. 2025. Russia equips decoy Gerbera drones with explosive payloads. Available: https://euromaidanpress.com/2025/02/03/russia-equips-decoy-gerbera-drones-with-explosive-payloads/.

Index

Note: Locators in *italics* represent figures and **bold** indicate tables in the text.